Lecture Notes in Artificial Intelligence 8754

Subseries of Lecture Notes in Computer Science

LNAI Series Editors

Randy Goebel
University of Alberta, Edmonton, Canada
Yuzuru Tanaka
Hokkaido University, Sapporo, Japan
Wolfgang Wahlster
DFKI and Saarland University, Saarbrücken, Germany

LNAI Founding Series Editor

Joerg Siekmann
DFKI and Saarland University, Saarbrücken, Germany

Linda C. van der Gaag   Ad J. Feelders (Eds.)

Probabilistic Graphical Models

7th European Workshop, PGM 2014
Utrecht, The Netherlands, September 17-19, 2014
Proceedings

 Springer

Volume Editors

Linda C. van der Gaag
Utrecht University
Faculty of Science
Department of Information and Computing Sciences
Princetonplein 5
3584 CC Utrecht, The Netherlands
E-mail: l.c.vandergaag@uu.nl

Ad J. Feelders
Utrecht University
Faculty of Science
Department of Information and Computing Sciences
Princetonplein 5
3584 CC Utrecht, The Netherlands
E-mail: a.j.feelders@uu.nl

ISSN 0302-9743   e-ISSN 1611-3349
ISBN 978-3-319-11432-3   e-ISBN 978-3-319-11433-0
DOI 10.1007/978-3-319-11433-0
Springer Cham Heidelberg New York Dordrecht London

Library of Congress Control Number: 2014948603

LNCS Sublibrary: SL 7 – Artificial Intelligence

Typesetting: Camera-ready by author, data conversion by Scientific Publishing Services, Chennai, India

Printed on acid-free paper

Springer is part of Springer Science+Business Media (www.springer.com)

Preface

The biennial European Workshop on Probabilistic Graphical Models (PGM) brings together researchers interested in all aspects of graphical models for probabilistic reasoning, decision making, and learning. It provides a forum for discussion of research developments, both theoretical and applied, in these fields.

Previous PGM meetings were held in Cuenca (2002), Leiden (2004), Prague (2006), Hirtshals (2008), Helsinki (2010), and Granada (2012). The 7th European Workshop on Probabilistic Graphical Models was held in Utrecht, The Netherlands, on September 17–19, 2014. This PGM meeting was the first to celebrate the publication of accepted papers in the *Lecture Notes in Artificial Intelligence* series by Springer. The 38 papers presented at the workshop were selected from 44 submitted manuscripts; these manuscripts involved a total of 92 authors working in 22 different countries. Each submission underwent rigorous reviewing by three members of the PGM Program Committee, with each PC member reviewing at most four papers. All papers were presented in a plenary session.

In addition to the presentations of the accepted papers, we were honored to have for our invited speaker

> Robert Cowell (City University London, UK):
> *Analysis of DNA Mixtures Using Bayesian Networks*

We are most grateful for his highly inspiring presentation.

To conclude, we would like to thank the 42 members of the Program Committee and the additional reviewer for their efforts, and for their punctual and high-quality reviews specifically; these reviews were most instrumental in selecting the best submissions for presentation during the workshop. And, last but not least, we are most indebted to our sponsors for their financial support.

August 2014

Linda C. van der Gaag
Ad J. Feelders

Organization

PGM 2014 was hosted by Utrecht University, The Netherlands, and organized by members of the Department of Information and Computing Sciences, Faculty of Science.

Executive Committee

Conference Chairs

Linda C. van der Gaag
Ad Feelders

Program Chairs

Ad Feelders
Linda C. van der Gaag
Silja Renooij

Organization Committee

Janneke H. Bolt
Arnoud Pastink

Silja Renooij
Steven P.D. Woudenberg (Chair)

Program Committee

Alessandro Antonucci (Switzerland)
Concha Bielza (Spain)
Cory J. Butz (Canada)
Andrés Cano (Spain)
Barry R. Cobb (USA)
Giorgio Corani (Switzerland)
Robert Cowell (UK)
Fabio G. Cozman (Brazil)
Adnan Darwiche (USA)
Luis M. De Campos (Spain)
Francisco Javier Díez (Spain)
Marek J. Druzdzel (USA, Poland)
M. Julia Flores (Spain)
José A. Gámez (Spain)

Helge Langseth (Norway)
Pedro Larrañaga (Spain)
Philippe Leray (France)
Jose A. Lozano (Spain)
Peter Lucas (The Netherlands)
Andres R. Masegosa (Spain)
Serafín Moral (Spain)
Petri Myllymäki (Finland)
Thomas D. Nielsen (Denmark)
Kristian G. Olesen (Denmark)
Thorsten Ottosen (Denmark)
Jose M. Peña (Sweden)
José M. Puerta (Spain)
David Rios (Spain)

Table of Contents

Structural Sensitivity for the Knowledge Engineering of Bayesian
Networks 1
   David Albrecht, Ann E. Nicholson, and Chris Whittle

A Pairwise Class Interaction Framework for Multilabel Classification . . . 17
   Jacinto Arias, José A. Gámez, Thomas D. Nielsen, and
   José M. Puerta

From Information to Evidence in a Bayesian Network 33
   Ali Ben Mrad, Véronique Delcroix, Sylvain Piechowiak, and
   Philip Leicester

Learning Gated Bayesian Networks for Algorithmic Trading 49
   Marcus Bendtsen and Jose M. Peña

Local Sensitivity of Bayesian Networks to Multiple Simultaneous
Parameter Shifts 65
   Janneke H. Bolt and Silja Renooij

Bayesian Network Inference Using Marginal Trees 81
   Cory J. Butz, Jhonatan de S. Oliveira, and Anders L. Madsen

On SPI-Lazy Evaluation of Influence Diagrams 97
   Rafael Cabañas, Andrés Cano, Manuel Gómez-Olmedo, and
   Anders L. Madsen

Extended Probability Trees for Probabilistic Graphical Models 113
   Andrés Cano, Manuel Gómez-Olmedo, Serafín Moral, and
   Cora B. Pérez-Ariza

Mixture of Polynomials Probability Distributions for Grouped Sample
Data 129
   Barry R. Cobb

Trading off Speed and Accuracy in Multilabel Classification 145
   Giorgio Corani, Alessandro Antonucci, Denis D. Mauá, and
   Sandra Gabaglio

Robustifying the Viterbi Algorithm 160
   Cedric De Boom, Jasper De Bock, Arthur Van Camp, and
   Gert de Cooman

Extended Tree Augmented Naive Classifier 176
  Cassio P. de Campos, Marco Cuccu, Giorgio Corani, and
  Marco Zaffalon

Evaluation of Rules for Coping with Insufficient Data in Constraint-
Based Search Algorithms 190
  Martijn de Jongh and Marek J. Druzdzel

Supervised Classification Using Hybrid Probabilistic Decision Graphs ... 206
  Antonio Fernández, Rafael Rumí, José del Sagrado, and
  Antonio Salmerón

Towards a Bayesian Decision Theoretic Analysis of Contextual Effect
Modifiers 222
  Gabor Hullam and Peter Antal

Discrete Bayesian Network Interpretation of the Cox's Proportional
Hazards Model 238
  Jidapa Kraisangka and Marek J. Druzdzel

Minimizing Relative Entropy in Hierarchical Predictive Coding 254
  Johan Kwisthout

Treewidth and the Computational Complexity of MAP
Approximations.. 271
  Johan Kwisthout

Bayesian Networks with Function Nodes 286
  Anders L. Madsen, Frank Jensen, Martin Karlsen, and
  Nicolaj Soendberg-Jeppesen

A New Method for Vertical Parallelisation of TAN Learning Based on
Balanced Incomplete Block Designs 302
  Anders L. Madsen, Frank Jensen, Antonio Salmerón,
  Martin Karlsen, Helge Langseth, and Thomas D. Nielsen

Equivalences between Maximum a Posteriori Inference in Bayesian
Networks and Maximum Expected Utility Computation in Influence
Diagrams 318
  Denis D. Mauá

Speeding Up k-Neighborhood Local Search in Limited Memory
Influence Diagrams.. 334
  Denis D. Mauá and Fabio G. Cozman

Inhibited Effects in CP-Logic 350
  Wannes Meert and Joost Vennekens

Learning Parameters in Canonical Models Using Weighted Least
Squares... 366
   Krzysztof Nowak and Marek J. Druzdzel

Learning Marginal AMP Chain Graphs under Faithfulness 382
   Jose M. Peña

Learning Maximum Weighted (k+1)-Order Decomposable Graphs by
Integer Linear Programming...................................... 396
   Aritz Pérez, Christian Blum, and Jose A. Lozano

Multi-label Classification for Tree and Directed Acyclic Graphs
Hierarchies.. 409
   *Mallinali Ramírez-Corona, L. Enrique Sucar, and
   Eduardo F. Morales*

Min-BDeu and Max-BDeu Scores for Learning Bayesian Networks...... 426
   Mauro Scanagatta, Cassio P. de Campos, and Marco Zaffalon

Causal Discovery from Databases with Discrete and Continuous
Variables 442
   Elena Sokolova, Perry Groot, Tom Claassen, and Tom Heskes

On Expressiveness of the AMP Chain Graph Interpretation 458
   Dag Sonntag

Learning Bayesian Network Structures When Discrete and Continuous
Variables Are Present 471
   Joe Suzuki

Learning Neighborhoods of High Confidence in Constraint-Based
Causal Discovery 487
   Sofia Triantafillou, Ioannis Tsamardinos, and Anna Roumpelaki

Causal Independence Models for Continuous Time Bayesian
Networks 503
   Maarten van der Heijden and Arjen Hommersom

Expressive Power of Binary Relevance and Chain Classifiers Based on
Bayesian Networks for Multi-label Classification 519
   Gherardo Varando, Concha Bielza, and Pedro Larrañaga

An Approximate Tensor-Based Inference Method Applied to the Game
of Minesweeper 535
   Jiří Vomlel and Petr Tichavský

Compression of Bayesian Networks with NIN-AND Tree Modeling 551
   Yang Xiang and Qing Liu

A Study of Recently Discovered Equalities about Latent Tree Models
Using Inverse Edges 567
    Nevin L. Zhang, Xiaofei Wang, and Peixian Chen

An Extended MPL-C Model for Bayesian Network Parameter Learning
with Exterior Constraints 581
    Yun Zhou, Norman Fenton, and Martin Neil

Author Index 597

Structural Sensitivity for the Knowledge Engineering of Bayesian Networks

David Albrecht, Ann E. Nicholson, and Chris Whittle

Monash University, Australia
{david.albrecht,ann.nicholson,chris.whittle}@monash.edu

Abstract. Whether a Bayesian Network (BN) is constructed through expert elicitation, from data, or a combination of both, evaluation of the resultant BN is a crucial part of the knowledge engineering process. One kind of evaluation is to analyze how sensitive the network is to changes in inputs, a form of sensitivity analysis commonly called "sensitivity to findings". The properties of d-separation can be used to determine whether or not evidence (or findings) about one variable may influence belief in a target variable, given the BN structure only. Once the network is parameterised, it is also possible to measure this influence, for example with mutual information or variance. Given such a metric of change, when evaluating a BN, it is common to rank nodes for either a maximum such effect or the average such effect. However this ranking tends to reflect the structural properties in the network: the longer the path from a node to the target node, the lower the influence, while the influence increases with the number of such paths. This raises the question: how useful is the ranking computed with the parameterised network, over and above what could be discerned from the structure alone? We propose a metric, *Distance Weighted Influence*, that ranks the influence of nodes based on the structure of the network alone. We show that not only does this ranking provide useful feedback on the structure in the early stages of the knowledge engineering process, after parameterisation the interest from an evaluation perspective is how much the ranking has changed. We illustrate the practical use of this on real-world networks from the literature.

Keywords: Bayesian Networks, Structure Sensitivity, Knowledge Engineering.

1 Introduction

Bayesian networks (BNs) [18,13] are being increasingly used for reasoning and decision making under uncertainty in many application domains, such as medicine, education, the military, environmental management and engineering. The proliferation of BN applications is due in part to the availability of both commercial and free BN software packages that combine the powerful inference algorithms developed in the AI research community with relatively easy-to-use graphical user interfaces (GUIs). But the uptake of the technology is occurring despite

L.C. van der Gaag and A.J. Feelders (Eds.): PGM 2014, LNAI 8754, pp. 1–16, 2014.

the fact that a lot of Bayesian network construction is done by hand, requiring painstaking involvement of domain experts, often combined with automated learning with small or noisy datasets, and a lot of time both building and validating models, creating a "knowledge-engineering bottleneck"[10].

There is a small, but growing literature on knowledge engineering (KE) Bayesian networks (reviewed in Section 2.2). Whether a BN is constructed through expert elicitation, from data, or a combination of both, evaluation of the resultant BN is a crucial part of the knowledge engineering process. One kind of evaluation is to analyze how sensitive the network is to changes in inputs, a form of sensitivity analysis commonly called "sensitivity to findings".[1] The properties of d-separation can be used to determine whether or not evidence (or findings) about one variable may influence belief in a target variable, given the BN structure only. Once the network is parameterised, it is also possible to measure this influence. Mutual information is the common measure of how much uncertainty is represented in a probability mass, hence it is one possible metric of change. A second measure of uncertainty sometimes used is the variance.

Given such a metric of change, when evaluating a BN, it is common to rank nodes for either a maximum such effect or the average such effect. There are many examples in the literature describing the use of rankings based on influence in the KE process including: ecological risk assessment [20,22], the spread of resistant bacteria [21], healthcare management [1]. It is also advocated as part of the evaluation process [9,2,19]. However this ranking tends to reflect the structural properties in the network: the longer the path from a node to the target node, the lower the influence of that node, while the influence increases with the number of such paths. This raises the question of how useful is the ranking computed with the parameterised network, over and above what could be discerned from the structure alone? Moreover, the sensitivity-to-findings ranking requires that the network be parameterised, which is expensive and time-consuming if done by expert elicitation.

In this paper we propose a new metric, so-called *Distance Weighted Influence*, that ranks the influence of query nodes based on the structure of the network alone. This ranking provides useful feedback on the structure to the knowledge engineer at an earlier stage in the knowledge-engineering process, *before* parameterisation, which can reduce wasted effort eliciting the wrong parameters.

In large complex networks, a list of nodes in ranked order of influence (whatever the measure) may not be very informative. Here we show how the influence measure can be used to generate a color intensity "heat-map" that provides a useful visualisation. When presenting the knowledge engineer with two rankings of relative influence – one provided by DWI and based on structure alone, and the second available after parameterisation – comparing heatmaps helps the knowledge engineer see where in the BN the parameterisation has *changed* the ranking. We also present a quantitative measure of the difference between two node orderings that represents the size of that change.

---

[1] The term used in the Netica BN software.

After presenting our approach in Section 3, we illustrate its practical use on three real-world networks from the literature in Section 4, before concluding and suggesting further work in Section 5.

2   Background

2.1   Bayesian Networks

Bayesian networks (BNs) [18,13] are an increasingly popular paradigm for reasoning under uncertainty. A Bayesian network is a directed, acyclic graph whose nodes represent the random variables in the problem. A set of directed arcs connect pairs of nodes, representing the direct dependencies (which are often causal connections) between variables. The set of nodes which have arcs pointing to X are called its parents, and is denoted $pa(X)$. The relationship between variables is quantified by conditional probability tables (CPTs) associated with each node, namely $P(X|pa(X))$. The CPTs together compactly represent the full joint distribution. Users can set the values of any combination of nodes in the network that they have observed. This evidence, e, propagates through the network, producing a new posterior probability distribution $P(X|e)$ for each node in the network. There are a number of efficient exact and approximate inference algorithms for performing this probabilistic updating, providing a powerful combination of predictive, diagnostic and explanatory reasoning.

2.2   Knowledge Engineering Bayesian Networks

Most approaches to knowledge engineering Bayesian networks (KEBN) are based on the inherent sequential stages reflecting the components of a BN: build the structure (the nodes, their states, and the arcs between nodes), then do the parameterisation, followed by evaluation (e.g. [16,8]). The notions of prototyping and spiral development from the software engineering literature (e.g. [7,3]) were first advocated for KEBN by Laskey & Mahoney [14], and later formalised as iterative, incremental development in [5]. A key aspect is the importance of evaluating at each stage and at each iteration. This is in order to identify and correct errors as early as possible in the development lifecycle.

When evaluating the structure, some key questions are [13, §10.3.10]: are the nodes in the BN the right ones? are the state space for each nodes correct? are the arcs right? And if the answer to any of these is 'no', the structure will change. If this changes is made after parameterisation, many of the parameters in the CPTs will have to be thrown away and the time spent learning or eliciting them will have been wasted.

Despite the importance of evaluating the structure, the methods available for this prior to parameterisation are relatively limited: elictation review and model walk-throughs [14], and exploring d-separation and other dependence relationships (e.g., in Matilda [6]).

2.3   Sensitivity Measures

Mutual Information is a commonly used measure of sensitivity when analysing Bayesian networks [18].[2]

Definition 1. *The mutual information, $I(X, Y)$, between two discrete variables, X and Y is defined as:*

$$I(X, Y) = \sum_{x,y} p(x, y) \ln(\frac{p(x, y)}{p(x)p(y)})$$

and has the following properties:

- $0 \leq I(X, Y) \leq H(X)$, where $H(X) = -\sum_x p(x) \ln(p(x))$.
- $I(X, Y) = 0$ iff X and Y independent
- In a Bayesian network which is a polytree, for any variables, X and Y, on a path, $I(X, Y)$ is non-increasing as the distance (number of edges) between X and Y increases.

The standard approach in sensitivity analysis of BNs is to take the variable you are interested in, say Y. Then compute $I(X, Y)$, for all the other variables, X, in the network. These measures then determine the influence on Y, namely the higher the measure the more influence the corresponding variable has on Y. This can be extended to the case where findings are available for a set of evidence nodes E, with a straightforward extension to $I(X, Y|E)$. However in this paper we will restrict our attention to the case where no evidence has been given.

3   Structural Sensitivity

3.1   Distance Weighted Influence

In general $I(X, Y)$ depends on all the ways that the variable X can influence variable Y, and vice-versa. In Bayesian networks, a node, X, can only influence another node, Y, if there exists an *unblocked path* [13] joining X and Y. In the case of a network without any given evidence, this means there is no node Z on the path where both path arcs lead into Z. So, to determine the influence of X on Y, we see that we only need to consider the set, $\mathcal{S}(X, Y)$, of *simple paths* (no nodes are visited twice) in the Bayesian network which are not blocked and join the nodes X and Y.

The above observations lead us to the following simple measure for influence, which we call *Distance Weighted Influence*. This measure depends upon the number of paths and for any path decreases as the length of the path increases.

---

[2] We note that mutual information has also been used as a measure of BN arc strength in other ways, e.g., for visualisation of arc strength based on the thickness of the arc [4], and for approximate inference [11].

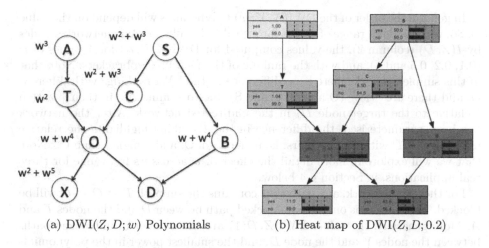

(a) DWI($Z, D; w$) Polynomials   (b) Heat map of DWI($Z, D; 0.2$)

Fig. 1. Asia Network (with target node D)

Definition 2. *The* Distance Weighted Influence *of X on Y, DWI($X, Y; w$), is defined as:*

$$DWI(X, Y; w) = \sum_{s \in \mathcal{S}(X,Y)} w^{|s|},$$

where $|s|$ is the length of the simple path s and $0 \leq w \leq 1$.

DWI($X, Y; w$) is a polynomial related to the reliability polynomial [17] and the weight, w, is related to the concept of arc weight defined in [11]. Although arc weights will in general depend on the arc, before parameterization we do not have any idea of their values. So, we treat all the arc weights as the same throughout the network.

The polynomial contains a wealth of information about the relationship between X and Y, e.g., the sum of the coefficients (also the value of the polynomial when $w = 1$) is the number of unblocked paths joining X and Y; and the smallest power is the smallest length of an unblocked path between X and Y. We have also extended this definition to handle the case where evidence has been given, however in this paper we will only consider examples where there is no evidence.

Consider the network in Figure 1(a), based on the Lauritzen and Spiegelhalter's Asia network [15], with D the designated target node.[3] Next to every node, Z, is the corresponding value DWI($Z, D; w$). In this case, for $0 \leq w \leq 1$, we have:

$$w + w^4 \geq w^2 + w^3 \geq w^2 + w^5 \geq w^2 \geq w^3$$
$$B, O \qquad S, C \qquad X \qquad T \qquad A$$

---

[3] Note that while it would be more realistic in a medical diagnostic BN such as Asia to make the target variable one of the disease nodes, e.g. T(uberculosis) or C(ancer), we chose the symptom $D(yspnea)$ because this provides the multiple paths to the target, and illustrates more fully how DWI works.

In general the order of the DWI$(X, Y; w)$ polynomials will depend on the value w. So, we consider a range of values for w. Table 1 ranks the Asia network's nodes by $I(Z, D)$ (column 2), the values computed for DWI$(Z, D; w)$ for 4 values of w (0.1, 0,2, 0.5 and 1), and with the ranking of the measures in brackets. Note that in this simple network, there is no difference in the DWI rankings with different w, and there are "ties" (O and B, C and S) due to symmetries in their position (relative to the target node D) in the undirected network. After the network has been parameterised, the difference in the I ranking highlights the relative influences on D within these pairs: B higher than O, and S higher than C.[4] Note that we will explore in more detail the effect of w across its full range for three real applications, in Section 4.4 below.

For the Asia network, any path that contains the subpath $T \rightarrow O \rightarrow C$ will be blocked. Hence there is only one unblocked path between D and the nodes T and A. More generally, the values DWI$(Z, D; 1)$ are the number of unblocked paths between the nodes Y and the node D, and the smallest power in the polynomials DWI$(Z, D; w)$ are the lengths of the smallest unblocked paths between the nodes Z and D.

Table 1. Structural sensitivity performed on the Asia network, with node D as the target node

	$I(Z,D)$	DWI$(Z, D; w)$ (rank)			
		$w = 0.1$	$w = 0.2$	$w = 0.5$	$w = 1$
D	0.98814	–	–	–	–
B	0.36156 (1)	0.10010 (1)	0.2016 (1)	0.5625 (1)	2 (1)
S	**0.04045 (2)**	**0.01100 (3)**	**0.0480 (3)**	**0.3750 (3)**	**2 (1)**
O	**0.02955 (3)**	**0.10010 (1)**	**0.2016 (1)**	**0.5625 (1)**	**2 (1)**
C	0.02538 (4)	0.01100 (3)	0.0480 (3)	0.3750 (3)	2 (1)
X	0.01517 (5)	0.01001 (5)	0.0403 (5)	0.2813 (5)	2 (1)
T	0.00397 (6)	0.01000 (6)	0.0400 (6)	0.2500 (6)	1 (6)
A	0.00001 (7)	0.00100 (7)	0.0003 (7)	0.0313 (7)	1 (6)
IC	-	2	2	2	0

3.2   Representing Difference in Influence Order

We now have two different influence rankings, one obtained by computing the mutual information on the parameterised network, the other generated by *Distance Weighted Influence* (for a particular influence weight, w) on the BN structure only. How can these rankings be presented to the knowledge engineer in a way that is useful?

First, we provide a visualisation of the influence using a so-called heatmap of the network, where the value computed using the influence metric (whether DWI or I) is mapped into a colour intensity (using a logarithmic scale to better depict

---

[4] In this case, S may have more influence through B.

for the human eye the relative sensitivity values due to the spread over several orders of magnitude). Figure 1(b) shows a heatmap based on the $DWI(Z, D; 0.2)$ measure, with blue denoting the target node and the intensity of the red on the remaining nodes being logarithmically proportional to the sensitivity, i.e., paler means less influence.

Next, we show how we can quantify the difference between two rank orderings by performing an *inversion count* [12] on the two ordered lists. In this algorithm, the nodes are firstly sorted by their I metric and are then replaced by their respective DWI value. A divide-and-conquer inversion counting algorithm is then applied to this list, whereby the list is recursively partitioned into two. Then the total number of inversions in the first partition, second partition, and occurring across the partitions are recursively calculated. In cases where nodes have equal influence values, the list order is chosen such that the inversion count is a minimum, making the inversion count a measure of the agreement between the orders. To achieve this minimum, when performing the initial sort of the nodes by I, nodes of equal I values are forced to adopt an order based on their relative DWI values. That is, they are already in order and thus no inversions occur amongst that group of nodes. Equal DWI values are accounted for by defining an inversion as being a strict inequality in the algorithm.

The inversion count (IC) provides a quantification of the agreement/difference between influence rankings generated by I for different parameter values, w, of DWI. The IC for the different w in our simple Asia experiment, are shown in the bottom row of Table 1. We developed IC while investigating the significance of the chosen weighting value w, for example the interesting regions, such as the values of w for which the agreement is at a maximum or is unchanging. In Section 4.4 below, we investigate the effect of changing w on the difference in the DWI and I rankings for our three application domains.

4   Case Studies

In order to assess the usefulness of the DWI method as a measure of structural sensitivity, it was calculated for a number of previously engineered networks reported in the literature and then used to rank node influences. In these examples, the knowledge engineers all made use of the network mutual information as a way of analysing node sensitivity and assessing its correctness.

4.1   The Goulburn Fish BN

Pollino et al. [20] developed a Bayesian network to investigate the decline of the native fish population in the Goulburn Catchment in the Murray-Darling Basin in Victoria, Australia. As part of the knowledge engineering process, the mutual information between the node describing the future abundance of fish (node FA) and other nodes was calculated. The resulting ranking of node influence was then compared with the expectations of domain experts as a form of validation.

We performed a structural sensitivity analysis on the Goulburn Fish BN using values of w ranging from 0 to 1 with steps of 0.05; results for $w = 0.1, 0.2, 0.5$ and 1 are given in Table 2, along with the rank order (in brackets), and the $I(Z, FA)$ values for each node Z.[5] We can see that there are many similarities in the DWI and I rankings, confirming that a lot of the sensitivity information can be ascertained from the structure alone (without parameterisation). For example Biological Potential (BP), Water Quality (WQ) and Overall Flow (OF) – all parents of FA – rank highly with all values of w. We note also that another parent, Time Scale (TS), has its influence reduced for $w=0.5$, which is in fact closer to the ranking based on I (after parameterisation), as does that of Future Diversity (FD), FA's child.

Table 2. Structural sensitivity performed on the Goulburn Catchment network with the future abundance (FA) as the target node

	$I(Z, FA)$	DWI(Z,FA; w) (rank)			
		$w = 0.1$	$w = 0.2$	$w = 0.5$	$w = 1$
FA	0.7599	–	–	–	–
FD	0.0879 (1)	0.1110 (2)	0.2511 (2)	1.9502 (13)	131 (4)
WQ	0.0563 (2)	0.1015 (4)	0.2275 (4)	3.1484 (4)	122 (6)
OF	0.0308 (3)	0.1007 (5)	0.2223 (5)	3.5312 (2)	187 (2)
BP	0.0304 (4)	0.1115 (1)	0.2649 (1)	3.6992 (1)	216 (1)
Si	0.0284 (5)	0.0151 (8)	0.1303 (7)	2.6094 (5)	32 (15)
Ba	**0.0221 (6)**	**0.0067 (26)**	**0.0678 (13)**	**1.8906 (14)**	**32 (15)**
Ty	**0.0221 (6)**	**0.0015 (31)**	**0.0261 (31)**	**1.3047 (30)**	**32 (15)**
Te	**0.0218 (8)**	**0.0107 (24)**	**0.0535 (25)**	**1.1797 (32)**	**32 (15)**
AS	0.0193 (9)	0.0120 (11)	0.0758 (11)	2.3125 (8)	61 (9)
		...MS, AW, LFS, MW, Ty, St, PS, Sa, Co ...			
Ri	0.0021 (19)	0.0114 (19)	0.0645 (19)	1.4922 (20)	32 (15)
SH	**0.0019 (20)**	**0.1004 (6)**	**0.2128 (6)**	**2.1758 (10)**	**88 (7)**
HS	0.0013 (21)	0.0114 (19)	0.0645 (21)	1.4922 (20)	32 (15)
Fi	0.0011 (22)	0.0025 (28)	0.0337 (28)	1.3984 (26)	32 (15)
TS	**0.0010 (23)**	**0.1100 (3)**	**0.2400 (3)**	**0.7500 (34)**	**2 (34)**
Al	0.0008 (24)	0.0025 (28)	0.0337 (28)	1.3984 (26)	32 (15)
		...NM, Fo, CD, Sn, pH, CC, Mi ...			
Fl	0.0001 (31)	0.0112 (23)	0.0550 (24)	1.3594 (29)	58 (12)
DO	**0.0000 (33)**	**0.0115 (16)**	**0.0660 (14)**	**1.5391 (15)**	**32 (15)**
To	**0.0000 (33)**	**0.0115 (13)**	**0.0660 (14)**	**1.5391 (15)**	**32 (15)**
IC	–	176	173	175	135

The data were graphically represented on a heatmap of the network, allowing qualitative visual comparisons between the $I(Z, FA)$ and DWI(Z,FA;w) values. Figure 2 shows the heatmaps for (a) DWI(Z, FA; 0.2) and (b) $I(Z, FA)$ for each node Z in the network.

---

[5] For reasons of space, some of the rows are omitted. The names of the removed nodes still appear in the table.

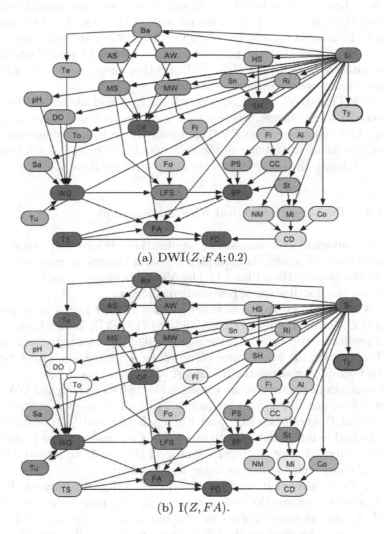

(a) DWI$(Z, FA; 0.2)$

(b) I(Z, FA).

Fig. 2. Heatmaps of the Goulburn Catchment network

In addition to obtaining useful information about relative node influence without using the parameterisation, we can use the I and DWI rankings to make observations about the parameterisation. Type (Ty), Temperature (Te), Barrier (Ba) and Connectivity (Co) are all ranked highly by I compared to their DWI values. This implies that their influence is significant and has contributions from beyond just their structural context. As expected, the conditional probability tables of all these nodes are largely deterministic (Ty is exactly deterministic). In contrast, Toxicants (To) and Dissolved Oxygen (DO) rank much more highly using DWI, indicating that the parameterisation results in some form of independence with the target node. Indeed, inspection shows that the conditional probability table has very little variation between rows.

Time Scale (TS) DWI values rank very high for small w, due to the node's close proximity to the target, whereas its I influence is much lower, indicating that it has little influence upon FA. This means there is little difference in the impact on the future abundance of the fish, whether a 1 year or a 5 year time frame is considered; the other factors are much stronger drivers.

4.2   Alpine Peatland Ecological Risk Assessment

In another ecological risk assessment application, White [22] engineered a Bayesian network to model the interplay between climate change and various systems in the Bogong High Plains in The Victorian Alps, as well as the resultant effect on the distribution of peatlands in the area.

Figure 3 shows the heatmaps of this Peatland BN for target node peatland condition (PC) for (a) DWI$(Z, PC; 0.2)$ and (b) $I(Z, PC)$. While Table 3 shows the set of results for this network – the nodes ranked according to the I values,[6] the DWI values for the same range of w, along with the associated ranking. Again, the bottom row shows the IC values.

These rankings again show many consistencies between the I and DWI rankings. The nodes describing Fire Frequency (FF), Hydrological characteristics (HC), Physical Disturbance (PD), and Entrenchment and Drainage (ED) are shown to be highly influential by both measures. Conversely, nodes Resort Pressures (RP), Subcatchment Size (Su), Ignitions (Ig), and Aqueducts Diverted (AD) are all consistently ranked as being less influential.

The Fire Probability (FP) and Increase in Very High to Extreme Weather Days (VI) are two noteworthy nodes, ranking significantly higher with I than with DWI. Again, this suggests a degree of determinism in the node CPTs, which was then confirmed by checking the CPTs presented in [22] (i.e. rows were made of up single high-valued cells and other low-probability results). Also, noteworthy is the low mutual information value for Slope Gradient (SG) which has a high DWI rank for low values of w. The low mutual information is due to the low mutual information between the Physical Suitability (PS) and SG and between Entrenchment Drainage (ED) and SG, and suggests that these arcs could be

---

[6] Here we use the I values reported in [22], as not all the CPTs are publically available.

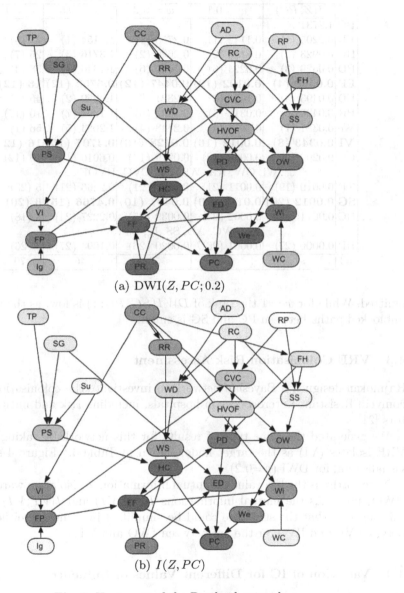

(a) DWI$(Z, PC; 0.2)$

(b) $I(Z, PC)$

Fig. 3. Heatmaps of the Peatland network

Table 3. Structural sensitivity performed on the BHP Peatland network with the Peatland Condition (PC) as the target node

	$I(Z, PC)$	DWI$(Z, PC; w)$ (rank)			
		$w = 0.1$	$w = 0.2$	$w = 0.5$	$w = 1$
PC	1.2731	–	–	–	–
FF	0.2205 (1)	0.1142 (1)	0.2758 (1)	1.5154 (1)	27 (6)
HC	0.0923 (2)	0.1123 (2)	0.2620 (2)	1.3716 (3)	25 (7)
PD	0.0529 (3)	0.0224 (6)	0.1027 (6)	1.2346 (4)	38 (4)
FP	**0.0520 (4)**	**0.0102 (12)**	**0.0447 (12)**	**0.5752 (12)**	**16 (12)**
ED	0.0493 (5)	0.1034 (3)	0.2321 (3)	1.4259 (2)	46 (3)
PR	0.0426 (6)	0.0312 (5)	0.1322 (5)	1.1094 (7)	10 (17)
We	0.0381 (7)	0.1014 (4)	0.2148 (4)	1.2084 (5)	56 (1)
VI	**0.0343 (8)**	**0.0012 (16)**	**0.0128 (16)**	**0.4707 (16)**	**16 (12)**
CC	0.0299 (9)	0.0023 (13)	0.0219 (13)	0.6016 (10)	16 (12)
	...Wi, OW, RR, WS, PS, CVC, HVOF ...				
TP	0.0019 (19)	0.0011 (21)	0.0097 (21)	0.2266 (21)	5 (23)
SG	**0.0012 (20)**	**0.0111 (10)**	**0.0497 (10)**	**0.4766 (15)**	**6 (20)**
RC	0.0011 (21)	0.0002 (25)	0.0039 (25)	0.2227 (24)	8 (18)
	...WC, AD, SS, Ig, Su ...				
RP	0.0000 (27)	0.0001 (26)	0.0020 (26)	0.1094 (27)	3 (26)
IC	–	54	53	46	79

omitted. While for $w = 1$ the rank of $DWI(SG, PC; 1)$ is low, as the number of unblocked paths between PC and SG is small.

4.3   VRE Colonisation Risk Assessment

Rajmokan designed a Bayesian network to investigate the colonisation of Vancomycin Resistant Enterococcus in hospitals, including risk and mitigating factors [21].

We generated the same type of results for this network by taking the node VRE isolates (VI) as the target node, shown in Table 4.[7] Figure 4 shows the visualisation for DWI (w=0.2).

Noteworthy is the low value of mutual information for isolation ward overflow (IWO) node. As the mutual informations $I(VT, VI)$ and $I(VP, VI)$ are high, this suggests that the structure could be simplified by removing either the arc between VP and IWO, or the arc between IWO and VT.

4.4   Variation of IC for Different Values of Influence

We observed that modifying w can change the order of the DWI ranking and we therefore produced a graph of IC against w to further investigate these variations. Figure 5 shows the IC for each of the studied networks as a function

---

[7] Here we use the I values reported in [21], as the parameterised BN is not publically available.

Table 4. Structural sensitivity performed on the VRE network with the VRE isolates (VI) as the target

	$I(Z,VI)$	DWI$(Z,VI;w)$ (rank)			
		$w = 0.1$	$w = 0.2$	$w = 0.5$	$w = 1$
VI	0.15497	–	–	–	
VT	0.02487 (1)	0.1010 (1)	0.2080 (1)	0.6250 (1)	2 (1)
VP	0.00860 (2)	0.1010 (1)	0.2080 (1)	0.6250 (1)	2 (1)
VU	0.00064 (3)	0.0101 (6)	0.0416 (6)	0.3125 (6)	2 (1)
Sc	**0.00042 (4)**	**0.0100 (9)**	**0.0400 (9)**	**0.2500 (9)**	**1 (12)**
Ha	**0.00035 (5)**	**0.0100 (9)**	**0.0400 (9)**	**0.2500 (9)**	**1 (12)**
CA	0.00032 (6)	0.0100 (9)	0.0400 (9)	0.2500 (9)	1 (12)
CU	0.00022 (7)	0.0101 (6)	0.0416 (6)	0.3125 (6)	2 (1)
VC	0.00020 (8)	0.0101 (6)	0.0416 (6)	0.3125 (6)	(1)
WO	**0.00015 (9)**	**0.0110 (4)**	**0.0480 (4)**	**0.3750 (4)**	**2 (1)**
Ov	**0.00007 (10)**	**0.0110 (4)**	**0.0480 (4)**	**0.3750 (4)**	**2 (1)**
St	0.00004 (11)	0.0100 (9)	0.0400 (9)	0.2500 (9)	1 (12)
IWO	**0.00003 (12)**	**0.0200 (3)**	**0.0800 (3)**	**0.5000 (3)**	**2 (1)**
KV	0.00001 (13)	0.0010 (13)	0.0083 (13)	0.1562 (13)	2 (1)
	...RP, SO, TP, MI, MP, PC, OTC, ED ...				
PBO	0.00000 (16)	0.0010 (16)	0.0080 (16)	0.1250 (16)	1 (12)
IC	–	28	28	28	29

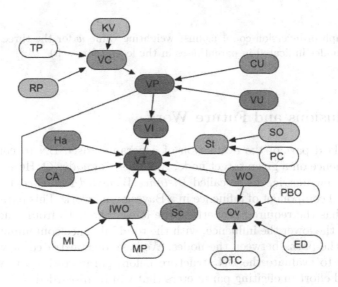

Fig. 4. A heatmap of the VRE network for $I(Z,VI)$

of w. We can see that the value of the inversion count, a quantification of the difference in ranking between the DWI and I rankings, not unexpectedly reflects the complexity of the BN, being highest for the Goulburn Fish BN (most nodes and highest connectivity), and lowest for the VRE (smallest and a simple tree structure). For the Goulburn Fish BN, the minimum inversion count was 161 at 0.25, whereas the minimum IC for the BHP Peatland BN was 45 at 0.55. The IC for the VRE BN is essentially unchanged, due to the tree-like structure of the BN, with very few multiple paths. In all cases IC is observed to dip for $w = 1$, due to an increased number of ties caused by nodes having the same number of unblocked paths.

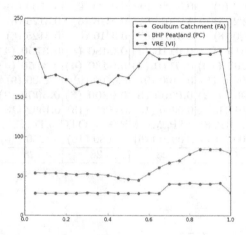

Fig. 5. A graph of inversion count against weighting value w for the three real-world BNs (target nodes indicated in parentheses in the legend for each)

5 Conclusions and Future Work

It is a standard part of the evaluation of a parameterised BN to consider the relative influence on a given target node, X, of a set of nodes Q. Here, we have a presented a new metric, the so-called *Distance Weighted Influence*, to represent the structural component of influence in a Bayesian network. This parameterised polynomial has the required properties: the longer the path from a node to the target node, the lower the influence, with the total influence combining the influences of all the paths between the nodes. We have demonstrated how the DWI can be used to evaluate the BN structure before parameterisation, which will avoid wasted effort in eliciting parameters that will be discarded if the structure changes later. Furthermore, we have shown the influence measure can be used to generate a heatmap of influence, which provides a useful visualisation of those influences. In addition, by explicitly separating out the structural component of the relative influence, the knowledge engineer can compare it to the relative influence after parameterisation, and focus on the areas of most change.

We have also shown how a heatmap provides a useful visual qualitative comparison, while our so-called inversion count provides a quantitative measure of the difference in the orderings. The IC also allows us to explore the effect of varying DWI's parameter w on the resultant ordering.

In the case studies explored in this paper, all the BNs were used in a predictive mode, with the designated target node a leaf, or near leaf node. In future work we intend to consider target nodes across the BN, for example root nodes in BNs being used for diagnostic purposes.

As mentioned above, we have extended our definition of DWI to include the incorporation of evidence, that is, $DWI(X, Y|E; w)$, as this changes the influences in the network, as well as changing the set of paths which are blocked. The next step is to evaluate the use of DWI with evidence on real case studies.

We have demonstrated the usefulness of structural sensitivity on real-world BNs described in the literature, i.e. retrospectively. The next step is to apply it in the knowledge engineering of a BN in a new domain.

Acknowledgements. We would like to thank Andrea White for providing a copy of her thesis, and Lucas Azzola, Michael Gill, Stuart Lloyd, and Ahmed Shifaz for their preliminarily work in structural sensitivity. This work has been partially funded by FEDER funds and the Spanish Government (MINECO) through project TIN2010-20900-C04-03.

References

1. Aktas, E., Ulengin, F., Sahin, S.: A decision support system to improve the efficiency of resource allocation in healthcare management. Socio-Economic Planning Sciences 41, 130–146 (2007)
2. Bednarski, M., Cholewa, W., Frid, W.: Identification of sensitivities in Bayesian networks. Engineering Applications of Artificial Intelligence 17, 327–335 (2004)
3. Boehm, B.W.: A spiral model of software development and enhancement. IEEE Computer, 61–72 (1988)
4. Boerlage, B.: Link strengths in Bayesian Networks. Master's thesis, Department of Computer Science, University of British Columbia (1995)
5. Boneh, T.: Ontology and Bayesian Decision Networks for Supporting the Meteorological Forecasting Process. PhD thesis, Clayton School of Information Technology, Monash University (2010)
6. Boneh, T., Nicholson, A., Sonenberg, L.: Matilda: A visual tool for modelling with Bayesian networks. International Journal of Intelligent Systems 21(11), 1127–1150 (2006)
7. Brooks, F.: The Mythical Man-Month: Essays on Software Engineering, 2nd edn. Addison-Wesley, Reading (1995)
8. Cain, J.: Planning improvements in natural resources management: Guidelines for using Bayesian networks to support the planning and management of development programmes in the water sector and beyond. Technical report, Centre for Ecology and Hydrology, Crowmarsh Gifford, Wallingford, Oxon, UK (2001)

9. Chen, S., Pollino, C.: Good practice in Bayesian network modelling. Environmental Modelling and Software 37, 134–145 (2012)
10. Feigenbaum, E.: The art of artificial intelligence: Themes and case studies of knowledge engineering. In: Fifth International Conference on Artificial Intelligence – IJCAI 1977, pp. 1014–1029. Morgan Kaufmann, San Mateo (1977)
11. Jitnah, N.: Using Mutual Information for Approximate Evaluation of Bayesian Networks. PhD thesis, Monash University, School of Computer Science and Software Engineering (2000)
12. Knuth, D.: The Art of Computer Programming, 2nd edn. Sorting and Searching, vol. 3. Addison Wesley, Longman (2000)
13. Korb, K., Nicholson, A.: Bayesian Artificial Intelligence, 2nd edn. CRC Press (2010)
14. Laskey, K., Mahoney, S.: Network engineering for agile belief network models. IEEE: Transactions on Knowledge and Data Engineering 12(4), 487–498 (2000)
15. Lauritzen, S.L., Spiegelhalter, D.J.: Local computations with probabilities on graphical structures and their application to expert systems. Journal of the Royal Statistical Society 50(2), 157–224 (1988)
16. Marcot, B.: A process for creating Bayesian belief network models of species-environment relations. Technical report, USDA Forest Service, Portland, Oregon (1999)
17. Moore, E., Shannon, C.: Reliable circuits using reliable relays. Journal of the Franklin Institute 262, 191–208 (1956)
18. Pearl, J.: Probabilistic Reasoning in Intelligent Systems. Morgan Kaufmann, San Mateo (1988)
19. Pitchforth, J., Mengersen, K.: A proposed validation framework for expert elicited Bayesian networks. Expert Systems with Applications 40, 162–167 (2013)
20. Pollino, C., Woodberry, O., Nicholson, A., Korb, K., Hart, B.T.: Parameterisation of a Bayesian network for use in an ecological risk management case study. Environmental Modelling and Software 22(8), 1140–1152 (2007)
21. Rajmokan, M., Morton, A., Mengersen, K., Hall, L., Waterhouse, M.: Using a Bayesian network to model colonisation with Vancomycin Resistant Enterococcus (VRE). In: Third Annual Conference of the Australasian Bayesian Network Modelling Society, ABNMS 2011 (2011)
22. White, A.: Modelling the impact of climate change on peatlands in the Bogong High Plains, Victoria. PhD thesis, The University of Melbourne, School of Botany (2009)

A Pairwise Class Interaction Framework for Multilabel Classification

Jacinto Arias[1], José A. Gámez[1], Thomas D. Nielsen[2], and José M. Puerta[1]

[1] Department of Computing Systems, University of Castilla-La Mancha,
Albacete, Spain
{jacinto.arias,jose.gamez,jose.puerta}@uclm.es
[2] Department of Computer Science, Aalborg University, Denmark
tdn@cs.aau.dk

Abstract. We present a general framework for multidimensional classification that captures the pairwise interactions between class variables. The pairwise class interactions are encoded using a collection of base classifiers (Phase 1), for which the class predictions are combined in a Markov random field that is subsequently used for multi-label inference (Phase 2); thus, the framework can be positioned between ensemble methods and label transformation-based approaches. Our proposal leads to a general framework supporting a wide range of base classifiers in the first phase as well as different inference methods in the second phase. We describe the basic framework and its main properties, including detailed experimental results based on a range of publicly available databases. By comparing the performance with other multilabel classifiers we see that the proposed classifier either outperforms or is competitive with the tested straw-men methods. We also analyse the scalability of our approach and discuss potential drawbacks and directions for future work.

Keywords: Multidimensional classification, probabilistic classifiers, Markov random fields.

1 Introduction

Supervised classification is the problem of assigning a value to a distinguished variable, the *class* C, for a given instance defined over a set of predictive attributes. In *multi-label* classification, several class variables are simultaneously considered and the task consists of assigning a configuration of values to all the class variables. In the multi-label setting, classes (or *labels*) are binary. *Multidimensional* classification is a generalization of multi-label classification that allows class variables to have more than two values. Recent literature, however, also uses the term multi-label when dealing with n-ary class variables, so we will use both names in an interchangeable way. A wide range of applications for multi-dimensional classification can be observed [21]: bio-informatics, document/music/movie categorization, semantic scene classification, multi-fault diagnosis, etc.

L.C. van der Gaag and A.J. Feelders (Eds.): PGM 2014, LNAI 8754, pp. 17–32, 2014.

One approach to solve a multi-dimensional problem is to first *transform* the problem into a set of single-class classification problems, and then combine the outputs into a joint configuration of class values. Managing the interaction among class variables is a key point when dealing with multi-dimensional problems. Transformation-based methods range from binary relevance, where no interaction is modeled, to brute-force label power set methods, where all the class variables are aggregated into a single compound class. In-between these two extremes, new and/or adapted algorithms have been developed to deal with the multi-dimensional problem, managing in a natural way the interactions between the variables. From this second family, probabilistic methods and, in particular, those based on Bayesian networks (BNs) [16] have demonstrated a convincing performance [2]. In this paper we focus on the probabilistic approach to multi-dimensional classification.

We propose a two-stage framework for multi-dimensional classification. The framework can be positioned between the transformation-based classifiers and the family of multi-dimensional PGM-based classifiers[1]. In the first stage we learn a single-class classifier for each pair of class variables in the domain, hence this stage of the framework follows a transformation-based approach. The framework does not prescribe a particular type of classifier, but only requires that the outcome of the classifier should be a weighted distribution over the (compound) class values. Standard probabilistic classifiers meet this criteria. In the second stage a Markov random field (MRF) is constructed based on the results from the first stage. The MRF thus models the dependencies between the class variables, and thereby connects the framework to the class of multi-dimensional PGM-based classifiers. Subsequent classification is achieved by performing inference in the induced MRF.

The proposed framework is flexible: (1) different types of classifiers can be applied in the first stage; (2) preprocessing can be done separately for each single-class classifier, thus allowing one to take advantage of state-of-the-art algorithms for supervised discretization and feature selection; and (3) different types of MRF-based inference algorithms can be used for the subsequent classification, the choice of which can therefore depend on the complexity of the model (exact or approximate inference) and the score to be maximized (calculation of marginal probabilities or a most probable configuration). Furthermore, since the first stage can be carried out in parallel, the method scales with the computational resources available. Nevertheless, naively dealing with all pairs of class variables imposes a strong limitation on the number of class variables the algorithm can handle. We therefore also outline strategies for scaling up the algorithm to datasets having a large number of class variables. Experiments carried out over a collection of benchmark datasets confirm the feasibility of the approach and show that the proposed method significantly outperforms or is comparable to the straw-men methods included in the comparison.

---

[1] We refer to probabilistic graphical models (PGMs) based classifiers in order to accommodate a wider range of graphical models [12] in addition to the more common approaches based on Bayesian networks.

We would like to remark that the proposal does not fit into the so-called *pair-wise multi-label approach*, which interprets the labels of the instances as preferences and whose goal is to obtain a *ranking* among the labels [15] and not a joint configuration of class values. Furthermore, although our approach trains several classifiers in the first phase, it does not strictly fall into the class of *ensemble* methods either, as each base-classifier only provides a partial answer to the multi-dimensional problem.

The rest of the paper is structured as follows: In Section 2 we introduce the notation and the required background in the field of multidimensional classification; in Section 3 we describe our framework and discuss its main properties; in Section 4 we evaluate our approach by carrying out several experiments with real world data, and finally, in Section 5, we summarize the obtained results and introduce ideas for future work.

2 Background

2.1 Notation and Problem Definition

We assume that the available dataset consists of a collection of instances $\mathbf{D} = \{(\mathbf{a}^{(1)}, \mathbf{c}^{(1)}), \ldots, (\mathbf{a}^{(t)}, \mathbf{c}^{(t)})\}$, where the first part of an instance, $\mathbf{a}^{(i)} = (a_1^{(i)}, \ldots, a_n^{(i)})$, is a configuration of values defined over a set $\mathbf{A} = \{A_1, \ldots, A_n\}$ of predictive attributes, while the second part, $\mathbf{c}^{(i)} = (c_1^{(i)}, \ldots, c_m^{(i)})$, is a configuration of values defined over a set $\mathbf{C} = \{C_1, \ldots, C_m\}$ of classes.[2] For ease of exposition, we will in this paper assume that all the variables are discrete (nominal), i.e., the state spaces $dom(A_i)$ and $dom(C_j)$ are finite sets of mutually exclusive and exhaustive states, $\forall i, 1 \le i \le n$ and $\forall j, 1 \le j \le m$.

Our goal is to induce a multidimensional classifier f that maps configurations of the predictive variables to configurations of the class variables:

$$f : \bigotimes_{i=1}^{n} dom(A_i) \longrightarrow \bigotimes_{j=1}^{m} dom(C_j),$$

$$(x_1, x_2, \ldots, x_n) \to (c_1, c_2, \ldots, c_m),$$

where $\bigotimes$ denotes the Cartesian product.

2.2 Evaluation

Although different evaluation metrics can be used to evaluate a multidimensional classifier (see e.g. [2, Sec. 5]), we resort in this paper to two of the more widely used once, both relating to *accuracy*. Given a dataset D consisting of t multidimensional instances $((a_1, \ldots, a_n), (c_1, \ldots, c_m))$ together with the predictions obtained by a multidimensional classifier H $f(a_1, \ldots, a_n) = (c_1', \ldots, c_m')$, we compute the:

---

[2] We will omit the superscript when no confusion is possible, just writing $(\mathbf{a}, \mathbf{c})$ and $((a_1, \ldots, a_n), (c_1, \ldots, c_m))$ instead of $(\mathbf{a}^{(i)}, \mathbf{c}^{(i)})$ and $((a_1^{(i)}, \ldots, a_n^{(i)}), (c_1^{(i)}, \ldots, c_m^{(i)}))$, respectively.

– *Exact match* (or *global accuracy*).

$$acc(D, H) = \frac{1}{t} \sum_{i=1}^{t} \delta(\mathbf{c}^{(i)}, \mathbf{c}'^{(i)}), \tag{1}$$

where δ is a Kronecker's delta function.
– *Hamming* (or *mean*) *accuracy*.

$$H_{acc}(D, H) = \frac{1}{t} \sum_{i=1}^{t} \frac{1}{m} \sum_{j=1}^{m} \delta(\mathbf{c}_j^{(i)}, \mathbf{c}_j'^{(i)}). \tag{2}$$

In both cases, the higher the value the better. Obviously, *acc* is a harder scoring criterion than H_{acc}.

From a probabilistic perspective, different inference tasks [16] can be used in order to maximize each of the two scores. Thus, computing the *most probable configuration* (MPE) for the class variables will maximize global accuracy, while maximizing the Hamming accuracy requires the computation of the most probable marginal assignment for each of the individual class variables.

2.3 Approaches to Multi-dimensional Classification

In the literature we can find several approaches to deal with multi-dimensional classification problems. Below, we briefly review some of them, detailing a bit more those approaches used for comparison in this paper and those that are more related to our proposal.

– *Transformation methods*. According to [21] this is a family of methods that transform the multi-dimensional classification problem into one or several single-class classification problems. Perhaps the two most well-known approaches coming from the multi-label domain are the classifiers based on *label power-sets* (LP) and *binary relevance* (BR). In their simplest form, LP-based classifiers construct a new (compound) single-class variable having as possible values all the different configurations of the class values (labels) included in the training set. This method implicitly considers the dependences between classes, but its obvious main drawback is that it is computationally tractable only for a relatively few number of class variables.

On the other hand, *Binary Relevance* (BR) methods learn a single-class classifier (or base classifier) for each class variable, $f_i : \bigotimes_{j=1}^{n} dom(A_j) \to dom(C_i)$. A solution to the multi-dimensional problem is then found by combining the single-class outputs from the base classifiers. This method does not consider any dependencies between class variables, but can be a good predictor according to Hamming accuracy.

In-between BR and brute-force LP-based methods other approaches have been developed. One example is RA*k*EL [23], which is based on training several single-class classifiers each using as class a compound variable constructed as the Cartesian product of k class variables. The number of single-class variable models is usually linear in the number of class variables, and random selection is used to choose the k variables that will form the compound variable. Later, from the $k-$tuples predicted, voting is used to obtain a joint configuration of values for all the class variables.

Chain classifiers (CC) [18] are an alternative to BR that incorporate dependences between class variables, while still maintaining the computational efficiency of BR. In CC an ordering σ is defined over the class variables. Let $C_{\sigma(i)}$ be the i-th class variable according to ordering σ. As in BR, m single-class classifiers are induced in a CC, but when learning the single-class classifier having $C_{\sigma(i)}$ as class, the variables $C_{\sigma(1)}, \ldots, C_{\sigma(i-1)}$ are also included as predictive attributes: $f_i : \bigotimes_{j=1}^{n} dom(A_j) \times \bigotimes_{k=1}^{i-1} dom(C_{\sigma(k)}) \to dom(C_{\sigma(i)})$. Therefore, the class variable in position i of σ, depends on the class variables appearing earlier in the ordering. As a consequence, inference over the single-class classifiers must be done sequentially by following the ordering imposed by σ.

- *Adaptation methods.* These are methods that directly modify/adapts existing single-class classification algorithms to accommodate multiple classes, e.g. based on decision trees, nearest neighbours, support vector machines, etc. See [21] for an overview.

- *Multi-dimensional Bayesian Networks classifiers (MBCs)* [10,24,2]. These are multi-dimensional classifiers that use the formalism of BNs to model the problem. However, as for single-class domains, instead of learning an unconstrained BN, the learning process is constrained by biasing the resulting graph. Thus, in MBCs three subgraphs are usually considered: (a) the class subgraph, which codifies dependence relations between classes; (b) the feature subgraph, which codifies dependence relations between features (predictive attributes); and (c) the bridge subgraph which codifies dependence relations from classes to features.

  Depending on the complexity of the types of graphs allowed in the class and feature subgraphs, several models/algorithms can arise: trees and/or polytrees [10,24,19], k-dependence limited models [19], general BN structures [2], etc. With respect to the search strategy used to guide the learning process, filter and wrapper approaches have been analyzed in [2], while skeleton-based ones are proposed in [3,4] based on Markov blankets and in [26] using mutual information.

- *Ensembles.* As in single-class classification, ensembles of multi-dimensional classifiers have shown potential to improve performance compared to single classifiers. This is, for example, the case of the ensemble of CC, where each member of the ensemble uses a different (usually random) ordering [18]. In the particular case of using MBCs as base classifier for the ensemble, recent studies [27,20,1] explore the idea of using as many members in the ensemble as there are class variables. In [27,20] an undirected tree structure is first learnt for the class variables, and then each class variable is set as root and the resulting topological ordering is used as a CC in the ensemble. The resulting configuration of class values is obtained by voting. In [1] no structural learning is done over the classes, and instead a naive Bayes structure among the class variables is used, with a different root for each member of the ensemble. In contrast to previous approaches, the resulting configuration is obtained by probabilistic inference.

3   A General Framework for Multi-label Classification

In this section we describe the proposed framework for multi-label classification as well as strategies for scaling up the framework and exploiting MapReduce architectures.

3.1   The Proposed Framework

The proposed framework for doing multi-label classification positions itself between ensemble classifiers [1] and transformation-based classifiers [21], by combining the results from a collection of classifiers learned for each possible pairwise interaction between the class variables. The legal types of classifiers are restricted to those classifiers that for a given instance $\mathbf{a}$ can provide a factor $\phi_{ij \mid \mathbf{a}} : dom(C_i, C_j) \to \mathbb{R}^+$ for each class pair C_i and C_j such that the greater the value the higher the compatibility between the class states. We shall refer to these classifiers as *base classifiers*. Given the class-pair factors produced by the base classifiers for an instance $\mathbf{a}$, we pose the problem of doing multi-label classification as an inference problem in the pairwise Markov random field (MRF) induced by the factors. Specifically, for m class variables, the pairwise Markov random field specified by the factors $\phi_{ij \mid \mathbf{a}}$ defines a joint distribution

$$P_{\mathbf{a}}(C_1, \ldots, C_m) = \frac{1}{Z} \prod_{i \neq j} \phi_{ij \mid \mathbf{a}}(C_i, C_j),$$

where

$$Z = \sum_{C_1, \ldots, C_m} \prod_{i \neq j} \phi_{ij \mid \mathbf{a}}(C_i, C_j)$$

is the partition function.[3] Based on this specification, we perform classification by doing inference in the MRF model. Thus, for global accuracy we look for the most probable explanation (MPE) in the MRF

$$\mathbf{c}^* = \arg \max_{\mathbf{c} = (c_1, \ldots, c_m)} \frac{1}{Z} \prod_{i \neq j} \phi_{ij \mid \mathbf{a}}(C_i, C_j)$$

$$= \arg \max_{\mathbf{c} = (c_1, \ldots, c_m)} \prod_{i \neq j} \phi_{ij \mid \mathbf{a}}(C_i, C_j),$$

---

[3] We abuse notation slightly and use variable summation to denote summation over states of a variable.

and for Hamming accuracy we consider the most probable class variable configurations separately:

$$c_k^* = \arg\max_{c_k} \sum_{C_l:l\neq k} \prod_{i\neq j} \phi_{ij\,|\,\mathbf{a}}(C_i, C_j).$$

In summary, given a collection of base classifiers, multi-label classification of an instance **a** consists of two steps:

1. For each pair of class variables, C_i and C_j, employ the corresponding base classifier to find a factor $\phi_{ij\,|\,\mathbf{a}}$ that for each configuration (c_i, c_j) encodes the affinity between c_i and c_j for instance **a**.
2. Establish the class-labels of **a** by performing inference in the pairwise Markov random field defined by the factors $\phi_{ij\,|\,\mathbf{a}}$ found in step 1.

The overall framework is flexible in the sense that it can accommodate several different types of base classifiers (e.g., probabilistic classifiers, neural networks, etc.). Hence, we say that a particular choice of base classifier instantiates the framework, and in what follows we shall refer to the instantiated framework as a *factor-based multi-labeled classifier (FMC)*. In the present paper, we focus on probabilistic base classifiers, and for ease of exposition we will in this section only consider naive Bayes classifiers: for each pair of class variables C_i and C_j we have a naive Bayes classifier (NBC), where the state space of the class variable consists of all combinations of class labels for C_i and C_j. With this type of base classifier, the factors $\phi_{ij\,|\,\mathbf{a}}$ in the FMC correspond to the posterior probabilities $P(C_i, C_j\,|\,\mathbf{a})$. The relationship between the NB base classifiers and the induced MRF is illustrated in Figure 1 for a domain with three attributes $\{A_1, A_2, A_3\}$ and three class variables $\{C_1, C_2, C_3\}$.

Comparing the proposed framework to multi-dimensional Bayesian network classifiers (MBCs) [24,2,3] we see that the induced MRF plays the role of the class-subgraph in the MBC. Similarly, the feature subgraph and the bridge subgraph are captured by the base classifiers that, in addition, also allow each class pair to employ different types of discretization and encode different dependency structures.

3.2 Scalable Learning and Inference

The complexity of learning and inference in an FMC is determined by

- the complexity of learning and doing inference in the base classifiers; considering all class variable pairs, we have $m(m-1)/2$ base classifiers.
- the complexity of performing inference in the generated MRF.

For learning the base classifiers, we first observe that for most types of base classifiers the computational complexity can be reduced by exploiting that the overall learning setting is trivially parallelizable and can easily be adapted to a MapReduce architecture [6]; although one could envision complex classifiers that

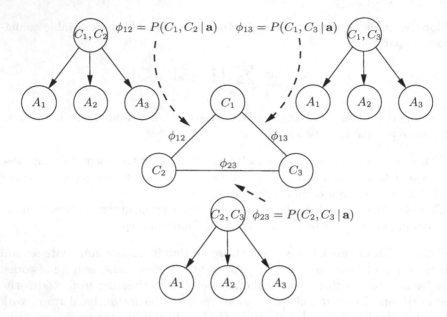

Fig. 1. An FMC is established for a domain consisting of three attributes and three class variables. The joint probabilities calculated using the NB base classifiers serve as factors in the MRF, which is in turn used for finding the class labels.

share substructures across class variable pairs, the more standard base classifiers can be learned independently.

However, even with a MapReduce architecture, the computational complexity of learning the base classifiers may still be demanding, since the number of classifiers is quadratic in the number of class variables. One immediate approach to overcome this difficulty is to only consider a restricted candidate subset of class variable pairs for which base classifiers should be learned. A simple strategy for selecting this candidate set is to greedily include the best k pairs of class variables according to a heuristic function that captures the dependence/affinity between these variables. In this paper we have used as heuristic, the empirical mutual information between class variable pairs

$$MI(C_i, C_j) = \sum_{C_i} \sum_{C_j} \hat{P}(C_i, C_j) \log \left(\frac{\hat{P}(C_i, C_j)}{\hat{P}(C_i)\hat{P}(C_j)} \right).$$

Given this strategy, we not only reduce the number of base classifiers to be learned, but we also reduce the complexity of the subsequently induced MRF, thereby also obtaining a reduction in the computational complexity when performing inference in the MRF. Note that with the described approach we may end up with a disconnected class structure with some components containing a single class variable only, our model can deal with this by learning a single class classifier for each one of these singleton variables and then set up the obtained distribution after the inference performed in the first phase as the node potential of the disconnected variables in the MRF. As a special case we can alternatively construct a Chow-Liu tree [5] over the variables, selecting in this case a subset k of $m - 1$ pairs and afterwards perform inference in this reduced model structure in which are variables are connected.

Alternatively, we could use a threshold function to select the subset of class variables pairs instead of using greedy selection. We propose a second scalable approach that selects pairs of variables by using a χ^2 statistical test to measure the dependency between the class variables. This new approach can be parametrized to allow the classifier to select a larger number of pairs by controlling the significance level of the test.

4   Experimental Evaluation

In the first part of this section we report on the results from the experimental evaluation performed for the proposed approach in its basic configuration. In the second part we evaluate the different scalability strategies described in the previous section.

4.1   Experiments

We evaluate our approach using two well-known base classifiers: Naive Bayes (NB), being a simple classifier with light computational and memory requirements; and A1DE [25], which is a more expressive classifier but requires larger amounts of resources. For comparison, we also include three state-of-the art multilabel classifiers in our experiments, which all have available public implementations: Binary-Relevance (BR), Ensembles of Classifier Chains (ECC), and RAkEL [23]; the three classifiers also require a base classifier and for the comparison we have used the same base classifiers as for our proposal. ECC has been configured to learn 10 different models for the ensemble, and in the case of RAkEL, we have considered two different configurations: One is the recommended configuration from the original paper [23], using $2m$ models with triplets of label combinations $k = 3$ (RAkEL$^{k=3}$). For the second one, we use pairwise label combinations, $k = 2$, considering all possible models (RAkEL$^{k=2}$), thus replicating the first stage of our approach in which all pairwise classifiers are built. For the subsequent predictions we use the voting scheme that RakEL implements.

Our implementation is still a prototype and is based on different platforms: The first step of the classifier has been implemented using the Mulan [22] library for multilabel dataset management and Weka [13] for the base classifiers. The second step performs inference over the Markov random field using the UGM[4] Matlab package. Due to the size of the MRF models, we perform approximate inference as exact inference is not feasible for all the data sets. Specifically,

---

[4] http://www.di.ens.fr/~mschmidt/Software/UGM.html

Table 1. Datasets used in the evaluation

Database	Birds	CAL500	Emotions	Enron	Genbase	Medical	Scene	Yeast
Classes	19	174	6	53	27	45	6	14
Features	260	68	72	1001	1186	1449	294	103
Instances	645	502	593	593	662	978	2407	2417

we have chosen the Loopy Belief Propagation algorithm [17] from the above mentioned Matlab package. Preliminary experiments over small datasets show that the approximate inference scheme obtains almost the same results as exact inference in terms of predictions, although this result should be checked for larger and more complex domains. For the experiments involving the BR, ECC and RAkEL classifiers we have used their open implementations in Mulan.

We have carried out several experiments using a collection of publicly available datasets taken from the Mulan repository[5]. Because of the high computational requirements of the basic version of our proposal, we have run the experiments using only those datasets with a moderate number of labels and attributes. The characteristics of the datasets used in the experiments can be found in Table 1.

We have discretized the numerical features using two different discretization methods. Due to the lack of standardized supervised multidimensional discretization algorithms, a simple unsupervised discretization has been performed using three equal width bins for each attribute, as it has proved to be the best configuration tested in preliminary experiments. However, the properties of the first step of our approach, and also that of the BR, ECC and RAkEL algorithms, allow us to use supervised discretization for each individual base classifier, as the feature variables are not used in the aggregation process. Thus, for each base classifier we have performed an MDL-based discretization [8] guided by the generated compound class label. We report on the results for both methods, showing the advantages of considering label information in the discretization process.

The experiments were conducted on a dedicated server with a Pentium Xeon 3.0 Ghz and 16GB of RAM running Linux, and for each dataset and classifier we have performed a 10 fold cross validation. We report both global accuracy (acc) and Hamming accuracy (H_{acc}) as performance indicators. Tables 2 and 3 show the results obtained for the unsupervised and supervised discretized datasets, respectively; the best results for the different datasets and measures are shown in boldface.

Empty cells correspond to unfinished executions, where the majority is due to memory limitations ($\leq$16GB RAM) as the A1DE algorithm is very inefficient when the number of features is high. We can observe a higher number of finished experiments when using supervised discretization, as this method removes some feature attributes which are left with only one state after the discretization process, and, in general, it obtains a dataset with attributes of lower cardinality, resulting in a considerable reduction of the number of parameters to be learned by the classifiers.

---

[5] http://mulan.sourceforge.net/datasets.html

We have not included results regarding execution time as it would be an unfair comparison as our prototype implementation is far from being optimized; tentative results show that the BR approach is the most efficient approach, followed by RAkEL, ECC and our proposed classifier being the least efficient one. Regarding base classifiers NB is more efficient than A1DE, whose complexity increases exponentially also in the number and the cardinality of the feature attributes.

We can observe that both supervised and unsupervised discretization methods obtains similar results and are superior to each other in some domains[6]. In any case, we have decided to continue the rest of the analysis by using the results obtained with supervised discretization, as it is a parameter free solution and, in addition, all algorithms obtain a good improvement from the aforementioned feature selection that it allows.

When comparing the different algorithms among themselves the results show that the proposed classifier has a clear advantage over the other approaches regarding both measures. To extend our evaluation, we have performed statistical tests for both global accuracy and Hamming accuracy [7,11]. In both cases, the tests have been performed based on the data underlying Table 3 obtained using the supervised discretization scheme. Regarding global accuracy, the Friedman [9] test, with a 5% significance level, rejects the hypothesis that all classifiers are equivalent with p-value $= 1.4484 \cdot 10^{-6}$. We also performed a post-hoc test using Holm's procedure [14], where the results can be found on Table 4 together with the ranking computed for the Friedman test; the tests compare all classifiers with the approach having the highest accuracy (MRF-A1DE) as control, and it rejects all the hypotheses, showing that our approach significantly outperforms the other ones. We have performed the same tests for Hamming accuracy, obtaining the p-value $= 9.9369 \cdot 10^{-11}$ for the Friedman test, and thus rejecting the hypothesis that all classifiers obtain equivalent results. Holm's procedure again rejects all the comparative hypotheses. The results can be found on Table 4.

This statistical analysis confirms that our approach has the overall better performance, even by using NB as base classifier.

4.2   Results on Scalability

We have replicated the previous experimental set-up to analyze the scalability approaches discussed in section 3. All the new experiments have been conducted using A1DE as base classifier and supervised discretization. We have run the pruned classifier by selecting the best subsets with size $k = 2m$ and $k = 4m$ of pair variables according to the empirical mutual information as well as the described approach of building a Chow-Liu tree between the variables. In addition, we have also tested the described method of selecting the class pairs by using a χ^2 tests, with confidence levels of 0.05 and 0.01.

---

[6] Note that for the *enron, genbase and medical* datasets the results obtained from using each discretization method are identical as there are not numerical attributes to discretize.

Table 2. Global accuracy and Hamming accuracy for each classifier and dataset using unsupervised discretization. Highlighted results show the best result for the corresponding dataset and empty cells refer to unfinished executions due to memory limitations.

Algorithm	Birds		CAL500		Emotions		Enron	
	acc	H_{acc}	acc	H_{acc}	acc	H_{acc}	acc	H_{acc}
BR-A1DE	**0.4720**	0.9491	**0.0000**	0.8350	0.3237	0.7979	-	-
BR-NB	0.3760	0.8944	**0.0000**	0.7895	0.2360	0.7686	0.0012	0.7832
ECC-A1DE	-	-	-	-	0.3338	**0.8038**	-	-
ECC-NB	0.3791	0.8959	**0.0000**	0.7261	0.2748	0.7714	0.0018	0.7991
RAkEL$^{k=3}$-A1DE	0.4689	**0.9495**	**0.0000**	**0.8378**	0.3337	0.7981	-	-
RAkEL$^{k=3}$-NB	0.3793	0.9206	**0.0000**	0.8046	0.2765	0.7821	0.0018	0.8262
RAkEL$^{k=2}$-A1DE	-	-	-	-	0.3370	0.8023	-	-
RAkEL$^{k=2}$-NB	0.3760	0.9091	**0.0000**	0.7933	0.2950	0.7841	0.0012	0.7835
FMC-A1DE	0.4689	0.9495	-	-	**0.3430**	0.7988	-	-
FMC-NB	0.3869	0.9196	**0.0000**	0.8130	0.2942	0.7744	**0.0037**	**0.9141**
	Genbase		Medical		Scene		Yeast	
	acc	H_{acc}	acc	H_{acc}	acc	H_{acc}	acc	H_{acc}
BR-A1DE	-	-	-	-	0.4292	0.8653	0.1692	0.7732
BR-NB	0.2749	0.9661	0.2586	0.9745	0.2069	0.8042	0.1088	0.7243
ECC-A1DE	-	-	-	-	0.4865	0.8781	0.2003	0.7779
ECC-NB	0.2416	0.9626	0.2249	0.9761	0.2127	0.8085	0.1245	0.7206
RAkEL$^{k=3}$-A1DE	-	-	-	-	0.5733	0.8990	0.1825	0.7782
RAkEL$^{k=3}$-NB	**0.2780**	**0.9662**	0.2607	0.9751	0.3444	0.8529	0.1221	0.7478
RAkEL$^{k=2}$-A1DE	-	-	-	-	0.4890	0.8826	0.1804	0.7796
RAkEL$^{k=2}$-NB	0.2749	0.9661	0.2597	0.9746	0.2638	0.8308	0.1167	0.7364
FMC-A1DE	-	-	-	-	**0.6277**	**0.9005**	**0.2106**	**0.7882**
FMC-NB	0.2779	0.9660	**0.2822**	**0.9784**	0.4865	0.8713	0.1423	0.7547

The results for all approaches regarding Global and Hamming accuracy can be found in Table 5. In order to check the effectiveness of the pruning approaches, we have included the number of variables selected for each dataset in Table 6, and a comparison of the execution time for each approach and each dataset in Table 7.

As we can observe, the pruned models clearly improve the efficiency of both stages of the algorithm. This improvement comes with a minimum loss of quality in the obtained predictions, which remain competitive when compared with the results obtained in the previous experiments, even being superior for some databases. This new approach reduces the model's computational complexity from $O(m^2)$ to $O(m)$ scaling-up the original classifier, so that it is capable of targeting more complex databases than the original brute force approach was unable to handle such as *CAL500, enron, genbase* and *medical*.

When comparing the different approaches between them we can observe that, although building a Chow-Liu tree is the most efficient method, selecting a larger number of class pairs by using a greedy strategy leads to better results in the majority of domains. Regarding the approach using the χ^2 test we can observe that a large amount of class pairs are selected, excessively decreasing the efficiency of the classifier without obtaining a significant improvement.

Table 3. Global accuracy and Hamming accuracy for each classifier and dataset using supervised discretization. Highlighted results show the best result for the corresponding dataset and empty cells refer to unfinished executions due to memory limitations.

Algorithm	Birds		CAL500		Emotions		Enron	
	acc	H_{acc}	acc	H_{acc}	acc	H_{acc}	acc	H_{acc}
BR-A1DE	**0.4751**	0.9233	0.0000	0.8520	0.1517	0.7745	-	-
BR-NB	0.2980	0.7758	0.0000	0.7671	0.2275	0.7542	0.0012	0.7832
ECC-A1DE	**0.4751**	0.9341	0.0000	0.8504	0.1736	0.7833	-	-
ECC-NB	0.3197	0.7869	0.0000	0.7143	0.2952	0.7570	0.0018	0.7991
RAkEL$^{k=3}$-A1DE	0.2797	0.8946	0.0000	0.7570	0.2361	0.7851	-	-
RAkEL$^{k=3}$-NB	0.3168	0.8448	0.0000	0.7888	0.2627	0.7679	0.0018	0.8262
RAkEL$^{k=2}$-A1DE	0.3045	0.8776	0.0000	0.5304	0.1923	0.7711	-	-
RAkEL$^{k=2}$-NB	0.3044	0.7947	0.0000	0.7761	0.2849	0.7718	0.0012	0.7835
FMC-A1DE	0.4473	**0.9408**	**0.0200**	0.8534	**0.3051**	**0.8007**	-	-
FMC-NB	0.3075	0.8772	0.0000	**0.8611**	0.2932	0.7841	**0.0037**	**0.9141**
	Genbase		Medical		Scene		Yeast	
	acc	H_{acc}	acc	H_{acc}	acc	H_{acc}	acc	H_{acc}
BR-A1DE	-	-	-	-	0.4761	0.8695	0.0364	0.7392
BR-NB	0.2749	0.9661	0.2586	0.9745	0.2863	0.7966	0.1026	0.7227
ECC-A1DE	-	-	-	-	0.5426	0.8821	0.0352	0.7440
ECC-NB	0.2416	0.9626	0.2249	0.9761	0.3008	0.8002	0.1374	0.6965
RAkEL$^{k=3}$-A1DE	-	-	-	-	0.5874	0.9093	0.0534	0.6817
RAkEL$^{k=3}$-NB	**0.2780**	**0.9662**	0.2607	0.9751	0.4400	0.8587	0.1188	0.7435
RAkEL$^{k=2}$-A1DE	-	-	-	-	0.4645	0.8896	0.0451	0.6794
RAkEL$^{k=2}$-NB	0.2749	0.9661	0.2597	0.9746	0.3523	0.8291	0.1262	0.7342
FMC-A1DE	-	-	-	-	**0.6465**	**0.9126**	**0.1858**	**0.7894**
FMC-NB	0.2779	0.9660	**0.2822**	**0.9784**	0.4487	0.8808	0.1432	0.7626

Table 4. Results from the statistical tests for both global accuracy and Hamming accuracy for the results obtained by using supervised discretization, showing the ranking computed for the Friedman test and the adjusted p-value using Holm's procedure. All hypotheses are rejected.

Global Accuracy (acc)			Hamming Accuracy (H_{acc})		
i Algorithm	Rank	p-value	i Algorithm	Rank	p-value
0 FMC-A1DE	1.9000	-	0 FMC-A1DE	1.3600	-
1 FMC-NB	4.8750	0.0500	1 FMC-NB	3.3200	0.05
2 ECC-NB	4.9249	0.0250	2 ECC-A1DE	3.4000	0.025
3 RAkEL$^{k=3}$-NB	5.8250	0.0166	3 BR-A1DE	4.6000	0.0166
4 ECC-A1DE	5.6250	0.0125	4 RAkEL$^{k=3}$-A1DE	5.7200	0.0125
5 RAkE$^{k=2}$-NB	5.3000	0.0100	5 RAkEL$^{k=3}$-NB	6.0000	0.01
6 RAkEL$^{k=3}$-A1DE	6.0249	0.0083	6 RAkEL$^{k=2}$-NB	6.5400	0.0083
7 BR-A1DE	6.1999	0.0071	7 RAkEL$^{k=2}$-A1DE	7.0400	0.0071
8 BR-NB	7.1250	0.0062	8 ECC-NB	8.4600	0.0062
9 RAkEL$^{k=2}$-A1DE	7.2000	0.0055	9 BR-NB	8.5600	0.0055

Table 5. Global accuracy and Hamming accuracy for each pruning method and dataset using A1DE as base classifier and supervised discretization. Highlighted results show the best result for the corresponding dataset and empty cells refer to unfinished executions due to memory limitations.

Algorithm	Birds		CAL500		Emotions		Enron	
	acc	H_{acc}	acc	H_{acc}	acc	H_{acc}	acc	H_{acc}
FMC-A1DE	**0.4689**	**0.9495**	**0.0200**	0.8534	**0.3430**	0.7988	-	-
best-mi-2	0.4382	0.9348	0.0000	**0.8598**	0.3051	**0.7990**	-	-
best-mi-4	0.4441	0.9382	0.0000	0.8580	0.3051	0.8001	-	-
test-chisq-0.01	0.4535	0.9418	**0.0199**	0.8534	0.2747	0.7878	-	-
test-chisq-0.05	0.4565	0.9415	**0.0199**	0.8534	0.2781	0.7898	-	-
tree-mi-0	0.4318	0.9297	**0.0199**	0.8534	0.2850	0.7904	**0.0488**	**0.9381**
	Genbase		Medical		Scene		Yeast	
	acc	H_{acc}	acc	H_{acc}	acc	H_{acc}	acc	H_{acc}
FMC-A1DE	-	-	-	-	0.6277	0.9005	**0.2106**	**0.7882**
best-mi-2	0.9320	0.9973	-	-	0.6365	0.9076	0.1854	0.7852
best-mi-4	**0.9351**	**0.9975**	-	-	**0.6465**	**0.9129**	0.1891	0.7894
test-chisq-0.01	0.9305	0.9973	-	-	0.4315	0.8696	0.1378	0.7794
test-chisq-0.05	0.9305	0.9973	-	-	0.4561	0.8695	0.1341	0.7786
tree-mi-0	0.9350	0.9974	0.3957	**0.9814**	0.6111	0.9032	0.1659	0.7787

Table 6. Number selected pairs and singleton variables respectively for each pruning approach expressed as the mean value among the 10 folds of the cross validation. Empty cells refer to unfinished executions due to memory limitations.

	birds	CAL500	emotions	enron	genbase	medical	scene	yeast
FMC-A1DE	171/0.0	15051/0	15/0.0	1378/0.0	351/0.0	990/0.0	15/0.0	91/0.0
best-mi-2	38/0.1	348/74.6	12/0.0	106/2.5	54/3.5	-	12/0.0	28/0.0
best-mi-4	76/0.0	696/43.3	15/0.0	-	108/1.2	-	15/0.0	56/0.0
test-chisq-0.01	162/0.0	12280/0.0	1/3.9	-	274/0.3	-	1/4.6	29/1.0
test-chisq-0.05	154/0.0	10766/0.0	1/4.0	-	245/0.3	-	0/6.0	23/1.0
tree-mi-0	18/0.0	173/0.0	5/0.0	52/0.0	26/0.0	44/0.0	5/0.0	13/0.0

Table 7. Execution time for each pruning approach and dataset expressed as the mean value among the 10 folds of the cross validation. Highlighted results show the best result for the corresponding dataset and empty cells refer to unfinished executions due to memory limitations.

	birds	CAL500	emotions	enron	genbase	medical	scene	yeast
FMC-A1DE	8.24	578.09	1.74	-	-	-	25.33	22.07
best-mi-2	2.72	11.20	1.87	1203.47	6.97	-	20.40	8.00
best-mi-4	3.93	19.41	1.65	-	12.70	-	25.74	14.14
test-chisq-0.01	7.63	374.01	**0.88**	-	23.09	-	7.54	6.72
test-chisq-0.05	7.60	336.46	0.91	-	29.81	-	7.17	5.32
tree-mi-0	**1.63**	**5.09**	0.93	**555.55**	**3.38**	484.39	4.41	**3.56**

5 Conclusions

We have introduced a new framework for multidimensional classification based on the construction of a collection of pairwise classifiers that in combination induce a Markov random field in which multi-label inference can be performed. In our experiments we have compared our proposal with a range of available state-of-the art classifiers and obtained favourable results. A potential concern about the framework is its scalability. We have outlined strategies to address this problem based on pruning class variable pairs or the induced MRF model. Preliminary experiments show that the complexity of our approach can be drastically reduced while maintaining competitive results.

In future work, we will conduct a more extensive experimentation. We plan to extend the comparison by including other related approaches to multi-dimensional classification and to extend the study by considering other evaluation metrics. Furthermore, we will study in greater detail the described procedures for pruning the models and will possibly refine them by exploring other metric and heuristics.

Acknowledgments. This work has been partially funded by FEDER funds and the Spanish Government (MINECO) through projects TIN2010-20900-C04-03 and TIN2013-46638-C3-3-P.

References

1. Antonucci, A., Corani, G., Mauá, D., Gabaglio, S.: An ensemble of Bayesian networks for multilabel classification. In: Proceedings of the Twenty-Third International Joint Conference on Artificial Intelligence, IJCAI 2013, pp. 1220–1225. AAAI Press (2013)
2. Bielza, C., Li, G., Larrañaga, P.: Multi-dimensional classification with Bayesian networks. International Journal of Approximate Reasoning 52(6), 705–727 (2011)
3. Borchani, H., Bielza, C., Martínez-Martín, P., Larrañaga, P.: Markov blanket-based approach for learning multi-dimensional Bayesian network classifiers: An application to predict the european quality of life-5 dimensions (EQ-5D) from the 39-item Parkinson's disease questionnaire (PDQ-39). Journal of Biomedical Informatics 45(6), 1175–1184 (2012)
4. Borchani, H., Bielza, C., Toro, C., Larrañaga, P.: Predicting human immunodeficiency virus inhibitors using multi-dimensional bayesian network classifiers. Artificial Intelligence in Medicine 57(3), 219–229 (2013)
5. Chow, C.K., Liu, C.: Approximating discrete probability distributions with dependence trees. IEEE Transactions on Information Theory 14, 462–467 (1968)
6. Dean, J., Ghemawat, S.: MapReduce: Simplified data processing on large clusters. In: Proceedings of the 6th Conference on Symposium on Operating Systems Design & Implementation, vol. 6, pp. 137–150. USENIX Association (2004)
7. Demšar, J.: Statistical comparisons of classifiers over multiple data sets. The Journal of Machine Learning Research 7, 1–30 (2006)
8. Fayyad, U.M., Irani, K.B.: Multi-interval discretization of continuous-valued attributes for classification learning. In: Proceedings of the 13th International Joint Conference on Artificial Intelligence (IJCAI 1993), pp. 1022–1029 (1993)

9. Friedman, M.: A comparison of alternative tests of significance for the problem of m rankings. The Annals of Mathematical Statistics, 86–92 (1940)
10. van der Gaag, L.C., de Waal, P.R.: Multi-dimensional Bayesian network classifiers. In: 3rd European Workshop on Probabilistic Graphical Models (PGM 2006), pp. 107–114 (2006)
11. García, S., Herrera, F.: An extension on statistical comparisons of classifiers over multiple data sets for all pairwise comparisons. Journal of Machine Learning Research 9(2677-2694), 66 (2008)
12. Guo, Y., Gu, S.: Multi-label classification using conditional dependency networks. In: IJCAI Proceedings-International Joint Conference on Artificial Intelligence, IJCAI 2011, vol. 22, pp. 1300–1305. AAAI Press (2011)
13. Hall, M., Frank, E., Holmes, G., Pfahringer, B., Reutemann, P., Witten, I.H.: The WEKA data mining software: an update. ACM SIGKDD Explorations Newsletter 11(1), 10–18 (2009)
14. Holm, S.: A simple sequentially rejective multiple test procedure. Scandinavian Journal of Statistics, 65–70 (1979)
15. Hüllermeier, E., Fürnkranz, J., Cheng, W., Brinker, K.: Label ranking by learning pairwise preferences. Artificial Intelligence 172(16-17), 1897–1916 (2008)
16. Jensen, F.V., Nielsen, T.D.: Bayesian Networks and Decision Graphs, 2nd edn. Springer Publishing Company, Incorporated (2007)
17. Murphy, K.P., Weiss, Y., Jordan, M.I.: Loopy belief propagation for approximate inference: An empirical study. In: Proceedings of the Fifteenth Conference on Uncertainty in Artificial Intelligence (UAI 1999), pp. 467–475. Morgan Kaufmann (1999)
18. Read, J., Pfahringer, B., Holmes, G., Frank, E.: Classifier chains for multi-label classification. Machine Learning 85(3), 333–359 (2011)
19. Rodríguez, J.D., Lozano, J.A.: Multi-objective learning of multi-dimensional Bayesian classifiers. In: 8th International Conference on Hybrid Intelligent Systems (HIS 2008), pp. 501–506 (2008)
20. Sucar, L.E., Bielza, C., Morales, E.F., Hernandez-Leal, P., Zaragoza, J.H., Larrañaga, P.: Multi-label classification with bayesian network-based chain classifiers. Pattern Recognition Letters 41, 14–22 (2014)
21. Tsoumakas, G., Katakis, I.: Multi-label classification: An overview. International Journal of Data Warehousing and Mining 3(3), 1–13 (2007)
22. Tsoumakas, G., Katakis, I., Vlahavas, I.: Mining multi-label data. In: Data Mining and Knowledge Discovery Handbook, pp. 667–685. Springer (2010)
23. Tsoumakas, G., Katakis, I., Vlahavas, I.P.: Random k-labelsets for multilabel classification. IEEE Transactions on Knowledge and Data Engineering 23(7), 1079–1089 (2011)
24. de Waal, P.R., van der Gaag, L.C.: Inference and learning in multi-dimensional Bayesian network classifiers. In: Mellouli, K. (ed.) ECSQARU 2007. LNCS (LNAI), vol. 4724, pp. 501–511. Springer, Heidelberg (2007)
25. Webb, G.I., Boughton, J.R., Wang, Z.: Not so naive Bayes: aggregating one-dependence estimators. Machine Learning 58(1), 5–24 (2005)
26. Zaragoza, J.C., Sucar, L.E., Morales, E.F.: A two-step method to learn multidimensional bayesian network classifiers based on mutual information measures. In: 24th International Florida Artificial Intelligence Research Society Conference, FLAIRS 2011 (2011)
27. Zaragoza, J.H., Sucar, L.E., Morales, E.F., Bielza, C., Larrañaga, P.: Bayesian chain classifiers for multidimensional classification. In: 22nd International Joint Conference on Artificial Intelligence (IJCAI 2011), pp. 2192–2197. AAAI Press (2011)

From Information to Evidence
in a Bayesian Network

Ali Ben Mrad[1,2], Véronique Delcroix[1],
Sylvain Piechowiak[1], and Philip Leicester[3]

[1] University Lille Nord de France, LAMIH UMR 8201, F-59313 Valenciennes, France
{Ali.BenMrad,Veronique.Delcroix}@univ-valenciennes.fr
[2] Université de Sfax, ENIS, CES Lab, 3038, Sfax, Tunisie
[3] Centre for Renewable Energy Systems Technology, Loughborough University

Abstract. Evidence in a Bayesian network comes from information based on the observation of one or more variables. A review of the terminology leads to the assessment that two main types of non-deterministic evidence have been defined, namely likelihood evidence and probabilistic evidence but the distinction between fixed probabilistic evidence and not fixed probabilistic evidence is not clear, and neither terminology nor concepts have been clearly defined. In particular, the term *soft evidence* is confusing. The article presents definitions and concepts related to the use of non-deterministic evidence in Bayesian networks, in terms of specification and propagation. Several examples help to understand how an initial piece of information can be specified as a finding in a Bayesian network.

Keywords: Non deterministic evidence, uncertain evidence, fixed probabilistic finding, likelihood finding, soft evidence, virtual evidence.

1 Introduction

Bayesian networks are probabilistic graphical models that provide a powerful way to embed knowledge and to update one's beliefs about target variables given new information about other variables. In a Bayesian network, prior knowledge is represented by a probability distribution P on the set of variables which define the system, whereas updated beliefs are represented by the posterior probability distribution $P(. \mid obs)$ where obs represents new information. Evidence is the starting point of inference methods and refers to new information in a Bayesian network, also called observations or findings. A finding on a variable commonly refers to an instantiation of the variable. This can be represented by a vector with one element equal to 1, corresponding to the state the variable is in, and all remaining elements equal to zero. This type of evidence is usually referred to as hard evidence or deterministic evidence [40] though other terms are sometimes used. This paper focuses on other types of evidence that cannot be represented by such vectors: non-deterministic evidence (or uncertain evidence). The objective of this paper is to clarify the terms of non deterministic evidence and

L.C. van der Gaag and A.J. Feelders (Eds.): PGM 2014, LNAI 8754, pp. 33–48, 2014.

their underlying concepts. Three types of non deterministic evidence are distinguished, namely likelihood evidence, fixed probabilistic evidence and not-fixed probabilistic evidence. A review of the terminology makes clear two main points: (1) the inconsistent use of terms for non-deterministic evidence is problematic, in particular the term *soft evidence*, the misuse of which causes real confusion. (2) two of the three types of non-deterministic evidence are not clearly distinguished. In order to counter this, we propose the use of the three following terms for non-deterministic evidence in a Bayesian network:

Likelihood evidence: defined by Pearl [43] also called virtual evidence;
Not-fixed probabilistic evidence: concept referred to in Jeffrey's rule [12];
Fixed-probabilistic evidence: concept of soft evidence [48].

The terms "likelihood" and "probabilistic" capture the way in which evidence is specified. Likelihood evidence is represented as a likelihood ratio whereas probabilistic evidence is specified by a probability distribution of one or more variables. The adjectives "fixed" and "not fixed" describe the posterior probability distribution after further evidence is obtained. The term "probabilistic evidence" is also inspired from two contributions: (1) in the context of Bayesian network revision, the input is named a *probabilistic constraint* [45]. The difference between such input and probabilistic evidence concerns only the life span of the input, leading to either inference (evidence propagation) or Bayesian network revision[1]. (2) in the Bayesian network engine BayesiaLab [26], the user may input several kinds of non-deterministic evidence that are referred to by the way they are specified : *probability distribution (fixed or not)*, and likelihood ratio. The rest of the paper is organized as follows. Section 2 is devoted to the notation and basics of Bayesian networks. Section 3 presents definitions and concepts related to likelihood evidence in Bayesian networks. Section 4 deals with both not-fixed probabilistic evidence and fixed probabilistic evidence. In this section, we present shared and specific properties of these two types of probabilistic evidence, together with several illustrative examples. Section 5 proposes a review of terminology about evidence in a Bayesian network that explains the proposed vocabulary.

2   Notations and Basics of Bayesian Network

Bayesian networks [5], [15], [24], [29], [43] are a class of probabilistic graphical models. A *Bayesian network* (BN) is a couple (G, P), where $G = (\mathbf{X}, \mathbf{E})$ is a directed acyclic graph (DAG) with nodes $\mathbf{X} = \{X_1, ..., X_n\}$ and directed edges E which represent conditional dependencies between nodes. The joint probability distribution for $X = \{X_1, ..., X_n\}$ is given by the chain rule: $P(X_1, ..., X_n) = \prod_{i=1}^{n} P(X_i \mid pa(X_i))$ where $pa(X_i)$ represents the parents of X_i as defined by the presence of directed edge from a parent node to X_i. In the following, capital letters are used to represent random variables, and lower-case letters represent their values. Bold capital letters correspond to sets of variables. In this paper

---

[1] This paper deals with evidence propagation and not with model revision.

we consider only discrete random variables. Here are some more notations used in the rest of the paper: $X \in \mathbf{X}$ denotes a BN node having its states (or values) in $\mathcal{D}_X = \{x_1, ..., x_m\}$, $P(\mathbf{X})$ denotes $P(X_1, ..., X_n)$, $P(x)$ denotes $P(X = x)$, $P(X)$ is the probability distribution $(P(X = x_1), ..., P(X = x_m))$. Once the BN is defined, algorithms of inference are used to propagate through the network some information based on the observation of one or more variables. The definitions of this paper concern different types of findings and naturally extend to evidence, which is a set of findings on variables of a Bayesian network.

Definition 1 (Deterministic finding or hard finding). *A deterministic finding e on a variable X in a Bayesian network with values in $\mathcal{D}_X$ is defined by an observation vector of size $m = |\mathcal{D}_X|$ containing a single 1, at the position corresponding to a state $x \in \mathcal{D}_X$ and 0 in the positions of the other states. This finding represents the instantiation of X to the value x and it is characterized by $P(X = x \mid e) = 1$.*

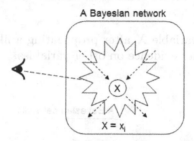

Fig. 1. Hard finding on X. The observation of the variable X is clear and direct.

In the literature, a deterministic finding on a variable is also called a hard finding, an observation, or a specific, regular or positive finding. Despite the variety of terminology in the literature, the definition of hard evidence is clear and presents no practical problems. We use terms deterministic finding (evidence) and hard finding (evidence) as synonyms since there is no ambiguity in the definition (instantiation of the variable, see definition1).

3 Likelihood Evidence: Definition and Characteristics

Likelihood evidence can be characterized as follows : "the uncertainty bears on the meaning of the input; the existence of the input itself is uncertain, due to, for instance, the unreliability of the source that supplies inputs" [19].

3.1 Definition, Properties and Examples

The definition of a likelihood finding in a Bayesian network is followed by two properties describing how likelihood evidence interacts with beliefs before and after its propagation.

Definition 2 (Likelihood finding or virtual finding). *A likelihood finding on a variable X of a Bayesian network is specified by a likelihood ratio $L(X) = (L(X = x_1) : \ldots : L(X = x_m))$, where the $L(X = x_i)$ are quantities relative to each other representing the relative strength of confidence toward the observed event Obs given X is in one state or another. The likelihood ratio $L(X)$ is defined by*

$$L(X) = (P(Obs \mid x_1) : \ldots : P(Obs \mid x_n))$$

where $P(Obs \mid x_i)$ is interpreted as the probability of the observed event given X is in the state x_i.

A likelihood finding specified by a vector of zeros and ones is sometimes referred to as a *negative finding*, meaning that the states of X corresponding to the zeros are impossible.

Property 1. Likelihood evidence is specified "without a prior", as a consequence, propagating likelihood evidence takes into account the beliefs on the variable before the evidence.

Property 2. Belief on a variable X after propagating a likelihood finding on X can be modified by further evidence on other variables.

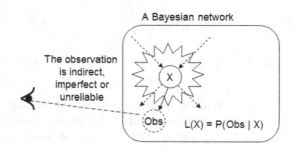

Fig. 2. Likelihood finding on X. The variable X is observed with uncertainty (e.g. via an imperfect sensor). The evidence is specified by the likelihood of the observed value *Obs* (which is not a variable of the model) with respect to each value of X.

Example 1 (Likelihood finding on a manuscript character). A Bayesian network includes a variable X representing a letter of the alphabet. The set of values of X is the set of letters of the alphabet. A piece of uncertain information on X is received from a character recognition technology. The input of this system is an image of a manuscript character and the output is a vector of similarity between the image of the manuscript character and each letter of the alphabet. Let o represents the observed image. Consider a case where, due to lack of clarity, o can be recognized as either the letter 'v' or 'u'. The character recognition technology provides the indices such that $P(o \mid X = \text{'v'}) = 0.8$, $P(o \mid X = \text{'u'}) = 0.4$, which means that there is twice as much chance of observing o if the writer had wanted to write the letter 'v' than if she had wanted to write the letter 'u'. Such a finding on X is a likelihood finding on X.

3.2 Propagation of Likelihood Evidence with Pearl's Method of Virtual Evidence

Virtual evidence refers to Pearl's (1988) idea of interpreting likelihood evidence on a set of events as hard evidence on some virtual events that only depend on this set of events. Pearl's method to propagate likelihood finding on X extends the given BN by adding a virtual node which is a child of X. The uncertain information on X is replaced by deterministic finding on the added node which is propagated using a classical inference algorithm in the augmented BN. The uncertainty of the information is specified in the conditional probability table of the added virtual node.

4 Fixed Probabilistic Evidence and Not-fixed Probabilistic Evidence

Probabilistic evidence can be characterized as follows : "the input is a partial description of a probability measure; the uncertainty is part of the input and is taken as a constraint on the final cognitive state. The input is then a correction to the prior cognitive state" [19]. This type of evidence has been studied in [1–3], [16].

4.1 Definition and Shared Properties

Definition 3 (Probabilistic finding). *A probabilistic finding on a variable $X \in \mathbf{X}$ is specified by a local probability distribution $R(X)$ that defines a constraint on the belief on X after this information has been propagated; it describes the state of beliefs on the variable X "all things considered". A probabilistic finding is* fixed *(or not) when the distribution $R(X)$ can not be (or can be) modified by the propagation of other findings.*

The next two properties describe how probabilistic evidence interacts with beliefs before and after its propagation. They concern both fixed and not-fixed probabilistic evidence.

Property 3. A probabilistic finding $R(X)$ on a variable X of a Bayesian network replaces any prior belief or knowledge on X. As a consequence, the prior $P(X)$ is not used in the propagation of $R(X)$, and any previous finding on X is lost.

Property 4. A probabilistic finding $R(X)$ on a variable X is preserved when updating belief. The beliefs after considering the probabilistic finding on X is represented by a probability distribution Q on $\mathbf{X}$ such that $Q(X) = R(X)$.

Probabilistic evidence behaves as hard evidence in that the specified evidence remains unchanged after its propagation. There are two main differences between probabilistic evidence and likelihood evidence. Firstly the specification: for probabilistic evidence the distribution is specified "all things considered" whereas for likelihood evidence the likelihood ratio is "without a prior". Secondly the propagation: while probabilistic evidence remains unchanged by updating the observed variables, likelihood evidence has to be combined with a previous belief in order to update the belief on the observed variable(s). The difference between fixed and not-fixed probabilistic evidence is only visible when several pieces of evidence are received and propagated (see Properties 5 and 6).

4.2 Not-fixed Probabilistic Evidence: Specific Properties and Examples

Not-fixed probabilistic evidence is the type of input considered in Jeffrey's rule [23] (see section 4.3).

Property 5. A not-fixed probabilistic finding on X can be modified by further evidence on any variable in the model, including likelihood evidence on X. As a consequence, the propagation of several not-fixed probabilistic finding does not commute.

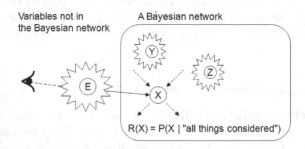

Fig. 3. Not-fixed probabilistic finding on X. The new information may include the observation of a variable outside of the BN model. The evidence on X is specified by a probability distribution $R(X)$ that includes the influence of all observations.

Example 2 (not-fixed probabilistic finding: example 1 continued). Consider for the variable X in example 1 that the language of the word from which the character comes, and the frequency of letters in that language are known. If the BN does not contain the variable "language of the text", this information can be applied as probabilistic evidence for the variable "manuscripts character": $R(X) = (R(X = \text{'a'}), R(X = \text{'b'}), \ldots, R(X = \text{'z'}))$. For example, given that the word containing the letter comes from English, where the frequency of the letter 'v' is 1%, provides the probabilistic evidence $R(X)$ with $R(X = \text{'v'}) = 0.01$. This has to replace the prior belief on the event $X = \text{'v'}$. However, since that information about X could be improved by further evidence such as the likelihood finding on X described in Example 1, it is a not-fixed probabilistic finding on X.

4.3   Propagating Not-fixed Probabilistic Evidence: Jeffrey Rule and Conversion in Likelihood Evidence

Propagating a probabilistic finding on $X \in \mathbf{X}$ requires a revision of the probability distribution P on $\mathbf{X}$ by a local probability distribution $R(X)$. The difficulty arises since Bayes' rule cannot be applied because $R(X)$ is not an event [44]. A probabilistic finding $R(X)$ requires a reconsideration of the joint probability distribution P because it replaces the existing prior on the variable X. The propagation of probabilistic evidence requires the replacement of the initial probability distribution P by another probability distribution Q that reflects the beliefs on the variables of the model after accepting the probabilistic evidence. In a Bayesian network, this replacement is not definitive: it lasts as long as the specific observed case holds, whereas the Bayesian network applies to a larger population. Jeffrey's rule [23] specifies evidence using posterior probabilities. This approach is known as "probability kinematics"; it is based on the requirements that (1) the posterior distribution $Q(X)$ is unchanged: $Q(X) = R(X)$, (2) the conditional probability distribution of other variables given X remains invariant under the observation: $Q(\mathbf{X} \setminus \{X\} \mid X) = P(\mathbf{X} \setminus \{X\} \mid X)$. In other words, even if P and Q disagree on X, they agree on the consequences of X on other variables. However, Jeffrey's rule cannot be directly applied to BNs, because their operations are defined on full joint probability distributions. Another way to propagate a not-fixed probabilistic finding is to convert it to a likelihood finding: $R(X)$ can be converted to a likelihood ratio

$$L(X) = \frac{R(x_1)}{P(x_1)} : \ldots : \frac{R(x_n)}{P(x_n)}. \tag{1}$$

Propagating the likelihood finding $L(X)$ with Pearl's method provides the same results as propagating $R(X)$ by Jeffrey's rule [12]. Thus, the posterior probability of X after propagating $L(X)$ by Pearl's method, is equal to $R(X)$. In case of several probabilistic findings, the method of converting probabilistic findings into likelihood findings does not preserve probabilistic findings. A simple example can be found in [12], [44]. It therefore holds that the inclusion of several pieces of probabilistic evidence with Jeffrey rule does not commute, meaning that those methods are relevant only for not-fixed probabilistic evidence. In case the commutation is required after propagating several pieces of probabilistic evidence, it means that the user has to consider fixed probabilistic evidence instead of not-fixed probabilistic evidence. This second type of probabilistic evidence is presented in the next section.

4.4   Fixed Probabilistic Evidence: Specific Properties and Examples

Fixed probabilistic evidence corresponds to the concept described as *soft evidence* in [4], [33], [42], [44], [48].

Property 6. A fixed probabilistic finding on X is not modified by further evidence on any other variables of the model, and a further finding on X is not possible, unless it overwrites the current evidence. Any kind of evidence received after fixed probabilistic evidence makes it necessary to re-propagate previous fixed probabilistic evidence together with the new evidence, in order to keep the former probabilistic evidence fixed. As a consequence, the propagation of several fixed probabilistic findings commutes: the result of propagation is independent of the order in which fixed probabilistic findings are received.

We present below three types of examples of the use of probabilistic evidence. The first example concerns the propagation of an observation on a continuous variable in a discrete Bayesian network (Figure 4 and Example 3). The second example is linked to the observation of a specific subset of cases (Figure 5 and Example 4). The third example is about using probabilistic evidence for distributed Bayesian networks (Figure 6).

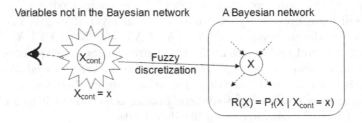

Fig. 4. Fixed probabilistic finding on X coming from the fuzzy discretization of a continuous variable. The model includes a discretized variable X; The information comes from the observation of the continuous variable. The evidence on X is specified by a probability distribution $R(X)$ obtained from a fuzzy discretization of the continuous variable.

Example 3 (Fixed probabilistic finding coming from the fuzzy discretization of a continuous variable). Let A be a variable representing the age of a person with value in $\mathcal{D}_A = \{child, adult, senior\}$. The information "she is 15 years old" can be specified by a fixed probabilistic finding on A, by adding the knowledge of the fuzzy membership function linking the real age to the values of $\mathcal{D}_A$. The result is for example: $R(A) = (0.7, 0.3, 0)$. This is a fixed probabilistic finding since it cannot be modified by other information.

A second type of example of fixed probabilistic evidence is given by the precise observation of a variable on a sub-population (Figure 5 and Example 4).

The use of probabilistic findings to propagate observations on a sub-population allow not to re-learning the parameters of the BN.

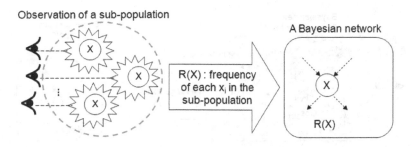

Fig. 5. Fixed probabilistic finding on X coming from the observation of a sub-population. The information is given by the observation of a variable X in a sub-population. The evidence on X is specified by a probability distribution $R(X)$ representing the frequency of each event on X in the observed sub-population.

Example 4 (Fixed probabilistic evidence coming from the observation of a sub-population). Consider the BN Asia [36] which contains eight binary nodes, among which there is a (root) node *Smoking* and a (leaf) node *Dyspnea*. Instead of having findings about a single person, consider findings coming from the data of a particular sub-population, such as the workers in a given factory. Observing that half of them have dyspnea and a tenth of them smoke constitutes fixed probabilistic findings on these variables such that: $R(Dyspnea) = (0.5, 0.5)$ and $R(Smoking) = (0.1, 0.9)$. When no more details are available, these findings cannot be considered as a single piece of probabilistic evidence on the two variables $R(Dyspnea, Smoking)$ as proposed in [48].

The third type of example of fixed probabilistic finding comes from the concept of Agent Encapsulated Bayesian Networks (AEBN) [8]. The information on a variable X is supplied by an expert on X, and her judgement on X cannot be improved by other evidence on any other variables of the model (see Figure 6). In an agent organization based on AEBN, fixed probabilistic evidence is considered by a receiver agent in order to take into account the information sent by a publisher agent, which is considered as an expert of its output variables.

4.5   Propagating Fixed Probabilistic Evidence

Propagating a single probabilistic finding can be done by its transformation into a likelihood finding as in Equation 1. This section concerns the propagation of several fixed probabilistic findings. Several algorithms were recently proposed to propagate fixed probabilistic evidence in a BN. Most of them are based on the *Iterative Proportional Fitting Procedure* (IPFP) algorithm [17], [25], [32] which is an iterative method of revising a probability distribution to respect a set of given probability constraints in the form of posterior marginal probability distributions over a subset of variables. However, the IPFP works on full joint distributions, and thus is not directly applicable to belief update in Bayesian networks. It could be applied only for very small Bayesian networks. Here is a list of several

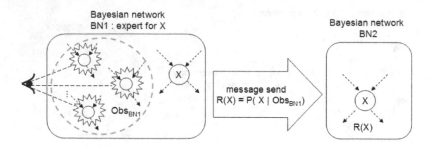

Fig. 6. Fixed probabilistic finding on X (AEBN) The variable X is shared by two Bayesian network. The model BN1 is an expert of the variable X and evaluates its value thanks to a set of observations Obs_{BN1} on its local variables. The model BN2 updates its local model with the received information specified as a fixed probabilistic finding.

algorithms for the propagating of fixed probabilistic evidence. These algorithms solutions are suitable for Bayesian networks.This synthesis is inspired by [44, 45]. The definition of probabilistic evidence (Definition 3) has been extended in [48]

Big-Clique [8], [48]	recoding of the junction tree algorithm based on IPFP
Soft updating [47]	proposed in the context of hybrid Bayesian network
Lazy big clique [33]	modifies the updating algorithm [37]
BN-IPFP (1 and 2)	combine IPFP and the conversion to likelihood evidence; and thus are independent of the inference algorithm
BN-IPFP-1 [44]	converts separately each probabilistic evidence
BN-IPFP-2 [44]	uses IPFP to calculate $R(X_1, \ldots, X_p)$ from the set $R(X_1), \ldots, R(X_p)$ then propagates by transforming to uncertain evidence
SMOOTH [44], [50]	propagation of inconsistent probabilistic evidence (proof of convergence for two pieces of evidence.)

in order to consider information about one or several variables of the model, specified by different forms of probabilistic description: (a) a joint probability distribution, (b) a conditional probability distribution, (c) probability assignments on arbitrary events, (d) probability assignments on arbitrary logic formulae. This extended notion of probabilistic evidence can be handled for evidence updating by the introduction of an observation node [48]. This technique allows the reformulation of extended probabilistic evidence into probabilistic evidence on a single new observation node. In case of evidence on a set of observation variables $E_1, \ldots, E_p$ that are independent in the BN, the propagation can be done by considering a single piece of evidence $R(E_1, \ldots, E_p) = R(E_1) \times \ldots \times R(E_p)$.

Table 1. Properties of the different types of evidence in a Bayesian network

	Hard evidence	Likelihood evidence	not fixed probabilistic evidence	fixed probabilistic evidence
The evidence defines a constraint on the posterior probability distribution. It is specified "all things considered".	Yes	No	Yes	Yes
Several observations can be combined on the **same** variable	No	Yes	Yes	No
Posterior probability distribution is modified by latter observations on **other** variables	No	Yes	Yes	No
The propagation requires the prior probability distribution	No	Yes	No	No
The propagation of several findings commutes	Yes	Yes	No	Yes

Table 2. Features of Bayesian network software about evidence updating

	Hard evidence	Likelihood evidence	not fixed probabilistic evidence	fixed probabilistic evidence
Elvira [20], Analytica[21], Samlam[14]	X			
AgenaRisk 6.1 [7], Bayes Server 5.5 [46] BNT 1.0.7 [39], Genie 2.0 [18], Hugin 8.0 [35], Infer.NET 2.5 [38], gRain 1.2-3 [22]	X	X		
Netica 5.12 [41]	X	X	X	
BayesiaLab 5.3 [26]	X	X	X	X

4.6 Synthesis

Table 1 summarizes the properties of the different types of evidence described above. Table 2 shows the features about the updating of non deterministic evidence among some available Bayesian network engine. The updating of fixed probabilistic evidence is available in BayesiaLab which is the only available BN engine to provide that feature[2]. This is done by using a specific, unpublished likelihood matching algorithm, that exhibits no dependency on the order in which the probabilistic evidence is entered. The user defines a probability distribution on the target node, and BayesiaLab calculates the likelihood distribution which permits this distribution. When the distribution is fixed, the likelihood distribution is updated

---

[2] In may 2014.

dynamically according to other observations. BayesiaLab allows a user-defined average to be specified for a target node.

5   Review of Terminology in Related Works

This section presents a review of terminology about evidence in Bayesian network, both in the literature and in Bayesian network software.

5.1   About the Definition of Findings

A generalization of the definition of findings is proposed in [24] in which a finding on a variable X is an m-dimensional table of zeros and ones. In the case where the observation vector has several ones, it has to be understood as a *negative finding*, which means that some states of the variable are impossible. For example, the vector $(1, 0, 0, \ldots, 0, 1)$ represents the statement that X can only be in state x_1 or in state x_m, and cannot be in any other state. We do not follow this proposition since a negative finding is indeed a likelihood finding.

5.2   About the Use of the Terms *Soft Evidence*

A review of the literature over the past ten years shows that the term *soft evidence* is sometimes used to refer to likelihood evidence [6], [10], [11], [13], [28], [31]. However, the use of the term soft evidence is abandoned in [12]. In another paper [9], soft evidence means that a variable X does not take a specific value x, that is to say $X \neq x$, which is usually called a negative finding. Most available software for non-deterministic evidence propagation in BN engines implement Pearl's method of virtual evidence. Table 3 illustrates the different terms used in Bayesian network software for non-deterministic evidence. It shows that the term "soft evidence" is misused in several Bayesian network engines since the term is used to refer to likelihood evidence. Moreover, the term "soft evidence" is not used by the only two Bayesian network engines that permit the propagation of probabilistic evidence. This review of terminology in literature and Bayesian network software leads to the assessment that the term *soft evidence* is confusing since it is used inconsistently. Fixed probabilistic evidence is named *soft evidence* by Valtorta and his co-authors [27], [33, 34], [42], [44, 45] [48] and also [47], [30].

5.3   About the Distinction between Fixed and Not-fixed Probabilistic Evidence, and the Question of Commutation

Apart from two exceptions [4], [26] the concepts of fixed and not-fixed probabilistic evidence are never distinguished. Specifically, several published articles clearly distinguish between likelihood evidence and probabilistic evidence —regrouped under the term *uncertain evidence*— without identifying the distinction between fixed and not-fixed probabilistic evidence [12], [15], [42], [44], [49]. In [12], the authors provide an interesting discussion about the specification of likelihood

Table 3. Terms used in Bayesian network software for non deterministic evidence (may 2014)

	Likelihood evidence	Probabilistic evidence
BayesiaLab 5.3 [26]	likelihoods	probability distribution (fixed or not-fixed)
Netica 5.12 [41]	likelihood finding	calibration
Hugin 8.0 [35] gRain 1.2-3 [22]	likelihood evidence	
Genie 2.0 [18] Infer.NET 2.5 [38]	virtual evidence	
BNT 1.0.7 [39] Bayes Server 5.5 [46] AgenaRisk 6.1 [7]	"soft evidence"	

evidence and not-fixed probabilistic evidence, but no terminology is proposed for this latter concept. In [44], which focuses on fixed probabilistic evidence, the analysis of [12] is referred to, but it doesn't lead to a clear distinction of both concepts of probabilistic evidence. This lack of distinction between fixed and not fixed probabilistic evidence gave rise to debates between authors, about the question of commutation: The question "Should, and do, iterated belief revisions commute?" [12] concerns the case of propagation in a Bayesian network with several pieces of evidence, of which some are not deterministic. Iterated revisions commute with Pearl's method, but not with Jeffrey's rule. Some authors claim that several pieces of evidence carrying the "All things considered" interpretation must not be commutative [12]. Others argue that probabilistic evidence is a true observation of the distributions of some events, and as such, they should all be preserved in the updated "posterior" distribution [44]. In the first case, the arriving information is susceptible to improvement by further evidence: it is not-fixed probabilistic evidence whereas in the second case the arriving information has to behaves as hard evidence and can not be influenced by any other information: it is fixed probabilistic evidence.

6   Conclusion

This paper contributes to the clarification and standardization of the definitions and properties of different types of evidence in a Bayesian network. We have set out the definitions and properties of deterministic and non-deterministic findings in a BN. Three kinds of non-deterministic evidence are distinguished. (1) Likelihood evidence is unreliable, imprecise, or vague evidence: it is evidence *with* uncertainty; it is specified by a likelihood ratio and propagated by Pearl's method of virtual evidence. (2) Probabilistic evidence expresses a constraint on the state of some variables after this information has been propagated in the BN. A probabilistic finding on X is specified by a probability distribution $R(X)$ that is given "all things considered", meaning that is replaces any former belief and knowledge on X. (2a) A not-fixed probabilistic finding can be propagated by converting it to likelihood evidence. It can be later modified by further evidence on any node of the BN, including X itself. (2b) Fixed probabilistic evidence is also known as soft evidence. It cannot be altered by any further information on variables in the model. Thus the propagation of several pieces of fixed probabilistic evidence is commutative. Fixed probabilistic evidence has to be propagated by specific algorithms such as BN-IPFP. We provide several examples where the application and propagation of probabilistic evidence in a BN is of interest. The problem addressed in this paper is one of belief updating and not the problem of model revision, which leads to a change of the probability distribution (or even the graph) of the BN. In the case of probabilistic evidence, information is probabilistic in nature and leads to the replacement (temporarily) of the prior distribution extant in the defined model. Currently, many BN engines allow the propagation of likelihood evidence, even though the terminology is not yet standardized. However, very few of them possess the ability to propagate fixed or not-fixed probabilistic evidence. This would be of a great interest to the BN application user community. At least three new features are required in most BN engines. The first is a requirement to enter an observation on a subgroup of instances as a single finding, as a fixed probabilistic finding. Second, software should facilitate the implementation of an agent organization based on AEBN. This requires a subscriber agent to integrate the information from a corresponding publisher agent by propagating probabilistic evidence in its local BN. The third feature of interest would allow the propagation of observations of continuous variable using a fuzzy discretization with degrees of membership of two or more course intervals. Another feature of interest is to allow to combination of several not deterministic findings on the same variable.

References

1. Ben Mrad, A., Delcroix, V., Maalej, M.A., Piechowiak, S., Abid, M.: Uncertain evidence in Bayesian networks: Presentation and comparison on a simple example. In: Greco, S., Bouchon-Meunier, B., Coletti, G., Fedrizzi, M., Matarazzo, B., Yager, R.R. (eds.) IPMU 2012, Part III. CCIS, vol. 299, pp. 39–48. Springer, Heidelberg (2012)
2. Ben Mrad, A., Maalej, M.A., Delcroix, V., Piechowiak, S., Abid, M.: Fuzzy evidence in Bayesian networks. In: Proc. of Soft Computing and Pattern Recognition, Dalian, China (2011)
3. Ben Mrad, A., Delcroix, V., Piechowiak, S., Maalej, M.A., Abid, M.: Understanding soft evidence as probabilistic evidence: Illustration with several use cases. In: 2013 5th International Conference on Modeling, Simulation and Applied Optimization (ICMSAO), pp. 1–6 (2013)
4. Benferhat, S., Tabia, K.: Inference in possibilistic network classifiers under uncertain observations. Annals of Mathematics and Artificial Intelligence 64(2-3), 269–309 (2012)
5. Bessière, P., Mazer, E., Ahuactzin, J.M., Mekhnacha, K.: Bayesian Programming. CRC Press (2013)

6. Bilmes, J.: On soft evidence in Bayesian networks. Tech. Rep. UWEETR-2004-00016, Department of Electrical Engineering University of Washington, Seattle (2004)
7. Birtles, N., Fenton, N., Neil, M., Tranham, E.: Agenarisk, http://www.agenarisk.com/
8. Bloemeke, M.: Agent encapsulated Bayesian networks. Ph.d. thesis, Department of Computer Science, University of South Carolina (1998)
9. Butz, C.J., Fang, F.: Incorporating evidence in Bayesian networks with the select operator. In: Kégl, B., Lee, H.-H. (eds.) AI 2005. LNCS (LNAI), vol. 3501, pp. 297–301. Springer, Heidelberg (2005)
10. Chan, H.: Sensitivity Analysis of Probabilistic Graphical Models. Ph.d. thesis, University of California, Los Angeles (2005)
11. Chan, H., Darwiche, A.: Sensitivity analysis in Bayesian networks: From single to multiple parameters. In: UAI, pp. 67–75 (2004)
12. Chan, H., Darwiche, A.: On the revision of probabilistic beliefs using uncertain evidence. Artificial Intelligence 163(1), 67–90 (2005)
13. D'Ambrosio, B., Takikawa, M., Upper, D.: Representation for dynamic situation modeling. Technical report, Information Extraction and Transport, Inc. (2000)
14. Darwiche, A.: Samlam, http://reasoning.cs.ucla.edu/samiam
15. Darwiche, A.: Modeling and Reasoning with Bayesian Networks. Cambridge University Press (2009)
16. Delcroix, V., Sedki, K., Lepoutre, F.X.: A Bayesian network for recurrent multi-criteria and multi-attribute decision problems: Choosing a manual wheelchair. Expert Systems with Applications 40(7), 2541–2551 (2013)
17. Deming, W.E., Stephan, F.F.: On a least square adjustment of a sampled frequency table when the expected marginal totals are known. Annals of Mathematical Statistics 11, 427–444 (1940)
18. Druzdzel, M.J.: Genie smile, http://genie.sis.pitt.edu
19. Dubois, D., Moral, S., Prade, H.: Belief change rules in ordinal and numerical uncertainty theories. In: Gabbay, D., Smets, P. (eds.) Belief Change, (D. Dubois, H. Prade, eds.). Handbook of Defeasible Reasoning and Uncertainty Management Systems, vol. 3, pp. 311–392. Kluwer Academic Publishers, Dordrecht (1998)
20. Elvira: Elvira project, http://leo.ugr.es/elvira/
21. Henrion, M.: Analytica, lumina decision systems, http://www.lumina.com/
22. Højsgaard, S.: gRain, http://people.math.aau.dk/~sorenh/software/gR/
23. Jeffrey, R.C.: The Logic of Decision, 2nd edn. 246 pages. University of Chicago Press (1990)
24. Jensen, F.V., Nielsen, T.D.: Bayesian Networks and Decision Graphs, 2nd edn. Springer Publishing Company, Incorporated (2007)
25. Jiroušek, R.: Solution of the marginal problem and decomposable distributions. Kybernetika 27, 403–412 (1991)
26. Jouffe, L., Munteanu, P.: Bayesialab, http://www.bayesia.com
27. Kim, Y.G., Valtorta, M., Vomlel, J.: A prototypical system for soft evidential update. Applied Intelligence 21(1), 81–97 (2004)
28. Kjaerulff, U., Madsen, A.: Bayesian Networks and Influence Diagrams: A Guide to Construction and Analysis. Information science and statistics, 2nd edn., vol. 22. Springer (2013)
29. Korb, K., Nicholson, A.: Bayesian Artificial Intelligence, 2nd edn. Chapman and Hall (2010)
30. Koski, T., Noble, J.: Bayesian Networks: An Introduction. Wiley Series in Probability and Statistics. Wiley (2009)

31. Krieg, M.L.: A tutorial on Bayesian belief networks. Tech. Rep. DSTO-TN-0403, Surveillance Systems Division, Electronics and Surveillance Research Laboratory, Defense science and technology organisation, Edinburgh, South Australia, Australia (2001)

32. Kruithof, R.: Telefoonverkeersrekening. De Ingenieur 52, 15–25 (1937)

33. Langevin, S., Valtorta, M.: Performance evaluation of algorithms for soft evidential update in Bayesian networks: First results. In: Greco, S., Lukasiewicz, T. (eds.) SUM 2008. LNCS (LNAI), vol. 5291, pp. 284–297. Springer, Heidelberg (2008)

34. Langevin, S., Valtorta, M., Bloemeke, M.: Agent-encapsulated Bayesian networks and the rumor problem. In: AAMAS 2010 Proceedings of the 9th International Conference on Autonomous Agents and Multiagent Systems, vol. 1, pp. 1553–1554 (2010)

35. Lauritzen, S.L.: Hugin, http://www.hugin.com

36. Lauritzen, S.L., Spiegelhalter, D.J.: Local computations with probabilities on graphical structures and their application to expert systems. Journal of the Royal Statistical Society, Series B 50, 157–224 (1988)

37. Madsen, A.L., Jensen, F.V.: Lazy propagation: A junction tree inference algorithm based on lazy evaluation. Artificial Intelligence 113(1-2), 203–245 (1999)

38. Minka, T., Winn, J.: Infer.net, http://research.microsoft.com/en-us/um/cambridge/projects/infernet/default.aspx

39. Murphy, K.: Bayesian network toolbox (bnt), http://www.cs.ubc.ca/~murphyk/Software/BNT/bnt.html

40. Naïm, P., Wuillemin, P.H., Leray, P., Pourret, O., Becker, A.: Réseaux bayésiens. Eyrolles, 3 edn. (2007)

41. Norsys: Netica application (1998), http://www.norsys.com

42. Pan, R., Peng, Y., Ding, Z.: Belief update in Bayesian networks using uncertain evidence. In: ICTAI, pp. 441–444 (2006)

43. Pearl, J.: Probabilistic reasoning in intelligent systems: networks of plausible inference. Morgan Kaufmann Publishers Inc., San Francisco (1988)

44. Peng, Y., Zhang, S., Pan, R.: Bayesian network reasoning with uncertain evidences. International Journal of Uncertainty, Fuzziness and Knowledge-Based Systems 18(5), 539–564 (2010)

45. Peng, Y., Ding, Z., Zhang, S., Pan, R.: Bayesian network revision with probabilistic constraints. International Journal of Uncertainty, Fuzziness and Knowledge-Based Systems 20(3), 317–337 (2012)

46. Sandiford, J.: Bayes server, http://www.bayesserver.com/

47. Tomaso, E.D., Baldwin, J.F.: An approach to hybrid probabilistic models. International Journal of Approximate Reasoning 47(2), 202–218 (2008)

48. Valtorta, M., Kim, Y.G., Vomlel, J.: Soft evidential update for probabilistic multiagent systems. International Journal of Approximate Reasoning 29(1), 71–106 (2002)

49. Vomlel, J.: Probabilistic reasoning with uncertain evidence. Neural Network World, International Journal on Neural and Mass-Parallel Computing and Information Systems 14(5), 453–465 (2004)

50. Zhang, S., Peng, Y., Wang, X.: An Efficient Method for Probabilistic Knowledge Integration. In: Proceedings of the 20th IEEE International Conference on Tools with Artificial Intelligence. IEEE Computer Society (November 2008)

Learning Gated Bayesian Networks
for Algorithmic Trading

Marcus Bendtsen and Jose M. Peña

Department of Computer and Information Science, Linköping University, Sweden
{marcus.bendtsen,jose.m.pena}@liu.se

Abstract. Gated Bayesian networks (GBNs) are a recently introduced extension of Bayesian networks that aims to model dynamical systems consisting of several distinct phases. In this paper, we present an algorithm for semi-automatic learning of GBNs. We use the algorithm to learn GBNs that output buy and sell decisions for use in algorithmic trading systems. We show how using the learnt GBNs can substantially lower risks towards invested capital, while at the same time generating similar or better rewards, compared to the benchmark investment strategy *buy-and-hold*.

Keywords: Probabilistic graphical models, Bayesian networks, algorithmic trading, decision support.

1 Introduction

Algorithmic trading can be viewed as a process of actively deciding when to own assets and when to not own assets, so as to get better risk and reward on invested capital compared to holding on to the assets over a long period of time. At the other end of the spectrum is the *buy-and-hold* strategy, where one owns assets continuously over a period of time without making any decisions of selling or buying during the period. This paper introduces a novel algorithm that can be used to learn *gated Bayesian networks* (GBNs, described in Sect. 2) for use as part of an algorithmic trading system. We also present a real-world application of this learning algorithm that shows that, compared to the benchmark *buy-and-hold* strategy, the expected risks and rewards are improved upon.

1.1 Algorithmic Trading

An algorithmic trading system contains several components, some which may be automated by a computer, and others that may be manually executed [1,2]. A schematic overview of the components of a general algorithmic trading system is shown in Fig. 1.

The type of data used at the research stage varies greatly, e.g. net profit, potential prospects, sentiment analysis, analysis of previous trades, or *technical analysis*, which will be the focus in the enclosed application (described in

L.C. van der Gaag and A.J. Feelders (Eds.): PGM 2014, LNAI 8754, pp. 49–64, 2014.

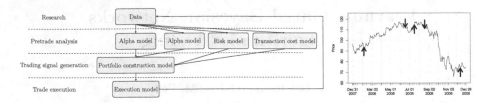

Fig. 1. Components of an algorithmic trading system   **Fig. 2.** Buy and sell signals

Sect. 5.1). The analysis of the data is split up into *alpha, risk* and *transaction cost* models. The alpha models are responsible for outputting decisions for buying and selling assets based on the data they are given. These decisions are known as buy and sell *signals*, examples of which are depicted in Fig. 2 (an arrow pointing upwards is a buy signal and a downwards facing arrow is a sell signal, the signals are drawn on top of the historical asset price). If followed, these buy and sell signals give rise to certain *risk* and *reward* on the initial investment (which will be described further in Sect. 3). The contribution of this paper is concerned with the use of GBNs as alpha models.

The risk and transaction cost models should be seen as strategies for managing risk and transaction costs in a system that has many alpha models. The output from these three types of models (alpha, risk and transaction) are in their turn the input to the *portfolio construction model* in the trading signal generation stage. Here the output of the previous components are combined to decide which signals to actually execute in order to create a portfolio that is based on a combination of alpha models. The final stage is the actual *execution* of the trading signals, which must be done in a manner that does not affect the price of the asset that is being bought. Although all components are important, we will not be addressing all of them in this paper, our focus will be on the alpha models.

The rest of the paper is organised as follows. We begin by giving a brief introduction to Bayesian networks (BN) and GBNs in Sect. 2, this is important to understand how GBNs can be used as alpha models. We continue by explaining how we can evaluate the performance of an alpha model in Sect. 3. We use this method of evaluation in Sect. 4, where we introduce a novel algorithm that can be used to learn GBNs. In Sect. 5 we make use of the learning algorithm in a real-world application, where we show how learnt GBNs can be used as alpha models. Finally we end the paper in Sect. 6 with a few words regarding our conclusions and future work.

2   Gated Bayesian Networks

BNs can be interpreted as models of causality at the macroscopic level, where unmodeled causes add uncertainty. Cause and effect are modelled using random variables that are placed in a directed acyclic graph (DAG). The causal model implies some probabilistic independencies among the variables, that can easily

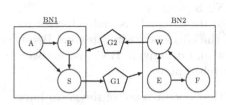
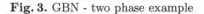

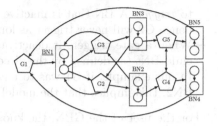

Fig. 3. GBN - two phase example   **Fig. 4.** GBN - multi phase example

be read off the DAG. Therefore, a BN does not only represent a causal model but also an independence model. The qualitative model can be quantified by specifying certain marginal and conditional probability distributions so as to specify a joint probability distribution, which can later be used to answer queries regarding posterior probabilities. The independencies represented in the DAG make it possible to compute these posteriors efficiently. See [3,4] for more details.

Although BNs have successfully been used in many domains, our interest is to model the process of buying and selling assets, and in this particular situation the BN model is not enough. This is the main motivation for us introducing GBNs [5], and the current paper builds upon this previous contribution. When trying to model the process of buying and selling assets, we want to model the continuous flow between looking for opportunities to buy and opportunities to sell. The model can be seen as being in one of two distinct phases: either looking for an opportunity to buy into the market, or an opportunity to sell and exit the market. These two phases can be very different and the random variables included in the BNs modelling them are not necessarily the same.

Switching between phases is done using so called *gates*. These gates are encoded with predefined logical expressions regarding posterior probabilities of random variables in the BNs. This allows activation and deactivation of BNs based on posterior probabilities. A GBN that uses two different BNs (BN1 and BN2) is shown in Fig. 3, follows does a brief explanation of this GBN and how it is used (for the full details we refer the reader to our previous publication [5]).

- A GBN consists of BNs and gates. BNs can be *active* or *inactive*. The label of BN1 is underlined, indicating that it is active at the initial state of the GBN. The BNs supply posterior probabilities to the gates via so called *trigger nodes*. The node S is a trigger node for gate G1 and W is a trigger node for G2. A gate can utilise more than one trigger node.
- Each gate is encoded with a predefined logical expression regarding its trigger nodes' posterior probability of a certain state, e.g. G1 may be encoded with $p(S = s1|e) > 0.7$. This expression is known as the *trigger logic* for gate G1.
- When evidence is supplied to the GBN an evidence handling algorithm updates posterior probabilities and checks if any of the logical statements in the gates are satisfied. If the trigger logic is satisfied for a gate it is said

to *trigger*. A BN that is inactive never supplies any posterior probabilities, hence G2 will never trigger as long as BN2 is inactive.

– When a gate triggers it deactivates all of its parent BNs and activates its child BNs (as defined by the direction of the edges between gates and BNs). In our example, if G1 was to trigger it would deactivate BN1 and activate BN2, this implies that the model has switched phase.

For the user of the GBN, the knowledge that one or more of the gates have triggered (i.e. the state of the GBN has changed), may be useful in a decision making process. As an example, if the GBN was used as an alpha model, knowing that the GBN has found a buying opportunity and has started modelling selling opportunities would suggest that a buy signal has been generated. Looking again at Fig. 2, each buy and sell signal is generated by the fact that the GBN switched back and forth between its states.

GBNs can consist of many phases, and the phases themselves can have sub-phases that are made up of several BNs. An example of a GBN with multiple phases is shown in Fig. 4.

3   Evaluating Alpha Models

Regression models can be evaluated by how well they minimise some error function or by their log predictive scores. For classification, the accuracy and precision of a model may be of greatest interest. Alpha models may rely on regression and classification, but can not be evaluated as either. An alpha model's performance needs to be based on its generated signals over a period of time, and the performance must be measured by the *risk* and *reward* of the model. This is known as *backtesting*.

3.1   Backtesting

The process of evaluating an alpha model on historic data is known as *backtesting*, and its penultimate goal is to produce metrics that describe the behaviour of a specific alpha model. These metrics can then be used for comparison between alpha models [6,7]. A time range, price data for assets traded and a set of signals are used as input. The backtester steps through the time range and executes signals that are associated with the current time (using the supplied price data) and computes an *equity curve* (which will be explained in Sect. 3.2). From the equity curve it is possible to compute metrics of risk and reward. To simulate potential transaction costs, often referred to as *commission*, every trade executed is usually charged a small percentage of the total value (0.06% is a common commission charge used in the enclosed application).

Alpha models are backtested separately from the other components of the algorithmic trading system, as the backtesting results are input to the other components. Therefore, we execute every signal from an alpha model during backtesting, whereas in a full algorithmic trading system we would have a portfolio construction model that would combine several alpha models and decide how to build a portfolio from their signals.

3.2   Alpha Model Metrics

What constitutes risk and reward is not necessarily the same for every investor, and investors may have their own personal preferences. However, there are a few that are common and often taken into consideration [7]. Here we will introduce a few metrics that we will use to evaluate the performance of our alpha models.

Equity Curve. Although not a metric on its own, the *equity curve* needs to be defined in order to define the following metrics. The equity curve represents the total value of a trading account at a given point in time. If a daily timescale is used, then it is created by plotting the value of the trading account day by day. If no assets are bought, then the equity curve will be flat at the same level as the initial investment. If assets are bought that increase in value, then the equity curve will rise. If the assets are sold at this higher value then the equity curve will again go flat at this new level. The equity curve summarises the value of the trading account including cash holdings and the value of all assets. We will use $\mathcal{E}_t$ to reference the value of the equity curve at point t.

Metric 1 (Return). The *return* of an investment is defined as the percentage difference between two points on the equity curve. If the timescale of the equity curve is daily, then $r_t = (\mathcal{E}_t - \mathcal{E}_{t-1})/|\mathcal{E}_{t-1}|$ would be the *daily* return between day t and $t-1$. We will use $\bar{r}$ and σ_r to denote the mean and standard deviation of a set of returns.

Metric 2 (Sharpe Ratio). One of the most well known metrics used is the so called *Sharpe ratio*. Named after its inventor Nobel laureate William F. Sharpe, this ratio is defined as: $(\bar{r} - \text{risk free rate})/\sigma_r$. The *risk free rate* is usually set to be a "safe" investment such as government bonds or the current interest rate, but is also sometimes removed from the equation [7]. The intuition behind the Sharpe ratio is that one would prefer a model that gives consistent returns (returns around the mean), rather than one that fluctuates. This is important since investors tend to trade *on margin* (borrowing money to take larger positions), and it is then more important to get consistent returns than returns that sometimes are large and sometimes small. This is why the Sharpe ratio is used as a reward metric rather than the return.

Drawdown Risks. Using the Sharpe ratio as a metric will ensure that the alpha models are evaluated on their risk adjusted return, however, there are other important alpha model behaviours that need to be measured. A family of these, that we will call *drawdown risks*, are presented here (please see Fig. 5 for examples of an equity curve and these metrics).

Metric 3 (Maximum Drawdown (MDD)). The percentage between the highest peak and the lowest trough of the equity curve during backtesting. The peak must come before the trough in time. The MDD is important from both a technical and psychological regard. It can be seen as a measure of the maximum risk that the investment will live through. Investors that use their existing

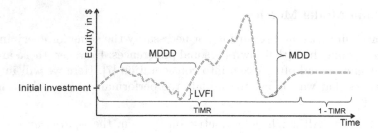

Fig. 5. Example of equity curve with drawdown risks

investments that have gained in value as safety for new investments may be put in a situation where they are forced to sell everything. Other risk management models may automatically sell investments that are loosing value sharply. For the individual who is not actively trading but rather placing money in a fund, the MDD is psychologically frustrating to the point where the individual may withdraw their investment at a loss in fear of loosing more money.

Metric 4 (Maximum Drawdown Duration (MDDD)). The longest it has taken from one peak of the equity curve to recover to the same value as that peak. Despite its unfortunate name it is not the duration of the MDD, but rather then longest drawdown period. There is an old adage amongst investors to "cut your losses early". In essence it means that it is better to take a loss straight away than to sit on an investments for months or years, hoping that it will come back to positive returns. During this time one could have re-invested the money elsewhere, rather then breaking-even much later (or taking a larger loss much later). Models that have long periods of drawdown lock resources when they could have been used better elsewhere.

Metric 5 (Lowest Value From Investment (LVFI)). The percentage between the initial investment and the lowest value of the equity curve. This is one of the most important metrics, and has a significant impact on technical and psychological factors. For investors trading on margin, a high LVFI will cause the lender to ask the investor for more safety capital (known as a *margin call*). This can be potentially devastating, as the investor may not have the capital required, and is then forced to sell the investment. The investor will then never enjoy the return the model could have produced. Individuals who are not investing actively, but instead are choosing between funds that invest in their place, should be aware of the LVFI as it is the worst case scenario if they need to retract their equity prematurely.

Metric 6 (Time In Market Ratio (TIMR)). The percentage of time of the investment period where the alpha model owned assets. This metric may seem odd to place within the same family as the other drawdown risks, however it fits naturally in this space. We can assume that the days the alpha model does not own any assets the drawdown risk is zero. If we are not invested, then there is no risk of loss. In fact, we can further assume that our equity is growing according to the risk free rate, as it is not bound in assets.

3.3 Buy and Hold Benchmark

At first the buy-and-hold strategy may seem naïve, however it has been shown that deciding when to own and not own assets requires consistent high accuracy of predictions in order to gain higher returns than the buy-and-hold strategy [8]. The buy-and-hold strategy has become a standard benchmark, not only because of the required accuracy, but also because it requires very little effort to execute (no complex computations and/or experts needed).

Now consider the family of metrics that we called drawdown risks. The buy-and-hold strategy holds assets over the entire backtesting period and so will be subject to the full force of these metrics. For instance, as an asset will be held throughout the period, the lowest point of the assets value will coincide with LVFI. Furthermore, the initial investment will always be locked in assets, not being able to make money from risk free rates during periods of decreasing value. These are serious risks of using buy-and-hold that algorithmic trading could improve upon, which we will explore in the enclosed application in Sect. 5.

4 Learning Algorithm

The algorithm proposed in this paper for semi-automatically learning the structure of a GBN consists of two parts: a *GBN template* and a novel combination of k-fold cross-validation and time series cross-validation (time series cross-validation is sometimes known as *rolling origin* [9] or *walk forward analysis* [6]).

4.1 Gated Bayesian Network Templates

A GBN template is a representation of the modelled phases, including the possible transitions between them. The template defines where BNs and gates can be placed. For each slot where a BN can be placed, there is a library of BNs to choose from, similarly so for gates (gates differ in their trigger logic, e.g. the thresholds may vary between them). A template with four slots and corresponding libraries is depicted in Fig. 6.

The only restrictions on the BNs and gates are the ones they place on each other, e.g. if the gates placed in G2 expect a particular node as trigger node, then the BNs placed in BN2 must contain that node. Except for these restrictions, the BNs and gates can be configured freely.

Selecting a BN and a gate from the libraries for each slot in the template creates a GBN (e.g. Fig. 3), we call this a *candidate* of the template. We use $\mathcal{C}_i$ to denote GBN candidate i of a GBN template.

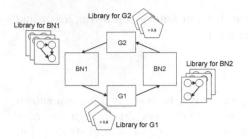

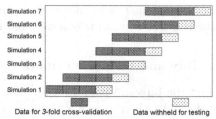

Fig. 6. GBN template

Fig. 7. Combined k-fold cross-validation and time series cross-validation using $n = 10$ blocks and $k = 3$ folds

4.2   K-Fold and Time Series Cross-Validation

In this section we will discuss how a GBN candidate, from a GBN template, is evaluated. In the domain of algorithmic trading it is natural for the test data to always come after the training data in temporal order. This is to ensure that we are not training on data that may carry information not available during the time the testing data was produced.

Splitting the Data. A data set $\mathcal{D}$ of consecutive evidence sets, e.g. observations over all or some of the random variables in the GBN, is divided into n equally sized blocks $(\mathcal{D}_1, ..., \mathcal{D}_n)$ such that they are mutually exclusive and exhaustive. Each block contains consecutive evidence sets and all evidence sets in block $\mathcal{D}_i$ come before all evidence sets in $\mathcal{D}_j$ for all $i < j$.

Depending on the amount of available data, k is chosen as the number of blocks used for training. These blocks will be used for k-fold cross-validation. Starting from index 1, blocks $1, .., k$ are used for training and $k + 1$ for testing, thus ensuring that the evidence sets in the testing data occurs after the training data (as in time series cross-validation). The procedure is then repeated starting from index 2 (i.e. blocks $2, .., k + 1$ are used for training and $k + 2$ for testing). By doing this we create repeated simulations, moving the testing data one block forward each time. An illustration of this procedure when $n = 10$ and $k = 3$ is show in Fig. 7.

4.3   Algorithm

Let $\mathcal{J}(\mathcal{C}_i, \mathcal{D}_j, \{\mathcal{D}\}_l^m)$ be the score, e.g. Sharpe ratio, LVFI, etc., for GBN candidate i when block j has been used for testing and the blocks $\mathcal{D}_l, ..., \mathcal{D}_m$ have been used for training. The algorithm then works in three steps (with an optional fourth):

1. For each simulation t, where (as discussed previously) $\mathcal{D}_{t+k}$ is the testing data and $\mathcal{D}_t...\mathcal{D}_{t+k-1}$ is the training data, find C^t that satisfies (1). This corresponds to finding the GBN candidate with the maximum mean score of

the k evaluations performed during k-fold cross-validation over the training data. This is done by taking into consideration every possible candidate, thus exhausting the search space.

$$\mathcal{C}^t = \arg\max_{\mathcal{C}_i} \; \frac{1}{k} \Sigma_{j=t}^{t+k-1} \mathcal{J}(\mathcal{C}_i, \mathcal{D}_j, \{\mathcal{D}\}_t^{t+k-1} \backslash \mathcal{D}_j) \; . \tag{1}$$

2. For each $\mathcal{C}^t$ calculate its score $\rho_{\mathcal{J}}^t$ on the testing set with respect to the scoring function $\mathcal{J}$ according to (2). This corresponds to training the found GBN candidate from (1) using all training data and evaluating the performance on the data withheld for testing.

$$\rho_{\mathcal{J}}^t = \mathcal{J}(\mathcal{C}^t, \mathcal{D}_{t+k}, \{\mathcal{D}\}_t^{t+k-1}) \; . \tag{2}$$

3. The expected performance $\bar{\rho}_{\mathcal{J}}$ of the algorithm, with respect to the score function $\mathcal{J}$, is then given by the average of the scores $\rho_{\mathcal{J}}^t$ (3).

$$\bar{\rho}_{\mathcal{J}} = \frac{1}{n-k} \Sigma_{t=1}^{n-k} \rho_{\mathcal{J}}^t \; . \tag{3}$$

4. (Optional) If the objective is to find the candidate to be used on future unseen data (i.e. block $\mathcal{D}_{n+1}$) then (1) is used once more to find $\mathcal{C}^{n-k+1}$. This candidate can then be used on $\mathcal{D}_{n+1}$ with an expected performance $\bar{\rho}_{\mathcal{J}}$.

It may seem unorthodox to use k-fold cross-validation with unordered data in step 1, i.e. the testing block may come before some training blocks. However, this step is only used to select a model to evaluate in step 2. The data used in step 2 is always ordered, i.e. the test block is always the immediate successor of the training blocks. This does give a fair evaluated performance on the testing data. Step 1 attempts to use the training data to its maximum, allowing for each candidate to be assessed on several data sets before selecting the one to move forward with.

In the description of the algorithm, one scoring function $\mathcal{J}$ has been used both for choosing a candidate in (1) and for evaluating the expected performance of the algorithm in (2). In Sect. 3.2 we have defined several metrics used to evaluate alpha models. The scoring function $\mathcal{J}$ used in (1) could internally use many of these metrics to come up with one score to compare the different candidates with. However, it is natural in the current setting to expose the actual values of these metrics during step 2, and so several scoring functions $\mathcal{J}$ can be used to get a vector of scores $[\rho_{\mathcal{J}_1}^t, ..., \rho_{\mathcal{J}_m}^t]$ and use a vector of means as the performance of the algorithm $[\bar{\rho}_{\mathcal{J}_1}, ..., \bar{\rho}_{\mathcal{J}_m}]$.

5 Application

In this section we show a real-world application where our proposed algorithm has been used to learn GBNs for use as alpha models, using backtesting to evaluate their performance. Following the discussion in Sect. 3.3, the aim is to generate buy and sell signals such that the drawdown risks defined in Sect. 3.2 are mitigated as compared to the buy-and-hold strategy, while at the same time maintaining similar or better rewards.

5.1 Methodology

The variables used in the BNs of our GBNs are all based on so called *technical analysis*. One of the major tenets in technical analysis is that the movement of the price of an asset repeats itself in recognisable patterns. *Indicators* are computations of price and volume that support the identification and confirmation of patterns used for forecasting. Many classical indicators exists, such as the *moving average* (MA), which is the average price over time, and the *relative strength index* (RSI) which compares the size of recent gains to the size of recent losses. Technical analysis is a topic that is being actively developed and researched [10]. In this application we will be using three indicators: the MA, the RSI and the relative difference between two MAs (MADIFF). Please see [11] for the full definition and calculations of these indicators.

GBN Template. A template with one BN per phase was created (see Fig. 6), along with eight BNs per BN slot (see Fig. 8) and four gates per gate slot, giving a total of 1024 candidates. The eight BNs used for BN1 are identical to those used in BN2, however the gates' trigger logic are different. The trigger logic for G1 asks for the posterior probability of a good buying opportunity (i.e. a predicted positive future climate) while the trigger logic for G2 asks for the posterior probability of a good selling opportunity (i.e. a predicted negative future climate).

The random variables in the BNs are discretizations of technical analysis indicators (RSI, MA and MADIFF) and their corresponding first and second order 1 and 5 day backward finite differences ($\nabla_1^1, \nabla_5^1, \nabla_1^2$ and ∇_5^2) which approximate the first and second order derivatives. The parameters used in the indicators are standard 14 day period for RSI [11] (written as RSI(14)), 20 day period for MA, representing 20 trading days in a month (written as MA(20)), and 5 and 20 day period for MADIFF, where 5 days represent the 5 trading days in a week (written as MADIFF(5,20) and calculated as $\frac{MA(5)-MA(20)}{MA(20)}$). We also consider the previous indicators but with an offset of 5 days in the past and 5 days into the future. The random variables that are offset into the future represent the future economical climate, one of which was involved in the trigger logic of the gates. The true values for these future random variables were naturally not part of the testing data sets. The BNs used for the BN slots are presented in Fig. 8. The node named S was used as the trigger node for all gates. The GBN generated trading signals as it transitioned between its two phases (as described in Sect. 2).

Data Sets. A set of actively traded stock shares where chosen for the evaluation of our learning algorithm: Apple Inc. (AAPL), Amazon.com Inc. (AMZN),

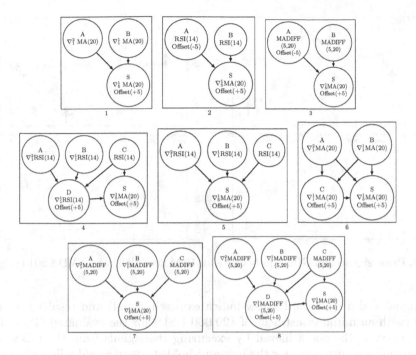

Fig. 8. BNs in GBN template libraries

International Business Machines Corporation (IBM), Microsoft Corporation (MSFT), NVIDIA Corporation (NVDA), General Electric Company (GE), Red Hat Inc. (RHT). The daily adjusted closing prices for these stocks between 2003-01-01 and 2012-12-31 were downloaded from Yahoo! Finance^{TM}. This gave a total of 10 years of price data for each stock, where each year was allocated to a block, and thus n = 10. For the learning algorithm, k was chosen to be 3, giving 7 simulations from which to calculate $[\bar{\rho}_{\mathcal{J}_1}, ..., \bar{\rho}_{\mathcal{J}_m}]$. The split of the data is visualised in Fig. 7.

Scoring Functions. The signals generated were backtested (see Sect. 3) in order to calculate the relevant metrics. For step 1 in the learning algorithm we used the Sharpe ratio. This choice was made as it combines both risk and reward into one score, which can then easily be compared between candidates. For step 2 we used the return and drawdown risks as described in Sect. 3.2 to create a score vector. For the buy-and-hold strategy the same metrics as in step 2 were calculated for the 7 simulations.

5.2 Results and Discussion

To visualise the backtesting that was done for each simulation, Fig. 9 gives two examples of stock price, generated signals (an upward arrow indicates a

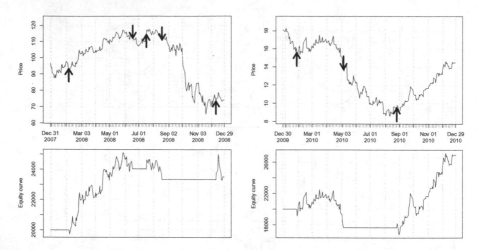

Fig. 9. Price, signals and GBN equity curve for IBM 2008 (left) and NVDA 2010 (right)

buy signal and a downward arrow indicates a sell signal) and resulting equity curve (with an initial investment of $20,000 USD) for the evaluated GBN. The equity curve is the one achieved by executing the signals from the GBN, the corresponding equity curve for the buy-and-hold strategy would follow the stock price exactly, as it holds shares over the entire period. The GBN equity curve grows in a more monotonic fashion, which is desirable because this decreases the drawdown risks, while at the same time generating positive returns. The buy-and-hold strategy would have made a loss in both these examples, because the final price is lower than the initial one, furthermore it would have displayed bad intermediate behaviour, reflected by the high drawdown risk values that would have been incurred. These are declining years for the shares, however the GBN does its best to get as much value as possible from the price movements.

Table 1 presents the score vectors from the learning algorithm versus the score vector of the buy-and-hold strategy over the 7 simulations. Rows named *min*, *max, mean* and *sd* (standard deviation) are based on (2) where *mean* corresponds to (3). As each block used by the learning algorithm had an approximate length of one year, the Sharpe ratio that is given by dividing the mean with the sd of the return column is a yearly Sharpe ratio based on seven years (where the risk-free rate has not been included). All values are ratios except for MDDD which is measured in number of days.

Analysis of Results. The Sharpe ratio is our measure of reward, premiered above the raw return for reasons discussed in Sect. 3.2. Our first concern is to ensure that the learnt GBNs are producing similar or better Sharpe ratios than the buy-and-hold strategy over the testing period. As can be seen in Table 1, this is the case except for NVDA and RHT. As we have previously discussed, it requires a very high accuracy of predictions to consistently beat the Sharpe ratio of buy-and-hold.

Table 1. Metric values comparing GBN with buy-and-hold

		GBN					Buy-and-hold				
		Return	MDD	MDDD	LVFI	TIMR	Return	MDD	MDDD	LVFI	TIMR
AAPL	min	-0.000	0.122	35.0	0.001	0.520	-0.559	0.129	28.0	0.001	1.000
	max	0.851	0.331	164.0	0.184	0.944	1.419	0.589	250.0	0.590	1.000
	mean	0.347	**0.206**	**95.0**	**0.055**	**0.723**	0.489	0.274	116.0	0.162	1.000
	sd	0.334	0.076	50.3	0.061	0.155	0.707	0.168	82.7	0.218	0.000
	Sharpe	**1.041**					0.691				
AMZN	min	-0.204	0.134	56.0	0.042	0.510	-0.466	0.157	45.0	0.001	1.000
	max	0.784	0.306	142.0	0.245	0.768	1.740	0.634	249.0	0.620	1.000
	mean	0.271	**0.218**	**101.7**	**0.109**	**0.630**	0.463	0.317	118.6	0.215	1.000
	sd	0.374	0.060	32.8	0.088	0.091	0.829	0.171	89.9	0.234	0.000
	Sharpe	**0.725**					0.559				
IBM	min	-0.022	0.062	53.0	0.013	0.494	-0.210	0.088	28.0	0.001	1.000
	max	0.238	0.176	176.0	0.121	0.944	0.596	0.442	190.0	0.302	1.000
	mean	0.125	**0.117**	112.3	**0.044**	**0.712**	0.170	0.174	**106.4**	0.086	1.000
	sd	0.094	0.042	45.4	0.042	0.173	0.245	0.120	59.7	0.101	0.000
	Sharpe	**1.332**					0.694				
MSFT	min	-0.256	0.099	88.0	0.001	0.365	-0.457	0.141	74.0	0.001	1.000
	max	0.381	0.305	197.0	0.279	0.741	0.659	0.498	250.0	0.498	1.000
	mean	0.056	**0.168**	**143.3**	**0.114**	**0.557**	0.069	0.249	168.6	0.200	1.000
	sd	0.202	0.068	41.9	0.091	0.156	0.338	0.119	67.8	0.155	0.000
	Sharpe	**0.278**					0.204				
NVDA	min	-0.420	0.182	64.0	0.032	0.241	-0.765	0.253	67.0	0.077	1.000
	max	0.342	0.541	227.0	0.467	0.700	1.230	0.820	249.0	0.821	1.000
	mean	0.016	**0.284**	148.1	**0.209**	**0.516**	0.202	0.458	172.3	0.311	1.000
	sd	0.284	0.120	62.1	0.140	0.171	0.701	0.195	76.6	0.268	0.000
	Sharpe	0.057					**0.288**				
GE	min	-0.302	0.049	60.0	0.015	0.404	-0.555	0.089	69.0	0.001	1.000
	max	0.461	0.465	217.0	0.438	0.570	0.222	0.657	217.00	0.642	1.000
	mean	0.040	**0.169**	**144.3**	**0.119**	**0.488**	-0.001	0.314	157.0	0.236	1.000
	sd	0.235	0.142	69.7	0.150	0.062	0.257	0.228	53.7	0.257	0.000
	Sharpe	**0.169**					-0.005				
RHT	min	-0.222	0.096	87.0	0.001	0.433	-0.370	0.143	40.0	0.001	1.000
	max	0.436	0.428	221.0	0.348	0.784	1.341	0.676	221.0	0.617	1.000
	mean	0.038	**0.254**	156.9	**0.136**	**0.613**	0.201	0.338	**133.0**	0.243	1.000
	sd	0.259	0.103	45.6	0.123	0.136	0.579	0.197	61.6	0.234	0.000
	Sharpe	0.145					**0.346**				

From this we can conclude that the GBNs do not get beaten consistently by the buy-and-hold strategy when considering the annual Sharpe ratio, even though it is considered a nearly optimal strategy. Furthermore, we should take into consideration TIMR. The GBNs are spending less time in the market, reducing risk to equity and possibly increasing equity value from risk free investments. Potential gain in equity from risk free rates have not been added to the Sharpe ratios presented in the table. Considering that the learnt GBNs consistently spend considerably less time in the market (shown by the low TIMR values), this could give a significant boost to the Sharpe ratios. An example of this can be seen for NVDA where the Sharpe ratio for GBN is lower than for buy-and-hold, but the GBN only spent on average 51.6% of the time in the market, risk free investments could potentially drive the Sharpe ratio for the GBN above that of the buy-and-hold strategy.

Turning our attention to the drawdown risks (as defined in Sec. 3.2) we first consider the MDD and MDDD. The difference of the MDD values are substantial, the MDD mean and sd are consistently smaller for the GBNs than they are for the buy-and-hold strategy. This signals that the equity we gain from our investments are at less risk when using the GBNs compared to the buy-and-hold strategy. For MDDD the means differ in favour of either approach, we would not prefer one in front of the other given only this metric.

The LVFI is a major threat to equity (see Sect. 3.2), and it is the one metric where buy-and-hold severely under-performs. Considering the max values we note that for NVDA the buy-and-hold strategy wiped out 82.1% of the equity at worst, while the GBNs did 46.7% at worst for NVDA. Considering the LVFI mean and sd for all stocks we note that they are consistently almost half for the GBNs compared to the buy-and-hold strategy. LVFI is important because it is the risk of the initial investment, loosing much of the initial investment may lead to premature withdrawal of funds and/or force liquidation by margin-calls.

All in all, the results above clearly indicate that GBNs are competitive with buy-and-hold in terms of Sharpe ratio, whereas they induce a more desirable behaviour in terms of MDD, LVFI and TIMR.

Post-Analysis. One of the benefits of using BNs is that we can get transparency as to why a particular signal was generated. Our aim here was to look at the non-discretized values of the variables at the time a signal was generated. We combined the signals from all simulations (regardless of which stock was traded) and then grouped the signals by which BN generated them and if they were buy or sell signals. We then did pair-wise combinations of the variables in each BN to create scatter plots with values of the variables along the axes and also added an approximated density using the frequency of signals. These scatter plots show when GBNs are generating signals. Examples of these plots for the BNs that generated the most signals are given in Fig. 10 (using 7 from Fig. 8) and Fig. 11 (using 5 from Fig. 8).

In Fig. 10 the BN is used to look for buying opportunities. In the first plot we see that most signals are generated when both $\nabla_1^1 MADIFF(5, 20)$ and $\nabla_1^2 MADIFF(5, 20)$ are positive, indicating that the difference between the two MAs is growing and increasing in speed, but not so positive so as to making it impossible to benefit from the trend. The second two plots in Fig. 10 plot $\nabla_1^2 MADIFF(5, 20)$ against $MADIFF(5, 20)$ and the $\nabla_1^1 MADIFF(5, 20)$ against $MADIFF(5, 20)$. Both these confirm what we knew about the first and second order difference, but also indicate that $MADIFF(5, 20)$ should be positive (so the short period MA should be above the long period MA). From a technical analysis perspective this kind of pattern is common, it indicates a trend change, as the shorter MA is moving above and away from the longer MA. It is noteworthy to mention that we have not set any priors on the BNs that would indicate that these are the kind of patterns we are interested in, so our learning algorithm is able to re-discover these human-like commonly used patterns. An example of selling signals is presented in Fig. 11, here we are using RSI which is bounded between 0 and 100. When RSI moves up towards 100 it indicates that the buying pressure is increasing, and should drive prices higher, the opposite is true when RSI moves towards 0. The first plot indicates that most selling signals are generated when $\nabla_1^1 RSI(14)$ is close to zero or negative (i.e. RSI has

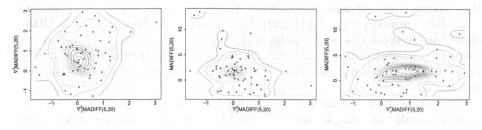

Fig. 10. Buy decisions using 7 from Fig. 8

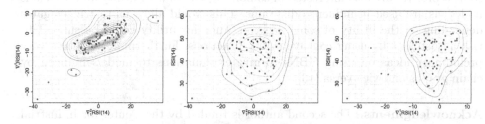

Fig. 11. Sell decisions using 5 from Fig. 8

started to decrease) and $\nabla_1^2 RSI(14)$ is bounded around ± 10. The two other plots in Fig. 11 represent $\nabla_1^2 RSI(14)$ against $RSI(14)$ and $\nabla_1^1 RSI(14)$ against $RSI(14)$. These last two figures confirm our findings in the first figure, but also indicates that the $RSI(14)$ should be below 50 (but not too much below 50 so as to miss the selling opportunity). This seems reasonable from a technical analysis perspective, as RSI goes below 50 and decreases, the selling pressure increases, indicating that the price will go lower, and so a selling signal is generated. We reemphasise that we did not set any prior in the BNs that would suggest that these are the type of signals we should be looking for.

Modelling Different Phases. The GBNs used herein do not attempt to switch between BNs to adapt to changes in non-stationary data, but instead they change when the decision being made has changed (i.e. first we are looking to buy, then to sell). GBNs in general model different phases in a process, albeit that data may be non-stationary in some or all phases. This makes GBNs different from formalisms that switch between models to adjust for shifts in non-stationary data, where it is common to take into consideration the performance of the models as part of the weighting or switching probability [12].

6   Conclusions and Future Work

We have introduced a novel algorithm for semi-automatic learning of GBNs, and shown how this algorithm can be used to learn GBNs for use as alpha models in algorithmic trading systems. We have applied the algorithm to evaluate the expected performance of the learnt GBNs as alpha models compared to using the benchmark buy-and-hold strategy. The results show that learnt GBNs consistently reduce risk with similar or better rewards and do so while at the same time staying out of the market for considerable amounts of time, during these non-invested days the equity is at zero risk and can gain value from risk free assets.

Our future work will include developing the learning algorithm to become more automatic, avoiding having to create a GBN template and rather allow the algorithm to place the phases, BNs and gates in such a way that it optimises some score. We are also interested in combining GBNs with utility and decisions nodes, as are used in influence diagrams. This would allow us to trigger gates depending on the utility of some decision, and this utility could be subject to risk adjustment by using concave utility functions. Furthermore, we have very preliminary ideas on using GBNs to give explanations to models induced by chain graphs and vice versa [13].

Acknowledgments. The second author is funded by the Center for Industrial Information Technology (CENIIT) and a so-called career contract at Linköping University, and by the Swedish Research Council (ref. 2010-4808).

References

1. Treleaven, P., Galas, M., Lalchand, V.: Algorithmic Trading Review. Commun. ACM. 56, 76–85 (2013)
2. Nuti, G., Mirghaemi, M., Treleaven, P., Yingsaeree, C.: Algorithmic Trading. Computer 44, 61–69 (2011)
3. Pearl, J.: Probabilistic Reasoning in Intelligent Systems: Networks of Plausible Inference. Morgan Kaufmann (1988)
4. Jensen, F.V., Nielsen, T.D.: Bayesian Networks and Decision Graphs. Springer (2007)
5. Bendtsen, M., Peña, J.M.: Gated Bayesian Networks. In: 12th Scandinavian Conference on Artificial Intelligence, pp. 35–44. IOS Press (2013)
6. Pardo, R.: The Evaluation and Optimization of Trading Strategies. John Wiley & Sons (2008)
7. Chan, E.P.: Quantitative Trading: How to Build Your Own Algorithmic Trading Business. John Wiley & Sons (2009)
8. Sharpe, W.F.: Likely Gains from Market Timing. Financial Analysts Journal 31, 60–69 (1975)
9. Tashman, L.J.: Out-of-Sample Tests of Forecasting Accuracy: An Analysis and Review. International Journal of Forecasting 11, 437–450 (2000)
10. Journal of Technical Analysis, http://www.mta.org
11. Murphy, J.J.: Technical Analysis of the Financial Markets. New York Institute of Finance (1999)
12. Liehr, S., Pawelzik, K., Kohlmorgen, J., Lemm, S., Müller, K.-R.: Hidden Markov Gating for Prediction of Change Points in Switching Dynamical Systems. In: ESANN, pp. 405–410 (1999)
13. Peña, J.M.: Every LWF and AMP Chain Graph Originates From a Set of Causal Models. ArXiv e-prints (2013)

Local Sensitivity of Bayesian Networks to Multiple Simultaneous Parameter Shifts

Janneke H. Bolt and Silja Renooij

Department of Information and Computing Sciences,
Utrecht University, Utrecht, The Netherlands

Abstract. The robustness of the performance of a Bayesian network to shifts in its parameters can be studied with a sensitivity analysis. For reasons of computational efficiency such an analysis is often limited to studying shifts in only one or two parameters at a time. The concept of sensitivity value, an important notion in sensitivity analysis, captures the effect of local changes in a single parameter. In this paper we generalise this concept to an *n-way sensitivity value* in order to capture the local effect of multiple simultaneous parameters changes. Moreover, we demonstrate that an n-way sensitivity value can be computed efficiently, even for large n. An n-way sensitivity value is direction dependent and its maximum, minimum, and direction of maximal change can be easily determined. The direction of maximal change can, for example, be exploited in network tuning. To this end, we introduce the concept of *sliced sensitivity function* for an n-way sensitivity function restricted to parameter shifts in a fixed direction. We moreover argue that such a function can be computed efficiently.

1 Introduction

The robustness of Bayesian networks to changes in their parameter probabilities can be studied with a sensitivity analysis. To this end, a function which describes the effect of varying one or more parameters on an output probability of interest can be established. From such a sensitivity function, various sensitivity properties can be derived that give insight into the effects of the parameter changes [6].

Most research has focused on one-way sensitivity analyses in which only a single parameter is varied at a time. These one-way analyses, however, do not provide full insight into the effects of multiple simultaneous parameter shifts; to study such effects, an n-way sensitivity analysis is required. To this end, we can establish an n-way sensitivity function. Unfortunately, the computation of multi-dimensional functions is generally expensive. Existing algorithms for n-way sensitivity analysis are only computationally feasible for larger n under certain conditions. For example, the efficient method for computing n-way sensitivity functions by Kjærulff and van der Gaag [8] assumes that the n parameters all belong to the same clique in the network's junction tree representation. Another example is the method introduced by Chan and Darwiche [2] for assessing which

L.C. van der Gaag and A.J. Feelders (Eds.): PGM 2014, LNAI 8754, pp. 65–80, 2014.

parameter shifts will enforce a given constraint with respect to some outcome probability: this method is feasible if all parameters concern the same CPT, and quickly becomes infeasible otherwise.

In this paper we are interested in studying the local effects of multiple simultaneous parameters changes in a Bayesian network. To this end we generalise the concept of *sensitivity value* [9] — well-known in the context of one-way sensitivity analysis — to an *n-way sensitivity value*. The sensitivity value captures the effect of local parameter changes by means of the derivative of the sensitivity function in the point corresponding with the original parameter assessment specified in the Bayesian network. We generalise this concept to multiple dimensions by using a directional derivative of the n-way sensitivity function. Moreover, we prove that computing this directional derivative can be done efficiently, due to the fact that we do not need the n-way sensitivity function: availability of the one-way sensitivity values of the parameters under consideration suffices. The n-way sensitivity value is direction dependent, but its maximum and minimum can be easily determined, together with the corresponding direction of maximal change. The latter information is not only useful for studying the robustness of a Bayesian network, but is also useful in the context of parameter tuning. In parameter tuning, network parameters are changed in order to fulfill constraints with respect to outcome probabilities. Assuming that small perturbations are preferred, we argue that tuning parameters by shifting them in the direction of maximal change will yield a good approximation of the optimal parameter change necessary to meet a given constraint. Moreover, since a fixed vector direction ties together all parameters linearly, we can efficiently establish the effect of such a combined parameter shift. To this end, we introduce the concept of *sliced sensitivity function*.

The remainder of the paper is organised as follows. In Section 2 we introduce our notational conventions, briefly review sensitivity analysis in Bayesian networks and review the mathematical notion of directional derivatives. In Section 3 we define the n-way sensitivity value and its bounds, and in Section 4 we address the question of how to compute an n-way sensitivity value efficiently. In Section 5 we discuss the use of our concepts in the context of parameter tuning and we conclude our paper with a discussion in Section 6.

2   Preliminaries

2.1   Bayesian Networks and Sensitivity Analysis

A Bayesian network compactly represents a joint probability distribution Pr over a set of stochastic variables $\mathbf{A}$ [7]. It combines an acyclic directed graph G, that captures the variables and their dependencies as nodes and arcs respectively, with conditional probability distributions for each variable A_i and its parents $\pi(A_i)$ in the graph, such that

$$\Pr(\mathbf{A}) = \prod_i \Pr(A_i \mid \pi(A_i))$$

Variables are denoted by capital letters, which are boldfaced in case of sets; specific values or instantiations are written in lower case. In examples we restrict ourselves to binary variables, writing a and $\bar{a}$ to denote the two possible instantiations of a variable A. We assume the conditional distributions are specified as tables (CPTs) and use the term *parameter* to refer to a CPT entry. The superscript 'o' is used to indicate that a probability is an original parameter value, or is computed from the network with parameter values as originally specified.

To investigate the effects of inaccuracies in its parameters, a Bayesian network can be subjected to a sensitivity analysis. In a sensitivity analysis, parameters of a network are varied and a probability of interest as a function of the varied parameters is computed.

General n-Way Analysis. In an n-way sensitivity analysis, simultaneous perturbations of multiple parameters are considered. The effect of varying the parameters $x_1, \ldots, x_n$ on a probability of interest $\Pr(y \mid e)$ is captured by a function of the form

$$f_{\Pr(y|e)}(x_1, \ldots, x_n) = \frac{f_{\Pr(y\,e)}(x_1, \ldots, x_n)}{f_{\Pr(e)}(x_1, \ldots, x_n)} = \frac{\sum_{\mathbf{X}_k \in \mathcal{P}(\{x_1, \ldots, x_n\})} c_k \cdot \prod_{x_i \in \mathbf{X}_k} x_i}{\sum_{\mathbf{X}_k \in \mathcal{P}(\{x_1, \ldots, x_n\})} d_k \cdot \prod_{x_i \in \mathbf{X}_k} x_i}$$

where $\mathcal{P}$ denotes the powerset, and c_k and d_k, $k = 0, \ldots, 2^n - 1$, are constants constructed from the non-varied network parameter [1]. A two-way function, for example, takes the following form:

$$f_{\Pr(y|e)}(x_1, \ldots, x_2) = \frac{c_0 + c_1 \cdot x_1 + c_2 \cdot x_2 + c_3 \cdot x_1 \cdot x_2}{d_0 + d_1 \cdot x_1 + d_2 \cdot x_2 + d_3 \cdot x_1 \cdot x_2}$$

The n parameters of an n-way sensitivity function are typically assumed to be independent, that is, parameters from the same CPT must come from different conditional distributions. Upon varying a parameter $x = \Pr(a_i \mid \boldsymbol{\pi})$, all probabilities $\Pr(a_j \mid \boldsymbol{\pi})$, $j \neq i$, pertaining to the same conditional distribution are assumed to co-vary proportionally.

An n-way sensitivity function in general requires the computation of 2^n constants and is thus computationally expensive; an algorithm to this end can be found in [8].

Single CPT Analysis. In the special case where all n parameters are independent parameters from the *same* CPT, the interaction terms in the n-way sensitivity function become zero and the function reduces to the following form [2]:

$$f_{\Pr(y|e)}(x_1, \ldots, x_n) = \frac{c_0 + \sum_i c_i \cdot x_i}{d_0 + \sum_i d_i \cdot x_i}$$

One-Way Analysis. Most research has focused on one-way sensitivity analysis, in which just a single parameter x is varied. In this case the sensitivity function becomes [4]:

$$f_{\Pr(y|e)}(x) = \frac{c_0 + c_1 \cdot x}{d_0 + d_1 \cdot x}$$

The constants of the one-way functions $f_{\Pr(y|e)}(x_i)$ for output probability $\Pr(y \mid e)$ can be established efficiently for *all* network parameters x_i simultaneously from just one inward and two outward propagations in the junction tree representation of the Bayesian network [8].

From the one-way sensitivity function, several sensitivity properties can be established [6]. The most well-known sensitivity property is the *sensitivity value* [9]. This value captures the sensitivity of an outcome probability of interest to small perturbations of the parameter under consideration. The sensitivity value of the one-way sensitivity function $f(x)$ for parameter x with original assessment x^o is defined as the absolute value of the first derivative of the function at $x = x^o$:

$$\left| \frac{df}{dx}(x^o) \right|$$

High sensitivity values indicate that the output probability of interest may change considerably as a result of small parameter changes. The one-way sensitivity function takes the form of a rectangular hyperbola, with its vertex (in which the derivative is $+1$ or -1) marking the transition from low to possibly high sensitivity.

2.2   Directional Derivatives

The sensitivity value is defined in terms of the first derivative of the one-way sensitivity function. For a one-dimensional function $f(x)$ we can refer to *the* derivative at $x = a$, since $\frac{df}{dx}(a)$ is a single value. The multi-dimensional analogue of the derivative is the *directional derivative*. The directional derivative of an n-dimensional function depends on the direction $\mathbf{v}$ and the specific point $\mathbf{x}$ of the function that is considered. To compute the directional derivative of a function $f(\mathbf{x})$ for $\mathbf{x} = (x_1, \ldots, x_n)$, we can use its *gradient* ∇f, that is, the vector of partial derivatives $(\frac{\partial f}{\partial x_1}, \ldots, \frac{\partial f}{\partial x_n})$ of f. The directional derivative at $\mathbf{x} = \mathbf{a}$ in the direction $\mathbf{v}$ now equals the following dot product

$$D_{\mathbf{u}}f(\mathbf{a}) = \nabla f(\mathbf{a}) \bullet \mathbf{u}$$

where unit vector $\mathbf{u}$ is the normalised vector of $\mathbf{v}$, that is, $\mathbf{u}$ is the vector in the direction of $\mathbf{v}$ that has length 1.

Although the directional derivative varies depending on the chosen direction, we can establish bounds on its value. The maximum directional derivative of f at $\mathbf{x} = \mathbf{a}$ is found in the direction of the gradient vector at $\mathbf{a}$, $\nabla f(\mathbf{a})$, and equals the length of the gradient vector at $\mathbf{a}$, $|\nabla f(\mathbf{a})|$. Similarly, the minimum directional derivative occurs in the opposite direction.

Example 1. Suppose we are interested in the directional derivative of $f(x, y) = x^2 + 4 \cdot x \cdot y$ at $(1, 2)$, in the direction $(-2, 1)$. We have that $\nabla f = (2 \cdot x + 4 \cdot y, 4 \cdot x)$ which yields $\nabla f(1, 2) = (10, 4)$. Vector $(-2, 1)$ has length $\sqrt{5}$ and is normalised to $\mathbf{u} = (\frac{-2}{\sqrt{5}}, \frac{1}{\sqrt{5}})$. The requested directional derivative thus equals $D_{\mathbf{u}}f(1, 2) = (10, 4) \bullet (\frac{-2}{\sqrt{5}}, \frac{1}{\sqrt{5}}) = \frac{-16}{\sqrt{5}}$. The maximum directional derivative at $(1, 2)$ occurs in the direction $(10, 4)$ and equals $\sqrt{10^2 + 4^2} \approx 10.77$; the minimum directional derivative at this point equals -10.77 and occurs in the direction $(-10, -4)$.

3   Defining an n-Way Sensitivity Value

The sensitivity value as defined in [9] reflects the local sensitivity of some outcome of interest to a single parameter shift. In this section we generalise the definition of sensitivity value in order to capture the local sensitivity given multiple simultaneous parameter shifts.

A (one-way) sensitivity value is defined in terms of the first derivative of a one-way sensitivity function. In mathematics, the notion of *first derivative* of a function with a single variable generalises to the notion of *directional derivative* for a function with multiple variables. We therefore define an n-way sensitivity value in terms of a directional derivative. In contrast to the definition of the one-way sensitivity value, we will not consider the n-way sensitivity value to be an absolute value. In our opinion, using the absolute value results in loss of useful information concerning the direction of change in the output of interest upon local perturbation of the parameters. For this reason, we also introduce a signed version of the one-way sensitivity value, which equals the sensitivity value prior to taking the absolute value.

Definition 1 (signed sensitivity value). *Let $f(x)$ be a one-way sensitivity function and x^o the original value for parameter x. The signed sensitivity value for $f(x)$, denoted sv^x, equals the first derivative of f at x^o:*

$$sv^x = \frac{df}{dx}(x^o)$$

We now generalise the concept of (signed) sensitivity value to multiple dimensions.

Definition 2 (n-way sensitivity value). *Let $f(\mathbf{x})$ be an n-way sensitivity function and let $\mathbf{x^o}$ be the vector of original parameter settings. Consider a shift of the parameters in the direction $\mathbf{v}$. The n-way sensitivity value for $f(\mathbf{x})$, denoted $sv_{\mathbf{v}}^{\mathbf{x}}$, equals the directional derivative of f at the original parameter assessments $\mathbf{x^o}$ in the direction $\mathbf{v}$:*

$$sv_{\mathbf{v}}^{\mathbf{x}} = D_{\mathbf{u}} f(\mathbf{x^o})$$

where unit vector $\mathbf{u}$ is the normalised vector of $\mathbf{v}$.

Note that sv^x is a special case of $sv_{\mathbf{v}}^{\mathbf{x}}$ for $n = 1$ and $\mathbf{u} = (1)$.

Whereas a single parameter can only be changed to lower or higher values, multiple simultaneous parameters shifts can occur in an infinite number of directions. Hence the dependence on $\mathbf{v}$ in our definition of n-way sensitivity value. Fortunately, the n-way sensitivity values have an upper- and lowerbound.

Definition 3 ($sv_{\max}^{\mathbf{x}}$). *Let $f(\mathbf{x})$ be an n-way sensitivity function and $\mathbf{x^o}$ the vector of original parameter settings. The maximum n-way sensitivity value, denoted $sv_{\max}^{\mathbf{x}}$, equals*

$$sv_{\max}^{\mathbf{x}} = \max_{\mathbf{v}} \; sv_{\mathbf{v}}^{\mathbf{x}} = \max_{\mathbf{u}} \; D_{\mathbf{u}} f(\mathbf{x^o})$$

where unit vector $\mathbf{u}$ is the normalised vector of $\mathbf{v}$.

Since the n-way sensitivity value is defined as a directional derivative, its maximum value in fact equals the length of the gradient vector of f at $\mathbf{x}^o$, that is, $sv_{max}^{\mathbf{x}} = |\nabla f(\mathbf{x}^o)|$; moreover, $sv_{max}^{\mathbf{x}}$ is obtained in the direction $\nabla f(\mathbf{x}^o)$. We can similarly define $sv_{min}^{\mathbf{x}} = -sv_{max}^{\mathbf{x}}$ as the minimum n-way sensitivity value, which occurs in the opposite direction $-\mathbf{1} \bullet \nabla f(\mathbf{x}^o)$.

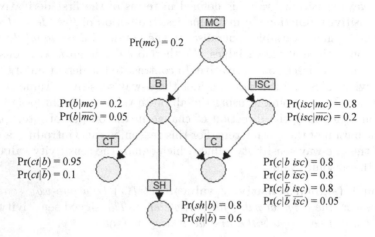

Fig. 1. An example Bayesian network

Example 2. Consider the example network from Fig. 1, representing some (fictitious) medical information. For a patient, the variables MC, B and SH represent the presence or absence of metastatic cancer, a brain tumour, and severe headaches, respectively. Variable ISC captures the presence or absence of an increased serum calcium level, variable C represents whether or not a patient is comatose, and CT whether or not the outcome of a CT scan is positive. Suppose that we are interested in the output probability of a brain tumour in a patient with a positive CT-scan, severe headaches, but who is not in a coma, that is, $\Pr(b \mid ct\,sh\,\bar{c})$. In addition, suppose that the assessments of the parameters $x = \Pr(mc)$ and $y = \Pr(sh \mid \bar{b})$ might be inaccurate. We now find the following sensitivity function, depicted in Fig. 2:

$$f_{\Pr(b|ct\,sh\,\bar{c})}(x, y) = \frac{0.76 + 2.28 \cdot x}{0.76 + 2.28 \cdot x + 7.6 \cdot y - 4.8 \cdot x \cdot y}$$

with gradient $\nabla f = (\frac{\partial f}{\partial x}, \frac{\partial f}{\partial y})$, where

$$\frac{\partial f}{\partial x} = \frac{20.976 \cdot y}{(0.76 + x \cdot (2.28 - 4.8 \cdot y) + 7.6 \cdot y)^2}$$

and

$$\frac{\partial f}{\partial y} = \frac{-5.776 - 13.68 \cdot x + 10.944 \cdot x^2}{(0.76 + x \cdot (2.28 - 4.8 \cdot y) + 7.6 \cdot y)^2}$$

The gradient at $(x^o, y^o) = (0.2, 0.6)$ then equals $\nabla f(0.2, 0.6) \approx (0.465, -0.299)$.

Now consider a parameter shift from $(x^o, y^o) = (0.2, 0.6)$ to $(0.1, 0.7)$, that is, a shift in the direction $(-0.1, 0.1)$. The directional derivative at $(0.2, 0.6)$ in this direction is $(0.465, -0.299) \cdot (\frac{-0.1}{\sqrt{0.02}}, \frac{0.1}{\sqrt{0.02}}) \approx -0.540$ and equals the sensitivity value $sv_{\vee}^{x,y}$ for this direction. The maximum sensitivity value $sv_{\max}^{x,y} = |(0.465, -0.299)| \approx 0.553$ and is found in the direction $(0.465, -0.299)$.

We can also compute the directional derivative at some other point (x, y) than the original parameter assessments. For example, the gradient at $(x, y) = (0.1, 0.1)$ equals $\nabla f(0.1, 0.1) \approx (0.581, -0.414)$. For a shift from this point, in the direction $(0.4, 0.2)$, we have a directional derivative of $(0.581, -0.414) \cdot (\frac{0.4}{\sqrt{0.2}}, \frac{0.2}{\sqrt{0.2}}) \approx -0.414$.

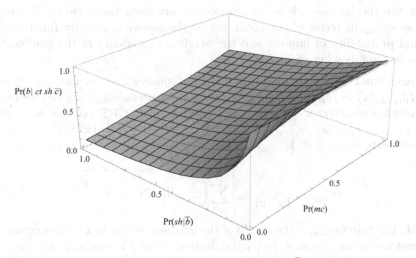

Fig. 2. $\Pr(b \mid ct\,sh\,\bar{c})$ as function of $\Pr(mc)$ and $\Pr(sh \mid \bar{b})$ given the network from Figure 1

4   Computing an n-Way Sensitivity Value

For the computation of an n-way sensitivity value, the partial derivatives $\frac{\partial f}{\partial x_i}$ of the n-way sensitivity function at $\mathbf{x}^o$ are required. These partial derivatives can be established in various ways. In Section 4.1 we will identify the possibilities and drawbacks of using various existing algorithms. In Section 4.2 we will subsequently point out a relation between the n-way sensitivity value and one-way sensitivity values that allows for efficiently computing the former.

4.1   Computing Partial Derivatives for Sensitivity Functions

There are basically two approaches that we can employ for computing our partial derivatives $\frac{\partial f}{\partial x_i}$ for sensitivity function $f(\mathbf{x})$: a direct and an indirect approach.

Indirect Approach. One approach is to establish the complete sensitivity function $f(\mathbf{x})$, using one of the available algorithms for computing its constants from the Bayesian network (see Section 2). Subsequently, the partial derivative with respect to x_i can be computed from the resulting function. This approach allows for computing partial derivatives at any value of x_i and not only at x_i^o. The major drawback of this approach, however, is that the currently most efficient algorithm to compute an n-way sensitivity function requires in the order of $2^n/n$ full junction tree propagations to establish 2^n equations from which the required constants can be solved; this can only be done more efficiently if all n parameters are in the same clique [8]. We note that in the special case where all n parameters are independent parameters from the same CPT, the n-way sensitivity function requires a linear rather than exponential number of constants. In that case, it is doable to compute the n-way sensitivity function.

For the special case where all parameters are from the same CPT, we can express $sv_{\max}^{\mathbf{x}}$ in terms of the constants of the n-way sensitivity function, the original probability of interest and the original probability of the evidence, as stated in the following proposition.

Proposition 1 ($sv_{\max}^{\mathbf{x}}$; x in single CPT). *Consider an n-way sensitivity function* $f_{\Pr(y|e)}(\mathbf{x}) = f_{\Pr(y\,e)}(\mathbf{x})/f_{\Pr(e)}(\mathbf{x})$ *for output probability* $\Pr(y \mid e)$ *and n independent parameters* $\mathbf{x} = (x_1, \ldots, x_n)$ *from a single CPT with original values* $\mathbf{x}^o = (x_1^o, \ldots, x_n^o)$. *Let* $f_{\Pr(y\,e)}(\mathbf{x}) = c_0 + \sum_i c_i \cdot x_i$ *and* $f_{\Pr(e)}(\mathbf{x}) = d_0 + \sum_i d_i \cdot x_i$ *for constants* c_i, d_i, $i = 1, \ldots, n$. *Then the maximum n-way sensitivity value equals*

$$sv_{\max}^{\mathbf{x}} = \frac{1}{\Pr^o(e)} \cdot \sqrt{\sum_{i=1}^{n} \left(c_i - d_i \cdot \Pr^o(y \mid e) \right)^2}$$

Proof. The value $sv_{\max}^{\mathbf{x}}$ is the length of the gradient vector in $\mathbf{x}^o$. To compute the gradient vector we compute the partial derivatives of f for each x_k, $k = 1, \ldots, n$:

$$\frac{\partial f}{\partial x_k} = \frac{c_k \cdot (d_0 + \sum_{i,i\neq k} d_i \cdot x_i) - d_k \cdot (c_0 + \sum_{i,i\neq k} c_i \cdot x_i)}{(d_0 + \sum_{i=1}^{n} d_i \cdot x_i)^2}$$

At $\mathbf{x}^o$ this partial derivative equals:

$$\frac{\partial f}{\partial x_k}(\mathbf{x}^o) = \frac{c_k \cdot (d_0 + \sum_{i,i\neq k} d_i \cdot x_i^o) - d_k \cdot (c_0 + \sum_{i,i\neq k} c_i \cdot x_i^o)}{(d_0 + \sum_{i=1}^{n} d_i \cdot x_i^o)^2}$$

$$= \frac{c_k \cdot (\Pr^o(e) - d_k \cdot x_k^o) - d_k \cdot (\Pr^o(y\,e) - c_k \cdot x_k^o)}{\Pr^o(e)^2}$$

$$= \frac{c_k \cdot \Pr^o(e) - d_k \cdot \Pr^o(y\,e)}{\Pr^o(e)^2} = \frac{c_k - d_k \cdot \Pr^o(y \mid e)}{\Pr^o(e)}$$

The result now follows directly. □

Direct Approach. The second approach is far more elegant for our purposes. A *differential* approach can be used to compute partial derivatives from the so-called *canonical network polynomial* $\mathcal{F}$. For $\frac{\partial f}{\partial x_i}$ a closed form in terms of first and second order partial derivatives of polynomial $\mathcal{F}$ exists; details are beyond the scope of this paper and can be found in [5]. As an alternative, for any parameter $x_i = \Pr(a \mid \boldsymbol{\pi})$ with $x_i^o \neq 0$, we can use the following probabilistic closed form, which is equivalent to the above-mentioned one based on partial derivatives [5]:

$$\frac{\partial f_{\Pr(y|e)}}{\partial x_i}(\mathbf{x}^o) = \frac{\Pr^o(y\, a\, \boldsymbol{\pi} \mid \mathbf{e}) - \Pr^o(y \mid \mathbf{e}) \cdot \Pr^o(a\, \boldsymbol{\pi} \mid \mathbf{e})}{x_i^o}$$

Since the closed forms allow for direct computation of partial derivatives, albeit at $x_i = x_i^o$ only, they are more efficient to compute than the approach using the n-way sensitivity function: rather than computing a number of constants that is exponential in n, we compute n partial derivatives. A drawback of using the partial-derivative-based closed form is that it cannot be computed using classical inference algorithms and requires the computation of both first and second order partial derivatives. A minor drawback of the probabilistic closed form is that the expression requires the computation of several probabilities per parameter for which it is not immediately clear what their relation to sensitivity analysis or sensitivity properties is.

4.2 n-Way Partial Derivatives From One-Way Functions

In the previous section we argued that the direct computation of partial derivatives is much more efficient than establishing them from an n-way sensitivity function. In this section we demonstrate that there is a simple correspondence between partial derivatives of n-way sensitivity functions and derivatives for one-way functions[1]. This provides us with an alternative way of efficiently establishing n-way sensitivity values during a sensitivity analysis.

Proposition 2. *Let $x_1, \ldots, x_n$ be $n > 1$ network parameters with original assessments x_i^o, $i = 1, \ldots, n$, and let P be an output probability of interest. Consider the n-way sensitivity function $f_P(x_1, \ldots, x_n)$ and the one-way sensitivity function $f_P^*(x_k)$, $k \in \{1, \ldots, n\}$. Then*

$$\frac{\partial f_P}{\partial x_k}(x_1^o, \ldots, x_n^o) = \frac{d\, f_P^*}{dx_k}(x_k^o)$$

Proof. Consider an output probability $P = \Pr(y \mid \mathbf{e}) = \frac{\Pr(y\, \mathbf{e})}{\Pr(\mathbf{e})}$. As a result of the factorisation defined by a Bayesian network, both numerator and denominator can be written as an expression of all network parameters consistent with y and/or $\mathbf{e}$ [1]. Suppose these expressions contain m independent parameters (the remaining ones will co-vary). A sensitivity function for $n < m$ of these independent parameters then basically is the m-dimensional sensitivity function with $m - n$ independent parameters fixed at their original value. This also holds for

---

[1] We note that this correspondence, formally stated in Proposition 2, has been implicitly exploited in, for example, [5]; to the best of our knowledge, however, it has not been formalised explicitly before.

$n = 1$. The partial derivative w.r.t x_k of an n-way function $f_P(x_1, \ldots, x_k, \ldots, x_n)$ with parameters $x_i \neq x_k$ kept at x_i^o, is therefore the same as the derivative of the one-way sensitivity function $f_P^*(x_k)$. This proves the proposition.   □

To assess the partial derivatives of an n-way sensitivity function given the original parameter assessments, we thus just need the appropriate one-way sensitivity functions. The above proposition thus gives a computationally feasible way of computing the n-way sensitivity value, since the constants of the one-way functions can be established efficiently. Note that if we are not interested in an n-way sensitivity value, but in the directional derivative at some other point than the original parameters assessments, then the one-way sensitivity functions will not suffice.

Example 3. Consider again the outcome of interest $\Pr(b \mid ct\,sh\,\bar{c})$ and the parameters $x = \Pr(mc)$ and $y = \Pr(sh \mid \bar{b})$ from *Example 2* and Fig. 1. The one-way sensitivity functions are given by

$$f^*_{\Pr(b|ct\,sh\,\bar{c})}(x) = \frac{0.76 + 2.28 \cdot x}{5.32 - 0.6 \cdot x} \quad \text{and} \quad f^{\Diamond}_{\Pr(b|ct\,sh\,\bar{c})}(y) = \frac{1.216}{1.216 + 6.64 \cdot y}$$

Their derivatives equal

$$\frac{df^*}{dx}(x^o) = \frac{12.586}{(5.32 - 0.6 \cdot x^o)^2} = 0.465, \quad \frac{df^{\Diamond}}{dy}(y^o) = \frac{-8.074}{(1.216 + 6.64 \cdot y^o)^2} = -0.299$$

at x^o and y^o, respectively. We observe that indeed

$$\left(\frac{\partial f}{\partial x}(x^o, y^o), \frac{\partial f}{\partial y}(x^o, y^o) \right) = \left(\frac{df^*}{dx}(x^o), \frac{df^{\Diamond}}{dy}(y^o) \right)$$

Using Proposition 2, we can express $sv^x_{\max}$ in terms of the constants of the one-way sensitivity functions and the original probability of the evidence.

Proposition 3 ($sv^x_{\max}$ in general). *Consider $n > 1$ network parameters* $\mathbf{x}$ $= (x_1, \ldots, x_n)$ *with original values* $\mathbf{x^o} = (x_1^o, \ldots, x_n^o)$, *and let* $\Pr(y \mid e)$ *be an output probability of interest. In addition, consider the n one-way sensitivity functions* $f^{(i)}_{\Pr(y|e)}(x_i) = f^{(i)}_{\Pr(y\,e)}(x_i) \,/\, f^{(i)}_{\Pr(e)}(x_i)$, $i = 1, \ldots, n$, *where* $f^{(i)}_{\Pr(y\,e)}(x_i) = c_0^i + c_1^i \cdot x_i$, *with constants* c_0^i, c_1^i, *and* $f^{(i)}_{\Pr(e)}(x_i) = d_0^i + d_1^i \cdot x_i$, *with constants* d_0^i, d_1^i. *Then the maximum n-way sensitivity value for the n-way function* $f_{\Pr(y|e)}(\mathbf{x})$ *equals*

$$sv^x_{\max} = \frac{1}{\Pr^o(e)^2} \cdot \sqrt{\sum_{i=1}^{n}(c_1^i \cdot d_0^i - c_0^i \cdot d_1^i)^2}$$

Proof. The derivative of the one-way sensitivity function $f^{(i)}_{\Pr(y|e)}(x_i)$ at the original parameter assessment x_i^o equals

$$\frac{df^{(i)}_{\Pr(y|e)}}{dx_i}(x_i^o) = \frac{c_1^i \cdot d_0^i - c_0^i \cdot d_1^i}{(d_0^i + d_1^i \cdot x_i^o)^2} = \frac{c_1^i \cdot d_0^i - c_0^i \cdot d_1^i}{\Pr^o(e)^2}$$

Since this derivative is equal to the partial derivative $\frac{\partial f_{\Pr(y|e)}}{\partial x_i}$ at $\mathbf{x}^o$ (Proposition 2), and $sv^{\mathbf{x}}_{\max}$ is the length of the gradient vector in $\mathbf{x}^o$, the proposition follows.   □

The following corollary states two convenient properties for $sv^{\mathbf{x}}_{\max}$. In addition, it states properties that can be used in case we are interested in the n-way sensitivity value in an arbitrary direction, rather than just in the maximum value.

Corollary 1. *Consider $n > 1$ network parameters $\mathbf{x} = (x_1, \ldots, x_n)$ with original values $\mathbf{x}^o = (x_1^o, \ldots, x_n^o)$, and let P be an output probability of interest. Consider the n-way sensitivity function $f_P(\mathbf{x})$ with n-way sensitivity value $sv^{\mathbf{x}}_{\mathbf{v}}$ in direction $\mathbf{v}$ of at most $sv^{\mathbf{x}}_{\max}$. In addition, consider the n one-way sensitivity functions $f_P^{(i)}(x_i)$, $i = 1, \ldots, n$, and let $\mathbf{s} = (sv^{x_1}, \ldots, sv^{x_n})$ be a vector of their one-way signed sensitivity values $sv^{x_i} = \frac{df_P^{(i)}}{dx_i}(x_i^o)$. Then*

1. $sv^{\mathbf{x}}_{\max} = |\mathbf{s}| = \sqrt{\sum_i (sv^{x_i})^2}$

2. $\mathbf{s} = \nabla f_P(\mathbf{x}^o)$

3. $sv^{\mathbf{x}}_{\mathbf{v}} = \mathbf{s} \bullet \mathbf{u}$, where unit vector $\mathbf{u}$ is the normalised vector of $\mathbf{v}$

4. if $|sv^{x_i}| < \frac{1}{\sqrt{n}}$ $\forall i$, then $sv^{\mathbf{x}}_{\max} < 1$ and $sv^{\mathbf{x}}_{\min} > -1$

Proof. Equalities *1.* and *2.* follow directly from the definition of the signed one-way sensitivity value and Proposition 2. Equality *3.* then follows directly from the definition of $sv^{\mathbf{x}}_{\mathbf{v}}$. Inequality *4.* follows directly from Equality *1.*   □

Note that the above corollary can be exploited both in the context of an indirect and a direct approach to computing partial derivatives. Moreover, inequality *4.* enables us to analyse what combinations of parameters may *not* be interesting enough to investigate further during a sensitivity analysis, allowing us to focus on more important parameters.

Example 4. We illustrate, using Fig. 3, the fact that some combinations of parameters may not be interesting enough for further investigation. This figure gives $sv^{x_1, x_2}_{\max}$ as a function of sv^{x_1} and sv^{x_2} for sensitivity values $|sv^{x_i}| < 1$. The figure in addition shows the plane $sv^{x_1, x_2}_{\max} = 1$. The fraction of combinations of sv^{x_1} and sv^{x_2} that result in $sv^{x_1, x_2}_{\max} < 1$ is found below the plane $sv^{x_1, x_2}_{\max} = 1$ and equals $\frac{\pi}{4} \approx 0.785$. From inequality *4.* of Corollary 1 it follows that, in order to result in a two-way sensitivity value ≥ 1, the absolute value of at least one of the individual values has to be $\geq \frac{1}{\sqrt{2}}$. Thus whenever both $|sv^{x_1}|$ and $|sv^{x_2}|$ are < 0.71, we can be sure that $sv^{x_1, x_2}_{\max} < 1$ and that $sv^{x_1, x_2}_{\min} > -1$, that is, $|sv^{x_1, x_2}_{\mathbf{v}}| < 1$ for any $\mathbf{v}$. If one of the two parameters, however, has a one-way sensitivity value ≥ 0.71, it depends on the sensitivity value of the other parameter whether $sv^{x_1, x_2}_{\max} \geq 1$ or not.

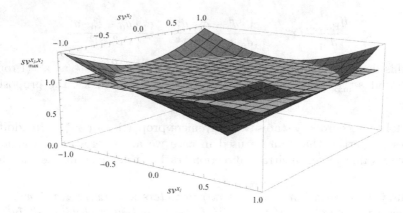

Fig. 3. $sv_{\max}^{x_1,x_2}$, as a function of $sv^{x_1}, sv^{x_2} \in \langle -1, 1 \rangle$ and the plane $sv_{\max}^{x_1,x_2} = 1$

4.3 Joint vs. Synergistic Effect in n-Way Analyses

We observe that the absolute value $|sv_{\mathbf{v}}^{\mathbf{x}}|$ may be higher (or lower) than each of the individual absolute values $|sv^{x_i}|$ of which it is composed. For example, in the network from Fig. 1 (see *Examples 2* and *3*), we had that $|sv^x| = 0.465$, $|sv^y| = 0.299$ and that a shift in the direction of $\mathbf{v} = (-0.1, 0.1)$ resulted in $|sv_{\mathbf{v}}^{x,y}| = 0.540$. A simultaneous shift thus may have a larger (or smaller) local effect on the outcome probability than each parameter shift separately.

We would like to note that this joint effect of multiple parameter changes is not the same as the *synergistic effect* of multiple parameter changes, as first described in [3]. A synergistic effect is caused by the fact that the exact form of the one-way sensitivity function of x_i, depends on the original values of the other parameters of the network, and thus may be different for different values of some other parameter x_j. Such a synergistic effect can only be present if the n-way sensitivity function of those parameters includes product terms of x_i and x_j. For a joint effect of multiple simultaneous changes, the presence of such product terms is not necessary. As mentioned in Section 2, given just parameters from a single CPT, a sensitivity function will not include product terms of its parameters. Given parameters from a single CPT, therefore, no synergistic effect will be observed; a joint effect, however, may be present.

5 Parameter Tuning

The theory we discussed in Sections 3 and 4 can be used to study the robustness of a network to small simultaneous parameter changes. Another area of application lies in parameter tuning. In building a network, we may want to adjust parameters in order to meet certain constraints. An example of such a constraint is $\Pr(y \mid \mathbf{e}) > t$ for some probability $\Pr(y \mid \mathbf{e})$ and a desired value t.

In [2], a method for parameter tuning is described in which the parameter adjustment is guided by the log-odd change of the varied parameters in order to keep the distance between the old and the new distribution as small as possible. The paper moreover describes a method to compute the solution space of all possible parameter changes that would fulfill the constraint. This method, although not mentioned as such by the authors, in essence provides for computing the constants of the sensitivity function of the varied parameters, and is exponential in the number of CPTs from which the parameters are chosen. The method thus is feasible only if the adjusted parameters come from a limited number of different CPTs.

We now propose another tuning approach, based on the theory introduced in Sections 3 and 4. In this approach, our goal is to satisfy some constraint by adjusting a given set of n parameters as little as possible, that is, by keeping the sum of the absolute values of the parameter changes as low as possible. In our method the direction of the maximal change is used to guide the parameter changes. Since the gradient $\mathbf{s} = (sv^{x_1}, \ldots, sv^{x_n})$ of an n-way sensitivity function at x^o gives the direction of local maximal increase of the outcome probability, we will simultaneously adjust the n parameters in or against the direction of $\mathbf{s}$ to achieve the desired value. As long as the changes needed are small, this adjustment will be a good approximation of the adjustments required to satisfy the desired constraint with a minimal change of the parameters.

The adjustments needed can be assessed using a sensitivity function in which the parameters are constrained to variation only in or against the direction of $\mathbf{s}$. Below we first define an n-way sensitivity function given parameter changes in or against a fixed direction $\mathbf{v}$ to be a *sliced sensitivity function* in the direction of $\mathbf{v}$.

Definition 4 (sliced sensitivity function). *Let $f_{\Pr(y|e)}(\mathbf{x})$ be an n-way sensitivity function. A sliced sensitivity function of f in the direction of $\mathbf{v}$, denoted $f^{\mathbf{v}}_{\Pr(y|e)}$, expresses $\Pr(y \mid e)$ as a function of the change of the parameters $\mathbf{x}$ in or against direction $\mathbf{v}$ only.*

The following proposition shows that a sliced sensitivity function can be expressed in a single parameter and takes the form of a fraction of two polynomial functions of degree at most the number of CPTs from which the parameters are chosen.

Proposition 4. *Consider an n-way sensitivity function $f_{\Pr(y|e)}(\mathbf{x})$ for an output probability $\Pr(y \mid e)$, and a change of its parameters $\mathbf{x} = (x_1, \ldots, x_n)$ in a fixed direction $\mathbf{v} = (v_1, \ldots, v_n)$. Then for any x_i, $i = 1, \ldots, n$, with $v_i \neq 0$ there exists a sliced sensitivity function $f^{\mathbf{v}}_{\Pr(y|e)}(x_i)$ of the form:*

$$f^{\mathbf{v}}_{\Pr(y|e)}(x_i) = \frac{c_0 + c_1 \cdot x_i^1 + \ldots + c_m \cdot x_i^m}{d_0 + d_1 \cdot x_i^1 + \ldots + d_m \cdot x_i^m}$$

where each x_i^k, $k = 1, \ldots, m$, is a polynomial term of degree k and m is the number of different CPTs from which $x_1, \ldots, x_n$ are chosen.

Proof. Given a change in or against a fixed direction $(v_1, \ldots, v_n)$, we can express all parameters x_j in parameter x_i, since $(x_j - x_j^o) = \frac{v_j}{v_i} \cdot (x_i - x_i^o) \Leftrightarrow x_j = \frac{v_j}{v_i} \cdot (x_i - x_i^o) + x_j^o$, which is linear in x_i. Product terms of parameters in the n-way function $f_{\Pr(y|e)}(\mathbf{x})$ will now result in polynomial terms in $f^{\mathbf{v}}_{\Pr(y|e)}(x_i)$, of which the degree is determined by the number of interacting parameters in $f_{\Pr(y|e)}(\mathbf{x})$. This number equals at most the number of CPTs from which the parameters $x_1, \ldots, x_n$ are chosen. $\qquad \square$

The n-way sensitivity function in any fixed vector direction $\mathbf{v}$ thus is a polynomial with as maximum degree the number of CPTs m from which the parameters are chosen and is determined by just $2 \cdot (m+1)$ constants. This observation also holds for the direction $\mathbf{s}$ of maximal increase. A constraint on $\Pr(y \mid e)$ now can be expressed in terms of a sliced sensitivity function in the direction of $\mathbf{s}$, and a feasible solution with minimal parameter change in or against the direction of $\mathbf{s}$ can be derived, if any. The solutions of a polynomial equation can be established analytically for polynomials up to degree 4; solutions for higher degree polynomials can be approximated.

In the above we assumed a *given* set of parameters $\{x_1, \ldots, x_n\}$. Note that a reasonable heuristic for choosing a set of parameters to adjust can be based on the one-way sensitivity values since $sv^{\mathbf{x}}_{\max} = \sqrt{\sum_i (sv^{x_i})^2}$; i.e. selecting the n parameters with highest one-way sensitivity value will allow for the largest possible local shift in the output upon their simultaneous perturbation.

Example 5. Consider again the example network from Fig. 1. Suppose we know that the outcome probability $\Pr(b \mid ct \, sh \, c)$ should be at least 0.80. In the network as it is we find that $\Pr(b \mid ct \, sh \, c) = 0.76$ so some parameter adjustment is required. First the signed one-way sensitivity values sv^x for all independent parameters x of the network are assessed[2]; these are given in Table 1. We observe that the parameters which will affect the outcome probability the most are $x = \Pr(b \mid \overline{mc})$ and $y = \Pr(ct \mid \overline{b})$. Suppose that we want to satisfy our constraint by adjusting those two parameters. The direction of maximal increase is $\mathbf{v} = (1.87, -1.81) \sim (1, -0.97)$. Tying y to x, the sliced sensitivity function in the direction of the maximal change now is

$$f^{\mathbf{v}}_{\Pr(b|ct \, sh \, c)}(x) = \frac{0.02432 + 0.4864 \cdot x}{0.0478424 + 0.318496 \cdot x + 0.09312 \cdot x^2}$$

Solutions to $f^{\mathbf{v}}_{\Pr(b|ct \, sh \, c)}(x) = 0.80$ are $x \approx 0.061$ and $x = 3.047$, where only the former is feasible. Parameter values that will satisfy $\Pr(b \mid ct \, sh \, c) \geq 0.80$ thus are $x = \Pr(b \mid \overline{mc}) = 0.061$ and $y = \Pr(ct \mid \overline{b}) = y^o - 0.97 \cdot (0.061 - x^o) = 0.10 - 0.97 \cdot (0.061 - 0.05) = 0.089$.

---

[2] Recall that dependent parameters are included in the analysis by covariation.

Table 1. Sensitivity values sv^x for independent parameters of the example network from Fig. 1

x	x^o	sv^x
$\Pr(mc)$	0.20	0.12
$\Pr(b \mid mc)$	0.20	0.56
$\Pr(b \mid \overline{mc})$	0.05	1.87
$\Pr(isc \mid mc)$	0.80	−0.09
$\Pr(isc \mid \overline{mc})$	0.20	−0.41
$\Pr(ct \mid b)$	0.95	0.19
$\Pr(ct \mid \bar{b})$	0.10	−1.81
$\Pr(c \mid b\ isc)$	0.80	0.11
$\Pr(c \mid b\ \overline{isc})$	0.80	0.11
$\Pr(c \mid \bar{b}\ isc)$	0.80	−0.20
$\Pr(c \mid \bar{b}\ \overline{isc})$	0.05	−0.46
$\Pr(sh \mid b)$	0.80	0.23
$\Pr(sh \mid \bar{b})$	0.60	−0.30

6 Discussion

The robustness of Bayesian networks to changes in their parameter probabilities can be studied with a sensitivity analysis. Since the study of multiple simultaneous parameter shifts is computationally expensive, most research has focused on one-way sensitivity analyses in which only a single parameter is varied at a time. An important notion in sensitivity analysis is the notion of *sensitivity value*, which captures the sensitivity of some outcome of the network to a small change in a single parameter under consideration. In this paper we generalised this concept to multiple dimensions and proved that the computation of such an n-way sensitivity value can be done efficiently from the one-way sensitivity values of the parameters under consideration.

In contrast to a one-way sensitivity value, an n-way sensitivity value varies depending on the direction of shift under consideration. We expressed the direction of maximal change in terms of one-way sensitivity values and provided bounds on the n-way sensitivity value. We argued that the maximal (minimal) sensitivity value and the corresponding direction of maximal change is not only useful for studying the robustness of a Bayesian network, but can also be used in the context of network tuning. For small parameter changes, a shift of the parameters in or against the direction of the maximal increase until some tuning constraint is met will yield a good approximation of the minimal parameter change necessary to meet this constraint. We also proved that, since a fixed vector direction of change ties all parameters linearly, the effect of a parameter shift in the direction of the maximal change on some outcome probability can be efficiently established. To this end we introduced the concept of *sliced sensitivity function* for a sensitivity function that captures such a linearly tied parameter shift.

In a sliced sensitivity function variables are tied linearly. Variables, however, can also be tied by some other meaningful relationship. In [2], for example, parameters are tied by their log-odds ratio changes. In future research, we would like to expand the notion of sliced sensitivity function to more general forms of constrained sensitivity functions and explore the use of these functions both within and outside the field of parameter tuning.

Acknowledgements. This research was supported by the Netherlands Organisation for Scientific Research.

References

1. Castillo, E., Gutiérrez, J.M., Hadi, A.S.: Parametric Structure of Probabilities in Bayesian Networks. In: Froidevaux, C., Kohlas, J. (eds.) ECSQARU 1995. LNCS (LNAI), vol. 946, pp. 89–98. Springer, Heidelberg (1995)
2. Chan, H., Darwiche, A.: Sensitivity Analysis in Bayesian Networks: From Single to Multiple Parameters. In: Chickering, M., Halpern, J. (eds.) Proceedings of the 20th Conference on Uncertainty in Artificial Intelligence, pp. 67–75. AUAI Press, Arlington (2004)
3. Coupé, V.M.H., Van der Gaag, L.C., Habbema, J.D.F.: Sensitivity Analysis: An Aid for Belief Network Quantification. The Knowledge Engineering Review 15, 215–232 (2000)
4. Coupé, V.M.H., Van der Gaag, L.C.: Properties of Sensitivity Analysis of Bayesian Belief Networks. Annals of Mathematics and Artificial Intelligence 36, 323–356 (2002)
5. Darwiche, A.: A Differential Approach to Inference in Bayesian Networks. Journal of the ACM 50, 280–305 (2003)
6. Van der Gaag, L.C., Renooij, S.: Analysing Sensitivity Data from Probabilistic Networks. In: Breese, J., Koller, D. (eds.) Proceedings of the 17th Conference on Uncertainty in Artificial Intelligence, pp. 530–537. Morgan Kaufmann, San Francisco (2001)
7. Jensen, F.V., Nielsen, T.D.: Bayesian Networks and Decision Graphs, 2nd edn. Springer (2007)
8. Kjærulff, U., Van der Gaag, L.C.: Making Sensitivity Analysis Computationally Efficient. In: Boutilier, C., Goldszmidt, M. (eds.) Proceedings of the 16th Conference on Uncertainty in Artificial Intelligence, pp. 317–325. Morgan Kaufmann, San Francisco (2000)
9. Laskey, K.B.: Sensitivity Analysis for Probability Assessments in Bayesian Networks. IEEE Transactions on Systems, Man and Cybernetics 25, 901–909 (1995)

Bayesian Network Inference
Using Marginal Trees[*],[**]

Cory J. Butz[1], Jhonatan de S. Oliveira[2], and Anders L. Madsen[3],[4]

[1] University of Regina, Department of Computer Science,
Regina, S4S 0A2, Canada
butz@cs.uregina.ca
[2] Federal University of Viçosa, Electrical Engineering Department,
Viçosa, 36570-000, Brazil
jhonatan.oliveira@gmail.com
[3] HUGIN EXPERT A/S, Aalborg, Denmark
Aalborg, DK-9000, Denmark
anders@hugin.com
[4] Aalborg University, Department of Computer Science,
Aalborg, DK-9000, Denmark

Abstract. *Variable Elimination* (VE) answers a query posed to a *Bayesian network* (BN) by manipulating the conditional probability tables of the BN. Each successive query is answered in the same manner. In this paper, we present an inference algorithm that is aimed at maximizing the reuse of past computation but does not involve precomputation. Compared to VE and a variant of VE incorporating precomputation, our approach fairs favourably in preliminary experimental results.

Keywords: Bayesian network, inference, marginal trees.

1 Introduction

Koller and Friedman [1] introduce readers to inference in *Bayesian networks* (BNs) [2] using the *Variable Elimination* (VE) [3] algorithm. The main step of VE is to iteratively eliminate all variables in the BN that are not mentioned in the query. Subsequent queries are also answered against the BN meaning that past computation is not reused. The consequence is that some computation may be repeated when answering a subsequent query.

Cozman [4] proposed a novel method attempting to reuse VE's past computation when answering subsequent queries. Besides the computation performed by VE to answer a given query, Cozman's method also performs precomputation that may be useful to answer subsequent queries. While Cozman's approach is meritorious in that it reduces VE's duplicate computation, one undesirable feature is that precomputation can build tables that are never used.

* Supported by NSERC Discovery Grant 238880.
** Supported by CNPq - Science Without Borders.

L.C. van der Gaag and A.J. Feelders (Eds.): PGM 2014, LNAI 8754, pp. 81–96, 2014.

In this paper, we introduce *marginal tree inference* (MTI) as a new exact inference algorithm in discrete BNs. MTI answers the first query the same way as VE does. MTI answers each subsequent query in a two-step procedure that can readily be performed in a new secondary structure, called a *marginal tree*. First, determine whether any computation can be reused. Second, only compute what is missing to answer the query. One salient feature of MTI is that it does not involve precomputation, meaning that every probability table built is necessarily used in answering a query. In preliminary experimental results, MTI fairs favourably when compared to VE and Cozman [4].

The remainder is organized as follows. Section 2 contains definitions. Marginal trees are introduced in Section 3. Section 4 presents MTI. Related work is discussed in Section 5. Section 6 describes advantages and give preliminary experimentals results. Conclusions are given in Section 7.

2   Definitions

2.1   Bayesian Network

Let U be a finite set of variables. Each variable $v_i \in U$ has a finite domain, denoted $dom(v_i)$. A *Bayesian network* (BN) [2] on U is a pair (B, C). B is a *directed acyclic graph* (DAG) with vertex set U and C is a set of *conditional probability tables* (CPTs) $\{p(v_i|P(v_i)) \mid v_i \in U\}$, where $P(v_i)$ denotes the parents (immediate predecessors) of $v_i \in B$. For example, Fig. 1 shows the *extended student Bayesian network* (ESBN) [1], where CPTs are not shown. The product of the CPTs in C is a joint probability distribution $p(U)$. For $X \subseteq U$, the *marginal distribution* $p(X)$ is $\sum_{U-X} p(U)$. Each element $x \in dom(X)$ is called a *row* (configuration) of X. Moreover, $X \cup Y$ may be written as XY. We call B a BN, if no confusion arises.

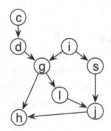

Fig. 1. The DAG of the ESBN [1]

2.2   Variable Elimination

Variable elimination (VE) [3] computes $p(X|E = e)$, where X and E are disjoint subsets of U, and E is observed taking value e. In VE (Algorithm 1), Φ is the set of CPTs for B, X is a list of query variables, E is a list of observed variables, e is the corresponding list of observed values, and σ is an elimination ordering for variables $U - (X \cup E)$. Evidence may not be denoted for simplified notation.

Algorithm 1. VE(Φ, X, E, e, σ)
Delete rows disagreeing with $E = e$ from $\phi \in \Phi$
While σ is not empty:
    Remove the first variable v from σ
    $\Phi = $ sum-out(v, Φ)
$p(X, E = e) = \prod_{\phi \in \Phi} \phi$
return $p(X, E = e) / \sum_X p(X, E = e)$

The sum-out algorithm eliminates v from a set Φ of *potentials* [1] by multiplying together all potentials involving v, then summing v out of the product.

Example 1. [1] Suppose $p(j|h = 0, i = 1)$ is the query issued to the ESBN in Fig. 1. One possible elimination order is $\sigma = (c, d, l, s, g)$. The evidence $h = 0$ and $i = 1$ are incorporated into $p(h|g, j)$, $p(i)$, $p(g|d, i)$, and $p(s|i)$. VE then computes:

$$p(d) = \sum_c p(c) \cdot p(d|c), \tag{1}$$

$$p(g|i) = \sum_d p(d) \cdot p(g|d, i), \tag{2}$$

$$p(j|g, s) = \sum_l p(l|g) \cdot p(j|l, s), \tag{3}$$

$$p(j|g, i) = \sum_s p(s|i) \cdot p(j|g, s), \tag{4}$$

$$p(h, j|i) = \sum_g p(g|i) \cdot p(h|g, j) \cdot p(j|g, i). \tag{5}$$

Next, the product of the remaining potentials is taken, $p(h, i, j) = p(h, j|i) \cdot p(i)$. VE answers the query by normalizing on the evidence variables, $p(j|h, i) = p(h, i, j) / \sum_j p(h, i, j)$.

3 Marginal Trees

We begin by motivating the introduction of marginal trees.

Example 2. Suppose $p(s|h = 0, i = 1)$ is the second query issued to VE. One possible elimination order is $\sigma = (c, d, l, j, g)$. VE performs:

$$p(d) = \sum_c p(c) \cdot p(d|c), \tag{6}$$

$$p(g|i) = \sum_d p(d) \cdot p(g|d, i), \tag{7}$$

$$p(j|g,s) = \sum_l p(l|g) \cdot p(j|l,s), \tag{8}$$

$$p(h|g,s) = \sum_j p(j|g,s) \cdot p(h|g,j),$$

$$p(h|i,s) = \sum_g p(h|g,s) \cdot p(g|i),$$

$$p(s,h,i) = p(s|i) \cdot p(i) \cdot p(h|i,s),$$

$$p(s|h,i) = p(s,h,i)/\sum_s p(s,h,i).$$

Note that VE's computation in (1)-(3) for the first query is repeated in (6)-(8) for the second query. We seek to avoid recomputation.

The second query $p(s|h = 0, i = 1)$ can be answered from the following factorization of the marginal $p(h,i,s)$:

$$p(h,i,s) = p(s|i) \cdot p(i) \cdot \sum_g p(g|i) \cdot \sum_j p(j|g,s) \cdot p(h|g,j),$$

where the past calculation of $p(g|i)$ in (2) and $p(j|g,s)$ in (3) are reused.

We introduce marginal trees as a representation of past computation. This secondary structure not only facilitates the identification of that computation which can be reused, but also enables the determination of what missing information needs to be constructed. It is based on the fact that VE can be seen as one-way propagation in a join tree [5]. A *join tree* [5] is a tree having sets of variables as nodes with the property that any variable in two nodes is also in any node on the path between the two.

Definition 1. *Given a Bayesian network B defining a joint probability distribution $p(U)$. A marginal tree M is a join tree on $X \subseteq U$ with CPTs of B assigned to nodes of M and showing constructed messages in one-way propagation to a chosen root node R of M yielding $p(R)$.*

The initial marginal tree has one node N, which has all variables from U. All the CPTs from the BN are assigned to N. For example, the initial marginal tree for the ESBN is depicted in Fig. 2 (i).

Each time a variable v is eliminated, a new marginal tree is uniquely formed by replacing one node with two nodes and the CPT (containing only those rows agreeing with the evidence) built during elimination is passed from one new node to the other. We say that a CPT from N_1 to N_2 is *outgoing* from N_1 and *incoming* to N_2.

Whenever v is eliminated, there exists a unique node N containing v without outgoing messages. Let Γ be the set of all assigned CPTs and incoming messages to N, Ψ is the set of all CPTs containing v, and τ the CPT produced by summing out v. Replace N by nodes N_1 and N_2. N_1 has Ψ assigned CPTs and N_2 has $\Gamma - \Psi$ assigned CPTs. The variables in nodes N_1 and N_2 are defined by the variables appearing in the CPTs assigned to N_1 and N_2, respectively. For each incoming

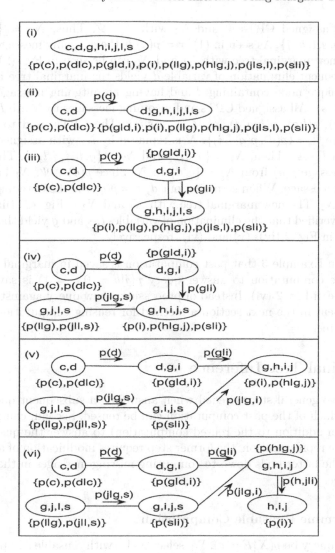

Fig. 2. In Example 3, the initial marginal tree is in (i). The respective marginal trees (ii)-(vi) formed by eliminating c,d,l,s and g are in (1)-(5), respectively.

message m from node N_i to N, if $m \in \Psi$, set m as incoming message from N_i to N_1; otherwise, set m as incoming message from N_i to N_2. The outgoing message from N_1 to N_2 is τ.

Example 3. In Example 1, the initial marginal tree is shown in Fig. 2 (i). Here, $\Gamma = \{p(c),\ p(d|c),\ p(g|d,i),\ p(i),\ p(l|g),\ p(h|g,j),\ p(j|s,l),\ p(s|i)\}$. The CPTs needed to eliminate variable c are $\Psi = \{p(c), p(d|c)\}$. A new marginal tree is uniquely formed as shown in Fig. 2 (ii). Node $N = \{c, d, g, h, i, j, l, s\}$ is replaced by N_1 with assigned CPTs Ψ and N_2 with $\Gamma - \Psi$. Thus, $N_1 = \{c, d\}$ and $N_2 = \{d, i, g, s, l, h, j\}$. As seen in (1), $\tau = p(d)$ is the outgoing message from N_1 to N_2. The new marginal tree with N_1 and N_2 can be seen in Fig. (2) (ii).

The subsequent elimination of variable d yields the marginal tree in Fig. (2) (iii). The unique node containing d and having no outgoing messages is $N = \{d, g, h, i, j, l, s\}$. All assigned CPTs and incoming messages to N are $\Gamma = \{p(d), p(g|d, i), p(i), p(l|g), p(h|g, j), p(j|s, l), p(s|i)\}$. The CPTs used to eliminate variable d are $\Psi = \{p(d), p(g|d, i)\}$. N is replaced by N_1 with assigned CPTs Ψ and N_2 with $\Gamma - \Psi$. Then, $N_1 = \{d, g, i\}$ and $N_2 = \{g, h, i, j, l, s\}$. There is one incoming message $p(d)$ from $N_i = \{c, d\}$ to N. Since $p(d) \in \Psi$, N_1 has $p(d)$ as an incoming message. When summing out d, $\tau = p(g|i)$ is the outgoing message from N_1 to N_2. The new marginal tree with N_1 and N_2 is Fig. (2) (iii).

It can be verified that the elimination of variables l, s and g yield the marginal trees shown in Fig. 2 (iv), (v) and (vi), respectively.

Observe in Example 3 that past computation is saved in marginal trees. For example, the computation to answer query $p(j|h = 0, i = 1)$ is saved in the marginal tree of Fig. 2 (vi). Instead of processing a new query against the given BN, we present in the next section a method for reusing computation saved in a marginal tree.

4   Marginal Tree Inference

There are two general steps needed when answering a subsequent query. First, determine which of the past computation can be reused. Second, compute what is missing (in addition to the reused computation) to answer the query. In our marginal tree representation, the former step requires modification of a marginal tree, while the latter boils down to computing missing messages in the modified marginal tree.

4.1   Determine Reusable Computation

Let the new query be $p(X|E = e)$. We select nodes with reusable computation in the marginal tree M with respect to the new query using the *selective reduction algorithm* (SRA) [6], described as follows. Mark variables XE in a copy M' of M. Repeatedly apply the following two operations until neither can be applied: (i) delete an unmarked variable that occurs in only one node; (ii) delete a node contained by another one.

Example 4. Let M be the marginal tree in Fig. 2 (vi) and M' be the copy in Fig. 3 (i). Let $N_1 = \{c, d\}$, $N_2 = \{d, g, i\}$, $N_3 = \{g, j, l, s\}$, $N_4 = \{g, i, j, s\}$, $N_5 = \{g, h, i, j\}$ and $N_6 = \{h, i, j\}$. Let the new query be $p(s|h = 0, i = 1)$. Variables s, h and i are first marked. Variable c can be deleted as it occurs only in N_1. Therefore, $N_1 = \{d\}$. Now N_1 can be deleted, since $N_1 \subseteq N_2$. It can be verified that after applying steps (i) and (ii) repeatedly, all that remains is $N_4 = \{g, i, j, s\}$ and $N_5 = \{g, h, i, j\}$, as highlighted in Fig. 3 (ii).

The SRA output is the portion of VE's past computation that can be reused. For instance, Example 4 indicates that the computation (1)-(3) for answering query $p(j|h = 0, i = 1)$ can be reused when subsequently answering the new query $p(s|h = 0, i = 1)$.

Now, we need to construct a marginal tree M' to be used to answer the new query $p(X|E = e)$ while at the same time reusing past computation saved in M.

Algorithm 3. Rebuild(M, XE)
Let M' be a copy of M
$M'' = \text{SRA}(M', XE)$
$N = \cup_{N_i \in M''} N_i$
Delete all messages between $N_i \in M''$ from M'
Delete all nodes $N_i \in M''$ from M'
Adjust messages as incoming to N accordingly
return M'

Example 5. Supposing that the new query is $p(s|h = 0, i = 1)$, then we call Rebuild$(M,\{s, h, i\})$, where M is the marginal tree in Fig. 2 (vi). Let M' be the copy in Fig. 3 (i) and $M'' = \{N_4 = \{g, i, j, s\}, N_5 = \{g, h, i, j\}\}$. In Rebuild, the next step sets $N = N_4 \cup N_5 = \{g, h, i, j, s\}$.

The message $p(j|g, i)$ from N_4 to N_5 in Fig. 3 (i) is ignored in Fig. 3 (iii). Moreover, all incoming messages to N_4 and N_5 remain as incoming messages to the new node N. The output from Rebuild is the modified marginal tree depicted in Fig. 3 (iv).

The key point is that the modified marginal tree has a node N containing all variables in the new query. Thus, the query could be answered by one-way probability propagation towards N in the modified marginal tree. For example, query $p(s|h = 0, i = 1)$ can be answered in the modified marginal tree in Fig. 4 (i) by propagating towards node $N = \{g, h, i, j, s\}$. However, some computation can be reused when answering a new query. For example, it can be seen in Fig. 4 (ii) that messages $p(d)$, $p(g|i)$ and $p(j|g, s)$ have already been built in (1)-(3) and do not have to be recomputed.

4.2   Determine Missing Computation

Given a marginal tree M constructed for a query and a new query $p(X|E = e)$ such that XE is contained within at least one node $N \in M$, the *partial-one-way-propagation* (POWP) algorithm determines which messages of M can be reused when answering $p(X|E = e)$, namely, it determines what missing computation is needed to answer $p(X|E = e)$. POWP works by determining the messages needed for one-way propagation to root node N in M [7,8], and then ignoring those messages that can be reused.

Example 6. Fig. 5 illustrates how POWP determines missing information to answer a new query $p(s|l = 0)$ from the marginal tree M in (i) previously built when answering query $p(j|h = 0, i = 1)$. In (ii), we consider node $\{g, j, l, s\}$ as

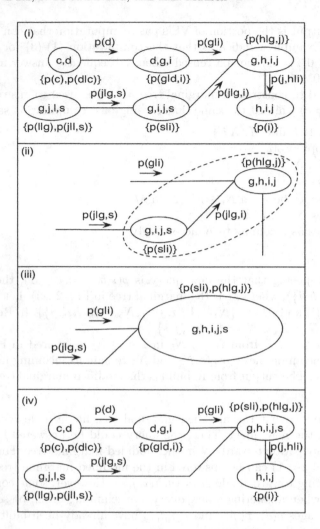

Fig. 3. (i) the marginal tree for Example 4. (ii) RSA outputs $N_4 = \{g, i, j, s\}$ and $N_5 = \{g, h, i, j\}$. (iii) $N = \{g, i, j, s\} \cup \{g, h, i, j\}$. (iv) the modified marginal tree built by RSA.

the new root node N, since the query variables s and l are contained in N. One-way propagation towards N is shown in Fig. 5 (ii). POWP determines that messages $p(d)$ and $p(g|i)$ can be reused from (1)-(2), and are hence ignored as shown in Fig. 5 (iii).

The important point is that the new query can be answered at the root node when POWP finishes.

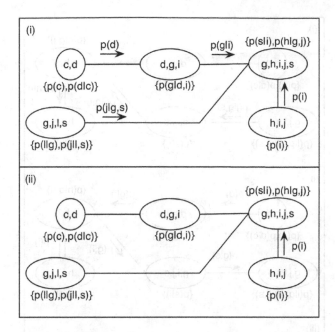

Fig. 4. (i) new messages needed to build a marginal factorization on new root $N = \{g, h, i, j, s\}$. (ii) messages $p(d)$, $p(g|i)$ and $p(j|g, s)$ built when answering a previous query in (1)-(3) can be reused.

Example 7. In Fig. 5 (iii), once POWP to the root node $N = \{g, j, l, s\}$ completes, $p(s|l = 0)$ can be answered from the marginal factorization of $p(N)$: $p(g, j, l, s) = p(l|g) \cdot p(j|l, s) \cdot p(g, s)$.

As another example, given the initial query $p(j|h = 0, i = 1)$ and marginal tree M in Fig. 3 (i), the following takes place to answer a new query $p(s|h = 0, i = 1)$. The modified marginal tree M' is depicted in Fig. 4 (i). The POWP algorithm determines the messages to propagate to the root node $N = \{g, h, i, j, s\}$, chosen as root since it contains the variables in the new query $p(s|h = 0, i = 1)$. POWP determines that messages $p(d)$, $p(g|i)$, and $p(j|g, s)$ in Fig. 3 (i) can be reused in Fig. 4 (i). Thus, only message $p(i)$ needs to be computed in Fig. 4 (ii). And, lastly, the query $p(s|h = 0, i = 1)$ can be answered from the marginal factorization of $p(N)$:

$$p(g, h, i, j, s) = p(g|i) \cdot p(j|g, s) \cdot p(i) \cdot p(s|i) \cdot p(h|g, j).$$

The culmination of the ideas put forth thus far is formalized as the *marginal tree inference* (MTI) algorithm.

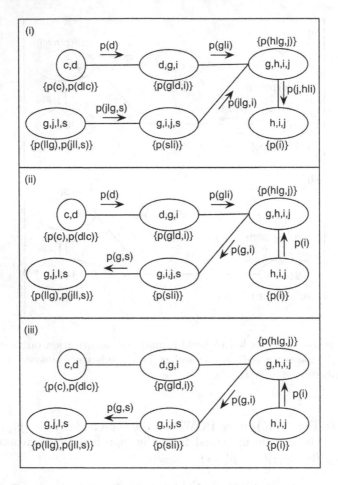

Fig. 5. (i) messages built answering the initial query $p(j|h = 0, i = 1)$ in node $\{h, i, j\}$. (ii) new root $N = \{g, j, l, s\}$. (iii) messages $p(d)$ and $p(g|i)$ can be reused while messages $p(g, s)$, $p(g, i)$ and $p(i)$ need to be built.

Algorithm 4. MTI($M, p(X|E = e)$)
$M' = $ SRA(M, XE)
Rebuild(M', XE)
POWP(M', XE)
Compute $p(X|E = e)$ at the root node of M'
Return $p(X|E = e)$

MTI takes as input a marginal tree M built from a previous query and a new query $p(X|E = e)$.

For instance, answering $p(j|h = 0, i = 1)$ in Example 1 builds the marginal tree M in Fig. 2 (vi), as described in Example 3. Let $p(s|l = 0)$ be the new query. SRA outputs two nodes highlighted in Fig. 3 (ii), as illustrated in Example 4.

In Example 5, Rebuild constructs M', as depicted in Fig. 3 (iv). Next, POWP determines reusable messages in Example 6, as illustrated in Fig. 4. Finally, $p(s|l = 0)$ can be calculated at the new root $N = \{g, j, l, s\}$, as discussed in Example 7.

5   VE with Precomputation

Cozman [4] gives a novel algorithm in an attempt to reuse VE's past computation. The crux of his algorithm can be understood as follows. As previously mentioned, Shafer [5] has shown that VE's computation to answer a query can be seen as one-way propagation to a root node in a join tree. It is also known [5] that conducting a subsequent outward pass from the root node to the leaves of the join tree, called two-way propagation, builds marginals $p(N)$ for every node N in the join tree. Cozman's method is to perform the outward second pass after VE has conducted the inward pass for a given query. The marginals built during the outward pass may be useful when answering a subsequent query.

As shown in [4], the VE algorithm can be described as follows. Given a query $p(X|E)$:

1. Compute the set of non-observed and non-query variables $N = U - XE$.
2. For each $v_i \in N$
   (a) Create a data structure B_i, called a *bucket*, containing:
       - the variable v_i, called the *bucket variable*;
       - all potentials that contain the bucket variable, called *bucket potentials*;
   (b) Multiply the potentials in B_i. Store the resulting potential in B_i; the constructed potential is called B_i's *cluster*.
   (c) Sum out v_i from B_i's cluster. Store the resulting potential in B_i; this potential is called B_i's *separator*.
3. Collect the potentials that contain the query variables in a bucket B_q. Multiply the potentials in B_q together and normalize the result.

Denote the bucket variable for B_i as v_i, the variables in B_i's separator by S_i, and the evidence contained in the sub-tree above and including bucket B_i by E_i. The outward pass is given by computing the marginal probability for every variable in a BN. In order to update buckets immediately above the root, denoted B_a, with the normalized potential containing v_a and some of the variables in X, compute:

$$p(v_a|E) = \sum_X p(v_a|X, E_a) \cdot p(X|E). \tag{9}$$

Similarly, to update the buckets away from the root, say B_b, compute:

$$p(v_b|E) = \sum_{S_b} p(v_b|S_b, E_b) \cdot p(S_b|E). \tag{10}$$

Example 8. Given query $p(j|h = 0, i = 1)$, Cozman's method runs VE to answer it. This inward pass generates a tree of buckets as shown in Fig. 6 (i). Next, Cozman performs an outward pass with the following computation. In order to use (9), first we need to compute $p(v_a|X, E_a)$:

$$p(g|h, i, j) = p(g, h, j|i)/ \sum_g p(g, h, j|i).$$ (11)

Now apply (9) to determine:

$$p(g|h, i) = \sum_j p(g|j, h, i) \cdot p(j|h, i).$$ (12)

Similarly, in order to use (10) we first compute $p(v_b|S_b, E_b)$:

$$p(d|g, i) = p(d, g|i)/ \sum_d p(d, g|i).$$ (13)

Now apply (10) to compute:

$$p(d|h, i) = \sum_g p(d, g|i) \cdot p(g|h, i).$$ (14)

The remainder of the example is as follows:

$$p(c|d) = p(c, d)/ \sum_c p(c, d),$$ (15)

$$p(c|h, i) = \sum_d p(c|d) \cdot p(d|h, i),$$ (16)

$$p(s|g, i, j) = p(j, s|g, i)/ \sum_s p(j, s|g, i),$$ (17)

$$p(s|h, i) = \sum_{g,j} p(s|g, i, j) \cdot (p(g|h, i) \cdot p(j|h, i)),$$ (18)

$$p(l|g, j, s) = p(l, j|g, s)/ \sum_l p(l, j|g, s),$$ (19)

$$p(l|h, i) = \sum_{g,j,s} p(l|g, j, s) \cdot (p(g|h, i) \cdot p(j|h, i) \cdot p(s|h, i)).$$ (20)

The outward pass can be illustrated in Fig. 6 (ii), where all buckets from (i) were updated.

Whereas VE's inward pass in Fig. 2 (vi) constructed $p(j|h = 0, i = 1)$ at the root node, Cozman's outward pass constructed posteriors for all nodes in Fig. 2 (vi), namely, $p(g|h = 0, i = 1)$ in (12), $p(d|h = 0, i = 1)$ in (14), $p(c|h = 0, i = 1)$ in (16), $p(s|h = 0, i = 1)$ in (18), and $p(l|h = 0, i = 1)$ in (20).

The precomputation performed during the outward pass in [4] can be exploited when answering subsequente queries as demonstrated in Example 9.

Example 9. Given a new query $p(s|h = 0, i = 1)$, Cozman's method can readily answer it using (18).

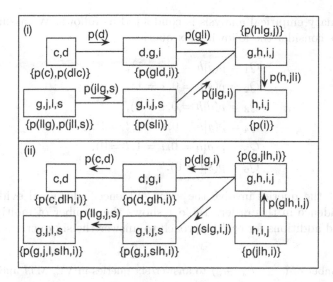

Fig. 6. In Example 8, (i) shows a tree of buckets for the inward pass and (ii) for the outward pass as described in [4]

6   Experimental Results

In the section, we compare and contrast VE, Cozman's approach, and marginal tree inference.

While VE is a simple approach to BN inference, repeatedly applying VE on the original BN can result in computation being duplicated, as previously discussed in Section 3.

Cozman [4] can alleviate some repeated computation. For example, as shown in Example 9, the computation in (17)-(18) for answering an initial query can be reused to answer a subsequent query in Example 9.

On the other hand, the price to pay is that precomputation can build tables that are never used. For instance, a second look at Example 9 reveals that the tables $p(l|g, j, s)$ in (19) and $p(l|h, i)$ in (20) are not reused. This means that the computation performed in (19)-(20) was wasteful.

We attempt here to stake out middle ground. One aim is to exploit VE as it is such a simple and clear approach to inference. Like Cozman, we seek to avoid repeated computation in VE. Unlike Cozman, however, our philosophy is to never build a probability table that goes unused.

MTI saves VE's past computation in marginal trees. Each time a new query is issued, MTI checks to see whether VE's past computation can be reused. If yes, MTI determines both that computation which can be reused and that computation which is missing. MTI proceeds to build only the missing computation. In this way, the query is answered and all tables built by MTI are used.

A preliminary empirical analysis is conducted as follows. We assume binary variables. We consider the following six queries:

$$Q_1 = p(j|h = 0, i = 1),$$
$$Q_2 = p(s|h = 0, i = 1),$$
$$Q_3 = p(d|h = 0, i = 1),$$
$$Q_4 = p(j|h = 0, i = 1, l = 0),$$
$$Q_5 = p(d|h = 0, i = 1, l = 0),$$
$$Q_6 = p(s|l = 0).$$

Observe that the queries involve the same evidence, additional evidence, and retracted evidence in this order. Table 1 shows the number of multiplications, divisions, and additions for each approach to answer the six queries.

Table 1. Number of ($\cdot$, / , +) to answer six queries in VE, MTI and Cozman's algorithm

$\dfrac{Algorithm}{Query}$	VE	MTI	Cozman
Q_1	(76,8,34)	(76,8,34)	(220,68,80)
Q_2	(72,8,34)	(44,8,20)	(0,0,24)
Q_3	(84,8,38)	(48,8,20)	(0,0,8)
Q_4	(84,16,38)	(80,16,32)	(212,68,100)
Q_5	(132,16,50)	(128,16,48)	(0,0,16)
Q_6	(56,4,24)	(44,4,22)	(136,48,74)
Total	(504,60,218)	(420,60,210)	(568, 184, 302)

Table 1 shows several interesting points. First, the number of multiplications, divisions, and additions for MTI will never exceed those for VE, respectively. MTI seems to show a small but noticeable improvement over VE. The usefulness of Cozman's approach is clearly evident from queries Q_2, Q_3 and Q_5, but is overshadowed by the wasteful precomputation for the other queries. Based on these encouraging results, future work includes a rigorous comparison on numerous real world and benchmark BNs, as well as establishing theoretical properties of marginal tree inference.

7   Conclusions

Applying VE against a given Bayesian network for each query can result in repeated computation. Cozman [4] alleviates some of this repeated computation, but at the expense of possibly building tables that remain unused. Our approach in this paper is to stake out middle ground.

Marginal tree inference seeks to maximize the reuse of VE's past computation, while at the same time ensuring that every table built is used to answer a query. This is consistent with [3], where it is emphasized that VE does not support precomputation.

Future work will investigate relationships and provide empirical comparisons between MTI and join tree propagation [5], including Lazy propagation [9,10,11,12] and various extensions such as using prioritized messages [13], message construction using multiple methods [14,15,16], and exploiting semantics [7,8].

References

1. Koller, D., Friedman, N.: Probabilistic Graphical Models: Principles and Techniques. MIT Press (2009)
2. Pearl, J.: Probabilistic Reasoning in Intelligent Systems: Networks of Plausible Inference. Morgan Kaufmann (1988)
3. Zhang, N.L., Poole, D.: A simple approach to Bayesian network computations. In: Proceedings of the Tenth Biennial Canadian Artificial Intelligence Conference, pp. 171–178 (1994)
4. Cozman, F.G.: Generalizing variable elimination in Bayesian networks. In: Workshop on Probabilistic Reasoning in Artificial Intelligence, Atibaia, Brazil (2000)
5. Shafer, G.: Probabilistic Expert Systems, vol. 67. Society for Industrial and Applied Mathematics, Philadelphia (1996)
6. Tarjan, R., Yannakakis, M.: Simple linear-time algorithms to test chordality of graphs, test acyclicity of hypergraphs, and selectively reduce acyclic hypergraphs. SIAM Journal on Computing 13(3), 566–579 (1984)
7. Butz, C.J., Yao, H., Hua, S.: A join tree probability propagation architecture for semantic modeling. Journal of Intelligent Information Systems 33(2), 145–178 (2009)
8. Butz, C.J., Yan, W.: The semantics of intermediate cpts in variable elimination. In: Fifth European Workshop on Probabilistic Graphical Models (2010)
9. Madsen, A.L., Butz, C.J.: Ordering arc-reversal operations when eliminating variables in Lazy AR propagation. International Journal of Approximate Reasoning 54(8), 1182–1196 (2013)
10. Madsen, A.L., Jensen, F.V.: Lazy propagation: A junction tree inference algorithm based on Lazy evaluation. Artificial Intelligence 113(1-2), 203–245 (1999)
11. Madsen, A.L.: Improvements to message computation in Lazy propagation. International Journal of Approximate Reasoning 51(5), 499–514 (2010)
12. Madsen, A.L., Butz, C.J.: On the importance of elimination heuristics in Lazy propagation. In: Sixth European Workshop on Probabilistic Graphical Models (PGM), pp. 227–234 (2012)
13. Butz, C.J., Hua, S., Konkel, K., Yao, H.: Join tree propagation with prioritized messages. Networks 55(4), 350–359 (2010)

14. Butz, C., Hua, S.: An improved Lazy-ar approach to Bayesian network inference. In: Nineteenth Canadian Conference on Artificial Intelligence (AI), pp. 183–194 (2006)
15. Butz, C.J., Konkel, K., Lingras, P.: Join tree propagation utilizing both arc reversal and variable elimination. International Journal of Approximate Reasoning 52(7), 948–959 (2011)
16. Butz, C.J., Chen, J., Konkel, K., Lingras, P.: A formal comparison of variable elimination and arc reversal in Bayesian network inference. Intelligent Decision Technologies 3(3), 173–180 (2009)

On SPI-Lazy Evaluation of Influence Diagrams

Rafael Cabañas[1], Andrés Cano[1], Manuel Gómez-Olmedo[1],
and Anders L. Madsen[2,3]

[1] Department of Computer Science and Artificial Intelligence,
CITIC, University of Granada, Spain
{rcabanas,acu,mgomez}@decsai.ugr.es
[2] HUGIN EXPERT A/S
Aalborg, Denmark
anders@hugin.com
[3] Department of Computer Science,
Aalborg University, Denmark

Abstract. Influence Diagrams are an effective modelling framework for analysis of Bayesian decision making under uncertainty. Improving the performance of the evaluation is an element of crucial importance as real-world decision problems are more and more complex. Lazy Evaluation is an algorithm used to evaluate Influence Diagrams based on message passing in a strong junction tree. This paper proposes the use of Symbolic Probabilistic Inference as an alternative to Variable Elimination for computing the clique-to-clique messages in Lazy Evaluation of Influence Diagrams.

Keywords: Influence Diagrams, Combinatorial Factorization Problem, Exact Evaluation, Heuristic Algorithm, Lazy Evaluation, Junction Tree.

1 Introduction

Influence Diagrams (IDs) [1] are a tool to represent and evaluate decision problems under uncertainty. A technique used to evaluate IDs is Lazy Evaluation (LE) [2,3]. Its basic idea is to maintain a decomposition of the potentials and postpone computations for as long as possible. Thus it is possible to exploit barren variables and independence induced by evidence.

LE is based on message passing in a strong junction tree, which is a representation of a decision problem represented as an ID. Computing the messages involves the removal of variables. In the original proposal, the method used is *Variable Elimination* (VE) [3]. An alternative method for removing a set of variables from a set of potentials is *Symbolic Probabilistic Inference* algorithm (SPI) [4,5,6], which considers the removal as a combinatorial factorization problem. That is, SPI tries to find the optimal order for the combinations and marginalizations (i.e. max-marginalization and sum-marginalization). In a previous paper [7], the basic version of the SPI algorithm was described for the direct evaluation of IDs. This algorithm was also proposed as an alternative for computing clique-to-clique messages in LE of Bayesian Networks (BNs) [8]. Our contribution is

L.C. van der Gaag and A.J. Feelders (Eds.): PGM 2014, LNAI 8754, pp. 97–112, 2014.

to describe how the SPI algorithm can be used for computing the messages in LE of IDs. This new method for evaluating IDs is called *SPI-Lazy Evaluation (SPI-LE)*. The differences between BNs and IDs must be considered: two kinds of potentials, the temporal order between decisions, etc. The experimental work shows how SPI can improve the efficiency of LE. In the experimental work we use a set of IDs present in the literature.

The paper is organized as follows: Section 2 introduces basic concepts about IDs, LE and the motivation of this work; Section 3 describes how SPI can be used for computing the messages in LE of IDs; Section 4 includes the experimental work and results; finally Section 5 details our conclusions and lines for future work.

2   Preliminaries

2.1   Influence Diagrams

An ID [1] is a Probabilistic Graphical Model for decision analysis under uncertainty which contains three kinds of nodes: *decision nodes* (squares) that correspond with the actions which the decision maker can control; *chance nodes* (circles) representing random variables; and *utility nodes* (diamonds) representing the decision maker preferences. Fig. 1 shows an example of an ID.

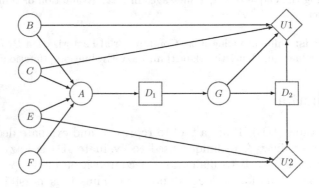

Fig. 1. An example of an ID with the partial order is $\{A\} \prec D_1 \prec \{G\} \prec D_2 \prec \{B, C, E, F\}$

We denote by $\mathcal{U}_C$ the set of chance nodes, by $\mathcal{U}_D$ the set of decision nodes, and by $\mathcal{U}_V$ the set of utility nodes. The decision nodes have a temporal order, $D_1, \ldots, D_n$, and the chance nodes are partitioned into a collection of disjoint sets according to when they are observed: $\mathcal{I}_0$ is the set of chance nodes observed before D_1, and $\mathcal{I}_i$ is the set of chance nodes observed after decision D_i is taken and before decision D_{i+1} is taken. Finally, $\mathcal{I}_n$ is the set of chance nodes observed after D_n. That is, there is a partial order: $\mathcal{I}_0 \prec D_1 \prec \mathcal{I}_1 \prec \cdots \prec D_n \prec \mathcal{I}_n$.

In the description of an ID, it is more convenient to think in terms of predecessors: the parents of a chance node X_i, denoted $pa(X_i)$, are also called *conditional predecessors*. The parents of a utility node V_i, denoted $pa(V_i)$, are also
called conditional predecessors. Similarly, the parents of a decision D_i are called
informational predecessors and are denoted $pa(D_i)$. Informational predecessors
of each decision D_i, must include previous decisions and their informational
predecessors (*no-forgetting assumption*).

The *universe* of the ID is $\mathcal{U} = \mathcal{U}_C \cup \mathcal{U}_D = \{X_1, \ldots, X_m\}$. Let us suppose
that each variable X_i is discrete and it takes values on a finite set $\Omega_{X_i} = \{x_1, \ldots, x_{|\Omega_{X_i}|}\}$. Each chance node X_i has a conditional probability distribu-
tion $P(X_i | pa(X_i))$ associated. In the same way, each utility node V_i has a utility
function $U(pa(V_i))$ associated. In general, we will talk about potentials (not necessarily normalized). The set of all variables involved in a potential ϕ is denoted
$dom(\phi)$, defined on $\Omega_{dom(\phi)} = \times \{\Omega_{X_i} | X_i \in dom(\phi)\}$. The elements of $\Omega_{dom(\phi)}$
are called configurations of ϕ. Therefore, a *probability potential* denoted by ϕ is
a mapping $\phi : \Omega_{dom(\phi)} \to [0, 1]$. A *utility potential* denoted by ψ is a mapping
$\psi : \Omega_{dom(\psi)} \to \mathbb{R}$. The set of probability potentials is denoted by Φ while the set
of utility potentials is denoted by Ψ.

An arc between an informational predecessor and a decision is redundant if it
is d-separated from the utility nodes given the rest of informational predecessors.
Any redundant arc can be removed [9]. If no-forgetting arcs have been added
and redundant arcs have been removed, the parents of a decision compose its
relevant past. A chance or decision node is a *barren node* if it is a sink, in other
words, it has no children or only barren descendants. Any barren node can be
directly removed from the ID.

The goal of evaluating an ID is to obtain an *optimal policy* δ_i for each decision
D_i, that is a function of a subset of its informational predecessors. The optimal
policy maximizes the *expected utility* for the decision. A strategy is an ordered
set of policies $\Delta = \{\delta_1, \ldots, \delta_n\}$, including a policy for each decision variable. An
optimal strategy $\widehat{\Delta}$ returns the optimal choice the decision maker should take
for each decision.

Optimal policy: *Let ID be an influence diagram over the universe* $\mathcal{U} = \mathcal{U}_C \cup \mathcal{U}_D$
and let $\mathcal{U}_V$ *be the set of utility nodes. Let the temporal order of the variables be
described as* $\mathcal{I}_0 \prec D_1 \prec \mathcal{I}_1 \prec \cdots \prec D_n \prec \mathcal{I}_n$. *Then, an optimal policy for* D_i *is*

$$\delta_{D_i}(\mathcal{I}_0, D_1, \ldots, \mathcal{I}_{i-1}) =$$

$$= \arg\max_{D_i} \sum_{\mathcal{I}_i} \max_{D_{i+1}} \cdots \max_{D_n} \sum_{\mathcal{I}_n} \prod_{X \in \mathcal{U}_C} P(X | pa(X)) \left(\sum_{V \in \mathcal{U}_V} U(pa(V)) \right) \quad (1)$$

2.2 Lazy Evaluation

Lazy Evaluation (LE) was already used for making inference in BNs [10], so it can
be adapted for evaluating IDs [2,3]. The basic idea of this method is to maintain
the decomposition of the potentials for as long as possible and to postpone

computations for as long as possible, as well as to exploit barren variables. LE is based on message passing in a *strong junction tree*, which is a representation of an ID built by moralization and by triangulating the graph using a strong elimination order [11].

Nodes in the strong junction trees correspond to *cliques* (maximal complete subgraphs) of the triangulated graph. Each clique is denoted by C_i where i is the index of the clique. The root of the strong junction tree is denoted by C_1. Two neighbour cliques are connected by a separator which contains the intersection of the variables in both cliques. The size of a clique C_i, denoted $|C_i|$, is the number of variables. The weight of a clique C_i, denoted $w(C_i)$, can be defined as $\prod_{X \in C_i} |\Omega_X|$. An example of a strong junction tree is shown in Fig. 2.

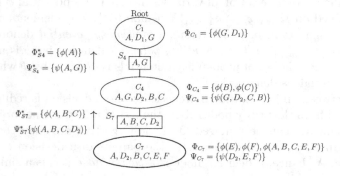

Fig. 2. Strong junction tree for the ID shown in Fig. 1 with the potentials associated to each clique (right) and messages stored at each separator (left)

Propagation is performed by message-passing. Initially, each potential is associated to the closest clique to the root containing all its variables. These potentials are not combined, so during propagation each clique and separator keeps two sets of potentials (one for probabilities and another for utilities). Sets of potentials stored in a clique C_j are denoted Φ_{C_j} and Ψ_{C_j}. Similarly, sets of potentials (or messages) stored in a separator S_j are denoted $\Phi^*_{S_j}$ and $\Psi^*_{S_j}$. Message propagation starts by invoking the *Collect Message* algorithm in the root (Algorithm 1).

Algorithm 1 (Collect Message) *Let C_j be a clique where Collect Message is invoked, then:*

1. *C_j invokes Collect Message in all its children.*
2. *The message to the clique parent of C_j is built and sent by absorption (Algorithm 2).*

A clique can send the message to its parent (*Absorption*) if it has received all the messages from its children. Consider a clique C_j and its parent separator S_j.

Absorption in C_j amounts to eliminating the variables of $C_j \backslash S_j$ from the list of probability and utility potentials associated with C_j and the separators of $ch(C_j)$ and then associating the obtained potentials with S_j. The original proposal [3] uses VE for removing the variables. Thus we will refer to this method as VE-Lazy Evaluation (VE-LE)

Algorithm 2 (Absorption) *Let C_j be a clique, S_j be the parent separator and $S' \in ch(C_j)$ be the child separators. If Absorption is invoked on C_j, then:*

1. *Let $\mathcal{R}_{S_j} = \Phi_{C_j} \cup \Psi_{C_j} \cup \bigcup_{S' \in ch(C_j)}(\Phi_{S'}^* \cup \Psi_{S'}^*)$.*
2. *Let $\mathbf{X} = \{X | X \in C_j, X \notin S_j\}$ the variables to be removed.*
3. *Choose an order to remove the variables in $\mathbf{X}$.*
4. *Marginalize out all variables in $\mathbf{X}$ from $\mathcal{R}_{S_j}$. Let $\Phi_{S_j}^*$ and $\Psi_{S_j}^*$ be the set of probability and utility potentials obtained.*
5. *Associate $\Phi_{S_j}^*$ and $\Psi_{S_j}^*$ to the parent separator S_j.*

The propagation finishes when the root clique has received all the messages. The utility potential from which each variable D_i is eliminated during the evaluation should be recorded as the *expected utility* for the decision D_i. The values of the decision that maximizes the expected utility is the policy for D_i. In case of decisions that are attached to the root node, the expected utility and policy is calculated by marginalizing out all variables in the root clique that do not belong to the relevant past of the decision.

2.3 Motivation

When Absorption is invoked on a clique, all variables in the clique not present in the parent separator are marginalized out from the set of relevant potentials. The original definition of LE proposes using *Variable Elimination* (VE) [3]. This algorithm chooses at each step a variable to remove based on any criteria or heuristic. This removal involves combining all potentials containing the chosen variable. Let us consider the strong junction tree shown in Fig. 2. If Absorption is invoked on C_7, variables E and F are marginalized out in order to compute the messages to S_7, denoted $\Phi_{S_7}^*$ and $\Psi_{S_7}^*$. The computations performed using VE are:

$$\Phi_{S_7}^* = \left\{ \sum_E P(E) \sum_F P(F)P(A|B,C,E,F) \right\} \tag{2}$$

$$\Psi_{S_7}^* = \left\{ \frac{\sum_E P(E)P(A|B,C,E)\frac{\sum_F P(F)P(A|B,C,E,F)U(D_2,E,F)}{P(A|B,C,E)}}{P(A|B,C)} \right\} \tag{3}$$

Assuming that all the variables are binary, the computation of $\Phi_{S_7}^*$ and $\Psi_{S_7}^*$ requires 144 multiplications and 72 additions and 48 divisions. Independently of the elimination ordering used to remove E and F, VE will always have to combine the marginal potentials with a large potential such as $P(A|B, C, E, F)$. However, with a re-order of the operations this situation can be avoided:

$$\Phi^*_{S_7} = \left\{ \sum_E \sum_F \left(P(A|B, C, E, F) \left(P(E)P(F) \right) \right) \right\} \tag{4}$$

$$\Psi^*_{S_7} = \left\{ \frac{\sum_E \sum_F \left(P(A|B, C, E, F) \left(P(E)P(F) \right) U(D_2, E, F) \right)}{P(A|B, C)} \right\} \tag{5}$$

Using Eq. (4) and (5) the computation of the messages requires 100 multiplications, 72 additions and 16 divisions. In some cases it could be better to combine small potentials even if they do not share any variable (e.g., $P(E)$ and $P(F)$). This combination will never be performed using VE since it is guided by the elimination ordering. Thus the efficiency of the message computation can be improved if an optimal ordering for the operations of marginalization and combination is found [5]. To overcome this, we propose for computing the clique-to-clique messages the SPI algorithm , which is more flexible than VE.

3   SPI Lazy Evaluation

3.1   Overview

SPI Lazy Evaluation (SPI-LE) uses SPI instead of VE in order to compute the messages. Thus the process for building the strong junction tree and *Collect Message* algorithm are the same. The general scheme of the *Absorption* algorithm is slightly different (see Algorithm 3). The main difference is that the set of variables to remove is partitioned into disjoint subsets of chance variables or single sets of decisions. Then, the removal of these subsets of variables is invoked in an order that respects the the temporal constraints. Notice that the removal of a subset of variables $\mathbf{X}_k$ is invoked only on the set of potentials containing any variable in $\mathbf{X}_k$.

Algorithm 3 (Absorption SPI) *Let C_j be a clique, S_j be the parent separator and $S' \in ch(C_j)$ be the child separators. If Absorption is invoked on C_j, then:*

1. *Set the relevant potential sets:*
   $\Phi^*_{S_j} := \Phi_{C_j} \cup \bigcup_{S' \in ch(C_j)} \Phi^*_{S'}$ $\Psi^*_{S_j} := \Psi_{C_j} \cup \bigcup_{S' \in ch(C_j)} \Psi^*_{S'}$
2. *Let $\mathbf{X} := \{X | X \in C_j, X \notin S_j\}$ the variables to remove.*
3. *Partition $\mathbf{X}$ into disjoint subsets of chance variables or single sets of decisions. Determine a partial order of the subsets that respects the temporal constraints: $\{\mathbf{X}_1 \prec \mathbf{X}_2 \prec \cdots \prec \mathbf{X}_n\}$*
4. *For $k := n$ to 1 do:*
   (a) *Set $\Phi_{\mathbf{X}_k} := \{\phi \in \Phi^*_S | \mathbf{X}_k \cap dom(\phi) \neq \emptyset\}$ and $\Psi_{\mathbf{X}_k} := \{\psi \in \Psi^*_S | \mathbf{X}_k \cap dom(\psi) \neq \emptyset\}$.*

(b) Remove $\mathbf{X}_k$ from $\Phi_{\mathbf{X}_k}$ and $\Psi_{\mathbf{X}_k}$. If $\mathbf{X}_k$ is a subset of chance variables use Algorithm 4, otherwise use Algorithm 6. The inputs to these algorithms are $\mathbf{X}_k$, $\Phi_{\mathbf{X}_k}$ and $\Psi_{\mathbf{X}_k}$ while $\Phi_{\mathbf{X}_k}^*$ and $\Psi_{\mathbf{X}_k}^*$ are the sets of potentials obtained.

(c) Update the relevant potential sets:

$$\Phi_{S_j}^* := (\Phi_{S_j}^* \backslash \Phi_{\mathbf{X}_k}) \cup \Phi_{\mathbf{X}_k}^* \quad \Psi_{S_j}^* := (\Psi_{S_j}^* \backslash \Psi_{\mathbf{X}_k}) \cup \Psi_{\mathbf{X}_k}^*$$

5. Associate $\Phi_{S_j}^*$ and $\Psi_{S_j}^*$ to the parent separator S_j.

3.2 Removal of Chance Variables

In order to remove a subset of chance variables $\mathbf{X}$ from $\Phi_{\mathbf{X}}$ and $\Psi_{\mathbf{X}}$, SPI considers probability and utility potentials separately: first, SPI tries to find the best order for combining all potentials in $\Phi_{\mathbf{X}}$. For that purpose, all possible pairwise combinations between the probability potentials are stored in the set combination candidate set B. Besides, B also contains those probability potentials that contain any variable of $\mathbf{X}$ which is not present in any other potential of $\Phi_{\mathbf{X}}$, that is a variable that can be directly removed (without performing any combination). At each iteration, an element of B is selected. If this element is a pair, both potentials are combined. The procedure stops when all variables have been removed. A variable can be removed in the moment it only appears in a single probability potential. Notice that this algorithm produces a factorization of potentials. This procedure is shown in Algorithm 4.

Algorithm 4 (Removal of a subset of chance variables) Let $\mathbf{X}$ be a set of chance variables, $\Phi_{\mathbf{X}}$ and $\Psi_{\mathbf{X}}$ be sets of probability and utility potentials relevant for removing $\mathbf{X}$. If the removal of $\mathbf{X}$ is invoked on $\Phi_{\mathbf{X}}$ and $\Psi_{\mathbf{X}}$, then:

1. Initialize the combination candidate set $B := \emptyset$.
2. Repeat:
    (a) Add all pairwise combinations of elements of $\Phi_{\mathbf{X}}$ to B which are not already in B.
    (b) Add to B all potentials in $\Phi_{\mathbf{X}}$ which are not already in B and that contain any variable of $\mathbf{X}$ which is not present in any other potential of $\Phi_{\mathbf{X}}$, that is a variable that can be removed.
    (c) Select a pair $p := \{\phi_i, \phi_j\}$ or a singleton $p := \{\phi_i\}$ from B according to some heuristic.
    (d) If p is a pair, then $\phi_{ij} := \phi_i \otimes \phi_j$. Otherwise, $\phi_{ij} := \phi_i$
    (e) Determine the set $\mathbf{W}$ of variables that can be sum-marginalized:

    $$\mathbf{W} := \{W \in dom(\phi_{ij}) \cap \mathbf{X} | \forall \phi \in \Phi_{\mathbf{X}} \backslash p : W \notin dom(\phi)\}$$

    (f) Select the utility potentials relevant for removing $\mathbf{W}$:

    $$\Psi_{\mathbf{W}} := \{\psi \in \Psi_{\mathbf{X}} | \mathbf{W} \cap dom(\psi) \neq \emptyset\}$$

(g) If $\mathbf{W} \neq \emptyset$, sum-marginalize variables in $\mathbf{W}$ from ϕ_{ij} and $\Psi_{\mathbf{W}}$. A probability potential $\phi_{ij}^{\downarrow \mathbf{W}}$ and a set of utility potentials $\Psi^{\downarrow \mathbf{W}}$ are obtained as
a result (Algorithm 5).

(h) Update:
  - $\mathbf{X} := \mathbf{X} \backslash \mathbf{W}$
  - If p is a pair, $\Phi_{\mathbf{X}} := \Phi_{\mathbf{X}} \backslash \{\phi_i, \phi_j\}$ and remove any element in B
    containing ϕ_i or ϕ_j. Otherwise, $\Phi_{\mathbf{X}} := \Phi_{\mathbf{X}} \backslash \{\phi_i\}$ and remove any
    element in B containing ϕ_i.
  - $\Phi_{\mathbf{X}} := \Phi_{\mathbf{X}} \cup \{\phi_{ij}^{\downarrow \mathbf{W}}\}$ $\Psi_{\mathbf{X}} := (\Psi_{\mathbf{X}} \backslash \Psi^{\mathbf{W}}) \cup \Psi^{\downarrow \mathbf{W}}$

  Until $\mathbf{X} = \emptyset$:

3. Return $\Phi_{\mathbf{X}}$ and $\Psi_{\mathbf{X}}$.

In Algorithm 4 only probability potentials are combined while utility potentials are not. The utility potentials must be combined with ϕ_{ij} which is the
resulting potential of combining all potentials containing X. For that reason,
the utilities can only be combined when a variable can be removed. That is the
moment when ϕ_{ij} has been calculated. The procedure for sum-marginalizing a
set of variables (Algorithm 5) involves finding good order for summing the utility
potentials. The procedure for that is quite similar to the procedure for combining probabilities, the main difference is that in the moment a variable can be
removed, the probability and utility potentials resulting of the marginalization
are computed. Notice that this procedure is invoked on $\Psi_{\mathbf{W}} \subseteq \Psi_{\mathbf{X}}$.

Algorithm 5 (Sum-marginalization) *Let ϕ be a probability potential and
$\Psi_{\mathbf{W}}$ a set of utility potentials relevant for removing the chance variables in $\mathbf{W}$.
Then, the procedure for sum-marginalizating $\mathbf{W}$ from ϕ and $\Psi_{\mathbf{W}}$ is:*

1. *Initialize the combination candidate set $B' := \emptyset$.*
2. *if $\Psi_{\mathbf{W}} = \emptyset$, then return $\sum_{\mathbf{W}} \phi$*
3. *Repeat:*
   (a) *Add all pairwise combinations of elements of $\Psi_{\mathbf{W}}$ to B' which are not
       already in B'.*
   (b) *Add to B' all potentials in $\Psi_{\mathbf{W}}$ which are not already in B' that contains
       any variable of $\mathbf{W}$ which is not present in any other potential of $\Psi_{\mathbf{W}}$,
       that is a variable that can be removed.*
   (c) *Select a pair $q := \{\psi_i, \psi_j\}$ or a singleton $q := \{\psi_i\}$ from B' according to
       some heuristic.*
   (d) *If q is a pair, then $\psi_{ij} := \psi_i + \psi_j$. Otherwise, $\psi_{ij} := \psi_i$*
   (e) *Determine the set $\mathbf{V}$ of variables that can be sum-marginalized:*

   $$\mathbf{V} := \{V \in dom(\psi_{ij}) \cap \mathbf{W} | \forall \psi \in \Psi_{\mathbf{W}} \backslash q : V \notin dom(\psi)\}$$

   (f) *If, $\mathbf{V} \neq \emptyset$, sum-marginalize $\mathbf{V}$, giving as a result:*

   $$\phi^{\downarrow \mathbf{V}} := \sum_{\mathbf{V}} \phi \qquad \psi^{\downarrow \mathbf{V}} := \sum_{\mathbf{V}} (\phi \otimes \psi_{ij}) / \phi^{\downarrow \mathbf{V}}$$

(g) *Update:*
- $\mathbf{W} := \mathbf{W} \backslash \mathbf{V}$
- *If q is a pair, $\Psi_{\mathbf{W}} := \Psi_{\mathbf{W}} \backslash \{\psi_i, \psi_j\}$ and remove any element in B' containing ψ_i or ψ_j. Otherwise, $\Psi_{\mathbf{W}} := \Psi_{\mathbf{W}} \backslash \{\psi_i\}$ and remove any element in B' containing ψ_i.*
- $\phi := \phi^{\downarrow \mathbf{V}}$ *and* $\Psi_{\mathbf{W}} := \Psi_{\mathbf{W}} \cup \{\psi^{\downarrow \mathbf{V}}\}$

Until $\mathbf{W} = \emptyset$

4. *Return ϕ and $\Psi^{\mathbf{W}}$.*

3.3   Removal of Decision Variables

The removal of a decision variable does not imply the combination of any probability potential since any decision is d-separated from its predecessors [12] and any successor has already been removed (the removal order of the disjoint subsets of variables must respect the temporal constraints). Thus, any probability potential $\phi(D_k, \mathbf{X})$ must be directly transform into $\phi(\mathbf{X})$ if D_k is a decision and $\mathbf{X}$ is a set of chance variables that belong to $\mathcal{I}_i$ with $i < k$. This property is used at step 2 of Algorithm 6.

Algorithm 6 (Removal of a decision) *Let D be a decision variable, Φ_D and Ψ_D be sets of probability and utility potentials relevant for removing D. If the removal of D is invoked on Φ_D and Ψ_D, then:*

1. *For each $\phi \in \Phi_D$, remove D by restricting ϕ to any of the values of D. The set of potentials $\Phi^{\downarrow D}$ is given as a result.*
2. *Max-marginalize variable D from Ψ_D. A new potential $\psi^{\downarrow D}$ is obtained as a result (Algorithm 7).*
3. *Return $\Phi^{\downarrow D}$ and $\psi^{\downarrow D}$*

Algorithm 7 shows the procedure for finding the best order for summing all utility potentials containing a decision D. Notice that the pairwise candidate set does not contain singletons and the sum-marginalization is performed once all utility potentials have been summed.

Algorithm 7 (Max-marginalization) *Let D be a decision variable and Ψ_D be a set of utility potentials containing D. Then, the procedure for max-marginalizating D from Ψ_D is:*

1. *Initialize the combination candidate set $B' := \emptyset$.*
2. *While $|\Psi_D| > 1$:*
   (a) *Add all pairwise combinations of elements of Ψ_D to B' which are not already in B'.*
   (b) *Select a pair $q := \{\psi_i, \psi_j\}$ according to some heuristic and sum both potentials giving as a result ψ_{ij}.*
   (c) *Update:*

      – *Delete all pairs p of B' where $\psi_i \in p$ or $\psi_j \in p$.*
      – $\Psi_D := \Psi_D \backslash \{\psi_i, \psi_j\} \cup \{\psi_{ij}\}$.

3. *Let ψ^D be the single potential in Ψ.*
4. *Max-marginalize D, giving as a result $\psi^{\downarrow D} := \max_D \psi^D$ and record the policy for D.*
5. *Return $\psi^{\downarrow D}$.*

3.4 Heuristics

During the removal of the chance variables, at each iteration a pair of probability potentials is selected to be combined (Definition 4, step 2.c). For that, any heuristic used with VE can be adapted for selecting a pair. Let $p := \{\phi_i, \phi_j\}$ be a candidate pair to be combined, let $\phi_{ij} = \phi_i \otimes \phi_j$ be the resulting potential of the combination. Then the heuristic *minimum size* [13] will select a pair minimizing Eq. (6). Thus *minimum size* heuristic chooses a pair that minimizes the number of variables in the resulting potential. This heuristic can also be used for selecting a pair of utility potentials at steps 3.c and 2.b of Definitions 5 and 7 respectively.

$$min\_size(p) = |dom(\phi_i) \cup dom(\phi_j)| = |dom(\phi_{ij})| \qquad (6)$$

    Let $\mathbf{W}$ be the set of variables that can be removed after combining potentials in p. Then the algorithm should check if any variable in $\mathbf{W}$ is a *probabilistic barren*, that is a barren node if only the set of probability potentials are considered. The removal of a probabilistic barren from a probability potential leads to an unity-potential. During the calculation of messages unity-potentials are not calculated and if the denominator of a division is a unity-potential, then the division is no not performed.

3.5 Example

To illustrate the computations of the messages using SPI-LE, let us consider the strong junction tree shown in Fig. 2 representing the ID in Fig. 1 with binary variables. To simplify the notation, $\phi(X_1, \ldots, X_n)$ will be denoted $\phi_{X_1, \ldots, X_n}$.

    Initially, *Collect Message* (Algorithm 1) is invoked on the root clique C_1 and recursively invoked on its children. Once *Collect Message* is invoked on the leaf clique C_7, the messages for the parent separator S_7 are computed (Algorithm 3). The relevant potentials are:

$$\Phi_S^* := \{\phi_E, \phi_F, \phi_{ABCEF}\} \qquad \Psi_S^* := \{\psi_{D_2EF}\}$$

    Variables $\{E, F\}$ must be removed for computing the messages. Both of them belong to $\mathcal{I}_2$, so they can be removed in any order. Then, the removal of $\{E, F\}$ is invoked on $\{\phi_E, \phi_F, \phi_{ABCEF}\} \cup \{\psi_{D_2EF}\}$ (Algorithm 4). The initial combination candidate set B is:

$$B := \{\{\phi_E, \phi_F\}, \{\phi_E, \phi_{ABCEF}\}, \{\phi_F, \phi_{ABCEF}\}\}$$

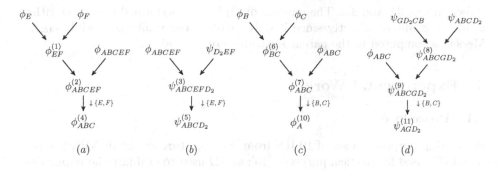

Fig. 3. Combination order of the probability and utility potentials obtained using SPI for removing chance variables when computing the messages sent from C_7 (a and b) and from C_4 (c and d) to their respective parent separators in the strong junction tree shown in Fig.2.

Notice that the set B does not contain any singleton because none of the variables appear only in one potential. If *minimum size* is the heuristic used, the element of B chosen is the pair $\{\phi_E, \phi_F\}$. Both potentials are combined giving as a result a new potential ϕ_{EF}. None of the variables can be marginalized as they are also contained in ϕ_{ABCEF}. In the second iteration, the combination candidate set is updated as follows:

$$B := \{\{\phi_{EF}, \phi_{ABCEF}\}\}$$

Now, the single pair of B is chosen and both potentials are combined giving as a result a new potential ϕ_{ABCEF} from which variables E and F can be sum-marginalized out. Using Algorithm 5 these variables are removed from the resulting probability potential and also from the utility potential ψ_{D_2EF}. For that, a combination candidate set B' with the relevant utility potentials is generated:

$$B' := \{\{\psi_{D_2EF}\}\}$$

Since B' contains only one element which is a singleton, the algorithm performs directly removal of variables E and F from ϕ_{ABCEF} and ψ_{D_2EF} (Algorithm 5 step 3.f). The resulting potentials are ϕ_{ABC} and ψ_{ABCD_2} which are, in this case, the messages for the parent separator.

The whole process for computing the messages from C_7 is shown in Fig. 3.a and 3.b in two factor graphs [14]. Nodes without any parent correspond to initial potentials while child nodes correspond to the resulting potentials of a combination. The numbers above each potentials indicate the combination ordering and arcs labels indicate the variables that are sum-marginalized.

When absorption is invoked on C_4, a similar procedure is followed for removing variables B, C and D_2. Now, these variables are partitioned into the disjoint subsets $\{\{D_2\} \prec \{B, C\}\}$. First, variables B and C are removed using Algorithm 4 giving as a result the potentials ϕ_A and ψ_{AGD_2}. The process for this removal is shown in Fig. 3.c and 3.d. The removal of D_2 is now performed using Algorithm 6 and it is almost directly since it only involves the utility potential ψ_{AGD_2}. Messages computed to the parent separator are ϕ_A and ψ_{AG}.

4   Experimental Work

4.1   Procedure

For testing purposes, a set of 10 IDs from the literature are used: NHL is a real world IDs used for medical purposes [15]; an ID used to evaluate the population population viability of wildlife species [16]; the oil wildcatter's problem[17]; the Chest Clinic ID [18] obtained from the Asia BN; an ID representing the decision problem in the poker game [12]; an ID used at agriculture for treating mildew [12]; an ID to model a simplified version of the dice game called *Think-box*[1]; finally, three synthetic IDs are used: the motivation example shown in Fig. 1 with binary variables and the ID used by Jensen et al. in [2]. The details of these IDs are shown in Table 1, which contains the number of nodes of each kind.

Table 1. Features of the IDs used for the experimental work

ID	Number of nodes		
	Chance	Decisions	Utility
NHL	17	3	1
Wildlife	9	1	1
Oil Wildcatter	2	2	2
ChestClinic	8	2	2
Poker	7	1	1
Mildew	7	2	2
Motivation ID	6	2	2
Jensen et al.	12	4	4
Thinkbox	5	2	4
Car buyer	3	3	1

From each ID a strong junction tree is built using the *minimum size* heuristic [13] for triangulating the graph. Table 2 shows, for each tree, the number of cliques, the minimum and maximum clique size $|C|$ and the minimum and maximum clique weight $w(C)$.

The message passing is performed using SPI-LE and VE-LE. In the first case, the heuristic used for selecting the next pair to combine is *minimum size* as described in Section 3.4. For the VE-LE the order used for removing the variables is the same than the one used during the triangulation. It must be noticed that in both cases a similar heuristic is used in order to obtained comparable results. For each evaluation, the number of operations involved is measured. The ratio

---

[1] http://www.hugin.com/technology/samples/think-box

Table 2. Features of the strong junction trees used for the experimental work obtained with *minimum size* heuristic

	Cliques	$|C|$		$w(C)$	
		min	max	min	max
NHL	8	5	12	32	$5.530 \cdot 10^5$
Wildlife	7	3	4	8	16
Oil Wildcatter	1	4	4	36	36
ChestClinic	5	3	6	8	64
Poker	5	3	3	32	324
Mildew	4	4	6	256	9408
Motivation ID	3	3	6	8	64
Jensen et al.	9	3	5	8	32
Thinkbox	2	4	6	8	384
Car buyer	1	6	6	384	384

between the number of operations using both methods can be computed using Eq. (7). A value lower than 1 means a better performance of SPI-LE.

$$ratio = \frac{SPI\text{-}LE \text{ operations}}{VE\text{-}LE \text{ operations}} \tag{7}$$

In order to check that SPI-LE does not have a large overhead, the CPU time needed for the evaluation of each ID is also measured. All the algorithms are implemented in Java with the Elvira Software[2]. The tests are run on a Intel Core i7-2600 (8 cores, 3.4 GHz). Each evaluation is repeated 10 times to avoid the effects of outliers. To analyze the reduction in the evaluation time, the *speed-up* can be computed using Eq. (8). In this case, a value higher than 1 means a better performance of SPI-LE.

$$speed\text{-}up = \frac{VE\text{-}LE \text{ CPU time}}{SPI\text{-}LE \text{ CPU time}} \tag{8}$$

4.2   Results

Table 3 shows the total number of operations needed for each evaluation and the ratio of the number of operations using SPI-LE to the number of operations using VE-LE. It can be observed that in most of the cases the number of operations required using SPI-LE is lower than using VE-LE. The ID representing the wildcatter's problem requires the same number of operations with both algorithms. The reason for that is that this ID is too simple to obtain any gain. By contrast the higher reduction (lower ratio) is obtained when evaluating the NHL ID, which is the largest ID used in the experimentation.

Table 4 shows the number of operations of each kind used for the evaluation using each method. That is the number of multiplications, divisions, additions

---

[2] http://leo.ugr.es/~elvira

Table 3. Total number of operations used for evaluating each ID using SPI-LE and VE-LE

ID	SPI-LE	VE-LE	ratio
NHL	$2.848 \cdot 10^6$	$4.660 \cdot 10^6$	0.611
Wildlife	163	172	0.948
Oil Wildcatter	121	121	1
ChestClinic	534	576	0.927
Poker	1781	1916	0.93
Mildew	$3.133 \cdot 10^4$	$3.218 \cdot 10^4$	0.974
Motivation ID	320	428	0.748
Jensen et al.	270	280	0.964
Thinkbox	1872	2308	0.811
Car buyer	1518	1535	0.989

and max-comparisons. It can be observed that the gain is mainly obtained due to the reduction in the number of multiplications. The number of multiplications required using SPI-LE is always lower or equal than using VE-LE. By contrast, in some cases, the rest of operations is higher if SPI-LE is used.

Table 4. Number of each kind operation used for evaluating each ID using SPI-LE and VE-LE

ID	Multiplications		Divisions		Additions		Max comparisons	
	SPI-LE	VE-LE	SPI-LE	VE-LE	SPI-LE	VE-LE	SPI-LE	VE-LE
NHL	$1.566 \cdot 10^6$	$2.674 \cdot 10^6$	$1.248 \cdot 10^4$	$5.794 \cdot 10^5$	$1.258 \cdot 10^6$	$1.386 \cdot 10^6$	$1.104 \cdot 10^4$	$2.120 \cdot 10^4$
Wildlife	104	110	8	8	50	52	1	2
Oil Wildcatter	60	60	12	12	42	42	7	7
ChestClinic	268	280	80	96	168	180	18	20
Poker	970	1186	18	0	792	729	1	1
Mildew	$1.552 \cdot 10^4$	$1.626 \cdot 10^4$	1408	1472	$1.346 \cdot 10^4$	$1.346 \cdot 10^4$	944	992
Motivation ID	152	212	24	72	138	138	6	6
Jensen et al.	152	156	16	16	92	96	10	12
Thinkbox	592	624	224	256	936	1296	120	132
Car buyer	736	864	224	192	440	364	118	115

The reduction in the number of operations needed for evaluating an ID should lead more efficient algorithms. Table 5 shows the CPU time needed for evaluating each ID and the speed-up obtained using SPI-LE with respect to VE-LE. For all the IDs, the CPU time is lower if SPI-LE is used instead for VE-LE. The speed-up obtained when evaluating NHL ID is not really high though there is a great difference in the number of operations. The reason for that is that SPI-LE is likely to have a larger overhead than VE-LE, specially with large ID where the algorithm considers a large number of pairwise combinations. However, the

Table 5. CPU time in milliseconds and speed-up needed for evaluating each ID using SPI-LE and VE-LE

ID	SPI-LE	VE-LE	speed-up
NHL	7944.2	$1.461 \cdot 10^4$	1.839
Wildlife	167.7	188	1.121
Oil Wildcatter	78.6	89.1	1.134
ChestClinic	42.7	204.3	4.785
Poker	16.1	187.6	11.652
Mildew	81.2	204.7	2.521
Motivation ID	63.5	159.2	2.507
Jensen et al.	54.3	311.2	5.731
Thinkbox	50.3	131.8	2.62
Car buyer	63.6	159.8	2.513

gain obtained with the reduction in the number of operations compensates this large overhead.

5 Conclusions and Future Work

In the present paper we have described how the SPI algorithm can be used for computing clique-to-clique messages in LE of IDs. This algorithm considers the removal of a set of variables as a combinatorial factorization problem. Thus SPI is more fine-grained than VE as it considers the order in which potentials are combined. Detailed methods for computing the messages using SPI-LE are given.

The experimentation have shown that using SPI for computing the messages is more efficient than using VE. The number of operations, particularly the number of multiplications, is reduced for most of the IDs. Even though SPI-LE usually have a larger overhead than VE-LE, experiments have shown that the CPU time needed for the evaluation is reduced as well: the reduction in the number of operations compensates this larger overhead. The experimentation has been performed using one of the heuristics *minimum size*. It could be interesting looking for alternative heuristics since the efficiency of the evaluation depends on this heuristic.

The SPI algorithm used only considers next pair of potentials to combine. Thus another line of future research could be studying the behaviour of the algorithm using a higher neighbourhood degree.

Acknowledgments. This research was supported by the Spanish Ministry of Economy and Competitiveness under project TIN2010-20900-C04-01, the European Regional Development Fund (FEDER), the FPI scholarship programme (BES-2011-050604). The authors have also been partially supported by "Junta de Andalucía" under projects TIC-06016 and P08-TIC-03717.

References

1. Howard, R., Matheson, J.: Influence diagram retrospective. Decision Analysis 2(3), 144–147 (2005)
2. Jensen, F., Jensen, F.V., Dittmer, S.L.: From Influence Diagrams to junction trees. In: Proceedings of the 10th International Conference on Uncertainty in AI, pp. 367–373. Morgan Kaufmann Publishers Inc. (1994)
3. Madsen, A., Jensen, F.: Lazy evaluation of symmetric Bayesian decision problems. In: Proceedings of the 15th Conference on Uncertainty in AI, pp. 382–390. Morgan Kaufmann Publishers Inc. (1999)
4. Shachter, R.D., D'Ambrosio, B., Del Favero, B.: Symbolic probabilistic inference in belief networks. In: AAAI, vol. 90, pp. 126–131 (1990)
5. Li, Z., d'Ambrosio, B.: Efficient inference in Bayes networks as a combinatorial optimization problem. IJAR 11(1), 55–81 (1994)
6. D'Ambrosio, B., Burgess, S.: Some experiments with real-time decision algorithms. In: Proceedings of the 12th International Conference on Uncertainty in AI, pp. 194–202. Morgan Kaufmann Publishers Inc. (1996)
7. Cabañas, R., Madsen, A.L., Cano, A., Gómez-Olmedo, M.: On SPI for evaluating influence diagrams. In: Laurent, A., Strauss, O., Bouchon-Meunier, B., Yager, R.R. (eds.) IPMU 2014, Part I. CCIS, vol. 442, pp. 506–516. Springer, Heidelberg (2014)
8. Madsen, A.: Variations over the message computation algorithm of lazy propagation. IEEE Transactions on Systems, Man, and Cybernetics, Part B: Cybernetics 36(3), 636–648 (2005)
9. Fagiuoli, E., Zaffalon, M.: A note about redundancy in influence diagrams. International Journal of Approximate Reasoning 19(3), 351–365 (1998)
10. Madsen, A., Jensen, F.: Lazy propagation: a junction tree inference algorithm based on lazy evaluation. Artificial Intelligence 113(1-2), 203–245 (2004)
11. Kjærulff, U.: Triangulation of graphs – algorithms giving small total state space. Research Report R-90-09, Department of Mathematics and Computer Science, Aalborg University, Denmark (1990)
12. Jensen, F., Nielsen, T.: Bayesian networks and decision graphs. Springer (2007)
13. Rose, D.: A graph-theoretic study of the numerical solution of sparse positive definite systems of linear equations. Graph Theory and Computing 183, 217 (1972)
14. Bloemeke, M., Valtorta, M.: A hybrid algorithm to compute marginal and joint beliefs in Bayesian networks and its complexity. In: Proceedings of the 14th Conference on Uncertainty in AI, pp. 16–23. Morgan Kaufmann Publishers Inc. (1998)
15. Lucas, P., Taal, B.: Computer-based decision support in the management of primary gastric non-Hodgkin lymphoma. UU-CS (1998-33) (1998)
16. Marcot, B., Holthausen, R., Raphael, M., Rowland, M., Wisdom, M.: Using Bayesian belief networks to evaluate fish and wildlife population viability under land management alternatives from an environmental impact statement. Forest Ecology and Management 153(1), 29–42 (2001)
17. Raiffa, H.: Decision analysis: Introductory lectures on choices under uncertainty (1968)
18. Goutis, C.: A graphical method for solving a decision analysis problem. IEEE Transactions on Systems, Man and Cybernetics 25(8), 1181–1193 (1995)

Extended Probability Trees for Probabilistic Graphical Models

Andrés Cano, Manuel Gómez-Olmedo, Serafín Moral, and Cora B. Pérez-Ariza

CITIC-UGR, University of Granada, Spain
{acu,mgomez,smc,cora}@decsai.ugr.es

Abstract. This paper proposes a flexible framework to work with probabilistic potentials in Probabilistic Graphical Models. The so-called Extended Probability Trees allow the representation of multiplicative and additive factorisations within the structure, along with context-specific independencies, with the aim of providing a way of representing and managing complex distributions. This work gives the details of the structure and develops the basic operations on potentials necessary to perform inference. The three basic operations, namely restriction, combination and marginalisation, are defined so they can take advantage of the defined factorisations within the structure, following a lazy methodology.

Keywords: Probability trees, recursive probability trees, lazy propagation.

1 Introduction

Probabilistic Graphical Models (PGMs) enable efficient representation of joint distributions exploiting independencies among the variables. The independencies are encoded by means of the *d-separation* criterion [1]. Therefore only explicit dependencies will be represented and quantified. The values measuring the dependencies can be stored using several data structures, being *Conditional Probability Tables* (CPTs) the most common and straightforward. A CPT encoding a potential defined over a set of variables can be seen as a grid with a cell for each combination of values of these variables. This implies an exponential growth of memory space requirements depending on the number of variables. *Probability Trees* (PTs) [2,3,4] try to improve CPTs allowing context-specific independencies and usually obtaining memory space savings as a consequence. *Recursive Probability Trees* (RPTs) [5] suppose another step in this direction and can be considered as a generalisation of PTs. With this data structure it is possible to cover the modelling capabilities of PTs and to represent proportionalities, multinets and mixtures of conditional probability distributions (CPDs) as well. Moreover, RPTs try to keep the information as factorised as possible. These features are used during inference. This paper proposes a refinement of RPTs, Extended Probability Trees (ePTs), that extends their modelling capabilities by managing additive factorisations in addition to all the features of RPTs, along the same lines as structures like NAND trees [6] and chain event graphs [7].

L.C. van der Gaag and A.J. Feelders (Eds.): PGM 2014, LNAI 8754, pp. 113–128, 2014.
© Springer International Publishing Switzerland 2014

The necessary operations for making inference in BNs are *restriction, combination* (product) and *marginalisation* and must be properly defined on ePTs. Therefore there are three main objectives in this paper: the description of the ePT data structure; the explanation about its modelling capability; and the explanation of how these basic operations operate on them. This paper focuses on operating with discrete distributions, however a similar idea has been applied to hybrid domains before [8].

The rest of this paper is structured as follows: Sec. 2 offers the basic notions on PTs and presents the notation to be used in the whole paper; Sec. 3 gives a review on the proposed data structure, specifying all its details; Sec. 4 explains how to perform inference on ePTs; Sec. 5 gives an example of a model that can be managed with ePTs but not with the usual data structures; Sec. 6 presents the application of ePTs to real-world Bayesian networks and Sec. 7 closes the paper with the main conclusions of this work and future lines of research.

2   Potentials, Probability Trees and Inference in Bayesian Networks

Let $\mathbf{X_I} = \{X_1, X_2, \ldots, X_n\}$ be a set of variables. Let us assume that each variable X_i takes values on a finite set of states Ω_{X_i} (the domain of X_i). We shall use x_i to denote the value of X_i, $x_i \in \Omega_{X_i}$. The Cartesian product $\times_{X_i \in \mathbf{X_I}} \Omega_{X_i}$ will be denoted by $\Omega_{\mathbf{X_I}}$. The elements of $\Omega_{\mathbf{X_I}}$ are called configurations of $\mathbf{X_I}$ and will be represented as $\mathbf{x_I}$. The projection of a configuration $\mathbf{x_I}$ onto the set of variables $\mathbf{X_J}$, $J \subseteq I$, is denoted by $\mathbf{x}^{\downarrow \mathbf{X_J}}$.

A potential ϕ for $\mathbf{X_I}$ is a mapping from $\Omega_{\mathbf{X_I}}$ into $\mathbb{R}_0^+$. Given ϕ, $s(\phi)$ denotes the set of variables for which ϕ is defined and $sum(\phi)$ the addition of its values. Therefore probability distributions and utility functions can be seen as potentials.

A BN is a DAG where each node represents a random variable X_i. The topology of the graph shows the independence relations between variables according to the *d-separation* criteria [1]. Each node X_i has a CPD encoded by a potential $\phi_i(X_i | \pi(X_i))$, where $\pi(X_i)$ refers to the parents of X_i. Therefore a BN with nodes $\mathbf{X_I} = \{X_1, X_2, \ldots X_n\}$ determines a joint probability distribution:

$$\phi(\mathbf{X}) = \prod_{X_i \in X} \phi_i(X_i | \pi(X_i)) \tag{1}$$

Let $\mathbf{X_E} \subset \mathbf{X_I}$ be a set of *observed* variables and $\mathbf{x_E} \in \Omega_{\mathbf{X_E}}$ their observed values. An algorithm focused on computing the posterior distribution $\phi(\mathbf{X_J} \subseteq \mathbf{X_I} | \mathbf{X_E} = \mathbf{x_E})$ is called propagation or inference algorithm. Regardless of the data structure employed for potentials (CPTs, PTs, etc), the process of inference in PGMs requires the definition of three operations:

- *Restriction:* If ϕ is a potential about variables $\mathbf{X}_I$, and $\mathbf{x}_J$ is a configuration for the set of variables $\mathbf{X}_J$, $K = I \cap J$ and $L = I - K$, then the *restriction* of ϕ to the configuration $\mathbf{x}_J$ is a potential $\phi^{R(\mathbf{x}_J)}$ defined for variables $\mathbf{X}_L$

and given by $\phi^{R(\mathbf{x}_J)}(\mathbf{x}_L) = \phi(\mathbf{x}_L, \mathbf{x}_J)$. The restriction of a potential to a configuration $\mathbf{x}_J$ consists of returning the part of the potential which is consistent with the configuration.

- *Combination:* If ϕ_1 and ϕ_2 are potentials with $s(\phi_1) = \mathbf{X}_I$ and $s(\phi_1) = \mathbf{X}_J$ then its combination is the potential $\phi_1 \otimes \phi_2$ defined on $\mathbf{X} = \mathbf{X}_I \cup \mathbf{X}_J$ and given by $\phi_1 \otimes \phi_2(\mathbf{x}) = \phi_1(\mathbf{x}^{\downarrow \mathbf{X}_I}).\phi_2(\mathbf{x}^{\downarrow \mathbf{X}_J})$.
- *Marginalisation:* If ϕ is a potential with $s(\phi) = \mathbf{X}_I$ and $\mathbf{X}_J \subseteq \mathbf{X}_I$ then the marginalisation of ϕ to $\mathbf{X}_J$ is the potential $\phi^{\downarrow \mathbf{X}_J}$, given by $\phi^{\downarrow \mathbf{X}_J}(\mathbf{x}_J) = \sum_{\mathbf{x}_I^{\downarrow \mathbf{X}_J} = \mathbf{x}_J} \phi(\mathbf{x}_I)$.

PTs are a flexible data structure providing exact and approximate representations of probability potentials and utility functions. A *PT* (denoted as $\mathcal{T}$) is a list of variables $s(\mathcal{T})$ and a directed labelled tree with two kind of nodes: *internal* nodes represent variables and *leaf* nodes represent real numbers. All the variables of the internal nodes must belong to the list $s(\mathcal{T})$. Internal nodes have outgoing arcs (one per state of the corresponding variable). The size of $\mathcal{T}$, denoted as $size(\mathcal{T})$, is defined as its leaf count.

The chance to capture context-specific independencies usually makes PTs a more compact representation of potentials than CPTs. Moreover, context-specific independencies can be exploited during computation avoiding unnecessary operations.

To be able to compute with PTs in PGMs, we need to define the three basic operations on potentials (restriction, combination, and marginalisation) in this data structure. The notation of the operations in a particular data structure will be completely analogous to the notation for general potentials. So, if $\mathcal{T}$ is a probability tree such that $s(\mathcal{T}) = \mathbf{X}_I$ and $\mathbf{X}_J \subseteq \mathbf{X}_I$ we use $\mathcal{T}^{R(\mathbf{x}_J)}$ to denote a probability tree representing the *restriction* of the potential associated to $\mathcal{T}$.

We say that a potential is *decomposed* into a set of factors if their product is equal to the original potential. For example, a BN specifies a joint probability distribution by means of a set of factors: the CPDs of its variables. In general, a factorised potential needs less space than the original one. There are several inference algorithms for Bayesian networks based on using lists of potentials (factorisation of potentials) with the objective of reducing inference computational complexity: *lazy propagation* [9], *lazy-penniless* [10], *variable elimination* [11] and *mini-bucket* [12]. Some inference algorithms assume that the factorisation of the potentials is given as input, but others obtain the decomposition as a previous step. In [13,14,15,16] PTs are used to represent each factor, and algorithms are given to decompose PTs into two or more factors in exact or approximate ways, using these favourable factorisations in inference algorithms for BNs.

3   Extended Probability Trees

Extended Probability Trees (ePTs), as a refinement of *Recursive Probability Trees* [5], are a generalisation of PTs. RPTs were developed with the aim of enhancing PTs' flexibility and so they are able to represent different kinds of patterns that so far were out of the scope of PTs, such as multiplicative factorisations. With ePTs we refine RPTs in order to capture other patterns that were left out, such as additive factorisations. An ePT is a directed tree, to be traversed from root to leaves, where the nodes (both inner nodes and leaves) play different roles depending on their nature. In the simplest case, an ePT is equivalent to a PT, where the inner nodes represent variables, and the leaf nodes are labelled by numbers. In the context of ePTs, we will call this type of inner nodes as *Split* nodes and this kind of leaves as *Value* nodes.

As RPTs do, ePTs include factorisations within the data structure by incorporating a type of inner node that lists together all the factors. Therefore, a *Multiply* node represents a multiplicative factorisation by listing all the factors making up the factorisation. If a *Multiply* node stores a factorisation of k factors of a potential ϕ defined on $\mathbf{X_J}$, and every factor i (an ePT as well) encodes ϕ_i for a subset of variables $\mathbf{X_{J_i}} \subseteq \mathbf{X_J}$, then ϕ corresponds to $\prod_{i=1}^{k} \phi_i(\mathbf{X_{J_i}})$.

Moreover, ePTs propose to include additive factorisations within the representation. This is done by incorporating a type of inner node that again lists together all the factors. Therefore, a *Sum* node represents an additive factorisation by listing all the factors making up the sum. If a *Sum* node stores a factorisation of k factors of a potential ϕ defined on $\mathbf{X_J}$, and every factor i (an ePT as well) encodes a potential ϕ_i for a subset of variables $\mathbf{X_{J_i}} \subseteq \mathbf{X_J}$, then ϕ corresponds to $\sum_{i=1}^{k} \phi_i(\mathbf{X_{J_i}})$.

When necessary, ePTs will include a fourth type of node denominated *Potential* node. This is a leaf node and its purpose is to encapsulate a full potential within the leaf in an internal structure. This internal structure usually will not be an ePT, but a PT or a CPT instead. In fact, as long as the internal structure of a *Potential* node supports the basic operations on potentials (namely marginalisation, restriction and combination), it is accepted within the ePT representation. The size of a *Potential* node is defined as the number of probability values the internal structure uses to represent the potential. We define the *size* of an ePT as the total number of probability values stored within it, which is the addition of all the *Value* nodes plus the sizes of all the *Potential* nodes in the ePT.

In summary, an ePT can have five kind of nodes in total: *Split*, *Multiply* or *Sum* as inner nodes, and *Value* or *Potential* nodes as leaves. We can combine them in different ways in order to find the structure that best fits the potential to be represented, making ePTs an extremely flexible framework to work with.

Therefore, an ePT ($e\mathcal{PT}$) defined on a set of variables $\mathbf{X_I}$ represents the potential $\phi_{e\mathcal{PT}}(\mathbf{X_I}) : \Omega_{\mathbf{X_I}} \rightarrow \mathbb{R}_0^+$ if for each $\mathbf{x_I} \in \Omega_{\mathbf{X_I}}$ the value $\phi_{e\mathcal{PT}}(\mathbf{x_I})$ is the number obtained with the recursive procedure explained in Alg. 1. The procedure consists of checking the ePT from root to leaves, applying a different action depending on the kind of node. If *root* is a *Value* node, the algorithm returns the corresponding value; If *root* is a *Potential* node, the algorithm gets the value for the selected configuration $\mathbf{x_I}$; If *root* is a *Split* node: the procedure is recursively applied to the child consistent with the given configuration; If *root* is a *Multiply* node: the algorithm multiplies the results obtained from new recursive calls for every child; If *root* is a *Sum* node: the algorithm adds the results obtained from new recursive calls for every child.

Algorithm 1. getValue($e\mathcal{PT}$, $\mathbf{x_I}$)

Input: $e\mathcal{PT}$: a ePT defined on $\mathbf{X_I}$
      $\mathbf{x_I}$: a configuration
Output: $\phi_{e\mathcal{PT}}(\mathbf{x_I})$: a value

1 // Procedure to obtain a probability
2 // value from an ePT ($e\mathcal{PT}$) and a certain configuration $\mathbf{x_I}$.
3 **begin**
4     $root \leftarrow$ root node of $e\mathcal{PT}$
5     $type \leftarrow$ type of $root$ (*Value*, *Potential*, *Split*, *Multiply* or *Sum*)

6   **switch** *type* **do**
7   **case** *Value*
8   **return** $root$
9   **case** *Potential*
10            $\phi(\mathbf{X_J} \subseteq \mathbf{X_I}) \leftarrow$ potential defined on $\mathbf{X_J}$ represented by $root$
11   **return** $\phi^{R(x_I)}(\mathbf{x_J})$
12   **case** *Split*
13   X_i, variable of the node
14   x_i, value of X_i in the configuration $\mathbf{x_I}$
15            $ch_i(e\mathcal{PT}) \leftarrow$ child of $root$ for x_i value
16   **return** getValue($ch_i(e\mathcal{PT})$, $\mathbf{x_I}$)
17   **case** *Multiply*
18            $ch_1(e\mathcal{PT}), ch_2(e\mathcal{PT}) \dots ch_n(e\mathcal{PT})$ children of $root$
19   **return** $\Pi_{i=1}^{n}$getValue($ch_i(e\mathcal{PT})$, $\mathbf{x_I}$)
20   **case** *Sum*
21            $ch_1(e\mathcal{PT}), ch_2(e\mathcal{PT}) \dots ch_n(e\mathcal{PT})$ children of $root$
22   **return** $\sum_{i=1}^{n}$ getValue($ch_i(e\mathcal{PT})$, $\mathbf{x_I}$)

For example, consider a variable X_1 that has three parents: X_2, X_3 and X_4. Let's assume that the CPD for X_1 is given by a convex combination of two terms, on the one hand a CPD of X_1 given X_2 and X_3, and on the other, a CPD of X_1 given X_3 and X_4. That is, the considered mixture of CPD is defined as:

$$\phi(X_1|X_2, X_3, X_4) = \alpha\phi(X_1|X_2, X_3) + (1 - \alpha)\phi(X_1|X_3, X_4). \qquad (2)$$

The representation of Eq. 2 as an ePT can be seen in Fig. 1: the root of the tree is a *Sum* node, that fathers the two terms of the mixture. Each factor is represented through a *Multiply* node having as children a *Value* node encapsulating the weight of the convex combination, and a *Potential* node enclosing each reduced CPD. Once a model is represented as an ePT, we can perform inference over it as described in the next section.

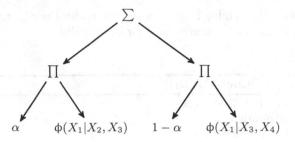

Fig. 1. A mixture of CPDs represented as an ePT

4   Inference with ePTs

The basic operations on potentials required to perform inference algorithms can be adapted to be supported by RPTs [5] and by ePTs, as explained in this Section. The operations are designed to deal with the multiplicative factorisations introduced within the ePTs by the *Multiply* nodes, and to manage the additive factorisations defined by means of the *Sum* nodes. The definition of the operations aims at postponing the actual computation of products or additions of numbers. In this sense, ePTs are compatible with inference schemes based on lazy propagation [9,17].

4.1   Restriction

Let ePT be an ePT and $\mathbf{x}_J$ a configuration for the variables in $\mathbf{X}_J$. The restriction operation of $ePT^{R(\mathbf{x}_J)}$ (ePT restricted to $\mathbf{x}_J$) will be performed recursively from root to leaf nodes acting on nodes according to their type: *Value* nodes will remain unaltered; *Multiply* and *Sum* nodes will transmit the operation to their children; potential nodes will produce the restriction of their potentials if needed (their domains include any of the variables in $\mathbf{X}_J$); on *Split* nodes the operation will be transmitted to children if the associated variable is not in $\mathbf{X}_J$, otherwise the leaf representing the value contained in $\mathbf{x}_J$ will be kept (and promoted to the place of its parent node) and the rest will be removed.

Observe that this operation is performed recursively from root to leaves, so the time consumed by the algorithm will be linear on the number of nodes of the ePT. The pseudocode for this procedure is shown in Alg. 2.

4.2   Combination

The *combination* (product) of two potentials can also be performed with ePTs in an easy way. The *combination* of two ePTs, ePT_1 and ePT_2, denoted by $ePT_1 \otimes ePT_2$, will be a new ePT representing the product of the potentials represented by ePT_1 and ePT_2. It is obtained with a *Multiply* node N_L including two children: $ePT1$ and $ePT2$. This idea is described in Alg. 3.

Algorithm 2. restrict($e\mathcal{PT}, \mathbf{x}_J$)

Input: An ePT $e\mathcal{PT}$, a configuration of variables $\mathbf{X}_J = \mathbf{x}_J$
Output: An ePT $e\mathcal{PT}^{R(\mathbf{x}_J)}$

1 **begin**
2   Let *root* be the root of $e\mathcal{PT}$;
3   **if** *root is a Value node labelled with* r **then**
4   **return** r;
5   **if** *root is a Potential node labelled with* P **then**
6   **return** $P^{R(\mathbf{x}_J)}$;
7   **if** *root is a Split node labelled with* X_i **then**
8   **if** $X_i \in \mathbf{X}_J$ **then**
9   Let $ch_i(e\mathcal{PT})$ be the child of *root* consistent with X_i in $\mathbf{x}_J$;
10   **return** restrict($ch_i(e\mathcal{PT}), \mathbf{x}_J$);
11   **else**
12   Make a new ePT $e\mathcal{PT}'$ with X_i as root;
13   **foreach** *child of the root of* $e\mathcal{PT}$, $ch_i(e\mathcal{PT})$ **do**
14              $e\mathcal{PT}'_i \leftarrow$ restrict($ch_i(e\mathcal{PT}), \mathbf{x}_J$);
15   Set $e\mathcal{PT}'_i$ as i-th child of $e\mathcal{PT}'$ root;
16   **return** $e\mathcal{PT}'$;
17   **if** *root is a Multiply or a Sum node* **then**
18   Make a new ePT $e\mathcal{PT}'$ with a *Multiply* node as root;
19   **foreach** *child of the root of* $e\mathcal{PT}$, $ch_i(e\mathcal{PT})$ **do**
20           $e\mathcal{PT}'_i \leftarrow$ restrict($ch_i(e\mathcal{PT}), \mathbf{x}_J$);
21   Set $e\mathcal{PT}'_i$ as i-th child of $e\mathcal{PT}'$ root;
22   **return** $e\mathcal{PT}'$;

Algorithm 3. combine($e\mathcal{PT}, f$)

Input: A ePT $e\mathcal{PT}$, a factor f (another ePT, a potential or a value).
Output: The result of combining $e\mathcal{PT}$ and f.

1 **begin**
2   Make a new ePT $e\mathcal{PT}'$ with a *Multiply* node as root;
3   Put $e\mathcal{PT}$ as child of $e\mathcal{PT}'$;
4   **if** f *is a numeric value* **then**
5   Make a new *Value* node labelled with f;
6   Put the *Value* node as child of $e\mathcal{PT}'$;
7   **if** f *is a potential* **then**
8   Make a new *Potential* node labelled with f;
9   Put the *Potential* node as child of $e\mathcal{PT}'$;
10   **if** f *is a ePT* **then**
11   Put $e\mathcal{PT}$ as child of $e\mathcal{PT}'$;
12   **return** $e\mathcal{PT}'$;

Note that Alg. 3 is formulated in a general way, so that it is designed for combining two ePTs but also for combining an ePT with another kind of potential, or an ePT with a value. In these last two cases, when the second factor is a constant, it is stored in a *Value* node (line 5, Alg. 3), whilst when the factor is a potential, it is enclosed within a *Potential* node (line 8, Alg. 3). These transformations are done prior to storing the factors as children of the *Multiply* node. An important feature of the combination process as described in Alg. 3 is that it does not really require the actual computation of any product of numbers.

4.3   Marginalisation

The *marginalisation* of a potential ϕ to a set of variables $\mathbf{X}_J$ can be performed directly over ePTs. Given an ePT $e\mathcal{PT}$ representing a potential ϕ defined for a set of variables $\mathbf{X}_I$, and $J \subseteq I$, the *marginalisation* of $e\mathcal{PT}$ over $\mathbf{X}_J$, denoted as $e\mathcal{PT}^{\downarrow \mathbf{X}_J}$, is a new ePT that represents the potential ϕ marginalised to $\mathbf{X}_J$, that is, $\phi^{\downarrow \mathbf{X}_J}(\mathbf{x}_J) = \sum_{\mathbf{x}_{I-J}} \phi(\mathbf{x}_J, \mathbf{x}_{I-J})$. The marginalisation is obtained by *deleting* from $e\mathcal{PT}$ all the variables $\{X_\iota\}$ where $\iota \in I - J$.

The deletion of variable $\{X_\iota\}$ is denoted by $e\mathcal{PT}^{\downarrow \mathbf{X}_I \setminus \{X_\iota\}}$ and represents the potential $\phi^{\downarrow \mathbf{X}_I \setminus \{X_\iota\}}$. The procedure to sum out a variable X_ι, which is explained in Alg. 4, is a recursive process that covers the ePT from root to leaves as follows:

- If the root is a *Value* node or a *Potential* node that does not contain X_ι in its domain, the result of the operation would be the node multiplied by the number of possible states of X_ι, denoted as $|\Omega_{X_\iota}|$ (lines 3 and 10, Alg. 4). If the *Potential* node has X_ι in its domain, then the marginalisation operation is performed according to the data structure of the potential enclosed in the node (line 8, Alg. 4).
- If the root is a *Split* node, then if X_ι labels it, the result of the operation will be a *Sum* node having as children all the children of the *Split* node (line 13, Alg. 4). Otherwise, the marginalisation operation is recursively applied to every child of the root (lines 15 to 19, Alg. 4).
- If the root is a *Multiply* node, then all its children are sorted into two sets, one containing the factors related to X_ι, and the other containing the remaining (line 21, Alg. 4). This first set must be combined pairwise (line 22, Alg. 4), and afterwards a recursive call to the algorithm is applied to the result of the multiplication (line 23, Alg. 4). The final ePT contains all the factors that were not related to X_ι plus the result of the recursive call (line 25, Alg. 4).
- If the root is a *Sum* node, then the marginalisation operation is propagated to every child of the root (lines 28 to 30, Alg. 4).

When dealing with *Multiply* nodes within the *marginalisation* of ePTs, we define a new operation that multiplies ePTs (line 22, Alg. 4). This operation is explained in Alg. 5, and differs from the *combination* or ePTs as defined in Alg. 3 in the sense that we now remove the *Multiply* node from the root of the result. The multiplication of ePTs starts considering the root of both ePTs to be multiplied (line 2, Alg. 5) and according to the type of node they are, performs in a different way:

Algorithm 4. sumOut$(X_\iota, e\mathcal{PT})$

Input: An ePT $e\mathcal{PT}$ defined for $\mathbf{X}_I$ and a variable $X_\iota \in \mathbf{X}_I$ (the variable to be summed out)
Output: An ePT for $e\mathcal{PT}^{\downarrow \mathbf{X}_I \setminus \{X_\iota\}}$

1 Let *root* be the root of $e\mathcal{PT}$;
2 **if** *root is a Value node labelled with a real number r* **then**
3 $\quad$ Make a new ePT $e\mathcal{PT}'$ with a *Value* node as root;
4 $\quad$ Put $r \cdot |\Omega_{X_\iota}|$ as the label of the root of $e\mathcal{PT}'$;
5 **else if** *root is a Potential node labelled with a potential* ϕ **then**
6 $\quad$ Make a new ePT $e\mathcal{PT}'$ with a *Potential* node as root;
7 $\quad$ **if** X_ι *is a variable in the domain of* ϕ **then**
8 $\quad\quad$ Put $\phi^{\downarrow \mathbf{X}_I \setminus \{X_\iota\}}$ as the label of the root of $e\mathcal{PT}'$ (this is an external operation that depends on the data structure that holds the potential) ;
9 $\quad$ **else**
10 $\quad\quad$ Put $\phi \cdot |\Omega_{X_\iota}|$ as the label of the root of $e\mathcal{PT}'$;
11 **else if** *root is a Split node labelled with* X_i **then**
12 $\quad$ **if** $X_i = X_\iota$ **then**
13 $\quad\quad$ Let $e\mathcal{PT}'$ be a new *Sum* node containing all the children of *root*;
14 $\quad$ **else**
15 $\quad\quad$ Make a new ePT $e\mathcal{PT}'$ with a *Split* node as root;
16 $\quad\quad$ Put X_i as the label of the root of $e\mathcal{PT}'$;
17 $\quad\quad$ **foreach** *child of root,* $ch_i(e\mathcal{PT})$ **do**
18 $\quad\quad\quad$ $e\mathcal{PT}'_i \leftarrow$ **sumOut**$(X_\iota, ch_i(e\mathcal{PT}))$;
19 $\quad\quad\quad$ Set $e\mathcal{PT}'_i$ as the ith child of the root of $e\mathcal{PT}'$;
20 **else if** *root is a Multiply node* **then**
21 $\quad$ Let *with* and *without* be the list of children of *root* containing and not containing X_ι respectively ;
22 $\quad$ Let $e\mathcal{PT}_1$ be the multiplication of all the factors in the *with* list (using Alg. 5);
23 $\quad$ $e\mathcal{PT}_2 \leftarrow$ **sumOut**$(X_\iota, e\mathcal{PT}_1)$;
24 $\quad$ Make a new ePT $e\mathcal{PT}'$ with a *Multiply* node as root;
25 $\quad$ Put $e\mathcal{PT}_2$ and all the factors in *without* list as children of the root of $e\mathcal{PT}'$;
26 **else if** *root is a Sum node* **then**
27 $\quad$ Make a new ePT $e\mathcal{PT}'$ with a *Sum* node as root;
28 $\quad$ **foreach** *child of root,* $ch_i(e\mathcal{PT})$ **do**
29 $\quad\quad$ $e\mathcal{PT}'_i \leftarrow$ **sumOut**$(X_\iota, ch_i(e\mathcal{PT}))$;
30 $\quad\quad$ Set $e\mathcal{PT}'_i$ as the ith child of the root of $e\mathcal{PT}'$;
31 **return** $e\mathcal{PT}'$;

- Multiply a *Multiply* node by any kind of node but a *Sum* node: the *Multiply* node will father the second factor, as shown in Alg. 3 (lines 7 to 14, Alg. 5).
- Multiply a *Sum* node by any kind of node but a *Multiply* node: the resulting structure will have a *Sum* node as root, and its children will be all the

---

Algorithm 5. multiply($e\mathcal{PT}_1, e\mathcal{PT}_2$)

Input: Two ePTs $e\mathcal{PT}_1$ and $e\mathcal{PT}_2$

Output: The multiplication of $e\mathcal{PT}_1$ and $e\mathcal{PT}_2$

1 **begin**
2   Let $root_1$ be the root of $e\mathcal{PT}_1$ and $root_2$ be the root of $e\mathcal{PT}_2$;
3   **if** $root_1$ *is a Multiply node* **then**
4   **if** $root_2$ *is a Sum node* **then**
5   **return** multiply($e\mathcal{PT}_2, e\mathcal{PT}_1$);
6   **else**
7   Make a new ePT $e\mathcal{PT}$ with a *Multiply* node as root;
8   Append all the children of $root_1$ as children of the root of $e\mathcal{PT}$;
9   **if** $root_2$ *is a Multiply node* **then**
10   Put all the children of $root_2$ as children of the root of $e\mathcal{PT}$;
11   **else**
12   Append the factor rooted by $root_2$ as a child of the root of $e\mathcal{PT}$;
13   // If $root_2$ is a `Multiply` node, we append its children instead.
14   **return** $e\mathcal{PT}$;
15   **else if** $root_1$ *is a Sum node* **then**
16   **if** $root_2$ *is a Multiply node* **then**
17   According to Eq. 3, either multiply or sum first;
18   **else**
19   Make a new ePT $e\mathcal{PT}$ with a *Sum* node as root;
20   **foreach** *child of* $root_1$, $ch_i(e\mathcal{PT}_1)$ **do**
21   Append multiply($ch_i(e\mathcal{PT}_1), e\mathcal{PT}_2$) as child of $e\mathcal{PT}$;
22   **return** $e\mathcal{PT}$;
23   **else if** $root_1$ *is a Split node* **then**
24   **if** $root_2$ *is a Sum or Multiply node* **then**
25   **return** multiply($e\mathcal{PT}_2, e\mathcal{PT}_1$);
26   **else**
27   Let X_i be the label of $root_1$;
28   Make a new ePT $e\mathcal{PT}$ with X_i labelling the root;
29   **foreach** *child of* $root_2$, $ch_i(e\mathcal{PT}_2)$ **do**
30   Make a new ePT $e\mathcal{PT}_i$ with a *Multiply* node as root;
31   Append $ch_i(e\mathcal{PT}_2)$ and $e\mathcal{PT}_1^{R(X_i=x_i)}$ as children of $e\mathcal{PT}_i$;
32   Put $e\mathcal{PT}_i$ as children of $e\mathcal{PT}$;
33   **return** $e\mathcal{PT}$;
34   Let f_1 and f_2 be the factors (*Value* or *Potential* nodes) in $e\mathcal{PT}_1$ and $e\mathcal{PT}_2$;
35   **return** $f_1 \cdot f_2$ (in this case f_1 and f_2 can be multiplied directly);

---

children of the *Split* node, every one of them multiplied by the other factor with Alg. 4 (lines 19 to 22, Alg. 5).

- Multiply a *Split* node by a *Value*, *Potential* or *Split* node: we change every child of the *Split* node by a *Multiply* node that fathers the original children and the other factor restricted to the context given by the branch (lines 27 to 33, Alg. 5).
- Multiply a *Potential* node by a *Value* node: the structures are directly multiplied (lines 34 and 35, Alg. 5).

Multiplying a *Sum* node N_S by a *Multiply* node N_M (line 17, Alg. 5) can be done in two ways, as illustrated in Fig. 2:

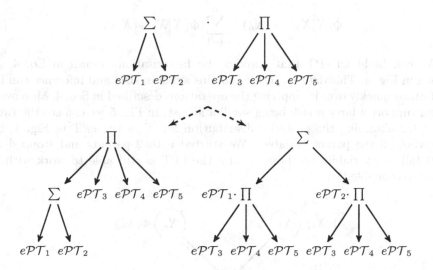

Fig. 2. Multiplying a *Sum* node by a *Multiply* node using Alg. 5

- Multiply first (bottom left part of Fig. 2). This implies creating a *Multiply* node that will father N_S and the children of N_M, as described in Alg. 3.
- Sum first (bottom right part of Fig. 2). In this case, the resulting structure will have a *Sum* node as root, and its children will be all the children of N_S, every one of them multiplied by N_M with Alg. 5.

To decide on how to perform the operation, we attend to the theoretical maximum size of the resulting structure for each possibility. We will multiply first if it holds that:

$$\Omega_{\mathbf{X}_{N_S} \cup \mathbf{X}_{N_M}} \;>=\; \Omega_{\mathbf{X}_{N_S}} + size(N_M) \tag{3}$$

where ch_{N_S} corresponds to the number of N_S' children, $size(N)$ corresponds to the number of probability values at N's leaves and $\mathbf{X}_{N_S}$ corresponds to the variables in the domain of N_S. This heuristic favours to postpone the the computation (by rooting the tree with a *Multiply* node) unless we are facing exceptional cases where the size gain is significant.

5   Example of Use

Consider a Bayesian network as defined in Fig. 3, where a variable X has a number n of parents. For high values of n, this structure can become unmanageable to most common data structures. One way of expressing the CPD for X could be decomposing it such as:

$$\phi(X|X_1, \cdots, X_n) = \sum_{i=1}^{n} (\phi(X|X_i)\phi(X_i)). \tag{4}$$

We can build an ePT that encodes the factorisations shown in Eq. 4, as shown in Fig. 4. This data structure remains small in size, and inference can be performed quickly over it, applying the operations described in Sec. 4. Moreover, it can support a large n still being small and fast. In Fig. 5 we can see the time spent for obtaining the posterior distribution for X in the ePT in Fig. 4, by removing all the parent variables. We started with 2 parents and stopped at 9900 (all the variables are binary), and the ePT is still able to work with it within reasonable time.

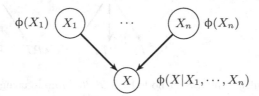

Fig. 3. A Bayesian network where a variable X has n parents

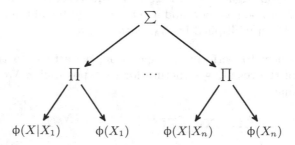

Fig. 4. An ePT encoding the factorisation given in Eq. 4.

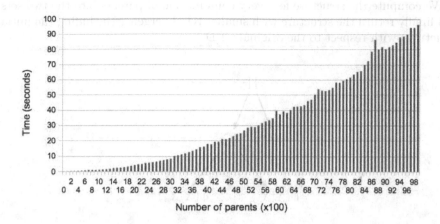

Fig. 5. Time spent for obtaining the posterior distribution for X, removing all the variables from the ePT in Fig. 4

6   Applying ePTs to Real Data

Applying ePTs to any problem is possible, as long as we can build the structure, as the basic operations on potentials are defined on ePTs. In this Section we present a way of approximating any CPD to a suitable ePT, afterwards we explain a heuristic to simplify an ePT in case it becomes too big, and finally we apply both methods to perform inference on two well known Bayesian networks.

6.1   Transforming CPDs into ePTs

To build an ePT from a CPD, we consider two possibilities. If the set of variables of the CPD contains 1 or 2 variables, we enclose it within a *Potential* node. To transform a bigger CPD into an ePT we follow an iterative algorithm that looks for the ePT that best fits, in terms of Kullback-Leibler divergence, the CPD. As the search space of different ePTs is huge, we limit the search as follows: we begin considering all the parents independently, so if we want to represent $\phi(X|\pi(X))$, n is the number of parents of X, and $X_1, \cdots, X_n \in \pi(X)$, we obtain:

$$\phi(X|\pi(X)) = \frac{1}{n} \sum_{i=1}^{n} \phi(X|X_i). \tag{5}$$

This structure is the starting point of the search, and the equivalent ePT can be seen in Fig. 6. We now try to compact further the representation by grouping the parents into two disjoint sets, where if $\pi_1(X) \cup \pi_2(X) = \pi(X)$ and $\pi_1(X) \cap \pi_2(X) = \emptyset$, we obtain:

$$\phi(X|\pi(X)) = \frac{1}{2}(\phi(X|\pi_1(X)) + \phi(X|\pi_2(X))) \tag{6}$$

We compute the structure for every combination of parents into the two sets, and finally return the structure with smaller KL divergence (including the initial structure) with respect to the original CPD.

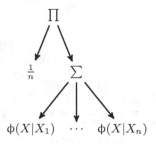

Fig. 6. The CPD $\phi(X|X_i, \cdots, X_n)$ encoded as an ePT

6.2   Simplifying ePTs

It may happen when working with large ePTs that after a number of operations, some branches become too factorised, that is, they contain small potentials of the same variables. When this happens we can simplify the tree. We have considered different heuristics to do so, and for the scope of this paper we present a recursive algorithm that traverses the ePT to the leaves, and then goes upwards simplifying the tree. The algorithm counts the number of variables of the potential defined by *Multiply* or *Sum* nodes, and exchanges them for equivalent *Potential* nodes in case the number of variables results bigger than a given threshold.

6.3   Inference Results

For this experiment we used two well-known medium-sized databases: Alarm[18] (37 nodes) and Water[19] (32 nodes). All the CPDs of both networks were approximated into ePTs using the algorithm described in Sec. 6.1. We applied the Variable Elimination algorithm to both networks, in order to remove all the variables, without evidence. We applied the simplification of ePTs, as described in Sec. 6.2 during the inference, after combining all the potentials related to the variable to be removed. For Alarm network, we used a simplification threshold of three variables, and for Water, as its relations are more complex, we increased the threshold to six variables. We computed the time needed to perform inference, the size of the maximum structure managed during the process and the average size of all the ePTs managed during the inference. We measured the size in terms of number of stored probability values. We compared the performance of ePTs with CPTs and slightly pruned [16] PTs (prune factor: 0.001).

The results are shown in Table 1, where we can see that while for Alarm ePTs work in a similar way as PTs, for Water ePTs do not obtain such good

Table 1. Inference results for two real-world databases

	Alarm network			Water network		
	CPT	PT	ePT	CPT	PT	ePT
time (ms)	195	174	**121**	16731	**578**	5635
max. size (values)	454	**211**	249	20492	**1239**	20498
avg. size (values)	137.1	**64.3**	83.2	2345.8	**117.7**	1450.8

results, but still perform better than CPTs in terms of time. The averages of the KL divergence of the posterior probabilities obtained with ePTs with respect to those obtained with PTs were 0.11 (s.d. 0.02) and 0.23 (s.d. 0.06), respectively.

7 Conclusions

This ongoing research presents a data structure to represent the probabilistic information in PGMs, along with the definition of the basic operations to perform inference on them, and some heuristics to build an ePT from a CPD, along with a way of compacting the structure in case it becomes too big.

We have seen that some models that are difficult or impossible to model with other structures are manageable using ePTs, and we have also tested the performance of ePTs against CPTs and PTs using real-world Bayesian networks. We conclude that for models that do not contain certain patterns, or are not factorised, the use of ePTs does not imply a benefit, as the operations are more complex on our data structure. On the other hand, ePTs increase the number of models that can be represented and efficiently managed. Learning ePTs is a key aspect to be developed further, along with an in-depth study to provide simplification heuristics that do not depend on thresholds. Moreover, a possible future use of ePTs is symbolic propagation.

Acknowledgements. This research was supported by the Spanish Ministry of Economy and Competitiveness under projects TIN2010-20900-C04-01 and P10-TIC-06016 and by the European Regional Development Fund (FEDER).

References

1. Geiger, D., Verma, T., Pearl, J.: Identifying independencies in Bayesian Networks. Networks 5, 507–534 (1990)
2. Cano, A., Moral, S., Salmerón, A.: Penniless propagation in join trees. International Journal of Intelligent Systems 15, 1027–1059 (2000)
3. Kozlov, D., Koller, D.: Nonuniform dynamic discretization in hybrid networks. In: Geiger, D., Shenoy, P. (eds.) Proceedings of the 13th Conference on Uncertainty in Artificial Intelligence, pp. 302–313. Morgan & Kaufmann (1997)
4. Boutilier, C., Friedman, N., Goldszmidt, M., Koller, D.: Context-specific independence in Bayesian networks. In: Horvitz, E., Jensen, F. (eds.) Proceedings of the 12th Conference on Uncertainty in Artificial Intelligence, pp. 115–123. Morgan & Kaufmann (1996)

5. Cano, A., Gómez-Olmedo, M., Moral, S., Pérez-Ariza, C., Salmerón, A.: Inference in Bayesian networks with recursive probability trees: data structure definition and operations. International Journal of Intelligent Systems 28 (2013)
6. Xiang, Y., Jia, N.: Modeling causal reinforcement and undermining for efficient CPT elicitation. IEEE Transactions on Knowledge and Data Engineering 19(12), 1708–1718 (2007)
7. Smith, J.Q., Anderson, P.E.: Conditional independence and chain event graphs. Artificial Intelligence 172, 42–68 (2008)
8. Langseth, H., Nielsen, T., Rumí, R., Salmerón, A.: Inference in hybrid Bayesian networks with mixtures of truncated basis functions. In: Cano, A., Gómez-Olmedo, M., Nielsen, T. (eds.) Proceedings of the 6th European Workshop on Probabilistic Graphical Models (PGM 2012), pp. 171–178 (2012)
9. Madsen, A., Jensen, F.: Lazy propagation: a junction tree inference algorithm based on lazy evaluation. Artificial Intelligence 113, 203–245 (1999)
10. Cano, A., Moral, S., Salmerón, A.: Lazy evaluation in Penniless propagation over join trees. Networks 39, 175–185 (2002)
11. Kohlas, J., Shenoy, P.P.: Computation in valuation algebras. In: Handbook of Defeasible Reasoning and Uncertainty Management Systems, pp. 5–39 (1997)
12. Dechter, R., Rish, I.: A scheme for approximating probabilistic inference. In: Geiger, D., Shenoy, P. (eds.) Proceedings of the 13th Conference on Uncertainty in Artificial Intelligence, pp. 132–141. Morgan & Kaufmann (1997)
13. Cano, A., Gómez-Olmedo, M., Pérez-Ariza, C., Salmerón, A.: Fast factorisation of probabilistic potentials and its application to approximate inference in Bayesian networks. International Journal of Uncertainty, Fuzziness and Knowledge-Based Systems 20(2), 223–243 (2012)
14. Martínez, I., Moral, S., Rodríguez, C., Salmerón, A.: Factorisation of probability trees and its application to inference in Bayesian networks. In: Gámez, J., Salmerón, A. (eds.) Proceedings of the First European Workshop on Probabilistic Graphical Models, pp. 127–134 (2002)
15. Martínez, I., Moral, S., Rodríguez, C., Salmerón, A.: Approximate factorisation of probability trees. In: Godo, L. (ed.) ECSQARU 2005. LNCS (LNAI), vol. 3571, pp. 51–62. Springer, Heidelberg (2005)
16. Martínez, I., Rodríguez, C., Salmerón, A.: Dynamic importance sampling in Bayesian networks using factorisation of probability trees. In: Proceedings of the Third European Workshop on Probabilistic Graphical Models, pp. 187–194 (2006)
17. Madsen, A.: Improvements to message computation in lazy propagation. International Journal of Approximate Reasoning 51, 499–514 (2012)
18. Beinlich, I., Suermondt, H.J., Chavez, R.M., Cooper, G.F.: The ALARM monitoring system: A case study with two probabilistic inference techniques for belief networks. In: Proceedings of the Second European Conference on Artificial Intelligence in Medicine, vol. 38, pp. 247–256 (1989)
19. Jensen, F.V., Kjærulff, U., Olesen, K.G., Pedersen, J.: Et forprojekt til et ekspertsystem for drift af spildevandsrensning (an expert system for control of waste water treatment - - a pilot project) (in danish). Technical report, Judex Datasystemer A/S, Aalborg, Denmark (1989)

Mixture of Polynomials Probability Distributions for Grouped Sample Data

Barry R. Cobb

Missouri State University, Department of Management,
Springfield, Missouri, USA
BarryCobb@MissouriState.edu

Abstract. This paper describes techniques for developing a mixture of polynomials (MOP) probability distribution from a frequency distribution (also termed grouped data) summarized from a large dataset. To accomplish this task, a temporary dataset is produced from the grouped data and the parameters for the MOP function are estimated using a Bspline interpolation technique. Guidance is provided regarding the composition of the temporary dataset, and the selection of split points and order of the MOP approximation. Good results are obtained when using grouped data as compared to the underlying dataset, and this can be a major advantage when using a decision support system to obtain information for estimating probability density functions for random variables of interest.

Keywords: Bayesian information criterion, B-spline interpolation, frequency distribution, grouped data, mixture of polynomials.

1 Introduction

This paper describes the construction of a mixture of polynomials (MOP) probability density function (PDF) from a frequency distribution developed from sample data, also termed here *grouped data*. In general, the MOP function can provide a method for approximating a PDF from data in a flexible form that can be readily manipulated for mathematical calculations.

Kernel density estimation is a well-known method for assigning a PDF to empirical data [1]. However, the functional form of many kernel density estimators is not amenable to use in probabilistic graphical models, and the sample data must be retained to reproduce the density function. To construct a hybrid Bayesian network or influence diagram with continuous variables that are not exclusively Gaussian, or to build models that cannot be represented in the conditional linear Gaussian framework, a functional form that permits closed-form addition, multiplication, and integration, and a form that maintains results in the same class of functions is desirable.

Mixtures of truncated exponentials [2], mixtures of polynomials [3], and mixtures of truncated basis functions [4] are methods suggested for overcoming the

L.C. van der Gaag and A.J. Feelders (Eds.): PGM 2014, LNAI 8754, pp. 129–144, 2014.

integration problem in hybrid Bayesian network models. To estimate PDFs accurately for such models, several methods have been developed to find parameters for PDFs of continuous random variables from data [5,6,7,8].

In many applications in practice, including a supply chain management problem discussed in a related working paper [9], raw data to estimate a PDF for a continuous random variable of interest is not readily available. Managing and transferring a dataset that includes all of the empirical data can become difficult because of its size. Furthermore, the data may have to be extracted directly from a database, which may be a more difficult task than simply accessing a report from a decision support system that includes frequency distributions calculated for grouped data as part of its standard output.

This paper examines whether an adequate mixture of polynomials PDF can be estimated from a frequency distribution of grouped data without resorting to accessing the entire dataset. The approach is to create a temporary dataset and use the B-spline interpolation approach suggested by López-Cruz et al. [10]. The issues that arise when using this approach are the number of values to include in the temporary dataset, the split points to use for the MOP functions, and the order of the polynomials in the approximation. We examine each of these issues in examples where we estimate MOP density functions from known distributions using a full dataset of sample data and grouped data summarized from the full dataset. The issue of using uniform split points versus equal probability split points with the B-spline technique is examined, and this discussion is relevant regardless of whether the full dataset or summary grouped data is used to create an MOP approximation. In general, we find that acceptable MOP approximations can be created from grouped data.

The remainder of the paper is structured as follows. The next section introduces notation and definitions used throughout the paper. Section 3 reviews a B-spline technique for estimating MOP functions from data developed by López-Cruz et al. [10]. Section 4 describes the process of estimating an MOP from grouped data using the B-spline method. Section 5 applies the technique to two examples where both grouped data and the simulated underlying dataset are available to allow comparison of results. Section 6 concludes the paper.

2   Notation and Definitions

This section reviews definitions and notation used throughout the paper.

2.1   Grouped Data

We suppose that an unobserved dataset $\mathcal{D} = \{x_1, \ldots, x_\mathcal{N}\}$ is summarized into $\mathcal{K}$ groups. A series of $\mathcal{K} + 1$ split points, $\mathbf{s} = \{\mathcal{S}_0, \mathcal{S}_1, \ldots, \mathcal{S}_{\mathcal{K}-1}, \infty\}$, defines the groups. The split points are defined without an upper bound because we often encounter cases in practice where the last group is not explicitly bounded, although if there is a finite upper bound this can be assigned as $\mathcal{S}_\mathcal{K}$.

A specific data point x_i is classified into group j if $\mathcal{S}_{j-1} \leq x_i < \mathcal{S}_j$. The frequency of observations in each group are denoted by $\mathbf{f} = \{\mathcal{F}_1, \ldots, \mathcal{F}_\mathcal{K}\}$ and the "mid-points" of the intervals are denoted by $\mathbf{m} = \{m_1, \ldots, m_\mathcal{K}\}$. For all examples in this paper, the mid-points are defined as $m_1 = \mathcal{S}_0 + (2/3) \cdot (\mathcal{S}_1 - \mathcal{S}_0)$, $m_\mathcal{K} = \mathcal{S}_{\mathcal{K}-1} + 0.5 \cdot (\mathcal{S}_{\mathcal{K}-1} - \mathcal{S}_{\mathcal{K}-2})$, and $m_j = (\mathcal{S}_j - \mathcal{S}_{j-1})/2$ for $j = 2, \ldots, \mathcal{K} - 1$. The assignment for m_1 (at 2/3 of the interval distance) assumes knowledge that a point in the first group is more likely to fall closer to the right end-point than the left end-point of the interval, as is the case for the motivating supply chain management problem (see [9]). If this knowledge is not available, we can simply define m_1 in the same way that m_2 is defined. Similarly, if $\mathcal{S}_\mathcal{K}$ is finite, $m_\mathcal{K}$ can be calculated in the same way as $m_{\mathcal{K}-1}$.

The mean and variance of the grouped data are calculated as

$$\overline{X}_G = \frac{1}{N} \sum_{j=1}^{\mathcal{K}} \mathcal{F}_j \cdot m_j \quad \text{and} \quad s_G^2 = \frac{1}{N-1} \sum_{j=1}^{\mathcal{K}} \mathcal{F}_j \cdot (m_j - \overline{X}_G)^2 . \qquad (1)$$

Example 1. Consider the grouped data shown in Table 1. The frequency distribution summarizes a dataset with 1000 observations into $\mathcal{K} = 6$ groups. The last group is unbounded on the right-hand side, but the mid-point $m_\mathcal{K}$ is calculated as if the width of the last interval is the same as the group defined on $[40, 50)$.

Table 1. Grouped dataset for Example 1

Interval $[\mathcal{S}_{j-1}, \mathcal{S}_j)$	$\mathcal{F}_j$	Probability	m_j	$\mathcal{F}_j \cdot m_j$	$\mathcal{F}_j \cdot (m_j - \overline{X}_G)^2$
0-10	88	0.088	6.67	586.67	23611
10-20	388	0.388	15	5820	25123
20-30	298	0.298	25	7450	1137
30-40	138	0.138	35	4830	19718
40-50	48	0.048	45	2160	23134
50+	40	0.040	55	2200	40841
TOTAL	$N = 1000$	1.000		23046.67	133562
				$\overline{X}_G = 23.05$	$s_G^2 = 133.7$

2.2 Mixtures of Polynomials

One method of modeling continuous PDFs in hybrid Bayesian networks is to approximate PDFs with MOP functions [3]. The MOP model provides a closed-form approximation that can be easily utilized in calculations involving addition, multiplication, and integration.

The definition of mixture of polynomials given here is used by Shenoy [11]. A one-dimensional function $f : \mathbb{R} \to \mathbb{R}$ is said to be a *mixture of polynomials* function if it is a piecewise function of the form:

$$f(x) = \begin{cases} a_{0i} + a_{1i}x + \cdots + a_{ni}x^n & \text{for } x \in A_i, i = 1, \ldots, k, \\ 0 & \text{otherwise} \end{cases} \qquad (2)$$

where $A_1, \ldots, A_k$ are disjoint intervals in $\mathbb{R}$ that do not depend on x, and $a_{0i}, \ldots, a_{ni}$ are constants for all i. We will say that f is a k-piece (ignoring the 0 piece), and n-degree (assuming $a_{ni} \neq 0$ for some i) MOP function. We will refer to the *order* of the MOP function as $n + 1$. Hereafter, MOP functions are assumed to be zero in undefined regions. MOP functions were first utilized in Bayesian network models by Shenoy and West [3].

Example 2. A sample of $\mathcal{N} = 71$ observations is taken from the $N(280, 210)$ distribution. A 2-piece, 2-degree (or third order) MOP density function that approximates this normal PDF is

$$\hat{f}(x) = \begin{cases} -3.3726 + 0.0241x - 0.000043x^2 & 236 \leq x < 271 \\ 0.7102 - 0.0061x + 0.000013x^2 & 271 \leq x \leq 306 \ . \end{cases} \tag{3}$$

The function above was constructed using B-spline functions [12] using the method suggested by López-Cruz et al. [10]. This will be discussed later in the paper. The MOP distribution is shown in Fig. 1 overlaid on the actual $N(280, 210)$ PDF (see the left panel). The right panel of Fig. 1 will be referred to in Section 3. Note that the MOP function was created from a small sample without knowledge of the actual underlying distribution.

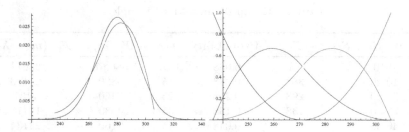

Fig. 1. MOP density function overlaid on the $N(280, 210)$ distribution (left) and B-spline functions used to construct the distribution (right)

2.3   Quality and Size of MOP Approximations

This section discusses four measures of the accuracy of a MOP approximation with respect to a PDF defined on the same domain: the Kullback-Leibler (KL) divergence, maximum absolute deviation (for the PDF and cumulative distribution function (CDF)), and maximum "right-tail" absolute deviation for the CDF. A measure of the "size" of the MOP function is also introduced.

If f is a PDF on the interval (a, b), and $\hat{f}$ is a PDF that is an approximation of f such that $\hat{f}(x) > 0$ for $x \in (a, b)$, then the *KL divergence* between f and $\hat{f}$, denoted by $KL(f, \hat{f})$, is defined as follows [13]:

$$KL(f, \hat{f}) = \int_a^b \ln \left(\frac{f(x)}{\hat{f}(x)} \right) f(x) \, dx. \tag{4}$$

$KL(f, \hat{f}) \geq 0$, and $KL(f, \hat{f}) = 0$ if and only if $\hat{f}(x) = f(x)$ for all $x \in (a, b)$. A closer approximation will produce a smaller KL divergence value.

The *maximum absolute deviation* between f and $\hat{f}$, denoted by $MAD(f, \hat{f})$, is given by:

$$MAD(f, \hat{f}) = sup\{|f(x) - \hat{f}(x)| : a < x < b\}$$

The maximum absolute deviation can also to applied to CDFs. Thus, if $F(\cdot)$ and $\hat{F}(\cdot)$ are the CDFs corresponding to $f(\cdot)$, and $\hat{f}(\cdot)$, respectively, then the maximum absolute deviation between F and $\hat{F}$, denoted by $MAD(F, \hat{F})$, is

$$MAD(F, \hat{F}) = sup\{|F(x) - \hat{F}(x)| : a < x < b\}$$

The value $MAD(F, \hat{F})$ is in units of probability, whereas the value $MAD(f, \hat{f})$ is in units of probability density, and the two values cannot be compared to each other.

For examples in this paper, we are particularly interested in values in the upper tail of the distribution. Thus, we will calculate a variation $MAD_{rt}(F, \hat{F})$ of mean absolute deviation that measures the maximum distance between the approximation and the target CDF in the region above the expected value. This is done as

$$MAD_{rt}(F, \hat{F}) = sup\{|F(x) - \hat{F}(x)| : E(X) < x < b\}$$

where $E(X) = \int_a^b x \cdot \hat{f}(x) \, dx$.

To illustrate these definitions, let $f(\cdot)$ denote the $N(280, 210)$ PDF truncated to $[236, 306]$. Consider $\hat{f}(\cdot)$, the 2-piece, 2-degree MOP approximation of $f(\cdot)$ as described in Eq. (1). Also, let $F(\cdot)$ and $\hat{F}(\cdot)$ denote the CDFs corresponding to f and $\hat{f}$, respectively. The goodness of fit statistics for $\hat{f}$ are as follows: $KL(f, \hat{f}) \approx 0.0181$, $MAD(f, \hat{f}) \approx 0.0034$, $MAD(F, \hat{F}) \approx 0.0371$, and $MAD_{rt}(F, \hat{F}) \approx 0.0031$. For clarification, recall that the comparison of the approximate distribution is made to the actual distribution truncated and normalized over the same interval.

The MOP models in this paper are developed using Mathematica 9.0 software. This package provides a function called `LeafCount` that gives the total "size" of an expression defined using the `Piecewise` representation, based on applying the `FullForm` function [14]. `LeafCount` (denoted by $\mathcal{L}$) will be used to measure the complexity of the resulting function. For additional information, please see the related working paper [9].

3   B-Spline Estimation of MOPs

Lopez-Cruz et al.[10] suggest using a linear combination of B-spline functions to construct MOP approximations from datasets where the parametric form of the underlying probability distribution is unknown. B-spline functions are piecewise polynomial functions defined by the number of control points, $n + 1$, and the degree of the polynomial, d. The control points define a knot vector $\mathbf{t} = \{t_0, t_1, t_2, \ldots, t_n\}$.

B-spline functions [12] have two definitions, one when $d = 1$ and another when $d > 1$. When $d = 1$, the functions are defined as

$$B_{j,1}(x) = \begin{cases} 1 & t_j \le x < t_{j+1} \\ 0 & \text{otherwise} . \end{cases}$$

For $d > 1$, the functions are calculated as

$$B_{j,k}(x) = \left(\frac{x - t_j}{t_{j+k-1} - t_j} \right) \cdot B_{j,k-1}(x) + \left(\frac{t_{j+k} - x}{t_{j+k} - t_{j+1}} \right) \cdot B_{j+1,k-1}(x) .$$

The control points are indexed by $j = 0, \ldots, n$ and the degree of the functions are indexed by $k = 1, \ldots, d$.

The B-spline functions are used to form a n-piece MOP density function

$$\hat{f}(x) = \sum_{i=1}^{m} \alpha_i \cdot B_{i,d}(x) \tag{5}$$

by selecting mixing coefficients α_i, $i = 1, \ldots, m$, where $m = n + d - 1$. Thus, the PDF for $\hat{f}$ will be a mixture of the m B-splines of order d.

Given a dataset $\mathcal{D} = \{x_1, \cdots, x_N\}$, Zong [12] suggests using the following iterative formula for determining the maximum likelihood estimators, $\hat{\alpha} = \{\hat{\alpha}_1, \ldots, \hat{\alpha}_m\}$, for the mixing coefficients in (5):

$$\hat{\alpha}_i^{(q)} = \frac{d}{N \cdot (t_i - t_{i-d})} \sum_{x \in \mathcal{D}} \frac{\hat{\alpha}_i^{q-1} B_{i,d}(x)}{\hat{f}\left(x; \hat{\alpha}^{(q-1)}\right)} . \tag{6}$$

Beginning with equivalent values for each α_i, the expression in (6) is used iteratively for $i = 1, \ldots, m$ until $\left| \dfrac{like^{(q)} - like^{(q-1)}}{like^{(q)}} \right| < \epsilon$, where $like^{(q)}$ is the log-likelihood of $\mathcal{D}$ given $\hat{f}\left(x; \hat{\alpha}^{(q)}\right)$ at iteration q in the optimization process. Using $\epsilon = 10^{-6}$ appears to be adequate for most applications [10]. To avoid placing constraints on the calculated values for the $\hat{\alpha}$ parameters, the function $\hat{f}_X\left(x; \hat{\alpha}^{(q)}\right)$ can simply be normalized to integrate to 1 at each iteration.

The goal is to develop a PDF $\hat{f}$ that is reasonably accurate; however, we would like the number of pieces and the degree of the polynomial functions comprising the MOP density to be as small as possible to avoid overfitting and speed up computation in subsequent applications. Thus, we will consider several possible values for d and n for each PDF and select the approximation that maximizes the Bayesian information criterion (BIC) calculated as

$$BIC\left(\hat{f}(x), \mathcal{D}\right) = \mathcal{L}\left(\mathcal{D}|\hat{f}(x)\right) - ((m - 1)\log N)/2 . \tag{7}$$

The second term in the BIC expression is a penalty for adding parameters to the model. Since the mixing coefficients must add to 1, $m - 1$ is the number of "free" mixing coefficients. In practice, once we settle on good values for d and n, this step could be avoided.

Example 3. Consider again the MOP density function $\hat{f}$ in Eq. 3 that approximates the $N(280, 210)$ PDF. Recall that the MOP is not fit to the normal PDF, but rather a small sample of data generated from the normal PDF. The four B-splines used to construct the MOP function are shown in the right panel of Figure 1. The mixing coefficients determined via 37 iterations of equation (6) are $\alpha_1 = 0.04$, $\alpha_2 = 0.07$, $\alpha_3 = 0.83$, and $\alpha_4 = 0.06$.

4   Grouped Data Estimation

Two issues arise in the process of fitting an MOP density function to grouped data: 1) creating a temporary dataset to approximate the unknown dataset used to compile the frequency distribution, and 2) determining the control points used to construct the B-splines and the MOP approximation. These two issues are further explored in this section.

4.1   Creating an Approximate Dataset

To fit a MOP density function to grouped data, such as shown in Table 1, we will produce a temporary dataset $\hat{\mathcal{D}}$ that approximates the unknown dataset $\mathcal{D}$ used to tabulate the frequency distribution for the grouped data. The number of observations in $\mathcal{D}$ is $\mathcal{N}$, and we will assign a multiple η that determines the approximate number of observations $\hat{\mathcal{N}} = \eta \cdot \mathcal{N}$ in $\hat{\mathcal{D}}$.

To construct $\hat{\mathcal{D}}$, we assign $2\mathcal{K} - 2$ different values to the $\hat{\mathcal{N}}$ observations in $\hat{\mathcal{D}}$. The mid-point m_1 is assigned approximately $\eta \cdot \mathcal{F}_1$ sample observations and the mid-point $m_{\mathcal{K}}$ is assigned approximately $\eta \cdot \mathcal{F}_{\mathcal{K}}$ sample observations. The intervals $j = 2, \ldots, \mathcal{K} - 1$ are each assigned two different sample values determined as

$$\hat{x}_{j,k} = \mathcal{S}_{j-1} + k \cdot \frac{\mathcal{S}_j - \mathcal{S}_{j-1}}{3} \quad \text{for} \quad k = 1, 2.$$

The approximate number of sample observations included in the dataset $\hat{\mathcal{D}}$ for $\hat{x}_{j,1}$ and $\hat{x}_{j,2}$, respectively, are

$$\hat{\mathcal{N}}_{j,1} = \eta \cdot \mathcal{F}_j \cdot \frac{\mathcal{F}_{j-1}}{\mathcal{F}_{j-1} + \mathcal{F}_{j+1}} \quad \text{and} \quad \hat{\mathcal{N}}_{j,2} = \eta \cdot \mathcal{F}_j \cdot \frac{\mathcal{F}_{j+1}}{\mathcal{F}_{j-1} + \mathcal{F}_{j+1}}.$$

The idea is to weight the number of sample observations included in the dataset for the two points in each interval by the relative number of points in the adjacent intervals. If the formulas above produce a decimal number, this is simply rounded to the nearest integer.

Example 4. Consider the grouped data in Table 1. With $\eta = 0.5$, the values shown in Table 2 are determined in order to create a dataset of $\hat{\mathcal{N}} = 502$ observations for use in the B-spline approximation algorithm. Points such as $\hat{x}_{2,1} = 13.33$ and $\hat{x}_{2,2} = 16.67$ simply divide the interval $[10, 20)$ into three equal sections. The point $\hat{x}_{2,1} = 13.33$ is assigned $\hat{\mathcal{N}}_{2,1} = 0.5 \cdot 388 \cdot 88/(298 + 88) \approx 44$ sample observations in $\hat{\mathcal{D}}$.

Table 2. Values required to reconstruct dataset $\hat{\mathcal{D}}$ for grouped data Table 1

$[\mathcal{S}_{j-1}, \mathcal{S}_j)$	$\mathcal{F}_j$	$\hat{\mathcal{N}}_{j,k}$	m_j	$\hat{x}_{j,k}$
$[0, 10)$	88	44	6.67	6.67
$[10, 20)$	388	44	15	13.33
$[10, 20)$	388	150	15	16.67
$[20, 30)$	298	110	25	23.33
$[20, 30)$	298	39	25	26.67
$[30, 40)$	138	60	35	33.33
$[30, 40)$	138	10	35	36.67
$[40, 50)$	48	19	45	43.33
$[40, 50)$	48	6	45	46.67
$[50, \infty)$	40	20	55	55
TOTAL	**1000**	502		

Creating the approximate dataset is a matter of assigning the data points in each interval defined for the grouped data, then assigning an adequate number of sample observations for each point. The two-point approximation described above has proven adequate in all of the example problems studied while conducting this research. Increasing the number of points beyond two has yielded results that are not significantly different than the two-point approximation when employed with the control points established in the next section.

4.2   Control Points

The B-spline functions are determined by the control points (in addition to the degree of the resulting polynomial) and define a knot vector $\mathbf{t} = \{t_0, t_1, t_2, \ldots, t_n\}$. We will consider three possibilities: uniform control points, equal probability control points, and control points determined by the intervals in the grouped data. The first two heuristics (uniform and equal probability) will be used when fitting MOP functions to sample data that is not grouped. The first and third (uniform and grouped data interval points) will be used to fit MOP functions to grouped data.

Since we consider the case where the last interval is unbounded, implementing the B-spline algorithm requires us to assign a maximum value as t_n. This is an important consideration, because the tails of probability distributions are important for understanding when an unusual event has occurred. In fact, for the supply chain management application in the related working paper [9], we are interested in knowing whether an observation that is outside the boundaries of all previous observations is likely to be observed from a process that is functioning correctly.

Chebyshev's inequality [15] states for a random variable X with mean μ and standard deviation σ that

$$P(|X - \mu| \geq z \cdot \sigma) \leq \frac{1}{z^2} \ .$$

We will target a probability that the maximum density not captured by the MOP approximation be 1% based on the grouped mean and variance as calculated in Eq. 1. This entails considering an upper bound on the domain of the MOP of $t_n = \overline{X}_G + 10 \cdot \sigma_G$. We also assume that $t_0 = S_0$. Thus, the points in $\mathbf{t}$ remaining to be assigned are $t_1, \ldots, t_{n-1}$.

Uniform Control Points. With t_0 and t_n assigned above, the remaining values in a set of uniform control points, denoted by UC, are determined as

$$t_i = t_0 + i \cdot \frac{t_n - t_0}{n} \quad \text{for} \quad i = 1, \ldots, n - 1.$$

These points are used to create *uniform* B-splines.

Equal Probability Control Points. While the primary topic of this paper is fitting MOP functions to grouped data, another interesting research question is whether or not using equal probability points (or percentiles) in a sample dataset produces a superior MOP approximation than using uniform control points. Here, since we will compare sample data and grouped data approximations, we will maintain the assignment of $t_0 = S_0$ and $t_n = \overline{X}_G + 10 \cdot \sigma_G$. In a situation where grouped data was not provided and access to the complete dataset is available, these could be assigned differently. The remainder of the equal probability control points, denoted by EP, are

$$t_i = P_{100i/n} \quad \text{for} \quad i = 1, \ldots, n - 1 \,.$$

P_k denotes the values in the sample dataset that have a greater value than at least k percent of all the elements in the set. The equal probability control points are not used to estimate PDFs from grouped data because it would be difficult to make an adequate assumption about the location of the percentiles in the pre-defined intervals.

Grouped Data Control Points. Another option for assigning the control points when developing a probability distribution for grouped sample data is to simply use the split points employed to tabulate the data. Due to the fact that the right tail of the MOP approximation is extended to 10 sample group standard deviations above the mean, we also fit one additional piece in the region corresponding with the last interval of the grouped data. Thus, with $t_0 = S_0$ and $t_n = \overline{X}_G + 10 \cdot \sigma_G$, the entire set of grouped data *plus one* control points, denoted by $GD1$, is defined as

$$\mathbf{t} = \{S_0, S_1, \ldots, S_{\mathcal{K}-1}, (S_{\mathcal{K}-1} + t_n)/2, t_n\}.$$

When the $GD1$ control points are used, the number of pieces in the MOP approximation is $n = \mathcal{K} + 1$.

We will also consider adding an additional point in the left-tail of the distribution because the shape of the PDF may change quickly in this region. This will be termed the grouped data *plus two* model (or $GD2$ for short) and is characterized by a $n = \mathcal{K} + 2$ piece MOP function defined by the control points

$$\mathbf{t} = \{\mathcal{S}_0, (\mathcal{S}_0 + \mathcal{S}_1)/2, \mathcal{S}_1, \ldots, \mathcal{S}_{\mathcal{K}-1}, (\mathcal{S}_{\mathcal{K}-1} + t_n)/2, t_n\}.$$

Example 5. For the grouped data shown in Table 1 with $\overline{X}_G = 23.05$ and $s_G^2 = 133.7$, we define $t_0 = 0$ and $t_n = 23.05 + 10 \cdot \sqrt{133.7} = 138.7$. The UC points for an $n = 7$ piece MOP approximation to the distribution for the grouped data are

$$\mathbf{t} = \{0, 19.8, 39.6, 59.4, 79.2, 99.1, 118.9, 138.7\}.$$

The $GD1$ control points for a 7-piece MOP approximation are

$$\mathbf{t} = \{0, 10, 20, 30, 40, 50, 94.35, 138.7\}.$$

The $GD2$ control points for an 8-piece MOP approximation are

$$\mathbf{t} = \{0, 5, 10, 20, 30, 40, 50, 94.35, 138.7\}.$$

Grouped data control points are not used to estimate MOP functions directly from the sample data, because the points would simply be arbitrary. When grouped data is available, we suppose that the intervals are intentionally selected for the frequency distribution because they are relevant for the application under consideration.

Note that the sample points in the dataset $\hat{\mathcal{D}}$ and the control points in $\mathbf{t}$ need not necessarily correspond. The control points define the B-splines that will be mixed to construct the MOP approximation and the sample points in $\hat{\mathcal{D}}$ are used to create the maximum likelihood estimators $\hat{\alpha}$ for the mixing coefficients.

Again, there are many potential heuristics that could be used to define the control points. We could potentially sub-divide the grouped data intervals further, or combine adjacent intervals. Several such variations were considered in the course of this research, but the methods described here were ultimately used because they produce good results, as demonstrated later in the paper.

5   Examples

This section presents two examples using the B-spline approach to construct an MOP approximation to the probability distribution that generated the grouped data. The approach will be to sample from a known density function. In practice, we would not assume any knowledge of the underlying probability distribution or require the raw data used to develop the frequency distribution; however, for the two examples in this section, using data simulated from a known probability distribution will help evaluate the resulting MOP distribution.

We will fit MOP approximations to the full (ungrouped) sample, then subsequently to the grouped data. These two approximations can be compared to to the true PDF using the measures of fit defined in Section 2.3. In terms of calculating the measures of fit, the MOP approximation $\hat{f}_S$ fit to the sample data and the MOP approximation $\hat{f}_G$ fit to the grouped data are both approximations of the true density f.

To facilitate calculation of the measures of fit, all MOP approximations will be defined over the same interval $[\mathcal{S}_0, \overline{X}_G + 10 \cdot \sqrt{s_G^2}]$. The underlying PDF will be truncated and normalized to integrate to 1 on this same interval. Obviously, if the full sample dataset was available in practice and we chose to use this dataset to construct the MOP approximation, we might choose endpoints for the first and last pieces of the MOP function differently.

Some issues we will consider are as follows:

1. The performance of uniform versus equal probability control points for estimating MOP functions from the full sample dataset.
2. The performance of uniform versus grouped control points for estimating MOP functions from grouped data.
3. The effect of varying the parameter η to adjust the size $\hat{\mathcal{N}}$ of the dataset $\hat{\mathcal{D}}$.

5.1 Example 1

In this section, we consider the grouped data shown in Table 1. The grouped data was summarized from 1000 random variates simulated from the $LN(3, 0.5^2)$ distribution. For the grouping represented in Table 1, the $GD2$ heuristic requires an 8-piece MOP function, so this is the largest number of pieces we will consider when using any heuristic. We will consider MOP functions with as few as 4 pieces for the UC and EP points. When selecting the best MOP function for a given control point heuristic, and for selecting the best MOP density function developed from the same dataset, we will use the BIC criterion defined earlier.

Results of several models estimated from the full sample dataset and the grouped sample dataset are shown in Table 3. For purposes of comparing computational time, the MOP estimated from the full dataset with UC points is considered the baseline model. The percentage change from this baseline is reported for the other models, with a negative percentage representing an improvement in required CPU time. Where relevant, the model selected for inclusion in Table 3 is the one that produced the highest BIC score. For example, the model developed from the full dataset with UC points is a fifth order MOP with 7 pieces. A similar function with 8 pieces had a BIC score of -3806.81.

The MOP functions in Table 3 can be compared as follows:

UC versus EP. For the full dataset, the UC points provided the MOP approximation with the highest BIC score, and this function is of smaller size and required far less CPU time to calculate. The UC function also had the lowest MAD measurements for the CDF. For this problem, using uniform control points seems to be preferable to EP points.

UC versus GD1/GD2. When comparing the three columns related to $\eta = 1$ in Table 3, we can see that the $GD1$ heuristic produced the MOP with the highest

Table 3. Results of MOP estimation for Example 1. The best measure of fit for each dataset type is shown in **bold**.

Dataset	Full	Full	Grouped	Grouped	Grouped	Grouped	Grouped
Control Points	UC	EP	UC	GD1	GD2	GD1	GD1
Pieces	7	6	7	7	8	7	7
Order	5	9	5	4	4	4	4
η	N/A	N/A	1	1	1	0.5	0.1
KL	0.0257	**0.0059**	0.0758	0.0469	0.0793	0.0465	**0.0461**
MAD PDF	0.0075	**0.0026**	**0.0081**	0.0103	0.0147	0.0104	0.0105
MAD CDF	**0.0303**	0.0317	**0.0343**	0.0566	0.0743	0.0574	0.0592
MAD_{rt} CDF	**0.0256**	0.0317	0.0334	**0.0286**	0.0358	0.0293	0.0326
BIC	**−3803.95**	−3807.64	−3766.54	**−3753.45**	−3754.6	−1895.27	−403.241
CPU	0%	359%	15%	−44%	2%	−69%	−89%
Leaf Count ($\mathcal{L}$)	**207**	298	194	**172**	195	**172**	**172**

BIC score, and this approximation required the least amount of CPU time to estimate. Its size (measured using the `LeafCount` function in Mathematica) is also the smallest among the competing models. Its KL statistic is the lowest among these three MOPs, and the MAD measurement for the right tail of the CDF is also the smallest among any of the models developed with grouped data. The grouped data points were preferable in this case to the UC points. Obviously, the grouped data points are established by an analyst or manager when creating the frequency distribution, so the relevance of these points to the application under consideration could affect these results.

Varying η Parameter. The BIC values shown in the last two columns of Table 3 are not comparable to those in any other column because they are based on different datasets. However, if we examine the goodness-of-fit statistics for the MOPs produce with datasets constructed from $\eta < 1$, we see very comparable to results to the case where $\eta = 1$; the CPU time required to produce these functions is much lower. In this example, it seems perfectly acceptable to reduce the size of the temporary dataset.

The PDF and CDF for the MOP estimated from $GD1$ points with $\eta = 0.1$ are shown in Fig. 2 overlaid on the corresponding functions for the $LN(3, 0.5^2)$ PDF. The CDF developed from the MOP PDF is also a MOP function. Recall that these MOP functions are constructed from a frequency distribution based on 1000 sample points from the $LN(3, 0.5^2)$ without knowledge of the underlying distribution.

How much can we reduce the η parameter to gain additional computational efficiency? This seems to depend partially on the size of the original dataset. For instance, if we reduce η to 0.05 in this example and use a dataset of 51 observations (and fit a 7-piece, 4th order MOP), all of the goodness-of-fit statistics are worse than for the $\eta = 0.1$ distribution. There is a critical mass of points needed in the dataset $\hat{\mathcal{D}}$.

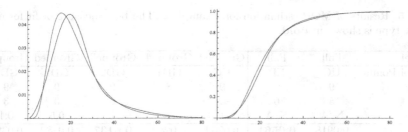

Fig. 2. MOP PDF estimated from grouped data with $GD1$ points and $\eta = 0.1$ (left) and CDF associated with the MOP function (right). Both functions are overlaid on the corresponding functions for the $LN(3, 0.5^2)$ PDF.

5.2 Example 2

The second example is based on a sample of 1000 points from a mixture distribution for a random variable X that is determined as $X = 0.6 \cdot X_1 + 0.4 \cdot X_2$, where $X_1 \sim LN(3, 0.5^2)$ and $X_2 \sim LN(4.5, 0.25^2)$. The grouped data for this example is shown in Table 4. The mean and variance of the grouped data are $\overline{X}_G = 52.56$ and $s_G^2 = 1394$, so the MOP functions for this example will be defined on $[0, 426]$. Use of the $GD2$ heuristic will require a 10-piece MOP approximation, so we will consider MOP approximations from 5-10 pieces, where applicable. The pieces in the $GD1$ and $GD2$ MOP functions are determined by the number of grouped data intervals.

The results for MOP models estimated from the full dataset for Example 2 and the grouped data shown in Table 4 are shown in Table 5. Some observations regarding these results are as follows:

Table 4. Grouped dataset for Example 2

Interval $[S_{j-1}, S_j)$	$\mathcal{F}_j$	Probability	m_j	$\mathcal{F}_j \cdot m_j$	$\mathcal{F}_j \cdot (m_j - \overline{X}_G)^2$
0-20	277	0.277	13.33	3693.33	426301
20-40	248	0.248	30	7440	126258
40-60	71	0.071	50	3550	467
60-80	112	0.112	70	7840	34052
80-100	154	0.154	90	13860	215832
100-120	88	0.088	110	9680	290309
120+	50	0.050	130	6500	299822
TOTAL	$\mathcal{N} = 1000$	1.000		52563.33	1393040
				$\overline{X}_G = 52.56$	$s_G^2 = 1394$

UC versus EP. When evaluating the UC and EP control point heuristics for use with the full dataset, the BIC score is highest for the model estimated using the EP points. Three of the measures of fit favor the EP model, although it does

Table 5. Results of MOP estimation for Example 2. The best measure of fit for each dataset type is shown in **bold**.

Dataset	Full	Full	Grouped	Grouped	Grouped	Grouped	Grouped
Control Points	UC	EP	UC	GD1	GD2	GD2	GD2
Pieces	9	9	9	8	9	9	9
Order	5	6	4	6	6	3	3
η	N/A	N/A	1	1	1	0.5	0.1
KL	0.0918	**0.0561**	0.0969	0.0280	**0.0137**	0.0382	0.0383
MAD PDF	0.0106	**0.0034**	0.0106	0.0058	0.0037	**0.0029**	0.0034
MAD CDF	0.0816	**0.0431**	0.0824	0.0351	0.0298	**0.0241**	0.0323
MAD_{rt} CDF	**0.0362**	0.0431	0.0484	0.0309	0.0298	**0.0241**	0.0321
BIC	−4806.92	**−4791.28**	−4780.75	−4720.84	**−4703.18**	−2370.66	−507.47
CPU	**0%**	143%	−15%	1%	−36%	−89%	−97%
Leaf Count ($\mathcal{L}$)	**226**	400	194	276	309	**174**	174

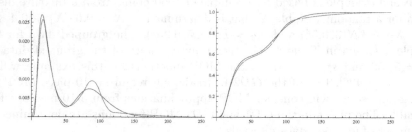

Fig. 3. MOP PDF estimated from grouped data with $GD2$ points and $\eta = 0.5$ (left) and CDF associated with the MOP function (right). Both functions are overlaid on the corresponding functions for the underlying mixture distribution.

require more CPU time and storage space. In the case of a bi-modal distribution, the EP heuristic deserves consideration when selecting the best MOP model.

UC versus GD1/GD2. For the three models estimated with $\eta = 1$ from the grouped data, the $GD2$ temporary dataset provided the result with the highest BIC score. The KL divergence of this model is the lowest of all the models, and the MAD statistics associated with the CDF are very competitive with other possible models. Again, using grouped data control points seems to work well. It may be possible to use a similar idea with the full dataset, i.e. we could divide points evenly over a portion of the domain, then define a wider distance between control points in the right-tail. This deserves some further investigation.

Varying η parameter. Although the grouped data is bi-model, reducing the number of elements in the temporary dataset used to construct MOP distributions still provides good results at a lower computational expense. For instance, the model estimated with $\eta = 0.5$ provides good results for the MAD measurements for the CDF at an 89% reduction in computational expense. The PDF and

CDF for this approximation are shown in Figure 3 overlaid on the actual PDF and CDF for the mixture distribution, respectively. As with the last example, we see that there may be a critical mass of points required in $\hat{\mathcal{D}}$, and this may be greater if the frequency distribution is multi-modal. Table 5 shows that all goodness-of-fit statistics for the $\eta = 0.1$ model are worse than those recorded for the $\eta = 0.5$ model.

6 Conclusions

This paper has described a method for estimating a mixture of polynomials PDF from a frequency distribution of grouped data. The approach described is an extension of the B-spline approximation technique implemented by López-Cruz et al. [10]. The motivation for adapting this method for use with grouped data is the construction of a cycle time distribution by using a standard report from a decision support system used by a manufacturer that utilizes refillable containers. A description of this application is available in a related working paper [9]. The approach described here is used to develop a PDF for the cycle time of refillable containers, which is then used to draw conclusions about the size of the container population (or *fleet*).

Some issues that arise when fitting MOP functions to grouped data are the size of the temporary dataset and the split points assigned for the MOP function. We find that using a fraction of the points contained in the original dataset was sufficient in both applications examined in this paper, although in the second example using too few observations produced results that were not as good as a trial with half the number of the original datasets points. The PDF in this example was bi-modal. When fitting an MOP to grouped data, utilizing the dividing points for the bins of the frequency distribution produced better results than the uniform control point heuristic. Some reasonable guidelines would be to reduce the number of points in the original dataset to 10% of the original dataset for uni-modal distributions and 50% of the original dataset for multimodal distributions. Use of the grouped data plus one control points may be adequate for uni-modal PDFs, whereas adding an additional control point at both ends of the domain may be necessary for multi-modal distributions.

Acknowledgments. Thank you to the reviewers for comments and suggestions which improved the paper.

References

1. Fryer, M.J.: A review of some non-parametric methods of density estimation. Journal of Applied Mathematics 20(3), 335–354 (1977)
2. Moral, S., Rumí, R., Salmerón, A.: Mixtures of truncated exponentials in hybrid Bayesian networks. In: Benferhat, S., Besnard, P. (eds.) ECSQARU 2001. LNCS (LNAI), vol. 2143, pp. 156–167. Springer, Heidelberg (2001)

3. Shenoy, P.P., West, J.C.: Inference in hybrid Bayesian networks using mixtures of polynomials. International Journal of Approximate Reasoning 52(5), 641–657 (2011)
4. Langseth, H., Nielsen, T., Rumí, R., Salmerón, A.: Mixtures of truncated basis functions. International Journal of Approximate Reasoning 53(2), 212–227 (2012)
5. Rumí, R., Salmerón, A., Moral, S.: Estimating mixtures of truncated exponentials in hybrid bayesian networks. Test 15(2), 397–421 (2006)
6. Romero, V., Rumí, R., Salmerón, A.: Learning hybrid bayesian networks using mixtures of truncated exponentials. International Journal of Approximate Reasoning 42(1-2), 54–68 (2006)
7. Langseth, H., Nielsen, T.D., Rumi, R., Salmerón, A.: Parameter estimation and model selection for mixtures of truncated exponentials. International Journal of Approximate Reasoning 51(5), 485–498 (2010)
8. Langseth, H., Nielsen, T.D., Rumí, R., Salmerón, A.: Learning mixtures of truncated basis functions from data. In: Cano, A., Gómez-Olmedo, M., Nielsen, T. (eds.) Proceedings of the Sixth European Conference on Probabilistic Graphical Models (PGM 2012), Granada, Spain, pp. 163–170 (2012)
9. Cobb, B.R.: Fleet management with cycle time distributions constructed from grouped sample data. Working paper, Missouri State University, Department of Management, Springfield, MO (2014)
10. López-Cruz, P.L., Bielza, C., Larrañaga, P.: Learning mixtures of polynomials of multidimensional probability densities from data using b-spline interpolation. International Journal of Approximate Reasoning 55(4), 989–1010 (2014)
11. Shenoy, P.P.: Two issues in using mixtures of polynomials for inference in hybrid Bayesian networks. International Journal of Approximate Reasoning 53(5), 847–866 (2012)
12. Zong, Z.: Information-Theoretic Methods for Estimating Complicated Probability Distributions. Elsevier, Amsterdam (2006)
13. Kullback, S., Leibler, R.A.: On information and sufficiency. Annals of Mathematical Statistics 22, 76–86 (1951)
14. Wolfram, S.: The Mathematica Book, 5th edn. Wolfram Media, Champaign (2003)
15. Larsen, R.J., Smith, A.F.M.: An Introduction to Mathematical Statistics and its Applications, 3rd edn. Prentice-Hall, Boston (2001)

Trading off Speed and Accuracy in Multilabel Classification

Giorgio Corani[1], Alessandro Antonucci[1],
Denis D. Mauá[2], and Sandra Gabaglio[3]

[1] Istituto Dalle Molle di Studi sull'Intelligenza Artificiale,
Lugano, Switzerland
{giorgio,alessandro}@idsia.ch
[2] Universidade de São Paulo,
São Paulo, Brazil
denis.maua@usp.br
[3] Institute for Information Systems and Networking,
Lugano, Switzerland
sandra.gabaglio@supsi.ch

Abstract. In previous work, we devised an approach for multilabel classification based on an ensemble of Bayesian networks. It was characterized by an efficient structural learning and by high accuracy. Its shortcoming was the high computational complexity of the MAP inference, necessary to identify the most probable joint configuration of all classes. In this work, we switch from the ensemble approach to the single model approach. This allows important computational savings. The reduction of inference times is exponential in the difference between the treewidth of the single model and the number of classes. We adopt moreover a more sophisticated approach for the structural learning of the class subgraph. The proposed single models outperforms alternative approaches for multilabel classification such as binary relevance and ensemble of classifier chains.

1 Introduction

In traditional classification each instance is assigned to a single class. *Multilabel classification* generalizes this idea by allowing each instance to be assigned to multiple *relevant classes*. Multilabel classification allows to deal with complex problems such as tagging news articles or videos.

A simple approach to deal with multilabel classification is *binary relevance* (BR), which decomposes the problem into a set of traditional (i.e., single label) classification problems. Given a problem with n classes, binary relevance trains n independent single-label classifiers. Each classifier predicts whether a specific class is relevant or not for the given instance. Binary relevance is attractive because of its simplicity. Yet, it ignores dependencies among the different class variables. This might result in sub-optimal accuracy, since the class variables are often correlated [8]. According to the global accuracy metric, a classification is accurate only if the relevance of *every* class is correctly predicted. A sound model

L.C. van der Gaag and A.J. Feelders (Eds.): PGM 2014, LNAI 8754, pp. 145–159, 2014.

of the joint probability of classes given the observed features is thus necessary (see [13] and the references therein). This requires identifying the *maximum a posteriori* (MAP) configuration of the class relevances.

The classifier chain [12] is a state-of-the-art approach to model dependencies among classes. It achieves good accuracy; however, it has no probabilistic interpretation.

Bayesian networks (BNs) are an appealing tool for probabilistic multilabel classification, as they compactly represent the joint distribution of class and feature variables. When dealing with multilabel classification they pose two main challenges: structural learning and predictive inference. As for structural learning, the graph is typically partitioned into three pieces [14, 3]: the *class subgraph*, namely the structure over the class variables; the *feature subgraph*, namely the structure over the features variables; the *bridge subgraph*, namely the structure linking the feature to the class variables [14, 3, 4].

In a previous work [1] we introduced the *multilabel naive* assumption, which provides an interesting trade-off between computational speed of structural learning and effectiveness of the learned dependencies. The assumption is that the features are independent *given the classes*, thus generalizing naive Bayes to the multilabel case. As a consequence, the feature subgraph is empty. The bridge subgraph is optimally learned by independently looking for the optimal parents set of each feature. It does not require iterative adjustments. This allows for efficient structural learning. We accompany the multilabel naive assumption with a simple but effective algorithm for feature selection.

Our previous approach [1] was based on an ensemble of different Bayesian networks. Under the multilabel assumption the different BNs had an empty feature subgraph and shared the same optimal bridge subgraph. Each model had a different naive Bayes class subgraph. The ensemble approach achieved good performance. Its main shortcoming was the high complexity of the MAP inference regarding the most probable joint configuration of all classes. The high cost of MAP inference in multilabel Bayesian network classifiers has been discussed previously in the literature. In [5], the authors limit the MAP inferential complexity by constraining the underlying graph to be a collection of small disjoint graphs. However, this severely limits the expressivity of the models.

In this work, we aim at largely decreasing the computational times of our previous approach while keeping accuracy as high as possible. To this end, we move from the ensemble to a single model. Single models are known to be less accurate than ensembles. To compensate this effect, we introduce a more sophisticated structural learning procedure for the class sub-graph. We allow the class subgraph to be either a naive Bayes or a forest-augmented naive Bayes (FAN). This yields two different multilabel classifiers, called in the following mNB and mFAN. Both models optimize in two steps the class subgraph. In the first step each class is considered as a possible root of mNB (or mFAN). For each possible root, we identify the optimal naive (or FAN) structure. We thus identify n different naive (or FAN) structures. In the second step we select the highest scoring naive (or FAN) structure.

Both mNB and mFAN are less accurate than the original ensemble. Yet, the accuracy gap is not huge. Moreover, mNB and mFAN outperform both binary relevance and the ensemble of classifier chains (implemented using naive Bayes as base classifier). The single model approach allows large computational savings compared to the ensemble. The saving is exponential in the difference between the number of classes and the treewidth of the single model which has been learned.

We then analyze the learned class sub-graphs. Define the relevance of a class as the percentage of instances for which it is relevant. We found a positive correlation between the relevance of the root class, the score of the class sub-graph and the number of non-root classes being to the root class. The explanation is as follows. Most classes have low relevance. A class which is labeled more often as relevant allows for better estimating the correlations with the remaining classes. This yields more non-root classes being connected to the root and also a higher score of the resulting graph. This might explain why our previous ensemble is only slightly more accurate than the single model. Many models of the ensemble had as a root of the class subgraph a class with low relevance. Such models were unlikely to convey helpful information when classifying the instances.

As a final contribution we discuss the need for preventing the *empty* prediction. A prediction is empty if all classes are predicted to be *not* relevant.

2 Probabilistic Multilabel Classification

We denote the array of class *relevances* as $\mathbf{C} := (C_1, \ldots, C_n)$; this is an array of Boolean variables, with variable C_i, $i = 1, \ldots, n$, expressing the relevance of the i-th class for the given instance. Thus, $\mathbf{C}$ takes its values in $\{0, 1\}^n$. We denote the set of features as $\mathbf{F} := (F_1, \ldots, F_m)$. We assume the availability of a set of complete training instances $\mathcal{D} = \{(\mathbf{c}, \mathbf{f})\}$, where $\mathbf{c} = (c_1, \ldots, c_n)$ and $\mathbf{f} = (f_1, \ldots, f_m)$ represent an *instantiation* of class relevances and features, respectively.

A *probabilistic multilabel classifier* estimates a joint probability distribution over the class relevances conditional on the features, $P(\mathbf{C}|\mathbf{F})$. Such model can predict the class relevances on new instances. We denote as $\mathbf{c}$ and $\hat{\mathbf{c}}$ respectively the set of actual and predicted class relevances. The actual and the predicted relevance of class C_i are denoted respectively by c_i and $\hat{c}_i$.

A common metric to evaluate multilabel classifiers on a given instance is *global accuracy* (also called *exact match*):

$$\text{acc} := I(\mathbf{c} = \hat{\mathbf{c}}), \tag{1}$$

where I is the indicator function.

Another measure of performance is *Hamming accuracy* (also called *mean label accuracy*):

$$\text{H}_{acc} := \frac{1}{n} \sum_{i=1}^{n} I(\hat{c}_i = c_i). \tag{2}$$

Commonly Hamming loss ($1\text{-}H_{acc}$) rather than Hamming accuracy is reported. We report Hamming accuracy to simplify results readability: both acc and H_{acc} are better when they are higher. Global accuracy is often zero on data sets with many classes. On such data sets Hamming accuracy is thus more meaningful than global accuracy.

When classifying an instance with features $\mathbf{f}$, two different inferences are performed depending on whether the objective is to maximize global accuracy or Hamming accuracy.

To maximize global accuracy we search for the most probable joint configuration of the class relevances (*joint query*):

$$\hat{\mathbf{c}} = \arg \max_{\mathbf{c} \in \{0,1\}^n} P(\mathbf{c}|\mathbf{f}) = \arg \max_{\mathbf{c} \in \{0,1\}^n} P(\mathbf{c}, \mathbf{f}) . \qquad (3)$$

In the context of Bayesian networks, the above problem is known as Maximum A Posteriori (MAP) or Most Probable Explanation (MPE) inference. Unlike in standard MAP inference problems, the prediction of all classes being non-relevant (*empty prediction*) is considered invalid. Each instance should be assigned to at least one class. If the most probable joint configuration is the empty prediction, we ignore it and return the second most probable configuration, which is necessarily non-empty. To our knowledge this issue has not yet been pointed out. For instance the algorithms implemented by MEKA[1] do not prevent the empty prediction. We show empirically in Section 5 that preventing the empty prediction can increase accuracy in some domains.

To maximize Hamming accuracy we select the most probable configuration (relevant or non-relevant) of each class C_i (*marginal query*):

$$\hat{c}_i = \arg \max_{c_i \in \{0,1\}} P(c_i|\mathbf{f}) = \arg \max_{c_i \in \{0,1\}} P(c_i, \mathbf{f}) , \qquad (4)$$

where $P(c_i|\mathbf{f}) = \sum_{\mathbf{C} \setminus \{C_i\}} P(\mathbf{c}|\mathbf{f})$. If all classes are predicted to be non-relevant, the empty prediction is avoided by predicting as relevant only the class C_i with the highest posterior probability of being relevant.

We model the joint distribution $P(\mathbf{C}, \mathbf{F})$ as a Bayesian network, and obtain classifications by running standard algorithms for either MAP inference or marginal inference in the network, according to the chosen performance measure. Once a structure (i.e., a directed acyclic graph over $\mathbf{C}, \mathbf{F}$) for the corresponding Bayesian network has been defined, the model parameters are efficiently computed using Bayesian estimation. Learning a good structure is a challenging problem, which we tackle by making a number of assumptions on the structure that enable fast and exact learning. We detail the assumptions in the next section.

3 Structural Learning

We address structural learning assuming the data set to be complete. The problem of how to efficiently learn the structure of a probabilistic multilabel classifier

---

[1] http://meka.sourceforge.net

has been studied in the past [14, 3]. The graph is typically partitioned into three pieces: the *class subgraph*, namely the structure over the class variables modeling class-to-class (in)dependences; the *feature subgraph*, namely the structure over the features variables modeling feature-to-feature (in)dependences; the *bridge subgraph*, namely the structure linking the feature to the class variables and modeling the (in)dependences between features and classes.

The approach of [14] constrains both the class subgraph and feature subgraph to be a tree-augmented naive Bayes (TAN). A bridge subgraph is proposed, and the TANs of the two subgraphs are correspondingly learned. The bridge subgraph is iteratively updated following a *wrapper* approach [11]. Every time the bridge subgraph is updated, the two TANs are re-learned. Also [3] adopts a similar approach, considering a wider variety of topologies for the class and the feature subgraphs. Such approaches can result in high computational times because the bridge subgraph is incrementally improved at each iteration, requiring to correspondingly update also the other subgraphs. Conversely, [4] keeps empty both the class subgraph and the feature subgraph. This approach is fast but cannot properly model correlated class variables.

Instead, we assume the features to be independent *given the classes* as in a previous work of ours [1]. Since the feature nodes cannot have children, the feature subgraph is empty. This allows us to optimally learn the bridge subgraph by independently looking for the optimal parents set of each feature.

Our procedure is based on maximizing the BDeu score, which decomposes, in the case of complete data, as the sum of the BDeu scores of each node:

$$\mathrm{BDeu}(\mathcal{G}) := \sum_{X_i \in \{\mathbf{C}, \mathbf{F}\}} \mathrm{BDeu}(X_i, \mathrm{Pa}(X_i)),$$

where $\mathcal{G}$ denotes the entire graph (i.e., the union of class, feature and bridge subgraphs), X_i a generic node and $\mathrm{Pa}(X_i)$ its parents set in $\mathcal{G}$. The number of joint configurations of the parents of X_i is denoted by q_i. The score function for a single node is:

$$\mathrm{BDeu}(X_i, \mathrm{Pa}(X_i)) := \sum_{j=1}^{q_i} \left[\log \frac{\Gamma(\alpha_{ij})}{\Gamma(\alpha_{ij} + n_{ij})} + \sum_{k=1}^{|X_i|} \log \frac{\Gamma(\alpha_{ijk} + n_{ijk})}{\Gamma(\alpha_{ijk})} \right], \quad (5)$$

where n_{ijk} is the number of records such that X_i is in its k-th state and its parents in their j-th configuration, while $n_{ij} = \sum_k n_{ijk}$. Finally, α_{ijk} is equal to the equivalent sample size α divided by the number of states of X_i and by the number of (joint) states of the parents, while $\alpha_{ij} = \sum_k \alpha_{ijk}$.

Class and feature nodes have only class nodes as parents (i.e., $\mathrm{Pa}(X_i) \subseteq \mathbf{C}$ for every node/variable X_i). Hence, the BDeu decomposes in two terms, referring respectively to the class and the bridge subgraphs:

$$\mathrm{BDeu}(\mathcal{G}) = \sum_{i=1}^{n} \mathrm{BDeu}(C_i, \mathrm{Pa}(C_i)) + \sum_{j=1}^{m} \mathrm{BDeu}(F_j, \mathrm{Pa}(F_j)).$$

Moreover, the two terms can be optimized separately, as the combined directed graph is necessarily acyclic. The optimizations of each term require different approaches:

Bridge Subgraph. We optimize the BDeu score of the bridge subgraph by independently searching for the optimal parents set of each feature. This strategy is optimal since: (i) the feature subgraph is empty, thus preventing the introduction of directed cycles; (ii) the BDeu scores decompose over the different feature variables.

Any subset of $\mathbf{C}$ is a candidate for the parents set of a feature, this reducing the problem to m independent local optimizations. The optimal parents set of F_j is found as follows:

$$\mathbf{C}_{F_j} := \arg \max_{\mathrm{Pa}(F_j) \subseteq \mathbf{C}} \mathrm{BDeu}(F_j, \mathrm{Pa}(F_j)),\qquad(6)$$

for each $j = 1, \ldots, m$. The optimization in Equation (6) becomes more efficient by considering the pruning techniques proposed in [6].

Feature Selection. Naive Bayes is surprisingly effective in traditional classification despite its simplicity. Moreover, careful feature selection allowed naive Bayes even to win data mining competitions [9].

We thus perform feature selection before learning the bridge graph. We rely on the correlation-based feature selection (CFS) [15, Chap. 7.1], which has been developed for traditional classification. We perform CFS n times, once for each different class variable. Eventually, we retain the *union* of the features selected in the different runs. This is a useful pre-processing step which reduces the number of features, removing the non-relevant ones (Table 2). It also helps from the computational viewpoint. Feature selection for multilabel classification is however an open problem, and more sophisticated approaches can be designed to this end.

Class Subgraph. Unlike the feature subgraph, the optimizations of the class variables cannot be carried out independently, as this might introduce directed cycles. Instead, we enable efficient structure learning by restricting the class of allowed structures. We allow the class subgraph to be either a naive Bayes or a forest-augmented naive Bayes (FAN). This yields two different multilabel classifiers, called in the following mNB and mFAN. The leading 'm' shows that such classifiers are designed for multilabel classification.

Let us assume that the class which serves as root node (C^{root}) of naive Bayes or FAN is given. We then search for the class subgraph which maximizes the BDeu score. As for mNB, the highest scoring naive Bayes is obtained by computing for each non-root class C_i the two scores $\mathrm{BDeu}(C_i, \emptyset)$ and $\mathrm{BDeu}(C_i, C^{\mathrm{root}})$. Class C_i is linked to the root class if

$$\mathrm{BDeu}(C_i, C^{\mathrm{root}}) > \mathrm{BDeu}(C_i, \emptyset)\qquad(7)$$

and unlinked otherwise. The above procedure is repeated n times. Every time a different class $C_1, \ldots, C_n$ is taken as root of the naive Bayes. This yields n different naive Bayes structures. The structure with maximum score among them is taken as class subgraph for the multilabel classifier. Summing up, we perform a two-steps optimization: first we compute the optimal naive structure for each possible root. Then we select the highest scoring structure among those identified in the first step.

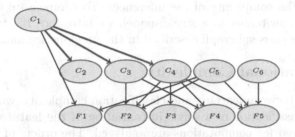

Fig. 1. Example of a mNB model. The class subgraph is a naive Bayes, with C_5 and C_6 unlinked from the root class. The arcs of the class subgraph are thicker and shown in blue.

An analogous two-step optimization is adopted when a FAN structure is looked for. The difference between FAN and TAN is that the former augments naive Bayes by a forest, while the latter augments naive Bayes by a tree (i.e., the underlying graph of a TAN is necessary connected). TAN is thus a special case of FAN, and therefore the latter can achieve a higher BDeu score than the former, as its structure has more degrees of freedom. To identify the optimal FAN we do not independently optimize the parents set of each non-root node, as this approach could introduce cycles. We instead solve an optimization problem characterized by the following constraints: cycles are not allowed; the root node has no parents; each non-root node C_i has three feasible configuration for its parents set: the empty set; the root class; the root class and another non-root class. Strategies for efficiently learning the optimal FAN structure are discussed for instance in [6, 7].

Figure 1 depicts an example of mNB, with the multilabel naive assumption for the bridge subgraph and a naive Bayes as class subgraph.

An Ensemble of Bayesian Networks. The previously described procedure yields a single Bayesian network model. Its bridge graph is optimally learned under the multilabel naive assumption. Its class graph is either an optimal FAN or an optimal naive Bayes.

In a previous work [1], we considered instead an ensemble approach. The ensemble was constituted by n different Bayesian networks (BNs). Each BN was based on naive multilabel assumption. We recall that, under the multilabel naive assumption, the optimal bridge subgraph is independent from the class subgraph. Thus, the BNs shared the same bridge subgraph.

The class subgraph was constituted by a different naive Bayes for each member of the ensemble. Each member of the ensemble used a different root for naive Bayes. The structure of naive Bayes was not optimized. The inferences produced by the different member of the ensemble were combined through logarithmic opinion pooling. The ensemble approach showed good performance, but also high computational times, especially for MAP inference (see the discussion at the end of Section 4).

In this paper we move from the ensemble to a single model. In this way, we largely reduce the complexity of the inferences. To compensate for the loss of accuracy due to switching to a single model, we introduce the two-layer optimization for the class subgraph described in the previous section.

4   Computational Complexity

We distinguish between learning and classification complexity, where the latter refers to the classification of a single instantiation of the features. Both space and time required for computations are analyzed. The orders of magnitude of these descriptors are reported as a function of the (training) dataset size d, the number of classes n, the number of features m, the average number of states for the features $f = m^{-1} \sum_{j=1}^{m} |F_j|$, and the maximum in-degree of the features $g = \max_{j=1,\dots,m} \|\mathbf{C}_{F_j}\|$ (where $\mathbf{C}_{F_j}$ is the optimal parents set of F_j computed as in Equation (6), $|\cdot|$ returns the cardinality of a variable, and $\|\cdot\|$ the number elements in a joint variable). As the class variables cannot have more than two parents in the mFAN and no more than one in the mNB, g is also the maximum in-degree of the network (provided that $g > 1$).

Regarding space, the conditional probability table (CPT) of the j-th feature, i.e., $P(F_j|\mathbf{C}_{F_j})$, needs space $O(|F_j| \cdot 2^{\|\mathbf{C}_{F_j}\|})$, while the n CPTs associated to the classes have size bounded by a small constant (8 numbers for the mFAN and 4 for the mNB). Overall, this means a space complexity $O(n + f2^g)$. These tables should be available during both learning and classification.

Regarding the time required by the learning, let us first note that, according to Equation (5), the computation of the BDeu score associated to a variable takes a time of the same order of the space required to store the corresponding CPT. The number of scores to be evaluated in order to determine the parents of F_j as in Equation (6) is of the same order of the binomial coefficient $\binom{n}{g}$, that means $O(n2^g)$. Then, we sum over the features and obtain $O(mn2^g)$ time. Regarding the class graph, for mNB we only have to evaluate the inequality in Equation (7), which only takes constant time, on the non-root classes. This means $O(n)$ time. Learning a FAN can be instead achieved in $O(n^2)$ [7]. Finally, the quantification of the network parameters requires the scan of the dataset, i.e., for the whole ensemble, $O((n + m)d)$. Such a learning procedure should be iterated over the n models of the ensemble, in order to select the one with the highest likelihood. This only affects the class subgraph, and the relative term should be therefore additionaly multiplied by n.

Concerning the classification time, both the MAP inference in Equation (3) and the computation of marginals in Equation (4) can be solved exactly by junction tree algorithms in time exponential in the *treewidth* of the network's moral graph.[2] The treewidth measures the connectivity of the network, which according to our learning procedure depends on unconditional class correlations, and the capability of the features to induce additional (conditional) correlations among classes. In the mFAN, the treewidth is at least 3, while in mNB it is at least 2. The experiments reported in the next section show that the treewidth of our models is usually much smaller than the number of classes, and often small enough to enable exact inference (by junction tree algorithms).

The situation of our previous ensemble was radically different [1]. For each pair of class variables, say C_i and C_j, there was at least a network in the ensemble such that C_i was parent of C_j (and vice-versa). Thus, when merging all the networks of the ensemble in a single Markov random field, there was a clique including all the classes, which made the treewidth equal to n, the number of classes. Such a treewidth made exact inference intractable for large n, and we were forced to resort to approximate methods such as max-product.

The ratio of the inference time of the ensemble to the inference time of the single model increases exponentially with the difference between the number of classes and the actual treewidth of the single model. Yet, there are no theoretical guarantees for the treewidth of the single model being small (i.e., bounded by a constant) [10]. In fact, as the inference problem is NP-hard even in structures as simple as the bridge graph alone, we expect the treewidth of the single models to be at least super logarithmic (i.e., greater than $\log(n)$) in the *worst case*. When this is the case, (i.e., when the treewidth is too high), approximate algorithms are used instead. In these cases the time complexity is $O(n2^g)$.

Summarizing, the maximum in-degree represents the bottleneck for space, learning time, and the classification time with approximate methods. Since, as proved by [6], $g = O(\log d)$, the overall complexity is polynomial. Regarding the classification time with exact inference, this is exponential in the treewidth.

5   Experiments

We compare mNB and mFAN against different alternative models: the ensemble of BNs we proposed in [1]; the binary relevance algorithm; the ensemble of chain classifiers (ECC). For both binary relevance and ECC we use naive Bayes as base classifier. Binary relevance thus runs n independent naive Bayes classifiers, where n is the number of classes.

ECC stands for 'ensemble of chain classifiers'. Each chain is characterized by a different order of the labels in the chain. We set to 20 the number of chains in the ensemble. Therefore, ECC runs $20 \cdot n$ naive Bayes, We use the implementation of these methods provided by MEKA.[3]

---

[2] The moral graph of a Bayesian network is the undirected graph obtained by linking nodes with a common child and dropping arc directions; its treewidth is the maximum size of a clique after being triangulated.

[3] http://meka.sourceforge.net

It has not been possible to include in our experiments other multi-label classifiers based on BNs [14, 3] because of the lack of public domain software.

Regarding the parameters of our model, we set the equivalent sample size for the structural learning to $\alpha = 5$. No other parameter needs to be specified.

We have implemented the high-level part of the algorithm in Python. For structural learning, we adopted the GOBNILP package.[4] We performed the inferences using the junction tree and belief propagation algorithms implemented in libDAI, a library for inference in probabilistic graphical models.[5]

We compare the classifiers on 8 different data sets, whose characteristics are given in Table 1. The *density* is the average number of relevant labels per instance.

Table 1. Datasets used for experiments

Data set	Classes	Features	Instances	Density
Emotions	6	72	593	.31
Scene	6	294	2407	.18
Yeast	14	103	2417	.30
Slashdot	22	1079	3782	.05
Genbase	27	1186	662	.04
Enron	53	1001	1702	.06
Cal500	174	68	502	.15
Medical	45	1449	978	.03

We validate the classifiers by a 5-folds cross-validation. We stratify training and test sets according to the least relevant label (i.e., the label which is less often annotated as relevant, and whose distribution among folds risks to be very uneven if not stratified).

Before training any classifier, we perform two pre-processing steps. First, we discretize numerical features into four bins. The bins are given by the 25-th, the 50-th and 75-th percentile of the value of the feature. Then we perform feature selection as described in Section 3. The effectiveness of feature selection can be appreciated from the third column of Table 2.

The results regarding global accuracy and Hamming accuracy are provided in Table 3 and 4 respectively. The Friedman test rejects the null hypothesis of all classifiers having the same median rank. This happens both for global accuracy ($p<0.01$) and for Hamming accuracy ($p<0.01$). The following rank is consistently found under both accuracies: the first ranked classifier is the ensemble, followed by mNB, mFAN, binary relevance and ECC.

We exclude from the subsequent analysis the mFAN model. It has lower rank than mNB on both Hamming accuracy and global accuracy, despite higher complexity. The reason of this phenomenon is not yet clear and it is worth further

---

[4] http://www.cs.york.ac.uk/aig/sw/gobnilp
[5] http://www.libdai.org

Table 2. Treewidth and feature selection on the benchmark data sets

Data set	Treewidth (mNB/ensemble)	Features (selected/original)
Emotions	5/6	22/72
Scene	6/6	184/294
Yeast	8/14	28/103
Slashdot	22/22	465/1079
Genbase	23/27	82/1186
Enron	32/53	220/1001
Cal500	10/174	66/68
Medical	35/45	436/1449

Table 3. Global accuracy. Classifiers are sorted according to their average rank. Lower rank is better.

Data set	ensemble	mNB	mFAN	ECC	binary rel.
Cal500	0.00	0.00	0.00	0.00	0.00
Emotions	0.28	0.22	0.22	0.25	0.25
Enron	0.12	0.06	0.11	0.02	0.02
Genbase	0.95	0.94	0.93	0.76	0.77
Medical	0.69	0.65	0.62	0.20	0.18
Scene	0.64	0.61	0.59	0.30	0.29
Slashdot	0.49	0.42	0.42	0.44	0.41
Yeast	0.14	0.07	0.09	0.13	0.11
Average rank	**1.2**	**3.1**	**3.2**	**3.4**	**4.0**

Table 4. Hamming accuracy. Classifiers are sorted according to their average rank. Lower rank is better.

Data set	ensemble	mNB	mFAN	ECC	binary rel.
Cal500	0.86	0.86	0.86	0.59	0.63
Emotions	0.79	0.77	0.77	0.76	0.75
Enron	0.95	0.94	0.94	0.77	0.87
Genbase	1.00	1.00	1.00	0.99	0.98
Medical	0.99	0.99	0.98	0.98	0.69
Scene	0.91	0.90	0.89	0.83	0.82
Slashdot	0.96	0.95	0.95	0.95	0.86
Yeast	0.78	0.77	0.76	0.76	0.75
Average rank	**1.3**	**2.2**	**2.8**	**4.0**	**4.8**

investigation. However, according to Occam razor, mNB should be preferred over mFAN.

We then perform the statistical multiple comparisons among ensemble, mNB, binary relevance and ECC. We adopt the Wilcoxon signed-rank test to perform the pairwise comparisons. Before declaring significance, we adjust the p-values according to the false discovery rate (FDR) correction [2]. FDR adjusts the p-values of multiple comparisons in a more powerful (i.e., less conservative) way than traditional methods such as Bonferroni. We report a significant difference when the *adjusted* p-value is smaller than 0.05.

The ensemble is significantly more accurate than binary relevance, ECC and mNB. This is verified both on global accuracy and Hamming accuracy. No significant difference can be detected between ECC and mNB. Moreover, both ECC and mNB have significantly higher Hamming accuracy than binary relevance. As for global accuracy, the ECC and mNB have higher rank than binary relevance, but the difference is not significant. A possible reason is that on the Cal500 data sets the global accuracy of all classifiers is zero because of the high number of classes. Thus, global accuracy provides less evidence than Hamming accuracy when analyzed by a hypothesis test.

Summing up, the ensemble is significantly more accurate than mNB, which however compares favorably to both binary relevance and ECC. However, mNB provides huge computational savings compared to the ensemble. The saving is linear in the number in classes when performing the marginal inference (Equation 4). The ensemble performs the marginal query n^2 times (n members of the ensemble multiplied by n classes). The mNB classifier performs this query n times (once for each class). As already discussed in Section 4, even larger computational savings are obtained on the query about the most probable joint configuration of the classes (Equation 3). Consider the ratio of the inference time of the ensemble to the inference time of mNB. This ratio varies between 3 and 280 depending on the data set. Figure 2 shows the relation between the logarithm of the ratio and the difference between the treewidth of the ensemble (equal to the number of classes, see Section 4) and the treewidth of mNB (see the values in Table 2). The relation is roughly linear as expected.

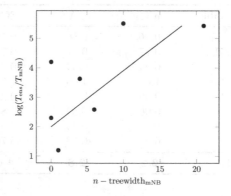

Fig. 2. Logarithm of the ratio of inference times (ensemble/mNB, in log scale) against difference between the number of classes and the treewidth of the mNB. A log ratio of 2.5 implies a 12-folds speed up. A log ratio of 5 implies a 150-folds speed up.

5.1   Insights on the mNB Structure

In this section we derive some insights analyzing the structure of the class subgraphs learned for the mNB model. Learning the class subgraph according to
the two steps procedure of Section 3 boils down to identify a) which non-root
classes should be connected to each possible root class and b) which is the best
structure, among the n characterized by different roots. Let us call *optimal root*
the root of the graph which is eventually chosen.

The optimal root is usually among the most relevant classes (i.e., those which
are more frequently tagged as relevant) available in the data set. This point
is illustrated by the following analysis. Each class has its own relevance (% of
instances in which it is relevant). Consider the relevance of the class selected
as optimal root. Sort the classes according to their relevance and compute the
percentile of the root class. This percentile is generally well above 0.8. Often the
root class is the most relevant one (Figure 3, left).

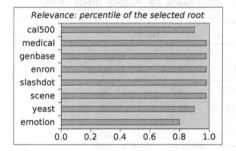
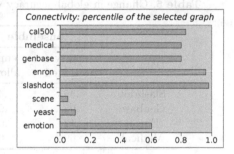

Fig. 3. Percentile of relevance of the optimal root (*left*) and connectivity of the optimal
naive Bayes (*right*)

We then analyze the connectivity of the optimal naive Bayes. If n is the number of classes, naive Bayes contains at most $n - 1$ arcs between the root and
the non-root classes. The *connectivity* of a naive Bayes is how many arcs are
instantiated out of the $n - 1$ possible ones. Consider the connectivity of the
naive Bayes selected as optimal. Sort the n different naive Bayes (each characterized by a different root) according to their connectivity. Take the percentile
of the optimal naive Bayes. Such percentile is usually well above 0.6 (Figure 3,
right). An exception is found on the Yeast and Scene data sets. On Scene, the
optimal naive Bayes has connectivity of 4/5 while the alternative naive Bayes
have connectivity 5/5. The optimal model has thus high connectivity, but this is
hidden when computing the percentile. Only on yeast the optimal naive Bayes
has limited connectivity (6/13).

Our explanation is as follows. Consider that most classes have low relevance:
the average relevance of a class is around 10%, with huge differences among
data sets. A class which is more often relevant allows to reliably estimate the
correlations with the relevance of the remaining classes. Using a class of this type

as root of the class subgraph results both in higher score and in higher number of arcs going from the root to the non-root classes.

5.2 Avoiding Empty Predictions

The least dense data sets are Slashdot, Genbase, Enron and Medical (see Table 1). On such data sets, *global accuracy* increases when empty predictions are prevented (Table 5). The effect is noteworthy on Slashdot. We did not find empty predictions on the other data sets.

Preventing empty prediction sometimes improves also *Hamming accuracy*, but the impact on this indicator is narrow. Preventing empty predictions should become common practice in multilabel classification because of the following reasons: it sometimes improves accuracy; it never worsens it. It is trivial to implement. Most important, it avoids returning a non-sensible prediction.

Table 5. Change in global accuracy when preventing the empty prediction

Data set	ensemble		mNB	
	empty prevented	empty allowed	empty prevented	empty allowed
Slashdot	.49	.40	.42	.34
Genbase	.95	.94	.94	.94
Enron	.12	.07	.06	.05
Medical	.69	.64	.65	.60

6 Conclusions

A new approach to multilabel classification based on Bayesian networks has been proposed. Some of the ideas of our previous work [1] are kept: the naive multiclass assumption (i.e., features are conditionally independent given the joint class), a singly connected subgraph over the classes, and no feature-to-class arcs. Yet, we consider a more sophisticated learning of the structure. We show empirically that the approximation largely decreases the computational burden while incurring only a small worsening of accuracy. We also provide an original analysis of the identified structures. We plan to make our code available in the near future.

Acknowledgements. We thank Cassio P. de Campos and David Huber for valuable discussions. Work supported by the Swiss NSF grant no. 200020_134759 / 1, by the Hasler foundation grant n. 10030 and by FAPESP grant no. 2013/23197-4.

References

[1] Antonucci, A., Corani, G., Mauá, D., Gabaglio, S.: An ensemble of Bayesian networks for multilabel classification. In: Rossi, F. (ed.) Proceedings of the 23rd International Joint Conference on Artificial Intelligence (IJCAI 2013), pp. 1220–1225 (2013)

[2] Benjamini, Y., Hochberg, Y.: Controlling the false discovery rate: a practical and powerful approach to multiple testing. Journal of the Royal Statistical Society Series B (Methodological), 289–300 (1995)

[3] Bielza, C., Li, G., Larrañaga, P.: Multi-dimensional classification with Bayesian networks. International Journal of Approximate Reasoning 52(6), 705–727 (2011)

[4] Bolt, J.H., van der Gaag, L.C.: Multi-dimensional classification with naive Bayesian network classifiers. In: Uiterwijk, J., Roos, N., Winands, M. (eds.) BNAIC 2012 the 24th Benelux Conference on Artificial Intelligence, pp. 27–34 (2012)

[5] Borchani, H., Bielza, C., Larrañaga, P.: Learning CB-decomposable multi-dimensional Bayesian network classifiers. In: Myllymaki, P., Roos, T., Jaakkola, T. (eds.) Proceedings of the 5th European Workshop on Probabilistic Graphical Models (PGM 2010), pp. 25–32 (2010)

[6] de Campos, C., Ji, Q.: Efficient structure learning of Bayesian networks using constraints. Journal of Machine Learning Research 12, 663–689 (2011)

[7] de Campos, C., Cuccu, M., Corani, G., Zaffalon, M.: The extended tree augmented naive classifier. In: Van Der Gaag, L., Feelders, A. (eds.) Proceedings PGM 2014 (2014)

[8] Dembczynski, K., Waegeman, W., Hüllermeier, E.: An analysis of chaining in multi-label classification. In: De Raedt, L., Bessiere, C., Dubois, D., Doherty, P., Frasconi, P., Heintz, F., Lucas, P. (eds.) Proceedings of the 20th European Conference on Artificial Intelligence (ECAI), pp. 294–299 (2012)

[9] Elkan, C.: Magical thinking in data mining: lessons from coil challenge 2000. In: Proc. KDD 2001: ACM SIGKDD International Conference on Knowledge Discovery and Data Mining, pp. 426–431 (2001)

[10] Karpas, E., Solomon, E., Beimel, A.: Approximate belief updating in max-2-connected Bayes networks is NP-hard. Artificial Intelligence 173(12-13), 1150–1153 (2009)

[11] Kohavi, R., John, G.: Wrappers for feature subset selection. Artificial Intelligence 97(1), 273–324 (1997)

[12] Read, J., Pfahringer, B., Holmes, G., Frank, E.: Classifier chains for multi-label classification. Machine Learning 85(3), 333–359 (2011)

[13] Read, J., Bielza, C., Larranaga, P.: Multi-dimensional classification with super-classes. IEEE Transactions on Knowledge and Data Engineering 26(7), 1720–1733 (2014)

[14] Van Der Gaag, L., De Waal, P.: Multi-dimensional Bayesian network classifiers. In: Studeny, M., Vomlel, J. (eds.) Proceedings of the Third European Workshop on Probabilistic Graphical Models, pp. 107–114 (2006)

[15] Witten, I., Frank, E., Hall, M.: Data Mining: Practical Machine Learning Tools and Techniques. Morgan Kaufmann (2011)

Robustifying the Viterbi Algorithm

Cedric De Boom, Jasper De Bock, Arthur Van Camp, and Gert de Cooman

SYSTeMS Research Group, Ghent University,
Technologiepark 914, 9052 Zwijnaarde, Belgium
{Cedric.DeBoom,Jasper.DeBock,Arthur.VanCamp,Gert.deCooman}@UGent.be

Abstract. We present an efficient algorithm for estimating hidden state sequences in imprecise hidden Markov models (iHMMs), based on observed output sequences. The main difference with classical HMMs is that the local models of an iHMM are not represented by a single mass function, but rather by a set of mass functions. We consider as estimates for the hidden state sequence those sequences that are maximal. In this way, we generalise the problem of finding a state sequence with highest posterior probability, as is commonly considered in HMMs, and solved efficiently by the Viterbi algorithm. An important feature of our approach is that there may be multiple maximal state sequences, typically for iHMMs that are highly imprecise. We show experimentally that the time complexity of our algorithm tends to be linear in this number of maximal sequences, and investigate how this number depends on the local models.

Keywords: Imprecise hidden Markov model, Viterbi algorithm, maximality, hidden state sequence, robustness.

1 Introduction

The popularity of Bayesian networks has increased rapidly over the last decades, and their power has been illustrated in numerous applications. Nevertheless, some of the assumptions they are based on are rather severe and, in some cases, even unreasonable. For example, in order to specify a Bayesian network, one has to quantify its local probability mass functions exactly. If limited data and/or expert knowledge is available, this is clearly an unrealistic requirement. By enforcing precision nevertheless, the resulting model and the inferences it produces are, although precise, not guaranteed to be supported by the evidence, thereby creating a false sense of correctness.

In order to avoid this problem, one can allow for local models that are represented by a set of mass functions instead of a single one, thereby obtaining a so-called *credal network* [1]. In this paper, we will consider the special case of an *imprecise hidden Markov model* (iHMM), which is the credal network version of an HMM. We explain how the problem of finding a state sequence with maximal posterior probability can be generalised to this framework, and present an algorithm that is capable of solving this new version of the problem in an efficient manner. In this way, we obtain a robust alternative to the *Viterbi algorithm* [5].

L.C. van der Gaag and A.J. Feelders (Eds.): PGM 2014, LNAI 8754, pp. 160–175, 2014.

A similar study has been conducted in Ref. [2] as well. However, the iHMM that was considered in that paper was of a completely different kind: instead of regarding an iHMM as a collection of HMMs—as we will do, and as is technically referred to as assuming 'strong independence'—the authors of Ref. [2] considered a so-called iHMM under epistemic irrelevance; see Ref. [1] for more information. In the conclusions of Ref. [2], the authors wondered whether or not it was possible to obtain similar results for our version of an iHMM. The present paper illustrates that this is indeed the case.

We start in Section 2 by introducing HMMs, also discussing the problem that is solved by the Viterbi algorithm. In Section 3, we generalise this problem towards *imprecise* hidden Markov models, which we introduce, and explain how it leads us to consider a set of maximal sequences as estimates for the hidden state sequence. We derive a manageable expression for this set in Section 4, and use it in Section 5 to derive an algorithm that is able to calculate the set of all maximal sequences in a recursive manner. In Section 6, we explain how for some common imprecise models, the parameters that are required to run our algorithm can be calculated easily. We end the paper in Section 7 by presenting a number of experiments, showing that the time complexity of our algorithm tends to be linear in the number of maximal sequences, and illustrating how this number depends on the local models of the iHMM.

2 Hidden Markov Models

A hidden Markov model (HMM) is a probabilistic graphical model that has a graphical structure of the form depicted in Figure 1.

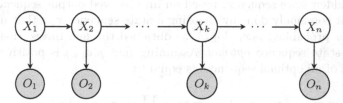

Fig. 1. Graphical structure of a hidden Markov model

It consists of $2n$ random variables that can be categorised into n hidden state variables $X_1, X_2, \ldots, X_n$ and n observable output variables $O_1, O_2, \ldots, O_n$. For any given k in $\{1, \ldots, n\}$, the variables X_k and O_k take values in their respective possibility space $\mathcal{X}_k$ and $\mathcal{O}_k$. We assume that every possibility space is finite.

2.1 Local Uncertainty Models

For the first state variable X_1, we have an *initial model* that, since $\mathcal{X}_1$ is assumed to be finite, can be characterised by a probability mass function p_1 on $\mathcal{X}_1$. For any x_1 in $\mathcal{X}_1$, $p_1(x_1)$ is the probability that X_1 takes the value x_1. For the subsequent state variables X_k, with k in $\{2, \ldots, n\}$, we have a *transition model* p_k. For any

x_k in $\mathcal{X}_k$ and x_{k-1} in $\mathcal{X}_{k-1}$, $p_k(x_k|x_{k-1})$ is the probability that X_k assumes the value x_k, conditional on X_{k-1} being equal to x_{k-1}. For notational convenience, we introduce a trivial state variable X_0 that assumes only a single value $\square$; hence, $\mathcal{X}_0 := \{\square\}$ and, whenever we write x_0, this is taken to be equal to $\square$. This trick allows us to regard p_1 as a conditional model as well, by defining $p_1(x_1|\square) := p_1(x_1)$. Finally, for every output variable O_k, with k in $\{1, \ldots, n\}$, we have an *emission model* q_k. For every o_k in $\mathcal{O}k$ and x_k in $\mathcal{X}_k$, it provides us with the conditional probability $q_k(o_k|x_k)$ that O_k assumes the value o_k, given that X_k is equal to x_k.

2.2   Constructing a Joint Model

By imposing the usual Markov condition for Bayesian networks, the global model of an HMM—a global mass function p—is completely determined by its local models; it suffices to multiply them. For all $x_{1:n}$ in $\mathcal{X}_{1:n} := \times_{k=1}^n \mathcal{X}_k$ and $o_{1:n}$ in $\mathcal{O}_{1:n} := \times_{k=1}^n \mathcal{O}_k$, we find that

$$p(x_{1:n}, o_{1:n}) = \prod_{k=1}^n p_k(x_k|x_{k-1})q_k(o_k|x_k),$$

where we use the shorthand notations $x_{1:n} := (x_1, \ldots, x_n)$ and $o_{1:n} := (o_1, \ldots, o_n)$ to refer to the state and output sequence, respectively.

2.3   The Viterbi Algorithm

One of the most important problems in an HMM is to try and estimate the unknown hidden state sequence, based on an observed output sequence $o_{1:n}$ in $\mathcal{O}_{1:n}$. This is commonly done by choosing a state sequence $x_{1:n}$ that maximises the posterior probability $p(x_{1:n}|o_{1:n})$, as obtained through Bayes' rule. We will call such a state sequence *optimal*. Assuming that $p(o_{1:n})$ is positive, we find that the set of all optimal sequences is equal to

$$\underset{x_{1:n} \in \mathcal{X}_{1:n}}{\arg\max}\, p(x_{1:n}, o_{1:n}) = \underset{x_{1:n} \in \mathcal{X}_{1:n}}{\arg\max} \prod_{k=1}^n p_k(x_k|x_{k-1})q_k(o_k|x_k). \tag{1}$$

A well-known method for finding an arbitrary element of this set—and hence an estimate for $x_{1:n}$—is to apply the Viterbi algorithm [5]; see Ref. [3] for a good introduction. By proceeding in a recursive fashion, this algorithm manages to be very efficient: its time complexity is only $O(nm^2)$, where m is the size of the biggest possibility space for the states: $m := \max\{|\mathcal{X}_k| : k \in \{1, \ldots, n\}\}$.

3   Imprecise Hidden Markov Models

In classical HMMs—the ones considered in the previous section—the local uncertainty models are mass functions, which have to be quantified exactly, say, with arbitrary precision. However, in many instances (e.g., if little data and/or expert knowledge is available), this requirement is clearly unreasonable. For example, what if there is some probability for which an expert is only able to provide an interval, rather than an exact value? In order to model such situations in a more flexible manner, one can use a so-called *imprecise hidden Markov model* (iHMM) which, basically, is just a set of HMMs. They all share the same graphical structure, but their local probability mass functions may differ.

3.1   Local Imprecise Uncertainty Models

The crucial difference between iHMMs and HMMs is that their local models are not required to consist of a single mass function. Instead, a set of mass functions may be used. The only restrictions we impose is that this set should be *compact*, and that it should assign positive probability to every event; for ease of reference, we call this last requirement the *positivity assumption*.

In order to make this more formal, we denote by $\Sigma_{\mathcal{X}}$ the set of all mass functions on some generic possibility space $\mathcal{X}$ and let $\mathrm{int}\Sigma_{\mathcal{X}}$ be its interior, as defined by $\mathrm{int}\Sigma_{\mathcal{X}} := \{p \in \Sigma_{\mathcal{X}} : (\forall x \in \mathcal{X})\, p(x) > 0\}$. We allow our local models to be any compact subset $\mathcal{M}$ of $\mathrm{int}\Sigma_{\mathcal{X}}$. Compactness is imposed to ensure that minima and maxima are well-defined, thereby allowing us to consider, for all x in $\mathcal{X}$, its so-called *lower and upper probability*, as defined by

$$\underline{p}(x) := \min\{p(x) : p \in \mathcal{M}\} \quad \text{and} \quad \overline{p}(x) := \max\{p(x) : p \in \mathcal{M}\}.$$

The positivity assumption is imposed for didactical reasons only, as it allows us to avoid a number of cumbersome technicalities. In its most general form—which, due to page limit constraints, we are unable to discuss here—the algorithm we are about to introduce works for compact subsets of $\Sigma_{\mathcal{X}}$ as well.

We introduce the following notation for the different local uncertainty models of an iHMM. For X_1, the initial model is denoted by $\mathcal{M}_1$ and can be any compact subset of $\mathrm{int}\Sigma_{\mathcal{X}_1}$. Similarly, the transition model at position k, conditional on $X_{k-1} = x_{k-1}$, is a compact subset $\mathcal{M}_{X_k|x_{k-1}}$ of $\mathrm{int}\Sigma_{\mathcal{X}_k}$. This set may be different for every x_{k-1} in $\mathcal{X}_{k-1}$. It will be convenient to refer to them collectively by means of the shorthand notation $\mathcal{M}_{X_k|X_{k-1}} := \bigtimes_{x_{k-1} \in \mathcal{X}_{k-1}} \mathcal{M}_{X_k|x_{k-1}}$. An element $p_k(\cdot|X_{k-1})$ of $\mathcal{M}_{X_k|X_{k-1}}$ is then a tuple, consisting of probability mass functions $p_k(\cdot|x_{k-1})$ in $\mathcal{M}_{X_k|x_{k-1}}$, one for every x_{k-1} in $\mathcal{X}_{k-1}$. For $k = 1$, we have that $\mathcal{M}_{X_1|X_0} = \mathcal{M}_{X_k|\square} := \mathcal{M}_1$. Finally, the emission model at position k, conditional on $X_k = x_k$, is a compact subset $\mathcal{M}_{O_k|x_k}$ of $\mathrm{int}\Sigma_{\mathcal{O}_k}$. We write $\mathcal{M}_{O_k|X_k} := \bigtimes_{x_k \in \mathcal{X}_k} \mathcal{M}_{O_k|x_k}$ to refer to all the different conditional models for O_k at once.

3.2   Constructing an Imprecise Joint Model

By specifying these imprecise, set-valued local models, we also specify, in a very natural way, a corresponding family of joint probability mass functions

$$\mathcal{M} := \Big\{ \prod_{k=1}^{n} p_k(X_k|X_{k-1})q_k(O_k|X_k) : \tag{2}$$
$$(\forall k \in \{1,\ldots,n\})\, p_k(\cdot|X_{k-1}) \in \mathcal{M}_{X_k|X_{k-1}},\, q_k(\cdot|X_k) \in \mathcal{M}_{O_k|X_k} \Big\}.$$

Every probability mass function p in $\mathcal{M}$ corresponds to a different HMM, whose local probability mass functions are selected from the imprecise, set-valued local models that were discussed in the previous section. Together, these HMMs—and their joint mass functions—constitute an iHMM. Note that, since multiplication is a continuous operation, the compactness of the local imprecise models guarantees that $\mathcal{M}$ is compact as well. Furthermore, since the local models satisfy the positivity assumption, $\mathcal{M}$ satisfies it too.

3.3 Generalising the Notion of Optimality

Since we are now working with a set $\mathcal{M}$ of joint mass functions rather than a single mass function p, the concept of 'maximising posterior probability' is no longer well-defined. Hence, we need to come up with some other way of estimating the hidden sequence $x_{1:n}$ based on an observed output sequence $o_{1:n}$; we need a new notion of optimality. Different imprecise-probabilistic decision criteria can be used for this purpose; see Ref. [4] for an overview. In the precise case—if $\mathcal{M}$ is a singleton—all these approaches coincide with the one that is adopted in Section 2.3.

The approach that we will use here is to adopt the decision criterion of maximality [6]. The idea is to introduce a strict preference relation $\succ$ between state sequences. For any two state sequences $x_{1:n}$ and $\hat{x}_{1:n}$ in $\mathcal{X}_{1:n}$, we say that $x_{1:n}$ is better than $\hat{x}_{1:n}$, and write $x_{1:n} \succ \hat{x}_{1:n}$, if

$$(\forall p \in \mathcal{M}) \; p(x_{1:n}|o_{1:n}) > p(\hat{x}_{1:n}|o_{1:n}), \tag{3}$$

This preference relation induces a strict partial order on the set of all state sequences $\mathcal{X}_{1:n}$, and we call a sequence $\hat{x}_{1:n}$ *maximal* if it is undominated in this partial order or, equivalently, if no other sequence is better. This leads us to consider as optimal sequences the elements of

$$\mathrm{opt}_{\max}(\mathcal{X}_{1:n}|o_{1:n}) := \big\{\hat{x}_{1:n} \in \mathcal{X}_{1:n} \colon (\forall x_{1:n} \in \mathcal{X}_{1:n}) \; x_{1:n} \nsucc \hat{x}_{1:n}\big\}. \tag{4}$$

4   A More Convenient Characterisation of Maximality

In its current form, our characterisation of maximality is—although intuitive—rather impractical. Therefore, as a first step in developing an efficient algorithm for calculating the maximal sequences, we set out to derive a more convenient expression for $\mathrm{opt}_{\max}(\mathcal{X}_{1:n}|o_{1:n})$.

4.1 Defining the Local Parameters

We start by introducing a number of important local parameters. As we will see, they are crucial to the developments in the remainder of this paper. For all k in $\{1, \ldots, n\}$, o_k in $\mathcal{O}_k$, x_k and $\hat{x}_k$ in $\mathcal{X}_k$, and x_{k-1} and $\hat{x}_{k-1}$ in $\mathcal{X}_{k-1}$, we define

$$\omega_k(x_k, \hat{x}_k, o_k) := \min_{q_k(\cdot|X_k) \in \mathcal{M}_{O_k|X_k}} \frac{q_k(o_k|x_k)}{q_k(o_k|\hat{x}_k)}$$

and

$$\chi_k(x_k, x_{k-1}, \hat{x}_k, \hat{x}_{k-1}) := \min_{p_k(\cdot \mid X_{k-1}) \in \mathcal{M}_{X_k \mid X_{k-1}}} \frac{p_k(x_k \mid x_{k-1})}{p_k(\hat{x}_k \mid \hat{x}_{k-1})}. \qquad (5)$$

The following result establishes that, for most values of x_k, $\hat{x}_k$, x_{k-1} and $\hat{x}_{k-1}$, these parameters can be calculated easily.

Proposition 1. *The parameters $\omega_k(x_k, \hat{x}_k, o_k)$ and $\chi_k(x_k, x_{k-1}, \hat{x}_k, \hat{x}_{k-1})$ can be calculated easily in most instances:*

$$\omega_k(x_k, \hat{x}_k, o_k) = \begin{cases} 1 & \text{if } x_k = \hat{x}_k, \\ \frac{q_k(o_k \mid x_k)}{\overline{q}_k(o_k \mid \hat{x}_k)} & \text{if } x_k \neq \hat{x}_k, \end{cases}$$

$$\chi_k(x_k, x_{k-1}, \hat{x}_k, \hat{x}_{k-1}) = \begin{cases} 1 & \text{if } x_k = \hat{x}_k \text{ and } x_{k-1} = \hat{x}_{k-1}, \\ \frac{p_k(x_k \mid x_{k-1})}{\overline{p}_k(x_k \mid \hat{x}_{k-1})} & \text{if } x_{k-1} \neq \hat{x}_{k-1}. \end{cases}$$

The only exception—which is the case that is not covered by Proposition 1—is $\chi_k(x_k, x_{k-1}, \hat{x}_k, \hat{x}_{k-1})$, with $x_k \neq \hat{x}_k$ and $x_{k-1} = \hat{x}_{k-1}$. In general, this parameter will have to be calculated by performing the actual minimisation in Eq. (5), for example, by fractional linear programming techniques. However, for many commonly used local models, closed-form expressions are available even in this case; we will come back to this in Section 6.

4.2 Rewriting the Solution Set

As we are about to show, the local parameters that were just introduced allow us to greatly simplify Eq. (4). As a first step, we rewrite Eq. (3) in the following manner:

$$x_{1:n} \succ \hat{x}_{1:n} \Leftrightarrow (\forall p \in \mathcal{M}) \; p(x_{1:n}, o_{1:n}) > p(\hat{x}_{1:n}, o_{1:n})$$

$$\Leftrightarrow (\forall p \in \mathcal{M}) \; \frac{p(x_{1:n}, o_{1:n})}{p(\hat{x}_{1:n}, o_{1:n})} > 1 \Leftrightarrow \min_{p \in \mathcal{M}} \frac{p(x_{1:n}, o_{1:n})}{p(\hat{x}_{1:n}, o_{1:n})} > 1, \qquad (6)$$

where the equivalences are a consequence of Bayes' rule, our positivity assumption, and the compactness of $\mathcal{M}$.

The nice thing about Eq. (6) is that the minimum at the right hand side can be easily calculated. Indeed, by exploiting the factorised form of the mass functions p in $\mathcal{M}$, splitting up the global minimum, and pushing the resulting individual minima inside, we find that

$$\min_{p \in \mathcal{M}} \frac{p(x_{1:n}, o_{1:n})}{p(\hat{x}_{1:n}, o_{1:n})} = \min_{p \in \mathcal{M}} \prod_{k=1}^{n} \frac{p_k(x_k \mid x_{k-1}) \, q_k(o_k \mid x_k)}{p_k(\hat{x}_k \mid \hat{x}_{k-1}) \, q_k(o_k \mid \hat{x}_k)}$$

$$= \prod_{k=1}^{n} \min_{p_k(\cdot \mid X_{k-1}) \in \mathcal{M}_{X_k \mid X_{k-1}}} \frac{p_k(x_k \mid x_{k-1})}{p_k(\hat{x}_k \mid \hat{x}_{k-1})} \min_{q_k(\cdot \mid X_k) \in \mathcal{M}_{O_k \mid X_k}} \frac{q_k(o_k \mid x_k)}{q_k(o_k \mid \hat{x}_k)}$$

$$= \prod_{k=1}^{n} \chi_k(x_k, x_{k-1}, \hat{x}_k, \hat{x}_{k-1}) \omega_k(x_k, \hat{x}_k, o_k). \qquad (7)$$

It is now but a small step to reformulate Eq. (4). By combining Eqs. (6) and (7), we easily find that

$$\hat{x}_{1:n} \in \mathrm{opt}_{\max}(\mathcal{X}_{1:n}|o_{1:n}) \Leftrightarrow \max_{x_{1:n}\in\mathcal{X}_{1:n}} \prod_{k=1}^{n} \chi_k(x_k, x_{k-1}, \hat{x}_k, \hat{x}_{k-1})\omega_k(x_k, \hat{x}_k, o_k) \leq 1,$$

(8)

where the maximum is trivially attained because $\mathcal{X}_{1:n}$ is a finite set.

For a given, fixed state sequence $\hat{x}_{1:n}$ in $\mathcal{X}_{1:n}$, calculating the maximum in Eq. (8) is a problem that is—formally—very closely related the one that is tackled by the Viterbi algorithm; compare Eqs. (1) and (8). Hence, it should not come as a surprise that, by Eq. (8), checking whether $\hat{x}_{1:n}$ is maximal can be done in an equally efficient way: $O(nm^2)$. It suffices to calculate the maximum in Eq. (8) in a recursive fashion.

5   A Recursive Algorithm

Although Eq. (8) already simplifies the problem of finding $\mathrm{opt}_{\max}(\mathcal{X}_{1:n}|o_{1:n})$, applying it directly is clearly not efficient enough. The main bottleneck is that it requires us to check the maximality of each individual state sequence separately. Since there are exponentially many such state sequences, this quickly becomes intractable. In order to avoid this exponential blow-up, we will now develop an algorithm that is able to rule out the maximality of many state sequences at once, without having to explicitly check the maximality of each of them individually.

5.1   Ruling out Multiple Sequences at Once

The central idea of our algorithm is to regard the set of all state sequences $\mathcal{X}_{1:n}$ as a search tree in which we can navigate while deciding whether a branch is useful or not to explore further. If we are able to infer that there is no maximal state sequence that starts with a given initial segment, then we can completely ignore all branches that start with this segment.

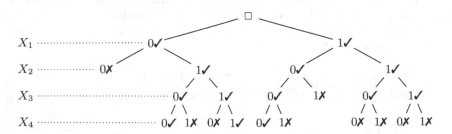

Fig. 2. Example of a search tree for binary sequences of length four.

Example 1. Consider the tree in Figure 2, which corresponds to $\mathcal{X}_{1:4}$, with binary local state spaces $\mathcal{X}_k := \{0,1\}$. In the leftmost part of the tree, the initial segment '00' is investigated. If we are able to infer that there is no maximal sequence that starts with '00'—for the sake of this example, we assume this is the case—then we can stop exploring the tree further in this direction and cut off the tree at the corresponding node. For the initial segment '101', a similar situation occurs.   ◆

In order to get this idea to work, it is crucial to have a simple check that allows us to conclude that none of the maximal state sequences in $\operatorname{opt}(\mathcal{X}_{1:n}|o_{1:n})$ starts with some given initial segment. In other words, for any k in $\{1,\ldots,n\}$ and $\hat{x}_{1:k}^*$ in $\mathcal{X}_{1:k}$, we need to be able to check if

$$(\forall \hat{x}_{1:n} \in \operatorname{opt}(\mathcal{X}_{1:n}|o_{1:n}))\ \hat{x}_{1:k} \neq \hat{x}_{1:k}^*. \tag{C1}$$

By Eq. (8), this is equivalent to checking whether

$$(\forall \hat{x}_{k+1:n} \in \mathcal{X}_{k+1:n})\ \max_{\substack{x_{1:n} \\ \in \mathcal{X}_{1:n}}} \prod_{i=1}^{k} \chi_i(x_i, x_{i-1}, \hat{x}_i^*, \hat{x}_{i-1}^*)\omega_i(x_i, \hat{x}_i^*, o_i)$$
$$\chi_{k+1}(x_{k+1}, x_k, \hat{x}_{k+1}, \hat{x}_k^*)\omega_{k+1}(x_{k+1}, \hat{x}_{k+1}, o_{k+1})$$
$$\prod_{j=k+2}^{n} \chi_j(x_j, x_{j-1}, \hat{x}_j, \hat{x}_{j-1})\omega_j(x_j, \hat{x}_j, o_j) > 1.$$

$$\Leftrightarrow \min_{\substack{\hat{x}_{k+1:n} \\ \in \mathcal{X}_{k+1:n}}} \max_{\substack{x_{1:n} \\ \in \mathcal{X}_{1:n}}} \prod_{i=1}^{k} \chi_i(x_i, x_{i-1}, \hat{x}_i^*, \hat{x}_{i-1}^*)\omega_i(x_i, \hat{x}_i^*, o_i)$$
$$\chi_{k+1}(x_{k+1}, x_k, \hat{x}_{k+1}, \hat{x}_k^*)\omega_{k+1}(x_{k+1}, \hat{x}_{k+1}, o_{k+1})$$
$$\prod_{j=k+2}^{n} \chi_j(x_j, x_{j-1}, \hat{x}_j, \hat{x}_{j-1})\omega_j(x_j, \hat{x}_j, o_j) > 1. \tag{C1'}$$

The problem with criterion (C1') is that it is very difficult to calculate the maximum and minimum because they run over exponentially large spaces. In order to circumvent this issue, one can split the global maximum into local maxima that run over the individual states x_i in $\mathcal{X}_i$, and push these maxima inside. This leads to the equivalent condition

$$\min_{\substack{\hat{x}_{k+1:n} \\ \in \mathcal{X}_{k+1:n}}} \max_{x_1 \in \mathcal{X}_1} \chi_1(x_1, \square, \hat{x}_1^*, \square)\omega_1(x_1, \hat{x}_1^*, o_1) \cdots$$
$$\max_{x_k \in \mathcal{X}_k} \chi_k(x_k, x_{k-1}, \hat{x}_k^*, \hat{x}_{k-1}^*)\omega_k(x_k, \hat{x}_k^*, o_k)$$
$$\max_{x_{k+1} \in \mathcal{X}_{k+1}} \chi_{k+1}(x_{k+1}, x_k, \hat{x}_{k+1}, \hat{x}_k^*)\omega_{k+1}(x_{k+1}, \hat{x}_{k+1}, o_{k+1})$$
$$\max_{x_{k+2} \in \mathcal{X}_{k+2}} \chi_{k+2}(x_{k+2}, x_{k+1}, \hat{x}_{k+2}, \hat{x}_{k+1})\omega_{k+2}(x_{k+2}, \hat{x}_{k+2}, o_{k+2})$$
$$\cdots \max_{x_n \in \mathcal{X}_n} \chi_n(x_n, x_{n-1}, \hat{x}_n, \hat{x}_{n-1})\omega_n(x_n, \hat{x}_n, o_n) > 1. \tag{C1''}$$

By applying a similar procedure to the global minimum, we finally obtain the following inequality:

$$\max_{x_1 \in \mathcal{X}_1} \chi_1(x_1, \Box, \hat{x}_1^*, \Box)\omega_1(x_1, \hat{x}_1^*, o_1) \cdots$$

$$\max_{x_k \in \mathcal{X}_k} \chi_k(x_k, x_{k-1}, \hat{x}_k^*, \hat{x}_{k-1}^*)\omega_k(x_k, \hat{x}_k^*, o_k)$$

$$\min_{\hat{x}_{k+1} \in \mathcal{X}_{k+1}} \max_{x_{k+1} \in \mathcal{X}_{k+1}} \chi_{k+1}(x_{k+1}, x_k, \hat{x}_{k+1}, \hat{x}_k^*)\omega_{k+1}(x_{k+1}, \hat{x}_{k+1}, o_{k+1})$$

$$\min_{\hat{x}_{k+2} \in \mathcal{X}_{k+2}} \max_{x_{k+2} \in \mathcal{X}_{k+2}} \chi_{k+2}(x_{k+2}, x_{k+1}, \hat{x}_{k+2}, \hat{x}_{k+1})\omega_{k+2}(x_{k+2}, \hat{x}_{k+2}, o_{k+2})$$

$$\cdots \min_{\hat{x}_n \in \mathcal{X}_n} \max_{x_n \in \mathcal{X}_n} \chi_n(x_n, x_{n-1}, \hat{x}_n, \hat{x}_{n-1})\omega_n(x_n, \hat{x}_n, o_n) > 1. \tag{C2}$$

However, this criterion is not equivalent to the previous ones. By pushing the local minima inside and beyond the maxima, we obtain a number that, although it is guaranteed never to be bigger, might be lower than the number we started out from. Hence, criterion (C2) is only a *sufficient* condition for (C1) to hold.

Nevertheless, we prefer criterion (C2) over (C1) because it is easier to check. Indeed, if for all k in $\{1, \ldots, n\}$, x_k in $\mathcal{X}_k$ and $\hat{x}_{1:k}$ in $\mathcal{X}_{1:k}$, we consider the parameters $\gamma_k(x_k, \hat{x}_k)$ and $\delta_k(x_k, \hat{x}_{1:k})$, as defined recursively by

$$\gamma_k(x_k, \hat{x}_k) := \min_{\hat{x}_{k+1} \in \mathcal{X}_{k+1}} \max_{x_{k+1} \in \mathcal{X}_{k+1}} \chi_{k+1}(x_{k+1}, x_k, \hat{x}_{k+1}, \hat{x}_k)$$
$$\omega_{k+1}(x_{k+1}, \hat{x}_{k+1}, o_{k+1})\gamma_{k+1}(x_{k+1}, \hat{x}_{k+1})$$

and

$$\delta_k(x_k, \hat{x}_{1:k}) := \max_{x_{k-1} \in \mathcal{X}_{k-1}} \chi_k(x_k, x_{k-1}, \hat{x}_k, \hat{x}_{k-1})\omega_k(x_k, \hat{x}_k, o_k)\delta_{k-1}(x_{k-1}, \hat{x}_{1:k-1}),$$

starting from $\gamma_n(x_n, \hat{x}_n) := 1$ and $\delta_1(x_1, \hat{x}_1) := \chi_1(x_1, \Box, \hat{x}_1, \Box)\omega_1(x_1, \hat{x}_1, o_1)$, then it is relatively easy to see that criterion (C2) reduces to the following simple inequality:

$$\max_{x_k \in \mathcal{X}_k} \delta_k(x_k, \hat{x}_{1:k}^*)\gamma_k(x_k, \hat{x}_k^*) > 1. \tag{C2'}$$

Whenever (C2') holds, we are guaranteed that (C1) holds as well and therefore, that there are no maximal state sequences that start with $\hat{x}_{1:k}^*$. As we will see, it is now but a small step to turn this into a working algorithm.

For $k = n$, criterion (C2') is even more powerful. In that case, the minima in expressions (C1'), (C1") and (C2) disappear, thereby making these conditions equivalent. Hence, we find that for $k = n$, (C1) and (C2') are equivalent. Furthermore, criterion (C2') now reduces to

$$\max_{x_n \in \mathcal{X}_n} \delta_n(x_n, \hat{x}_{1:n}^*) > 1 \tag{C2*}$$

and (C1) and therefore also (C2*) serves as a necessary as well as sufficient condition for $\hat{x}_{1:n}^*$ *not* to be maximal: $\hat{x}_{1:n}^*$ is a maximal state sequence if and only if criterion (C2*) *fails*.

5.2   Turning it into an Algorithm

It is relatively easy to turn the ideas and formulas of the previous section into a working algorithm that is able to construct the set $\mathrm{opt}_{\max}(\mathcal{X}_{1:n}|o_{1:n})$ in an efficient manner. Algorithm 1 provides a pseudo-code version. As input data, it requires an output sequence $o_{1:n}$, the local parameters χ_k and ω_k and the global parameters γ_k. All these parameters can—and should—be calculated beforehand. Note that this is not the case for the parameters δ_k; it is not feasible to calculate these beforehand, as there are simply too many.

Algorithm 1. MaxiHMM

Data: the local parameters χ_k and ω_k, an output sequence $o_{1:n}$, and the corresponding global parameters γ_k

Result: the set $\mathrm{opt}(\mathcal{X}_{1:n}|o_{1:n})$ of all maximal state sequences

1 $\mathrm{opt}(\mathcal{X}_{1:n}|o_{1:n}) \leftarrow \emptyset$

2 **for** $\hat{x}_1 \in \mathcal{X}_1$ **do**

3   **for** $x_1 \in \mathcal{X}_1$ **do**

4         $\delta_1(x_1,\hat{x}_1) \leftarrow \chi_1(x_1,\hat{x}_1)\omega_1(x_1,\hat{x}_1,o_1)$

5   **if** $\max\limits_{x_1\in\mathcal{X}_1}\delta_1(x_1,\hat{x}_1)\gamma_1(x_1,\hat{x}_1) \leq 1$ **then** Recur$(1, \hat{x}_1, \delta_1(\cdot,\hat{x}_1))$

6 **return** $\mathrm{opt}(\mathcal{X}_{1:n}|o_{1:n})$

Procedure Recur$(k,\hat{x}_{1:k},\delta_k(\cdot,\hat{x}_{1:k}))$

---

1 **if** $k = n$ **then**

2   add $\hat{x}_{1:n}$ to $\mathrm{opt}(\mathcal{X}_{1:n}|o_{1:n})$ ▷ We found a solution!

3 **else**

4   **for** $\hat{x}_{k+1} \in \mathcal{X}_{k+n}$ **do**

5        $\hat{x}_{1:k+1} \leftarrow (\hat{x}_{1:k},\hat{x}_{k+1})$ ▷ Append $\hat{x}_{k+1}$ to the end of $\hat{x}_{1:k}$

6   **for** $x_{k+1} \in \mathcal{X}_{k+1}$ **do**

7           $\delta_{k+1}(x_{k+1},\hat{x}_{1:k+1}) \leftarrow \max\limits_{x_k\in\mathcal{X}_k} \chi_{k+1}(x_{k+1},x_k,\hat{x}_{k+1},\hat{x}_k)$

8                        $\omega_{k+1}(x_{k+1},\hat{x}_{k+1},o_{k+1})$

9                        $\delta_k(x_k,\hat{x}_{1:k})$

10   **if** $\max\limits_{x_{k+1}\in\mathcal{X}_{k+1}}\delta_{k+1}(x_{k+1},\hat{x}_{1:k+1})\gamma_{k+1}(x_{k+1},\hat{x}_{k+1}) \leq 1$ **then**

11   Recur$(k+1, \hat{x}_{1:k+1}, \delta_1(\cdot,\hat{x}_{1:k+1}))$

The Procedure Recur implements the recursive nature of our algorithm. In it, we traverse the search tree that corresponds to $\mathcal{X}_{1:n}$ in depth-first order. That is, if the algorithm is unable to infer that there are no maximal sequences starting with $\hat{x}_{1:k}$—if criterion (C2') fails—it presumes there are and immediately descends to depth $k+1$ to check criterion (C2') again. In order to be able to perform this check for $k+1$, we need the parameters $\delta_{k+1}(x_{k+1}, \hat{x}_{1:k+1})$, which—as said before—have not been calculated beforehand. However, luckily, these parameters can easily be calculated while running the algorithm, based on the parameters $\delta_k(x_k, \hat{x}_{1:k})$ that were used in the previous step; see Lines 7–9 of the Procedure `Recur`.

When we arrive at depth n, we check criterion (C2')—which is now equivalent to (C2*)—and, if it fails, we add the current sequence $\hat{x}_{1:n}$ to the solution set. After all, if criterion (C2*) fails, we are guaranteed to have found a maximal solution. Since, while running the algorithm, we have only "ignored" sequences that were definitely not maximal—because criterion (C2') was true—this means that the MaxiHMM algorithm does indeed succeed in constructing the set $\text{opt}_{\max}(\mathcal{X}_{1:n}|o_{1:n})$ correctly.

Example 2. In Figure 2, the MaxiHMM algorithm starts by checking criterion (C2') for $\hat{x}_1 = 0$. The criterion fails, and therefore, the algorithm descends to depth 2, now checking criterion (C2') for $\hat{x}_{1:2} = 00$. This time, the criterion is true, allowing us to "ignore" all the sequences that start with '00'. Next, the algorithm checks criterion (C2') for $\hat{x}_{1:2} = 01$, which turns out to be false. By proceeding in this way, we eventually find that in this case, $\text{opt}_{\max}(\mathcal{X}_{1:4}|o_{1:4})$ consists of three maximal sequences: '0100', '0111' and '1000'.   ♦

5.3   Complexity Analysis

The time complexity of the MaxiHMM algorithm depends on a number of factors. First of all, we have to take into account the size S of the set $\text{opt}_{\max}(\mathcal{X}_{1:n}|o_{1:n})$ we are looking for. After all, if all state sequences in $\mathcal{X}_{1:n}$ are maximal, then no single branch can be pruned from the search tree. In that case, the complete tree has to be traversed, which clearly has a time complexity that is exponential in n. Note that this is far from surprising: in this case, even simply printing all the maximal sequences has such a complexity.

In general, our algorithm is linear in the number of times criterion (C2') fails or, equivalently, the number of times we execute Line 5 of the MaxiHMM algorithm or Line 11 of the Procedure `Recur`. For ease of reference, let us denote this number by C. For example, in Figure 2, C is the number of ✓-signs. By taking a closer look at the pseudo-code of our algorithm, we find that it has a time complexity of the order $\text{O}(Cm^2)$.

Let us now assume that criterion (C2') is equivalent to (C1). Then every time criterion (C2')—and hence (C1)—fails, the current node in the search tree is guaranteed to be part of a maximal state sequence. Since there are S maximal state sequences, each of which consists of n nodes, we find that C is bounded above by Sn. Hence, under the assumption that (C2') is equivalent to (C1), the time complexity of our algorithm is $\text{O}(Snm^2)$. Interestingly, this is linear in the number of maximal sequences S. It is also comparable to the complexity of the Viterbi algorithm, since in that particular case, $S = 1$.

Of course, as explained in Section 5.1, the criteria (C1) and (C2') are not equivalent and therefore, from a theoretical point of view, the aforementioned complexity cannot be guaranteed. For example, it might occur that $C > Sn$; in Figure 2, we see that $13 = C > Sn = 12$. Nevertheless, it turns out that in practice—we illustrate this in Section 7.1—the time complexity of our algorithm tends to increase linearly in S. This suggests that—despite the fact that criteria (C2') and (C1) are not guaranteed to be identical—the aforementioned time complexity of $O(Snm^2)$ might be a good approximation of reality.

6   Common Local Models and Their Parameters

In order to apply the above algorithm, all that is needed are the local parameters ω_k and χ_k. For general compact local models, these can be obtained by applying the formulas in Section 4.1. However, for some specific classes of local models, closed-form expressions for these parameters are available as well. In the following, we discuss two such instances: local models that are obtained by means of ϵ-contamination and local models that are derived from data using Walley's Imprecise Dirichlet Model (IDM) [7].

6.1   Frequently Used Imprecise-Probabilistic Models

The perhaps simplest way to obtain an imprecise local model is to ϵ-contaminate a mass function p in $\Sigma_\mathcal{X}$. For any ϵ in $[0, 1]$, the corresponding ϵ-contaminated model is defined as

$$\mathcal{M}_p^\epsilon := \{(1 - \epsilon)p + \epsilon q \colon q \in \Sigma_\mathcal{X}\}.$$

It is a closed, bounded and therefore also compact subset of $\Sigma_\mathcal{X}$. For $\epsilon = 0$, we find that $\mathcal{M}_p^0 = \{p\}$, thereby recovering the precise-probabilistic case. As ϵ increases, additional mass functions are added. For $\epsilon = 1$, $\mathcal{M}_p^1$ is equal to $\Sigma_\mathcal{X}$, thereby representing complete model uncertainty. In order to satisfy our positivity assumption, we require that p is an element of $\mathrm{int}\Sigma_\mathcal{X}$ and that $\epsilon < 1$.

The corresponding lower and upper probabilities are easily calculated. For example, if $|\mathcal{X}| \geq 2$, then for any singleton $x \in \mathcal{X}$, we find that

$$\overline{p}_\epsilon(x) := \max\{\tilde{p}(x) \colon \tilde{p} \in \mathcal{M}_p^\epsilon\} = (1 - \epsilon)p(x) + \epsilon, \tag{9}$$

$$\underline{p}_\epsilon(x) := \min\{\tilde{p}(x) \colon \tilde{p} \in \mathcal{M}_p^\epsilon\} = (1 - \epsilon)p(x). \tag{10}$$

Example 3. Consider a ternary sample space $\mathcal{X} = \{a, b, c\}$. Then any probability mass function p on $\mathcal{X}$ can be identified with a point in an equilateral triangle with height one, which represents the simplex $\Sigma_\mathcal{X}$; see Figure 3. The ϵ-contaminated model $\mathcal{M}_p^\epsilon$ is represented by an equilateral triangle with height ϵ, which 'grows' around p as ϵ increases.   ♦

The following lemma establishes a technical property of ϵ-contaminated models that will enable us to obtain closed-form expressions for the local parameters ω_k and χ_k.

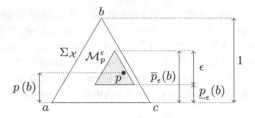

Fig. 3. Constructing an ϵ-contaminated model

Proposition 2. *Consider any $p \in \text{int}(\Sigma_\mathcal{X})$, with $|\mathcal{X}| \geq 2$, and any $\epsilon \in [0,1)$. Then $\mathcal{M}_p^\epsilon \subseteq \text{int}\,\Sigma_\mathcal{X}$ and, for all $x, \hat{x} \in \mathcal{X}$ such that $x \neq \hat{x}$:*

$$\min_{p' \in \mathcal{M}_p^\epsilon} \frac{p'(x)}{p'(\hat{x})} = \frac{\underline{p}_\epsilon(x)}{\overline{p}_\epsilon(\hat{x})}.$$

Besides ϵ-contamination, another popular method for constructing imprecise local models is to derive them from data by means of Walley's IDM; see Ref. [7] for a thorough discussion. For our presents purposes, it suffices to mention that the resulting predictive model on $\mathcal{X}$ coincides with an ϵ-contaminated model $\mathcal{M}_p^\epsilon$, with p and ϵ constructed as follows. For a data set of n experiments, n_x of which are equal to x, we let $p(x) := {}^{n_x}/n$. Furthermore, $\epsilon := {}^s/n+s$, where $s > 0$ is a parameter of the IDM that can be interpreted as a degree of cautiousness. In order for our positivity assumption to be satisfied, we require that, for all $x \in \mathcal{X}$, $n_x > 0$. Using this connection between the IDM and ϵ-contamination, we find that, with $|\mathcal{X}| \geq 2$, for any $x \in \mathcal{X}$:

$$\underline{p}_{\text{IDM}}(x) := \frac{n_x}{n+s} \text{ and } \overline{p}_{\text{IDM}}(x) := \frac{n_x + s}{n+s}.$$

6.2 Local Parameters for an ϵ-Contaminated Model

An important advantage of working with ϵ-contaminated local models (and therefore also the IDM) is that the corresponding local parameters ω_k^ϵ and χ_k^ϵ are extremely easy to calculate.

For $\omega_k^\epsilon(x_k, \hat{x}_k, o_k)$, it suffices to plug the local lower and upper probabilities, as obtained by Eqs. (9) and (10), into the expression that is provided by Proposition 1. We can proceed in much the same way to calculate $\chi_k^\epsilon(x_k, x_{k-1}, \hat{x}_k, \hat{x}_{k-1})$, except if $x_k \neq \hat{x}_k$ and $x_{k-1} = \hat{x}_{k-1}$. However, luckily, in this case, Proposition 2 is applicable, which allows us to optimise the numerator and denominator in Eq. (5) separately anyway, as in the case $x_{k-1} \neq \hat{x}_{k-1}$. Hence, we find that

$$\chi_k^\epsilon(x_k, x_{k-1}, \hat{x}_k, \hat{x}_{k-1}) = \begin{cases} 1 & \text{if } x_k = \hat{x}_k \text{ and } x_{k-1} = \hat{x}_{k-1}, \\ \dfrac{\underline{p}_k^\epsilon(x_k|x_{k-1})}{\overline{p}_k^\epsilon(x_k|\hat{x}_{k-1})} & \text{if } x_k \neq \hat{x}_k \text{ or } x_{k-1} \neq \hat{x}_{k-1}. \end{cases}$$

7   Experiments

We conclude this paper with a number of experiments. In order to allow us to visualise them easily, we focus on binary (i)HMMs. Hence, for all k in $\{1, \ldots, n\}$: $\mathcal{X}_k = \mathcal{O}_k := \{0, 1\}$. We start from a precise stationary HMM. By stationarity, and since binary mass functions can be specified by means of a single number, the local models of this HMM are completely characterised by five numbers: $q(0|0) = 0.9$, $q(0|1) = 0.1$, $p_1(0) = 0.5$, $l := p(0|0)$ and $m := p(0|1)$. We turn this precise HMM into an imprecise one by ϵ-contaminating all of its local models with the same ϵ. In this way, we obtain a stationary iHMM, meaning that the *imprecise* local models do not depend on k. Note however that, by Eq. (2), the corresponding family of *precise* HMMs contains stationary as well as non-stationary ones.

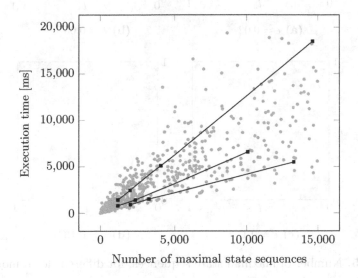

Fig. 4. Scatterplot of $760 + 10$ complexity experiments

7.1   Computational Complexity Experiments

We begin by corroborating the statement that was made in Section 5.3: that in practice, the time complexity of our algorithm tends to increase linearly in the number of maximal state sequences. Figure 4 illustrates the correlation between the execution time of the algorithm and the number of maximal sequences it produces. For these experiments, we chose $l = 0.9$ and $m = 0.1$. The grey dots correspond to 760 randomly generated output sequences of length $n = 100$, with values of ϵ ranging from 0.01 to 0.1. We clearly recognise some kind of cone, which already suggests that the execution time increases linearly in the number of maximal sequences. In black, we plot the results for three additional random but fixed sequences, for different values of ϵ; experiments that correspond to the same output sequence have been connected. This time, the observed linearity is rather striking.

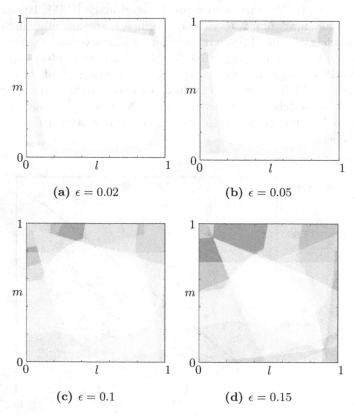

(a) $\epsilon = 0.02$ (b) $\epsilon = 0.05$

(c) $\epsilon = 0.1$ (d) $\epsilon = 0.15$

Fig. 5. Number of maximal state sequences, for different local models

7.2   A Closer Look at the Number of Maximal State Sequences

Since the complexity of our algorithm depends so crucially on the number of maximal sequences it returns, we will now take a closer look at this number and investigate the extent to which it depends on the local models of our iHMM. We do this by visualising the number of maximal sequences as a function of the transition probabilities l and m in four heat plots, each of which corresponds to a different value for ϵ. The output sequence is '1100110011', with $n = 10$. The results are depicted in Figure 5. White corresponds to a single maximal sequences, whereas pitch black corresponds to 200 sequences being maximal.

As expected, the number of maximal state sequences increases with ϵ. That is, the regions that correspond to a higher number of maximal sequences become wider as ϵ increases. The maximum number of maximal sequences that can be observed in these plots is about 160—for $\epsilon = 0.15$, the tiny black dot near the upper left corner of the heat plot ($l = 0.1$ and $m = 0.9$). Note that this is only 16% of the maximum number possible, which is $2^{10} = 1024$. The large, (dark) gray regions correspond to 77 maximal sequences. Finally, and this is rather remarkable, we observe that there are fairly large regions in which—even for $\epsilon = 0.15$—there is only one maximal state sequence.

8 Conclusions and Future Work

The main contribution of this paper is an algorithm that can construct the set of all maximal state sequences of an iHMM in an efficient manner, thereby providing a robust version of the Viterbi algorithm. Our experiments show that the time complexity of this algorithm tends to increase linearly in the number of maximal state sequences. Finally, we have illustrated how this number depends upon the parameters of the iHMM.

We see a number of interesting avenues for future research, the most important of which is perhaps to apply our algorithm to a real-life problem, and to compare the results with those of the Viterbi algorithm. Another, more theoretically oriented line of research is to develop the algorithm without the positivity assumption. Finally, we would like to see whether the ideas in this paper can be used to develop similarly efficient algorithms for credal networks whose graphical structure is more complicated than that of an HMM.

Acknowledgements Jasper De Bock is a PhD Fellow of the Research Foundation Flanders (FWO) and wishes to acknowledge its financial support. The authors would also like to thank Alessandro Antonucci and Cassio P. de Campos for a number of stimulating discussions on the topic of this paper.

References

1. Cozman, F.G.: Credal networks. Artificial Intelligence 120, 199–233 (2000)
2. De Bock, J., De Cooman, G.: State sequence prediction in imprecise hidden Markov models. In: 7th ISIPTA (2011)
3. Rabiner, L.: A tutorial on hidden markov models and selected applications in speech recognition. Proceedings of the IEEE 77(2), 257–286 (1989)
4. Troffaes, M.C.M.: Decision making under uncertainty using imprecise probabilities. International Journal of Approximate Reasoning 45, 17–29 (2007)
5. Viterbi, A.J.: Error bounds for convolutional codes and an asymptotically optimum decoding algorithm. IEEE Transactions on Information Theory 13(2), 260–269 (1967)
6. Walley, P.: Statistical Reasoning with Imprecise Probabilities. Chapman & Hall (1991)
7. Walley, P.: Inferences from Multinomial Data: Learning about a Bag of Marbles. Journal of the Royal Statistical Society Series B 58(1), 3–57 (1996)

Extended Tree Augmented Naive Classifier

Cassio P. de Campos[1], Marco Cuccu[2], Giorgio Corani[1], and Marco Zaffalon[1]

[1] Istituto Dalle Molle di Studi sull'Intelligenza Artificiale (IDSIA), Switzerland
[2] Università della Svizzera italiana (USI), Switzerland
{cassio,giorgio,zaffalon}@idsia.ch, marco.cuccu@usi.ch

Abstract. This work proposes an extended version of the well-known tree-augmented naive Bayes (TAN) classifier where the structure learning step is performed without requiring features to be connected to the class. Based on a modification of Edmonds' algorithm, our structure learning procedure explores a superset of the structures that are considered by TAN, yet achieves global optimality of the learning score function in a very efficient way (quadratic in the number of features, the same complexity as learning TANs). A range of experiments show that we obtain models with better accuracy than TAN and comparable to the accuracy of the state-of-the-art classifier averaged one-dependence estimator.

1 Introduction

Classification is the problem of predicting the *class* of a given object on the basis of some attributes (*features*) of it. A classical example is the iris problem by Fisher: the goal is to correctly predict the *class*, that is, the species of iris on the basis of four features (sepal and petal length and width). In the Bayesian framework, classification is accomplished by updating a prior density (representing the beliefs before analyzing the data) with the likelihood (modeling the evidence coming from the data), in order to compute a posterior density, which is then used to select the most probable class.

The naive Bayes classifier [1] is based on the assumption of stochastic independence of the features given the class; since the real data generation mechanism usually does not satisfy such a condition, this introduces a bias in the estimated probabilities. Yet, at least under the zero-one accuracy, the naive Bayes classifier performs surprisingly well [1, 2]. Reasons for this phenomenon have been provided, among others, by Friedman [3], who proposed an approach to decompose the misclassification error into bias error and variance error; the bias error represents how closely the classifier approximates the target function, while the variance error reflects the sensitivity of the parameters of the classifier to the training sample. Low bias and low variance are two conflicting objectives; for instance, the naive Bayes classifier has high bias (because of the unrealistic independence assumption) but low variance, since it requires to estimate only a few parameters. A way to reduce the naive Bayes bias is to relax the independence assumption using a more complex graph, like a tree-augmented naive Bayes (TAN) [4]. In particular, TAN can be seen as a Bayesian network where

L.C. van der Gaag and A.J. Feelders (Eds.): PGM 2014, LNAI 8754, pp. 176–189, 2014.
© Springer International Publishing Switzerland 2014

each feature has the class as parent, and possibly also a feature as second parent. In fact, TAN is a compromise between general Bayesian networks, whose structure is learned without constraints, and the naive Bayes, whose structure is determined in advance to be naive (that is, each feature has the class as the only parent). TAN has been shown to outperform both general Bayesian networks and naive Bayes in a range of experiments [5, 4, 6].

In this paper we develop an extension of TAN that allows it to have (i) features without the class as parent, (ii) multiple features with only the class as parent (that is, building a forest), (iii) features completely disconnected (that is, automatic feature selection). The most common usage of this model is traditional classification. However it can also be used as a component of a graphical model suitable for multi-label classification [7].

Extended TAN (or simply ETAN) can be learned in quadratic time in the number of features, which is essentially the same computational complexity as that of TAN. The goodness of each (E)TAN structure is assessed through the Bayesian Dirichlet likelihood equivalent uniform (BDeu) score [8, 9, 10]. Because ETAN's search space of structures includes that of TAN, the BDeu score of the best ETAN is equal or superior to that of the best TAN. ETAN than provides a better fit: a higher score means that the model better fits the joint probability distribution of the variables. However, it is well known that this fit does not necessarily imply higher classification accuracy [11]. To inspect that, we perform extensive experiments with these classifiers. We empirically show that ETAN yields in general better zero-one accuracy and log loss than TAN and naive Bayes (where log loss is computed from the posterior distribution of the class given features). Log loss is relevant in cases of cost-sensitive classification [12, 13]. We also study the possibility of optimizing the equivalent sample size of TAN, which makes its accuracy become closer to that of ETAN (although still slightly inferior).

This paper is divided as follows. Section 2 introduces notation and defines the problem of learning Bayesian networks and the classification problem. Section 3 presents our new classifier and an efficient algorithm to learn it from data. Section 4 describes our experimental setting and discusses on empirical results. Finally, Section 5 concludes the paper and suggests possible future work.

2 Classification and Learning TANs

The classifiers that we discuss in this paper are all subcases of a Bayesian network. A Bayesian network represents a joint probability distribution over a collection of categorical random variables. It can be defined as a triple $(\mathcal{G}, \mathcal{X}, \mathcal{P})$, where $\mathcal{G} = (V_{\mathcal{G}}, E_{\mathcal{G}})$ is a directed acyclic graph (DAG) with $V_{\mathcal{G}}$ a collection of nodes associated to random variables $\mathcal{X}$ (a node per variable), and $E_{\mathcal{G}}$ a collection of arcs; $\mathcal{P}$ is a collection of conditional mass functions $p(X_i|\Pi_i)$ (one for each instantiation of Π_i), where Π_i denotes the parents of X_i in the graph (Π_i may be empty), respecting the relations of $E_{\mathcal{G}}$. In a Bayesian network every variable is conditionally independent of its non-descendant non-parents given its parents (Markov condition). Because of the Markov condition, the Bayesian network represents a joint probability distribution by the expression $p(\mathbf{x}) = p(x_0, \ldots, x_n) = \prod_i p(x_i | \boldsymbol{\pi}_i)$, for every $\mathbf{x} \in \Omega_{\mathcal{X}}$ (space of joint configurations of variables), where every x_i and $\boldsymbol{\pi}_i$ are consistent with $\mathbf{x}$.

In the particular case of classification, the *class* variable X_0 has a special importance, as we are interested in its posterior probability which is used to predict unseen values; there are then several feature variables $\mathcal{Y} = \mathcal{X} \setminus \{X_0\}$. The supervised classification problem using probabilistic models is based on the computation of the posterior density, which can then be used to take decisions. The goal is to compute $p(X_0|\mathbf{y})$, that is, the posterior probability of the classes given the values $\mathbf{y}$ of the features in a *test* instance. In this computation, p is defined by the model that has been learned from labeled data, that is, past data where class and features are all observed have been used to infer p. In order to do that, we are given a complete data set $D = \{D_1, \ldots, D_N\}$ with N instances, where $D_u = \mathbf{x}_u \in \Omega_{\mathcal{X}}$ is an instantiation of all the variables, the first learning task is to find a DAG $\mathcal{G}$ that maximizes a given score function, that is, we look for $\mathcal{G}^* = \mathrm{argmax}_{\mathcal{G} \in \mathbf{\mathcal{G}}} \, s_D(\mathcal{G})$, with $\mathbf{\mathcal{G}}$ the set of all DAGs with nodes $\mathcal{X}$, for a given score function s_D (the dependency on data is indicated by the subscript D).[1]

In this work we only need to assume that the employed score is decomposable and respects likelihood equivalence. Decomposable means it can be written in terms of the local nodes of the graph, that is, $s_D(\mathcal{G}) = \sum_{i=0}^{n} s_D(X_i, \Pi_i)$. Likelihood equivalence means that if $\mathcal{G}_1 \neq \mathcal{G}_2$ are two arbitrary graphs over $\mathcal{X}$ such that both encode the very same conditional independences among variables, then s_D is likelihood equivalent if and only if $s_D(\mathcal{G}_1) = s_D(\mathcal{G}_2)$.

The naive Bayes structure is defined as the network where the class variable X_0 has no parents and every feature (the other variables) has X_0 as sole parent. Figure 1(b) illustrates the situation. In this case, there is nothing to be learned, as the structure is fully defined by the restrictions of naive Bayes. Nevertheless, we can define $\mathcal{G}^*_{\mathrm{naive}}$ as being its (fixed) optimal graph.

The class X_0 has also no parents in a TAN structure, and every feature must have the class as parent (as in the naive Bayes). However, they are allowed to have at most one other feature as parent too. Figure 1(c) illustrates a TAN structure, where X_1 has only X_0 as parent, while both X_2 and X_3 have X_0 and X_1 as parents. By ignoring X_0 and its connections, we have a tree structure, and that is the reason for the name TAN. Based on the BDeu score function, an efficient algorithm for TAN can be devised. Because of the likelihood equivalence of BDeu and the fact that every feature has X_0 as parent, the same score is obtained whether a feature X_i has X_0 and X_j as parent (with $i \neq j$), or X_j has X_0 and X_i, that is,

$$s_D(X_i, \{X_0, X_j\}) + s_D(X_j, \{X_0\}) = s_D(X_j, \{X_0, X_i\}) + s_D(X_i, \{X_0\}) \ . \quad (1)$$

This symmetry allows for a very simple and efficient algorithm [14] that is proven to find the TAN structure which maximizes any score that respects likelihood

---

[1] In case of many optimal DAGs, then we assume to have no preference and argmax returns one of them.

equivalence, that is, to find

$$\mathcal{G}^*_{\mathrm{TAN}} = \operatorname*{argmax}_{\mathcal{G} \in \mathcal{G}_{\mathrm{TAN}}} s_D(\mathcal{G}) \ , \tag{2}$$

where $\mathcal{G}_{\mathrm{TAN}}$ is the set of all TAN structures with nodes $\mathcal{X}$. The idea is to find the minimum spanning tree in an undirected graph defined over $\mathcal{Y}$ such that the weight of each edge (X_i, X_j) is defined by $w(X_i, X_j) = -(s_D(X_i, \{X_0, X_j\}) - s_D(X_i, \{X_0\}))$. Note that $w(X_i, X_j) = w(X_j, X_i)$. Without loss of generality, let X_1 be the only node without a feature as parent (one could rename the nodes and apply the same reasoning). Now,

$$\max_{\mathcal{G} \in \mathcal{G}_{\mathrm{TAN}}} s_D(\mathcal{G}) = \max_{\Pi'_i : \forall i > 1} \left(\sum_{i=2}^{n} s_D(X_i, \{X_0, X_{\Pi'_i}\}) + s_D(X_1, \{X_0\}) \right)$$

$$= s_D(X_1, \{X_0\}) - \min_{\Pi'_i : \forall i > 1} \left(-\sum_{i=2}^{n} s_D(X_i, \{X_0, X_{\Pi'_i}\}) \right)$$

$$= \sum_{i=1}^{n} s_D(X_i, \{X_0\}) - \min_{\Pi'_i : \forall i > 1} \sum_{i=2}^{n} w(X_i, X_{\Pi'_i}) \ . \tag{3}$$

This last minimization is exactly the minimum spanning tree problem, and the argument that minimizes it is the same as the argument that maximizes (2). Because this algorithm has to initialize the $\Theta(n^2)$ edges between every pair of features and then to solve the minimum spanning tree (e.g. using Prim's algorithm), its overall complexity time is $O(n^2)$, if one assumes that the score function is given as an oracle whose queries take time $O(1)$. In fact, because we only consider at most one or two parents for each node (two only if we include the class), the computation of the whole score function can be done in time $O(Nn^2)$ and stored for later use. As a comparison, naive Bayes can be implemented in time $O(Nn)$, while the averaged one-dependence estimator (AODE) [15] needs $\Theta(Nn^2)$, just as TAN does.

2.1   Improving Learning of TANs

A simple extension of this algorithm can already learn a forest of tree-augmented naive Bayes structures. One can simply define the edges of the graph over $\mathcal{Y}$ as in the algorithm for TAN, and then remove those edges (X_i, X_j) such that $s_D(X_i, \{X_0, X_j\}) \leq s_D(X_i, \{X_0\})$, that is, when $w(X_i, X_j) \geq 0$, and then run the minimum spanning tree algorithm over this reduced graph. The optimality of such an idea can be easily proven by the following lemma, which guarantees that we should use only X_0 as parent of X_i every time such choice is better than using $\{X_0, X_j\}$. It is a straightforward generalization of Lemma 1 in [16].

Lemma 1. *Let X_i be a node of $\mathcal{G}$, a candidate DAG where the parent set of X_i is Π'_i. Suppose $\Pi_i \subset \Pi'_i$ is such that $s_D(X_i, \Pi_i) \geq s_D(X_i, \Pi'_i)$, where s_D is a decomposable score function. If Π'_i is the parent set of X_i in an optimal DAG, then the same DAG but having Π_i as parent of X_i is also optimal.*

Using a forest as structure of the classifier is not new, but to the best of our knowledge previous attempts to learn a forest (in this context) did not globally optimize the structure, they instead selected a priori the number of arcs to include in the forest [17].

We want to go even further and allow situations as in Figs. 1(a) and 1(d). The former would automatically disconnect a feature if such feature is not important to predict X_0, that is, if $s_D(X_i, \emptyset) \geq s_D(X_i, \Pi_i)$ for every Π_i. The latter case allows some features to have another feature as parent without the need of having also the class. For this purpose, we define the set of structures named Extended TAN (or ETAN for short), as DAGs such that X_0 has no parents and X_i ($i \neq 0$) is allowed to have the class and at most one feature as parent (but it is not obliged to having any of them), that is, the parent set Π_i is such that $|\Pi_i| \leq 1$, or $|\Pi_i| = 2$ and $\Pi_i \supseteq \{X_0\}$.

$$\mathcal{G}^*_{\text{ETAN}} = \underset{\mathcal{G} \in \mathcal{G}_{\text{ETAN}}}{\text{argmax}} \, s_D(\mathcal{G}) \; . \tag{4}$$

This is clearly a generalization of TAN, of the forest of TANs, and of naive Bayes in the sense that they are all subcases of ETAN. Note that TAN is not a generalization of naive Bayes in this sense, as TAN forces arcs among features even if these arcs were not useful. Because of that, we have the following result. The next section discusses how to efficiently learn ETANs.

Lemma 2. *The following relations among subsets of DAGs hold.*

$$s_D(\mathcal{G}^*_{ETAN}) \geq s_D(\mathcal{G}^*_{TAN}) \quad and \quad s_D(\mathcal{G}^*_{ETAN}) \geq s_D(\mathcal{G}^*_{naive}) \; .$$

3   Learning Extended TANs

The goal of this section is to present an efficient algorithm to find the DAG defined in (4). Unfortunately the undirected version of the minimum spanning tree problem is not enough, because (1) does not hold anymore. To see that, take the example in Fig. 1(d). The arc from X_1 to X_2 cannot be reversed without changing the overall score (unless we connect X_0 to X_2). In other words, every node in a TAN has the class as parent, which makes possible to use the minimum spanning tree algorithm for undirected graphs by realizing that any orientation of the arcs between features will produce the same overall score (as long as the weights of the edges are defined as in the previous section).

Edmonds' algorithm [18] (also attributed to Chu and Liu [19]) for finding minimum spanning arborescence in directed graphs comes to our rescue. Its application is however not immediate, and its implementation is not as simple as the minimum spanning tree algorithm for TAN. Our algorithm to learn ETANs is presented in Algorithm 1. It is composed of a preprocessing of the data to create the arcs of the graph that will be given to Edmonds' algorithm for directed

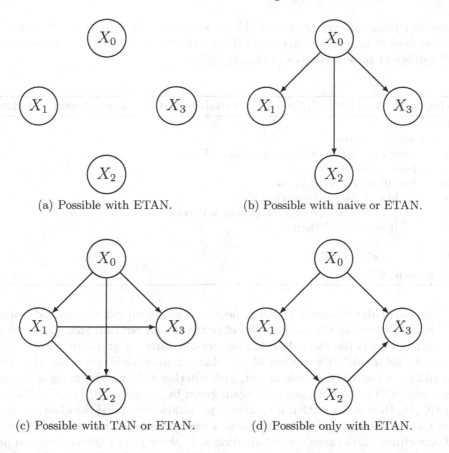

(a) Possible with ETAN.   (b) Possible with naive or ETAN.

(c) Possible with TAN or ETAN.   (d) Possible only with ETAN.

Fig. 1. Some examples of structures allowed by the different classifiers. The labels indicate which classifier allows them as part of their whole structure.

minimum spanning tree (in fact, we assume that Edmonds' algorithm computes the directed maximum spanning tree, which can be done trivially by negating all weights). EdmondsContract and EdmondsExpand are the two main steps of that algorithm, and we refer the reader to the description in Zwick's lecture notes [20] or to the work of Tarjan [21] and Camerini et al. [22] or Gabow et al. [23] for further details on the implementation of Edmonds' idea. In fact, we have not been able to find a stable and reliable implementation of such algorithm, so our own implementation of Edmonds' algorithm has been developed based on the description in [20], even though some fixes had to be applied. Because Edmonds' algorithm finds the best spanning tree for a given "root" node (that is, a node that is constrained not to have features as parents), Algorithm 1 loops over the possible roots and extract from Edmonds' the best parent for each node given that fixed root node (line 6), and then stores the best solution over all such possible root nodes. At each loop, Algorithm 3 is called and builds a graph using the information from the result of Edmonds'. Algorithm 1 also loops over a set of score functions that are given to it. This is used later on to optimize the value of the equivalent sample size in each of the learning steps by giving a list of scores with different prior strengths to the algorithm.

Algorithm 1. ETAN($\mathcal{X}, S$): $\mathcal{X}$ are variables and S is a set of score functions

---
1: $s^* \leftarrow -\infty$
2: **for all** $s_D \in S$ **do**
3:      $(\text{arcs}, \text{classAsParent}) \leftarrow \text{ArcsCreation}(\mathcal{X}, s_D)$
4:      EdmondsContract(arcs)
5:      **for all** root $\in \mathcal{X} \setminus \{X_0\}$ **do**
6:          in $\leftarrow$ EdmondsExpand(root)
7:          $\mathcal{G} \leftarrow$ buildGraph($\mathcal{X}$, root, in, classAsParent)
8:          **if** $s_D(\mathcal{G}) > s^*$ **then**
9:              $\mathcal{G}^* \leftarrow \mathcal{G}$
10:             $s^* \leftarrow s_D(\mathcal{G})$
11: **return** $\mathcal{G}^*$

The particular differences with respect to a standard call of Edmonds' algorithm are defined by the methods `ArcsCreation` and `buildGraph`. The method `ArcsCreation` is the algorithm that creates the directed graph that is given as input to Edmonds'. The overall idea is that we must decide whether the class should be a parent of a node or not, and whether it is worth having a feature as a parent. The core argument is again given by Lemma 1. If $s_D(X_i, \{X_0\}) \leq s_D(X_i, \emptyset)$, then we know that no parent is preferable to having the class as a parent for X_i. We store this information in a matrix called `classAsParent` (line 2 of Algorithm 2). Because this information is kept for later reference, we can use from that point onwards the value $\max(s_D(X_i, \emptyset), s_D(X_i, \{X_0\}))$ as the weight of having X_i with only the class as parent (having or not the class as parent cannot create a cycle in the graph, so we can safely use this max value). After that, we loop over every possible arc $X_j \to X_i$ between features, and define its weight as the maximum between having X_0 also as parent of X_i or not, minus the value that we would achieve for X_i if we did not include X_j as its parent (line 8). This is essentially the same idea as done in the algorithm of TAN, but here we must consider both $X_j \to X_i$ and $X_i \to X_j$, as they are not necessarily equivalent (this happens for instance if for one of the two features the class is included in its parent set and for the other it is not, depending on the maximization, so scores defining the weight of each arc direction might be different). After that, we also keep track of whether the class was included in the definition of the weight of the arc or not, storing the information in `classAsParent` for later recall. In case the weight is not positive (line 9), we do not even include this arc in the graph that will be given to Edmonds' (recall we are using the maximization version of Edmonds'), because at this early stage we already know that either no parents for X_i or only the class as parent of X_i (which one of the two is the best can be recalled in `classAsParent`) are better than the score obtained by including X_j as parent, and using once more the arguments of Lemma 1 and the fact that the class as parent never creates a cycle, we can safely disregard X_j as parent of X_i. All these cases can be seen in Fig. 1 by considering that the variable X_2 shown in the figure is our X_i. There are four options for X_i: no parents (a), only X_0 as parent (b), only X_j as parent (d), and both X_j and X_0 (c). The trick is that Lemma 1 allows us to reduce these four options to two: best between (a) and (b), and best between (c) and (d). After the arcs with positive weight are inserted in a list of arcs that will be given to Edmonds' and classAsParent is built, the algorithm ends returning both of them.

Algorithm 2. ArcsCreation$(\mathcal{X}, s_D)$

1: **for all** $X_i \in \mathcal{X} \setminus \{X_0\}$ **do**
2:      classAsParent$[X_i] \leftarrow s_D(X_i, \{X_0\}) > s_D(X_i, \emptyset)$
3: arcs $\leftarrow \emptyset$
4: **for all** $X_i \in \mathcal{Y}$ **do**
5:      **for all** $X_j \in \mathcal{Y}$ **do**
6:          twoParents $\leftarrow s_D(X_i, \{X_0, X_j\})$
7:          onlyFeature $\leftarrow s_D(X_i, \{X_j\})$
8:          $w \leftarrow$ max(twoParents, onlyFeature) $-$ max$(s_D(X_i, \emptyset), s_D(X_i, \{X_0\}))$
9:          **if** $w > 0$ **then**
10:             Add $X_j \to X_i$ with weight w into arcs
11:             classAsParent$[X_j \to X_i] \leftarrow$ twoParents $>$ onlyFeature
12:          **else**
13:             classAsParent$[X_j \to X_i] \leftarrow$ classAsParent$[X_i]$
14: **return** (arcs, classAsParent)

Finally, Algorithm 3 is responsible for building back the best graph from the result obtained by Edmonds'. Inside in is stored the best parent for each node, and root indicates a node that shall have no other feature as parent. The goal is to recover whether the class shall be included as parent of each node, and for that we use the information in classAsParent. The algorithm is quite straightforward: for each node that is not the root and has a parent chosen by Edmonds', include it as parent each check if that arc was associated to having or not the class (if it had, include also the class); for each node that has no parent as given by Edmonds' (including the root node), simply check whether it is better to have the class as parent.

Somewhat surprisingly, learning ETANs can be accomplish in time $O(n^2)$ (assuming that the score function is given as an oracle, as discussed before), the same complexity for learning TANs. Algorithm 2 takes $O(n^2)$, because it loops over every pair of nodes and only performs constant time operations inside the loop. EdmondsContract can be implemented in time $O(n^2)$ and EdmondsExpand in time $O(n)$ [21, 22]. Finally, buildGraph takes time $O(n)$ because of its loop over nodes, and the comparison between scores of two ETANs as well as the copy of the structure of an ETANs takes time $O(n)$. So the overall time of the loop in Algorithm 1 takes time $O(n^2)$. Our current implementation can be found at http://ipg.idsia.ch/software.

Algorithm 3. buildGraph($\mathcal{X}$, root, in, classAsParent)

1: $\mathcal{G} \leftarrow (\mathcal{X}, \emptyset)$
2: **for all** node $\in \mathcal{X} \setminus \{X_0\}$ **do**
3:    $\Pi_{\text{node}} \leftarrow \emptyset$
4:    **if** node $\neq$ root **and** in[node] $\neq$ null **then**
5:       $\Pi_{\text{node}} \leftarrow \Pi_{\text{node}} \cup \{\text{in[node]}\}$
6:       **if** classAsParent[in[node] $\rightarrow$ node] **then**
7:          $\Pi_{\text{node}} \leftarrow \Pi_{\text{node}} \cup \{X_0\}$
8:    **else if** classAsParent[node] **then**
9:       $\Pi_{\text{node}} \leftarrow \Pi_{\text{node}} \cup \{X_0\}$
10: **return** $\mathcal{G}$

---

4   Experiments

This section presents results with naive Bayes, TAN and ETAN using 49 data sets from the UCI machine learning repository [24]. Data sets with many different characteristics have been used. Data sets containing continuous variables have been discretized in two bins, using the median as cut-off. Our empirical results are obtained out of 20 runs of 5-fold cross-validation (each run splits the data into folds randomly and in a stratified way), so the learning procedure of each classifier is called 100 times per data set. For learning the classifiers we use the Bayesian Dirichlet equivalent uniform (BDeu) and assume parameter independence and modularity [10]. The BDeu score computes a function based on the posterior probability of the structure $p(\mathcal{G}|D)$. For that purpose, the following function is used:

$$s_D(\mathcal{G}) = \log \left(p(\mathcal{G}) \cdot \int p(D|\mathcal{G}, \boldsymbol{\theta}) \cdot p(\boldsymbol{\theta}|\mathcal{G}) d\boldsymbol{\theta} \right) \ ,$$

where the logarithm is used to simplify computations, $p(\boldsymbol{\theta}|\mathcal{G})$ is the prior of $\boldsymbol{\theta}$ (vector of parameters of the Bayesian network) for a given graph $\mathcal{G}$, assumed to be a symmetric Dirichlet. BDeu respects likelihood equivalence and its function is decomposable. The only free parameter is the prior strength α (assuming $p(\mathcal{G})$ is uniform), also known as the equivalent sample size (ESS). We make comparisons using different values of α. We implemented ETAN such that $\alpha \in \{1, 2, 10, 20, 50\}$ is chosen according to the value that achieves the highest BDeu for each learning call, that is, we give to ETAN five BDeu score functions with different values of α. Whenever omitted, the default value for α is two.

As previously demonstrated, ETAN always obtains better BDeu score than its competitors. TAN is usually better than naive Bayes, but there is no theoretical guarantee it will always be the case. Table 1 shows the comparisons of BDeu scores achieved by different classifiers. It presents the median value of the difference between averaged BDeu of the classifiers on the 49 datasets, followed by the number of wins, ties and losses of ETAN against the competitors, and finally the p-value from the Wilcoxon signed rank test (one-sided in the direction of the median value). We note that naive Bayes and TAN might win against ETAN

Table 1. Median value of the BDeu difference between ETAN and competitor (positive means ETAN is better), followed by number of wins, ties and losses of ETAN over competitors, and p-values using the Wilcoxon signed rank test on 49 data sets (one-sided in the direction of the median difference). Names of competitors indicate their equivalent sample size (ESS), and *All* means that ESS has been optimized at each learning call.

| Competitor | BDeu | | |
vs. ETAN	Median	W/T/L	p-value
Naive(1)	454	49/0/0	2e-15
Naive(2)	340	48/0/1	3e-15
Naive(10)	276	48/0/1	4e-15
TAN(1)	182	49/0/0	1e-15
TAN(2)	129	46/0/3	8e-13
TAN(10)	23.6	40/1/8	3e-9
TAN(All)	128	46/0/3	2e-8

(as it does happen in Tab. 1) because the values of α used by the classifiers in different learning problems are not necessarily the same (the learning method is called 100 for each dataset over different data folds). Nevertheless, the statistical test indicates that the score achieved by ETAN is significantly superior than scores of the other methods.

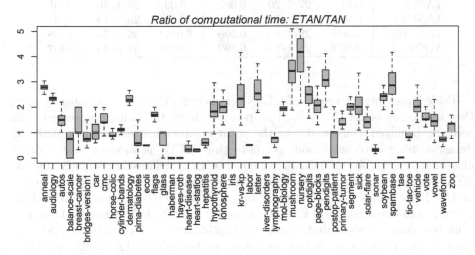

Fig. 2. Computational time to learn the classifier as a ratio ETAN time divided by TAN time, so higher values mean ETAN is slower.

Figure 2 shows the computational time cost to run the learning in the 100 executions per data set for ETAN and TAN (both optimizing α as described before), that is, each of the 20 times 5-fold cross-validation executions. We can see in the graph that learning ETAN has been less than five times slower than learning

TAN in all situations (usually less than three times) and has performed better than TAN in a reasonable amount of instances. We recall that both classifiers can be run in quadratic time in the number of features and linear in the sample size, which is asymptotically as efficient as other state-of-the-art classifiers, such as the averaged one-dependence estimator (AODE) [15].

We measure the accuracy of classifiers using zero-one accuracy and log loss. Zero-one accuracy is the number of correctly classified instances divided by the total number of instances, while log loss equals minus the sum (over the testing instances) of the log-probability of the class given the instance's features.

Table 2. Median value of the difference between ETAN and competitor (positive means ETAN is better), followed by number of wins, ties and losses of ETAN over competitors, and p-values using the Wilcoxon signed rank test on 49 data sets (one sided in the direction of the median difference). Names of competitors indicate their equivalent sample size.

Competitor	Zero-one accuracy			Log loss		
vs. ETAN	Median	W/T/L	p-value	Median	W/T/L	p-value
Naive(1)	0.74%	35/3/21	1e-5	0.17	45/0/4	3e-12
Naive(2)	0.64%	36/1/12	4e-5	0.13	46/0/3	5e-12
Naive(10)	1.35%	38/1/10	8e-6	0.12	38/0/11	8e-8
TAN(1)	0.13%	29/1/19	0.022	0.05	43/0/6	2e-8
TAN(2)	0.01%	27/2/20	0.087	0.03	38/1/10	3e-6
TAN(10)	0.01%	28/3/18	0.261	0.01	29/1/19	0.047
TAN(All)	0.06%	29/1/19	0.096	0.0004	26/0/23	0.418
AODE	-0.07%	21/1/27	0.192	-0.005	24/0/25	0.437

Table 2 presents the results of ETAN versus other classifiers. Number of wins, ties and losses of ETAN, as well as p-values from the Wilcoxon signed rank test are displayed, computed over the point results obtained for each of the 49 datasets using cross-validation. We note that ETAN is superior to the other classifiers, except for AODE, in which case the medians are slightly against ETAN and the difference is not significant (pvalues of 0.192 for zero-one accuracy and 0.437 for log loss, in both cases testing whether AODE is superior to ETAN). Median zero-one accuracy of ETAN is superior to TAN(All), although the signed rank test does not show that results are significant at 5% confidence level. The same is true for log loss. In fact, we must emphasize that TAN with optimized choice of α could also be considered as a novel classifier (even if it is only a minor variation of TAN, we are not aware of implementations of TAN that optimize the equivalent sample size).

Figures 3 and 4 show the performance of ETAN versus AODE in terms of zero-one accuracy and log loss, respectively. Each boxplot regards one data set and considers 100 points defined by the runs of cross-validation. In Fig. 3, the values are the zero-one accuracy of ETAN divided by the zero-one accuracy of its competitor in each of the 100 executions. In Fig. 4, it is presented the difference in log loss between AODE and ETAN. In both figures we can see cases where ETAN performed better, as well as cases where AODE did.

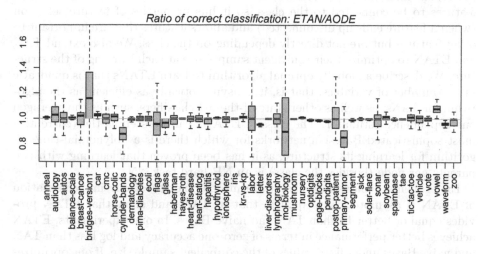

Fig. 3. Comparison of zero-one loss with AODE. Values are ratios of the accuracy of ETAN divided by the competitor, so higher values mean ETAN is better.

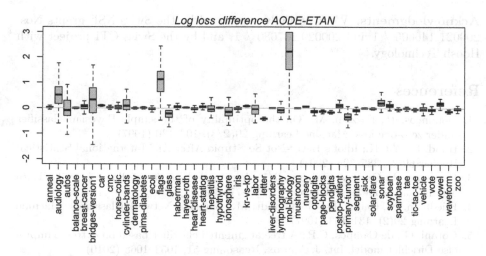

Fig. 4. Comparison of log loss with AODE. Values are differences in the log loss of the competitor minus ETAN, so higher values mean ETAN is better.

5   Conclusions

We presented an extended version of the well-known tree-augmented naive Bayes (TAN) classifier, namely the extended TAN (or ETAN). ETAN does not demand features to be connected to the class, so it has properties of feature selection (when a feature ends up disconnected) and allows features that are important to other features but are not directly depending on the class. We also extend TAN and ETAN to optimize their equivalent sample size at each learning of the structure. We describe a globally optimal algorithm to learn ETANs that is quadratic in the number of variables, that is, it is asymptotically as efficient as the algorithm for TANs, as well as other state-of-the-art classifiers, such as the averaged one-dependence estimator. The class of ETANs can be seen as the (currently) most sophisticated Bayesian networks for which there is a polynomial-time algorithm for learning its structure, as it has been proven that learning with two parents per node (besides the class) is an NP-hard task [25].

Experiments demonstrate that the time complexity of our implementation of ETAN is asymptotically equal to that of TAN, and show that ETAN provides equal or better fit than TAN and naive Bayes. In our experiments, ETAN achieves better performance in terms of zero-one accuracy and log loss than TAN and naive Bayes under fixed values of the equivalent sample size. If one optimizes the equivalent sample size of TAN, then ETAN has performed in general slightly better than TAN (even though not significant in a statistical test). Future work will investigate further the relation between BDeu and classification accuracy, as well as scenarios where ETAN might be preferable, and will study additional structures beyond ETAN that could be useful in building classifiers.

Acknowledgments. Work partially supported by the Swiss NSF grants Nos. 200021_146606 / 1 and 200020_137680 / 1, and by the Swiss CTI project with Hoosh Technology.

References

1. Domingos, P., Pazzani, M.: On the optimality of the simple Bayesian classifier under zero-one loss. Machine Learning 29(2/3), 103–130 (1997)
2. Hand, D., Yu, K.: Idiot's Bayes-Not So Stupid After All? International Statistical Review 69(3), 385–398 (2001)
3. Friedman, J.: On bias, variance, 0/1 - loss, and the curse-of-dimensionality. Data Mining and Knowledge Discovery 1, 55–77 (1997)
4. Friedman, N., Geiger, D., Goldszmidt, M.: Bayesian Network Classifiers. Machine Learning 29(2), 131–163 (1997)
5. Corani, G., de Campos, C.P.: A tree augmented classifier based on extreme imprecise Dirichlet model. Int. J. Approx. Reasoning 51, 1053–1068 (2010)
6. Madden, M.: On the classification performance of TAN and general Bayesian networks. Knowledge-Based Systems 22(7), 489–495 (2009)
7. Corani, G., Antonucci, A., Maua, D., Gabaglio, S.: Trading off speed and accuracy in multilabel classification. In: Proceedings of the 7th European Workshop on Probabilistic Graphical Models (2014)

8. Buntine, W.: Theory refinement on Bayesian networks. In: Proceedings of the 8th Conference on Uncertainty in Artificial Intelligence, UAI 1992, pp. 52–60. Morgan Kaufmann, San Francisco (1991)
9. Cooper, G.F., Herskovits, E.: A Bayesian method for the induction of probabilistic networks from data. Machine Learning 9, 309–347 (1992)
10. Heckerman, D., Geiger, D., Chickering, D.M.: Learning Bayesian networks: the combination of knowledge and statistical data. Machine Learning 20, 197–243 (1995)
11. Kontkanen, P., Myllymäki, P., Silander, T., Tirri, H.: On supervised selection of Bayesian networks. In: Proceedings of the 15th Conference on Uncertainty in Artificial Intelligence, pp. 334–342. Morgan Kaufmann Publishers Inc. (1999)
12. Elkan, C.: The foundations of cost-sensitive learning. In: Proceedings of the Seventeenth International Joint Conference on Artificial Intelligence (IJCAI 2001), pp. 973–978. Morgan Kaufmann (2001)
13. Turney, P.D.: Cost-sensitive classification: Empirical evaluation of a hybrid genetic decision tree induction algorithm. Journal of Artificial Intelligence Research 2, 369–409 (1995)
14. Chow, C.K., Liu, C.N.: Approximating discrete probability distributions with dependence trees. IEEE Transactions on Information Theory IT-14(3), 462–467 (1968)
15. Webb, G.I., Boughton, J.R., Wang, Z.: Not so naive Bayes: Aggregating one-dependence estimators. Machine Learning 58(1), 5–24 (2005)
16. de Campos, C.P., Ji, Q.: Efficient structure learning of Bayesian networks using constraints. Journal of Machine Learning Research 12, 663–689 (2011)
17. Lucas, P.J.F.: Restricted Bayesian network structure learning. In: Gámez, J.A., Moral, S., Salmerón, A. (eds.) Advances in Bayesian Networks. STUDFUZZ, vol. 146, pp. 217–232. Springer, Berlin (2004)
18. Edmonds, J.: Optimum branchings. Journal of Research of the National Bureau of Standards B 71B(4), 233–240 (1967)
19. Chu, Y.J., Liu, T.H.: On the shortest arborescence of a directed graph. Science Sinica 14, 1396–1400 (1965)
20. Zwick, U.: Lecture notes on Analysis of Algorithms: Directed Minimum Spanning Trees (April 22, 2013)
21. Tarjan, R.E.: Finding optimum branchings. Networks 7, 25–35 (1977)
22. Camerini, P.M., Fratta, L., Maffioli, F.: A note on finding optimum branchings. Networks 9, 309–312 (1979)
23. Gabow, H.N., Galil, Z., Spencer, T., Tarjan, R.E.: Efficient algorithms for finding minimum spanning trees in undirected and directed graphs. Combinatorica 6(2), 109–122 (1986)
24. Asuncion, A., Newman, D.: UCI machine learning repository (2007), http://www.ics.uci.edu/~mlearn/MLRepository.html
25. Dasgupta, S.: Learning polytrees. In: Proceedings of the Conference on Uncertainty in Artificial Intelligence, pp. 134–141. Morgan Kaufmann, San Francisco (1999)

Evaluation of Rules for Coping with Insufficient Data in Constraint-Based Search Algorithms

Martijn de Jongh[1] and Marek J. Druzdzel[1,2]

[1] Decision System Laboratory, School of Information Sciences and
Intelligent Systems Program, University of Pittsburgh,Pittsburgh, PA, 15260, USA
[2] Faculty of Computer Science, Białystok University of Technology, Wiejska 45A,
15-351 Białystok, Poland
mad159@pitt.edu,marek@sis.pitt.edu

Abstract. A fundamental step in the PC causal discovery algorithm
consists of testing for (conditional) independence. When the number of
data records is very small, a classical statistical independence test is typically unable to reject the (null) independence hypothesis. In this paper,
we are comparing two conflicting pieces of advice in the literature that
in case of too few data records recommend (1) assuming dependence and
(2) assuming independence. Our results show that assuming independence is a safer strategy in minimizing the structural distance between
the causal structure that has generated the data and the discovered structure. We also propose a simple improvement on the PC algorithm that
we call *blacklisting*. We demonstrate that blacklisting can lead to orders
of magnitude savings in computation by avoiding unnecessary independence tests.

Keywords: Causal discovery, PC algorithm, independence testing,
limited data.

1 Introduction

Even a superficial examination of the constraint-based search algorithms for
causal discovery, such as the PC algorithm [13], shows that the first phase of the
algorithm, consisting of a series of tests of independence, is both computationally intensive and crucial in terms of accuracy of the results. The results of a
combinatorial number of conditional independence tests determine the skeleton
of the causal graph. In case of discrete variables, for each edge (X, Y) under
consideration and for each conditioning set $\mathbf{S}$, we construct a contingency table.
Using this table, we calculate a test statistic, typically the G^2 statistic, and,
using a χ^2 distribution,[1] we determine the p-value for the independence hypothesis, which we subsequently compare against our preset significance level α. If
the p-value for any of the tests is larger than α, the algorithm removes the edge

---

[1] In our evaluation we apply Fienberg's method [8] of calculating degrees of freedom,
which corrects for empty rows, columns, and cells.

L.C. van der Gaag and A.J. Feelders (Eds.): PGM 2014, LNAI 8754, pp. 190–205, 2014.
© Springer International Publishing Switzerland 2014

and stores the conditioning set **S**, which may influence the result of the edge orientation phase.

The quality of a Bayesian network produced by the PC algorithm depends mostly on this phase, as it will determine the skeleton of the network. In turn, the quality of this phase hinges on the quality of the independence tests. One of the factors impacting the quality of an instance of the independence test is the number of data records that the test is based on. When there are insufficient data available, contingency tables will contain many zeros, which means in practice that co-occurrences of the variable states were not recorded in the data. This reduces the value of the G^2 statistic and the number of degrees of freedom used for the χ^2 test. A smaller value for G^2 can make it seem that the two variables of the edge are independent of each other. Consequently, an independence test for which we have insufficient data available will typically result in the test's inability to reject the (null) independence hypothesis and a Type II error.

In their influential book, Spirtes et al. [13] suggest a minimum ratio of ten to one of the number of records (samples) to the number of cells in the contingency table, which ensures a minimum level of reliability for the statistical tests. The advice that they put forward is that when this ratio is not satisfied, the test should not be performed and the variables should be considered dependent:

> "In testing the conditional independence of two variables given a set of other variables, if the sample size is less than ten times the number of cells to be fitted we assume the variables are conditionally dependent." [13, page 95]

Interestingly, we have come across a statement by Tsamardinos et al. [15], in which the authors state that they follow the advice of Spirtes et al. in assuming independence (!) if there are too few samples:

> "Following the practice used in [13] in our experiments we do not perform an independence test (i.e., we assume independence) unless there are at least five training instances on average per parameter (count) to be estimated." [15, page 43]

At this point, we are uncertain whether this conflicting advice is due to misreading, a typographical error, or any other reason. It is not obvious which of the two mutually exclusive rules is superior. Because assuming direct dependence between two variables amounts to keeping an arc between them in the discovered causal graph and assuming independence amounts to removing the arc, in the remainder of this paper we will call the two approaches the *keep rule* and the *remove rule* respectively.

The focus of this paper is an empirical evaluation of the impact of choosing between the *keep* and *remove* rules. In other words, we examine the question: Should we assume conditional independence (absence of an edge) or dependence in case there are too few data records to perform a reliable independence test? We found that from the point of view of distance metrics (i.e., the distance between the gold standard and the discovered graph) it is typically beneficial to follow the interpretation of [15]. Removing edges makes discovered causal graphs more sparse (and, hence, more efficient for the purpose of later inference) and typically closer to the original causal graphs. For those researchers, who prefer to assume dependence, we propose an improvement on the PC algorithm that we call *blacklisting*. Blacklisting avoids performing independence tests that are provably unable to reject the H_0 hypothesis of independence and, by this, may lead to orders of magnitude savings in computation.

The remainder of this paper is structured as follows. We first introduce black-listing in Section 2. We follow this up by two experiments: (1) comparing the effectiveness of the retrieval of gold standard causal graphs using the *keep* and *remove* rules (Section 4), and (2) comparing the classification accuracy of models based on the two rules (Section 5). We finish with a brief discussion of the two approaches (Section 6).

2   The Blacklisting Rule

Once they make a decision that two nodes X and Y are independent given a set of nodes $\mathbf{Z}$, constraint search-based causal discovery algorithms stop testing for independence of X and Y using different sets of conditioning nodes. When dealing with small data sets, the remove rule leads to a significant reduction of computation: The moment the algorithm realizes that there are not enough data to test for independence, it will assume independence and, effectively, remove the edge and stop further tests. It turns out that it is possible to avoid useless testing when using the keep rule as well.

We will first define the concept of *data to cells ratio*, i.e., the ratio of samples in the data set to the number of cells in the contingency table.

Definition 1 (data to cells ratio). *The* data to cells ratio, *$R(\mathbf{D}, X, Y, \mathbf{Z})$, in an independence test $I(X, Y|\mathbf{Z})$ is the ratio between the number of records $|\,\mathbf{D}\,|$ in the data set $\mathbf{D}$ and the number of cells in the contingency table for the variables X, Y, and the variables in the conditioning set $\mathbf{Z}$.*

An independence test $I(X, Y \mid \mathbf{Z})$ requires a contingency table with N elements, where N is the product of the variable cardinalities:

$$N = Card(X)\, Card(Y) \prod_{i=1}^{n} Card(Z_i)\,.$$

The data to cells ratio can be computed in the following way:

$$R(\mathbf{D}, X, Y, \mathbf{Z}) = \frac{|\,\mathbf{D}\,|}{N} = \frac{|\,\mathbf{D}\,|}{Card(X)\, Card(Y) \prod_{i=1}^{n} Card(Z_i)}\,. \tag{1}$$

Now we will prove a theorem that binds the *data to cells ratio* to increasing the size of the conditioning set.

Theorem 1. *Let r be a predefined minimum data to cells ratio. If in a given data set $\mathbf{D}$ an independence test $I(X, Y|\mathbf{Z})$ fails to satisfy r, then an independence test $I(X, Y|\mathbf{Z} \cup W)$ fails to satisfy r as well.*

Proof. Let the size of the conditioning set be $\mid \mathbf{Z} \mid = n$. By (1), we have

$$R(\mathbf{D}, X, Y, \mathbf{Z}) = \frac{\mid \mathbf{D} \mid}{N} = \frac{\mid \mathbf{D} \mid}{Card(X)\,Card(Y)\,\prod_{i=1}^{n} Card(Z_i)}\,.$$

If $W \in \mathbf{Z}$, we have $R(\mathbf{D}, X, Y, \mathbf{Z}) = R(\mathbf{D}, X, Y, \mathbf{Z} \cup W)$ and the theorem is true by assumption. Otherwise, we have

$$R(\mathbf{D}, X, Y, \mathbf{Z} \cup W) = \frac{\mid \mathbf{D} \mid}{Card(X)\,Card(Y)\,Card(W)\,\prod_{i=1}^{n} Card(Z_i)}$$

$$= \frac{R(\mathbf{D}, X, Y, \mathbf{Z})}{Card(W)}\,.$$

Because $Card(W) \geq 1$, we have

$$R(\mathbf{D}, X, Y, \mathbf{Z} \cup W) \leq R(\mathbf{D}, X, Y, \mathbf{Z}) < r\,,$$

which proves the theorem. ∎

Once we find out that an independence test of any pair of variables X and Y conditional on a set of variables $\mathbf{Z}$ does not satisfy the minimum ratio, Theorem 1 allows us to skip all tests of independence of X and Y conditional on any superset of $\mathbf{Z}$. We call the application of this rule in a constraint search-based causal discovery algorithm the *Blacklisting rule*.

Definition 2 (Blacklisting Rule). *If for any edge none of the conditional tests can be performed due to an insufficient sample to cell ratio, this edge is blacklisted and no longer checked in the subsequent iterations of the algorithm.*

Blacklisting can lead to substantial savings in computation.

In practice, we can enhance blacklisting by putting an upper bound on the number of variables in the conditioning set $\mathbf{Z}$. This allows us to skip calculation of the data to cells ratio for a given data set when the size of the conditioning set gets large enough. The upper bound on the number of conditioning variables depends on the minimum variable cardinality and the number of records of the data set:

$$n \leq \left\lfloor \frac{\log\left(\frac{S}{5}\right)}{\log\left(\min_i Card\left(i\right)\right)} \right\rfloor - 2\,.$$

S is the sample size and $Card(X_i)$ is the cardinality of a node X_i. We found at least one case where this upper bound optimization was applied [14] and we expect that others may have done so as well.

We did not apply this upper bound heuristic in the experiments described in this paper because we wanted to ensure that our keep and remove rules get applied whenever appropriate. Once we reach the upper bound, independence testing stops, leaving edges behind that otherwise would be subjected to one of the rules. Although it should not matter when applying the keep rule, combining this heuristic with the remove rule could possibly give results different from applying only the remove rule. We consider it beyond the scope of this paper to quantify the additional effect of the upper bound heuristic on the quality of patterns (also known as CPDAGS) produced by the PC algorithm.

3   Measure of Data Set Size for the Experiments

Constraint search-based causal discovery algorithms perform independence tests with an increasing number of conditioning variables, starting with unconditional independence tests, then one, two, three, etc., conditioning variables, all in an attempt to detect whether a pair of variables (nodes in a causal graph) are directly connected. Rather than the number of records, it is of more importance how many conditioning variables are feasible given a data set.

In our experiments, we chose to express the results as a function of an index C, which denotes the size of the conditioning set in the independence tests in the constraint search-based algorithm, such that rules are guaranteed to kick in. The following equation expresses the minimum data set size to guarantee at least 5 records in every cell of the contingency table when we test independence with C conditioning variables:

$$S = 5 \cdot 2^{C+2} . \tag{2}$$

Base 2 in the equation assumes binary variables, so when the data set size is smaller than S, the rules are guaranteed to kick in, even in the best case, when all variables are binary. The data set size derived this way guarantees the activation of either the keep or remove rule for all cases involving C or more conditioning variables. We present all our plots with results with the index C on the x axis. Our index is fairly easy to translate into the number of records. For example, $C = 5$ means that the experiment was based on a data set consisting of $5 \cdot 2^{5+2} = 640$ records.

Rather than generating for each model data sets of different sizes from scratch each time, we generated a single data set of size 10K records (please note that this data set is sufficiently large for $C = 9$, which corresponds to 10,240 records. Subsequently, we limited the size during the experiments to obtain the number of records required by the index C.

4   Experiment 1: Keep vs. Remove Rules in Structure Retrieval

Our first experiment focused on a comparison of the keep and remove rules in their ability to retrieve the gold standard structure that has generated the data.

4.1   Methodology

We selected nine different Bayesian networks, listed in Table 1, for our gold standard networks.

The experiment consisted of 100 repetitions of the following steps:

1. Generate a data set (see Section 3).
2. For each of the conditioning size limits $C = 0, 1, 2, \ldots, 9$:

Table 1. Gold Standard Evaluation Networks

Network	Nodes	Edges	Reference
Asia	8	8	[9]
Nursery[2]	9	10	[12]
Adult[2]	15	18	[12]
Bank[2]	17	28	[12]
Chess[2]	37	100	[12]
Alarm	37	46	[4]
Hailfinder	56	66	[1]
Hepar	70	123	[10]
CPCS179	179	239	[11]

(a) Reduce the sample size of the data to S that corresponds to C (see Section 3, Equation 2).

(b) Learn a structure (a pattern) from the data using the PC algorithm using (a) the keep rule and (b) the remove rule.

(c) Calculate distance statistics:

    i. Hamming distance from the learned structures to the gold standard[2]

    ii. Hamming distance between the two learned structures.

    iii. Skeletal distance from the learned structures to the gold standard.

    iv. Skeletal distance between the learned structures.

(d) Count the number of edge orientation mistakes between the learned structures and the gold standard.

(e) Count the number of edge orientation mistakes between the learned structures.

(f) Store the the running time of the algorithm and the number of times the rules were applied.

The Hamming distance [2,15] between (CP)DAG structures A and B counts the number of changes that need to made to transform A into B (and vice versa, because Hamming distance is a symmetric measure). It consists of the sum of three components: (1) the number of edges that are present in A but missing in B, (2) the number of edges that are present in B but missing in A, and (3) the number of edges in A that are present in B but oriented differently. The skeletal distance, also measured in our experiments, consists of just the first two components of the Hamming distance.

4.2 Results

Figure 1 shows the gold standard recovery performance of the two rules, i.e., Hamming distance between the recovered patterns and the patterns embedded

---

[2] We convert the original DAG structure of the gold standard into a pattern by applying Chickering's algorithm [5]

[2] These networks were created by Ratnapinda and Druzdzel in an unrelated experiment [12] using data from the UCI Machine Learning Repository [3]

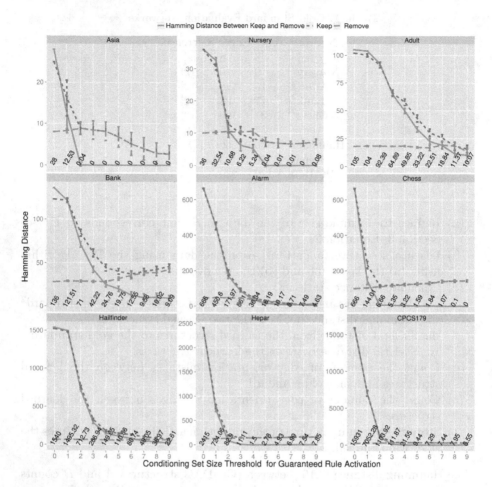

Fig. 1. Hamming distance between the keep and remove rules and the gold standard networks as a function of data set size. Numbers indicate difference between keep and remove rules.

in the gold standard networks, and Hamming distance between the recovered patterns. The difference between the two rules is most apparent for data sets with small number of data records. In the extreme case, the keep rule results in a completely connected graph, and the remove rule results in an completely disconnected graph. The worst case scenario is for the remove rule much closer to the gold standard than the worst case scenario of the keep rule. The reason for this is that causal graphs seem to be naturally sparse. When more data are available, the algorithm has enough data to perform independence tests and the difference between the two rules disappears. The remove rule performs in that case at least as well as the keep rule and sometimes better.

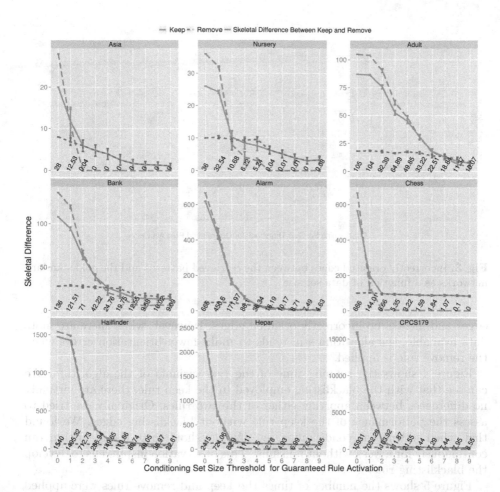

Fig. 2. Skeletal differences between patterns for keep and remove rules and the gold standard as a function of data set size. Numbers indicate differences between keep and remove rules.

Figure 2 shows skeletal differences between the recovered patterns and the patterns derived from the gold standard networks. In most cases, the skeletons of the retrieved graphs tend to converge when there are enough data. We can see that the discrepancies between the keep and remove rules in Figure 1 most likely result from errors in edge orientation.

Figure 3 illustrates the edge orientation errors made by both algorithm variants and shows how many edges differed between the patterns. Overall, the remove rule makes fewer mistakes, but it is important to note that this measure counts only cases where in both patterns (result and gold standard, or both results) an edge exists, which is why in the first few conditioning limits the number is low, starting with 0 in the first step. In that case, the remove rule results in no

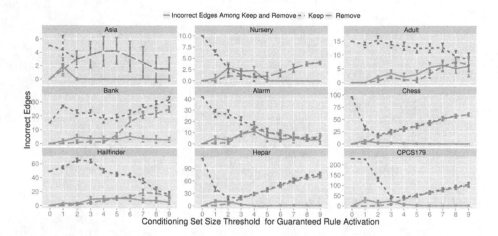

Fig. 3. Incorrectly oriented edges between the retrieved patterns and the gold standard networks as a function of data set size

edges and, hence, no incorrectly oriented edges. However, when there are more data available, the algorithm still tends to make fewer orientation errors when the remove rule is applied.

Figure 4 shows the running time of the two variants of the algorithm. We can see that with the blacklisting employed in the keep rule, there are virtually no differences between applying either of the two rules. Originally, we tried to assess the performance of the keep rule without blacklisting edges. We found that when setting the conditioning set size threshold to 0, the algorithm ran continuously for a week without finishing. This observation inspired us to develop the blacklisting rule.

Figure 5 shows the number of times the keep and remove rules were applied (please note the logarithmic scale on the y-axis). The keep rule clearly required a much larger number of independence tests and applications of the keep rule due to insufficient data. This has to do with the size of the conditioning sets, which are composed of nodes that are still possibly connected to the tested pairs of nodes. Removing edges decreases the number of edges that need to be tested in subsequent iterations of the algorithm, limits the number of possible future conditioning sets, and reduces the possibility of not having sufficient data to run an independence test.

To summarize, the remove rule performs at least as well as the keep rule, when recovering the structure of a gold standard network, and better when fewer records data are available. However, with blacklisting implemented in the keep rule, the remove rule does not necessarily improve the running time of the algorithm.

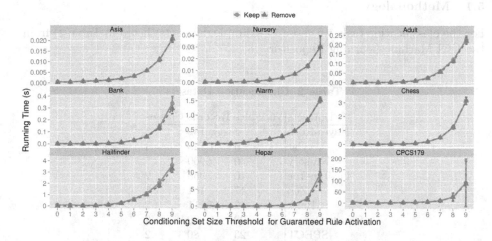

Fig. 4. Average running times for the keep and remove rules

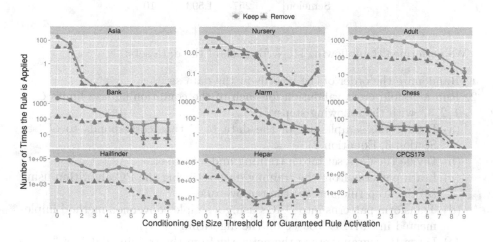

Fig. 5. The average number of rule applications

5 Experiment 2: Keep Rule vs. Remove Rule in Classification Accuracy

Keeping vs. removing edges in a graph has an effect on the ability of the network to represent the joint distribution over the modeled variables and should have an effect on the overall precision of the networks when they are used in practice. In Experiment 2, we compared classification accuracy of networks retrieved using the keep and remove rules. Classification is a typical machine learning task that requires modeling accuracy.

5.1   Methodology

For the classification task, we chose 12 example data sets (listed in Table 2) from the UCI Machine Learning Repository [3].

Table 2. Classification Datasets

Datasets	Variables	Records	Classes
Hayes-Roth	4	132	3
Car	6	1,728	4
Nursery	9	12,960	5
Tic-Tac-Toe	10	958	2
Zoo	17	101	7
Bank	17	45,211	2
Breast-Cancer	22	80	2
SPECT	23	80	2
Soybean	36	266	15
Chess	37	3,196	2
Connect-4	43	67,557	3
Semeion	257	1,593	10

We performed the following steps for each of the data sets:

1. Randomize the order of the records in the data set.
2. Run a 10-fold cross validation for each of the conditioning size limits $C = 1, 2, 3$,[3] with the following steps:
   (a) Reduce the sample size of the data to S that corresponds to C (see Section 3, Equation 2).
   (b) Divide the data set into a training and a test set.
   (c) Learn patterns from the training data set using the PC algorithm using (a) the keep rule and (b) the remove rule.
   (d) Transform patterns into DAGs using a simple conversion method implemented in SMILE©(based on the algorithm by [7]).
   (e) Learn the parameters of the networks from the training data set.[4]
   (f) Determine the classification accuracy of each network by predicting the class variable using samples from the test data set, and verify the predictions by means of the actual class values.
   (g) Record:
      i. The number of times the rules were applied in both cases.
      ii. The running time of the algorithm in both cases.

---

[3] We used $C = 1, 2, 3$ rather than 1 through 9 in this experiment, because this is the range of data set sizes for which, as discovered in Experiment 1, there were differences between the keep and remove rules.

[4] We used SMILE©'s implementation of the EM algorithm to learn the parameters, even though our data was complete. In this case it is reduced to plain maximum likelihood estimation.

iii. The time required to run inference on the Bayesian network in both cases.

iv. The total clique size of the junction tree used for inference.

5.2 Results

Three of the data sets (Bank, Soybean and Semeion) failed to produce tractable Bayesian networks when applying the keep rule. Although the PC algorithm produced a pattern applying either rule, and the patterns were successfully converted into DAGs, a common problem were very large parent sets in some of the nodes. This, in combination with nodes having a considerable number of variable states, resulted in the conditional probability tables (CPTs) becoming too large, causing the experiment to stall in the conversion process.

For nine of the twelve data sets, the experiment completed successfully. Figures 6 through 9 show the results for these datasets.

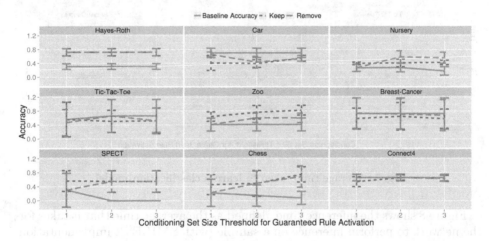

Fig. 6. Average classification accuracy as a function of the data set size for both keep and remove rules

Following the procedure described by [6], we performed Wilcoxon signed rank tests [16] after ranking the performance of the algorithms to determine if either of the rules significantly outperformed the other. Additionally, we tested if the rules outperformed the baseline accuracy model, which simply bets on the most likely class a-priori. For the Wilcoxon test, we considered 27 paired data points (9 datasets, 3 constrained versions). These points were the ranks of the algorithms after comparing their classification accuracy, averaged over the ten folds. We found that there were no significant differences between the two rules from the point of view of classification accuracy. Both rules did better than the baseline performance, although the result was barely significant (see Table 3).

Table 3. p-values for the Wilcoxon signed rank test comparing the classification accuracy of models recovered using the two rules

	Remove	Baseline
Keep	0.8563	**0.03345**
Remove		**0.03694**

Figure 7 shows the running time needed by the algorithms to learn the models. For the data sets tested (specifically, their three restricted versions), neither of the rules drastically influenced the running time. No significant difference was found between the algorithms (Wilcoxon test, V = 139, p = 0.2386).

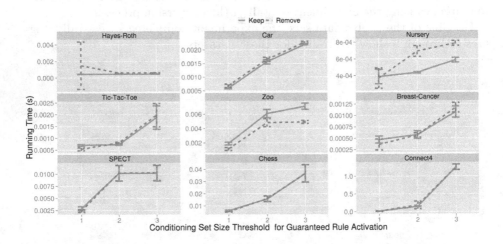

Fig. 7. Average time taken to learn a classification model

Figure 8 shows the inference time, defined as the average time that it takes for the network to perform inference on a sample (with SMILE$^©$'s implementation of the clustering algorithm). Remove rule outperformed the keep rule (one-sided Wilcoxon test, $V = 349.5$, $p = 5.247\text{e-}06$). From the practical point of view, the performance differences are only noticeable when the models are learned from very small data sets (resulting in dense graphs learned with the keep rule).

The networks learned from very small data sets were much denser. Denser networks result in larger cliques. The total clique size, i.e, the sum of all elements of all the clique potentials, gives an indication of how much time inference will take. Figure 9 shows the total clique sizes for the retrieved networks. Here also we see that the remove rule outperforms the keep rule in terms of learning smaller networks.

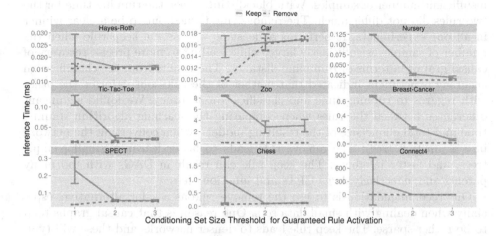

Fig. 8. Average classification inference times

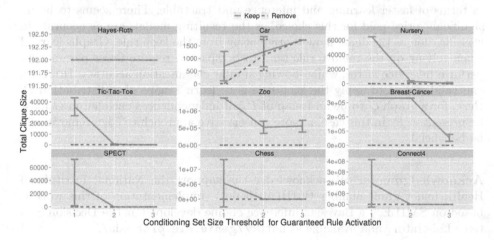

Fig. 9. Average Total Size of All Cliques in the Network

6   Discussion

We tested two approaches suggested in the literature for dealing with insufficient number of samples when performing independence test in the context of constrain search-based causal discovery algorithms: (1) assuming dependence (keep rule) and (2) assuming independence (remove rule). We performed two experiments: (1) recovery of gold standard networks from data sets of varying size, and (2) comparison of computational and spatial complexity and classification accuracy of models learned using the two approaches.

To reduce the run time of the keep rule, we introduced blacklisting, which amounts to reducing independence tests that will predictably fail because of insufficient number of samples. With blacklisting edges, the running time for the two rules did not differ much. The remove rule turned out to be a clear winner in all our experiments. It performs at least as well as the keep rule when the number of samples is sufficiently large, while leading to more precise recovery of causal graphs in terms of Hamming distance. The results from the classification accuracy experiment indicated that neither of the rules outperforms the other with regards to running time and classification accuracy. We found significant differences between the rules when comparing the learning algorithm running time and total clique size of the resulting models (when converting the network into a junction tree): The remove rule leads to faster learning, smaller clique trees, and faster inference. The keep rule can result in DAGs with too many parents and, effectively, huge CPTs and junction trees.

Our recommendation is to use the remove rule as a safer alternative, especially when dealing with small data sets. One reason is that causal graphs tend to be rather sparse. The keep rule leads to denser networks and these will typically differ more from the pursued causal graphs. Remove rule leads to networks that are not only closer to the pursued causal graphs. They are more practical in terms of faster learning and inference and tractable. There seems to be no significant price paid for this choice, as the resulting models are as accurate in classification as the dense models originating from the keep rule. Graphs learned when relying on the remove rule tend to be sparser, require fewer parameters, perform inference faster, while showing similar classification accuracy. However, one valid reason to apply the keep rule is to learn I-maps of a distribution P. Here, for a graph G to be an I-map of P, all independences encoded in G must be present in P. In this case, a more conservative approach to edge removal will be beneficial.

Acknowledgments. We acknowledge the support the National Institute of Health under grant number U01HL101066-01. Implementation of this work is based on SMILE©, a Bayesian inference engine developed at the Decision Systems Laboratory and available at http://genie.sis.pitt.edu/.

References

1. Abramson, B., Brown, J., Edwards, W., Winkler, R., Murphy, A.: Hailfinder: A Bayesian system for forecasting severe weather. International Journal of Forecasting 12(1), 57–72 (1996)
2. Acid, S., de Campos, L.M.: Searching for Bayesian network structures in the space of restricted acyclic partially directed graphs. Journal of Artificial Intelligence Research 18, 445–490 (2003)
3. Bache, K., Lichman, M.: UCI Machine Learning Repository (2013)
4. Beinlich, I., Suermondt, J., Chavez, M., Cooper, G.: The ALARM Monitoring System: A Case Study with Two Probablistic Inference Techniques for Belief Networks. In: Second European Conference on Artificial Intelligence in Medicine, London, pp. 247–256 (1989)

5. Chickering, D.M.: A transformational characterization of equivalent Bayesian network structures. In: Proceedings of the 11th Annual Conference on Uncertainty in Artificial Intelligence (UAI 1995), pp. 87–98. Morgan Kaufmann, San Francisco (1995)
6. Demšar, J.: Statistical comparisons of classifiers over multiple data sets. J. Mach. Learn. Res. 7, 1–30 (2006)
7. Dor, D., Tarsi, M.: A simple algorithm to construct a consistent extension of a partially oriented graph. Technicial Report R-185, Cognitive Systems Laboratory, UCLA (1992)
8. Fienberg, S.E., Holland, P.W.: Methods for eliminating zero counts in contingency tables. In: Random Counts on Models and Structures, pp. 233–260 (1970)
9. Lauritzen, S.L., Spiegelhalter, D.J.: Local computations with probabilities on graphical structures and their application to expert systems. Journal of the Royal Statistical Society 50, 157–224 (1988)
10. Onisko, A.: Probabilistic Causal Models in Medicine: Application to Diagnosis of Liver Disorders. PhD thesis, Institute of Biocybernetics and Biomedical Engineering, Polish Academy of Science, Warsaw (March 2003)
11. Pradhan, M., Provan, G., Middleton, B., Henrion, M.: Knowledge engineering for large belief networks. In: Proceedings of the Tenth International Conference on Uncertainty in Artificial Intelligence, pp. 484–490. Morgan Kaufmann Publishers Inc. (1994)
12. Ratnapinda, P., Druzdzel, M.J.: An empirical comparison of Bayesian network parameter learning algorithms for continuous data streams. In: Recent Advances in Artificial Intelligence: Proceedings of the Nineteenth International Florida Artificial Intelligence Research Society Conference (FLAIRS–2013), pp. 627–632 (2013)
13. Spirtes, P., Glymour, C., Scheines, R.: Causation, Prediction, and Search, 2nd edn. MIT Press (2000)
14. Stojnic, R., Fu, A.Q., Adryan, B.: A graphical modelling approach to the dissection of highly correlated transcription factor binding site profiles. PLoS Computational Biology 8(11), e1002725 (2012)
15. Tsamardinos, I., Brown, L.E., Aliferis, C.F.: The max-min hill-climbing Bayesian network structure learning algorithm. Mach. Learn. 65(1), 31–78 (2006)
16. Wilcoxon, F.: Individual comparisons by ranking methods. Biometrics 1(6), 80–83 (1945)

Supervised Classification Using Hybrid Probabilistic Decision Graphs

Antonio Fernández[1], Rafael Rumí[1], José del Sagrado[2], and Antonio Salmerón[1]

[1] Dept. of Mathematics, University of Almería,
Ctra. Sacramento s/n, 04120 Almería, Spain
[2] Dept. of Computer Science, University of Almería,
Ctra. Sacramento s/n, 04120 Almería, Spain
{afalvarez,rrumi,jsagrado,antonio.salmeron}@ual.es
http://elvira.ual.es/programo

Abstract. In this paper we analyse the use of probabilistic decision graphs in supervised classification problems. We enhance existing models with the ability of operating in hybrid domains, where discrete and continuous variables coexist. Our proposal is based in the use of mixtures of truncated basis functions. We first introduce a new type of probabilistic graphical model, namely probabilistic decision graphs with mixture of truncated basis functions distribution, and then present an initial experimental evaluation where our proposal is compared with state-of-the-art Bayesian classifiers, showing a promising behaviour.

Keywords: Supervised classification, Probabilistic decision graphs, Mixtures of truncated basis functions, Mixtures of polynomials, Mixtures of truncated exponentials.

1 Introduction

The Probabilistic Decision Graph (PDG) model was introduced in [2] as an efficient representation of probabilistic transition systems. In this study, we consider the more general version of PDGs proposed in [8].

PDGs are probabilistic graphical models that can represent some context specific independencies that are not efficiently captured by conventional graphical models as Bayesian Networks (BNs). In addition, probabilistic inference can be carried out directly over the PDG structure in a time linear in the size of the PDG model.

PDGs have mainly been studied as representations of joint distributions over *discrete* random variables, showing a competitive performance when compared to BNs and Latent class Naive BNs [15]. The discrete PDG model has also been successfully applied to supervised classification problems [16] and unsupervised clustering [4].

The need to handle discrete and continuous variables simultaneously has motivated the development of new probabilistic graphical models incorporating that feature, mainly hybrid Bayesian networks [3, 14, 9, 10, 17, 18]. Also, PDGs have

L.C. van der Gaag and A.J. Feelders (Eds.): PGM 2014, LNAI 8754, pp. 206–221, 2014.

been recently extended in order to allow the inclusion of *continuous* variables
when the joint distribution of the model is a *conditional Gaussian* (CG) [6].

In this paper we extend the PDG classifiers introduced in [16] in order to incorporate continuous variables. We rely on the *Mixture of truncated basis functions*
(MoTBFs) model [10]. Unlike CG models, MoTBFs do not rely on the normality
assumption, and do not impose any restriction on the structure of the conditional
distributions involved in the PDG, so that discrete variables can be conditioned
on continuous ones and vice versa.

2   Notation and Preliminaries

We will use uppercase letters to denote random variables, and boldfaced uppercase letters to denote random vectors, e.g. $\mathbf{X} = \{X_0, X_1, \ldots, X_N\}$. By $R(X)$ we
denote the set of possible states of variable X, and similarly for random vectors,
$R(\mathbf{X}) = \times_{X_i \in \mathbf{X}} R(X_i)$. By lowercase letters x (or $\mathbf{x}$) we denote some element
of $R(X)$ (or $R(\mathbf{X})$). When $\mathbf{x} \in R(\mathbf{X})$ and $\mathbf{Y} \subseteq \mathbf{X}$, we denote by $\mathbf{x}[\mathbf{Y}]$ the projection of $\mathbf{x}$ onto coordinates $\mathbf{Y}$. Throughout this document we will consider a
set $\mathbf{W}$ of discrete variables and a set $\mathbf{Z}$ of continuous variables, and we will use
$\mathbf{X} = \mathbf{W} \cup \mathbf{Z}$.

In what concerns the structure of the PDG model, we will make use of the
following notation. Let G be a directed graph over nodes $\mathbf{V}$. Let $\nu \in \mathbf{V}$, we
then denote by $pa_G(\nu)$ the set of parents of node ν in G, by $ch_G(\nu)$ the set of
children of ν in G, by $de_G(\nu)$ the set of descendants of ν in G, that is recursively
defined as $de_G(\nu) = \{\nu' : \nu' \in ch_G(\nu) \vee [\nu' \in ch_G(\nu'') \wedge \nu'' \in de_G(\nu)]\}$, and
we use as shorthand notation $de_G^*(\nu) = de_G(\nu) \cup \nu$. By $an_G(\nu)$ we understand
the set of ancestors (or predecessors) of ν in G, that is recursively defined as
$an_G(\nu) = \{\nu' : \nu' \in pa_G(\nu) \vee [\nu' \in pa_G(\nu'') \wedge \nu'' \in an_G(\nu)]\}$.

2.1   Discrete PDGs

The discrete PDG model was introduced in [8] as representation of joint distributions over discrete random variables. The structure is formally defined as
follows:

Definition 1 (PDG Structure [8]). *Let F be a forest of directed tree struc-
tures over a set of discrete random variables $\mathbf{W}$. A PDG structure $G = \langle \mathbf{V}, \mathbf{E} \rangle$
for $\mathbf{W}$ w.r.t. F is a set of* rooted *DAGs, such that:*

1. *Each node $\nu \in \mathbf{V}$ is labelled with exactly one $W \in \mathbf{W}$. By $\mathbf{V}_W$, we will refer
   to the set of all nodes in a PDG structure labelled with the same variable
   W. For every variable W, $\mathbf{V}_W \neq \emptyset$, we will say that ν represents W when
   $\nu \in \mathbf{V}_W$.*
2. *For each node $\nu \in \mathbf{V}_W$, each possible state $w \in R(W)$ and each successor
   $Y \in ch_F(W)$ there exists exactly one edge labelled with w from ν to some
   node ν' representing Y. Let $Y \in ch_F(W)$, $\nu \in \mathbf{V}_W$ and $w \in R(W)$. By
   $succ(\nu, Y, w)$ we will then refer to the unique node $\nu' \in \mathbf{V}_Y$ that is reached
   from ν by an edge with label w.*

An example, taken from [6], of a PDG structure and its corresponding variable forest are depicted in Figs. 1(b) and (a) respectively. A PDG structure consists of two layers, one for variable and one for nodes. The variable layer conforms a directed forest over the variables F and the node layer is a one-root directed acyclic graph structure. Children, parents, descendants or ancestors of a variable in a PDG structure G, are located according to structure F. So, using Fig. 1(b) as an example, on the variable layer we have: $pa_G(W_1) = \{W_0\}$, $ch_G(W_0) = \{W_1, W_2\}$, $de_G(W_0) = \{W_1, W_2, W_3\}$ and $an_G(W_3) = \{W_1, W_0\}$. On the node layer, we have: $succ(\nu_0, W_1, 0) = \nu_1$, $succ(\nu_0, W_2, 1) = \nu_4$ and $succ(\nu_1, W_3, 0) = \nu_6$.

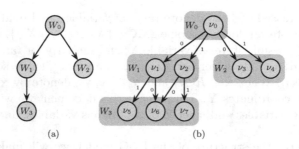

(a)   (b)

Fig. 1. (a) A forest structure F formed by a single tree over binary variables $\mathbf{W} = \{W_0, W_1, W_2, W_3\}$. (b) A PDG structure over $\mathbf{W}$ with underlying forest F.

A PDG structure G is *instantiated* by assigning a real function $f^\nu : R(W_i) \to \mathbb{R}_0^+$, with $\nu \in \mathbf{V}_{W_i}$ to every node ν in the structure. It represents the global real function $f_G : R(\mathbf{W}) \to \mathbb{R}_0^+$, recursively defined as follows:

$$f_G^\nu(\mathbf{w}) := f^\nu(\mathbf{w}[W]) \prod_{Y \in ch_F(W)} f_G^{succ(\nu, Y, \mathbf{w}[W])}(\mathbf{w}), \tag{1}$$

for all $\mathbf{w} \in R(\mathbf{W})$. f_G is then defined on $R(\mathbf{W})$ as:

$$f_G(\mathbf{w}) := \prod_{\nu:\nu \text{ is a root}} f_G^\nu(\mathbf{w}). \tag{2}$$

Equation (1) defines a factorisation with one factor f^ν for each $W \in \mathbf{W}$. The set of nodes in a PDG structure associated with a given element $\mathbf{w} \in R(\mathbf{W})$ is characterised by function *reach*.

Definition 2 (Reach). *A node ν representing variable W_i in G is reached by* $\mathbf{w} \in R(\mathbf{W})$ *if*

1. *ν is a root in G, or*
2. *$W_j = pa_F(W_i)$, node ν' representing variable W_j is reached by $\mathbf{w}$ and $\nu = succ(\nu', W_i, \mathbf{w}[W_j])$.*

By $reach_G(W_i, \mathbf{w})$ we denote the unique node representing W_i reached by $\mathbf{w}$ in PDG structure G.

As an example [6], consider again the PDG structure of Fig. 1(b) and let $\mathbf{w} = \{W_0 = 0, W_1 = 1, W_2 = 1, W_3 = 1\}$. Then $reach_G(W_0, \mathbf{w}) = \nu_0$, $reach_G(W_1, \mathbf{w}) = \nu_1$, $reach_G(W_2, \mathbf{w}) = \nu_3$ and $reach_G(W_3, \mathbf{w}) = \nu_5$. Function f_G in (2) can be reformulated as

$$f_G(\mathbf{w}) := \prod_{W_i \in \mathbf{W}} f^{reach_G(W_i, \mathbf{w})}(\mathbf{w}[W_i]). \tag{3}$$

When all the local functions f^ν in an instantiated PDG structure G over $\mathbf{W}$ are probability distributions, f_G defines a joint multinomial probability distribution over $\mathbf{W}$ [8]. In fact, f_G^ν in (1) defines a multinomial distribution over variables $W \cup ch_F(W)$. We will refer to such instantiated PDG structures as PDG models.

Definition 3 (PDG model [8]). *A PDG model $\mathcal{G}$ is a pair $\mathcal{G} = \langle G, \theta \rangle$, where $G = \langle \mathbf{V}, \mathbf{E} \rangle$ is a valid PDG structure (Definition 1) over some set $\mathbf{W}$ of discrete random variables and $\theta = \{f^\nu : \nu \in \mathbf{V}\}$ is a set of real functions, each of which defines a discrete probability distribution.*

Example 1 (taken from [6]). Consider the PDG structure in Fig. 1. It encodes a factorisation of the joint distribution of $\mathbf{W} = \{W_0, W_1, W_2, W_3\}$, with $f^{\nu_0} = P(W_0)$, $f^{\nu_1} = P(W_1|W_0 = 0)$, $f^{\nu_2} = P(W_1|W_0 = 1)$, $f^{\nu_3} = P(W_2|W_0 = 0)$, $f^{\nu_4} = P(W_2|W_0 = 1)$, $f^{\nu_5} = P(W_3|W_0 = 0, W_1 = 1)$, $f^{\nu_6} = P(W_3|W_1 = 0, \{W_0 = 0 \vee W_0 = 1\})$, $f^{\nu_7} = P(W_3|W_0 = 1, W_1 = 1)$.

The PDG structure plus the set of conditional distributions given above constitute a PDG model over the set of variables $\mathbf{W} = \{W_0, W_1, W_2, W_3\}$. Assume that we want to evaluate the PDG model for a given configuration of $\mathbf{W}$, for instance, $(0, 1, 1, 1)$. According to (1), the returned value is

$$\begin{aligned} f_G(0, 1, 1, 1) &= f^{\nu_0}(0) f^{\nu_1}(1) f^{\nu_3}(1) f^{\nu_5}(1) \\ &= P(W_0 = 0) P(W_1 = 1|W_0 = 0) P(W_2 = 1|W_0 = 0) \\ &\quad P(W_3 = 1|W_0 = 0, W_1 = 1). \end{aligned}$$

2.2 Conditional Gaussian PDGs

The first attempt to include continuous variables in PDGs came along with the so-called *conditional Gaussian PDGs* [6]. These models represent joint distributions of discrete and continuous variables simultaneously, conforming a conditional Gaussian (CG) distribution [12]. It means that the joint over the continuous variables is assumed to be a mixture of multivariate Gaussians, and the joint over the discrete variables is a multinomial.

Formally, a CG-PDG is a PDG with forest F where variables can be discrete and continuous. Discrete variables are treated as in PDGs, and the continuous ones follow the next requirements:

- Every continuous variable $Z \in \mathbf{Z}$ is only allowed to have continuous children in F.
- A node ν representing a continuous variable $Z \in \mathbf{Z}$ has exactly one outgoing edge for each child of Z in F.
- A node ν representing $Z \in \mathbf{Z}$ with predecessors in F $\{Z_1, \ldots, Z_n\}$ defines the conditional density $f^\nu = \mathcal{N}(z; \alpha^\nu + \sum_{i=1}^{n} \beta_i^\nu z_i, \sigma^{2^\nu})$.

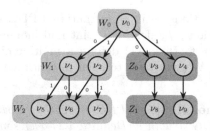

Fig. 2. An example of a structure of a CG-PDG, with discrete variables W_0, W_1, W_2 and continuous variables Z_0 and Z_1

Example 2. The structure depicted in Fig. 2 is compatible with a CG-PDG with discrete variables W_0, W_1, W_2 and continuous variables Z_0 and Z_1. The structure is instantiated as described in Table 1. Note that nodes corresponding to discrete variables contain probability tables, while nodes corresponding to continuous variables have densities instead. For instance, $f^{\nu_3} = \rho(Z_0|W_0 = 0)$ is a Gaussian density with fixed mean and variance. If a continuous variable has at least one continuous predecessor, then the density in each of its nodes is also Gaussian with fixed variance, but the mean is not constant, but rather a linear function of the continuous predecessors. That is the case of nodes ν_8 and ν_9.

Table 1. Instantiation of the structure in Fig. 2

$f^{\nu_0} = P(W_0)$	$f^{\nu_5} = P(W_2\|W_0 = 0, W_1 = 1)$
$f^{\nu_1} = P(W_1\|W_0 = 0)$	$f^{\nu_6} = P(W_2\|W_1 = 0)$
$f^{\nu_2} = P(W_1\|W_0 = 1)$	$f^{\nu_7} = P(W_2\|W_0 = 1, W_1 = 1)$
$f^{\nu_3} = \rho(Z_0\|W_0 = 0)$	$f^{\nu_8} = \rho(Z_1\|Z_0, W_0 = 0)$
$f^{\nu_4} = \rho(Z_0\|W_0 = 1)$	$f^{\nu_9} = \rho(Z_1\|Z_0, W_0 = 1)$

2.3   Mixtures of Truncated Basis Functions

The MoTBF framework is based on the abstract notion of real-valued *basis functions* $\psi(\cdot)$, which includes both polynomial and exponential functions as special cases. Let X be a continuous variable with domain $R(X) \subseteq \mathbb{R}$ and let $\psi_i : \mathbb{R} \to \mathbb{R}$, for $i = 0, \ldots, k$, define a collection of real basis functions. We say that a function $g_k : R(X) \mapsto \mathbb{R}_0^+$ is an MoTBF potential of level k wrt. $\Psi = \{\psi_0, \psi_1, \ldots, \psi_k\}$ if g_k can be written as [10]

$$g_k(x) = \sum_{i=0}^{k} a_i \, \psi_i(x),$$
(4)

where a_i are real numbers. The potential is a density if $\int_{R(X)} g_k(x)\, dx = 1$.

Example 3. By letting the basis functions correspond to polynomial functions, $\psi_i(x) = x^i$ for $i = 0, 1, \ldots$, the MoTBF model reduces to an MOP model [18] for univariate distributions. Similarly, if we define the basis functions as $\psi_i(x) = \{1, \exp(-x), \exp(x), \exp(-2x), \exp(2x), \ldots\}$, the MoTBF model corresponds to an MTE model [14] with the exception that the parameters in the exponential functions are fixed.

In a conditional MoTBF density, the influence a set of continuous parent variables $\mathbf{Z}$ has on their child variable X is encoded only through the partitioning of the domain of $\mathbf{Z}$, denoted as $R(\mathbf{Z})$, into hyper-cubes, and not directly in the functional form of $g_k(x|\mathbf{z})$ inside each hyper-cube. More precisely, for a partitioning $\mathcal{P} = \{R(\mathbf{Z})^1, \ldots, R(\mathbf{Z})^m\}$ of $R(\mathbf{Z})$, the conditional MoTBF is defined for $\mathbf{z} \in R(\mathbf{Z})^j$, $1 \leq j \leq m$, as

$$g_k^{(j)}(x|\mathbf{z} \in R(\mathbf{Z})^j) = \sum_{i=0}^{k} a_i^{(j)} \, \psi_i^{(j)}(x).$$
(5)

Similarily, MoTBFs can be defined for discrete variables, in which case each potential value $g(x)$ represents the value $P(X = x)$ with $\sum_x g(x) = 1$. Conditional distributions of discrete variables given continuous and/or discrete variables can be defined analogously to (5). See [10, 11] for more details.

3   Hybrid PDGs Based on MoTBFs

CG-PDGs have two limitations. One is the normality assumption, that may not hold in applications with real data. The other one is the structural restriction that forbids discrete variables to have continuous parents in the structure. For instance, the structure in Fig. 3 is not valid for a CG-PDG since discrete variable W_1 has a continuous predecessor, Z_0.

Our proposal to sidestep the above-mentioned restrictions is to adopt the MoTBF framework (see Sect. 2.3) within the PDG model. The formal definition is as follows.

Definition 4 (MoTBF-PDG). *We define an MoTBF-PDG as a PDG with forest F where variables can be discrete and continuous. Discrete variables are treated as in PDGs, and the continuous ones follow the next requirements:*

- *Continuous variables are allowed to have discrete and continuous successors and predecessors.*

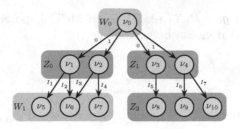

Fig. 3. An example of a PDG structure that is not compatible with a CG-PDG, since discrete variable W_1 has a continuous predecessor, Z_0. Labels $I_1, \ldots, I_7$ indicate intervals of the domain of the corresponding continuous variable.

- A node ν representing $Z \in \mathbf{Z}$ can have one or more *outgoing edges for each Z_i child of Z in F. Each edge represents a interval the domain of Z, and all of them constitute a* partition *of it.*
- A node ν representing $Z \in \mathbf{Z}$ with continuous predecessors $\{Z_1, \ldots, Z_n\}$ defines an MoTBF density $f^\nu(z)$ conditional *on the branch that leads from the root to ν.*

Example 4. Consider the PDG structure in Fig. 3. Assume W_0 and W_1 are binary variables and Z_0, Z_1 and Z_2 are continuous variables with domain $[0, 1]$. An instantiation of the PDG structure is given in Table 2, where the labels $I_1, \ldots, I_7$ are, respectively, intervals $[0, 0.5), [0.5, 1], [0.5, 1], [0, 0.5), [0, 1], [0, 0.3), [0.3, 1]$.

Table 2. An instantiation of the PDG structure in Fig. 3. Potentials denoted as ρ correspond to conditional MoTBF densities as in (5).

$$
\begin{aligned}
f^{\nu_0} &= P(W_0) & f^{\nu_5} &= P(W_1 | W_0 = 0, Z_0 \in [0, 0.5)) \\
f^{\nu_1} &= \rho(Z_0 | W_0 = 0) & f^{\nu_6} &= P(W_1 | Z_0 \in [0.5, 1]) \\
f^{\nu_2} &= \rho(Z_0 | W_0 = 1) & f^{\nu_7} &= P(W_1 | W_0 = 1, Z_0 \in [0, 0.5)) \\
f^{\nu_3} &= \rho(Z_1 | W_0 = 0) & f^{\nu_8} &= \rho(Z_2 | W_0 = 0) \\
f^{\nu_4} &= \rho(Z_1 | W_0 = 1) & f^{\nu_9} &= \rho(Z_2 | W_0 = 1, Z_1 \in [0, 0.3)) \\
& & f^{\nu_{10}} &= \rho(Z_2 | W_0 = 1, Z_1 \in [0.3, 1])
\end{aligned}
$$

The next proposition shows that an MoTBF-PDG actually represents a joint distribution of class MoTBF over the variables it contains.

Proposition 1. *Let G be a MoTBF-PDG over variables $\mathbf{X} = \mathbf{W} \cup \mathbf{Z}$. Function*

$$
f_G(\mathbf{x}) = \prod_{X \in \mathbf{X}} f^{reach_G(X, \mathbf{x})}(\mathbf{x}[X])
$$

represents an MoTBF distribution over $\mathbf{X}$.

Proof. According to (3), $f^{reach_G(X,\mathbf{x})}$ is the function stored in the unique parameter node ν of variable X that is reached by the path in G determined by configuration $\mathbf{X} = \mathbf{x}$. Let us index the variables in $\mathbf{X}$ as $X_1, \ldots, X_n$ and denote by $\nu_{\mathbf{x}[X_i]}$ the unique parameter node of variable X_i that is reached by the path in G determined by configuration $\mathbf{X} = \mathbf{x}$. Then,

$$f_G(\mathbf{x}) = \prod_{i=1}^{n} f^{reach_G(X_i, \mathbf{x})}(\mathbf{x}[X_i]) = \prod_{i=1}^{n} f^{\nu_{\mathbf{x}[X_i]}}(\mathbf{x}[X_i])$$

$$= \prod_{i=1}^{n} f^{\nu_{\mathbf{x}[X_i]}}(\mathbf{x}[X_i]|\mathbf{x}[X_1, \ldots, X_{i-1}]) \tag{6}$$

where, according to Definition 4, $f^{\nu_{\mathbf{x}[X_i]}}$ is a conditional probability function of X_i given the configuration in the branch upwards the root. Furthermore, also according to Definition 4, $f^{\nu_{\mathbf{x}[X_i]}}$ is of class MoTBF. As the product of MoTBF functions is known to be an MoTBF as well (see [10]) we can conclude, by applying the chain rule, that the factorisation in (6) is a joint MoTBF distribution over $X_1, \ldots, X_n$, i.e. over $\mathbf{X}$. □

In the next section we will study the problem of supervised classification and how MoTBF-PDGs can be used in that context.

4   PDG Classifiers

A classification problem can be described in terms of a set of *feature variables* $\mathbf{X} = \{X_1, \ldots, X_n\}$, that describes an individual, and a *class variable*, C, that indicates the class to which that individual belongs. A *classifier* is a model oriented to predict the value of variable C given that the values of the features $\mathbf{X}$ are known. If the joint probability distribution of C and $\mathbf{X}$ is known, it can be used to solve the classification problem by assigning to any individual with observed features $x_1, \ldots, x_n$ the class c^* such that

$$c^* = \arg\max_{c \in R(C)} P(C = c | \mathbf{X} = x_1, \ldots, x_n). \tag{7}$$

By *supervised classification* we understand the problem of learning a classifier from a set of labeled examples, i.e., from a database with variables $X_1, \ldots, X_n, C$ where the value of C is known in all the records in the database.

Probabilistic graphical models, and more precisely Bayesian networks, have been used as classifiers, as they provide compact representations of joint probability distributions. Usually, the structure of the network is restricted in such a way that the class variable is set as root and the feature variables are connected to the class [5]. Similarly, PDGs have been used as classifiers by imposing certain structural restriction, ensuring that all the features are connected to the class. The formal definition is as follows.

Definition 5 (PDG Classifier [16]). *A PDG classifier C is a PDG model that, in addition to the structural constraints of Definition 1, satisfies the following two structural constraints:*

1. *G defines a forest containing a single tree over the variables* $\mathbf{C} = \{C\} \cup \mathbf{X}$,
2. *C is the root of this tree.*

In a PDG classifier, the forest is restricted to contain a single tree in order to guarantee that all the feature variables are connected to C by a path in G. By forcing C to be placed at the root, typical Bayesian network classifiers structures can be easily replicated, as for instance, the Naïve Bayes (NB) model.

Definition 6 (MoTBF-PDG Classifier). *An* MoTBF-PDG classifier $\mathcal{C}$ *is an MoTBF-PDG that satisfies the structural constraints in Definition 5.*

In a classification problem with class variable C and features $X_1, \ldots, X_n$ (discrete or continuous), if we denote by $\mathbf{C}$ the set $\{C, X_1, \ldots, X_n\}$, an MoTBF-PDG classifier G would be used to represent the joint distribution

$$f_G(\mathbf{c}) = f_G(c, x_1, \ldots, x_n).$$

According to (7), we need to determine the value

$$c^* = \arg\max_{c \in R(C)} f_G(c|x_1, \ldots, x_n) = \arg\max_{c \in R(C)} \frac{f_G(c, x_1, \ldots, x_n)}{\sum_{c \in R(C)} f_G(c, x_1, \ldots, x_n)}.$$

As $\sum_{c \in R(C)} f_G(c, x_1, \ldots, x_n)$ does not depend on c, solving the classification problem is equivalent to finding the value

$$c^* = \arg\max_{c \in R(C)} f_G(c, x_1, \ldots, x_n).$$

Hence, in order to classify an item with observed features $\mathbf{x} = (x_1, \ldots, x_n)$, we just have to compute, for each $c \in R(C)$, the value $f_G(c, x_1, \ldots, x_n)$, which amounts to evaluate the conditional MoTBF functions in the parameter nodes reached by $(c, x_1, \ldots, x_n)$ as described in (6).

4.1   Learning MoTBF-PDG Classifiers from Data

Learning an MoTBF-PDG classifier from data consists of determining the structure of the PDG and estimating the conditional MoTBF distributions in the parameter nodes. Assuming a fixed PDG structure (see Definition 1) the MoTBF probability function corresponding to each parameter node ν is estimated by first determining the elements in the data sample that reach ν, and then learning a univariate MoTBF using those data points following the method described in [11]. We refer the reader to that reference for further details on the estimation procedure for univariate MoTBF densities.

For determining the structure, we have considered three basic approaches:

1. Fix a *Naïve Bayes-like* structure, so that all the features are directly connected to the class variable. We denote this approach as NB. An example of a PDG with NB structure is depicted in Fig. 4.

2. Rank the feature variables according to their mutual information with the class variable and connect the variables conforming a chain rooted by the class variable followed by the features in a sequence according to their rank. The mutual information is estimated from data, discretising the continuous variables. We will refer to this approach by the term `ranked`. An example of a structure obtained in this way is found in the left panel of Fig. 5.

3. Rank the feature variables according to their mutual information with the class variable, and include the feature variables in the PDG one by one according to the rank (the class variable is always included on top of the structure). Unlike in the `ranked` approach, here each variable is inserted below any previously inserted variable. Among all possible insertion points, the one resulting in a better classification rate (CR) is chosen. The CR is computed in a validation set randomly drawn form the training database. In this paper, we have used an 80% of the training database for learning and a 20% for validation. We denote this approach by `rankedCR`. An example of an structure compatible with this approach is displayed in Fig. 5 (right).

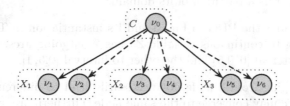

Fig. 4. A PDG structure compatible with the `NB` approach

During the process of inserting new variables in a PDG structure, it is necessary to decide the number of nodes to store at each variable, and the connections among them. In general, the maximum number of nodes at each variable depends on the number of nodes at its parent variable, and of the number of outgoing edge of each node at the parent.

Example 5. Consider the PDG at the right of Fig. 5. The maximum number of nodes at X_3 is 4, as its parent, X_2 has 2 nodes and each one has 2 outgoing arcs. However, in the PDG at the left of Fig. 5, the maximum number of nodes for X_3 is 8, as in this case X_2 has 4 nodes with 2 outgoing arcs each one. Note that, even though the maximum allowed is 8, in this example X_3 actually has 6 nodes, which indicates the presence of context specific independencies.

The number of outgoing arcs of a node corresponding to a discrete variable is equal to the number of possible values of the variable, which means that there are at least 2 outgoing arcs in such case. For a node corresponding to a continuous variable, the number of outgoing arcs may vary from 1 to any positive integer. In practice, it is necessary to establish a maximum number of arcs when learning

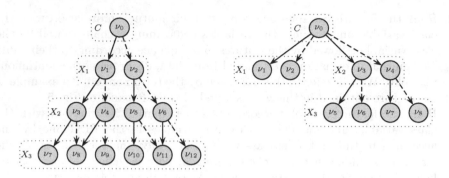

Fig. 5. Examples of PDG structures corresponding to the `ranked` (left) and `rankedCR` (right) approaches. It is assumed that the ranking of variables with respect to their mutual information with the class variable is X_1, X_2, X_3.

the PDG from data. Each outgoing arc corresponds to a subset of the domain of the continuous variable, so that the collection of the subsets associated with all the edges conforms a partition of its domain.

Example 6. Consider the PDG in Fig. 3 and its instantiation in Table 2. Node ν_1, corresponding to continuous variable Z_0 has 2 outgoing arcs. One of them corresponds to interval $[0, 0.5)$ and the other to interval $[0.5, 1]$.

The arcs emerging from a node not necessarily lead to different nodes, i.e., more than one arc may converge to the same node. Furthermore, arcs emerging from different nodes may converge to the same one. In this paper, we have considered a fixed number of intervals for every continuous variable, which is given as an argument to the classifier learning algorithm. The borders of the intervals are obtained following an equal frequency binning process.

Note that each node in a PDG represents a portion of the training data, namely those items that correspond to configurations that reach the parameter node. Therefore, as the PDG is expanded during its construction, the amount of data available for estimating the MoTBFs distributions in each node goes down. In order to avoid estimating the MoTBF functions from tiny samples, whenever a new variable is inserted during the process of constructing the PDG, we generate all the nodes for that variable and carry out a *collapse* operation [16], so that nodes that are not reached by a given minimum number of training items are collapsed. More precisely, the collapse operation involves the following steps:

1. Establish a *collapse threshold* $r_c > 0$.
2. Let $\nu_1, \ldots, \nu_k$ be the nodes for the current variable.
3. Let $size(\nu_i)$, $i = 1, \ldots, n$ denote the number of parameters estimated from data for node ν_i. This is the number of possible values of the variable minus 1, if the variable is discrete, and the parameters of the MoTBF density plus the number of intervals minus 1 if it is continuous.

4. Pairs of nodes (ν_i, ν_j) that are reached by a number of training items lower than $r_c \times size(\nu_i)$ and $r_c \times size(\nu_j)$ respectively, are chosen in increasing order of the previously mentioned thresholds and collapsed into a single new node, for which the corresponding probability function is re-estimated using the union of both training samples. When collapsing two nodes, the incoming and outgoing arcs are re-arranged appropriately.

The process is repeated until no nodes below the threshold remain, or until there is only one, in which case it is coupled with the smallest (in terms of $r_c \times size(\nu)$) parameter node for the same variable, and collapsed into a single one.

Once the MoTBF-PDG classifier has been constructed by any of the three approaches mentioned above, we carry out an operation aimed at reducing the chances of overfitting for the learnt model. The operation is called *merge*. It traverses the PDG structure bottom-up. For each variable, we explore a fraction of its parameter nodes, determined by a rate $r_m \in [0, 1]$ that we call *merge rate*. Then, the sampled parameter nodes are collapsed into a single one, similarly to the case of the collapse operation, but in this case *only if the classification rate*, computed from the validation set, *is increased* respect to the current model.

5   Experimental Evaluation

We have carried out an initial experimental evaluation aimed at evaluating the performance of MoTBF classifiers over a set of benchmark databases taken from the UCI (http://archive.ics.uci.edu/ml) and KEEL [1] repositories. A description of the datasets used in the experiments is given in Table 3.

In the experiments, we have induced MoTBF-PDG classifiers for each dataset using the three approaches explained in Sect. 4.1, i.e. NB, ranked and rankedCR, and several combinations of the parameters described there. More precisely, we have tested 2 and 3 intervals for the domain of the continuous parent variables, collapse thresholds of $r_c = 3, 5$ and 7, merge rates of $r_m = 0.25$ and 0.5. For the densities in the parameter nodes, we have used mixtures of polynomials (MOPs), which are one of the possible types of MoTBFs together with MTEs [10]. The polynomials have been learnt using the procedure described in [11] with limits on the degree of the polynomials equal to 4, 6, 8 and 10. We also tested discrete PDG classifiers as well as four Bayesian classifiers available in software Weka, namely Gaussian Naïve Bayes (called NB Simple in Weka), kernel NB, discrete NB and discrete TAN. We tested each algorithm measuring the classification rate (CR) using 5-fold cross validation. Our implementation of PDGs has been done using the R statistical package.

We found that the best combination of parameter values for the MoTBF-PDG classifiers was 3 intervals for the domain of the continuous parent variables, collapse threshold $r_c = 3$, merge rate $r_m = 0.25$, and maximum degree for polynomials equal to 6. We chose this combination of parameters by trying each possible combination of them an counting how many times each one was the winner in terms of classification rate of the constructed model. The results of

Table 3. Description of the datasets used in the experiments

	instances	#features	#continuous	#categorical	#classStates
appendicitis	106	7	7	0	2
banknote	1372	4	4	0	2
fourclass	862	2	2	0	2
haberman	306	3	3	0	2
iris	150	4	4	0	3
liver	345	6	6	0	2
newthyroid	215	5	5	0	3
phoneme	5404	5	5	0	2
pima	768	8	8	0	2
seeds	209	7	7	0	3
teaching	151	5	3	2	3
vertebral	309	6	6	0	2
wine	178	13	13	0	3

the experiments in terms of CR attained by each classifier are shown in Table 4. The sizes of the obtained models, measured as the number of parameters they contain, are displayed in Table 5. The values shown for MoTBF-PDG classifiers correspond to the configuration of parameters described above.

Table 4. Classification rates attained by the tested classifiers

Database	MoTBF-PDG			Discrete-PDG			Weka			
	NB	ranked	rankedCR	NB	ranked	rankedCR	Discrete-NB	Kernel-NB	Gaussian-NB	Discrete-TAN
appendicitis	0.8403	0.8307	0.8597	0.7935	0.8121	0.8117	0.8022	0.8771	0.868	0.8489
banknote	0.8586	0.9752	0.9738	0.8571	0.9425	0.9388	0.6348	0.9227	0.8397	0.9344
fourclass	0.7553	0.8365	0.8202	0.725	0.7982	0.7982	0.6439	0.8677	0.7506	0.8411
haberman	0.7482	0.7384	0.7287	0.706	0.7125	0.7222	0.7353	0.7418	0.7483	0.7255
iris	0.94	0.94	0.92	0.9333	0.92	0.9333	0.7867	0.9667	0.96	0.9267
liver	0.5768	0.5739	0.6232	0.658	0.5884	0.6	0.5797	0.658	0.5623	0.5768
newthyroid	0.907	0.893	0.8791	0.9023	0.8698	0.8884	0.707	0.9581	0.9674	0.9442
phoneme	0.779	0.7613	0.7846	0.7435	0.8116	0.795	0.7065	0.7841	0.7606	0.805
pima	0.7474	0.7045	0.7383	0.7318	0.7358	0.7267	0.651	0.7422	0.7591	0.7448
seeds	0.8949	0.8854	0.8854	0.8854	0.8707	0.8806	0.8801	0.8994	0.9138	0.8994
teaching	0.411	0.4239	0.4566	0.5101	0.5034	0.5166	0.4297	0.5428	0.5295	0.4503
vertebral	0.7411	0.738	0.8412	0.7572	0.7604	0.7605	0.6764	0.7668	0.7766	0.8057
wine	0.9611	0.8873	0.9611	0.9552	0.8817	0.9497	0.944	0.9775	0.966	0.9662

5.1 Discussion

In order to determine the significance of the results in Table 4, we run Friedman's test with maxT statistic [7], reporting significant differences among the tested classifiers (p-value below 0.05) in terms of accuracy (classification rate). Then we carried out a *post hoc* analysis following Wilcoxon-Nemenyi-McDonald-Thompson's procedure for pairwise comparisons. The result of the *post hoc* analysis is shown in Fig. 6, where a box plot for the differences in classification rate between every pair of algorithms is displayed. Green boxes are used to highlight the cases where statistically significant differences were found, which are, from left to right, discrete TAN vs. discrete NB, Gaussian NB vs. discrete NB, kernel NB vs. discrete NB, MoTBF-PDG (rankedCR) vs. discrete NB, kernel NB vs. discrete PDG (NB) and kernel NB vs. discrete PDG (ranked).

Table 5. Sizes of the learnt classifiers computed as the number of parameters they contain

Database	NB	MoTBF-PDG ranked	rankedCR	Discrete-PDG NB	ranked	rankedCR	Weka Discrete-NB	Kernel-NB	Gaussian-NB	Discrete-TAN
appendicitis	38.6	120	35	29	77.4	35.4	127	743	27	139
banknote	36	307.8	216.6	17	83.4	40.2	73	5489	19	79
fourclass	9	22	17.4	9	15.8	15.4	37	1725	9	39
haberman	19.6	37.4	21	13	42.6	14.6	55	919	13	59
iris	45.6	85.8	51.6	26	32.8	26	110	602	26	119
liver	56.2	275	117.6	25	165.8	63.4	109	2071	25	119
newthyroid	59.4	169.6	71.8	32	82	36.4	137	1077	32	149
phoneme	53.2	988.6	105.8	21	309.8	49	91	27021	21	99
pima	66.8	686.6	165.2	33	460.2	114.6	145	6145	29	159
seeds	80.6	192.4	100.6	44	90.8	47.6	191	1465	44	209
teaching	19.6	90.8	42.4	26	58.2	28.2	87	459	16	151
vertebral	50.6	242.2	151.6	25	128.6	56.6	109	1855	25	119
wine	146	381.4	189.2	80	238.4	95.2	353	2316	80	389

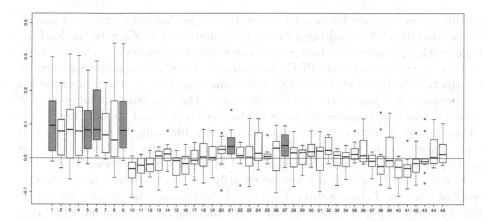

Fig. 6. Results of the pairwise comparison between the tested methods. Boxes in green indicate statistically significant differences

MoTBF-PDG classifiers are in general heavier in terms of parameters, as shown in Table 5, except kernel NB which always returns the largest model. However, it must be pointed out that we have not considered variable selection during the construction of any of the tested classifiers.

Even though defining a variable selection strategy for learning MoTBF-PDG classifiers is a matter of future research, we carried out a simple experiment in order to have a glimpse on the impact of variable selection both on size and accuracy of the learnt PDGs. The experiment consisted of inducing a classification tree and then learning the PDG but using only the variables actually included in the tree. Table 6 shows a comparison of the results with and without variable selection. The results suggest that variable selection can have a remarkable impact on the obtained models. Note that the model sizes dramatically decrease while the CR is not seriously deteriorated and even improved in some cases.

Table 6. Results of the preliminary experiment with variable selection. Columns Old CR, Old size, New CR and New size contain the results of MoTBF-PDG classifier with rankedCR strategy and the same configuration of parameters as in the main experiment, where 'old' indicates including all the variables and 'new' only with the same variables as in the induced classification tree. #Vars indicates the ratio of included variables.

Database	Old CR	New CR	Old size	New size	#Vars
appendicitis	0.8597	0.8403	35	13	2/7
haberman	0.7287	0.7318	21	18.6	2/3
iris	0.92	0.94	51.6	28.4	2/4
pima	0.7383	0.7435	165.2	107.4	6/8
seeds	0.8854	0.9139	100.6	53.6	4/7
wine	0.9611	0.9216	189.2	40.8	3/13

6   Concluding Remarks

In this paper we have introduced a new hybrid probabilistic graphical model, called MoTBF-PDG, resulting from the combination of PDGs with the MoTBF framework. We have done that within the context of supervised classification, motivated by the fact that PDGs had already been successfully employed as classifiers. The initial experimental evaluation suggests that the new classifiers are potentially competitive with existing ones. The analysis also suggests that variable selection is necessary in order to obtain compact models less prone to overfitting. A more extensive experimentation, including larger datasets and more algorithms as the one presented in [13] is necessary for determining the impact of the different parameters involved in the learning process.

Finally, MoTBF-PDGs are not necessarily models restricted to the framework of supervised classification. They could, for instance, be used for regression problems, where the variable to predict is continuous. As a general model for representing a joint probability distribution, the necessary operations for carrying out probabilistic inference still remain to be developed for MoTBF-PDGs.

Acknowledgments. This work has been supported by the Spanish Ministry of Economy and Competitiveness, grant TIN2010-20900-C04-02, by Junta de Andalucía grants P11-TIC-7821, P12-TIC-2541, and by ERDF (FEDER) funds.

References

[1] Alcalá-Fdez, J., Fernández, A., Luengo, J., Derrac, J., García, S., Sánchez, L., Herrera, F.: KEEL data-mining software tool: Data set repository, integration of algorithms and experimental analysis framework. J. Mult.-Valued Log. Soft Comput. 17, 255–287 (2011)

[2] Bozga, M., Maler, O.: On the Representation of Probabilities over Structured Domains. In: Halbwachs, N., Peled, D.A. (eds.) CAV 1999. LNCS, vol. 1633, pp. 261–273. Springer, Heidelberg (1999)

[3] Cobb, B.R., Shenoy, P.P.: Inference in Hybrid Bayesian Networks with Mixtures of Truncated Exponentials. Int. J. Approximate Reasoning 41, 257–286 (2006)

[4] Flores, M.J., Gámez, J.A., Nielsen, J.D.: The PDG-mixture Model for Clustering. In: 11th Int. Conf. on Data Warehousing & Knowledge Discovery, pp. 378–389 (2009)

[5] Friedman, N., Geiger, D., Goldszmidt, M.: Bayesian Network Classifiers. Machine Learning 29, 131–163 (1997)

[6] Gámez, J.A., Nielsen, J.D., Salmerón, A.: Modelling and Inference with Conditional Gaussian Probabilistic Decision Graphs. Int. J. Approximate Reasoning 53, 929–945 (2012)

[7] Horthorn, T., Hornik, K., van de Wiel, M.A., Zeileis, A.: Implementing a Class of Permutation Tests: The coin Package. J. Stat. Soft. 28, 1–23 (2008)

[8] Jaeger, M.: Probabilistic Decision Graphs - Combining Verification and AI Techniques for Probabilistic Inference. Int. J. Uncertainty Fuzziness Knowledge Based Syst. 12, 19–42 (2004)

[9] Langseth, H., Nielsen, T.D., Rumí, R., Salmerón, A.: Parameter Estimation and Model Selection for Mixtures of Truncated Exponentials. Int. J. Approximate Reasoning 51, 485–498 (2010)

[10] Langseth, H., Nielsen, T.D., Rumí, R., Salmerón, A.: Mixtures of Truncated Basis Functions. Int. J. Approximate Reasoning 53, 212–227 (2012)

[11] Langseth, H., Nielsen, T.D., Pérez-Bernabé, I., Salmerón, A.: Learning mixtures of truncated basis functions from data. Int. J. Approximate Reasoning 55, 940–956 (2014)

[12] Lauritzen, S.L., Wermuth, N.: Graphical Models For Associations Between Variables, Some of Which Are Qualitative and Some Quantitative. The Annals of Statistics 17, 31–57 (1989)

[13] López-Cruz, P.L., Bielza, C., Larrañaga, P.: Learning mixtures of polynomials of multidimensional probability densities from data using B-spline interpolation. Int. J. Approximate Reasoning 55, 989–1010 (2014)

[14] Moral, S., Rumí, R., Salmerón, A.: Mixtures of Truncated Exponentials in Hybrid Bayesian Networks. In: Benferhat, S., Besnard, P. (eds.) ECSQARU 2001. LNCS (LNAI), vol. 2143, pp. 135–143. Springer, Heidelberg (2001)

[15] Nielsen, J.D., Jaeger, M.: An Empirical Study of Efficiency and Accuracy of Probabilistic Graphical Models. In: Third European Workshop on Probabilistic Graphical Models, pp. 215–222 (2006)

[16] Nielsen, J.D., Rumí, R., Salmerón, A.: Supervised Classification Using Probabilistic Decision Graphs. Comput. Stat. Data Anal. 53, 1299–1311 (2009)

[17] Romero, V., Rumí, R., Salmerón, A.: Learning Hybrid Bayesian Networks Using Mixtures of Truncated Exponentials. Int. J. Approximate Reasoning 42, 54–68 (2006)

[18] Shenoy, P., West, J.: Inference in Hybrid Bayesian Networks Using Mixtures of Polynomials. Int. J. Approximate Reasoning 52, 641–657 (2011)

Towards a Bayesian Decision Theoretic Analysis of Contextual Effect Modifiers

Gabor Hullam[1,2] and Peter Antal[1]

[1] Department of Measurement and Information Systems,
Budapest University of Technology and Economics,
Magyar tudosok krt. 2, 1117 Budapest, Hungary
[2] MTA-SE Neuropsychopharmacology and Neurochemistry
Research Group, Hungarian Academy of Sciences,
Nagyvarad ter 4, 1089 Budapest, Hungary
{gabor.hullam,peter.antal}@mit.bme.hu

Abstract. Relevance measures based on parametric and structural properties of Bayesian networks can be utilized to characterize predictors and their interactions. The advantage of the Bayesian framework is that it allows a detailed view of parametric and structural aspects of relevance for domain experts. We discuss two particularly challenging scenarios from psycho-genetic studies, (1) the analysis of weak effects, and (2) the analysis of contextual relevance, where a factor has a negligible main effect and it modifies an effect of another factor only in a given subpopulation. To cope with this challenge, we investigate the formalization of expert intuitions and preferences from the exploratory data analysis phase. We propose formal losses for these two scenarios. We introduce and evaluate a Bayesian effect size measure using an artificial data set related to a genetic association study, and real data from a psycho-genetic study.

Keywords: Bayesian association measures, effect size, contextual relevance, decision theoretic discovery.

1 Introduction

Probabilistic graphical models, especially Bayesian networks, are intensively used in the field of genomics and biomedicine as they allow the modeling of complex dependency structures between various clinical, genetic and environmental factors [20]. The capability of modeling complex dependency relationships is vital for understanding the mechanisms of multifactorial diseases (e.g. depression, asthma). In recent years, genetic factors related to such illnesses were investigated by genome-wide association studies (GWAS) and candidate gene association studies (CGAS) [19]. It became apparent, that standard statistical analyses relying on the hypothesis testing framework posed considerable limitations that hindered the interpretation and communication of results. These phenomenon induced an extensive research for applicable univariate Bayesian methods [25]. Furthermore, Bayesian network based multivariate methods also attracted much attention, as they perform normative correction for multiple hypothesis testing [21],[24],[28].

L.C. van der Gaag and A.J. Feelders (Eds.): PGM 2014, LNAI 8754, pp. 222–237, 2014.

Bayesian network based methods not only allow the detection of possibly complex dependency relationships, but also provide a rich tool set for the detailed characterization of associations [3]. Various association measures can be derived based on parametric and structural properties of Bayesian networks. Several methods focus on the learning of structural properties (or the whole structure) from data, which provides valuable information on the relationships between factors [8]. The main goal of such structure based learning methods is to enable the selection of relevant factors with respect to a specified target. However parametric properties, such as effect size, are typically neglected in those contexts. On the other hand, there are other methods which investigate the parametric aspect of dependency relationships. For example, effect size measures can be used to characterize the nature of associations, i.e. whether a factor increases the risk of a disease or it has a protective effect [4]. Both the structural and the parametric aspects appear to be relevant, and provide a distinct view on dependency relationships [13].

In domains with complex dependency patterns the analysis of factors involving effect modifiers poses a further challenge. Let us consider an example from a psycho-genetic study, in which the effect of genetic factors was assessed with respect to depression [15]. Figure 1.A shows the distribution of a Bayesian association measure (odds ratio, see section 3) between a single nucleotide polymorphism (SNP) and depression based on the whole data set. As the distribution is centered around the neutral effect size of 1 (i.e. no effect) one might conclude that the factor is non-relevant. However, Figure 1.B shows that in case we consider an environmental factor related to recent negative life events (RLE), and investigate the subpopulations defined by its subtypes (RLE-0: 0 or 1 RLE, RLE-1: 2 RLEs, RLE-2: 3 or more RLEs), then in one subpopulation (RLE-2) a relevant effect (odds ratio above 2) can be observed.

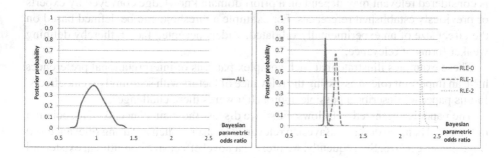

Fig. 1. Posterior distribution of Bayesian parametric odds ratio for a selected SNP based on the whole (ALL) data set (left) and on subpopulations (right) defined by recent negative life events (RLE) subtypes.

Figure 2 shows a more complex example from the same study concerning the effect of a SNP on depression involving multiple environmental factors. In case of certain configurations of environmental factors relevant odds ratios can be observed for patients with age lower than 30 (Figure 2.A), whereas in case of patients above 30 most odds ratios are negligible (Figure 2.B). Without preferences provided by an expert, the investigation of such scenarios can be overlooked and not reported. Such cases motivate the application of a decision theoretic framework based on formalized loss functions (see section 4).

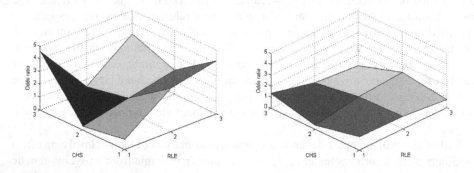

Fig. 2. Odds ratios for an effect modifier genetic factor with respect to depression based on subpopulations according to the subtypes of environmental factors with strong main effects. *RLE* denotes recent negative life events and *CHA* denotes childhood adversity. Odds ratios are displayed separately for patients below age 30 (left) and above (right).

Even though Bayesian network based structural relevance measures such as strong relevance (see section 3) quantify the relevance of factors, they do not provide insight on effect size, which is generally required by experts. Furthermore, whether an effect size is considered relevant may depend on a priori domain knowledge conveyed by experts or previously established references. For example a threshold can be defined based on the effect size of an experimentally validated, widely accepted factor, thereby defining weaker forms of relevance.

These examples illustrate that such complex patterns of interpretational preferences have an important role in assessing the relevance of factors with seemingly weak effects. In this paper we describe a possible approach towards these challenges.

The paper is organized as follows: First, we discuss the main concepts of association measures. Then, we describe Bayesian relevance measures such as strong relevance and Bayesian effect size. Subsequently, we present a novel effect size conditional existential relevance measure (ECER). We also provide interpretation, reporting and publication preferences using loss functions and outline a corresponding decision theoretic framework. In summary, the paper demonstrates the application of Bayesian relevance measures on an artificial data set related to a genetic association study, and on real data from a psycho-genetic study.

2   Association Measures

Association measures are the basic building blocks of scientific investigation aiming to reveal dependency relationships between measured factors. Some provide an existential statement, that a dependency relationship exists between two (or more) factors, while others quantitatively describe the nature and the strength of the effect of one factor on another. We can distinguish at least three main approaches: parametric, existential (structural), and causal approaches.

2.1 Effect Size

The most frequently used parametric approach utilizes effect size measures that do not take structural aspects of dependency relationships into account. Thus multivariate relationships are also neglected. In several fields, such as biomedicine and genetics, the most frequently used effect size measure is the odds ratio (OR) [4]. In the frequentist framework an odds is defined as the ratio of conditional probabilities, which can be used to compute the odds ratio.

Definition 1 (Odds). *Let $X_1, X_2, ..., X_n$ denote discrete variables that have $r_1, r_2, ..., r_n$ states respectively, and let Y denote the target variable with $y_1, ..., y_q$ possible states. Then $X_i = x_{i_1}$ denotes variable X_i in state x_{i_1}, and an* odds *is defined as*

$$\text{Odds}(X_i^{(j)}, Y^{(m,n)}) = \frac{p(Y = y_m | X_i = x_{i_j})}{p(Y = y_n | X_i = x_{i_j})}. \tag{1}$$

Definition 2 (Odds ratio). *An* odds-ratio *for variable X_i between states x_{i_j} and x_{i_k} is given as*

$$\text{OR}(X_i^{(k,j)}, Y^{(m,n)}) = \frac{Odds(X_i^{(k)}, Y^{(m,n)})}{Odds(X_i^{(j)}, Y^{(m,n)})} \tag{2}$$

If the target variable Y is binary then it typically serves as a disease indicator such that $Y = 0$: non-affected (control), $Y = 1$: affected (case). Assuming that variables $X_1, X_2, ..., X_n$ represent single nucleotide polymorphisms (SNP), then $r_1, r_2, ..., r_n$ encode states $\{0, 1, 2\}$ that refer to common(wild) homozygote, heterozygote, rare (mutant) homozygote genotypes respectively. In that case $\text{OR}(X_i^{(1,0)})$ denotes an odds ratio of heterozygous (1) versus common homozygous (0) genotypes of SNP X_i.

2.2 Strong Relevance

The basic concept of the existential approach is structural uncertainty, that is whether a dependency relationship exists between two factors. Bayesian networks are used extensively in this approach, which allow a detailed analysis of relationships, such as whether a relationship is direct or it is mediated by other factors. Structural uncertainty is described by a probability measure based on the data. As in most cases the aim of these methods is to identify relevant factors with respect to a target, the probability measure is related to (or conditioned on) the selected target.

The underlying concept of structural relevance can be defined formally in multiple ways. On one hand, it can be stated using conditional probability distributions without being specific to the applied model class used as a predictor, the optimization algorithm, the data set, and the loss function [16].

Definition 3 (Strong and weak relevance). *A feature X_i is strongly relevant to Y, if there exist some $X_i = x_i, Y = y$ and $s_i = x_1, \ldots, x_{i-1}, x_{i+1}, \ldots, x_n$ for which $p(x_i, s_i) > 0$ such that $p(y|x_i, s_i) \neq p(y|s_i)$. A feature X_i is weakly relevant, if it is not strongly relevant, and there exists a subset of features S_i' of S_i for which there exist some x_i, y and s_i' for which $p(x_i, s_i') > 0$ such that $p(y|x_i, s_i') \neq p(y|s_i')$. A feature is relevant, if it is either weakly or strongly relevant; otherwise it is irrelevant.*

Assuming a Bayesian framework, a high posterior probability for strong relevance indicates a close structural connection between X_i and Y [8].

Note that strong relevance does not imply a large effect size, i.e. parametric relevance, and vice versa. A high posterior for the strong relevance of X_i with respect to Y is not necessarily accompanied by a high odds ratio, as it is possible that X_i is only an interaction term and only has a joint effect with another factor X_j on Y. In such a case the individual odds ratio of X_i can be close to 1 [7]. Similarly, a high odds ratio does not entail strong relevance, as this effect size measure does not impose strong criteria on an underlying (Bayesian network) structure; in fact the exact structural relation between X_i and Y is unknown. For example, a strong effect can be transitive, i.e. the effect of X_i on Y is mediated by several other factors, which means that X_i then cannot be strongly relevant (but it can be weakly relevant).

In summary, the structural and the parametric aspects are separate dimensions of relevance [13].

3 Bayesian Network Based Association Measures

Bayesian networks provide a graph based language for encoding relevance and representing dependency relationships. The advantage of a Bayesian network based framework is that both parametric and structural association measures can be investigated. The structural aspect is closely related to the learning of structural properties of Bayesian networks, whereas the parametric aspect is connected to the parameter prior and the parameterization learned from the data.

In case of structure learning of a Bayesian network from data, the identification of model properties is essential, since in realistic cases the identification of the whole network is typically not possible from a statistical point of view [9]. Instead, learning smaller substructures or structural properties is a possible solution [5],[6],[17],[18],[23].

Previously, we investigated various structural properties of Bayesian networks related to relevance, and we demonstrated the Bayesian application (i.e. in a Bayesian statistical framework) of Bayesian networks in relevance analysis [1]. Subsequently, we proposed a Bayesian network based Bayesian multilevel analysis of relevance (BN-BMLA) [2], and carried out a comparative study of BN-BMLA against other methods in [14]. Then we applied the BN-BMLA method in a candidate gene association study of asthma [27].

3.1 Strong Relevance in Bayesian Networks

The Bayesian interpretation of strong relevance involves Markov blanket sets, which are special structural properties of Bayesian networks. A model-based, probabilistic definition of Markov blankets can be stated as [22]:

Definition 4 (Markov blanket). *A set of variables* $\mathbf{X}' \subseteq V$ *is called a* Markov blanket *set of* X_i *with respect to the distribution* $p(V)$, *if* $(X_i \perp\!\!\!\perp V \setminus \mathbf{X}'|\mathbf{X}')_p$, *where* $\perp\!\!\!\perp$ *denotes conditional independence.*

In [26] the conditions for the unambiguous Bayesian network representation of the relevant structural properties were derived. Based on these results, given a distribution p defined by a Bayesian network(G,θ) we refer to the Markov blanket set of Y as MBS(Y,G) by the implicit assumption that p is Markov compatible with G (i.e. each variable in p is conditionally independent of the set of all its non-descendants given the set of all its parents).

This means, that for a given G all the strongly relevant variables X_j with respect to Y are in MBS(Y,G). In other words, MBS(Y,G) is a strongly relevant set of variables upon which a pairwise relation can be defined.

Definition 5 (Markov blanket membership). *The pairwise relation* MBM(X_j,Y) *indicating whether* X_j *is a member of* MBS(Y,G) *is called Markov blanket membership.*

The posterior probability for any structural property $p(f|D)$, e.g. the MBM posterior $p(MBM(X_i,Y))$, can be computed by evaluating the corresponding expression $f(G)$ for all possible structures G given data set D as

$$p(f|D) = \sum_G p(G|D)f(G). \tag{3}$$

However, since averaging over all possible structures is generally intractable, an alternative estimation method is required. For this purpose we used a Markov chain Monte Carlo (MCMC) method in order to facilitate a random walk in the space of directed acyclic graph structures by applying operators for inserting, deleting, and inverting edges [2]. The probability of applying these operators in the proposal distribution was uniform [11]. In each MCMC step, the Markov blanket set (with respect to a selected target) corresponding to the directed acyclic graph in the current step is determined and the relative frequency of this Markov blanket set is updated. In the final step, a normalized posterior is produced for each Markov blanket set. An MBM posterior $p(MBM(X_i,Y))$ is computed for each X_i by summing the posteriors of those Markov blanket sets that contained X_i.

3.2 Bayesian Effect Size

A possible Bayesian approach to effect size estimation is to utilize the underlying BN(G,θ) that is a graph structure G and its parametrization θ for odds ratio computation OR$(X_i,Y|\theta,G)$. This OR can be treated as a random variable induced by the distribution $p(\Theta|G,D)$, where D denotes the data set, and Θ is a random variable of possible parameterizations. However, properties such as mean and credible interval for the posterior of odds ratios cannot be analytically derived, thus they have to be estimated.

In previous works, we utilized the strong relevance property of Markov blankets to guide the estimation process, and created a structure conditional Bayesian effect size measure [12,13]. This hybrid effect size measure encompasses both the structural and the parametric aspects of relevance.

Following the Bayesian univariate association analysis framework, here we investigate a purely parametric Bayesian effect size measure in which the parametrization θ_k is sampled from a Dirichlet distribution $Dir(\mathbf{U_n}, \alpha'_n)$, defined by the applied Bayesian Dirichlet prior

$$Dir(U_1, ..., U_{n-1}, \alpha_1, ..., \alpha_n) = \frac{1}{\beta(\alpha)} \cdot \prod_{i=1}^{n} U_i^{\alpha_i - 1}, \tag{4}$$

where $U_1, ..., U_n$ denote discrete variables defined by data D, and $\alpha_1, ..., \alpha_n$ denote corresponding hyperparameters. The sampling is performed based on the posterior $Dir(\mathbf{U_n}, \alpha'_n)$ using updated hyperparameters $\alpha'_i = \alpha_i + N$, where N is the sample size. Note that the Bayesian Dirichlet prior is also used for structure learning to estimate the posterior probability $p(G|D)$. Based on θ_k odds ratios are then estimated as

$$\hat{OR}(X_i, Y|D) = \frac{1}{q} \sum_{k=1}^{q} OR(X_i, Y|\theta_k) \tag{5}$$

where q is the number of samples taken from $Dir(\mathbf{U_n}, \alpha'_n)$. Apart from computing the effect size estimate, the distribution $p(OR(X_i, Y|\Theta))$ can be used to define a credible interval $(CR_{0.9})$, which satisfies $p(OR(X_i, Y|\Theta) \in CR) = 0.9$ [10]. The credible interval is essentially the Bayesian equivalent of a confidence interval in standard (frequentist) statistics. Later on we assume that the credible interval is the smallest such interval, i.e. a high probability density region.

Figure 3 shows the posterior distribution of Bayesian odds ratios for a selected variable from the investigated artificial data set (for details see Section 5) in case of various sample sizes. Each curve corresponds to the histogram of a posterior distribution, and its endpoints mark the borders of the related credible interval. The effect of growing sample size, i.e. more evidence, can be clearly seen on the Bayesian odds ratio, as the credible interval becomes smaller and the distribution becomes more peaked.

3.3 Effect Size Conditional Existential Relevance

The distribution $p(OR(X_i, Y|\Theta))$ can also be used to devise an existential relevance measure which is based on effect size parameters and reflects experts' preferences. Note that this is an opposite approach compared to our previous hybrid association measure [12].

This measure can be formalized by defining an interval of negligible effect size C_ε around the neutral odds ratio of 1, e.g. in a symmetric case $\varepsilon = 0.2$ means that $OR \in C_\varepsilon$ if $0.9 < OR < 1.1$. Note that this corresponds to an ε-insensitive $0-1$ loss-function. If the credible interval for $OR(X_i, Y|\Theta)$ intersects with interval C_ε then we can state that based on the effect size distribution $p(OR(X_i, Y|\Theta))$ the variable X_i is partially non-relevant in terms of parametric relevance. More specifically the distribution mass in C_ε quantifies the parametric irrelevance of variable X_i. Straightforwardly, the greater the mass of $p(OR(X_i, Y|\Theta) \in C_\varepsilon)$ the less parametrically relevant X_i is.

If the credible interval of $p(OR(X_i, Y|\Theta))$ does not intersect with C_ε that means that the effect size distribution of X_i only contains non-negligible values, that is X_i is parametrically relevant. Another interpretation is that there is no parametrically encoded

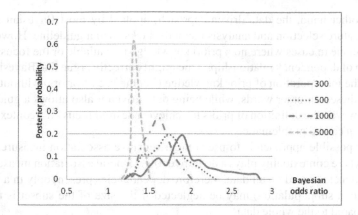

Fig. 3. Posterior distribution of Bayesian parametric odds ratio for a selected variable in case of various sample sizes. The horizontal axis displays Bayesian parametric odds ratio, whereas the vertical axis displays posterior probabilities.

independence between X_i and Y. This means that X_i and Y are dependent not only on the parametric level, but also on the structural level. In other words, based on a statement of parametric relevance we can form a statement for structural relevance.

Definition 6 (Effect size conditional existential relevance - ECER). *Given an interval of negligible effect size C_ε with size $\varepsilon \geq 0$, and a distribution of effect size $p(OR(X_i,Y|\Theta)$ for variable X_i with respect to a selected Y, let $I_{\{ECER_\varepsilon(X_i,Y)\}}$ denote that $OR(X_i,Y|\Theta) \notin C_\varepsilon$ that is the effect size of variable X_i is relevant as it is outside the C_ε interval. In the Bayesian framework the posterior of effect size conditional existential relevance $ECER_\varepsilon(X_i,Y)$ can be defined as $p(I_{\{ECER_\varepsilon(X_i,Y)\}})$.*

Note that the optimal selection of ε is problem specific, and depends on multiple parameters such as sample size and effect strength. We evaluated the performance of ECER on an artificial data set detailed in section 5. This measure also ensures comparability with earlier structural posteriors, such as the MBM posterior $p(MBM(X_i,Y))$.

4   Contextual Relevance and Effect Modifiers

The hypothesis driven approach of traditional frequentist statistical methods has several weaknesses, from which the multiple hypothesis testing problem is the most severe. However when sufficient a priori knowledge is present that can be formed into consistent hypotheses, then having such a focus can remarkably enhance data analysis efforts. In biomedical domains a viable hypothesis can not be neglected as it may have a considerable impact on the success of the data analysis. For example in case of gene-environment interactions, the knowledge of relevant environmental factors, and more precisely the relevant environmental state that is known to modify the effect of genetic factors may allow the detection of otherwise negligible associations.

On the other hand, the data driven approach, applied by most Bayesian and non-Bayesian feature selection and analysis methods, lacks such a guideline. However, it is non-replaceable in cases where no a priori knowledge is available or the focus is on the exploration of dependency relationships. One might correctly argue that Bayesian methods allow the incorporation of prior knowledge (i.e. the use of priors) without defining exact hypotheses. In other words, while being data driven it also allows a guideline. In practice however, the definition of priors to achieve a desired focus on a context specific phenomenon can be problematic.

Another possible approach is to use context sensitive association measures in such scenarios where contextuality plays a definitive role. General association measures evaluate dependencies based on all data, therefore dependencies present only in a subset of the data (i.e. a subpopulation) may be neglected if the size of the subset is relatively small compared to the whole data.

The concept of having a context $C = c$ in which a variable X_i is independent from the target Y can be described formally by contextual irrelevance [2].

Definition 7 (Contextual Irrelevance). *Assume that* $X' = X_i \cup C$ *is relevant for* Y, *that is* $(Y \not\perp\!\!\!\perp (X_i \cup C))$, *and* $X_i \cap C) = \emptyset$). *We say that* X_i *is contextually irrelevant if there exists some* $C = c$ *for which* $(Y \perp\!\!\!\perp X_i | c)$.

Note that this definition is given from a conditional independence perspective. However, in genetic association studies we are typically interested in the complementary case, that is whether there exists such a context $C' = c'$ in which variable X_i is not independent from target Y. More specifically, given r possible value configurations $c'_1, c'_2, \ldots, c'_r$ of context C' if there is at least one c'^* in which X_i is not independent from target Y then X_i can be considered as contextually relevant. The biomedical motivation for such an approach is that even if X_i is not relevant with respect to Y given the whole data D, if there is a subset $D' \subset D$ for which it is relevant then it can be further investigated in a more focused study.

Related research preferences can be formalized in the Bayesian decision theoretic framework.

4.1   Bayesian Decision Theoretic Approach for Research Preferences

The practical scenarios described in section 1 can be summarized as a clear need for concepts and tools to explore and report weak dependencies, particularly effect modifying contextual dependencies. Bayesian decision theory provides an appropriate framework to formalize these intuitive expectations using informative losses. The main reason to apply such a framework is that some aspects of background knowledge cannot be expressed (or are difficult to express) as priors because they belong to a different phase of scientific discovery (e.g. reporting instead of exploratory analysis). For example the effect size of an experimentally validated genetic factor cannot be directly translated into a parameter prior, but it can be utilized as an overall expectation regarding relevant effect sizes. Loss functions allow an alternate way beyond priors to incorporate such a priori knowledge.

In case of parameter estimation, $L(\hat{\theta}|\theta)$ denotes the loss function for selecting (e.g. reporting) $\hat{\theta}$ instead of θ. Let D denote observations, and $p(\theta)$ the prior, then the optimal estimate $\hat{\theta}$ minimizes the posterior expected loss

$$\rho(p(\theta),\hat{\theta}|D) = \int L(\hat{\theta}|\theta)p(\theta|D)\,d\theta. \qquad (6)$$

This framework allows a detailed, expert-driven analysis of weak relevances using informative losses, which are demonstrated in the following four examples. For ease of explanation, let us assume that the parameters θ_i correspond to some effect size measure for predictors X_i, e.g. odds ratios for a binary outcome. In case of a well-known threshold τ corresponding to a validated factor with a weak effect, a threshold loss function can exactly quantify whether θ_i is less than this reference:

$$L(a_\tau|\theta_i) = \begin{cases} 0 \text{ if } \theta_i < \tau \\ \text{else } 1. \end{cases}$$

where a_τ denotes the action of reporting that the parameter is below the threshold τ. If this reference X_r with parameter θ_r can be included in the analysis, and interactions between this reference and other factors are neglected in a multivariate model, then a comparative loss can also quantify that a candidate is weaker than the reference:

$$L(a_{X_r \prec X_i}|\theta_i, \theta_r) = \begin{cases} 0 \text{ if } \theta_i < \theta_r \\ \text{else } 1. \end{cases}$$

The *effect size conditional existential relevance* can also be seen as an expected loss for reporting independence (i.e. for an action that $\hat{\theta} = 1$):

$$L_\varepsilon(a_{\hat{\theta}=1}|\theta_i) = \begin{cases} 0 \text{ if } |\hat{\theta} - \theta_i| < \varepsilon \\ \text{else } 1. \end{cases}$$

Note that this loss function can also be interpreted as an expected utility (confirmation) for reporting dependence.

Furthermore, loss functions can be also used to represent the preference to highlight special weak dependencies, e.g. contextual effect modifiers described in the second data exploratory scenario in section 1. Losses for contextual effect modifiers on the one hand should allow the specification of domain specific relevance from experts and on the other hand it should be neutral for independence in certain stratas. Let as assume that Y denotes the outcome, X denotes a predictor with strong main effect (i.e. $(Y \not\perp X)$), e.g. X is a well-known environmental variable , and Z denotes a contextual effect modifier, e.g. a weak genetic factor. For simplicity, we also assume the existence of a weak main effect for Z, i.e. $(Y \not\perp Z)$, but it is not necessary [7]. Furthermore, we assume the contextual independence of Z, i.e. that there exist(s) some value(s) x_i that $(Y \perp Z|X = x_i)$. In this case the loss of reporting dependence between Y and Z could express the dominance of the losses in those strata with dependence. This scenario is even more realistic, if X is an exogenous variable with a varying distribution under different circumstances or if its distribution depends from study design. In this case, for example, a contextual loss function can be defined as follows:

$$L(a_\tau|\theta_Z) = \min_i L(a_\tau|\theta_Z, x_i),$$

where $L(a_\tau|\theta_Z, x_i)$ denotes the threshold loss under the condition $X_i = x_i$.

From a practical point of view, the presented informative loss functions can be used to define the C_ε interval of negligible effect size. This allows the ECER measure to express experts' knowledge in the form of τ thresholds that could not be applied directly otherwise. In case there is no a priori knowledge, the selection of C_ε is more challenging, although in most cases there is at least an expected effect size related to the investigated domain. A possible solution is to use the effect size of relevant variables identified by auxiliary measures e.g. strong relevance.

Concerning the decision that a factor is existentially relevant based on its effect size the selection of the C_ε interval is the primary task. Defining the decision threshold (i.e. the required proportion of posterior probability distribution outside the C_ε interval) is only secondary. The reason behind this is that selecting a C_ε interval defines the relevant effect sizes (i.e. those that are outside C_ε). That is if related a priori knowledge is available then it can be directly applied in interval selection, thus rendering the selection of a decision threshold (other than e.g. 0.9) an additional option.

Note, that it is possible to use multiple references X_r within the same analysis. For example it is plausible to use two distinct X_r for protective and risk increasing factors. In such cases a combination of loss functions can be applied (e.g. using various weights or logic functions).

5 Results

We compared the discussed relevance measures on an artificial data set containing 115 variables and 5,000 samples. The data set was generated based on a model learned from real-world data of a candidate gene association study of asthma [14]. Data sets of various sizes (300, 500, 1000, 5000) were created by truncating the original data set of 5,000 samples. Strong relevance posteriors $MBM(X_i, Y)$ were estimated with the BN-BMLA method [2].

We investigated the performance of ECER with respect to the a priori known reference set of relevant variables. Standard performance measures of accuracy, sensitivity and specificity are shown in Table 1. The posterior of ECER was investigated for three different intervals of negligible effect size $C_{\varepsilon_1} : (0.9 - 0.1), C_{\varepsilon_2} : (0.66 - 1.5)$, and $C_{\varepsilon_3} : (0.5 - 2.0)$ in case of data sets of various sample sizes (300, 500, 1000, 5000). A variable X_i is only considered ECER relevant if it is true that $p(\text{ECER}_\varepsilon(X_i, Y)) > 0.95$, which means that the mass of its effect size distribution within C_ε is negligible.

Results indicate a high specificity (> 0.9) of ECER in case of C_{ε_2} and C_{ε_3} for all sample sizes, whereas the sensitivity is low for C_{ε_2} and even lower for C_{ε_3}. In contrast, the sensitivity related to C_{ε_1} is relatively higher for all sample sizes, though it comes at the price of a relatively lower specificity.

The reason of low sensitivity scores is that in the reference set a large portion of relevant variables are interaction terms, i.e. they do not have an individual effect, instead they have a joint effect on the target together with additional variables. Although interaction terms are strongly relevant (by definition), they are theoretically not detectable

Table 1. The performance of ECER for various sample sizes (Size) and for multiple negligible effect size intervals (C_{ε}). $C_{\varepsilon_1}, C_{\varepsilon_2}, C_{\varepsilon_3}$ denote negligible effect size intervals (0.9-0.1), (0.66-1.5), and (0.5-2.0) respectively. For each case the sensitivity (SENS), specificity(SPEC) and accuracy(ACC) measures are displayed. Partial measures (SENS*, SPEC*, ACC*) are also displayed in which interaction terms are not considered as relevant.

Size	C_{ε}	SENS	SPEC	ACC	SENS*	SPEC*	ACC*
300	C_{ε_1}	0.38	0.56	0.53	0.71	0.68	0.59
	C_{ε_2}	0.31	0.91	0.83	0.71	0.92	0.90
	C_{ε_3}	0.25	1	0.90	0.57	1	0.97
500	C_{ε_1}	0.56	0.51	0.51	0.86	0.52	0.54
	C_{ε_2}	0.31	0.96	0.87	0.71	0.97	0.95
	C_{ε_3}	0.25	1	0.90	0.57	1	0.97
1000	C_{ε_1}	0.44	0.65	0.62	0.86	0.67	0.68
	C_{ε_2}	0.31	0.97	0.88	0.71	0.97	0.96
	C_{ε_3}	0.25	0.99	0.89	0.57	0.99	0.97
5000	C_{ε_1}	0.50	0.92	0.86	0.86	0.91	0.90
	C_{ε_2}	0.44	1	0.92	0.71	0.98	0.97
	C_{ε_3}	0.31	1	0.90	0.57	0.99	0.97

by univariate measures, such as ECER. In contrast, $MBM(X_i, Y)$ is a univariate aggregate of a multivariate relevance measure $MBS(Y, G)$, and as such it correctly detects interaction terms. Therefore, a set of partial measures are also shown in table 1, where interaction terms are ignored.

The partial sensitivity measures show that ECER correctly detects the majority of relevant variables in case of C_{ε_1} and C_{ε_2}. In case of C_{ε_3} the sensitivity is still considerably lower, which indicates that interval $C_{\varepsilon_3} : (0.5 - 2.0)$ is possibly too large, and prohibits the detection of relevant variables. However, due to the high specificity related to C_{ε_3}, in terms of overall accuracy it remains a possible choice for some data sets. Based on the results $C_{\varepsilon_2} : (0.66 - 1.5)$ has the best trade-off between sensitivity and specificity.

We also compared the ECER posteriors with MBM posteriors to investigate the relationship between Bayesian network based 'pure' structural relevance and the effect size based structural relevance. The former relies on structural properties, whereas the latter is based on parametric relevance upon which a statement on structural relevance is formed. Figures 4 and 5 provide an overview on these relevance measures for negligible effect size intervals $C_{\varepsilon_2} : (0.66 - 1.5)$ and $C_{\varepsilon_3} : (0.5 - 2.0)$ in case of data sets with 500 and 5000 samples. Strongly relevant variables that are in a direct relationship with the target and strongly relevant variables that are interaction terms are marked separately on the figure. Non strongly relevant variables are either irrelevant or only transitively relevant, i.e. their effect is mediated by other factors.

The case of 500 samples shows that several non strongly relevant variables, i.e. variables with low MBM posterior, have a reasonably high ECER posterior. This is partially due to the transitively relevant elements, and because the insufficiency of the data at this sample size to discriminate between relevant and non-relevant variables based on their effect size. The insufficiency of the data is confirmed by the fact, that for some strongly relevant variables (based on reference) even the MBM posterior is low, as these

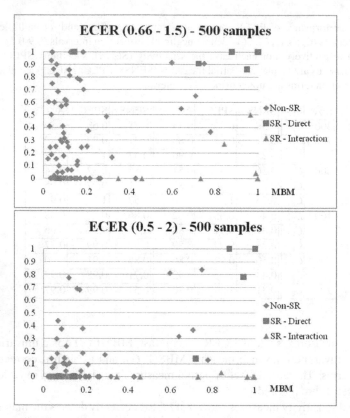

Fig. 4. The relation of MBM posteriors $p(\text{MBM}(X_i, Y))$ and ECER posteriors $p(\text{ECER}(X_i, \varepsilon))$ in case of 500 samples. Results are displayed for two different intervals of negligible effect size, $C_\varepsilon : 0.66 - 1.5$ and $C_\varepsilon : 0.5 - 2.0$. The horizontal axis displays MBM posteriors, whereas the vertical axis displays ECER posteriors. Variables were divided into three groups according to their strong relevance with respect to Y. 'Non-SR' denotes variables that are non strongly relevant, that is they are either irrelevant or only transitively relevant. 'SR - Direct' denotes strongly relevant variables that have a direct relationship with the target Y, and 'SR - Interaction' denotes variables that are strongly relevant as interaction terms.

variables are only detectable as relevant at a higher sample size. The majority of the strongly relevant variables that are in direct relationship with the target are correctly identified by both measures (upper right corner), as both $p(\text{MBM}(X_i, Y))$ and $p(\text{ECER}_\varepsilon(X_i, Y))$ posteriors are high. Interaction terms typically have a low or moderate individual effect size, and consequently a low ECER posterior, whereas they have a relatively high MBM posterior (lower right corner). The effect of the larger interval of C_{ε_3} is that fewer non strongly relevant variables have high ECER posteriors.

In case of 5000 samples the separation between strongly relevant and non strongly relevant variables based on their MBM and ECER posteriors can be clearly seen on Figure 5. The majority of non strongly relevant variables correctly have low ECER

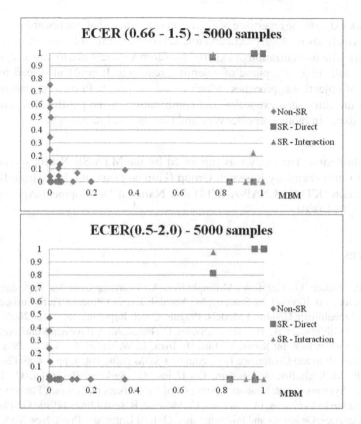

Fig. 5. The relation of MBM posteriors $p(\text{MBM}(X_i,Y))$ and ECER posteriors $p(\text{ECER}(X_i,\varepsilon))$ in case of 5000 samples. Results are displayed for two different intervals of negligible effect size, $C_\varepsilon : 0.66 - 1.5$ and $C_\varepsilon : 0.5 - 2.0$. The horizontal axis displays MBM posteriors, whereas the vertical axis displays ECER posteriors.

posteriors, although some strongly relevant variables that have a moderate effect size also have low ECER posteriors especially in case of $C_{\varepsilon_3} : (0.5 - 2.0)$.

6 Conclusions

There are several machine learning approaches for the incorporation of prior knowledge and for the fusion of heterogeneous data and knowledge, but to our knowledge there is no previous study which investigated the problem of research preferences that are relevant for evaluating and reporting the results. We outlined two paradigmatic biomedical scenarios illustrating such post-analysis, evaluation and communication oriented research preferences, we introduced related concepts, and investigated a corresponding Bayesian decision theoretic framework. Furthermore, we introduced the effect size conditional existential relevance measure, which allows the formal incorporation of experts' preferences for the evaluation of results.

This work extends the repertoire of Bayesian relevance analysis towards the decision theoretic formalization of the evaluation and scientific communication of the results of data analysis. The formalization of experts' preferences could clarify this last, currently rather informal, subjective phase of scientific research. It could also lead to the development of objective approaches, which could support both the repeatability of the results and the automated extraction and combination of the published results, which are critical issues in biomarker discovery and translational research.

Acknowledgments. The study was supported by the MTA-SE Neuropsychopharmacology and Neurochemistry Research Group (Hungary) and the Hungarian Brain Research Program (KTIA_13_NAP-A-II/14) and National Development Agency (KTIA _NAP_13-1-2013-0001).

References

1. Antal, P., Hullam, G., Gezsi, A., Millinghoffer, A.: Learning Complex Bayesian Network Features for Classification. In: Studeny, M., Vomlel, J. (eds.) Proc. of Third European Workshop on Probabilistic Graphical Models, Prague, Czech Republic, pp. 9–16 (2006)
2. Antal, P., Millinghoffer, A., Hullam, G., Szalai, C., Falus, A.: A Bayesian Multilevel Analysis of Feature Relevance. In: Saeys, Y., Liu, H., Inza, I., Wehenkel, L., Van de Peer, Y. (eds.) JMLR Workshop and Conference Proceedings: FSDM 2008, vol. 4, pp. 74–89 (2008)
3. Antal, P., Millinghoffer, A., Hullam, G., Hajos, G., Sarkozy, P., Szalai, C., Falus, A.: Bayesian, Systems-based, Multilevel Analysis of Biomarkers of Complex Phenotypes: From Interpretation to Decisions. In: Sinoquet, C., Mourad, R. (eds.) Probabilistic Graphical Models for Genetics, Genomics and Postgenomics. Oxford University Press, New York (in press)
4. Balding, D.J.: A Tutorial on Statistical Methods for Population Association Studies. Nature 7, 781–791 (2006)
5. Buntine, W.L.: Theory Refinement of Bayesian Networks. In: D'Ambrosio, B., Smets, P. (eds.) Proc. of the 7th Conf. on Uncertainty in Artificial Intelligence, Los Angeles, CA, USA, pp. 52–60. Morgan Kaufmann, San Mateo (1991)
6. Cooper, G.F., Herskovits, E.: A Bayesian Method for the Induction of Probabilistic Networks from Data. Machine Learning 9, 309–347 (1992)
7. Culverhouse, R., Suarez, B.K., Lin, J., Reich, T.: A Perspective on Epistasis: Limits of Models Displaying No Main Effect. Am. J. Hum. Genet. 70, 461–471 (2002)
8. Friedman, N., Koller, D.: Being Bayesian about Network Structure. Machine Learning 50, 95–125 (2003)
9. Friedman, N., Yakhini, Z.: On the Sample Complexity of Learning Bayesian Networks. In: Horvitz, E., Jensen, F.V. (eds.) Proc. of the Twelfth Conference on Uncertainty in Artificial Intelligence, Portland, Oregon, USA, pp. 274–282. Morgan Kaufman (1996)
10. Gelman, A., Carlin, J.B., Stern, H.S., Rubin, D.B.: Bayesian Data Analysis. Chapman & Hall, London (1995)
11. Giudici, P., Castelo, R.: Improving Markov Chain Monte Carlo Model Search for Data Mining. Machine Learning 50, 127–158 (2003)
12. Hullam, G., Antal, P.: Estimation of Effect Size Posterior Using Model Averaging Over Bayesian Network Structures and Parameters. In: Cano, A., Gomez-Olmedo, M., Nielsen, T.D. (eds.) Proceedings of the 6th European Workshop on Probabilistic Graphical Models, Granada, Spain, pp. 147–154 (2012)

13. Hullam, G., Antal, P.: The Effect of Parameter Priors on Bayesian Relevance and Effect Size Measures. Period. Polytech. Elec. 57, 35–48 (2013)
14. Hullam, G., Antal, P., Szalai, C., Falus, A.: Evaluation of a Bayesian Model-based Approach in GA Studies. In: Dzeroski, S., Geurts, P., Rousu, J. (eds.) JMLR Workshop and Conference Proceedings: MLSB 2009, vol. 8, pp. 30–43 (2010)
15. Juhasz, G., Hullam, G., Eszlari, N., Gonda, X., Antal, P., Anderson, I.M., Hokfelt, T.G., Deakin, J.F., Bagdy, G.: Brain Galanin System Genes Interact with Life Stresses in Depression-related Phenotypes. Proc. Natl. Acad. Sci. USA 111, E1666–E1673 (2014)
16. Kohavi, R., John, G.H.: Wrappers for Feature Subset Selection. Artificial Intelligence 97, 273–324 (1997)
17. Koivisto, M., Sood, K.: Exact Bayesian Structure Discovery in Bayesian Networks. Journal of Machine Learning Research 5, 549–573 (2004)
18. Madigan, D., Andersson, S.A., Perlman, M., Volinsky, C.T.: Bayesian Model Averaging and Model Selection for Markov Equivalence Classes of Acyclic Digraphs. Communications in Statistics: Theory Methods 25, 2493–2520 (1996)
19. Maher, B.: Personal Genomes: The Case of the Missing Heritability. Nature 456, 18–21 (2008)
20. Moreau, Y., Antal, P., Fannes, G., De Moor, B.: Probabilistic Graphical Models for Computational Biomedicine. Methods of Information in Medicine 42, 161–168 (2003)
21. Mourad, R., Sinoquet, C., Leray, P.: Probabilistic Graphical Models for Genetic Association Studies. Briefings in Bioinformatics 13, 20–33 (2012)
22. Pearl, J.: Probabilistic Reasoning in Intelligent Systems. Morgan Kaufman, San Francisco (1988)
23. Pena, J.M., Nilsson, R., Bjorkegren, J., Tegner, J.: Towards Scalable and Data Efficient Learning of Markov Boundaries. Int. J. Approx. Reason. 45, 211–232 (2007)
24. Rodin, A.S., Boerwinkle, E.: Mining Genetic Epidemiology Data with Bayesian Networks I: Bayesian Networks and Example Application (Plasma apoE Levels). Bioinformatics 21, 3273–3278 (2005)
25. Storey, J.D.: The Positive False Discovery Rate: A Bayesian Interpretation and the q-value. Ann. Statist. 31, 2013–2035 (2003)
26. Tsamardinos, I., Aliferis, C.: Towards Principled Feature Selection: Relevancy, Filters, and Wrappers. In: Proc. of the Ninth International Workshop on Artificial Intelligence and Statistics, Key West, Florida, USA, pp. 334–342. Morgan Kaufmann (2003)
27. Ungvari, I., Hullam, G., Antal, P., Kiszel, P.S., Gezsi, A., Hadadi, E., Virag, V., Hajos, G., Millinghoffer, A., Nagy, A., Kiss, A., Semsei, A.F., Temesi, G., Melegh, B., Kisfali, P., Szell, M., Bikov, A., Galffy, G., Tamasi, L., Falus, A., Szalai, C.: Evaluation of a Partial Genome Screening of Two Asthma Susceptibility Regions Using Bayesian Network Based Bayesian Multilevel Analysis of Relevance. PLoS ONE 7, e33573 (2012)
28. Verzilli, C.J., Stallard, N., Whittaker, J.C.: Bayesian Graphical Models for Genomewide Association Studies. Am. J. Hum. Genet. 79, 100–112 (2006)

Discrete Bayesian Network Interpretation of the Cox's Proportional Hazards Model

Jidapa Kraisangka[1] and Marek J. Druzdzel[1,2]

[1] Decision System Laboratory, School of Information Sciences and
Intelligent Systems Program, University of Pittsburgh,Pittsburgh, PA, 15260, USA
[2] Faculty of Computer Science, Białystok University of Technology, Wiejska 45A,
15-351 Białystok, Poland
{jik41,druzdzel}@pitt.edu

Abstract. Cox's Proportional Hazards (CPH) model is quite likely the
most popular modeling technique in survival analysis. While the CPH
model is able to represent relationships between a collection of risks and
their common effect, Bayesian networks have become an attractive alternative with far broader applications. Our paper focuses on a Bayesian
network interpretation of the CPH model. We provide a method of encoding knowledge from existing CPH models in the process of knowledge
engineering for Bayesian networks. We compare the accuracy of the resulting Bayesian network to the CPH model, Kaplan-Meier estimate, and
Bayesian network learned from data using the EM algorithm. Bayesian
networks constructed from CPH model lead to much higher accuracy
than other approaches, especially when the number of data records is
very small.

Keywords: Bayesian network, Cox's proportional hazard model, survival analysis.

1 Introduction

Survival analysis is a set of statistical methods that aim at modeling the relationship between a set of predictor variables and an outcome variable and, in
particular, prediction of the time when an event occurs [1]. In medical sciences,
survival analysis is primarily used to predict events such as death, relapse, or
development of a new disease. Several methods have been used in survival analysis. The simplest of these is the Kaplan-Meier (K-M) estimator [2]. The plot of
the K-M estimator depicts the probability of survival at a given point in time
for a group of subjects with particular characteristics.

An alternative and most popular method used in survival analysis is called
the Cox's Proportional Hazards (CPH) model [3]. The CPH model is similar
to a multiple linear regression technique that explores the relationship between
a hazard and related independent explanatory variables over a period of time.
It describes the impact of a risk factor or the effect of a treatment on patients
through a parameter called hazard ratio [4]. The hazard ratio of two groups, e.g.,

L.C. van der Gaag and A.J. Feelders (Eds.): PGM 2014, LNAI 8754, pp. 238–253, 2014.

treatment and control group in a clinical trial, represents the relative likelihood of survival at any time in the study. Usually, the hazard ratio is assumed to be constant over time.

The CPH model has been widely used for predicting patient survival rate. For example, the Seattle Heart Failure Model [5] uses the CPH model to predict 1-, 2-, and 3-year survival of heart failure patients. The Registry to Evaluate Early and Long-Term Pulmonary Arterial Hypertension (PAH) Disease Management (REVEAL) [6] also uses the CPH model to derive the Risk Score Calculator to determine probability of a PAH patient survival within an enrolled year.

While the CPH model has been popular in survival analysis, several researchers tried to find alternative models with comparable predictive ability. Compared to the CPH model and various other Artificial Intelligence and machine learning techniques, a Bayesian network can model explicitly the structure of the relationships among explanatory variables with their probability [7]. Researchers can intuitively design and build a Bayesian network from expert knowledge or available data. The networks can depict a complex structure of a problem and provide a way to infer probability distributions which are suitable for prognosis and diagnosis, particularly in medical decision support systems [8]. However, building Bayesian networks by obtaining their numerical parameters can be a time-consuming and costly task. We are motivated to build Bayesian networks for survival analysis by utilizing existing classical survival models. Our paper focuses on a Bayesian network interpretation of the CPH model. The application of our work is using the CPH models as data sources in the process of parameter estimation for Bayesian networks.

The remainder of our paper is structured as follows. First, we provide necessary background knowledge on survival analysis, the CPH model, the K-M estimators, and Bayesian networks. We follow this up by a Bayesian network interpretation of the CPH model along with an example. Finally, we report the result of a study comparing our Bayesian network interpretation of the CPH model to the original CPH models, K-M estimates, and Bayesian networks learned from data.

2   Background Knowledge

Survival analysis basically focuses on modeling time-to-event occurrences. For example, we may focus on time-to-death of patients with a specific disease, failure time of machines, or time to rearrest of prisoners who have been released. Survival analysis can be conducted to estimate time-to-event for a group, compare time-to-event between several groups, or just to study the relationship between variables and the predicted events.

The probability of an individual surviving beyond a given time t, i.e., the survivor function, is defined as

$$S(t) = Pr(T > t) . \tag{1}$$

T is a variable denoting the time of occurrence of an event of interest. The survival probability at the beginning, i.e., t_0, may be equal to 1 or to some baseline survival probability, which will drop down to zero over time. While survivor function represents the probability of survival, the hazard function represents the risk of event occurrence at time t. The hazard function is given by

$$\lambda(t) = \lim_{\Delta t \to 0} \frac{Pr(t \leq T < t + \Delta t \mid T \geq t)}{\Delta t}, \tag{2}$$

where T is also a time variable. The hazard is a measure of risk at a small time interval Δt which can be considered as a rate [1]. The hazard function can be an exponential distribution, where the rate is constant over time, or by Weibull distribution, where the rate can be increasing or decreasing over time.

The relationship between the hazard function and the survivor function (see more details in Allison's textbook [1]) is described as

$$\lambda(t) = -\frac{d}{dt} \log S(t) \tag{3}$$

or as

$$S(t) = \exp \int_0^t \lambda(u)\, du. \tag{4}$$

Hence, we can estimate the survival probability from the hazard function. There are several techniques used to model the hazard function or the survivor function, e.g., parametric regression techniques, non-parametric estimates, or semi-parametric models. In this paper, we focus only on the semi-parametric model, which is the CPH model, described in the next section.

2.1 Cox's Proportional Hazard Model

The Cox proportional hazard model [3] is a set of regression methods used in the assessment of survival based on its risk factors or explanatory variables. The risk factors can be time-independent (e.g., race or sex) or time-dependent, which can change throughout the study (e.g., blood pressure at different points of study time). In this paper, we focus only on the CPH model with time-independent risk factors. This model allows researchers to evaluate and control factors that affects the time to event [9].

As defined originally by Cox [3], the hazard regression model is expressed as,

$$\lambda(t) = \lambda_0(t) \exp^{\beta' \cdot \mathbf{X}}. \tag{5}$$

This hazard model is composed of two main parts: the baseline hazard function, $\lambda_0(t)$, and the set of effect parameters, $\beta' \cdot \mathbf{X} = \beta_1 X_1 + \beta_2 X_2 + ... + \beta_n X_n$. The baseline hazard function determines the risks at an underlying level of explanatory variables, i.e., when all explanatory variables are absent. According to Cox [3], this $\lambda_0(t)$ can be unspecified or follow any distribution, which makes the CPH model a semi-parametric model. The βs are the coefficients corresponding to the risk factors, $\mathbf{X}$.

CPH models can handle both continuous and discrete variables [1]. The CPH model treats these risk factors as numerical variables, so that the model can estimate the parameter coefficients, β. Researchers can treat risk factors as they are defined in the data set or do some data preprocessing. For example, in case of categorical variables with n categories, researchers need to create a set of dummy binary variables capturing $n-1$ categories, e.g., we can code a variable *color* having values as red, green, blue, as two binary variables (e.g., *color-red* and *color-green*). Some continuous variables, e.g., number of days in a hospital, can also be discretized. Once all risk factors have been established, β parameters are estimated by means of the Maximum Partial Likelihood technique.

The application of the CPH model relies on the assumption that the hazard ratio of two observations, e.g., treatment and control group in a clinical trial, is constant over time [3]. Given the hazard at time t_1 and the hazard at time t_2, the hazard at these two points in time are $\lambda(t_1)$ and $\lambda(t_2)$ respectively. The ratio of two hazards is a constant, defined as γ:

$$\gamma = \frac{\lambda(t_2)}{\lambda(t_1)} = \frac{\exp(\beta'X_2)}{\exp(\beta'X_1)} . \tag{6}$$

This hazard ratio is an estimate of the relative risk of the two groups. For example, given the hazard ratio of 2, the treatment group may have twice the risk of death from the treatment relative to the control group. Hence, the CPH model estimates relative rather than absolute risks.

If the explanatory variables are propositional, their value at time t_i could be expressed as *presence* $(X = 1)$ or as *absence* $(X = 0)$ of the risk factor. We assess the hazard ratio at time t_1 and t_2. The risk of the event at t_2 when the factor X_i is present can be compared to the risk of the event at t_1 where the factor X_i is absent [10]. Since the hazard ratio γ is assumed to be constant, we can use it to estimate the survival probability [11]. The relationship is expressed as

$$S(t) = S_0(t)^\gamma , \tag{7}$$

where $S_0(t)$ is the baseline survival probability. Hence, in case of dichotomous variables, we can assess the survival probability given a set of hazards from presence or absence of each individual hazard variable.

Example 1. A classical example application of the CPH model is an experimental study of recidivism of prisoners [12]. The data set has been made available to researchers and was used as an illustration in survival analysis examples by Allison using SAS [1] and Fox using R [13]. The data in the Recidivism data set describe 432 male prisoners who were under one year observation after being released from prison. The event of interest in this analysis is re-arrest, i.e., whether the prisoner is re-arrested during the period of study or not.

For the purpose of simplicity, we selected only four variables (or risk factors) from the 10 variables in the original Recidivism data set.

- **fin** : Financial aid status when arrested (No-FinancialAid = 0, Has-FinancialAid = 1)

- *race* : Prisoner's race (Other = 0, Black = 1).

- *wexp* : Status of having prior full-time working experience (No-Exp = 0, Has-Exp = 1).

- *prio* : Number of prior convictions, i.e., 0, 1, 2, 3, etc. We discretized this variable into two classes (Five-and-Below = 0, Above-Five = 1).

The time variable in this data set is **week**, which is the week when a prisoner was rearrested during the observation period of one year after having been released from prison. Its domain is 1 to 52 weeks. The event of interest (the survival variable) is **arrest** indicating the rearrest status of a prisoner (Rearrested = 1, Never Rearreasted = 0).

We used R as our tool to create the CPH model based on the Recidivism data set. We used the package *survival*, which provides a set of functions to create CPH models and other survival analysis methods. We used the *coxph* method to model the survival variable *arrest* with *race* and *wexp* variables based on each *week*. Table 1 shows the parameters of the constructed CPH model.

Table 1. Selected variables in the Recidivism data set example

Variables	β	$\exp(\beta)$	lower .95	upper .95
fin	-0.3899	0.6771	0.4664	0.9829
race	0.2591	1.2958	0.7110	2.3617
wexp	-0.5249	0.5916	0.4038	0.8667
prio	0.3330	1.3951	0.8462	2.3001

The β of each variable represents the coefficient in the model while the $\exp(\beta)$ is the multiplicative effect of the hazard [13]. The lower and upper bounds of the 95% confidence interval are in the third and fourth columns respectively. We show the survival curve along with the curves for the 95% confidence interval estimated from the generated CPH model in Figure 1.

The baseline survival probability in Figure 1 is estimated from the beginning of the observation period until the end of the 52nd week. The baseline survival probability, $S_0(t)$, is the probability measured when all risk factors are absent ($fin = 0, race = 0, wexp = 0, and prio = 0$) at time t. When other cases than the baseline are analyzed, we can estimate the survival probability based on this baseline and respective hazard ratios. For example, all things being equal, the hazard ratio of black prisoners (race = 1) compared to other races (race = 0) is 1.29. Therefore, the hazard (rearrest) at every time will be 1.29 times of its baseline. The survival probability of black prisoner group relative to the other-race prisoner group at any time t can be calculated from

$$S\left(t\right) = S_0(t)^{(1.29)} . \tag{8}$$

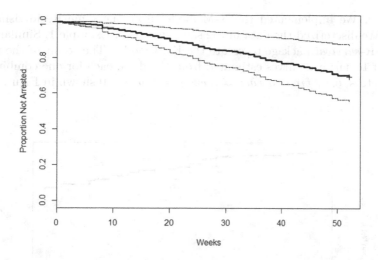

Fig. 1. The non-rearrest (survival) probability estimation based on the CPH model. The thick black line shows the baseline survival probability when all risk factors are absent, while the two grey lines show the upper and lower 95% confidence interval.

The baseline survival probability at the first week, $S_0(1)$, is 0.9973. If we want to assess the survival probability of the black prisoner group in the first week, we get $S(t_1) = 0.9973^{(1.29)} = 0.9965$. We can obtain survival probabilities for each week relative to the baseline by repeating the same steps. ∎

The CPH model provides a way of calculating the survival probabilities conditional on every combination of risk present. Although, in practice, not every combination of risk factors is present in the data set, the CPH model is still able to derive the entire probability distribution and provide reasonable predicted survival probabilities. This survival probability resulting from the calculation equals simply to the conditional probability distribution of survival given these risk factors. The survival probability reported in the CPH model can thus be mapped directly to the conditional probability in Bayesian networks.

2.2 Kaplan-Meier Estimator

The Kaplan-Meier (K-M) estimator [2] is an alternative method of modeling the survival curve. It amounts simply to calculating the survival probability for each time interval t based on the event occurrences at that time. From the data, the survival probabilities are estimated as follows,

$$S(t) = \prod_{t_i \leq t} (1 - \frac{d_i}{n_i}),$$

(9)

where n_i is the number of subjects at risk at the beginning of the time interval t_i and d_i is the number of subjects who have not survived during the time interval t_i.

Example 2. We implemented the K-M model from the same Recidivism data set [12]. We discretized the data in the same way as in Example 1. Similarly, we used the R-*survival* package to create the K-M model. The result of the model is a set of 16 survival curves estimated from the data, each for one combination of risk factors, e.g., $fin = 0, race = 1, wexp = 1, prio = 0$ shown in Figure 2. ∎

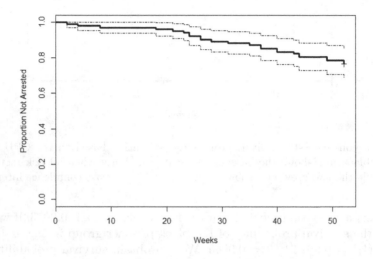

Fig. 2. The survival curve along with its 95% confidence interval from the K-M model produced by R shows the survival probability of a group of prisoners when $fin = 0, race = 1, wexp = 1, prio = 0$.

Unlike the CPH model, the K-M is learned directly from a data set. When there are enough data records to learn from, the K-M estimates provide good predicted survival probabilities. However, there could be few data records for each combination of risk factors. When there are not enough data records to learn from, the K-M estimates provide poor quality of survival prediction. In those cases, the CPH model is preferable.

3   Bayesian Networks

Bayesian networks [14] are probabilistic graphical models capable of modeling the joint probability distribution over a finite set of random variables. The model is static, i.e., it shows a snapshot of the system at a given time. The structure of a Bayesian network is an acyclic directed graph in which nodes are variables and directed arcs denote dependencies among them. A conditional probability table (CPT) of a variable X contains probability distributions over the states of X conditional on all combinations of states of X's parents. The joint probability distribution over all variables of the network can be calculated by taking the product of all prior and conditional probability distributions:

$$\Pr(\mathbf{X}) = \Pr(X_1, \ldots, X_n) = \prod_{i=1}^{n} \Pr(X_i | Pa(X_i)) . \tag{10}$$

The structure of a Bayesian network and all numerical probabilities can be obtained from experts or learned from data. Although Bayesian networks may take significant effort to construct, they are widely used in many areas, such as medical and engineering diagnosis or prognosis.

Bayesian network has become an alternative approach for survival analysis. It is well-structured, intuitive, and while also being theoretically sound [8]. Also, it has the ability to capture expert knowledge, handle model complexity, and offers more flexibility in model interpretation [7]. Bayesian networks are used in the prognostic models, e.g., prediction of metastasis in breast cancer [15], prognosis of patients with a carcinoid tumor [16] etc.

4   Bayesian Network Interpretation of the Cox Proportional Hazard Model

In this section, we provide an interpretation of the CPH model in the Bayesian network framework. We assume the CPH model's assumptions are not violated when making the estimation. The risk factors or random variables $\mathbf{X_i}$ are binary variables and time-independent. The Recidivism data set meets the CPH assumptions, thus we use the data set in our interpretation.

We first create the structure of a Bayesian network. From a given CPH model, we convert each of the risk factors and the survival variable into a random variable in a Bayesian network (in our example, *fin*, *race*, *wexp*, *prio*, and *arrest*). We make random variables representing risk factors parents of the survival node, *arrest*.

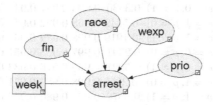

Fig. 3. A Bayesian network representing the interaction among variables in the Recidivism example

Note that, unlike the CPH model, Bayesian networks capture a snapshot in time of a system. We need, thus, to represent time explicitly; we achieve this by adding an indexing variable (in our example, *week*). For the purpose of simplicity, we reduce the number of states for the variable *week* from 52 to 13, which amounts to analyzing the system at 4-week steps. Other random variables (risk factors) have the same states as in the CPH model. We show the resulting structure of the Bayesian network in Figure 3.

In the next step, we create the conditional probability table for the survival node (*arrest*). Recall that we can obtain the survival probabilities from Equation 7. For each time snapshot captured in the variable *week*, we assess a set of survival probabilities, $S(t)$ from the CPH model. A set of survival probabilities here means that we configure the hazard ratio γ according to the combination of the parent states. γ is equal to the ratio of hazard of the conditioning case X_i to the baseline case X_b, i.e., case in which all risk variables are absent, i.e.,

$$\gamma = \frac{exp(\beta'X_i)}{exp(\beta'X_b)} = -0.3899fin + 0.2591race - 0.5249wexp + 0.3330prio. \quad (11)$$

Equation 11 allows us to assess the survival probabilities directly from the parameters of the CPH model. First, we configure all risk factor cases in Equation 11 to find all hazard ratio values. Then, we obtain the baseline survival probability at the 1st week from the CPH model ($S_0(t = 1) = 0.9973$) and use Equation 7 to find the survival probability. The survival probability calculated for each combination of risk factors corresponds to the conditional probability of survival. We show examples of conditional probabilities of survival for all combinations of states of the risk variables in Table 2.

In summary, we construct the CPT by deriving the survival probabilities using hazard ratios for each time step. $\mathbf{X_i}$, the set of risk factors in the CPH model, are

Table 2. Conditional probabilities of survival for all cases at each snapshot of time. γ is calculated from Equation 11 and $S(1), S(2), S(3), ..., S(13)$ are calculated from Equation 7. s is the survival variable *arrest*.

$Pr(s \mid X_i)$	γ	$S(1)$	$S(2)$	$S(3)$	$S(4)$	...	$S(12)$	$S(13)$
$\mathbf{Pr}(s \mid f = 0, r = 0, w = 0, p = 0)$	0.0000	0.997	0.987	0.962	0.946	...	0.729	0.692
$\mathbf{Pr}(s \mid f = 0, r = 0, w = 0, p = 1)$	0.3330	0.996	0.981	0.948	0.926	...	0.644	0.598
$\mathbf{Pr}(s \mid f = 0, r = 0, w = 1, p = 0)$	-0.5249	0.998	0.992	0.977	0.968	...	0.829	0.804
$\mathbf{Pr}(s \mid f = 0, r = 0, w = 1, p = 1)$	-0.1919	0.998	0.989	0.969	0.955	...	0.771	0.738
$\mathbf{Pr}(s \mid f = 0, r = 1, w = 0, p = 0)$	0.2591	0.997	0.983	0.951	0.931	...	0.664	0.621
$\mathbf{Pr}(s \mid f = 0, r = 1, w = 0, p = 1)$	0.5921	0.995	0.976	0.933	0.904	...	0.565	0.514
$\mathbf{Pr}(s \mid f = 0, r = 1, w = 1, p = 0)$	-0.2658	0.998	0.989	0.971	0.958	...	0.785	0.754
$\mathbf{Pr}(s \mid f = 0, r = 1, w = 1, p = 1)$	0.0672	0.997	0.986	0.959	0.942	...	0.714	0.675
$\mathbf{Pr}(s \mid f = 1, r = 0, w = 0, p = 0)$	-0.3899	0.998	0.991	0.974	0.963	...	0.808	0.779
$\mathbf{Pr}(s \mid f = 1, r = 0, w = 0, p = 1)$	-0.0569	0.997	0.987	0.964	0.949	...	0.742	0.706
$\mathbf{Pr}(s \mid f = 1, r = 0, w = 1, p = 0)$	-0.9148	0.999	0.996	0.985	0.978	...	0.881	0.863
$\mathbf{Pr}(s \mid f = 1, r = 0, w = 1, p = 1)$	-0.5818	0.999	0.992	0.979	0.969	...	0.838	0.814
$\mathbf{Pr}(s \mid f = 1, r = 1, w = 0, p = 0)$	-0.1308	0.998	0.988	0.967	0.952	...	0.758	0.724
$\mathbf{Pr}(s \mid f = 1, r = 1, w = 0, p = 1)$	0.2022	0.997	0.984	0.954	0.934	...	0.679	0.637
$\mathbf{Pr}(s \mid f = 1, r = 1, w = 1, p = 0)$	-0.6557	0.997	0.993	0.980	0.972	...	0.849	0.826
$\mathbf{Pr}(s \mid f = 1, r = 1, w = 1, p = 1)$	-0.3227	0.998	0.990	0.972	0.961	...	0.796	0.766

random variables in the Bayesian network. $S_0(t)$ is the baseline survival probability estimated from CPH at time t, while β is a set of regression coefficients corresponding to each risk factor. T is the time of interest or the time variable in the Bayesian network. The conditional probability to be encoded in the CPT can be estimated by

$$Pr(s \mid X_i, T = t) = S_0\left(t\right)^{e^{(\beta' x_i)}} . \tag{12}$$

5   Empirical Evaluation

Although, the Recidivism data is admittedly small, it is quite likely the most widely used example data for survival analysis, especially for the CPH model. It is not atypical for survival data to be satisfied by the CPH assumptions. Hence, we use the Recidivism data set to compare the precision of our Bayesian-Cox network (BNCox), the CPH model, the K-M model, and the Bayesian network learning directly from data (BNLearn). We used the R programming environment with the *survival* library [13] to implement the CPH model and the K-M model. For BNCox, we obtained all survival probabilities from the CPH model and used GeNIe [1] to implement its structure. We built the BNLearn model in GeNIe using the same structure as in the BNCox model. The BNLearn model learned the numerical parameters from data using the EM algorithm. We created all models with 4 risk factors: *fin, race, wexp,* and *prio.* These four risk factors are binary variables resulting in $2^4 = 16$ combinations of risk factors. In addition to the simplified, four-risk-factor model, we also created a complete Recidivism model with all eight risk factors (see Rossi's work [12]), seven of which are binary and one is categorical. We will provide the results of each evaluation in the following two sections.

5.1   Recidivism Prediction with Four Risk Factors

As mentioned previously, the Recidivism model with four risk factors produces 16 combinations of risk factors. We plotted the distribution of the number of records corresponding to these 16 cases in Figure 4. We compared the prediction of models for all 16 cases. However, in this paper, due to space limitations, we selected only four cases as samples, including one with the highest number of records (102 records), one medium-to-high number of records (61 records), one medium-to-small number of records (9 records), and one small number of records (2 records). We marked the selected cases as dark grey in Figure 4.

Figure 5 shows the survival probabilities predicted by all four models: the CPH model, the BNCox model, the K-M model, and the BNLearn model. We observe an almost perfect match between the CPH model and the BNCox model in all 16 cases. Both K-M model and BNLearn are close, although departing from the CPH model significantly as the number of records gets smaller. We provide

---

[1] http://genie.sis.pitt.edu/

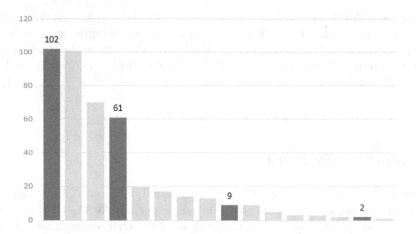

Fig. 4. Distribution of the number of records in the Recidivism data set with four risk factors for each of the 16 combination of risk factors (sorted in descending order). We selected four of the combination of risk factors (marked as dark grey) as examples for our comparison.

four examples of combinations of risk factors in Figure 5. The dark grey square-dotted line represents the CPH model prediction while the black diamond point represents the 13-snapshot survival probability produced by the BNCox model.

Both BNCox and BNLearn model produced 13 survival probabilities for each case while the K-M model produced more data according to the event occurrence in the Recidivism data set. Figure 5 shows the predicted survival curve for all four models: black-diamond/BNCox, hollow-triangle/BNLearn, dark grey square-dotted line/CPH model, and light grey round-dotted line/K-M model. We found that when we have enough data to learn, e.g., more than a hundred records, there is a remarkable agreement among all four models. However, when we lack data, we found that the predicted curve produced by the K-M estimate and the Bayesian network learned from data (BNLearn) are in some agreement with one another, while the BNCox model and the CPH model, which agree perfectly, are smoothing the curve.

5.2 Recidivism Prediction with All Risk Factors

The total number of combinations of states of all risk factors in the full version of Recidivism models is 512. We constructed all four models for this case as well. We found that the distribution of the cases in terms of the number of records is extremely skewed. As shown in Figure 6, the case best represented in the data has only 32 records, while more than 70 percent of cases (392 cases of the total of 512 cases) have zero records. We selected four cases, with 32, 27, 5, and 0 records respectively, for the purpose of the comparison.

Similarly to the simplified model presented in the previous section, we compared the predicted survival probability of the CPH model, the BNCox model,

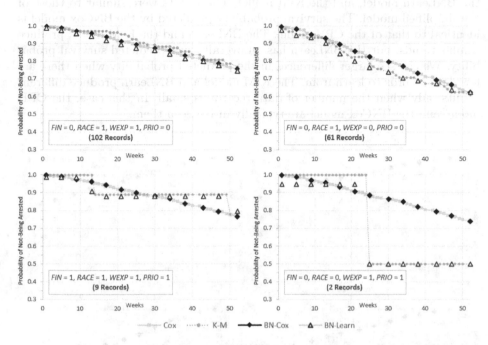

Fig. 5. The predicted survival curves generated by each model: CPH model (Cox), K-M model (K-M), BNCox model (BN-Cox) and BNLearn (BN-Learn) for four cases with different number of records.

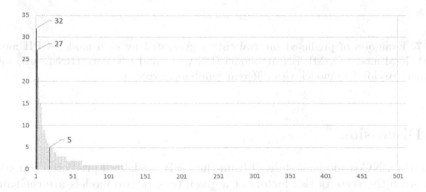

Fig. 6. Distribution of the number of records in the Recidivism data set for each combination of the risk factors (sorted in descending order). There are 392 cases with zero records in the data set. The marked bar indicates the selected combination of risk factors for further comparison.

the BNLearn model, and the K-M model. The results were similar to those of
the simplified model. The survival probability predicted by the BNCox model is
identical to that of the CPH model. The BNLearn and the K-M model produce
similar trends, but the BNLearn has an overall lower predicted survival probability. We can see larger differences in the predicted probability when there are
few data records to learn from. The K-M model and BNLearn produce different
results only when the number of data records is small. In this case, the CPH
model and the BNCox model are typically in-between them.

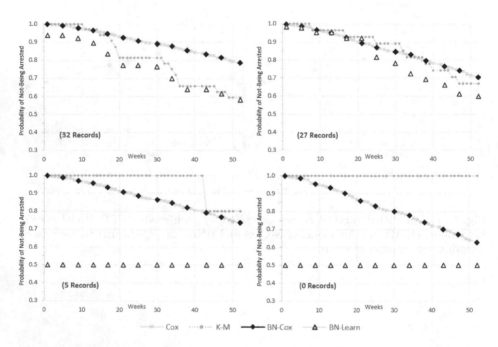

Fig. 7. Examples of predicted survival curves generated by each model: CPH model
(Cox), K-M model (K-M), BNCox model (BN-Cox), and BNLearn (BN-Learn) for a
complete Recidivism model with different numbers of records.

6   Discussion

Since the BNCox model is derived from the CPH model, all predicted probabilities given the state of risk factors at a given time of two models are the same.
However, in our experiment, the BNCox model captured part of the CPH model,
not the whole model. The complexity and predictive capability of the BNCox
model are a tradeoff.

The CPH model, the Bayesian network learning from data, and the K-M
estimates have a similar predictive ability when there are enough data to learn.
In those cases, the K-M estimate and the Bayesian network learning from data

are in agreement. The predicted curves from both models depart from the CPH survival curve when there are few data records. In practice, data can include many combinations of risk factors with very small number of records per case. In this case, the additional assumption of the CPH model are able to smoothen out the survival curve.

When there are few data records available, Bayesian network predictive quality tends to drop [17]. In those cases, the CPH model serves better in terms of smoothing out the distribution. The CPH model behaves similar to Bayesian networks with ICI gates, which smoothen out the distribution when there are not enough records to learn. Hence, the Bayesian network interpreted from the CPH model may more useful in practice.

We could consider using other special types of Bayesian networks as possible extensions of our BNCox model. One aspect could be using the Continuous Time Bayesian networks(CTBNs) [18] to handle survival time as continuous variable. Also, the CTBNs could be used for handling the CPH model with time-dependent risk factors, which have not been included in the current BN-Cox model. Another interesting model is to Bayesian networks with generalized Noisy-OR [19]. This could be an alternative model to capture survival probability from the CPH model.

7   Conclusion

Bayesian networks are a viable alternative to the CPH model and the K-M model for survival analysis. However, the process of building Bayesian networks can take a significant effort. The focus of this paper was a Bayesian network interpretation of the CPH model. The main application of our work is in knowledge engineering for Bayesian networks. The CPH model can be used as a data source in the process of parameter estimation for Bayesian networks.

We mapped survival probabilities from the CPH model to the conditional probability distributions for each combination of risk factors and built survival probabilities in the Bayesian network. The resulting Bayesian network produces survival probabilities which are almost identical to those of the CPH model.

We evaluated the proposed Bayesian-Cox model by comparing its predictive precision to the CPH model, the Bayesian network learning from data, and the K-M estimate on the same data set (Recidivism data from Rossi). We found that the Bayesian network interpreted from the CPH model produces exactly the same probabilities as the CPH model. All modeling methods make similar predictions when there are sufficient number of data records to learn from. However, when the number of records is very low, Bayesian network and the K-M estimate are unable to learn reliable models and produce inferior model, while the CPH model and the Bayesian network interpreted from the CPH model remain close to each other and presumably more accurate. In summary, we found that the interpreted Bayesian-Cox model assembles the advantages of Bayesian network and the CPH model: an intuitive structure and more reliable predictive quality.

Acknowledgments. We acknowledge the support the National Institute of Health under grant number U01HL101066-01 and the Faculty of Information and Communication Technology, Mahidol University, Thailand. Implementation of this work is based on GeNIe and SMILE, a Bayesian inference engine developed at the Decision Systems Laboratory and available at http://genie.sis.pitt.edu/. We would like to thank anonymous reviewers for valuable suggestions that improved the clarity of the paper.

References

1. Allison, P.D.: Survival Analysis Using SAS: A Practical Guide, 2nd edn. SAS Institute Inc., Cary (2010)
2. Kaplan, E.L., Meier, P.: Nonparametric estimation from incomplete observations. Journal of the American Statistical Association 53(282), 457–481 (1958)
3. Cox, D.R.: Regression models and life-tables. Journal of the Royal Statistical Society. Series B (Methodological) 34(2), 187–220 (1972)
4. Spruance, S.L., Reid, J.E., Grace, M., Samore, M.: Hazard ratio in clinical trials. Antimicrobial Agents and Chemotherapy 48(8), 2787–2892 (2004)
5. Levy, W.C., Mozaffarian, D., Linker, D.T., Sutradhar, S.C., Anker, S.D., Croppan, A.B., Anand, I., Maggionl, A., Burton, P., Sullivan, M.D., Pitt, B., Poole-Wilson, P.A., Mann, D.L., Packer, M.: The Seattle Heart Failure Model: Prediction of survival in heart failure. Circulation 113(11), 1424–1433 (2006)
6. Benza, R.L., Miller, D.P., Gomberg-Maitland, M., Frantz, R.P., Foreman, A.J., Coffey, C.S., Frost, A., Barst, R.J., Badesch, D.B., Elliott, C.G., Liou, T.G., McGoon, M.D.: Predicting survival in pulmonary arterial hypertension: Insights from the registry to evaluate early and long-term pulmonary arterial hypertension disease management (REVEAL). Circulation 122(2), 164–172 (2010)
7. Hanna, A.A., Lucas, P.J.: Prognostic models in medicine- AI and statistical approaches. Method Inform. Med. 40, 1–5 (2001)
8. Husmeier, D., Dybowski, R., Roberts, S.: Probabilistic modeling in bioinformatics and medical informatics. Springer (2005)
9. Klein, J.P., Moeschberger, M.L.: Survival Analysis: Censored and Truncated Data, 2nd edn. Springer-Verlag New York, Inc., New York (2003)
10. Christensen, E.: Multivariate survival analysis using Cox's regression model. Hepatology 7, 1346–1358 (1987)
11. Casea, L.D., Kimmickb, G., Pasketta, E.D., Lohmana, K., Tucker, R.: Interpreting measures of treatment effect in cancer clinical trials. The Oncologist 7(3), 181–187 (2002)
12. Rossi, P.H., Berk, R.A., Lenihan, K.J.: Money, Work, and Crime - Experimental Evidence. Academic Press, Inc., San Diego (1980)
13. Fox, J.: An R and S-Plus Companion to Applied Regression. Sage Publication Inc., CA (2002)
14. Pearl, J.: Probabilistic reasoning in intelligent systems: networks of plausible inference. Morgan Kaufmann Publishers Inc., San Francisco (1988)
15. Gevaert, O., Smet, F.D., Timmerman, D., Moreau, Y., Moor, B.D.: Predicting the prognosis of breast cancer by integrating clinical and microarray data with bayesian networks. Bioinformatics 22(14), e184–e190 (2006)

16. van Gerven, M.A., Taal, B.G., Lucas, P.J.: Dynamic Bayesian networks as prognostic models for clinical patient management. Journal of Biomedical Informatics 41(4), 515–529 (2007)
17. Oniśko, A., Druzdzel, M.J., Wasyluk, H.: Learning Bayesian network parameters from small data sets: Application of Noisy-OR gates. In: Working Notes on the European Conference on Artificial Intelligence (ECAI) Workshop Bayesian and Causal Networks: From Inference to Data Mining (August 22, 2000)
18. Nodelman, U., Shelton, C.R., Koller, D.: Continuous Time Bayesian Networks. In: Proceedings of the Eighteenth Conference on Uncertainty in Artificial Intelligence, pp. 378–387. Morgan Kaufmann Publishers Inc. (2002)
19. Srinivas, S.: A generalization of the Noisy-Or model. In: Proceedings of the Ninth International Conference on Uncertainty in Artificial Intelligence, pp. 208–215. Morgan Kaufmann Publishers Inc. (1993)

Minimizing Relative Entropy
in Hierarchical Predictive Coding

Johan Kwisthout

Radboud University Nijmegen, Donders Institute for Brain, Cognition and
Behaviour, Montessorilaan 3, 6525 HR Nijmegen, The Netherlands
j.kwisthout@donders.ru.nl

Abstract. The recent Hierarchical Predictive Coding theory is a very
influential theory in neuroscience that postulates that the brain continuously makes (Bayesian) predictions about sensory inputs using a generative model. The Bayesian inferences (making predictions about sensory
states, estimating errors between prediction and observation, and lowering the prediction error by revising hypotheses) are assumed to allow
for efficient approximate inferences in the brain. We investigate this assumption by making the conceptual ideas of how the brain may minimize
prediction error computationally precise and by studying the computational complexity of these computational problems. We show that each
problem is intractable in general and discuss the parameterized complexity of the problems.

1 Introduction

The assumption that the brain in essence is a Bayesian inferential machine,
integrating prior knowledge with sensory information such as to infer the most
probable explanation for the phenomena we observe, is quite wide spread in neuroscience [19]. Recently, this 'Bayesian brain' hypothesis has merged with the
hypothesis that the brain is a prediction machine that continuously makes predictions about future sensory inputs, based on a generative model of the causes
of these inputs [17] and with the free energy principle as a driving force of prediction error minimization [13]; the resulting theory has been called Hierarchical
Predictive Coding or Predictive Processing [7]. It is assumed to explain and unify
all cortical processes, spanning all of cognition [6]. Apart from being one of the
most influential current unifying theories of the *modus operandi* of the brain, it
has inspired researchers in domains such as developmental neurorobotics [23],
human-robot interaction [25], and conscious presence in virtual reality [26].

At the very heart of Hierarchical Predictive Coding (hereafter HPC) are the
Bayesian predictions, error estimations, and hypothesis revisions that are assumed to allow for efficient approximate Bayesian inferences in the brain [7]. As
Bayesian inferences are intractable in general, even to approximate [1,9], this
invites the question to what extent the HPC mechanism indeed renders these
inferences tractable [3,22]. In essence, minimizing prediction errors boils down
to minimizing the relative entropy or Kullback-Leibler divergence between the

L.C. van der Gaag and A.J. Feelders (Eds.): PGM 2014, LNAI 8754, pp. 254–270, 2014.
© Springer International Publishing Switzerland 2014

predicted and observed distributions [14]. Lowering the relative entropy between prediction and observation can be done in many ways: we can revise the hypothesized causes that generated the prediction; alternatively, we may adjust the probabilistic dependences that modulate how predictions are generated from hypotheses, or we might want to seek and include additional observations into the model in order to adjust the posterior distribution over the predictions. In contrast, we might also bring prediction and observation closer to each other by *intervention* in the world, thus hopefully manipulating the observation to better match what we predicted or expected. This is referred to as *active inference* in the HPC literature [15].

The contribution of this paper is to make these informal notions explicit and to study the computational complexity of minimizing relative entropy using these notions. We show that each conceptualization of prediction error minimization yields an intractable (i.e., NP-hard) computational problem. However, we can clearly identify where the border between tractable and intractable lies by giving fixed-parameter tractability results for all discussed problems. The remainder of this paper is structured as follows. In Section 2 we formally define HPC in the context of discrete Bayesian networks. We recall some needed preliminaries from computational complexity and discuss related work. In Section 3 we discuss the complexity of computing entropy and relative entropy in Bayesian networks. In Sections 4 and 5 we discuss *belief revision* and *model revision*, respectively, and in Section 6 we investigate the complexity of deciding which observation to make in order to decrease prediction error. In Section 7 we turn to the complexity of *active inference*, i.e., deciding which possible action to perform to decrease prediction error. We switch to the parameterized complexity of these problems in Section 8. In Section 9 we conclude this paper and sketch possible future work.

2   Preliminaries

A Bayesian network $\mathcal{B} = (\mathbf{G}_\mathcal{B}, \Pr_\mathcal{B})$ is a graphical structure that models a set of stochastic variables, the conditional independences among these variables, and a joint probability distribution over these variables. $\mathcal{B}$ includes a directed acyclic graph $\mathbf{G}_\mathcal{B} = (\mathbf{V}, \mathbf{A})$, modeling the variables and conditional independences in the network, and a set of conditional probability tables (CPTs) $\Pr_\mathcal{B}$ capturing the stochastic dependences between the variables. The network models a joint probability distribution $\Pr(\mathbf{V}) = \prod_{i=1}^{n} \Pr(V_i \mid \pi(V_i))$ over its variables, where $\pi(V_i)$ denotes the parents of V_i in $\mathbf{G}_\mathcal{B}$. By convention, we use upper case letters to denote individual nodes in the network, upper case bold letters to denote sets of nodes, lower case letters to denote value assignments to nodes, and lower case bold letters to denote joint value assignments to sets of nodes. We use the notation $\Omega(V_i)$ to denote the set of values that V_i can take. Likewise, $\Omega(\mathbf{V})$ denotes the set of joint value assignments to $\mathbf{V}$.

HPC can be understood as a cascading hierarchy of increasingly abstract hypotheses about the world, where the predictions on one level of the hierarchy are identified with the hypotheses at the subordinate level. At any particular level,

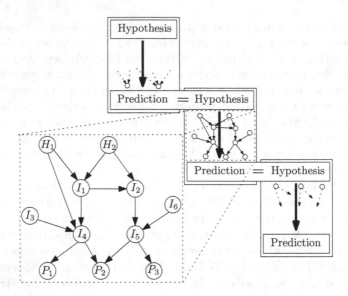

Fig. 1. An example level L of the HPC hierarchy, with hypothesis variables Hyp $= \{H_1, H_2\}$, prediction variables Pred $= \{P_1, P_2, P_3\}$, and intermediate variables Int $= \{I_1, \ldots, I_6\}$.

making a prediction based on the current hypothesis in any of the assumed levels corresponds to computing a posterior probability distribution $\Pr(\Pr_{\text{Pred}} \mid \Pr_{\text{Hyp}})$ over the space of candidate predictions, given the current estimated probability distribution over the space of hypotheses, modulated by contextual dependences. We can thus describe each level L of the HPC hierarchy as a Bayesian network $\mathcal{B}_L$, where the variables are partitioned into a set of hypothesis variables Hyp, a set of prediction variables Pred, and a set of intermediate variables Int, describing contextual dependences and (possibly complicated) structural dependences between hypotheses and predictions. We assume that all variables in Hyp are source variables, all variables in Pred are sink variables, and that the Pred variables in $\mathcal{B}_L$ are identified with the Hyp variables in $\mathcal{B}_{L+1}$ for all levels of the hierarchy save the lowest one (See Figure 1). As HPC is claimed to be a unifying mechanism describing all cortical processes [6], we do not impose additional *a priori* constraints on the structure of the network describing the stochastic relationships [16]. Motivated by the assumption that global prediction errors are minimized by local minimization [18], we will focus on the computations in a single level of the network.

Computing the prediction error at any level of the hierarchy corresponds to computing the relative entropy or Kullback-Leibler divergence

$$D_{\text{KL}}(\Pr_{(\text{Pred})} \| \Pr_{(\text{Obs})}) = \sum_{\mathbf{p} \in \Omega(\text{Pred})} \Pr_{\text{Pred}}(\mathbf{p}) \log \left(\frac{\Pr_{\text{Pred}}(\mathbf{p})}{\Pr_{\text{Obs}}(\mathbf{p})} \right)$$

between the probability distributions over the prediction Pred and the (possibly inferred) observation Obs[1]. In the remainder of this paper, to improve readability we abbreviate $D_{KL}(\Pr_{(Pred)} \| \Pr_{(Obs)})$ to simply D_{KL} when the divergence is computed between $\Pr_{(Pred)}$ and $\Pr_{(Obs)}$; we sometimes include brackets $D_{KL[\psi]}$ to refer to the divergence under some particular value assignment, parameter setting, or observation ψ.

The computed prediction error is used to bring prediction and observation closer to each other; either by belief revision, model revision, or by passive or active intervention. In belief revision, we lower prediction error by revising the probability distribution over the space of hypotheses $\Pr_{Hyp}$; by model revision by revising some parameters in $\Pr_B$; by passive intervention by observing the values of some of the intermediate variables; by active intervention by setting the values of some of the intermediate variables. These notions will be developed further in the remainder of the paper when we discuss the computational complexity of these mechanisms of lowering prediction error.

2.1   Computational Complexity

In the remainder, we assume that the reader is familiar with basic concepts of computational complexity theory, in particular Turing Machines, the complexity classes P and NP, and NP-completeness proofs. In addition to these basic concepts, to describe the complexity of various problems we will use the *probabilistic* class PP, oracle machines, and some basic principles from parameterized complexity theory. The interested reader is referred to [10] for more background on complexity issues in Bayesian networks, and to [12] for an introduction in parameterized complexity theory.

The class PP contains languages L that are accepted in polynomial time by a *Probabilistic Turing Machine*. This is a Turing Machine that augments the more traditional non-deterministic Turing Machine with a probability distribution associated with each state transition. Acceptance of an input x is defined as follows: the probability of arriving in an *accept state* is strictly larger than $1/2$ if and only if $x \in L$. This probability of acceptance, however, is not fixed and may (exponentially) depend on the input, e.g., a problem in PP may accept 'yes'-instances with size $|x|$ with probability $1/2 + 1/2^{|x|}$. This means that the probability of acceptance cannot in general be amplified by repeating the computation a polynomial number of times and making a decision based on a majority count, ruling out efficient randomized algorithms. Therefore, PP-complete problems are considered to be intractable. The canonical PP-complete problem is MAJSAT: given a Boolean formula ϕ, does the majority of the truth assignments satisfy ϕ? In Bayesian networks, the canonical problem of determining whether $\Pr(\mathbf{h} \mid \mathbf{e}) > q$ for a given rational q and joint variable assignments $\mathbf{h}$ and $\mathbf{e}$ (known as the INFERENCE problem) is PP-complete.

---

[1] Conform the definition of the Kullback-Leibler divergence, we will interpret the term $0 \log 0$ as 0 when appearing in this formula, as $\lim_{x \to 0} x \log x = 0$. The KL divergence is undefined if for any $\mathbf{p}$, $\Pr_{Obs}(\mathbf{p}) = 0$ while $\Pr_{Pred}(\mathbf{p}) \neq 0$.

A Turing Machine $\mathcal{M}$ has *oracle access* to languages in the class C, denoted as $\mathcal{M}^C$, if it can decide membership queries in C ("consult the oracle") in a single state transition. For example, NP^{PP} is defined as the class of languages which are decidable in polynomial time on a non-deterministic Turing Machine with access to an oracle deciding problems in PP.

Sometimes problems are intractable (i.e., NP-hard) in general, but become tractable if some *parameters* of the problem can be assumed to be small. Informally, a problem is called fixed-parameter tractable for a parameter k (or a set $\{k_1, \ldots, k_n\}$ of parameters) if it can be solved in time, exponential *only* in k and polynomial in the input size $|x|$, i.e., in time $\mathcal{O}(f(k) \cdot |x|^c)$ for a constant c and an arbitrary function f. In practice, this means that problem instances can be solved efficiently, even when the problem is NP-hard in general, if k is known to be small.

Finally, a word on the representation of numerical values. In the complexity proofs we assume that all parameter probabilities are rational numbers (rather than reals), and we assume that logarithmic functions are approximated when needed with sufficient precision, yet polynomial in the length of the problem instance. All logarithms in this paper have base 2.

2.2 Previous Work

The computational complexity of various problems in Bayesian networks is well studied. Interestingly, such problems tend to be complete for complexity classes with few other "real-life" complete problems. For example, deciding upon the MAP distribution is NP^{PP}-complete [24], as well as deciding whether the parameters in a network can be tuned to satisfy particular constraints [21]. Deciding whether a network is monotone is $co - NP^{PP}$-complete [27], and computing the same-decision probability of a network has a PP^{PP}-complete decision variant [11]. Some results are known on the complexity of entropy computations: In [8] it was established #P-hardness of computing the (total) entropy of a Bayesian network; computing the relative entropy between two arbitrary probability distributions is PP-hard [20]. In [2] it was proved that no approximation algorithm can compute a bounded approximation on the entropy of arbitrary distributions using a polynomial amount of samples.

While concerns with respect to the computational complexity of inferences in (unconstrained) HPC models have been raised in [3] and [22], and acknowledged in [6], this paper is (to the best of our knowledge) the first to explicitly address the complexity of minimizing relative entropy in the context of HPC.

3 The Complexity of Computing Relative Entropy in HPC

The first computational problem we will discuss is the computation of the entropy of a prediction, and the relative entropy between a prediction and an observation. While complexity results are known for the computation of the entropy of an entire network [8], respectively the relative entropy between two arbitrary distributions [20], we will here show that decision variants of both problems remain PP-complete even for *singleton* and *binary* hypothesis, prediction, and observation variables. The proof construct we introduce in this proof will be reused, with slight modifications, in subsequent proofs.

We start with defining a decision variant of ENTROPY.

ENTROPY
Instance: A Bayesian network $\mathcal{B}$ with designated variable subsets Pred and Hyp; rational number q.
Question: Is the entropy $E(\text{Pred}) = - \sum_{\mathbf{p} \in \Omega(\text{Pred})} \Pr(\mathbf{p}) \log \Pr(\mathbf{p}) < q$?

We will reduce ENTROPY from MINSAT, defined as follows:

MINSAT
Instance: A Boolean formula ϕ with n variables.
Question: Does the *minority* of truth assignments to ϕ satisfy ϕ?

Note that MINSAT is the complement problem of the PP-complete MAJSAT problem; as PP is closed under complement, MINSAT is PP-complete by a trivial reduction. In order to change as little as possible to the construct in subsequent proofs, we will sometimes reduce from MINSAT and sometimes from MAJSAT.

We will illustrate the reduction form MINSAT to ENTROPY using the example Boolean formula $\phi_{\text{ex}} = \neg x_1 \wedge (x_2 \vee \neg x_3)$; note that this is a 'yes'-instance to MINSAT as three out of eight truth assignments satisfy ϕ_{ex}. We construct a Bayesian network $\mathcal{B}_\phi$ from ϕ as follows. For every variable x_i in ϕ, we construct a binary variable X_i in $\mathcal{B}_\phi$, with values t and f and uniform probability distribution. The set of all variables $X_1, \ldots, X_n$ is denoted with $\mathbf{X}$. For each logical operator in ϕ, we create an additional variable in the network $\mathcal{B}_\phi$. The parents of this variable are the variables that correspond with the sub-formulas joined by the operator; its conditional probability table mimics the truth table of the operator. The variable associated with the top-level operator of ϕ will be denoted by V_ϕ. In addition, we include a binary hypothesis variable H, with uniformly distributed values t and f, and a binary prediction variable P, with values t and f. The parents of this variable are V_ϕ and H, and the conditional probability table of this variable mimics an *and*-operator, i.e., $\Pr(P = t \mid V_\phi, H) = 1$ if and only if both V_ϕ and H are set to t. In Figure 2 we illustrate how $\mathcal{B}_{\phi_{\text{ex}}}$ is thus constructed from ϕ_{ex}. We set Pred $= P$, Hyp $= H$, and $q = 1/2 - 3/4 \log 3/4$.

Theorem 1. ENTROPY *is* PP-*complete, even for singleton binary variables* Pred *and* Hyp.

Proof. Membership proof in PP follows from a trivial modification of the proof that computing the Kullback-Leibler divergence between two distributions is in PP, such as presented in [20].

To prove PP-hardness, we will reduce MINSAT to ENTROPY. Let ϕ be an instance of MINSAT and let $\mathcal{B}_\phi$ be the Bayesian network constructed from ϕ as

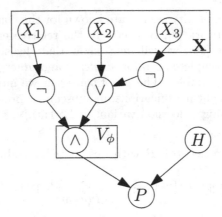

Fig. 2. The Bayesian network $\mathcal{B}_{\phi_{\mathrm{ex}}}$ that is constructed from the MINSAT example ϕ_{ex}. Note that we here have a single hypothesis node H (a source node) and a single prediction node P (a sink node).

described above. Observe that in $\mathcal{B}_\phi$, the posterior probability $\Pr(V_\phi = t \mid \mathbf{X} = \mathbf{x}) = 1$ if and only if the truth assignment corresponding with the joint value assignment $\mathbf{x}$ satisfies ϕ, and 0 otherwise. In particular, if exactly half of the truth assignments satisfy ϕ, then $\Pr(V_\phi = t) = 1/2$ and consequently $\Pr(P = t) = 1/4$. The entropy then equals $E(P) = -(\Pr(P = t) \log \Pr(P = t) + \Pr(P = f) \log \Pr(P = f)) = -(1/4 \log 1/4 + 3/4 \log 3/4) = 1/2 - 3/4 \log 3/4$. The entropy ranges from $E(P) = 0$ in case ϕ is not satisfiable (and hence $\Pr(P = t) = 0$) and $E(P) = 1$ in case ϕ is a tautology (and hence $\Pr(P = t) = 1/2$). In particular, if and only if the minority of truth assignments to ϕ satisfies ϕ, then $E(P) < 1/2 - 3/4 \log 3/4 = \mathrm{q}$. Note that the reduction can be done in polynomial time, given our assumptions on the tractable approximation of the logarithms involved; hence, ENTROPY is PP-complete.   □

Computing the *relative* entropy between a prediction and an observation is defined as a decision problem as follows.

RELATIVEENTROPY
Instance: A Bayesian network $\mathcal{B}$ with designated variable subset Pred, where $\Pr_{(\mathrm{Pred})}$ denotes the posterior distribution over Pred; an observed distribution $\Pr_{(\mathrm{Obs})}$ over Pred; a rational number q.
Question: Is the relative entropy $D_{\mathrm{KL}} < q$?

To prove PP-completeness, we use the same construction as above, but now we set $q = 3/4 \log 3/2 - 1/4$. In addition, we set $\Pr_{(\mathrm{Obs})}$ to $\Pr(P = t) = 1/2$.

Theorem 2. RELATIVEENTROPY *is PP-complete, even for singleton binary variables* Pred *and* Hyp.

Proof. Membership in PP of the more general problem of computing the Kullback-Leibler divergence between two arbitrary probability distributions was established in [20]. To prove PP-hardness, we reduce MAJSAT to RELATIVEEN-TROPY. Let ϕ be an instance of MAJSAT and let $\mathcal{B}_\phi$ be the Bayesian network constructed from ϕ as described above. Observe that in $\mathcal{B}_\phi$ D_{KL} decreases when $\Pr(V_\phi = t)$ increases; in particular, when $\Pr(V_\phi = t) = p$ (and hence $\Pr(P = t) = p/2$), $D_{\mathrm{KL}} = p/2 \log(\frac{p/2}{1/2}) + (2-p)/2 \log(\frac{(2-p)/2}{1/2})$. Note that $\Pr(V_\phi = t) = 1/2$ if exactly half of the truth assignments to ϕ satisfy ϕ. Hence, if and only if a *majority* of truth assignments to ϕ satisfies ϕ, then $D_{\mathrm{KL}} < 1/4 \log(\frac{1/4}{1/2}) + 3/4 \log(\frac{3/4}{1/2}) = 3/4 \log 3/2 - 1/4 = q$. As the reduction can be done in polynomial time, this proves that RELATIVEENTROPY is PP-complete.   □

In subsequent sections we will discuss the complexity of *lowering* the relative entropy D_{KL} by means of belief revision, model revision, or by passive or active intervention.

4   Revision of Beliefs

In this section we discuss *belief revision*, i.e., changing the probability distribution over the hypothesis variables, as a means to reduce relative entropy. We formulate two decision problems that capture this concept; the first one focuses on lowering the relative entropy *to* some threshold, the second one on lowering the relative entropy *by* some amount.

BELIEFREVISION1
Instance: A Bayesian network $\mathcal{B}$ with designated variable subsets Hyp and Pred, where $\Pr_{(\mathrm{Hyp})}$ denotes the prior distribution over Hyp, and $\Pr_{(\mathrm{Pred})}$ denotes the posterior distribution over Pred; an observed distribution $\Pr_{(\mathrm{Obs})}$ over Pred; a rational number q.
Question: Is there a (revised) prior probability distribution $\Pr_{(\mathrm{Hyp})'}$ over Hyp such that $D_{\mathrm{KL}[\mathrm{Hyp}']} < q$?

BELIEFREVISION2
Instance: As in BELIEFREVISION1.
Question: Is there a (revised) prior probability distribution $\Pr_{(\mathrm{Hyp})'}$ over Hyp such that $D_{\mathrm{KL}[\mathrm{Hyp}]} - D_{\mathrm{KL}[\mathrm{Hyp}']} > q$?

We prove that both problems are PP-hard via a reduction from MAJSAT, again using the construct that we used in the proof of Theorem 1, but we redefine the conditional probability distribution $\Pr(P \mid V_\phi, H)$ and we redefine $\Pr_{(\mathrm{Hyp})}$, $\Pr_{(\mathrm{Obs})}$, and q. Let $\Pr(P \mid V_\phi, H)$ be defined as follows:

$$\Pr(P = t \mid V_\phi, H) = \begin{cases} 3/8 & \text{if } V_\phi = t, H = t \\ 0 & \text{if } V_\phi = t, H = f \\ 1/8 & \text{if } V_\phi = f, H = t \\ 0 & \text{if } V_\phi = f, H = f \end{cases}$$

We set $\Pr_{(\text{Hyp})}$ to $\Pr(H = t) = 0$ and $\Pr_{(\text{Obs})}$ to $\Pr(P = t) = {}^{15}\!/_{16}$. For BE-LIEFREVISION1, we redefine $q = q_1 = {}^{1}\!/_{4}\log(\frac{1/4}{15/16}) + {}^{3}\!/_{4}\log(\frac{3/4}{1/16})$. For BELIEFRE-VISION2, we redefine $q = q_2 = 4 - {}^{1}\!/_{4}\log(\frac{1/4}{15/16}) - {}^{3}\!/_{4}\log(\frac{3/4}{1/16})$. We now claim the following.

Theorem 3. BELIEFREVISION1 *and* BELIEFREVISION2 *are* PP-*hard, even for singleton binary variables* Pred *and* Hyp.

Proof. To prove PP-hardness, we reduce BELIEFREVISION from MAJSAT. Let ϕ be an instance of MAJSAT and let $\mathcal{B}_\phi$ be the Bayesian network constructed from ϕ as described above. Observe that in $\mathcal{B}_\phi$ $D_{\text{KL}}[\text{Hyp}]$ is independent of $\Pr(V_\phi)$ as $\Pr(P = t \mid V_\phi, H) = 0$ (as $\Pr(H = t) = 0$) and thus $D_{\text{KL}}[\text{Hyp}] = 0 + \log(\frac{1}{1/16}) = 4$.

We now investigate the effect of revising the hypothesis distribution $\Pr_{(\text{Hyp})}$ to $\Pr_{(\text{Hyp})'}$. For every probability distribution $\Pr(V_\phi)$, D_{KL} increases when $\Pr(H = t)$ goes to 0, and decreases when $\Pr(H = t)$ goes to 1. That is, $D_{\text{KL}}[\text{Hyp}']$ is minimal for $\Pr_{(\text{Hyp})'} = \Pr(H = t) = 1$. In general, for $\Pr(H = t) = 1$ and $\Pr(V_\phi) = p$, $\Pr(P = t \mid V_\phi, H) = {}^{(2p+1)}\!/_{8}$ and $D_{\text{KL}}[\text{Hyp}'] = {}^{(2p+1)}\!/_{8}\log(\frac{(2p+1)/8}{15/16}) + {}^{(7-2p)}\!/_{8}\log(\frac{(7-2p)/8}{1/16})$. For $\Pr(V_\phi) = {}^{1}\!/_{2}$ and $\Pr_{(\text{Hyp})'} = \Pr(H = t) = 1$, $\Pr(P = t \mid V_\phi, H) = {}^{1}\!/_{4}$ and $D_{\text{KL}}[\text{Hyp}'] = {}^{1}\!/_{4}\log(\frac{1/4}{15/16}) + {}^{3}\!/_{4}\log(\frac{3/4}{1/16})$. We have in that case that $D_{\text{KL}}[\text{Hyp}] - D_{\text{KL}}[\text{Hyp}'] = 4 - {}^{1}\!/_{4}\log(\frac{1/4}{15/16}) - {}^{3}\!/_{4}\log(\frac{3/4}{1/16})$.

In particular if and only if $\Pr(V_\phi) > {}^{1}\!/_{2}$ there exists a revised hypothesis distribution $\Pr_{(\text{Hyp})'}$ (i.e., $\Pr(H = t) = 1$) such that $D_{\text{KL}}[\text{Hyp}'] < q_1$ and that $D_{\text{KL}}[\text{Hyp}] - D_{\text{KL}}[\text{Hyp}'] > q_2$. Now, $\Pr(V_\phi) > {}^{1}\!/_{2}$ if and only if there is a majority of truth assignments to ϕ that satisfies ϕ. Given that the reduction can be done in polynomial time, this proves PP-hardness of both BELIEFREVISION1 and BELIEFREVISION2.   □

Note that these problems are not known or believed to be in PP, as we need to determine a revised probability distribution $\Pr_{(\text{Hyp})'}$ as well as computing the relative entropy. In case Hyp is a singleton binary variable (as in our constrained proofs), the probability $\Pr_{(\text{Pred})}$ depends linearly on this distribution [4], but the complexity of this dependency grows when the distribution spans multiple variables. This makes a polynomial sub-computation of $\Pr_{(\text{Hyp})'}$, and thus membership in PP, unlikely. However, we can non-deterministically *guess* the value of $\Pr_{(\text{Hyp})'}$ and then decide the problem using an oracle for RELATIVEENTROPY; for this reason, the problems are certainly in the complexity class NP^{PP}.

5   Revision of Models

In the previous section we defined belief revision as the revision of the prior distribution over Hyp. We can also revise the stochastic dependences in the model, i.e., how Pred depends on Hyp (and Int). However, a naive formulation of *model revision* will give us a trivial algorithm for solving it, yet unwanted side effects.

NAIVEMODELREVISION
Instance: A Bayesian network $\mathcal{B}$ with designated variable subsets Hyp and Pred, where $\text{Pr}_{(\text{Pred})}$ denotes the posterior distribution over Pred; an observed distribution $\text{Pr}_{(\text{Obs})}$ over Pred; a rational number q.
Question: Is there a probability distribution Pr_{new} over the variables in $\mathcal{B}$ such that $D_{\text{KL}[\text{new}]} < q$?

Note that this problem can be solved rather trivially by reconfiguring the CPTs such that $\text{Pr}(\text{Pred}) = \text{Pr}(\text{Obs})$ and thus $D_{\text{KL}[\text{new}]} = 0$. This has of course consequences for previous experiences—we are likely to induce unexplained past prediction errors. However, we cannot assume that we have access to (all) previous predictions and observations, making it close to impossible to minimize joint prediction error over all previous predictions and observations. As we do want to constrain the revisions in some way or another, we propose to revise the current model by allowing modification only of a *designated subset* of parameters in the model. So, we reformulate model revision to decide whether we can decrease D_{KL} by a change in a subset **p** of parameter probabilities in the network.[2] As in belief revision, we define two variants of the decision problem.

MODELREVISION1
Instance: A Bayesian network $\mathcal{B}$ with designated variables Hyp and Pred, where $\text{Pr}_{(\text{Pred})}$ denotes the posterior distribution over Pred; an observed distribution $\text{Pr}_{(\text{Obs})}$ over Pred; a subset **P** of the parameter probabilities represented by $\text{Pr}_{\mathcal{B}}$; a rational number q.
Question: Is there a combination of values **p** to **P** such that $D_{\text{KL}[\mathbf{p}]} < q$?

MODELREVISION2
Instance: As in MODELREVISION1.
Question: Is there a combination of values **p** to **P** such that $D_{\text{KL}} - D_{\text{KL}[\mathbf{p}]} > q$?

We will show that these problems are NP^{PP}-complete, that is, as least as hard as PARTIAL MAP [24] and PARAMETER TUNING [21]. To prove NP^{PP}-hardness, we reduce from the following NP^{PP}-complete problem:

E-MAJSAT
Instance: A Boolean formula ϕ with n variables, partitioned into sets $\mathbf{X_E} = x_1, \ldots, x_k$ and $\mathbf{X_M} = x_{k+1}, \ldots, x_n$ for $1 \leq k \leq n$.
Question: Is there a truth assignment $\mathbf{x_E}$ to $\mathbf{X_E}$ such that the majority of truth assignments to $\mathbf{X_M}$ together with $\mathbf{x_E}$ satisfy ϕ?

---

[2] One of the anonymous made the interesting observation that *changing the network structure* (i.e., removing or adding arcs) can also be seen as model revision. We do not address that aspect here.

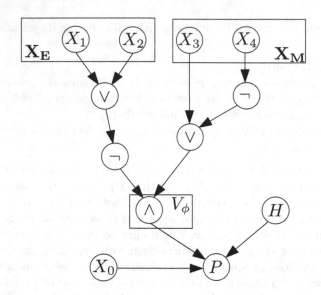

Fig. 3. The Bayesian network $\mathcal{B}_{\phi_{ex}}$ that is constructed from the E-MAJSAT example ϕ_{ex}

We will use the following E-MAJSAT instance $(\phi_{ex}, \mathbf{X_E}, \mathbf{X_M})$ as a running example in the construction: $\phi_{ex} = (\neg(x_1 \lor x_2)) \land (\neg(x_3 \lor \neg x_4))$, $\mathbf{X_E} = \{x_1, x_2\}$, $\mathbf{X_M} = \{x_3, x_4\}$; note that this is a 'yes'-instance to E-MAJSAT: for $x_1 = x_2 = f$, three out of four truth assignments to $\mathbf{X_M}$ satisfy ϕ_{ex}.

We construct $\mathcal{B}_{\phi_{ex}}$ from ϕ_{ex} in a similar way as in the proof of Theorem 3, but we add another binary variable X_0 as an additional parent of P, with prior probability distribution $\Pr(X_0 = t) = 0$ (Figure 3). We define $\Pr(P \mid V_\phi, H, X_0)$ as follows:

$$\Pr(P = t \mid V_\phi, H, X_0) = \begin{cases} 3/8 & \text{if } V_\phi = t, H = X_0 = t \\ 1/8 & \text{if } V_\phi = f, H = X_0 = t \\ 0 & \text{otherwise} \end{cases}$$

We redefine $\Pr_{(Hyp)}$ to $\Pr(H = t) = 1/2$ and $\Pr_{(Obs)}$ to $\Pr(P = t) = 31/32$. In addition, we designate the sets of variables $\mathbf{X_E}$ and $\mathbf{X_M}$ in the network, and we set $\mathbf{P} = \mathbf{X_E} \cup \{X_0\}$. We set $q = q_1 = 1/8 \log(\frac{1/8}{31/32}) + 7/8 \log(\frac{7/8}{1/32})$ and $q = q_2 = 5 - 1/8 \log(\frac{1/8}{31/32}) - 7/8 \log(\frac{7/8}{1/32})$.
We now claim the following.

Theorem 4. MODELREVISION1 *and* MODELREVISION2 *are* NP^{PP}-*complete, even for singleton binary variables* Pred *and* Hyp.

Proof. Membership follows from the following algorithm: non-deterministically guess a combination of values **p** and compute the (change in) relative entropy.

This can be done in polynomial time using a non-deterministic Turing Machine with access to an oracle for problems in PP.

To prove $\mathsf{NP}^{\mathsf{PP}}$-hardness, we reduce MODELREVISION from E-MAJSAT. Let $(\phi, \mathbf{X_E}, \mathbf{X_M})$ be an instance of MAJSAT and let $\mathcal{B}_\phi$ be the Bayesian network constructed from ϕ as described above. Observe that in $\mathcal{B}_\phi$, given the prior probability distribution of X_0, we have that $\Pr(P = t \mid V_\phi, H, X_0) = 0$ independent of the probability distribution of V_ϕ, and thus $D_{\mathrm{KL}} = 0 + \log(1/32) = 5$. If we revise the prior probability distribution of X_0, we observe that D_{KL} decreases when $\Pr(X_0 = t)$ goes to 1; $D_{\mathrm{KL}[\Pr(X_0=t)]}$ is minimal for $\Pr(X_0 = t) = 1$. In that case, for $\Pr(V_\phi) = p$, $\Pr(P = t \mid V_\phi, H, X_0) = (2p+1)/16$ and $D_{\mathrm{KL}[\Pr(X_0=t)=1]} = (2p+1)/16 \log(\frac{(2p+1)/16}{31/32}) + (15-2p)/16 \log(\frac{(15-2p)/16}{1/32})$.

For $\Pr(V_\phi) = 1/2$, $\Pr(P = t \mid V_\phi, H, X_0) = 1/8$ and $D_{\mathrm{KL}[\Pr(X_0=t)=1]} = 1/8 \log(\frac{1/8}{31/32}) + 7/8 \log(\frac{7/8}{1/32})$. We have in that case that $D_{\mathrm{KL}} - D_{\mathrm{KL}[\Pr(X_0=t)=1]} = 5 - 1/8 \log(\frac{1/8}{31/32}) - 7/8 \log(\frac{7/8}{1/32})$.

If there exists a truth assignment $\mathbf{x_E}$ to $\mathbf{X_E}$ such that the majority of truth assignments to $\mathbf{X_M}$ satisfies ϕ, then there exists a combination of values $\mathbf{p}$ to $\mathbf{P} = \mathbf{X_E} \cup \{X_0\}$ such that $\Pr(V_\phi) > 1/2$ and thus $D_{\mathrm{KL}[\Pr(X_0=t)=1]} < q_1$ and $D_{\mathrm{KL}} - D_{\mathrm{KL}[\Pr(X_0=t)=1]} > q_2$; namely, the combination of values to $\mathbf{X_E}$ that sets $\Pr(X_i = t)$ to 1 if $X_i \in \mathbf{X_E}$ is set to t, and $\Pr(X_i = t)$ to 0 if $X_i \in \mathbf{X_E}$ is set to f, together with setting $\Pr(X_0 = t)$ to 1. Vice versa, if we can revise $\mathbf{P}$ such that $D_{\mathrm{KL}[\Pr(X_0=t)=1]} < q_1$ and that $D_{\mathrm{KL}} - D_{\mathrm{KL}[\Pr(X_0=t)=1]} > q_2$, then there exists a truth assignment $\mathbf{x_E}$ to $\mathbf{X_E}$ such that the majority of truth assignments to $\mathbf{X_M}$ satisfies ϕ, namely, the truth assignment that sets $X_i \in \mathbf{X_E}$ to t if $\Pr(X_i = t) \geq 1/2$ and to f otherwise.

Given that the reduction can be done in polynomial time, this proves $\mathsf{NP}^{\mathsf{PP}}$-completeness of both MODELREVISION1 and MODELREVISION2. $\square$

6 Adding Additional Observations to the Model

Apart from revising the probability distribution of the hypotheses and from revising the parameters in the model, we can also lower relative entropy by some action that influences either the outside world or our perception of it. By observing previously unobserved variables in the model (i.e., changing our perception of the world), the posterior probability of the *prediction* can be influenced; similarly, we can *intervene* in the outside world, thus influencing the posterior probability over the *observation*. In both cases, we will need to decide on *which* observations to gather, respectively *which* variables to intervene on. Again we assume that the set of allowed observations, respectively interventions, is designated. We will first focus on the question which candidate observations to make. As in the previous two problems, we formulate two decision problems that capture this question.

ADDOBSERVATION1
Instance: A Bayesian network $\mathcal{B}$ with designated variables Hyp and Pred, where $\Pr_{(\mathrm{Pred})}$ denotes the posterior distribution over Pred; an observed distribution

$\Pr_{(Obs)}$ over Pred; and rational number q. Let $\mathbf{O} \subseteq$ Int denote the set of observable variables in $\mathcal{B}$.

Question: Is there a joint value assignment $\mathbf{o}$ to $\mathbf{O}$ such that $D_{KL}[\mathbf{o}] < q$?

ADDOBSERVATION2

Instance: As in ADDOBSERVATION1.

Question: Is there a joint value assignment $\mathbf{o}$ to $\mathbf{O}$ such that $D_{KL} - D_{KL}[\mathbf{o}] > q$?

While these problems are *conceptually* different from the MODELREVISION problems, from a *complexity* point of view they are very similar: the effect of setting a prior probability of a variable X_i in the proof construct to 1, and observing its value to be t, are identical; the same holds for setting it to 0, respectively observing its value to be f. This allows us to prove NP^{PP}-completeness of ADDOBSERVATION using essentially the same construct as in the proof of Theorem 4; however, we must take care that the prior probability distribution of X_0 is such that no inconsistencies in the network emerge as a result of observing its value to t. In particular, if $\Pr(X_0 = t) = 0$, then we cannot observe X_0 to be t without creating an inconsistency in the network.

So, we redefine $\Pr(X_0 = t) = 1/2$; now, $\Pr(P = t \mid V_\phi, H, X_0)$ (and thus also D_{KL}) becomes dependent of the probability distribution of V_ϕ. In particular, for $\Pr(V_\phi) = p$ we have that $\Pr(P = t \mid V_\phi, H, X_0) = (2p+1)/32$ and consequently, $D_{KL} = (2p+1)/32 \log(\frac{(2p+1)/32}{31/32}) + (31 - 2p)/32 \log(\frac{(31-2p)/32}{1/32})$. We therefore redefine $q_2 = 1/16 \log(\frac{1/16}{31/32}) + 15/16 \log(\frac{15/16}{1/32}) - q_1 = 1/16 \log(\frac{1/16}{31/32}) + 15/16 \log(\frac{15/16}{1/32}) - 1/8 \log(\frac{1/8}{31/32}) - 7/8 \log(\frac{7/8}{1/32})$. We set $\mathbf{O} = \mathbf{X_E} \cup \{X_0\}$.

Theorem 5. ADDOBSERVATION1 *and* ADDOBSERVATION2 *are* NP^{PP}-*complete*.

Proof. Membership follows from a similar argument as for MODELREVISION. To prove NP^{PP}-hardness, we again reduce from E-MAJSAT. Let $(\phi, \mathbf{X_E}, \mathbf{X_M})$ be an instance of E-MAJSAT and let $\mathcal{B}_\phi$ be the Bayesian network constructed from ϕ as described above. The probability distribution $\Pr(P = t \mid V_\phi, H, X_0)$ depends as follows on the observed value of X_0: $\Pr(P = t \mid V_\phi, H, X_0 = t) = (2p+1)/16$ and $\Pr(P = t \mid V_\phi, H, X_0 = f) = 0$. In particular, if $\Pr(V_\phi) > 1/2$, then $\Pr(P = t \mid V_\phi, H, X_0 = t) > 1/8$ and hence $D_{KL[X_0=t]} < 1/8 \log(\frac{1/8}{31/32}) + 7/8 \log(\frac{7/8}{1/32})$. Similarly, $\Pr(P = t \mid V_\phi, H, X_0 = f) = 0$ and hence $D_{KL[X_0=f]} = 5$. So, only if X_0 is observed to be t and $\Pr(V_\phi) > 1/2$ we have that $D_{KL[X_0=t]} < q_1$ and $D_{KL} - D_{KL[X_0=t]} > q_2$.

If there exists a truth assignment $\mathbf{x_E}$ to $\mathbf{X_E}$ such that the majority of truth assignments to $\mathbf{X_M}$ satisfies ϕ, then there exists a joint value assignment to $\mathbf{O} = \mathbf{X_E} \cup \{X_0\}$ such that $\Pr(V_\phi) > 1/2$ and $D_{KL}[\mathbf{o}] < q_1$ and that $D_{KL} - D_{KL}[\mathbf{o}] > q_2$. Namely, the joint value assignment that sets X_0 to t and sets the variables in $\mathbf{X_E}$ according to $\mathbf{x_E}$. And vice versa, if there exists a joint value assignment $\mathbf{o}$ to $\mathbf{O}$ such that $D_{KL}[\mathbf{o}] < q_1$ and $D_{KL} - D_{KL}[\mathbf{o}] > q_2$, then there is a truth assignment to $\mathbf{X_E}$ such that the majority of truth assignments to $\mathbf{X_M}$ satisfy ϕ, namely, the truth assignment that sets $X_i \in \mathbf{X_E}$ to t if $X_i \in \mathbf{o}$ is observed as t, and to f

otherwise. As this reduction can be done in polynomial time, this proves that ADDOBSERVATION1 and ADDOBSERVATION2 are $\mathsf{NP}^{\mathsf{PP}}$-complete. $\qquad\square$

7 Intervention in the Model

We can bring prediction and observation closer to each other by changing our prediction (by influencing the posterior distribution of the prediction by revision of beliefs, parameters, or observing variables), but also by what in the HPC framework is called *active inference*: actively changing the causes of the observation to let the observation ("the real world") match the prediction ("the model of the world"). This is a fundamental aspect of the theory, which is used to explain how a desire of moving one's arm—i.e., the expectation or prediction that one's arm will be in a different position two seconds from now—can yield actual motor acts that establish the desired movement. We implement this as *intervention* in the Bayesian framework, and the problem that needs to be resolved is to decide *how* to intervene.

The predicted result of an action of course follows from the generative model, which represents how (hypothesized) causes generate (predicted) effects, for example, how motor commands sent to the arm will change the perception of the arm. So, from a computational point of view, the decision variants of the INTERVENTION problem are identical to the decision variants of the OBSERVATION problem:

INTERVENTION1
Instance: A Bayesian network $\mathcal{B}$ with designated variables Hyp and Pred, where $\mathrm{Pr}_{(\mathrm{Pred})}$ denotes the posterior distribution over Pred; an observed distribution $\mathrm{Pr}_{(\mathrm{Obs})}$ over Pred; and rational number q. Let $\mathbf{A} \subseteq$ Int denote the set of intervenable variables in $\mathcal{B}$.
Question: Is there a joint value assignment $\mathbf{a}$ to $\mathbf{A}$ such that $D_{\mathrm{KL}[\mathbf{a}]} < q$?

INTERVENTION2
Instance: As in INTERVENTION1.
Question: Is there a joint value assignment $\mathbf{a}$ to $\mathbf{A}$ such that $D_{\mathrm{KL}} - D_{\mathrm{KL}[\mathbf{a}]} > q$?

Corollary 1. INTERVENTION1 *and* INTERVENTION2 *are* $\mathsf{NP}^{\mathsf{PP}}$-*complete.*

8 Parameterized Complexity

What situational constraints can render the computations tractable? From the intractability proofs above we can already infer what *does not* make prediction error minimization tractable. Even for binary variables, singleton hypothesis and prediction nodes, and at most three incoming arcs per variable, all problems remain intractable. It is easy to show that MODELREVISION, ADDOBSERVATION, and INTERVENTION remain PP-hard when there is just a single designated parameter, observable or intervenable variable. The complexity of these problems is basically in the *context* that modulates the relation between hypothesis and prediction.

ADDOBSERVATION and INTERVENTION are fixed-parameter tractable for the parameter set {treewidth of the network, cardinality of the variables, size of Pred} plus the size of **O**, respectively **A**. In that case, the computation of D_{KL} is tractable, and we can search joint value assignments to **O**, respectively **A** exhaustively. Similarly, when the computation of D_{KL} is tractable, one can use parameter tuning algorithms to decide MODELREVISION and BELIEFREVISION; these problems are fixed-parameter tractable for the parameter set {treewidth of the network, cardinality of the variables, size of Pred} plus the size of **P**, respectively Hyp [5].

9   Conclusion

Hierarchical Predictive Coding (HPC) is an influential unifying theory in theoretical neuroscience, proposing that the brain continuously makes Bayesian predictions about future states and uses the prediction error between prediction and observation to update the hypotheses that drove the predictions. In this paper we studied HPC from a computational perspective, formalizing the conceptual ideas behind hypothesis updating, model revision, and active inference, and studying the computational complexity of these problems. Despite rather explicit claims on the contrary (e.g., [7, p.191]), we show that the Bayesian computations that underlie the error minimization mechanisms in HPC are *not* computationally tractable in general, even when hypotheses and predictions are constrained to binary singleton variables. Even in this situation, rich contextual modulation of the dependences between hypothesis and prediction may render successful updating intractable. Further constraints on the structure of the dependences (such as small treewidth and limited choice in which parameters to observe or observations to make) are required.

In this paper, we focused on computations within a particular level of the hierarchy and on error minimization. There is more to say about the computations that are postulated within HPC, for example how increasingly rich and complex knowledge structures are *learned* from prediction errors. We leave that for further research.

References

1. Abdelbar, A.M., Hedetniemi, S.M.: Approximating MAPs for belief networks is NP-hard and other theorems. Artificial Intelligence 102, 21–38 (1998)
2. Batu, T., Dasgupta, S., Kumar, R., Rubinfeld, R.: The complexity of approximating the entropy. SIAM Journal on Computing 35(1), 132–150 (2005)
3. Blokpoel, M., Kwisthout, J., van Rooij, I.: When can predictive brains be truly Bayesian? Frontiers in Theoretical and Philosophical Psychology 3, 406 (2012)

4. Castillo, E., Gutiérrez, J.M., Hadi, A.S.: Sensitivity analysis in discrete Bayesian networks. IEEE Transactions on Systems, Man, and Cybernetics 27, 412–423 (1997)
5. Chan, H., Darwiche, A.: Sensitivity analysis in Bayesian networks: From single to multiple parameters. In: Proceedings of the 20th Conference in Uncertainty in Artificial Intelligence, pp. 67–75 (2004)
6. Clark, A.: The many faces of precision (Replies to commentaries on "Whatever next? Neural prediction, situated agents, and the future of cognitive science"). Frontiers in Theoretical and Philosophical Psychology 4, e270 (2013)
7. Clark, A.: Whatever next? Predictive brains, situated agents, and the future of cognitive science. Behavioral and Brain Sciences 36(3), 181–204 (2013)
8. Cooper, G.F., Herskovitz, E.: Determination of the entropy of a belief network is NP-hard. Technical Report KSL-90-21, Stanford University. Computer Science Deptu. Knowledge Systems Laboratory (March 1990)
9. Dagum, P., Luby, M.: Approximating probabilistic inference in Bayesian belief networks is NP-hard. Artificial Intelligence 60(1), 141–153 (1993)
10. Darwiche, A.: Modeling and Reasoning with Bayesian Networks. CU Press, Cambridge (2009)
11. Darwiche, A., Choi, A.: Same-decision probability: A confidence measure for threshold-based decisions under noisy sensors. In: 5th European Workshop on Probabilistic Graphical Models (2010)
12. Downey, R.G., Fellows, M.R.: Parameterized Complexity. Springer, Berlin (1999)
13. Friston, K.J.: The free-energy principle: A rough guide to the brain? Trends in Cognitive Sciences 13(7), 293–301 (2009)
14. Friston, K.J.: The free-energy principle: A unified brain theory? Nature Reviews Neuroscience 11(2), 127–138 (2010)
15. Friston, K.J., Daunizeau, J., Kilner, J., Kiebel, S.J.: Action and behavior: A free-energy formulation. Biological Cybernetics 102(3), 227–260 (2010)
16. Griffiths, T.L., Chater, N., Kemp, C., Perfors, A., Tenenbaum, J.B.: Probabilistic models of cognition: Exploring representations and inductive biases. Trends in Cognitive Sciences 14(8), 357–364 (2010)
17. Hohwy, J.: The Predictive Mind. Oxford University Press (2013)
18. Kilner, J.M., Friston, K.J., Frith, C.D.: The mirror-neuron system: A Bayesian perspective. Neuroreport 18, 619–623 (2007)
19. Knill, D., Pouget, A.: The Bayesian brain: The role of uncertainty in neural coding and computation. Trends in Neuroscience 27(12), 712–719 (2004)
20. Kwisthout, J.: The Computational Complexity of Probabilistic Networks. PhD thesis, Faculty of Science, Utrecht University, The Netherlands (2009)
21. Kwisthout, J., van der Gaag, L.C.: The computational complexity of sensitivity analysis and parameter tuning. In: Chickering, D.M., Halpern, J.Y. (eds.) Proceedings of the 24th Conference on Uncertainty in Artificial Intelligence, pp. 349–356. AUAI Press (2008)
22. Kwisthout, J., Van Rooij, I.: Predictive coding: Intractability hurdles that are yet to overcome [abstract]. In: Knauff, M., Pauen, M., Sebanz, N., Wachsmuth, I. (eds.) Proceedings of the 35th Annual Conference of the Cognitive Science Society. Cognitive Science Society, Austin (2013)
23. Park, J.-C., Lim, J.H., Choi, H., Kim, D.-S.: Predictive coding strategies for developmental neurorobotics. Frontiers in Psychology 3, 134 (2012)
24. Park, J.D., Darwiche, A.: Complexity results and approximation settings for MAP explanations. Journal of Artificial Intelligence Research 21, 101–133 (2004)

25. Saygin, A.P., Chaminade, T., Ishiguro, H., Driver, J., Frith, C.: The thing that should not be: Predictive coding and the uncanny valley in perceiving human and humanoid robot actions. Social Cognitive and Affective Neuroscience 7(4), 413–422 (2012)
26. Seth, A.K., Suzuki, K., Critchley, H.D.: An interoceptive predictive coding model of conscious presence. Frontiers in Psychology 2, e395 (2011)
27. van der Gaag, L.C., Bodlaender, H.L., Feelders, A.J.: Monotonicity in Bayesian networks. In: Chickering, M., Halpern, J. (eds.) Proceedings of the Twentieth Conference on Uncertainty in Artificial Intelligence, pp. 569–576. AUAI Press, Arlington (2004)

Treewidth and the Computational Complexity of MAP Approximations

Johan Kwisthout

Radboud University Nijmegen, Donders Institute for Brain,
Cognition and Behaviour, Montessorilaan 3, 6525 HR Nijmegen, The Netherlands
j.kwisthout@donders.ru.nl

Abstract. The problem of finding the most probable explanation to a designated set of variables (the MAP problem) is a notoriously intractable problem in Bayesian networks, both to compute exactly and to approximate. It is known, both from theoretical considerations and from practical experiences, that low treewidth is typically an essential prerequisite to efficient exact computations in Bayesian networks. In this paper we investigate whether the same holds for approximating MAP. We define four notions of approximating MAP (by value, structure, rank, and expectation) and argue that all of them are intractable in general. We prove that efficient value-, structure-, and rank-approximations of MAP instances with high treewidth will violate the Exponential Time Hypothesis. In contrast, we hint that expectation-approximation can be done efficiently, even in MAP instances with high treewidth, if the most probable explanation has a high probability.

1 Introduction

One of the most important computational problems in Bayesian networks is the MAP problem, i.e., the problem of finding the joint value assignment to a designated set of variables (the MAP variables) with the maximum posterior probability. The MAP problem is notably intractable; as it is NP^{PP}-hard, it is strictly harder (given usual assumptions in computational complexity theory) than the PP-hard inference problem [17]. In a sense, it can be seen as combining an *optimization* problem with an *inference* problem, both of which potentially contribute to the problem's complexity [17, p. 113]. Even when all variables in the network are binary and the network has the (very restricted) polytree topology, MAP remains NP-hard [5]. Only when both the optimization *and* the inference part of the problem can be computed tractably (for example, if both the treewidth of the network and the cardinality of the variables are small *and* the most probable joint value assignment has a high probability) MAP can be computed tractably [11]. It is known that, for arbitrary probability distributions and under the assumption of the Exponential Time Hypothesis, a small treewidth of the moralized graph of a Bayesian network is a necessary condition for the inference problem to be tractable [13]; this result can easily be extended to MAP.

L.C. van der Gaag and A.J. Feelders (Eds.): PGM 2014, LNAI 8754, pp. 271–285, 2014.

MAP is also intractable to approximate [1,11,12,17]. While it is obviously the case that a particular instance to the MAP problem can be approximated efficiently when it can be efficiently computed exactly, it is as yet unclear whether approximate MAP computations can be rendered tractable under *different* conditions than exact MAP computations. Crucial here is the question *what we mean* with a statement as 'algorithm A approximates the MAP problem'. Typically, in computer science, approximation algorithms guarantee that the output of the algorithm has a value that is within some bound of the value of the optimal solution. For example, the canonical approximation algorithm to the VERTEX COVER problem selects an edge at random, puts both endpoints in the vertex cover, and removes these nodes from the instance. This algorithm is guaranteed to get a solution that has at most twice the number of nodes in the vertex cover as the optimal vertex set. However, typical Bayesian approximation algorithms have no such guarantee; in contrast, they may converge to the optimal value given enough time (such as the Metropolis-Hastings algorithm), or they may find an optimal solution with a high probability of success (such as repeated local search strategies).

In this paper we assess different notions of approximation as relevant for the MAP problem; in particular value-approximation, structure-approximation, rank-approximation, and expectation-approximation of MAP. After introducing notation and providing some preliminaries (Section 2), we show that each of these approximations is intractable under the assumption that $P \neq NP$, respectively $NP \not\subseteq BPP$ (Section 3). Building on the result in [13] we show in Section 4 that bounded treewidth is indeed a necessary condition for efficient value-, structure-, and rank-approximation of MAP; however, we show that MAP can sometimes be efficiently expectation-approximated, even on networks where the moralized graph has a high treewidth, if the most probable joint value assignment to the MAP variables has a high probability. We conclude the paper in Section 5.

2   Preliminaries

In this section, we introduce our notational conventions and provide some preliminaries on Bayesian networks, graph theory, and complexity theory; in particular definitions of the MAP problem, treewidth, parameterized complexity theory, and the Exponential Time Hypothesis. For a more thorough discussion of these concepts, the reader is referred to textbooks such as [4], [3], and [6].

2.1   Bayesian Networks

A Bayesian network $\mathcal{B} = (\mathbf{G}_\mathcal{B}, \Pr)$ is a graphical structure that models a joint probability distribution over a set of stochastic variables. $\mathcal{B}$ includes a directed acyclic graph $\mathbf{G}_\mathcal{B} = (\mathbf{V}, \mathbf{A})$, where $\mathbf{V}$ models the variables and $\mathbf{A}$ models the conditional (in)dependences between them, and a set of parameter probabilities $\Pr$ in the form of conditional probability tables (CPTs), capturing the strengths of the relationships between the variables. The network models a joint probability distribution $\Pr(\mathbf{V}) = \prod_{i=1}^{n} \Pr(V_i \mid \pi(V_i))$ over its variables; here, $\pi(V_i)$ denotes the parents of V_i in $\mathbf{G}_{\mathcal{B}}$. We will use upper case letters to denote individual nodes in the network, upper case bold letters to denote sets of nodes, lower case letters to denote value assignments to nodes, and lower case bold letters to denote joint value assignments to sets of nodes.

One of the key computational problems in Bayesian networks is the problem to find the most probable explanation for a set of observations, i.e., the joint value assignment to a designated set of variables (the explanation set) that has highest posterior probability given the observed variables (the joint value assignment to the evidence set) in the network. If the network is bi-partitioned into explanation variables and evidence variables this problem is known as MOST PROBABLE EXPLANATION (MPE). The more general problem, where the network also includes variables that are neither observed nor to be explained is known as (PARTIAL or MARGINAL) MAP. This problem is typically defined formally as follows:

MAP
Instance: A Bayesian network $\mathcal{B} = (\mathbf{G}_{\mathcal{B}}, \Pr)$, where $\mathbf{V}$ is partitioned into a set of evidence nodes $\mathbf{E}$ with a joint value assignment $\mathbf{e}$, a set of intermediate nodes $\mathbf{I}$, and an explanation set $\mathbf{H}$.
Output: A joint value assignment $\mathbf{h}$ to $\mathbf{H}$ such that for all joint value assignments $\mathbf{h}'$ to $\mathbf{H}$, $\Pr(\mathbf{h} \mid \mathbf{e}) \geq \Pr(\mathbf{h}' \mid \mathbf{e})$.

In the remainder, we use the following definitions. For an arbitrary MAP instance $\{\mathcal{B}, \mathbf{H}, \mathbf{E}, \mathbf{e}\}$, let $cansol_{\mathcal{B}}$ denote a function returning candidate solutions to $\{\mathcal{B}, \mathbf{H}, \mathbf{E}, \mathbf{e}\}$, with $optsol_{\mathcal{B}}$ denoting a function returning the *optimal* solution (or, in case of a draw, one of the optimal solutions) to the MAP instance.

2.2   Treewidth

An important structural property of a Bayesian network $\mathcal{B}$ is its *treewidth*, which can be defined as the minimum width over all tree-decompositions of triangulations of the moralization $\mathbf{G}_{\mathcal{B}}^{\mathrm{M}}$ of the network. Treewidth plays an important role in the complexity analysis of Bayesian networks, as many otherwise intractable computational problems can be rendered tractable, provided that the treewidth of the network is small. The moralization (or 'moralized graph') $\mathbf{G}_{\mathcal{B}}^{\mathrm{M}}$ is the undirected graph that is obtained from $\mathbf{G}_{\mathcal{B}}$ by adding arcs so as to connect all pairs of parents of a variable, and then dropping all directions. A triangulation of $\mathbf{G}_{\mathcal{B}}^{\mathrm{M}}$ is any chordal graph $\mathbf{G}_{\mathbf{T}}$ that embeds $\mathbf{G}_{\mathcal{B}}^{\mathrm{M}}$ as a subgraph. A chordal graph is a graph that does not include loops of more than three variables without any pair being adjacent.

A tree-decomposition [18] of a triangulation $\mathbf{G}_{\mathbf{T}}$ now is a tree $\mathbf{T}_{\mathbf{G}}$ such that each node $\mathbf{X}_{\mathbf{i}}$ in $\mathbf{T}_{\mathbf{G}}$ is a bag of nodes which constitute a clique in $\mathbf{G}_{\mathbf{T}}$; and for every i, j, k, if $\mathbf{X}_{\mathbf{j}}$ lies on the path from $\mathbf{X}_{\mathbf{i}}$ to $\mathbf{X}_{\mathbf{k}}$ in $\mathbf{T}_{\mathbf{G}}$, then $\mathbf{X}_{\mathbf{i}} \cap \mathbf{X}_{\mathbf{k}} \subseteq \mathbf{X}_{\mathbf{j}}$. The width of the tree-decomposition $\mathbf{T}_{\mathbf{G}}$ of the graph $\mathbf{G}_{\mathbf{T}}$ is defined as the size of the largest bag in $\mathbf{T}_{\mathbf{G}}$ minus 1, i.e., $\max_i(|\mathbf{X}_{\mathbf{i}}| - 1)$. The treewidth tw of a Bayesian network $\mathcal{B}$ now is the minimum width over all possible tree-decompositions of triangulations of $\mathbf{G}_{\mathcal{B}}^{\mathrm{M}}$.

2.3 Complexity Theory

We assume that the reader is familiar with basic notions from complexity theory, such as intractability proofs, the computational complexity classes P, NP, and polynomial-time reductions. In this section we shortly review some additional concepts that we use throughout the paper, namely the complexity classes PP and BPP, the Exponential Time Hypothesis and some basic principles from parameterized complexity theory.

The complexity classes PP and BPP are defined as classes of decision problems that are decidable by a probabilistic Turing machine (i.e., a Turing machine that makes stochastic state transitions) in polynomial time with a particular (two-sided) probability of error. The difference between these two classes is in the bound on the error probability. *Yes*-instances for problems in PP are accepted with probability $1/2 + \epsilon$, where ϵ may depend exponentially on the input size (i.e., $\epsilon = 1/c^n$). *Yes*-instances for problems in BPP are accepted with a probability that is polynomially bounded away from $1/2$, i.e., (i.e., $\epsilon = 1/n^c$). PP-complete problems, such as the problem of determining whether the *majority* of truth assignments to a Boolean formula ϕ satisfies ϕ, are considered to be intractable; indeed, it can be shown that NP $\subseteq$ PP. In contrast, problems in BPP are considered to be tractable. Informally, a decision problem Π is in BPP if there exists an efficient randomized (Monte Carlo) algorithm that decides Π with high probability of correctness; given that the error is polynomially bounded away from $1/2$, the probability of answering correctly can be boosted to be arbitrarily close to 1. While obviously BPP $\subseteq$ PP, the reverse is unlikely; in particular, it is conjectured that BPP = P.

The *Exponential Time Hypothesis* (ETH), introduced by [8], states that there exists a constant $c > 1$ such that deciding any 3SAT instance with n variables takes at least $\Omega(c^n)$ time. Note that the ETH is a stronger assumption than the assumption that P $\neq$ NP. A sub-exponential but not polynomial-time algorithm for 3SAT, such as an algorithm running in $O(2^{\sqrt[3]{n}})$, would contradict the ETH but would not imply that P = NP. We will assume the ETH in our proofs that show the necessity of low treewidth for efficient approximation of MAP.

Sometimes problems are intractable (i.e., NP-hard) in general, but become tractable if some *parameters* of the problem can be assumed to be small. Informally, a problem is called fixed-parameter tractable for a parameter k (or a set $\{k_1, \ldots, k_n\}$ of parameters) if it can be solved in time, exponential (or even worse) *only* in k and polynomial in the input size $|x|$, i.e., in time $\mathcal{O}(f(k) \cdot |x|^c)$ for a constant c and an arbitrary function f. In practice, this means that problem instances can be solved efficiently, even when the problem is NP-hard in general, if k is known to be small. In contrast, if a problem is NP-hard even when k is small, the problem is denoted as para-NP-hard for k.

3 Approximating MAP

It is widely known, both from practical experiences and from theoretical results, that 'small treewidth' is often a necessary constraint to render exact Bayesian inferences tractable.[1] However, it is often assumed that such intractable computations can be efficiently *approximated* using inexact algorithms; this assumption appears to be warranted by the observation that in many cases approximation algorithms seem to do a reasonable job in, e.g., estimating posterior distributions. Whether this observation has a firm theoretical basis, i.e., whether approximation algorithms can or cannot in principle perform well even in situations where treewidth can grow large, is to date not known.

Crucial in answering this question is to make precise what *efficiently approximated* actually pertains to. The on-line Merriam-Webster dictionary lists as one of its entries for *approximate* 'to be very similar to but not exactly like (something)'. In computer science, this similarity is typically defined in terms of *value*: 'approximate solution A has a value that is close to the value of the optimal solution'. However, other notions of approximation can be relevant. One can think of approximating not the *value* of the optimal solution, but the *appearance*: 'approximate solution A' closely resembles the optimal solution'. Also, one can define an approximate solution as one that ranks close to the optimal solution: 'approximate solution A'' ranks within the top-k solutions'. Note that these notions can refer to completely different solutions. One can have situations where the second-best solution does not resemble at all the optimal solution, whereas solutions that look almost the same have a very low value as compared to the optimal solution [12]. Similarly, the second-best solution may either have a value that is almost as good as the optimal solution, or much worse.

In many practical applications, in particular of Bayesian inferences, these definitions of 'approximation' do not (fully) capture the actual notion we are interested in. For example, when trying to approximate a distribution using some sampling method we have no guarantee on how well the approximate distribution matches the original distribution (e.g., in terms of the Kullback-Leibler divergence); likely, we will (need to) settle for 'probably approximately correct' (PAC) approximations [19]. The added notion of approximation here, induced by the use of randomized computations, is the allowance of a bounded amount of error.

In the remainder of this section we will elaborate on these notions of approximation when applied to the MAP problem. We will give formal definitions of these approximate problems and show why all of them are intractable in general. For MAP-approximation by value and by structure we will interpret known results in the literature. For MAP-approximation by rank we give a formal proof of intractability; for MAP-approximation using randomized algorithms we give an argument from complexity theory.

3.1   Value-Approximation

Value-approximating MAP is the problem of finding an explanation that has a value, close to the value of the optimal solution. This problem is intractable in

---

[1] An exception to this general observation might be algorithms that employ specific local structures, such as context-specific dependences, in the network, as one of the anonymous reviewers noted.

general, even if the variables of the network are bi-partitioned into explanation and evidence variables (i.e., when we approximate an MPE problem). Abdelbar and Hedetniemi proved that it is NP-hard in general to find an explanation $\mathbf{h} \in cansol_\mathcal{B}$ with a constant ratio bound $\frac{\Pr(optsol_\mathcal{B} \mid \mathbf{e})}{\Pr(\mathbf{h} \mid \mathbf{e})} \leq \rho$ for any constant $\rho \geq 1$ [1]. In addition, it can be shown that it is NP-hard in general to find an explanation $\mathbf{h} \in cansol_\mathcal{B}$ with $\Pr(\mathbf{h}, \mathbf{e}) > \epsilon$ for any constant $\epsilon > 0$ [11]. The latter result holds even for networks with only binary variables and at most two incoming arcs per variable.

3.2   Structure-Approximation

Structure-approximating MAP is the problem of finding an explanation that structurally resembles the optimal solution. This is captured using a *solution distance function*, a metric associated with each optimization problem relating candidate solutions with the optimal solution [7]. For MAP, the typical structure distance function $d_H(\mathbf{h} \in cansol_\mathcal{B}, optsol_\mathcal{B})$ is the Hamming distance between explanation $\mathbf{h} \in cansol_\mathcal{B}$ and the most probable explanation $optsol_\mathcal{B}$. It has been shown in [12] that no algorithm can calculate the value of even a single variable in the most probable explanation in polynomial time, unless $\mathsf{P} = \mathsf{NP}$; that is, it is NP-hard to find an explanation with $d_H(\mathbf{h} \in cansol_\mathcal{B}, optsol_\mathcal{B}) \leq |optsol_\mathcal{B}| - 1$, even if the variables of the network are bi-partitioned into explanation and evidence variables.

3.3   Rank-Approximation

Apart from allowing an explanation that resembles, or has a probability close to, the most probable explanation, we can also define an approximate solution as an explanation which is one of the k best explanations, for a constant k. Note that this explanation may not resemble the most probable explanation nor needs to have a relatively high probability, only that it is *ranked* within the k most probable explanations. We will denote this approximation as a rank-approximation, and we will prove that it is NP-hard to approximate MAP using a rank-approximation for any constant k. We do so by a reduction from a variant of LexSat, based on the reduction in [14]. LexSat is defined as follows:

LexSAT
Instance: A Boolean formula ϕ with n variables $X_1, \ldots, X_n$.
Output: The lexicographically largest truth assignment $\mathbf{x}$ to $\mathbf{X} = \{X_1, \ldots, X_n\}$ that satisfies ϕ; the output is $\bot$ if ϕ is not satisfiable.

Here, the lexicographical order of truth assignments maps a truth assignment $\mathbf{x} = x_1, \ldots, x_n$ to a string $\{0, 1\}^n$, with $\{0\}^n$ (all variables set to FALSE) is the lexicographically *smallest*, and $\{1\}^n$ (all variables set to TRUE) is the lexicographically *largest* truth assignment. LexSat is NP-hard; in particular, LexSat has been proven to be complete for the class $\mathsf{FP^{NP}}$ [9]. In our proofs we will use the following variant that always returns a truth assignment (rather than $\bot$, in case ϕ is unsatisfiable):

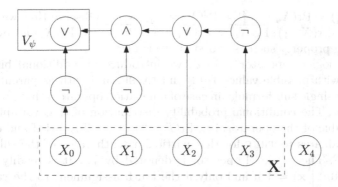

Fig. 1. Example construction of $\mathcal{B}_{\phi_{\mathrm{ex}}}$ from LexSat' instance ϕ_{ex}

LexSAT'
Instance: A Boolean formula ϕ with n variables $X_1, \ldots, X_n$.
Output: The lexicographically largest satisfying truth assignment $\mathbf{x}$ to $\psi = (\neg X_0) \vee \phi$ that satisfies ψ.

Note that if ϕ is satisfiable, then X_0 is never set to FALSE in the lexicographically largest satisfying truth assignment to ψ, yet X_0 is necessarily set to FALSE if ϕ is not satisfiable; hence, unsatisfying truth assignments to ϕ are always ordered after satisfying truth assignments in the lexicographical ordering. Note that LexSat trivially reduces to LexSat' using a simple transformation. We claim the following.

Theorem 1. *No algorithm can k-rank-approximate MAP, for any constant k, in polynomial time, unless* P = NP.

In our proof we describe a polynomial-time Turing reduction from LexSat' to k-rank-approximated-MAP for an arbitrary constant k. The reduction largely follows the reduction as presented in [14] with some additions. We will take the following LexSat'-instance as running example in the proof: $\phi_{\mathrm{ex}} = \neg X_1 \wedge (X_2 \vee \neg X_3)$; correspondingly, $\psi_{\mathrm{ex}} = (\neg X_0) \vee (\neg X_1 \wedge (X_2 \vee \neg X_3))$ in this example. We set $k = 3$ in the example construct. We now construct a Bayesian network $\mathcal{B}_\phi$ from ψ as follows (Figure 1).

For each variable X_i in ψ, we introduce a binary root variable X_i in $\mathcal{B}_\phi$ with possible values TRUE and FALSE. We set the prior probability distribution of these variables to $\Pr(X_i = \mathrm{TRUE}) = 1/2 - \frac{2^{i+1}-1}{2^{n+2}}$. In addition, we include a uniformly distributed variable X_{n+1} in $\mathcal{B}_\phi$ with k values $x_{n+1}^1, \ldots, x_{n+1}^k$. The variables $X_0, \ldots, X_n$ together form the set $\mathbf{X}$. Note that the prior probability of a joint value assignment $\mathbf{x}$ to $\mathbf{X}$ is higher than the prior probability of a different joint value assignment $\mathbf{x}'$ to $\mathbf{X}$, if and only if the corresponding truth assignment $\mathbf{x}$ to the LexSat' instance has a lexicographically larger truth assignment than $\mathbf{x}'$. In the running example, we have that $\Pr(X_0 = \mathrm{TRUE}) = 15/32$, $\Pr(X_1 = \mathrm{TRUE}) = 13/32$, $\Pr(X_2 = \mathrm{TRUE}) = 9/32$, and $\Pr(X_3 = \mathrm{TRUE}) = 1/32$, and

$\Pr(X_4 = x_4^1) = \Pr(X_4 = x_4^2) = \Pr(X_4 = x_4^3) = 1/3$. Observe that we have that $\Pr(X_1) \cdot \ldots \cdot \Pr(X_{i-1}) \cdot \overline{\Pr(X_i)} > \overline{\Pr(X_1)} \cdot \ldots \cdot \overline{\Pr(X_{i-1})} \cdot \Pr(X_i)$ for every i, i.e., the ordering property such as stated above is attained.

For each logical operator T in ψ, we introduce an additional binary variable in $\mathcal{B}_\phi$ with possible values TRUE and FALSE, and with as parents the subformulas (or single sub-formula, in case of a negation operator) that are bound by the operator. The conditional probability distribution of that variable matches the truth table of the operator, i.e., $\Pr(T = \text{TRUE} \mid \pi(T)) = 1$ if and only if the operator evaluates to TRUE for that particular truth value of the sub-formulas bound by T. The top-level operator is denoted by V_ψ. It is readily seen that $\Pr(V_\psi = \text{TRUE} \mid \mathbf{x}) = 1$ if and only if the truth assignment to the variables in ψ that matches $\mathbf{x}$ satisfies ψ. Observe that the k-valued variable X_{n+1} is independent of every other variable in $\mathcal{B}_\phi$. Further note that the network, including all prior and conditional probabilities, can be described using a number of bits which is polynomial in the size of ϕ. In the MAP instance constructed from ϕ, we set V_ψ as evidence set with $V_\psi = \text{TRUE}$ as observation and we set $\mathbf{X} \cup \{X_{n+1}\}$ as explanation set.

Proof. Let ϕ be an instance of LEXSAT', and let $\mathcal{B}_\phi$ be the network constructed from ϕ as described above. We have for any joint value assignment $\mathbf{x}$ to $\mathbf{X}$ that $\Pr(\mathbf{X} = \mathbf{x} \mid V_\psi = \text{TRUE}) = \alpha \cdot \Pr(\mathbf{X} = \mathbf{x})$ for a normalization constant α if $\mathbf{x}$ corresponds to a satisfying truth assignment to ψ, and $\Pr(\mathbf{X} = \mathbf{x} \mid V_\psi = \text{TRUE}) = 0$ if $\mathbf{x}$ corresponds to a non-satisfying truth assignment to ψ. Given the prior probability distribution of the variables in $\mathbf{X}$, we have that all satisfying joint assignments $\mathbf{x}$ to $\mathbf{X}$ are ordered by the posterior probability $\Pr(\mathbf{x} \mid V_\psi = \text{TRUE}) > 0$, where all non-satisfying joint value assignments have probability $\Pr(\mathbf{x} \mid V_\psi = \text{TRUE}) = 0$ and thus are ordered after satisfying assignments. The joint value assignment that has the highest posterior probability thus is the lexicographically largest satisfying truth assignment to ψ.

If we take the k-th valued variable X_{n+1} into account, we have that for every $\mathbf{x}$, the k joint value assignments to $\Pr(\mathbf{x}, X_{n+1} \mid V_\psi = \text{TRUE})$ have the same probability since $\Pr(\mathbf{x}, X_{n+1} \mid V_\psi = \text{TRUE}) = \Pr(\mathbf{x} \mid V_\psi = \text{TRUE}) \cdot \Pr(X_{n+1})$. But then, the k joint value assignments $\mathbf{x}^\mathbf{k}$ to $\mathbf{X} \cup \{X_{n+1}\}$ that correspond to the lexicographically largest satisfying truth assignment $\mathbf{x}$ to ψ all have the same posterior probability $\Pr(\mathbf{x}^\mathbf{k} \mid V_\psi = \text{TRUE})$. Thus, any algorithm that returns one of the k-th ranked joint value assignments to the explanation set $\mathbf{X} \cup \{X_{n+1}\}$ with evidence $V_\psi = \text{TRUE}$ can be transformed in polynomial time to an algorithm that solves LEXSAT'. We conclude that no algorithm can k-rank-approximate MAP, for any constant k, in polynomial time, unless P = NP. $\qquad\square$

Note that, technically speaking, our result is even stronger: as LEXSAT' is FP^{NP}-complete and the reduction described above actually is a one-Turing reduction from LEXSAT' to k-rank-approximation-MAP, the latter problem is FP^{NP}-hard. We can strengthen the result further by observing that all variables (minus V_ψ) that mimic operators deterministically depend on their parents and thus can be added to the explanation set without substantially changing the proof above. This implies that k-rank-approximation-MPE is also FP^{NP}-hard.

3.4   Expectation-Approximation

The last notion of MAP approximation we will discuss here returns in polynomial time an explanation that are likely to be the most probable explanation, but allows for a small margin of error; i.e., there is a small probability that the answer is not the optimal solution, and then no guarantees are given on the quality of that solution. These approximations are closely related to randomized algorithms that run in polynomial time but whose output has a small probability of error, viz., Monte Carlo algorithms. This notion of approximation–which we will refer to as *expectation-approximation* [15]–is particularly relevant for typical Bayesian approximation methods, such as Monte Carlo sampling and repeated local search algorithms.

In order to be of practical relevance, we want the error to be *small*, i.e., when casted as a decision problem, we want the probability of answering correctly to be bounded away from $1/2$. In that case, we can amplify the probability of answering correctly arbitrarily close to 1 in polynomial time, by repeated evocation of the algorithm. Otherwise, e.g., if the error depends exponentially on the size of the input, we need an exponential number of repetitions to achieve such a result. Monte Carlo randomized algorithms are in the complexity class BPP; randomized algorithms that may need exponential time to reduce the probability of error arbitrarily close to 0 are in the complexity class PP.

As MAP is NP-hard, an efficient randomized algorithm solving MAP in polynomial time with a bounded probability of error, would imply that NP $\subseteq$ BPP. This is considered to be highly unlikely, as almost every problem that enjoys an efficient randomized algorithm has been proven to be in P, i.e., be decidable in deterministic polynomial time.[2] On various grounds it is believed that P = BPP, and thus an efficient randomized algorithm for MAP would (under that assumption) establish P = NP. Therefore, no algorithm can expectation-approximate MAP in polynomial time with bounded margin of error unless NP $\subseteq$ BPP. This result holds also for MPE, which is in itself already NP-hard.

4   The Necessity of Low Treewidth for Efficient Approximation of MAP

In the previous section we have shown that for four notions of approximating MAP, no efficient general approximation algorithm can be constructed unless either P = NP or NP $\subseteq$ BPP. However, MAP is *fixed-parameter tractable* for a number of problem parameters; for example, $\{tw, c, 1 - p\}$-MAP is in FPT for parameters treewidth (tw), cardinality of the variables (c), and probability of the most probable solution $1 - p$. Surely, if we can compute $\{k_1, \ldots, k_m\}$-MAP exactly in FPT time, we can also approximate $\{k_1, \ldots, k_m\}$-MAP in FPT time.

---

[2] The most dramatic example of such a problem is PRIMES: given a natural number, decide whether it is prime. While efficient randomized algorithms for PRIMES have been around quite some time (establishing that PRIMES $\in$ BPP), only fairly recently it has been proven that PRIMES is in P [2].

A question remains, however, whether approximate MAP can be fixed-parameter tractable for a *different* set of parameters than exact MAP.

Treewidth has been shown to be a *necessary* parameter for efficient exact computation of the INFERENCE problem (and, by a trivial adjustment, also of MAP), under the assumption that the ETH holds [13]. In this section, we will show that low treewidth is also a necessary parameter for efficient *approximate* computation for value-, structure-, and rank-approximations. We also show that it is *not* a necessary parameter for efficient expectation-approximation. In the next sub-section we will review so-called treewidth-preserving reductions (tw-reductions), a special kind of polynomial many-one reduction that preserves treewidth of the instances [13]. In Subsection 4.2 we sketch how this notion can be used to tw-reduce CONSTRAINT SATISFACTION to INFERENCE. Together with the known result that CONSTRAINT SATISFACTION instances with high treewidth cannot have sub-exponential algorithms, unless the ETH fails [16], it was established in [13] that there cannot be a polynomial-time algorithm that decides INFERENCE on instances with high treewidth in sub-exponential time, unless the ETH fails; the reader is referred to [13] for the full proof.

Subsequently, we will show how this proof can be augmented to establish similar results for MAP, value-approximate MAP, structure-approximate MAP, and rank-approximate MAP (Sub-sections 4.3 and 4.4). In the last sub-section we will give a small example where a simple forward-sampling algorithm can efficiently expectation-approximate MAP despite high treewidth; we will elaborate on the constraints needed to render such algorithms provably fixed-parameter tractable and give pointers for future work.

4.1   Treewidth-Preserving Reductions

Treewidth-preserving reductions are defined in [13] as a means to reduce CONSTRAINT SATISFACTION to INFERENCE while ensuring that treewidth is preserved between instances in the reduction, modulo a linear factor.

Definition 1 ([13]). *Let A and B be computational problems such that treewidth is defined on instances of both A and B. We say that A is polynomial-time treewidth-preserving reducible, or tw-reducible, to B if there exists a polynomial-time computable function g and a linear function l such that $x \in A$ if and only if $g(x) \in B$ and $\mathrm{tw}(g(x)) = l(\mathrm{tw}(x))$. The pair (g, l) is called a tw-reduction.*

We will use this notion to show that CONSTRAINT SATISFACTION also tw-reduces to MAP, value-approximate MAP, structure-approximate MAP, and rank-approximate MAP.

4.2   Proof Sketch

The tw-reduction from (binary) CONSTRAINT SATISFACTION to INFERENCE, as presented in [13], constructs a Bayesian network $\mathcal{B}_\mathcal{I}$ from an instance $\mathcal{I} = (\mathbf{V}, \mathbf{D}, \mathbf{C})$ of CONSTRAINT SATISFACTION, where $\mathbf{V}$ denotes the set of variables of $\mathcal{I}$, $\mathbf{D}$ denotes the set of values of these variables, and $\mathbf{C}$ denotes the set of

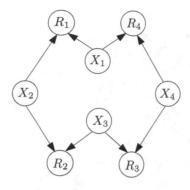

Fig. 2. Example construction of $\mathcal{B}_\mathcal{I}$ from example CSP instance $\mathcal{I}$

binary constraints defined over $\mathbf{V} \times \mathbf{V}$. The constructed network $\mathcal{B}_\mathcal{I}$ includes uniformly distributed variables X_i, corresponding with the variables in $\mathbf{V}$, and binary variables R_j, corresponding with the constraints in $\mathbf{C}$. The parents of the variables R_j are the variables X_i that are bound by the constraints; their conditional probability distributions match the imposed constraints on the variables (i.e., $\Pr(R_j = \text{TRUE} \mid \mathbf{x} \in \Omega(\pi(R_j))) = 1$ if and only if the joint value assignment $\mathbf{x}$ to the variables bound by R_j matches the constraints imposed on them by R_j. Figure 2, taken from [13], shows the result of the construction so far for an example CONSTRAINT SATISFACTION instance with four variables X_1 to X_4, where $\mathbf{C}$ contains four constraints that bind respectively (X_1, X_2), (X_1, X_4), (X_2, X_3), and (X_3, X_4).

The treewidth of the thus obtained network equals $\max(2, \text{tw}(\mathbf{G}_\mathcal{I}))$, where $\mathbf{G}_\mathcal{I}$ is the primal graph of $\mathcal{I}$; note that the treewidth of $\mathcal{B}_\mathcal{I}$ at most increases the treewidth of $\mathbf{G}_\mathcal{I}$ by 1. In order to enforce that *all* constraints are simultaneously enforced, the constraint nodes R_j need to be connected by extra nodes mimicking 'and' operators. A crucial aspect of the tw-reduction is the topography of this connection of the nodes R_j: care most be taken not to blow up treewidth by arbitrarily connecting the nodes, e.g., by a log-deep binary tree. The original proof uses a minimal tree-decomposition of the moralization of $\mathcal{B}_\mathcal{I}$ and describes a procedure to select which nodes need to be connected such that the treewidth of the resulting graph is at most the treewidth of $\mathbf{G}_\mathcal{I}$ plus 3. The conditional probability distribution of the nodes A_k is defined as follows.

$$\Pr(A_k = \text{TRUE} \mid \mathbf{x}) = \begin{cases} 1 \text{ if } \mathbf{x} = \bigwedge_{V \in \pi(A_k)}(V = \text{TRUE}) \\ 0 \text{ otherwise} \end{cases}$$

For a node A_k without any parents, $\Pr(A_k = \text{TRUE}) = 1$. The graph that results from applying this procedure to the example is given in Figure 3 (also taken from [13]). Now, $\Pr(A_1 = \text{TRUE} \mid \mathbf{x}) = 1$ if $\mathbf{x}$ corresponds to a satisfying value assignment to $\mathbf{V}$ and 0 otherwise; correspondingly, $\Pr(A_1 = \text{TRUE}) > 0$ if and only if the CONSTRAINT SATISFACTION instance is satisfiable.

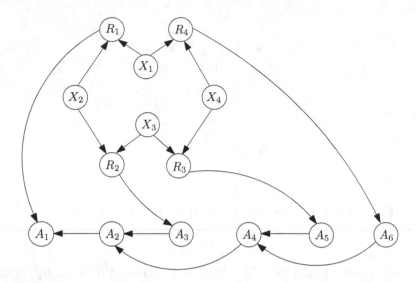

Fig. 3. Resulting graph $\mathcal{B}_\mathcal{I}$ after adding nodes A_k and appropriate arcs

4.3 MAP Result

The tw-reduction described in the previous sub-section can be easily be modified to a tw-reduction from CONSTRAINT SATISFACTION to MAP. We do this by adding a binary node $V_\mathcal{I}$ to the thus obtained graph, with A_1 as its only parent and with conditional probability $\Pr(V_\mathcal{I} = \text{TRUE} \mid A_1 = \text{TRUE}) = 1$ and $\Pr(V_\mathcal{I} = \text{TRUE} \mid A_1 = \text{FALSE}) = 1/2 - \epsilon$, where ϵ is a number, smaller than $1/|\mathbf{D}|^{|\mathbf{V}|}$. Consequently, we have that $\Pr(V_\mathcal{I} = \text{TRUE}) > 1/2$ if $\mathcal{I}$ is satisfiable, and $\Pr(V_\mathcal{I} = \text{TRUE}) < 1/2$ if $\mathcal{I}$ is not satisfiable; hence, a MAP query with explanation set $\mathbf{H} = V_\mathcal{I}$ will return $V_\mathcal{I} = \text{TRUE}$ if and only if $\mathcal{I}$ is satisfiable. We added a single node to $\mathcal{B}_\mathcal{I}$, with A_1 as only parent, thus increasing the treewidth of $\mathcal{B}_\mathcal{I}$ by at most 1. Hence, CONSTRAINT SATISFACTION tw-reduces to MAP.

4.4 Approximation Intractability Results

In a similar way we can modify the reduction from Sub-section 4.2 to show that value-, structure-, and rank-approximations can be tw-reduced from CONSTRAINT SATISFACTION, as sketched below.

Value-Approximation. We add a binary node $V_\mathcal{I}$, with A_1 as its only parent, and with conditional probability $\Pr(V_\mathcal{I} = \text{TRUE} \mid A_1 = \text{TRUE}) = 1$ and $\Pr(V_\mathcal{I} = \text{TRUE} \mid A_1 = \text{FALSE}) = 0$. We observe this variable to be set to TRUE. This enforces that $\Pr(A_1 = \text{TRUE} \mid V_\mathcal{I} = \text{TRUE})$ has a non-zero probability (i.e., $\mathcal{I}$ is solvable) since otherwise there is conflicting evidence in the thus constructed network. Thus, any value-approximation algorithm with with explanation set $\mathbf{H} = A_1$ and evidence $\mathbf{e} = V_\mathcal{I} = \text{TRUE}$ that can return a solution $\mathbf{h} \in cansol_\mathcal{B}$

with $\Pr(\mathbf{h}, \mathbf{e}) > \epsilon$ for any constant $\epsilon > 0$, effectively solves CONSTRAINT SATIS-
FACTION. Given that we added a single node to $\mathcal{B}_\mathcal{I}$, with A_1 as only parent, this
increases the treewidth of $\mathcal{B}_\mathcal{I}$ by at most 1. Hence, CONSTRAINT SATISFACTION
tw-reduces to value-approximate MAP.

Structure-Approximation. Observe from the tw-reduction to MAP in Sub-
section 4.3 that, since $\mathbf{H}$ consists of a singleton binary variable, we trivially have
that no algorithm can find an explanation with $d_H(\mathbf{h} \in cansol_\mathcal{B}, optsol_\mathcal{B}) \leq$
$|optsol_\mathcal{B}| - 1 = 0$ since that would solve the MAP query. We can extend this
result to hold for explanation sets with size k for any constant k, i.e., no structure-
approximation algorithm can guarantee to return the correct value of *one* of the
k variables in $\mathbf{H}$ in polynomial time in instances of high treewidth, unless the
ETH fails.

Instead of adding a single binary node $V_\mathcal{I}$ as in the tw-reduction to MAP,
we add k binary nodes $V_\mathcal{I}^1 \ldots V_\mathcal{I}^k$, all with A_1 as their only parent and with
$\Pr(V_\mathcal{I}^j = \text{TRUE} \mid A_1 = \text{TRUE}) = 1$ and $\Pr(V_\mathcal{I}^j = \text{TRUE} \mid A_1 = \text{FALSE}) = 1/2 - \epsilon$
for $1 \leq j \leq k$ and with ϵ as described in Sub-section 4.3. A MAP query with
explanation set $\mathbf{H} = \bigcup_{1 \leq j \leq k} V_\mathcal{I}^j$ will then return $\forall_{1 \leq j \leq k} V_\mathcal{I}^k = \text{TRUE}$ if and only if
$\mathcal{I}$ is satisfiable; if $\mathcal{I}$ is not satisfiable, a MAP query will return $\forall_{1 \leq j \leq k} V_\mathcal{I}^k = \text{FALSE}$
as most probable explanation. Hence, any structure-approximation algorithm
that can correctly return the value of one of the variables in $\mathbf{H}$, effectively solves
CONSTRAINT SATISFACTION. As we added k nodes to $\mathcal{B}_\mathcal{I}$, with A_1 as their
only parent, the treewidth of $\mathcal{B}_\mathcal{I}$ increases by at most k. Hence, CONSTRAINT
SATISFACTION tw-reduces to structure-approximate MAP.

Rank-Approximation. We modify the proof of Sub-section 4.3 as follows.
In addition to adding a binary node $V_\mathcal{I}$ as specified in that section, we also
add a uniformly distributed unconnected node $K_\mathcal{I}$ with k values to $\mathbf{H}$; a k-
rank-approximate MAP query with explanation set $\mathbf{H} = \{V_\mathcal{I}, K_\mathcal{I}\}$ will return
$V_\mathcal{I} = \text{TRUE}$ (and $K_\mathcal{I}$ set to an arbitrary value) if and only if $\mathcal{I}$ is satisfiable. The
addition of $K_\mathcal{I}$ does not increase treewidth, hence, CONSTRAINT SATISFACTION
tw-reduces to k-rank-approximate MAP.

4.5   Expectation-Approximation

In the previous section we showed that we cannot value-, structure-, or rank-
approximate MAP on instances with high treewith, unless the ETH fails. Now
what about expectation-approximation? We will argue that there are MAP instances with high treewidth that *can* be efficiently expectation-approximated,
provided that the probability $\Pr(optsol_\mathcal{B} \mid \mathbf{e})$ is high. Note that it remains NP-
hard (to be precise: para-PP-hard) to decide INFERENCE, even if the probability
of interest is arbitrarily close to 1 [10]; as INFERENCE is a degenerate special
case of MAP, it follows that computing MAP exactly is also NP-hard in this
case. While this sketchy argument is not a fully worked-out proof, it hints that
efficient expectation-approximation of MAP indeed depends on a *different* set of
parameters than the other notions of approximation discussed above.

The argument goes as follows. In order to generate MAP instances with high treewidth, we construct them from SAT instances in a similar way as described in Section 3.3. We can generate SAT instances with high treewidth by, e.g., picking an arbitrary formula ϕ and then boosting the treewidth by "inserting" tautologies $\wedge(x_i \vee \neg x_i)$ at strategic places in ϕ. We then construct a Bayesian network $\mathcal{B}_\phi$ from ϕ by including binary root *truth-setting* variables X_i for all variables X_i in ϕ, and adding binary *operator* variables T_j for all logical operators in ϕ, and connecting them as described in Section 3.3. We again denote the top-level operator as V_ϕ and we observe that $\Pr(V_\phi = \text{TRUE}) = \#SAT/2^n$, i.e., the probability distribution over $\Pr(V_\phi)$ corresponds to the number of satisfying truth assignments to ϕ. If a majority of truth assignments satisfy ϕ, a MAP query with $\mathbf{H} = V_\phi$ will return TRUE, if a minority of truth assignments satisfy ϕ, the same MAP query will return FALSE. Now, if the probability p of the most probable joint value assignment is bounded away from $1/2$, i.e., is guaranteed to be $1/2 + 1/n^c$ for a constant c, a simple forward sampling strategy (assigning random joint value assignments to the variables X_i and propagating the assignments according to the CPTs of the operator variables T_j) can decide this MAP query with a bounded degree of error. To be precise, using the Chernoff bound we can compute than the number of samples needed to have a degree of error lower than δ is $1/(p - 1/2)^2 \ln 1/\sqrt{\delta} = n^{c^2} \ln 1/\sqrt{\delta}$.

5   Conclusion

In this paper we analysed whether low treewidth is a prerequisite for approximating MAP in Bayesian networks. We formalized four distinct notions of approximating MAP (by value, structure, rank, or expectation) and argued that approximate MAP is intractable in general using either of these notions. In case of value-, structure-, and rank-approximation we showed that MAP cannot be approximated using these notions in instances with high treewidth, if the ETH holds. We argued that expectation-approximation, in contrast, may be rendered fixed-parameter tractable, even in instances with high treewidth, if the probability q of the most probable explanation is high (and the cardinality c of the variables is bounded). As INFERENCE (and thus also MAP) is intractable even when the probability of the most probable explanation is high, this result may indeed lead to a $\{q, c\}$-fixed parameter tractable expectation-approximation algorithm for MAP. We leave the proof of existence and the actual development and analysis of such an algorithm for future work.

References

1. Abdelbar, A.M., Hedetniemi, S.M.: Approximating MAPs for belief networks is NP-hard and other theorems. Artificial Intelligence 102, 21–38 (1998)
2. Agrawal, M., Kayal, N., Saxena, N.: PRIMES is in P. Annals of Mathematics 160(2), 781–793 (2004)
3. Arora, S., Barak, B.: Computational Complexity: A Modern Approach. Cambridge University Press (2009)

4. Darwiche, A.: Modeling and Reasoning with Bayesian Networks. Cambridge University Press (2009)
5. De Campos, C.P.: New complexity results for MAP in Bayesian networks. In: Proceedings of the Twenty-Second International Joint Conference on Artificial Intelligence, pp. 2100–2106 (2011)
6. Downey, R.G., Fellows, M.R.: Parameterized Complexity. Springer, Berlin (1999)
7. Hamilton, M., Müller, M., van Rooij, I., Wareham, H.T.: Approximating solution structure. In: Demaine, E., Gutin, G.Z., Marx, D., Stege, U. (eds.) Structure Theory and FPT Algorithmics for Graphs, Digraphs and Hypergraphs. Dagstuhl Seminar Proceedings, vol. (07281) (2007)
8. Impagliazzo, R., Paturi, R.: On the complexity of k-SAT. Journal of Computer and System Sciences 62(2), 367–375 (2001)
9. Krentel, M.W.: The complexity of optimization problems. Journal of Computer and System Sciences 36, 490–509 (1988)
10. Kwisthout, J.: The computational complexity of probabilistic inference. Technical Report ICIS–R11003, Radboud University Nijmegen (2011)
11. Kwisthout, J.: Most probable explanations in Bayesian networks: Complexity and tractability. International Journal of Approximate Reasoning 52(9), 1452–1469 (2011)
12. Kwisthout, J.: Structure approximation of most probable explanations in Bayesian networks. In: van der Gaag, L.C. (ed.) ECSQARU 2013. LNCS (LNAI), vol. 7958, pp. 340–351. Springer, Heidelberg (2013)
13. Kwisthout, J., Bodlaender, H.L., van der Gaag, L.C.: The necessity of bounded treewidth for efficient inference in Bayesian networks. In: Coelho, H., Studer, R., Wooldridge, M. (eds.) Proceedings of the 19th European Conference on Artificial Intelligence (ECAI 2010), pp. 237–242. IOS Press (2010)
14. Kwisthout, J.H.P., Bodlaender, H.L., van der Gaag, L.C.: The complexity of finding kth most probable explanations in probabilistic networks. In: Černá, I., Gyimóthy, T., Hromkovič, J., Jefferey, K., Královič, R., Vukolić, M., Wolf, S. (eds.) SOFSEM 2011. LNCS, vol. 6543, pp. 356–367. Springer, Heidelberg (2011)
15. Kwisthout, J., van Rooij, I.: Bridging the gap between theory and practice of approximate Bayesian inference. Cognitive Systems Research 24, 2–8 (2013)
16. Marx, D.: Can you beat treewidth? In: Proceedings of the 48th Annual IEEE Symposium on Foundations of Computer Science (FOCS 2007), pp. 169–179 (2007)
17. Park, J.D., Darwiche, A.: Complexity results and approximation settings for MAP explanations. Journal of Artificial Intelligence Research 21, 101–133 (2004)
18. Robertson, N., Seymour, P.D.: Graph minors II: Algorithmic aspects of tree-width. Journal of Algorithms 7, 309–322 (1986)
19. Valiant, L.G.: A theory of the learnable. Communications of the ACM 27(11), 1134–1142 (1984)

Bayesian Networks with Function Nodes

Anders L. Madsen[1,2], Frank Jensen[1], Martin Karlsen[1],
and Nicolaj Soendberg-Jeppesen[1]

[1] HUGIN EXPERT A/S, Aalborg, Denmark
[2] Department of Computer Science, Aalborg University, Denmark

Abstract. This paper introduces the notion of Bayesian networks augmented with function nodes. Two types of function nodes are considered. A real-valued function node represents a real value either used to parameterise one or more conditional probability distributions of the Bayesian network or a real value computed after a successful belief update or Monte Carlo simulation. On the other hand, a discrete function node represents a discrete marginal distribution. The paper includes four real-world examples that illustrate how function nodes have improved the flexibility and efficiency of utilizing Bayesian networks for reasoning with uncertainty in different domains.

Keywords: Bayesian networks, function nodes, belief update.

1 Introduction

A *Bayesian network* [18,3,10] is a powerful and popular model for probabilistic inference. Its graphical nature makes it well-suited for representing complex problems where the interactions between entities represented as variables are described using *conditional probability distributions* (CPDs). One of the main reasons for the popularity of Bayesian networks is the access to development tools and tools for integrating Bayesian networks into other applications.

Bayesian networks are often used in applications where belief update is only part of the calculations performed to produce the end result. That is the results of belief update, i.e., posterior probabilities, are used as input for further calculations outside the framework of Bayesian networks. Also, the results of belief update in one Bayesian network model may be used to parameterise another Bayesian network. Bayesian networks do not readily support post belief update calculations nor the possibility to link results of belief update in one model to parameters of a different model. Gated Bayesian networks [1] are a new formalism to combine several Bayesian networks such that they may be active or inactive during the belief update process depending on predefined logical statements. This is different from how function nodes as presented in this paper are used to link Bayesian networks where, for instance, information is passed between CPDs in different Bayesian networks.

This paper introduces the notion of Bayesian networks augmented with function nodes. A real-valued function node represents a real value whereas a discrete

L.C. van der Gaag and A.J. Feelders (Eds.): PGM 2014, LNAI 8754, pp. 286–301, 2014.

function node represents a marginal probability distribution. Either is computed from the results of a belief update or simulation operation. The main motivation for introducing Bayesian networks augmented with function nodes is to enable knowledge engineers and analysts to combine Bayesian networks with other techniques, specify post belief update calculations, link Bayesian network models and link parameterisation of a Bayesian network to user input. Bayesian networks augmented with function nodes enable the knowledge engineer to achieve this in a single integrated knowledge representation. Two types of function nodes are considered, where type refers to the value a function node can take, i.e., either a real value or a probability distribution. A real-valued function node specifies a mathematical expression (can be a constant) evaluated as part of the belief update or simulation process. A discrete function node can be used to compute distributions derived from other distributions, i.e., a distribution computed from the result of belief update.

This paper is organised as follows. Section 2 introduces preliminaries and notation. Section 3 introduces the notion of networks with function nodes and Section 4 describes how inference is performed in such networks. Section 5 describes four real-world examples showing how function nodes have improved the flexibility and efficiency of utilizing Bayesian networks for reasoning with uncertainty in different domains while Section 6 concludes the paper with a discussion and concluding remarks.

2   Preliminaries and Notation

Let $\mathcal{X} = \{X_1, \ldots, X_n\}$ be a set of discrete random variables such that $\mathrm{dom}(X)$ is the state space of X and $||X|| = |\mathrm{dom}(X)|$. A discrete Bayesian network $\mathcal{N} = (\mathcal{X}, G, \mathcal{P})$ over $\mathcal{X}$ consists of an acyclic directed graph (DAG) $G = (V, E)$ with vertices V and edges E and a set of CPDs $\mathcal{P} = \{P(X \,|\, \mathrm{pa}(X)) : X \in \mathcal{X}\}$, where $\mathrm{pa}(X)$ denotes the parents of X in G [18,3,10]. The discrete Bayesian network $\mathcal{N}$ specifies a joint probability distribution over $\mathcal{X}$

$$P(\mathcal{X}) = \prod_{i=1}^{n} P(X_i \,|\, \mathrm{pa}(X_i)).$$

A Conditional Linear Gaussian (CLG) Bayesian network $\mathcal{N} = (\mathcal{X}, G, \mathcal{P}, \mathcal{D})$ is similar to a discrete Bayesian network where $\mathcal{X} = \mathcal{X}_\Gamma \cup \mathcal{X}_\Delta$ is a partition of the random variables into discrete random variables $\mathcal{X}_\Delta$ and continuous random variables $\mathcal{X}_\Gamma$ and $\mathcal{D}$ is a set of CLG density functions [13]. The CLG Bayesian network $\mathcal{N} = (\mathcal{X}, G, \mathcal{P}, \mathcal{D})$ specifies a mixture distribution over $\mathcal{X}_\Gamma \cup \mathcal{X}_\Delta$

$$P(\mathcal{X}_\Delta) \cdot f(\mathcal{X}_\Gamma \,|\, \mathcal{X}_\Delta) = \prod_{X \in \mathcal{X}_\Delta} P(X \,|\, \mathrm{pa}(X)) \cdot \prod_{Y \in \mathcal{X}_\Gamma} f(Y \,|\, \mathrm{pa}(Y)).$$

If $\mathcal{X}_\Gamma = \emptyset$, then the CLG Bayesian network is a discrete Bayesian network.

Belief update in $\mathcal{N}$ is defined as the task of computing the posterior marginal $P(X \mid \epsilon)$, for each non-evidence variable $X \in \mathcal{X} \setminus \mathcal{X}_\epsilon$ given a set of variable instantiations ϵ, where $\mathcal{X}_\epsilon \subseteq \mathcal{X}$ is the set of variables instantiated by ϵ. A belief update operation is *successful* when no errors are produced, which, for instance, would be the result of propagating inconsistent evidence, i.e., evidence that has zero probability in $\mathcal{N}$. A simulation is the process of performing a Monte Carlo simulation over $\mathcal{N}$ given ϵ.

Each vertex $v \in V$ represents a random variable $X \in \mathcal{X}$ and each X is represented by a $v \in V$. This means that we will refer to a vertex and its random variable interchangeably. Let $\mathcal{N}$ be a Bayesian network with DAG $G = (V, E)$, then $G(V_i)$ is the subgraph induced by $V_i \subseteq V$. The moral graph G^m of a DAG G is produced by adding an edge between vertices with a common child and removing direction on the edges.

A *trail* π is a sequence of vertices $\pi = (v_1, \dots, v_n)$ such that there is a link $v_i \to v_{i+1}$ or $v_i \leftarrow v_{i+1}$, i.e., either $(v_i, v_{i+1}) \in E$ or $(v_{i+1}, v_i) \in E$, between each pair of consecutive vertices in the sequence. If each link between v_i and v_{i+1} is directed as (v_i, v_{i+1}) where v_i is parent and v_{i+1} is child, then π is a directed trail and v_1 is the source and v_n is the destination.

The ancestral set of a (vertex representing) variable X in G is denoted as $\mathrm{an}(X)$ and the ancestral set of set of variables $\mathcal{X}_i \subseteq \mathcal{X}$ is denoted as $\mathrm{an}(\mathcal{X}_i)$. Similarly, the sets of descendants are denoted as $\mathrm{de}(X)$ and $\mathrm{de}(\mathcal{X}_i)$, respectively.

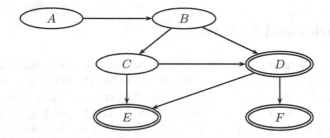

Fig. 1. A CLG Bayesian network with $\mathcal{X}_\Delta = \{A, B, C\}$ and $\mathcal{X}_\Gamma = \{D, E, F\}$

Example 1. Figure 1 shows an example of a CLG Bayesian network over the random variables A, B, C, D, E and F. Single-lined oval vertices represent discrete random variables while double-lined oval nodes represent continuous random variables. Thus, we have $\mathcal{X}_\Delta = \{A, B, C\}$ and $\mathcal{X}_\Gamma = \{D, E, F\}$.

3   Bayesian Networks with Function Nodes

A function node F expresses a real value or probability distribution calculated after a successful belief update or simulation or a constant computed before belief update over $\mathcal{N}$. This enables the knowledge engineer to create one integrated knowledge representation for belief update and calculations using the results of a belief update. This is a significant contribution to representing, testing and validating systems using Bayesian networks.

3.1   Function Nodes

This section introduces real-valued and discrete function nodes.

Definition 1 (Real-Valued Function Node). *A real-valued function node F represents a single real value.*

Definition 2 (Discrete Function Node). *A discrete function node F represents a discrete marginal distribution.*

In both cases, the entity is a function of the values or distributions of the parents. A real-valued function node is not (directly) involved in the belief update process. It cannot be instantiated by evidence, but its function can be evaluated using the results of the belief update or a simulation. A discrete function node can be instantiated by evidence with the same properties as a random variable except it will not induce a likelihood over its parents, i.e., it will have the same properties as a variable without parents (a root variable).

The set of real-valued function nodes is denoted $\mathcal{F}_{\mathbb{R}}$ and the set of discrete function nodes is denoted $\mathcal{F}_{\Delta}$.

Example 2. Figure 2 shows an example of a Bayesian network model augmented with discrete function nodes F_1 and F_2 (hexagon) and a real-valued function node p (hexagon with double lines). The real-valued function node p could denote a constant in an expression for defining the content of the conditional probability distribution for X_1, e.g., the p parameter of a Binomial distribution, while F_2 could denote a probability distribution computed from the posterior probability distributions of X_1 and X_3, e.g., if $P(X_3 \mid \epsilon)$ is an impact distribution and $P(X_1 \mid \epsilon)$ is a frequency distribution, then F_2 could specify a total impact distribution as the convolution of frequency and impact.

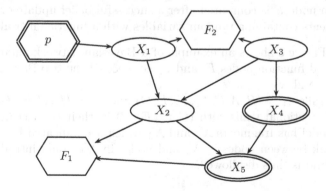

Fig. 2. A Bayesian network augmented with function nodes

3.2   Graphical Model Structure

A discrete Bayesian network augmented with function nodes $\mathcal{N} = (\mathcal{X}, G, \mathcal{P}, \mathcal{F})$ consists of a set of random variables $\mathcal{X}$, DAG $G = (V, E)$, a set of CPDs $\mathcal{P}$ and a set of function nodes $\mathcal{F}$. Each function node $F \in \mathcal{F}$ corresponds one-to-one with a vertex in G and $G(\mathcal{F}_i)$ is the subgraph induced by $\mathcal{F}_i \subseteq \mathcal{F}$. This means that G is a graph over $\mathcal{X} \cup \mathcal{F}$.

Definition 3 (Functional Link). *A functional link is a directed link $(v_i, v_j) \in E$ in $G = (V, E)$ that points to a vertex representing a function node or originates from a vertex representing a real-valued function node.*

Definition 4 (Functional Trail). *A trail $\pi = (v_1, \dots, v_n)$ of vertices in a graph G that contains a functional link which must be directed as (v_i, v_{i+1}) is called a functional trail.*

A trail that contains no functional link is a non-functional trail. A trail $\pi = (v_1, \dots, v_n)$ of only function nodes is called a *pure* functional trail. A functional trail $\pi = (v_1, \dots, v_n)$ where $v_1, v_n \in \mathcal{X}$ and $v_2, \dots, v_{n-1} \in \mathcal{F}$ is called a *connecting* functional trail. Notice that if X and Y are connected by a functional trail from X to Y, then Y is functionally dependent on X. On the other hand, if there exists a non-functional trail between X and Y, then the dependence relation is probabilistic.

The DAG G must not contain cyclic functional trails as a cyclic dependency in the evaluation of function nodes is not allowed. The following example has a cyclic dependency in the evaluation of function nodes. Let $X_1, X_2 \in \mathcal{X}$ and $F \in \mathcal{F}$ such that $\text{pa}(X_2) = \{X_1, F\}$ and $\text{pa}(F) = \{X_1\}$.

A maximal set of connected random variables is referred to as a *Bayesian network fragment* or *fragment*. Fragments are connected using connecting functional trails. A connecting functional trail represents the operation of transferring the results of belief update in the network fragment of the source variable to the network fragment of the destination variable.

A function node F is computed after a successful belief update or simulation in the fragments containing random variables with a functional trail to F.

Example 3. Figure 3 shows an example of a Bayesian network augmented with two real-valued function nodes F_1 and F_2 connecting the two Bayesian network fragments $\mathcal{N}_1$ and $\mathcal{N}_2$.

The two trails $\pi_1 = (B, A, C, F_1, F_2, E, G)$ and $\pi_2 = (D, C, F_1, F_2, E, H)$ are examples of functional trails with (C, F_1, F_2, E) in their connecting functional trail. The model has fragments $\mathcal{N}_1$ and $\mathcal{N}_2$ which are indicated by clouds. Any additional link between nodes in $\mathcal{N}_1$ and nodes in $\mathcal{N}_2$ would introduce a functional cycle and is thus disallowed.

3.3   Function Node Models

It is common for many implementations of Bayesian networks to support the specification of CPDs using a compact representation such as a mathematical equation

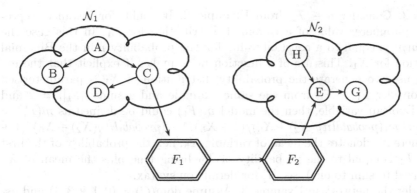

Fig. 3. Example of a Bayesian network augmented with function nodes connecting two Bayesian network fragments

or expression [20,15][1]. Here we adopt the framework introduced by [20,15]. This framework introduces the notion of subtype of discrete random variables. A discrete function node $F \in \mathcal{F}_\Delta$ has a subtype, which is either Labelled, Boolean, Numbered or Interval defining how the states of F are interpreted when evaluating expressions containing F or specifying the CPD for F. A real-valued function node expresses a calculation producing a real value and has no subtype.

Definition 5 (Function Node Model). *Let F be a function node in a Bayesian network $\mathcal{N}$ with DAG G. The model $m(F)$ specifies how the value or distribution of F is computed from the values or distributions of* pa(F) *in G.*

Function nodes only support *forward* reasoning, i.e., the value or distribution of a function node $F \in \mathcal{F}$ is computed using its model $m(F)$ and the values or distributions of pa(F).

For function node $F \in \mathcal{F}_\mathbb{R}$, the calculation defined by the model $m(F)$ should evaluate to a real value. Otherwise, the value of F is undefined and models of child nodes referring to the value of F cannot be evaluated. Similarly, for function node $F \in \mathcal{F}_\Delta$, the calculation defined by the model $m(F)$ should evaluate to a probability distribution. Otherwise, the distribution of F is undefined and models of child nodes referring to the distribution of F cannot be evaluated. Notice that $m(F)$ may, for instance, define a deterministic function that is translated into a probability distribution for a random variable when $m(F)$ is evaluated.

Different mathematical functions can be used to define $m(F)$ for $F \in \mathcal{F}_\mathbb{R}$ while additional operators can be used to define $m(F)$ for $F \in \mathcal{F}_\Delta$. This includes probability operators such as, for instance, the *probability*$(X = x)$-operator that returns the probability that variable $X \in \mathcal{X}_\Delta$ is in a specific state x. For $X \in \mathcal{X}_\Gamma$, the mean $\mu(X)$ of X is used as the value of X when evaluating models referring to X.

---

[1] E.g., Genie, Netica and HUGIN have this facility.

Example 4. Consider $p \in \mathcal{F}_\mathbb{R}$ from Example 2. It could, for instance, represent the parameter value of a Binomial distribution for X_1. In this case the model $m(p)$ is equal to a constant value for the probability p in the Binomial distribution for X_1. This use of a function node makes it explicit that the expression used to generate the probability distribution of X_1 is parameterised by p. Consider $F_1 \in \mathcal{F}_\Delta$ from the same example and assume $\|F_1\| = 2$ and X_2 is a Boolean variable, then the model $m(F_1)$ could be defined as $m(F_1) = Distribution((probability(X_2) + X_5)/(1 + X_5), 1 - (probability(X_2) + X_5)/(1 + X_5))$, where X_5 denotes the mean of variable X_5, i.e., the probability of the first state of F_1 is equal to the probability of X_2 being true plus the mean of X_5 (normalized to sum to one), see [7] for details on syntax.

Consider the network in Example 3. Assume $\text{dom}(C) = (0, 1, 2, 3, 4)$ and assume that the function node $F_1 \in \mathcal{F}_\mathbb{R}$ should represent the probability that $C \geq 2$. In this case, we set $m(F_1) = probability(C \geq 2)$.

4   Inference

Identifying dependence and independence properties between random variables using d-separation [6] is an important element of inference and the algorithm for determining d-separation needs to reflect the use of function nodes to link Bayesian networks.

4.1 d-Separation

When the network contains function nodes, trails containing functional links become directed. This means that d-separation is no longer a symmetric relation as it is necessary to distinguish between source and destination of the trails. The properties of d-separation for variables within the same network fragment are unchanged. Notice that if variables X and Y are connected by a functional trail, then they cannot be connected by a non-functional trail and vice versa.

[12] describes a criterion equivalent to d-separation for determining the independence properties in a Bayesian network:

> Given three sets of vertices: V, W and S. Construct the induced graph $G(V \cup W \cup S \cup \text{an}(V \cup W \cup S))$, and then form the moral graph G^m. If S separates $V \setminus S$ and $W \setminus S$ in G^m, then V and W are d-separated by S.

To determine the independence properties in a Bayesian network augmented with function nodes, the criterion is extended as follows. For functional links information must follow the direction of the links. This means that functional links are ignored when the ancestral graph is constructed, i.e., the ancestral graph does not contain real-valued function nodes as V, W and S do not contain real-valued function nodes. Subsequently, a depth first traversal is performed to identify reachable descendants. This means function nodes can be reached if they are descendants of a reachable vertex in the ancestral graph and that the relation is no longer symmetric.

Notice that a discrete function node should be considered equivalent to a root random variable.

4.2   Belief Update

Belief update in a Bayesian network augmented with function nodes is defined as the task of computing posterior marginals given evidence ϵ and calculating all function nodes. The order in which this is performed is important.

There exist a number of types of algorithms for belief update in a Bayesian network $\mathcal{N} = (\mathcal{X}, G, \mathcal{P})$. We describe how belief update in a Bayesian network augmented with function nodes is performed using the junction tree based algorithm of [8] and [13] referred to as HUGIN propagation. In HUGIN propagation over a discrete Bayesian network, a junction tree $T = (\mathcal{C}, \mathcal{S})$ is created from the moral graph G^m and triangulation of G^m producing G^T. The cliques $\mathcal{C}$ of T are identified from G^T and the cliques are connected by separators $\mathcal{S}$. Each CPD $P \in \mathcal{P}$ is assigned to a clique C such that $\text{dom}(P) \subseteq C$. Each clique $C \in \mathcal{C}$ has a probability potential ϕ_C and initially $\phi_C = \prod_{P \in \mathcal{P}_C} P$ where $\mathcal{P}_C$ is the set of CPDs assigned to C. Belief update is performed by two rounds of message passing over the separators $\mathcal{S}$ of T where messages are passed from the leaves of T to the root R and subsequently from R to the leaves, see [8] for details.

Let $\mathcal{N} = (\mathcal{X}, G, \mathcal{P}, \mathcal{F})$ be a Bayesian network augmented with function nodes containing a single fragment. Once function nodes in $\text{an}(\mathcal{X}_i) \cap \mathcal{F}$ are computed, $\mathcal{N}$ can be considered a self-contained Bayesian network (without function nodes) and belief update is performed as described above. After belief update, $\text{de}(\mathcal{X}) \cap \mathcal{F}$ are computed.

If $\mathcal{N} = (\mathcal{X}, G, \mathcal{P}, \mathcal{F})$ has more than one network fragment linked by connecting functional trails, then it is necessary to make sure that computations are performed in the right order. Let $\mathcal{X}_1, \ldots, \mathcal{X}_n$ be the sets of variables in the fragments $\mathcal{N}_1, \ldots, \mathcal{N}_n$ of $\mathcal{N}$. Before belief update in a fragment $\mathcal{X}_i$ can be performed, it is necessary to calculate functional trails with destination in $\mathcal{X}_i$. This means calculating $\mathcal{F} \cap \text{an}(\mathcal{X}_i)$. Belief update in each network fragment and calculation of function nodes can be organised as a DAG H where each node represents a network fragment and links represent pure functional trails between fragments. Belief update should be performed using a traversal scheme of H where parents are evaluated before children.

Proposition 1 (Correctness). *Belief update in a Bayesian network with function nodes is correct.*

Proof. Each Bayesian network fragment $\mathcal{X}_i$ is a self-contained Bayesian network once the values and distributions of function nodes in $\text{an}(\mathcal{X}_i) \cap \mathcal{F}$ are calculated. As there are no cyclic functional trails in $\mathcal{N}$, each network fragment can be compiled into a junction tree and belief update can be performed in the junction forest using a junction tree propagation algorithm by considering the trees according to the linking between network fragments.

Example 5. Figure 4 shows an example where belief update should be performed in $\mathcal{N}_1$, before $\mathcal{N}_2$ and $\mathcal{N}_3$ followed by $\mathcal{N}_4$.

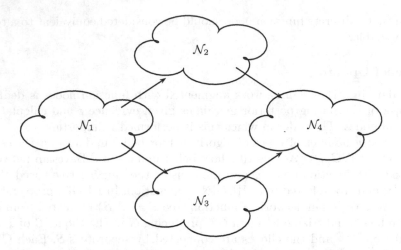

Fig. 4. Belief update in a model with fragments $\mathcal{N}_1$, $\mathcal{N}_2$, $\mathcal{N}_3$ and $\mathcal{N}_4$

The belief update algorithm for evaluating function nodes in $G(\mathcal{F}_i)$ performs a traversal from the root vertices in $G(\mathcal{F}_i)$ where $\mathcal{F}_i \subseteq \mathcal{F}$ is a maximal connected subset of $\mathcal{F}$ in G. In some cases it is not possible to compute the value of a function node F with model $m(F)$. This is the case when the model $m(F)$ depends on the value of a random variable for which the value cannot be computed, e.g., a non-instantiated random variable of subtype Labelled.

5   Real-World Application

Here four real-world applications of function nodes are described.

5.1   Credit Scoring

Many real-world applications of Bayesian networks in the domain of financial services require some kind of post-processing of the results of belief update. One of the first real-world applications of Bayesian networks where we experienced the need for post belief update calculations is BayesCredit described by [5] launched in 2001. In [5] the authors describe how Bayesian networks combined with logistic regression are used by Nykredit, the largest Danish mortgage provider, to predict the probability that a customer will default. The probability of default $P(D = 1)$ is, in principle, computed as:

$$P(D = 1) = \frac{\exp(\alpha_{EA} + \beta_{EA} \cdot H = j)}{1 + \exp(\alpha_{EA} + \beta_{EA} \cdot H = j)},$$

where α_{EA} and β_{EA} are real valued coefficients conditional on earlier arrears (EA) and H is a random variable representing the health state of the company.

The network in Fig. 5 illustrates the principle structure of the BayesCredit model (adopted from [5] where a hexagon with double border represents a real-valued function node). With a function node representation of D it is easy to include additional calculations into the model, e.g., adjusting the value in high risk periods or combining risk calculations in separate models into a single score.

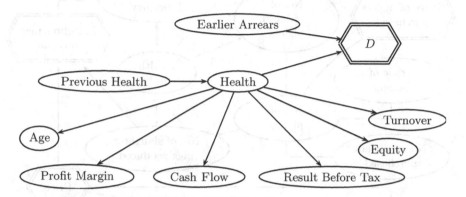

Fig. 5. Structure of the BayesCredit model using function nodes

Bayesian networks augmented with function nodes make it possible to specify the entire credit risk evaluation as a single integrated knowledge representation. This significantly improves the efficiency of evaluating and comparing different credit risk policies at the individual customer level and at the portfolio level. Policies can easily be adjusted and evaluated using one integrated knowledge representation. This significantly improves the evaluation of models and reduces the risk of errors in implementing the credit risk policy.

5.2   Risk Management in Pigs Production

RiBAY is a system for risk management in pigs production [19] launched in 2013. The core part of RiBAY is a Bayesian network augmented with function nodes $\mathcal{N} = (\mathcal{X}, G, \mathcal{P}, \mathcal{F})$. With $||\mathcal{X}|| = 28$ and $||\mathcal{F}|| = 162$, it is clear that function nodes play a significant and important role in the specification of $\mathcal{N}$. The Bayesian network is as an integrated modelling approach for representing uncertainty and analysing risk management in agriculture.

The Bayesian network consists of three types of network fragments representing uncertain prices, uncertain production of slaughter pigs and uncertain yields of rape seed and grain. The model covers one year of production with a focus on risk at the gross marginal level referred to at economic outcome. It is augmented with function nodes in part to support parameterisation of the model to the properties of a particular farm and in part to link uncertainties of different aspects of pigs production to the economic outcome of pigs production.

Function nodes are, for instance, used to encode farm specific information in the CPD expressions such as risk mitigation measures, budget numbers, system variables (i.e., the type of production at the specific farm) and forecast variables, e.g., prices on futures.

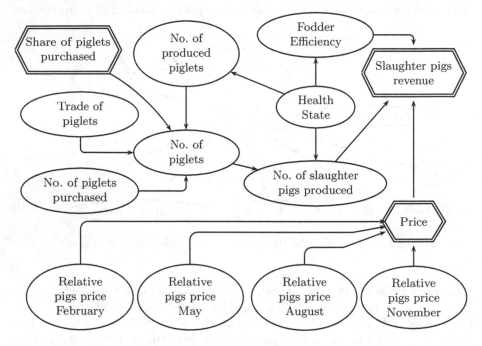

Fig. 6. Fragments representing uncertainty in the number of produced piglets and slaughter pigs

Figure 6 shows a number of (simplified) network fragments from RiBAY encoding uncertainty on the number of produced piglets and slaughter pigs [19]. Notice that *Share of piglets purchased* is defined as a real-valued function node and it is specified as an input parameter to a CPD. The random variable *No. of slaughter pigs produced* is parent of a function node *Slaughter pigs revenue*, which depends on random variables representing uncertainty on selling price.

The user interface to RiBAY is based on a web service implementation using the architecture introduced by [14]. This makes it simple to link a function node to a real value input widget on a web-site. This is used to link user input to a specific parameter of a function node model or a CPD. This is used extensively in the RiBAY application. See Fig. 7 for parts of the user interface for specifying system variables (in Danish).

With a complete specification of farm specific properties including risk mitigation measures, Monte Carlo simulation in the network fragments is used to create an empirical probability distribution for each function node including the function node representing economic outcome. This makes it possible to compute different statistics to support risk management at the farm level.

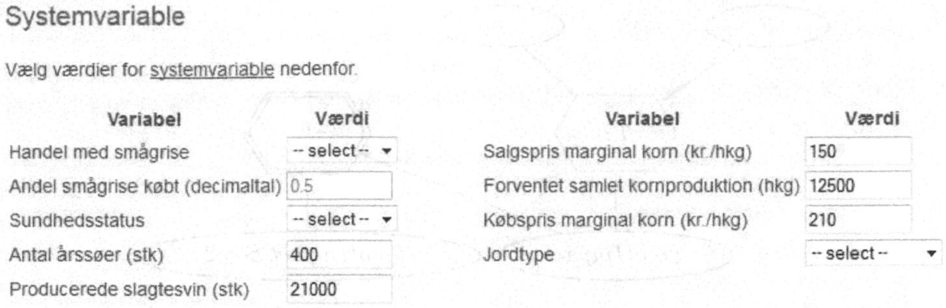

Systemvariable

Vælg værdier for systemvariable nedenfor.

Variabel	Værdi	Variabel	Værdi
Handel med smågrise	-- select -- ▾	Salgspris marginal korn (kr./hkg)	150
Andel smågrise købt (decimaltal)	0.5	Forventet samlet kornproduktion (hkg)	12500
Sundhedsstatus	-- select -- ▾	Købspris marginal korn (kr./hkg)	210
Antal årssøer (stk)	400	Jordtype	-- select -- ▾
Producerede slagtesvin (stk)	21000		

Fig. 7. Input interface of Ribay (in Danish)

5.3 Operational Risk Management

Operational risk is often defined as the risk of loss resulting from inadequate or failed internal processes, people and systems, or from external events, and includes risks such as fraud, system failures, legal risks, environmental risks and terrorism. Efficient handling of uncertainty is an important element of operational risk management (ORM) [16] and [2]. This means that a Bayesian network can be a critical tool in ORM as it is well suited for modeling risk on the basis of little or uncertain data.

In ORM it is common to compute a total impact of events distribution as the aggregation of a frequency and an impact of event distribution. The frequency distribution is a distribution for the number of events while the impact distribution is a distribution for severity of losses.

The posterior probability distribution for total impact, the aggregate loss distribution, is calculated based on the convolution of the frequency and a impact distribution. The frequency variable is of subtype Numbered and the impact variable is of subtype Interval. The total impact distribution is computed as $P(F = 1) \cdot I + P(F = 2) \cdot (I + I) + P(F = 3) \cdot (I + I + I) \cdots$ where the I variables are independent identically distributed random variables. This calculation cannot be specified in a Bayesian network (without function nodes) as the aggregated distribution is created from the two posterior distributions for frequency and impact. The calculation is represented as a discrete function node T of subtype Interval in a Bayesian network augmented with function nodes and the distribution is specified using the *aggregate*-operator in the model $m(T)$. After belief update the *aggregate*-operator computes the distribution of T from posterior distributions of I and F.

Figure 8, where the hexagon with single border represents the discrete function node TI of subtype Interval, shows an example of aggregation of two distributions and the calculation of $P(Total\ impact \geq 2)$. Assuming $\text{dom}(F) = (0, 1, 2)$, $\text{dom}(I) = ([0; 1[, [1; 2[)$ and $\text{dom}(TI) = ([0; 1[, [1; 2[, [2; 3[, [3; 4[, [4; inf[)$ with uniform probability distributions on F and I, the aggregated distribution is $P(TI) = (0.542, 0.292, 0.125, 0.042, 0)$. The real-valued function node p represents the probability that *Total impact* ≥ 2 equal to 0.167, i.e., $P(Total\ impact \geq$

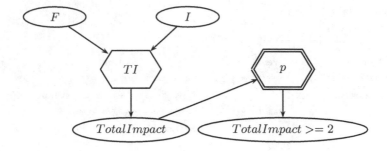

Fig. 8. Aggregation of distributions of F and I into the distribution for TI and calculation of $P(\text{Total impact} \geq 2)$

2) $= 0.167$. This value is used as the prior probability of the child variable of p. In this way, results of belief update in one Bayesian network fragment is transferred to other fragments using functional trails. The main advantage of the framework of networks with function nodes is this support for performing all computations in a single model linking Bayesian network fragments.

5.4   Geographic Information System

The combination of a Geographic Information System (GIS) and Bayesian networks is a powerful tool for decision analysis. It can, for instance, be used to evaluate the impact of geospatial management decisions, see, e.g., [21] for a classification application and, e.g., [17,22] for prediction using dynamic and object-oriented models. The geospatial area of interest is divided into a set of regions defined as placemarks on a GIS map. In a common approach to linking a GIS with Bayesian networks, the GIS provides placemark specific data to be analysed by one or more Bayesian networks and the results of the analysis are displayed on the map using color coding for visual interpretation of the consequences of management decisions.

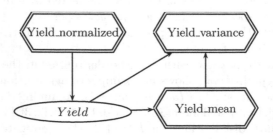

Fig. 9. Normalization of the expected value of a variable

One approach to mapping the results of belief update on a GIS map using color coding is to display the probability of a particular state of a specific variable. Alternatively, if the variable is of a numeric subtype, then the expected value of the variable can be described. This requires that the value is normalized. This can be integrated into the Bayesian network by augmenting the network with function nodes. This creates one single knowledge base representing the entire analysis and supports offline generation of the map for test and validation. Figure 9 shows a simplified example where the *Yield* variable of subtype Numbered is normalized and its variance is computed using the *probability*-operator.

The web service architecture [14] supports the use of Bayesian networks in combination with Google Maps, see Fig. 10. This is used in the OpenNESS project to operationalize Ecosystem Services[2]. In this work some output variables of Bayesian networks are of subtype Numbered. In this case, normalization of the expected value (as opposed to the most likely value) can be used for color coding the placemarks. This is readily supported within a single integrated knowledge representation by the use of function nodes adding significant additional value to the analysis using Bayesian network in combination with GIS.

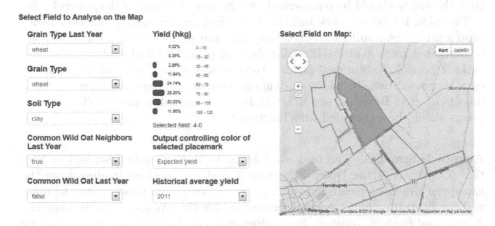

Fig. 10. An example of linking Bayesian networks and Google Maps

6   Discussion and Conclusion

We have introduced the notion of a Bayesian network augmented with function nodes. Both discrete and real-valued function nodes are considered. A real-valued function node expresses a calculation that should be performed using the results of a (successful) belief update or simulation. Similarly, a discrete function node can be used to compute distributions derived from other distributions.

---

[2] EU FP7 308428 — www.openness-project.eu

The use of Bayesian networks with function nodes is illustrated using four real-world applications in the domains of finance, insurance, risk management and geospatial decision making using GIS. In each case, Bayesian networks augmented with function nodes have made it possible to create a single knowledge representation of the properties of the problem domain under consideration and the calculations necessary to compute or specify parameter values as well as to calculate real values or distributions based on the results of belief update or simulation. This is extremely useful for the knowledge specification as well as for testing and validating the performance of systems based on Bayesian networks.

In some domains such as ORM it is common to use multiple Bayesian networks for computing a total impact distribution for each business line. The total impact distribution is computed using discrete function nodes while connecting functional trails can be used to link the networks into a single knowledge representation for risk management at the organisational level.

It is not possible to specify evidence on a real-valued function node and evidence on a discrete function node F does not induce a likelihood function over $\text{pa}(F)$. These are natural limitations from the definition of function nodes. If evidence should be entered or should induce a likelihood function over the parents, then the entity should be represented as a random variable in the network.

The plan for future work includes extending $\mathcal{F}_{\mathbb{R}}$ to include function nodes representing the result of a simulation, i.e., the real value of a function node is simulated from a probability distribution prior to belief update. This will make it possible to represent and use the result of a simulation within a knowledge representation. In addition, dynamic time-sliced Bayesian networks [4,9], object-oriented Bayesian networks [11,15] and limited-memory influence diagrams should be augmented with function nodes.

Acknowledgments. The research leading to the example described in Section 5.4 is part of the OpenNESS project and has received funding from the European Union's Seventh Programme for research, technological development and demonstration under grant agreement no 308428. We also want to thank the *Norma and Frode S. Jacobsen Foundation* and *The Nordea Bank Foundation* for their financial support of the research producing the example in Section 5.2.

References

1. Bendtsen, M., Peña, J.M.: Gated Bayesian Networks. In: Frontiers in Artificial Intelligence and Applications, Twelfth Scandinavian Conference on Artificial Intelligence, vol. 257, pp. 35–44 (2013)
2. Condamin, L., Louisot, J.-P., Naim, P.: Risk Quantification: Management, Diagnosis and Hedging. Wiley, New York (2006)
3. Cowell, R.G., Dawid, A.P., Lauritzen, S.L., Spiegelhalter, D.J.: Probabilistic Networks and Expert Systems. Springer (1999)
4. Dean, T., Kanazawa, K.: A model for reasoning about persistence and causation. Computational Intelligence 5, 142–150 (1989)

5. Ejsing, E., Vastrup, P., Madsen, A.L.: Probability of default for large corporates. In: Bayesian Networks: A Practical Guide to Applications, ch. 19, pp. 329–344. Wiley, New York (2008)
6. Geiger, D., Verma, T., Pearl, J.: Identifying independence in Bayesian networks. Networks 20(5), 507–534 (1990); Special Issue on Influence Diagrams
7. Jensen, F.: HUGIN API Reference Manual. Version 8.0 (2014), http://www.hugin.com
8. Jensen, F.V., Lauritzen, S.L., Olesen, K.G.: Bayesian updating in causal probabilistic networks by local computations. Computational Statistics Quarterly 4, 269–282 (1990)
9. Kjærulff, U.B.: dHugin: A computational system for dynamic time-sliced Bayesian networks. International Journal of Forecasting 11, 89–111 (1995)
10. Kjærulff, U.B., Madsen, A.L.: Bayesian Networks and Influence Diagrams: A Guide to Construction and Analysis, 2nd edn. Springer (2013)
11. Koller, D., Pfeffer, A.: Object-oriented Bayesian networks. In: Proc. of UAI, pp. 302–313 (1997)
12. Lauritzen, S.L., Dawid, A.P., Larsen, B.N., Leimer, H.G.: Independence properties of directed Markov fields. Networks 20(5), 491–505 (1990)
13. Lauritzen, S.L., Jensen, F.: Stable local computation with mixed Gaussian distributions. Statistics and Computing 11(2), 191–203 (2001)
14. Madsen, A.L., Karlsen, M., Barker, G.B., Garcia, A.B., Hoorfar, J., Jensen, F., Vigre, H.: A Software Package for Web Deployment of Probabilistic Graphical Models. In: Frontiers in Artificial Intelligence and Applications, Twelfth Scandinavian Conference on Artificial Intelligence, vol. 257, pp. 175–184 (2013)
15. Neil, M., Fenton, N., Nielsen, L.M.: Building Large-Scale Bayesian Networks. The Knowledge Engineering Review 15(3), 257–284 (2000)
16. Neil, M., Häger, D., Andersen, L.B.: Modeling operational risk in financial institutions using hybrid dynamic Bayesian networks. The Journal of Operational Risk 4(1), 3–33 (2009)
17. Nicholson, A.E., Flores, M.J.: Combining state and transition models with dynamic Bayesian networks. Ecological Modelling 222(3), 555–566 (2011)
18. Pearl, J.: Probabilistic Reasoning in Intelligent Systems: Networks of Plausible Inference. Series in Representation and Reasoning. Morgan Kaufmann Publishers, San Mateo (1988)
19. Rasmussen, S., Madsen, A.L., Lund, M.: Bayesian network as a modelling tool for risk management in agriculture. IFRO Working Paper 2013/12, University of Copenhagen, Department of Food and Resource Economics (2013)
20. SERENE. The SERENE Method Manual Version 1.0 (F). EC Project No. 22187. Project Doc. Number SERENE/5.3/CSR/3053/R/1 (1999)
21. Stassopoulou, A., Petrou, M., Kittler, J.: Application of a Bayesian network in a GIS based decision making system. International Journal Geographical Information Science 12, 23–45 (1998)
22. Wilkinson, L.A.T., Chee, Y.E., Nicholson, A., Quintana-Ascencio, P.: An Object-oriented Spatial and Temporal Bayesian Network for Managing Willows in an American Heritage River Catchment. In: UAI Applications Workshops: Big Data meet Complex Models and Models for Spatial, Temporal and Network Data, pp. 77–86 (2013)

A New Method for Vertical Parallelisation of TAN Learning Based on Balanced Incomplete Block Designs

Anders L. Madsen[1,2], Frank Jensen[1], Antonio Salmerón[3], Martin Karlsen[1], Helge Langseth[4], and Thomas D. Nielsen[2]

[1] HUGIN EXPERT A/S, Aalborg, Denmark
[2] Department of Computer Science, Aalborg University, Denmark
[3] Department of Mathematics, University of Almería, Spain
[4] Department of Computer and Information Science,
Norwegian University of Science and Technology, Norway

Abstract. The framework of Bayesian networks is a widely popular formalism for performing belief update under uncertainty. Structure restricted Bayesian network models such as the Naive Bayes Model and Tree-Augmented Naive Bayes (TAN) Model have shown impressive performance for solving classification tasks. However, if the number of variables or the amount of data is large, then learning a TAN model from data can be a time consuming task. In this paper, we introduce a new method for parallel learning of a TAN model from large data sets. The method is based on computing the mutual information scores between pairs of variables given the class variable in parallel. The computations are organised in parallel using balanced incomplete block designs. The results of a preliminary empirical evaluation of the proposed method on large data sets show that a significant performance improvement is possible through parallelisation using the method presented in this paper.

Keywords: Bayesian networks, TAN, parallel learning.

1 Introduction

A *Bayesian network* (BN) [5,14,15,16,19] is a powerful and popular model for probabilistic inference. Its graphical nature makes it well-suited for representing complex problems where the interactions between entities represented as variables are described using *conditional probability distributions*.

Structure restricted Bayesian network models such as the Naive Bayes (NB) Model [7] and Tree-Augmented Naive Bayes (TAN) Model [11] have shown impressive performance for solving classification tasks [7,20,24]. Data sets to be analysed using NB and TAN models are ever increasing in number and size. The size increases both with respect to the number of variables in the data sets and the number of cases in each data set. Large data sets may challenge the efficiency of pure sequential algorithms for constructing NB and TAN models

L.C. van der Gaag and A.J. Feelders (Eds.): PGM 2014, LNAI 8754, pp. 302–317, 2014.
© Springer International Publishing Switzerland 2014

from data. On the other hand, the computational power of computers is increasing and access to computers supporting parallel processing is improving. This includes improved access to computers with multiple CPUs and multicore CPUs as well as high performance computers such as supercomputers. Therefore, there is an increasing need for algorithms supporting parallel processing. In this paper, we present a new method for parallel computing when learning a TAN model from large data sets, where variables are scored pairwise in parallel. Hence, this method focuses mainly on learning from data sets, where the number of feature variables is large. The challenge is to distribute the pairwise scoring of subsets of variables onto a set of *compute nodes*.

Balanced Incomplete Block (BIB) diagrams [23] are related to the statistical issue of design of experiments. In [23] the author writes *"Combinatorial design theory concerns questions about whether it is possible to arrange elements of a finite set into subsets so that certain balance properties are satisfied."* When learning a TAN model from data with many feature variables, where the scoring is distributed onto a number of compute nodes, we need to arrange the features into subsets such that all pairs of variables are scored (at least) once. Design theory can provide one solution to this challenge. This paper describes how.

In [8] the authors describe a MapReduce-based method for learning Bayesian networks from massive data using a search & score algorithm while [4] describes a MapReduce-based method for machine learning on multicore computers. Also, [21] presents the R package **bnlearn** which provides implementations of some structure learning algorithms including support for parallel computing. [2] introduces a method for accelerating Bayesian network parameter learning using Hadoop and MapReduce. In this paper, we consider parallelisation of TAN learning on distributed memory concurrent computers using the standardized and portable message-passing system referred to as the *Message Passing Interface* (MPI) [10]. We employ the SPMD (Single Program, Multiple Data) technique to achieve parallelism in the learning of the TAN model from data through a MPI implementation. The implementation has a *master* process and a number of *worker* processes where the master process will also be a worker process. Tasks are divided into subtasks and run simultaneously on multiple processors with different input. The results of the subtasks are communicated to a master process, which collects the results and produces the final outcome.

This paper is organised as follows. Section 2 presents preliminaries and notation including an introduction to BIB designs. Section 3 describes the details of the proposed method for parallel TAN learning while Section 4 presents the results of a preliminary empirical evaluation. Finally, Section 5, Section 6 and Section 7 give a discussion, conclusions and outline future work, respectively.

2   Preliminaries and Notation

2.1   Bayesian Networks

Let $\mathcal{X} = \{X_1, \ldots, X_n\}$ be a set of discrete random variables such that $\mathrm{dom}(X)$ is the state space of X and $||X|| = |\mathrm{dom}(X)|$. A discrete BN $\mathcal{N} = (\mathcal{X}, G, \mathcal{P})$ over

$\mathcal{X}$ consists of an acyclic directed graph (DAG) $G = (V, E)$ with vertices V and edges E and a set of CPDs $\mathcal{P} = \{P(X \,|\, \mathrm{pa}(X)) : X \in \mathcal{X}\}$, where $\mathrm{pa}(X)$ denotes the parents of X in G [5,15,19]. The discrete BN $\mathcal{N}$ specifies a joint probability distribution over $\mathcal{X}$ as

$$P(\mathcal{X}) = \prod_{i=1}^{n} P(X_i \,|\, \mathrm{pa}(X_i)).$$

We only consider discrete Bayesian networks in this paper. We use upper case letters, e.g., X_i and Y, to denote variables and lower case letters, e.g., x_j and y, to denote states of variables. Sets of variables are denoted using calligraphic letters, e.g., $\mathcal{X}$ and $\mathcal{F}$.

A TAN model $\mathcal{T} = (\mathcal{X}, G, \mathcal{P})$ is a restricted type of BN with $\mathcal{X} = \{C\} \cup \mathcal{F}$ where C is a class variable and $\mathcal{F}$ is a set of feature variables and $G(\mathcal{F})$ is a tree where $G(\mathcal{X}')$ is the subgraph of G induced by $\mathcal{X}'$ and C is parent of each $F \in \mathcal{F}$.

Example 1. Figure 1 shows the graph G of a TAN model with $|\mathcal{F}| = n$ features where $\mathrm{pa}(F_i) = \{F_{i-1}, C\}$ for $i = 2, \ldots, n$, $\mathrm{pa}(F_1) = \{C\}$ and $\mathrm{pa}(C) = \emptyset$.

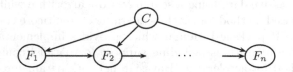

Fig. 1. A TAN model with n features

The class variable C is a parent of each $F \in \mathcal{F}$ and each $F \in \mathcal{F}$ has at most one other parent. If C is removed from G, a tree is obtained over the remaining variables, i.e., $\mathcal{F}$.

2.2 Learning a TAN from Complete Data

Let $\mathcal{D} = (c_1, \ldots, c_N)$ denote a data set of N complete cases over variables $\mathcal{X} = \{C\} \cup \mathcal{F}$ where C is the classification variable and $\mathcal{F}$ is a set of n features. The task of constructing a TAN model over $\mathcal{X}$ from $\mathcal{D}$ basically amounts to finding a maximal weighted spanning tree over $\mathcal{F}$, directing edges such that each vertex has at most one parent and adding C as a parent of each $F \in \mathcal{F}$. The algorithm of [11] based on [3] is basically:

1. Compute mutual information $I(F_i, F_j \,|\, C)$ for each pair, $i \neq j$.
2. Build a complete graph G over $\mathcal{F}$ with edges annotated by $I(F_i, F_j \,|\, C)$.
3. Build a maximal spanning tree T from G.
4. Select a vertex and recursively direct edges outward from it.
5. Add C as parent of each $F \in \mathcal{F}$.

In order to build the complete graph G over $\mathcal{F}$, we need to compute $I(F_i, F_j \mid C)$ for $\binom{n}{2} = n(n-1)/2$ pairs where $n = |\mathcal{F}|$, i.e., there are n feature variables. When using multiple processes we need to determine how to distribute the scoring between processors to avoid computing the score of any pair more than once. The ultimate level of parallelisation would be to create one process for each pair to score. However, this may not be the most efficient approach in practice.

Once the structure G has been determined, the parameters of $\mathcal{P}$ are estimated. We assume data $\mathcal{D}$ is complete and do not consider the process of estimating $\mathcal{P}$ from $\mathcal{D}$. Thus, we focus only on determining the structure of G.

2.3   Balanced Incomplete Block Designs

The use of block designs dates back to the statistical theory of design of experiments [9], highly motivated in its origin by agricultural experiments. In such scenarios, the goal was to compare the yield of different plant varieties, considering that the yield could be significantly affected by the environment, i.e., the conditions under which the plants are grown. The idea was to remove the effect of the environment by setting up *blocks* of uniform environmental conditions, and distribute the plants among the blocks, as testing every plant in each block might potentially have an unaffordable cost. The term *balanced design* refers to the fact of keeping the probability that two varieties are compared (i.e., that they fall inside the same block) constant for every pair. BIB designs are used to distribute the pairwise scoring to obtain the highest level of parallelism making sure that all pairs are scored and no pair is scored more than once.

The work of testing for independence between pairs of variables should be distributed evenly among the processes (or processors) available. At the same time, we want each process to access as little data as possible (in order to minimize IO and memory usage). If we have n variables, p processes, and each process reads data for k variables, then the following inequality must always be satisfied in order to cover all pairs of variables

$$p\binom{k}{2} \geq \binom{n}{2}.$$

Solving for k produces $k \geq n/\sqrt{p}$, and equality holds only when $p = 1$.

In order to come as close as possible to this theoretical minimum, we must distribute the data among the processes in such a way that each pair of variables is assigned to exactly one process. We use BIB designs to achieve this. In Design Theory, a design is defined as:

Definition 1 (Design [23]). *A design is a pair $(X, \mathcal{A})$ s. t. the following properties are satisfied:*

1. *X is a set of elements called points, and*
2. *$\mathcal{A}$ is a collection of nonempty subsets of X called blocks.*

We only consider cases where each block is a set (and not a multiset) and each point will correspond to a subset of variables. A BIB design is defined as:

Definition 2 (BIB design [23]). *Let v, k and λ be positive integers s. t. $v > k \geq 2$. A (v, k, λ)-BIB design is a design $(X, \mathcal{A})$ s. t. the following properties are satisfied:*

1. *$|X| = v$,*
2. *each block contains exactly k points, and*
3. *every pair of distinct points is contained in exactly λ blocks.*

We use BIB designs to control the process of scoring pairs of feature variables. A point corresponds to a subset of feature variables and a process is created for each block. The number of blocks in a design is denoted b and r denotes the *replication number*, i.e., how often each point appears in a block. Property 3 in the definition is the *balance* property that we need. We only want to score each pair once and therefore require $\lambda = 1$. A BIB design is called incomplete as $k < v$. A BIB design where $v = b$ or $r = k$ is symmetric, i.e., the number of points equals the number of blocks or the replication number equals the block size. In a (v, k, λ)-BIB design, every point occurs in r blocks where $r = \lambda(v - 1) / (k - 1)$ and the number of blocks is $b = vr/k$ [23].

Example 2. Consider the $(7, 3, 1)$-BIB design. In this design, $b = 7 \cdot 3/3 = 7$ and $r = 1 \cdot (7 - 1)/(3 - 1) = 3$. Hence, each point appears in three blocks and there are seven blocks. The blocks are (one out of a number of possibilities):

$$\{123\}, \{145\}, \{167\}, \{246\}, \{257\}, \{347\}, \{356\}, \tag{1}$$

where $\{abc\}$ is shorthand notation for $\{a, b, c\}$. This BIB design is *symmetric* as the number of blocks equals the number of points. This will not be the case in general.

Examples of other designs that are known to exist include $(16, 20, 5, 4, 1)$, $(91, 91, 10, 10, 1)$ and $(871, 871, 30, 30, 1)$, using the notation (v, b, r, k, λ) for each BIB design [6].

There is no single method to construct all BIB designs. However, a difference set can be used to generate some symmetric BIB designs.

Definition 3 (Difference Set[23]). *Assume $(G, +)$ is a finite group of order v in which the identity element is 0. Let k and λ be positive integers such that $2 \leq k < v$. A (v, k, λ)-difference set in $(G, +)$ is a subset $D \subseteq G$ that satisfies the following properties:*

1. *$|D| = k$,*
2. *the multiset $[x - y : x, y \in D, x \neq y]$ contains every element in $G \setminus \{0\}$ exactly λ times.*

In our case, we are restricted to using $(\mathbb{Z}_v, +)$, the integers modulo v.

If $D \subseteq \mathbb{Z}_v$ is a difference set in group $(G, +)$, then $D + g = \{x + g | x \in D\}$ is a translate of D for any $g \in G$. The multiset of all v translates of D is denoted $Dev(D)$ and called the development of D [23], page 42.

Theorem 1 ([23], Theorem 3.8 p. 43). *Let D be a (v, k, λ)-difference set in an Abelian group $(G, +)$. Then $(G, Dev(D))$ is a symmetric (v, k, λ)-BIB design.*

Example 3. The set $D = \{0, 1, 3\}$ is a $(7, 3, 1)$-difference set in $(\mathbb{Z}_7, +)$. The blocks constructed by iteratively adding one to each element of D (modulo 7) are:

$$\{013\}, \{124\}, \{235\}, \{346\}, \{450\}, \{561\}, \{602\}. \tag{2}$$

Notice that the ith element of each block is unique across all blocks. This property is used to assign blocks to processes.

Table 1 [12] shows difference sets for a set of symmetric BIB designs. The corresponding BIB design blocks are constructed as illustrated in Example 3. Notice that the first element of each difference set is unique. This means that the first element of each block can be used to associate process ranks and blocks.

Table 1. Difference sets for a set of symmetric BIB designs

BIB design	Difference set	k/v
(3,2,1)	(0,1)	0.67
(7,3,1)	(0,1,3)	0.43
(13,4,1)	(0,1,3,9)	0.31
(21,5,1)	(0,1,4,14,16)	0.24
(31,6,1)	(0,1,3,8,12,18)	0.19
(57,8,1)	(0,1,3,13,32,36,43,52)	0.14
(73,9,1)	(0,1,3,7,15,31,36,54,63)	0.12
(91,10,1)	(0,1,3,9,27,49,56,61,77,81)	0.11
(133,12,1)	(0,9,10,12,26,30,67,74,82,109,114,120)	0.09
(183,14,1)	(0,12,19,20,22,43,60,71,76,85,89,115,121,168)	0.08

Notice how the block size k increases and that some values are *missing* in the sequence. That is, there is no symmetric BIB design for $k = 7$ with $\lambda = 1$. We know that a symmetric BIB design exists when $k - 1$ is a *prime power* and a conjecture states that a symmetric BIB design exists only when this is the case. For instance, $8 - 1 = 7^1$ and $9 - 1 = 2^3$ whereas $7 - 1 = 6 = 2 * 3$ [13,12].

3 Parallelisation of TAN Learning

There are two obvious approaches to parallelise the TAN learning algorithm described in Section 2.2. One approach is to assign the same number of cases to each process. Each process would then count the configurations of all pairs of variables (together with the class variable) in the data assigned to the process. The counts from all processes are combined and used to perform the pairwise scoring. We refer to this as *horizontal* parallelisation. This approach to horizontal parallelization is *embarrassingly* parallel, i.e., it requires little effort to separate the problem into a number of parallel tasks.

A second approach (and the one investigated in this paper) is to distribute the scoring to the processes. The idea is to assign a set of variables to each process such that each pair of variables is assigned to exactly one process as we need to score each pair of variables at least once. This is only possible for certain combinations of numbers of variables and processes. We refer to this as vertical parallelisation.

Horizontal parallelisation mainly addresses learning from data sets with many cases whereas vertical parallelization mainly addresses learning from data sets with many feature variables. Horizontal and vertical parallelization can be combined to cope with data sets where both N and $|\mathcal{F}|$ are large. In this paper, we focus only on vertical parallelisation.

3.1   Parallel Scoring Using BIB Designs

In learning the structure of a TAN model, each pair of variables $X_i, X_j \in \mathcal{F}$ should be scored for mutual information given C (at least once). The task of calculating these scores in parallel can be solved using BIB designs. That is, BIB designs are used to control the process of scoring all pairs of features in Step 1 of the algorithm in Section 2.2, i.e., computing the mutual information between $F_i, F_j \in \mathcal{F}$ given the class variable C. This means that BIB designs are used to divide $\mathcal{F}$ into subsets to be assigned to each process. Each process will score pairs of features assigned to it.

Fisher's inequality states that $b \geq v$ [23] (who cites [9]). That is, no design with $b < v$ is possible. This means that the number of blocks b is larger than or equal to the number of points v. On the other hand, $|\mathcal{F}|$ is usually much larger than the number of processors available. This means that each point should represent a subset of variables, i.e., each point p represents a set of variables $\mathcal{F}_p \subseteq \mathcal{F}$. We do not include the class variable C in the set of points. As we need to score pairs exactly once, we are only interested in designs with $\lambda = 1$.

A separate process with a unique rank is created for each block where the rank is a number from zero to the number of processes minus one. Each process computes the pairwise scores represented by the block as described below. This means that ideally the number of blocks should match the number of processes and each point in all blocks should represent the same number of features. This may not be possible in practice as BIB designs for any combination of v and k do not necessarily exist. Instead either some processors will be idle, more blocks than processes can be created or idle processors can be used for other tasks such as horizontal parallelisation.

The process of computing the scores in Step 1 of the algorithm in Section 2.2 can be organized and distributed using a BIB design. Each process computes the score for each pair of features from different points. This is referred to as inter-point scoring. This means that all variables in different points are scored. In addition, each process computes the score for each pair of features in a unique point. This is referred to as intra-point scoring and ensures that all pairs are scored exactly once. This is demonstrated by the next example continuing Example 2.

Example 4. In the $(7, 3, 1)$-BIB design, each point $p = 1, \dots, 7$ represents a subset $\mathcal{F}_p \subseteq \mathcal{F}$. If we assume $|\mathcal{F}| = 140$, then $|\mathcal{F}_p| = 20$, i.e., each point represents 20 features. As $k = 3$ each process is assigned 60 features, but each process does not score all pairs as described below. The seven blocks ($b = 7$) of the $(7, 3, 1)$-BIB design are shown in (1).

Example 5. Consider again the $(7, 3, 1)$-BIB design and assume $|\mathcal{F}| = 14$. This means that each point represents two features. Each process is assigned six features. The seven blocks ($b = 7$) of the $(7, 3, 1)$-BIB design are shown in (1).

Figure 2 is a graphical illustration of how BIB designs are used to calculate the scores in parallel using seven processes and assuming $|\mathcal{F}| = 14$. Each process is assigned a block containing three points. Each point represents two variables.

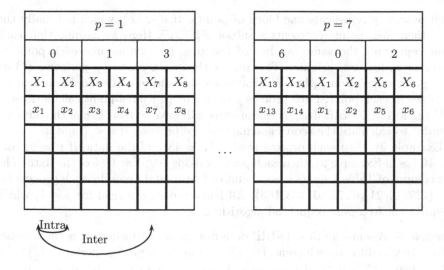

Fig. 2. Illustration of how BIB designs are used to parallelise the pairwise scoring

The figure illustrates how process $p = 1$ performs both inter-point and intra-point calculations. The block assigned to process $p = 1$ is $\{013\}$. Each point represents a unique pair of variables, e.g., point 0 represents the set $\{X_1, X_2\}$. Calculating the score for X_1 and X_2 is an intra-point operation whereas calculating the score for X_1 and X_3 is an inter-point score as X_1 and X_3 are in different points. The challenge is to make sure that all processes perform the same number of computations.

Notice that each process reads $k/v = 3/7 = 43\%$ of the feature data in addition to the class data.

Each process p computes the score for all pairs of feature variables X_i, X_j where X_i and X_j belongs to subsets represented by different points in the block represented by p. These are the inter-point operations. In addition, each process p computes the score for all pairs of feature variables X_i, X_j where X_i and X_j belong to the subset represented by the first point in the block represented by p. These are the intra-point operations. In this way, all pairs are scored exactly once.

3.2 Generating Symmetric BIB Designs with $\lambda = 1$

Symmetric BIB designs with $\lambda = 1$ can be generated using difference sets as described in Section 2.3 where Table 1 shows the difference sets needed to generate symmetric BIB designs with $\lambda = 1$ for $k \leq 14$. Each process generates its corresponding block using its unique rank by adding the rank to each element of the difference set modulus the number of processes created.

3.3 Theoretical Performance Improvement

Each process p represents one block of points. If $v < |\mathcal{F}|$, which is usually the case, then each point represents a subset $\mathcal{F}_p \subseteq \mathcal{F}$. Here we assume that each point represents the same number of features, i.e., we assume each point to represent m feature variables. This means that each process p performs $\binom{k}{2} m^2$ inter-point calculations, where k is the block size and m is the number of feature variables in each point. Each process p performs $\binom{m}{2}$ intra-point calculations.

Using a (v, k, λ)-BIB design each process will have to read k/v of the data set in order to calculate the scores assigned to the process. If $v = 7$ and $k = 3$ (as in Example 2), then each process reads $3/7 = 43\%$ of the data. If $v = 91$ and $k = 10$ (as in Example 2), then each process reads $10/91 = 11\%$ of the data. The last column of Table 1 shows the amount of feature data read by each process for $p \in \{3, 7, 13, 21, 57, 73, 91, 133, 183\}$. All feature data are read for $p = 1$, which is equivalent to a pure sequential algorithm.

Example 6. Assume a $(13, 4, 1)$-BIB design with $m = 10$ where m is the number of feature variables in each point, i.e., $|\mathcal{F}| = 130$. Each process performs $\binom{4}{2} 10^2 + \binom{10}{2} = 600 + 45 = 645$ calculations and reads 40 data files (in addition to the data file containing the class variable) out of 130 data files, i.e. $41/131 = 31\%$ of $\mathcal{D}$. In total $13 \cdot 41 = 533$ files are read by the 13 processes.

A pure sequential implementation will read all data, i.e., 130 feature data files and one class data file, and score $\binom{130}{2} = 8385$ pairs of feature variables. That is, each process in the parallel program computes 8% of the scores computed by the single process in a pure sequential application.

Although the proposed method achieves a linear (in the number of processes) speed-up in the number of calculations (mutual information scores) performed by a single process, it only achieves a speed-up of the square root of the number of processes in the amount of data needed by a single process. This is optimal for vertical parallelisation as explained in Section 2.3.

4   Empirical Evaluation

This section reports on a preliminary empirical evaluation of the proposed parallel TAN learning algorithm.

4.1   Data Sets

Three different sources of data sets were considered in the empirical evaluation. Random samples were generated from two real-world Bayesian networks of different sizes, i.e., the Munin1 [1] and Munin2 [1] networks. For each of these networks a variable was arbitrarily chosen as class variable. The third source of data was a sample generated from a real-world financial data set.

Table 2. Data sets used in the experiments

data set	$|\mathcal{X}|$	N
Munin1	189	750,000
Munin2	1,003	750,000
Bank	1,823	1,140,000

Table 2 describes properties of the data sets used in the experiments. Munin1 and Munin2 are data sets of 750,000 cases generated from the Munin1 and Munin2 networks, respectively, while Bank is a data set with 1,140,000 cases over financial data. Bank is an artificial data set generated from a real-world data set maintaining some of the statistical properties of the original data. Variable names and values have been anonymised. Continuous variables were discretized into five bins. All data sets used in the empirical evaluation are complete, i.e., there are no missing values in the data.

4.2   Hardware

The empirical evaluation was performed on three different computer systems. One Linux server and two supercomputers Fyrkat and Vilje both running Linux:

1. A linux server running Ubuntu (kernel 2.6.38-11-server) with a four-core Intel Xeon(TM) E3-1270 Processor and 32 GB RAM.
2. Fyrkat[1] is a computer cluster where each worker node used has 2 Intel Xeon (TM) X5260 Processors and 16GB RAM. It has a total of 80 such nodes. This cluster system uses SLURM (simple Linux Utility for Resource Management) for resource management.

---

[1] http://fyrkat.grid.aau.dk

3. Vilje[2] is a computer cluster where each worker node has dual eight-core Xeon E5-2670 Processors and 32GB. It has a total of 1404 such nodes. This cluster system uses PBS (Portable batch System) for resource management.

The algorithms were implemented using HUGIN software version 8.0 [17,18] and MPI. The HUGIN software does not have any special features necessary for implementing the ideas presented in this paper. It is important to notice that the experiments were performed when the system was being used by other users and running other applications. This is likely to impact performance and produce a higher variance in execution times than if the experiments were performed on a dedicated system.

4.3   Scoring Function

In the implementation, the score $I(X_i, X_j | C)$ was computed from the *likelihood* test statistics $G = 2 * \sum_i O_i \cdot \log(\frac{O_i}{E_i})$, where O_i is the observed frequency and E_i is the expected frequency under the null hypothesis. Since $G = 2 \cdot N \cdot I(X_i, X_j)$, the mutual information score can be computed as $I(X_i, X_j) = G/(2 \cdot N)$.

If the counts computed by each process are stored and communicated to the master process, then (assuming complete data) the parameters of the conditional probability distribution can be estimated.

4.4   Evaluations

The parallel algorithm was implemented employing the SPMD model. The system has a master process and a number of worker processes. Each process has a unique identifier referred to as its rank. The process of rank zero is referred to as the master process. Each process, including the master process, reads data for the feature variables assigned to it and the class variable. Each variable is stored in a single file. This means that each process only reads data for its assigned features and the class variable. The block assigned to a process is uniquely identified using the rank of the process and the difference set (each process knows the number of processes created). The unique block of process p is calculated using the rank of p and the difference set (see Theorem 1). All processes including the process of rank zero compute the mutual information for its assigned pairs and communicate the results back to the process of rank zero. The process of rank zero collects the results and creates the maximum spanning tree (Step 3 - 5). These last steps are very fast to perform compared to data reading and scoring.

The average computation time was calculated over ten runs with the same data. The computation time was measured as the elapsed (wall-clock) time between two specific points in the program. Time was measured in the master process from before it started reading data for its assigned features until the scoring was completed and all results communicated to the master process. We also report the time used for reading data and performing the scoring for the process of rank zero to get an indication of the division of work between reading data and computing scores as well as to verify the speed-up obtained.

---

[2] https://www.hpc.ntnu.no/display/hpc/Vilje

4.5   Results

This section reports on the results of the empirical evaluation of the proposed method for vertical parallelisation of learning the structure of a TAN. Experiments were performed using the three data sets described in Section 4.1 and three systems described in Section 4.2.

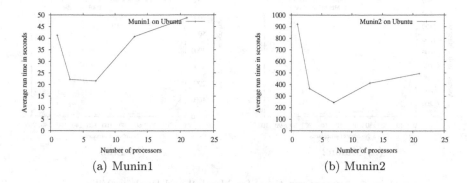

Fig. 3. Average run times for Munin1 and Munin2 on Ubuntu

Figure 3 (left) shows the average run time in seconds for Munin1 on Ubuntu while Fig. 3 (right) shows the average run time in seconds for Munin2 on Ubuntu. The figure shows that performance improved up to seven processes. For 13 and 21 processes performance deteriorated. This is expected as Ubuntu has only four physical cores (and eight logical cores).

On Fyrkat data files were assumed mounted on the compute nodes before executing the application. Figure 4 (left) shows the average running time for Munin2 on Fyrkat while Fig. 4 (right) shows the average running time for Bank on Fyrkat. It is clear that the average running time improved as the number of

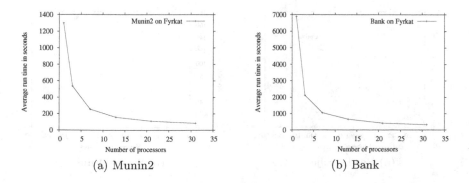

Fig. 4. Average run times for Munin2 and Bank on Fyrkat

blocks, i.e., processors used, increased. The performance should be expected to deteriorate if the number of blocks is higher than the number of processors used.

Figure 5 shows the average running time on Vilje for Munin2 (left) and Bank (right). It is clear that the average running time improved as the number of blocks, i.e., processors used, increased.

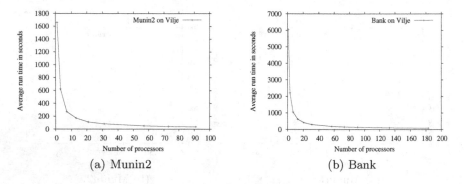

Fig. 5. Average run times for Munin2 and Bank on Vilje

Figure 6 (left) shows the speed-up factors for reading data (*Time*) and the *theoretical* number of files read by each process (*Files*) relative to the case of one processor as well as the square root of the number of processors (*Optimal*). The *theoretical* number of files read by each process is computed as $k/v \cdot |\mathcal{X}|$. Figure 6 (right) shows the speed-up factor for scoring as a function of the number of processors used.

It is clear from Fig. 6 that the theoretical number of files read by each process as expected follows the square root of the number of processes. Similarly, the speed-up factor for time used on testing is approximately a linear function of the number of processors used. It is more surprising that the measured time

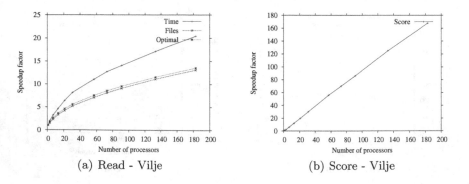

Fig. 6. Speed up as a function of the number of processors (relative to one processor)

performance shows a higher speed-up that the square root of the number of processes. The number of times a file was read by different processes increased as the number of processors used increased. This means that the impact of *caching* files increased with the number of processors used. We believe that this explains the unexpected observed behaviour.

5 Discussion

This paper has introduced a new method for vertical parallelisation of TAN learning based on using symmetric BIB designs with $\lambda = 1$. The use of BIB designs makes it possible to achieve a work *balance* between processes such that all processes score (almost) an equal number of pairs of feature variables. Symmetric BIB designs with $\lambda = 1$ do not exist for all values of v and k. Table 3 shows symmetric BIB designs with $\lambda = 1$ for $k \leq 14$. Difference sets for symmetric BIB designs with $\lambda = 1$ for much higher k are known, see e.g.,[12]. If the number of processors (or cores) does not match with a block size for which a symmetric BIB design with $\lambda = 1$ is known to exist, then idle processors can be used for horizontal parallelisation or parallelisation of the counting process. Difference sets for b up to 1893 are implemented and this can be increased as needed taking the prime power conjecture into account.

The results of the experimental evaluation show a clear time performance improvement as the number of blocks, i.e., processors used, is increased. Notice that even on a single CPU machine with multiple cores, a performance improvement is achieved. For this system performance deteriorates when the number of blocks is higher than the number of logical cores in the CPU. This should be expected. There is some variance in the run time measured. This should also be expected as the evaluation is performed on systems serving other users, i.e., the experiments have not been performed on isolated systems.

Notice that the performance evaluation has been performed on three different systems. From a personal computer running as a Linux server to powerful supercomputers using different types of resource management systems. The results of the experiments show that a performance improvement can be realised on each of these types of systems taking advantage of parallel computation. The empirical evaluation has been performed using data sets of different complexity both with respect to the number of variables and the number of cases.

6 Conclusion

This paper introduces a new method to vertical parallelisation of learning the structure of a TAN model from data. The approach is based on the use of BIB designs to distribute computing pairwise mutual information between features given the class variable.

The results of an empirical evaluation of the proposed method on desktop as well as supercomputers show a significant time performance improvement over the pure sequential method.

7   Future Work

The principle of vertical parallelisation of pairwise scoring introduced in this paper can be applied to the process of structure learning of a Bayesian network using, e.g., the PC algorithm [22]. In the PC algorithm all variables are initially tested for pairwise independence, which is similar to the scoring of all pairs of $\mathcal{F}$. In addition, a set of conditional independence tests are performed. The principles of vertical parallelisation can be applied in both cases. The plan for future work includes investigating how BIB designs can be applied to perform the conditional independence tests in parallel.

Parallelising the counting process (horizontal parallelisation) can be considered as orthogonal to the pairwise scoring (vertical parallelisation). This means that the methods can be combined to achieve even further performance improvements when data sets are extremely large, i.e., do not fit into main memory of the computer. Furthermore, we plan to investigate options for taking advantage of multithreaded programming solving the tasks assigned to each process. This may include both data reading for horizontal parallelisation of the counting scheme as well as inter and intra pairwise conditional independence testing. Furthermore, BIB designs can also be applied to improve the performance of the pairwise scoring by a single process.

Acknowledgments. This work was performed as part of the AMIDST project. AMIDST has received funding from the European Union's Seventh Framework Programme for research, technological development and demonstration under grant agreement no 619209. The data set Bank has kindly been provided for testing by Cajamar.

References

1. Andreassen, S., Jensen, F.V., Andersen, S.K., Falck, B., Kjærulff, U., Woldbye, M., Sørensen, A.R., Rosenfalck, A., Jensen, F.: MUNIN — an expert EMG assistant. In: Desmedt, J.E. (ed.) Computer-Aided Electromyography and Expert Systems, ch. 21. Elsevier Science Publishers, Amsterdam (1989)
2. Basak, A., Brinster, I., Ma, X., Mengshoel, O.J.: Accelerating Bayesian network parameter learning using Hadoop and MapReduce. In: Proceedings of the 1st International Workshop on Big Data, Streams and Heterogeneous Source Mining: Algorithms, Systems, Programming Models and Applications, pp. 101–108 (2012)
3. Chow, C.K., Liu, C.N.: Approximating discrete probability distributions with dependence trees. IEEE Transactions on Information Theory, IT 14(3), 462–467 (1968)
4. Chu, C.-T., Kim, S.K., Lin, Y.-A., Yu, Y., Bradski, G., Ng, A.Y., Olukotun, K.: Map-reduce for machine learning on multicore. In: NIPS, pp. 281–288 (2006)
5. Cowell, R.G., Dawid, A.P., Lauritzen, S.L., Spiegelhalter, D.J.: Probabilistic Networks and Expert Systems. Springer (1999)
6. Di Paola, J.W., Wallis, J.S., Wallis, W.D.: A list of (v,b,r,k,λ) designs for $r \leq 30$. In: Proc. 4th S-E Cont. Combinatorics, Graph Theory and Computing, pp. 249–258 (1973)

 7. Domingos, P., Pazzani, M.: On the optimality of the simple Bayesian classifier under zero-one loss. Machine Learning, 103–130 (1997)
 8. Fang, Q., Yue, K., Fu, X., Wu, H., Liu, W.: MapReduce-Based Method for Learning Bayesian Network from Massive Data. In: Ishikawa, Y., Li, J., Wang, W., Zhang, R., Zhang, W. (eds.) APWeb 2013. LNCS, vol. 7808, pp. 697–708. Springer, Heidelberg (2013)
 9. Fisher, R.A.: An examination of the different possible solutions of a problem in incomplete blocks. Annals of Eugenics, 52–75 (1940)
10. The MPI Forum. MPI: A Message Passing Interface (1993)
11. Friedman, N., Geiger, D., Goldszmidt, M.: Bayesian network classifiers. Machine Learning, 1–37 (1997)
12. Gordon, D.M.: La Jolla Difference Set Repository, http://www.ccrwest.org/diffsets/diff_sets/ (accessed May 15, 2014)
13. Gordon, D.M.: The Prime Power Conjecture is True for $n < 2000000$. Electronic J. Combinatorics 1(1, R6), 1–7 (1994)
14. Jensen, F.V., Nielsen, T.D.: Bayesian Networks and Decision Graphs, 2nd edn. Springer (2007)
15. Kjærulff, U.B., Madsen, A.L.: Bayesian Networks and Influence Diagrams: A Guide to Construction and Analysis, 2nd edn. Springer (2013)
16. Koller, D., Friedman, N.: Probabilistic Graphical Models — Principles and Techniques. MIT Press (2009)
17. Madsen, A.L., Jensen, F., Kjærulff, U.B., Lang, M.: HUGIN - The Tool for Bayesian Networks and Influence Diagrams. International Journal on Artificial Intelligence Tools 14(3), 507–543 (2005)
18. Madsen, A.L., Lang, M., Kjærulff, U.B., Jensen, F.: The Hugin Tool for Learning Bayesian Networks. In: Nielsen, T.D., Zhang, N.L. (eds.) ECSQARU 2003. LNCS (LNAI), vol. 2711, pp. 594–605. Springer, Heidelberg (2003)
19. Pearl, J.: Probabilistic Reasoning in Intelligent Systems: Networks of Plausible Inference. Series in Representation and Reasoning. Morgan Kaufmann Publishers, San Mateo (1988)
20. Rish, I.: An empirical study of the naive Bayes classifier. In: IJCAI Workshop on Empirical Methods in AI, pp. 41–46 (2001)
21. Scutari, M.: Learning Bayesian Networks with the bnlearn R Package. Journal of Statistical Software 35(3), 1–22 (2010)
22. Spirtes, P., Glymour, C., Scheines, R.: Causation, Prediction, and Search, Adaptive Computation and Machine Learning, 2nd edn. MIT Press (2000)
23. Stinson, D.: Combinatorial Designs — Constructions and Analysis. Springer (2003)
24. Zhang, N.L.: Hierarchical latent class models for cluster analysis. Journal of Machine Learning Research 5, 697–723 (2004)

Equivalences between
Maximum a Posteriori Inference
in Bayesian Networks and
Maximum Expected Utility
Computation in Influence Diagrams

Denis D. Mauá

Universidade de São Paulo,
São Paulo, Brazil
denis.maua@usp.br

Abstract. Two important tasks in probabilistic reasoning are the computation of the maximum posterior probability of a given subset of the variables in a Bayesian network (MAP), and the computation of the maximum expected utility of a strategy in an influence diagram (MEU). Despite their similarities, research on both problems have largely been conducted independently, with algorithmic solutions and insights designed for one problem not (trivially) transferable to the other one. In this work, we show constructively that these two problems are equivalent in the sense that any algorithm designed for one problem can be used to solve the other with small overhead. These equivalences extend the toolbox of either problem, and shall foster new insights into their solution.

Keywords: Bayesian networks, maximum a posteriori inference, influence diagrams, maximum expected utility.

1 Introduction

Maximum a posteriori inference (MAP) consists in finding a configuration of a certain subset of the variables that maximizes the posterior probability distribution induced by a Bayesian network [30]. MAP has applications, for example, in diagnostic systems and classification of relational and sequential data [15]. Solving MAP is computationally difficult, and the literature contains a plethora of approximate solutions, a few examples being the works in [4,29,17,18,13,21,26]

Influence diagrams extend Bayesian networks with preferences and actions to cope with decision making situations [10,12]. The maximum expected utility problem (MEU) is to select a mapping from observations to actions that maximizes the expected utility as defined by an influence diagram. MEU appears, for example, in troubleshooting and active sensing [12]. Although MEU is computationally difficult to solve, it counts with a large number of approximate solutions, for example, the works in [32,14,20,5,24,19,16,9,7,6].

L.C. van der Gaag and A.J. Feelders (Eds.): PGM 2014, LNAI 8754, pp. 318–333, 2014.

The MAP and MEU problems are closed for the complexity class NP^{PP} [30,5], which implies that any algorithm designed to solve one problem can *in principle* be used to solve the other.[1] Moreover, both problems are NP-complete when the treewidth of the underlying diagram is assumed bounded [3,22,23].[2] In practice, however, these two problems have been investigated independently, with a few similarities arising in the design of algorithms such as the use of clique-tree structures and message-passing for fast probabilistic inference [20,21].

In this work we provide *constructive* proofs of the equivalences between these two problems. We start by presenting background knowledge on graphs (Sec. 2), Bayesian networks (Sec. 3) and influence diagrams (Sec. 4), and formalizing the MAP and MEU problems. Then, we design a polynomial-time reduction that maps MAP problems into MEU problems (Sec. 5). We show that the reduction increases the treewidth by at most four, which makes the reduction closed in NP. We proceed to build a polynomial-time reduction of MEU into MAP problems (Sec. 6). The reduction increases treewidth by at most five, being also closed in NP. These reductions enlarge the algorithmic toolbox of either problem, and shall bring new insights into the design of new algorithms. We conclude with an overview of the results and a brief discussion on some shortcomings of the reductions developed here (Sec. 7).

2   Some Useful Concepts from Graph Theory

Consider a directed graph with nodes X and Y. A node X is a *parent* of a Y if there is an arc going from X to Y, in which case we say that Y is a *child* of X. The *in-degree* of a node is the number of its parents. We denote the parents of a node X by $pa(X)$ and its children by $ch(X)$. The family of a node comprises the node itself and its parents. A *polytree* is a directed acyclic graph (DAG) which contains no undirected cycles. A DAG is *loopy* if it is not a polytree. Polytrees are important, as they are among the simplest structures, and probabilistic inference can be performed efficiently in some polytree-shaped Bayesian networks.

The *moral graph* of a DAG is the undirected graph obtained by connecting nodes with a common child and dropping arc directions. The moral graph of a DAG might contain (undirected) cycles even when the DAG itself does not (e.g., any polytree with maximum in-degree greater than one).

A *tree decomposition* of an undirected graph G is a tree T such that

1. each node i associated to a subset $\mathcal{X}_i$ of nodes in G;
2. for every edge X-Y of G there is a node i of T whose associated node set $\mathcal{X}_i$ contains both X and Y;
3. for any node X in G the subgraph of T obtained by considering only nodes whose associate set contain X is a tree.

---

[1] We assume here that the number of incoming arcs into any decision node in an influence diagram is logarithmically bounded by the number of variables, which limits the size of strategies to a polynomial in the input size.

[2] The treewidth of a graph is a measure of its similarity to a tree.

The third property is known as the running intersection property. A *clique* is a set of pairwise connected nodes of an undirected graph. Any tree decomposition of a graph contains every clique of it included in some of the associated node sets [2]. The *treewidth* of a tree decomposition is the maximum cardinality of a node set $\mathcal{X}_i$ associated to a node of it minus one. The treewidth of a graph G is the minimum treewidth over all tree decompositions of it. The treewidth of a directed graph is the treewidth of its corresponding moral graph. Polytrees have treewidth given by the maximum in-degree of a node.

The elimination of a node X from a graph G produces a graph G' by removing X (and its incident arcs) and pairwise connecting all its neighbors. A node is *simplicial* if all its neighbors are pairwise connected. Eliminating a simplicial node is the same as simply removing it (and its incident arcs) from the graph. Let G be a graph of treewidth κ, and G' be the graph of treewidth κ' obtained from G by eliminating a node X of degree d. Then κ is at most $\max\{\kappa', d\}$, being exactly that when X is simplicial [2]. By removing arcs or nodes we generate graphs whose treewidth are not larger than the original graph.

3   Bayesian Networks and the MAP Problem

A Bayesian network consists of a DAG G over a set of variables $\mathbf{X}$ and a set of conditional probability assessments $P(X = x|\mathrm{pa}(X) = \pi)$, one assessment for every variable X in $\mathbf{X}$ and configurations x and π of X and $\mathrm{pa}(X)$, respectively. The DAG encodes a set of Markov conditions: every variable is independent of its non-descendant non-parents given its parents. These conditions induce a joint probability distribution over the variables that factorizes as $P(\mathbf{X}) = \prod_{X \in \mathbf{X}} P(X|\mathrm{pa}(X))$.

The treewidth of a Bayesian network is defined as the treewidth of its underlying DAG. When using tree decompositions of Bayesian networks we refer to the sets associated to nodes of the tree as variable sets, since every node is identified with a variable. Probabilistic inference can be performed in time at most exponential in the treewidth of the network, hence in polynomial-time if treewidth is bounded [15].

Let $(\mathbf{M}, \mathbf{E}, \mathbf{H})$ denote a partition of $\mathbf{X}$ and $\hat{\mathbf{e}}$ be an assignment to $\mathbf{E}$. The set $\mathbf{M}$ contains *MAP variables*, whose values we would like to infer; the set $\mathbf{E}$ contains *evidence variables*, whose values are known to be (i.e., they are fixed at) $\hat{\mathbf{e}}$; at last, the set $\mathbf{H}$ contains *hidden* variables, whose values we ignore (i.e., they are marginalized out). The MAP problem consists in computing the *value*

$$\max_{\mathbf{m}} P(\mathbf{M} = \mathbf{m}, \mathbf{E} = \hat{\mathbf{e}}) = \max_{\mathbf{m}} \sum_{\mathbf{H}} P(\mathbf{M} = \mathbf{m}, \mathbf{E} = \hat{\mathbf{e}}, \mathbf{H}) \,. \tag{1}$$

A configuration $\mathbf{m}^*$ which maximizes the equation above is known as a maximum a posteriori configuration or posterior mode, as it also maximizes the posterior probability distribution $P(\mathbf{M}|\mathbf{E} = \hat{\mathbf{e}})$. We can compute $\mathbf{m}^*$ by recursively solving MAP problems as follows. First, solve the MAP problem (call this problem unconstrained). Label all MAP variables free and repeat the following procedure until no free variables remain: Select a free variable M_i and clamp it at a value m_i^* such that the MAP problem with $M_i = m_i^*$ returns the same value as the unconstrained problem; label this variable fixed. Note however that most algorithms for the MAP problem are able to provide a configuration $\mathbf{m}^*$ without resorting to the procedure described (and with much less overhead).

MAP was shown to be NP-hard to approximate even in polytree-shaped networks [30]. Specifically, it was shown that the decision version of MAP is NP^{PP}-complete on loopy networks, and NP-complete on networks of bounded treewidth. More recently, de Campos [3] showed that the problem is NP-hard to solve even in polytree-shaped networks with ternary variables, but admits a fully polynomial-time approximation scheme in networks of bounded treewidth with variables taking on a bounded number of values. A large number of approximate algorithms have been designed to cope with such a computational difficulty, including search-based methods [29], branch-and-bound techniques[17], dynamic programming [12,21], message passing [18,13], function approximation [8,4,26], and knowledge compilation [11].

4   Influence Diagrams and the MEU Problem

An influence diagram extends a Bayesian network with preferences and actions in order to represent decision making situations. Formally, a influence diagram consists of a DAG over a set of *chance variables* $\mathbf{C}$, *decision variables* $\mathbf{D}$, and *value variables* $\mathbf{V}$. The sets $\mathbf{C}$, $\mathbf{D}$ and $\mathbf{V}$ are disjoint. A chance variable C represents quantities over which the decision maker has no control, and is associated with conditional probability assessments $P(C|\mathrm{pa}(C))$ as in a Bayesian network. The restriction of an influence diagram to its chance variables characterizes a Bayesian network. A decision variable D represents possible actions available to the decision maker conditional on the observation of the values of $\mathrm{pa}(D)$. Decision variables are not (initially) associated to any function. A value variable V represents costs or rewards of taking actions $\mathrm{pa}(V) \cap \mathbf{D}$ given an instantiation of $\mathrm{pa}(V) \cap \mathbf{C}$. Every value variable V is associated with utility functions $U(\mathrm{pa}(V)))$, which encode additive terms of the overall utility. The treewidth of an influence diagram is the treewidth of the corresponding moral graph after deleting value nodes.

A *decision rule* (a.k.a policy) for a decision variable D is a conditional distribution $P(D|\mathrm{pa}(D))$ specifying the probability of executing action $D = d$ upon observing $\mathrm{pa}(D) = \pi$. A decision rule prescribes an agent behavior, which is not necessarily deterministic (i.e., the agent might take different actions d when in scenario π according to $P(D|\pi)$). When $P(D|\mathrm{pa}(D))$ is degenerate for every π, we can identify a policy with a function mapping configurations π into actions d. Moreover, if D has no parents, then we can associated a decision rule for D with an assignment of a value of D. We will often refer to degenerate policies as functions or assignments (of root decision variables). A *strategy* is a set containing exactly one decision rule for each decision variable. The expected utility of a strategy $\mathcal{S} = \{P(D|\mathrm{pa}(D)) : D \in \mathbf{D}\}$ is given by

$$E(\mathcal{S}) = \sum_{V \in \mathbf{V}} \sum_{\mathrm{pa}(V)} U(\mathrm{pa}(V)) P(\mathrm{pa}(V)|\mathcal{S}) \qquad (2)$$

$$= \sum_{V \in \mathbf{V}} \sum_{\mathbf{C},\mathbf{D}} U(\mathrm{pa}(V)) \prod_{X \in \mathbf{C} \cup \mathbf{D}} P(X|\mathrm{pa}(X)) . \qquad (3)$$

The MEU problem is to compute the *value* of the maximum expected utility of a strategy, that is, to compute $\max_{\mathcal{S}} E(\mathcal{S})$. A strategy $\mathcal{S}^*$ whose expected utility equals that value is called an *optimal strategy*. We can obtain an optimal strategy by recursively solving MEU problems, in a similar fashion to the computation of a maximum a posteriori configuration of the MAP problem. Although in this work we state the results of the reductions in terms of the MAP and MEU problems (hence, problems whose output are numbers), the same results could be stated with respect to maximum a posteriori configurations of MAP and optimal strategies of MEU. It is well-known that the maximum expected utility can be attained by a strategy containing only degenerate policies. Hence, in what concerns the MEU problem there is no loss in allowing only deterministic policies.

The *perfect recall* condition characterizes a non-forgetting agent, and translates graphically to the property that the parents of any decision variable are also parents of any of its children. Perfect recall is a consequence of rationality when the decision problem involves a single agent with unlimited resources, as it equates with every known information being considered when making a decision. This is not the case when multiple agents are involved or resources such as memory and computing power are limited. A related concept is that of *regularity*, which requires a temporal order over the decision variables. Together, perfect recall and regularity enable the solution of MEU by dynamic programming due to Bellman's principle of optimality. In our definition, we do not require or assume perfect recall or regularity, although we do allow them to be present by explicit specification in the graph. Influence diagrams that do not enforce perfect recall and regularity are often called *limited memory influence diagrams* [16] or *decision networks* [33], although there is some ambiguity about the use of the latter.

De Campos and Ji [5] showed that the decision version of MEU is NP^{PP}-complete in loopy diagrams, and NP-complete in diagrams of bounded treewidth. Mauá et al. strengthened those results by showing the problem to be NP-hard even in polytree-shaped diagrams with ternary variables and a single value variable [25], and even in polytree-shaped diagrams with binary variables and arbitrarily many value variables [23]. They also showed that it is NP-hard to approximately solve the problem, even in polytree-shaped diagrams when variables can take on arbitrarily many values [25], but that there is a fully polynomial-time approximation scheme when both the diagram's treewidth and the maximum variable cardinality are bounded [22]. The problem was also shown to be polynomial-time computable in polytree-shaped diagrams with binary variables [23], and in diagrams that satisfy perfect recall and whose minimal diagram has bounded treewidth [16]. As with MAP, the computational hardness of the problem motivated the development of a large number of approximate algorithms.

Some of the approaches include branch-and-bound [28,27,32,14], dynamic programming [20,12,24], integer programming [5], message passing [19], combinatorial search [16,9], and function approximation [7,6].

5   Reducing MAP To MEU

Consider a MAP problem with Bayesian network $\mathcal{N} = (G, \mathbf{X}, \{P(X|\text{pa}(X))\})$, MAP variables $\mathbf{M} \subseteq \mathbf{X}$ and evidence $\mathbf{E} = \hat{\mathbf{e}}$. Assume w.l.o.g. that the variables in $\mathbf{E}$ have no children [1]. Consider also an ordering $M_1, \ldots, M_n$ of the variables in $\mathbf{M}$ consistent with the partial ordering defined by G (i.e., if there is a directed path from M_i to M_j in G then $j > i$), and an ordering $E_1, \ldots, E_m$ of the variables in $\mathbf{E}$ also consistent with G. Let $\hat{e}_i$ denote the assignment in $\mathbf{e}$ corresponding to E_j, $j = 1, \ldots, m$. Obtain an influence diagram $\mathcal{I}$ by augmenting the Bayesian network $\mathcal{N}$ in the following way.

1. Label every variable in $\mathbf{X}$ as chance variable;
2. Add root chance variables S_0 and T_0 with values t and f, and specify $P(S_0) = P(T_0) = 1/2$;
3. For $i = 1, \ldots, n$ add a decision variable D_i taking the same values as M_i;
4. For $i = 1, \ldots, n$ add a chance variable S_i with values t and f, and parents S_{i-1}, M_i and D_i, and specify

$$P(S_i = 1 | S_{i-1}, M_i, D_i) = \begin{cases} 1, & \text{if } S_{i-1} = t \text{ and } M_i = D_i; \\ 0, & \text{otherwise.} \end{cases}$$

5. For $j = 1, \ldots, m$ add a variable T_j with values t and f, parents T_{j-1} and E_j, and specify

$$P(T_j = 1 | T_{j-1}, E_j) = \begin{cases} 1, & \text{if } T_{j-1} = t \text{ and } E_j = \hat{e}_j; \\ 0, & \text{otherwise.} \end{cases}$$

6. Add a value variable V with parents S_n and T_m and utility function

$$U(S_n, T_m) = \begin{cases} 1, & \text{if } S_n = T_m = t; \\ 0, & \text{otherwise.} \end{cases}$$

Figure 1 illustrates the influence diagram obtained in reduction above.

Remark 1. The above reduction takes time polynomial in the size of the input Bayesian network.

Remark 2. The reduction might introduce (undirected) loops, that is, the reduction (potentially) maps a polytree-shaped Bayesian network into a loopy influence diagram.

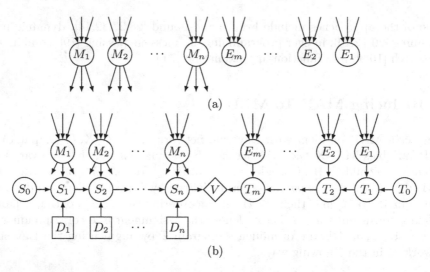

Fig. 1. Fragments of (a) of a Bayesian network and (b) its equivalent influence diagram produced by the procedure described

Lemma 1. *Let P be the probability measure induced by $\mathcal{I}$. Then,*

$$P(S_n = t, S_0, \ldots, S_{n-1}|\mathbf{M}, \mathbf{D}) = \begin{cases} 1/2, & \text{if } S_i = t \text{ and } M_i = D_i, i = 1, \ldots, n; \\ 0, & \text{otherwise.} \end{cases}$$

Proof. By induction in n. The base case for $n = 1$ follows from simple application of the Chain Rule in Bayesian networks:

$$P(S_1 = t, S_0|M_1, D_1) = P(S_0)P(S_1 = t|M_1, D_1) = \begin{cases} 1/2, & \text{if } M_1 = D_1; \\ 0, & \text{otherwise.} \end{cases}$$

Assume the result holds for some n. Applying the Chain Rule and using the conditional independences represented by the graph we obtain

$$P(S_{n+1} = t, S_0, \ldots, S_n|M_1, \ldots, M_{n+1}, D_1, \ldots, D_{n+1}) =$$
$$P(S_{n+1} = t|S_n, M_{n+1}, D_{n+1})P(S_0, \ldots, S_n|M_1, \ldots, M_n, D_1, \ldots, D_n).$$

By design $P(S_{n+1} = t|S_n, M_{n+1}, D_{n+1})$ vanishes unless $S_n = t$ and $M_{n+1} = D_{n+1}$, in which case the above equality equals

$$P(S_n = t, S_0, \ldots, S_{n-1}|M_1, \ldots, M_n, D_1, \ldots, D_n).$$

Hence, the induction hypothesis holds also for $n + 1$, and the result follows. □

Lemma 2. *Let P be the probability measure induced by $\mathcal{I}$. Then,*

$$P(T_m = t, T_1, \ldots, T_{m-1}|\mathbf{E}) = \begin{cases} 1, & \text{if } T_j = t \text{ and } E_j = \hat{e}_j, j = 1, \ldots, m; \\ 0, & \text{otherwise.} \end{cases}$$

Proof. The proof is similar to the proof of Lemma 1.   □

Theorem 1. *The MEU problem obtains the same value as the MAP problem.*

Proof. Consider an arbitrary strategy $\mathcal{S} = \{d_1, \ldots, d_n\}$, and let $\mathbf{S} = \{S_0, \ldots, S_{n-1}\}$ and $\mathbf{T} = \{T_0, \ldots, T_{m-1}\}$. By using the independences stated in $\mathcal{I}$ and the Total and Chain Rules we derive

$$E(\mathcal{S}) = \sum_{S_n, T_m} U(S_n, T_m) P(S_n, T_m | \mathcal{S}) = P(S_n = t, T_m = t | \mathcal{S})$$

$$= \sum_{\mathbf{X}, \mathbf{S}, \mathbf{T}} P(S_n = t, T_m = t, \mathbf{X}, \mathbf{S}, \mathbf{T} | \mathbf{D} = \mathcal{S})$$

$$= \sum_{\mathbf{X}, \mathbf{S}, \mathbf{T}} P(S_n = t, \mathbf{S} | \mathbf{M}, \mathbf{D} = \mathcal{S}) P(T_m = t, \mathbf{T} | \mathbf{E}) P(\mathbf{X}).$$

According to Lemmas 1 and 2, the product in the sum above vanishes whenever $S_i \neq t$, for some $i = 1, \ldots, n-1$, $M_i \neq d_i$, for some $i = 1, \ldots, n$, $T_i \neq t$, for $j = 1, \ldots, n-1$, or $E_j \neq \hat{e}_j$, for $j = 1, \ldots, m$. Whence,

$$E(\mathcal{S}) = \sum_{\mathbf{H}, S_0, T_0} P(S_n = t | \mathbf{M} = \mathcal{S}) P(T_m = t | \mathbf{E} = \hat{e}) P(\mathbf{M} = \mathcal{S}, \mathbf{E} = \hat{e}, \mathbf{H}) P(S_0) P(T_0)$$

$$= P(\mathbf{M} = \mathcal{S}, \mathbf{E} = \hat{e}, \mathbf{H}).$$

It follows from the above that $\max_{\mathcal{S}} E(\mathcal{S}) = \max_{\mathbf{h}} P(\mathbf{M} = \mathbf{h}, \mathbf{E} = \hat{e}, \mathbf{H})$, which proves the result.   □

The next result shows that the reduction devised maintains at least part of the structure of the reduced problem.

Theorem 2. *Let κ denote the treewidth of the Bayesian network $\mathcal{N}$. Then the diagram $\mathcal{I}$ has treewidth at most $\kappa + 4$.*

Proof. Let $\mathcal{T}$ be an optimal tree decomposition (i.e., one with minimum treewidth) for the DAG of $\mathcal{N}$ after deleting the arcs leaving variables in $\mathbf{E}$ (removing arcs leaving evidence nodes does not alter the result of MAP inference [1]). We obtain a tree decomposition for $\mathcal{I}$ whose treewidth is at most the treewidth of $\mathcal{T}$ plus three as follows. For $i = 1, \ldots, n$ find a node whose associated variable set includes $\{M_i\} \cup \mathrm{pa}(M_i)$, add a leaf node ℓ_i as its neighbor and associate ℓ_i with $\{M_i\} \cup \mathrm{pa}(M_i)$. Similarly, for $j = 1, \ldots, m$ find a node associated with a superset of $\{E_j\} \cup \mathrm{pa}(E_j)$, add a leaf node ℓ_{n+j} as its neighbor, and associate ℓ_{n+j} with $\{E_j\} \cup \mathrm{pa}(E_j)$. Transform the resulting structure such that it becomes binary, and denote the result by $\mathcal{T}_1$.[3] Root $\mathcal{T}_1$ at a node r (by orienting arcs away from r) such that $\ell_1, \ldots, \ell_{n+m}$ are visited in-order, that is, in a depth-first tree traversal of $\mathcal{T}_1$ rooted at r, ℓ_i is visited before ℓ_j if

---

[3] Any tree decomposition can be turned into a binary tree decomposition (i.e., one in which each node has at most three neighbors) of same treewidth [31].

and only if $i < j$. Obtain a structure $\mathcal{T}_2$ from $\mathcal{T}_1$ as follows. For every node ℓ_i, $i = 1, \ldots, n$, add a node ℓ'_i as a child of ℓ_i and associate it to $\{S_i, S_{i-1}, D_i, M_i\}$. Similarly, for every node ℓ_i, $i = n + 1, \ldots, m$, add a child node ℓ'_i associated to $\{T_i, T_{i-1}, E_i\}$. The structure $\mathcal{T}_3$ is a not a valid tree-decomposition, as it violates the running intersection property: e.g. the variable set associated to a node ℓ'_i, with $i = 1, \ldots, n$, contains the variable S_i, which is also in the variable set associated to ℓ'_{i+1} but not in the variable set associated to any other node in the path between them (as S_i does not appear in $\mathcal{T}$). We obtain a valid tree-decomposition $\mathcal{T}_3$ from $\mathcal{T}_2$ by walking around $\mathcal{T}_2$ in a Euler tour tree traversal where each edge is visited exactly twice and enforcing the running intersection property: for each node that appears after ℓ'_{i-1} and before ℓ'_i during the walk, we include S_{i-1} if $i < n$ and T_{i-1} if $i > n$ in its associated variable set. Since the Euler tour tree traversal visits each leaf once and each internal node at most three times, the procedure inserts at most three new variables in any sets associated to a node of $\mathcal{T}_3$. The treewidth of $\mathcal{T}_3$ thus exceeds the treewidth of $\mathcal{T}_2$ by at most 3. The last step is to obtain $\mathcal{T}'$ from $\mathcal{T}_3$ by covering $pa(V) = \{S_n, E_m\}$ while respecting the running intersection property. To this end, we include E_m in the variable set associate with every node in the path from ℓ'_n to ℓ'_m. This increases the treewidth by at most one, and guarantees that the treewidth of $\mathcal{T}'$ is in the worst case the treewidth of $\mathcal{T}$ plus four.   □

The above result implies that applying the reduction on the class of bounded treewidth Bayesian networks produces a class of bounded treewidth influence diagrams. Thus, (the decision version of) MAP problems that are NP-complete are mapped into (the decision version of) MEU problems which are also NP-complete.

6   Reducing MEU To MAP

Consider a MEU problem with influence diagram $\mathcal{I}$. In order to obtain a Bayesian network $\mathcal{N}$ we first need to apply a sequence of transformations that obtains an MEU-equivalent influence diagram where decision variables have no parents and there is a single value variable. The following transformation substitutes a decision variable with multiple parents by multiple parentless decision variables and preserves the value of the MEU.

Transformation 1 *Select a decision variable D with at least one parent, and let $\pi_1, \ldots, \pi_r$ be the configurations of $pa(D)$.*

1. *Remove D;*
2. *Add parentless decision variables $D_1, \ldots, D_r$ taking the same values as D;*
3. *Add variables $X_1, \ldots, X_r$ taking the same values as D; set $pa(X_1) = pa(D) \cup \{D_1\}$ and $pa(X_i) = pa(D) \cup \{M_i, X_{i-1}\}$ for $i = 2, \ldots, r$; specify*

$$\Pr(X_1 | D_1, pa(D)) = \begin{cases} 1, & \text{if } pa(D) = \pi_1 \text{ and } X_1 = D_1, \\ 0, & \text{if } pa(D) = \pi_1 \text{ and } X_1 \neq D_1, \\ 1/m & \text{if } pa(D) \neq \pi_1; \end{cases}$$

for $i = 2, \ldots, r$, specify

$$\Pr(X_i|X_{i-1}, D_i, \mathrm{pa}(D)) = \begin{cases} 1, & \text{if } \mathrm{pa}(D) \neq \pi_i \text{ and } X_i = X_{i-1}, \\ 0, & \text{if } \mathrm{pa}(D) \neq \pi_i \text{ and } X_i \neq X_{i-1}, \\ 1, & \text{if } \mathrm{pa}(D) = \pi_i \text{ and } X_i = D_i, \\ 0, & \text{if } \mathrm{pa}(D) = \pi_k \text{ and } X_i \neq D_i; \end{cases}$$

4. *Substitute D by X_r in $\mathrm{pa}(C)$ for every C in $\mathrm{ch}(D)$, and modify the conditional probability functions $\Pr(C|\mathrm{pa}(C))$ accordingly.*

Figure 2 depicts the result of applying Transformation 1 on a decision node. The bottleneck of the computational performance of the transformation is the specification of the $O(r^2v^3)$ probability values $\Pr(X_i = x_i|X_{i-1} = x_{i-1}, D_i = d_i, \mathrm{pa}(D) = \pi_k)$, where v is the cardinality of D.

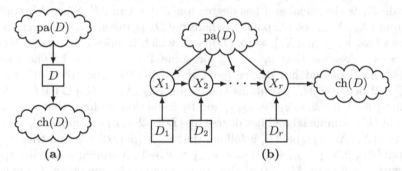

Fig. 2. A piece of a diagram before (a) and after (b) Transformation 1

Remark 3. Let c be the maximum cardinality of a variable in the family of D and $w = |\mathrm{pa}(D)|$. Then the transformation takes time $O(c^{2w+3})$. If we assume that the in-degree of decision variables are bounded, then w is a constant and the transformation takes time polynomial in the input size.

If the in-degree of decision variables is not bounded then the specification of an optimal strategy might take space exponential in the input. Thus, assuming that w is bounded is reasonable.

Remark 4. The transformation might create loops in polytree-shaped diagrams.

The following two results were proved in [25, Proposition 7].

Lemma 3. *Let $\mathcal{I}'$ be the result of applying Transformation 1 on a decision variable D in a diagram $\mathcal{I}$. There is a polynomial-time computable bijection between strategies of $\mathcal{I}$ and $\mathcal{I}'$ that preserves expected utility.*

Corollary 1. *Let $\mathcal{I}'$ be the result of applying Transformation 1 on a decision variable D in a diagram $\mathcal{I}$. The MEU of $\mathcal{I}'$ and $\mathcal{I}$ are equal.*

Transformation 1 might increase the treewidth of the graph. To see this, consider an influence diagram $\mathcal{I}$ containing one chance variable C, one decision variable D and one value variable V, with graph structure $C \to D \to V$. The treewidth of the transformed diagram is three while the treewidth of original graph is one. The following result shows that the increase in treewidth is small.

Lemma 4. *Transformation 1 increases the treewidth by at most two.*

Proof. Let $\mathcal{I}'$ be the result of applying the transformation in a diagram $\mathcal{I}$ of treewidth κ. Also, let M and M' be the moral graphs of $\mathcal{I}$ and $\mathcal{I}'$, respectively. We can obtain M from M' by sequentially eliminating nodes $D_1, \ldots, D_r$ and $X_1, \ldots, X_{r-1}$, in this order, and replacing X_r with D. Let $M_1, \ldots, M_{2r}$ be the graphs obtained by applying each of these operations. Thus, M_1 is the graph obtained by removing D_1 from M', and M_{2r} equals M. Let $\kappa_1, \ldots, \kappa_{2r}$ be the treewidth of the graphs $M_1, \ldots, M_{2r}$, respectively, and κ' be the treewidth of M'. The node D_1 is simplicial and has degree $|\mathrm{pa}(D)| + 1$ in M'. Since M_1 contains the clique $\{X_r, X_{r-1}, D_r\} \cup \mathrm{pa}(D)$ (where $\mathrm{pa}(D)$ is taken with respect to M), it follows that $\kappa_1 \geq |\mathrm{pa}(X_r)| = |\mathrm{pa}(D)| + 2$, which implies $\kappa = \max\{|\mathrm{pa}(D)| + 1, \kappa_1\} = \kappa_1$. Assume that $\kappa_\ell = \kappa_0$, for some $1 \leq \ell < r - 1$. The variable $D_{\ell+1}$ is simplicial and has degree $|\mathrm{pa}(D)| + 2$ in M_ℓ. The treewidth $\kappa_{\ell+1} \geq |\mathrm{pa}(D)| + 2$ because $M_{\ell+1}$ contains the clique $\{X_r, X_{r-1}, D_r\} \cup \mathrm{pa}(D)$. Hence, $\kappa_\ell = \max\{|\mathrm{pa}(D)| + 2, \kappa_{\ell+1}\} = \kappa_{\ell+1}$, and by induction we have that $\kappa' = \kappa_{r-1}$. The node D_r is simplicial and has degree $|\mathrm{pa}(D)| + 2$ in M_{r-1}. Since M_r contains the clique $\{X_r, X_{r-1}\} \cup \mathrm{pa}(D)$, it follows that $\kappa_r \geq |\mathrm{pa}(D)| + 1$, and thus $\kappa_{r-1} = \max\{|\mathrm{pa}(D)| + 2, \kappa_r\} \leq \kappa_r + 1$. Hence, $\kappa_{r-1} \leq \kappa_r + 1$. A similar reasoning applies for κ_ℓ with $r < \ell < 2r$. M_{r+1} (i.e., the graph obtained by removing X_1) contains a clique of size $|\{X_r, X_{r-1}\} \cup \mathrm{pa}(D)| = |\mathrm{pa}(D)| + 2$, and the node X_1 is simplicial and has degree $|\mathrm{pa}(D)| + 1$ in M_r. Hence, $\kappa_r = \max\{|\mathrm{pa}(D)| + 1, \kappa_{r+1}\} = \kappa_{r+1}$. Assume $\kappa_\ell = \kappa_m$ for $r < \ell < 2r - 2$. Then $X_{\ell-m+1}$ is simplicial and has degree $|\mathrm{pa}(D)| + 1$ in M_ℓ. Since $M_{\ell+1}$ contains the clique $\{X_r, X_{r-1}\} \cup \mathrm{pa}(D)$, it follows that $\kappa_\ell = \max\{|\mathrm{pa}(D)| + 1, \kappa_{\ell+1}\} = \kappa_{\ell+1}$. Thus, by induction, $\kappa_r = \kappa_{2r-2}$. Finally, the graph M_{2r-1} (obtained by removing X_{r-1} from M_{2r-2}) contains the clique $\{X_r\} \cup \mathrm{pa}(D)$ (so that $\kappa_{2r-1} \geq |\mathrm{pa}(D)|$), and X_{r-1} is simplicial and has degree $|\mathrm{pa}(D)| + 1$ in M_{2r-2}. Therefore, $\kappa_{2r-2} = \max\{|\mathrm{pa}(D)| + 1, k_{2r-1}\} \leq k_{2r-1} + 1$. Since the replacement of X_r with D used to generate $M_{2r} = M$ from M_{2r-1} does not change the treewidth (i.e., $\kappa = \kappa_{2r} = \kappa_{2r-1}$), we have that

$$\kappa' = \kappa_{r-1} \leq \kappa_r + 1 = \kappa_{2r-2} + 1 \leq \kappa_{2r-1} + 2 = \kappa + 2,$$

and the result follows. □

The previous result can be generalized to recurrent applications of Transformation 1:

Corollary 2. *Let $\mathcal{I}'$ be the result of applying Transformation 1 in a diagram $\mathcal{I}$ of treewidth κ repeatedly until all decision variables are parentless. Then the treewidth of $\mathcal{I}'$ is at most $\kappa + 2$.*

Proof. Applying the transformation on two different decision variables affect different parts of the graph of the original diagram. Hence, the new variables introduced by the repeated applications can be eliminated in parallel, which shows that the increase in treewidth remains bounded by two.   □

A final issue to circumvent in order to devise a mapping from MEU to MAP problems is the treatment of multiple value variables. The following transformation maps diagrams with multiple value variables into MEU-equivalent diagrams with a single value variable.

Transformation 2 *Take an influence diagram $\mathcal{I}$ with value variables $V_1, \ldots, V_n$, and let $\underline{U} = \min_{i,\pi_i} U(pa(V_i) = \pi_i)$ and $\overline{U} = \max_{i,\pi_i} U(pa(V_i) = \pi_i)$ denote, respectively, the minimum and maximum utility value associated to any value variable.*

1. *Substitute each value variables V_i by a binary chance variable W_i taking values t and f and with probability distribution given by*

$$P(W_i = t \mid pa(V_i)) = \frac{U(pa(V_i)) - \underline{U}}{\overline{U} - \underline{U}}.$$

2. *Add variables $O_1, \ldots, O_n$, each taking values t and f, with $pa(O_1) = \{W_1\}$, and $pa(O_i) = \{O_{i-1}, W_i\}$, $i = 2, \ldots, n$; specify $P(O_1 = t \mid W_1 = t) = 1$, $P(O_1 = t \mid W_1 = f) = 0$ and*

$$P(O_i = t \mid O_{i-1}, W_i) = \begin{cases} 1, & \text{if } O_{i-1} = W_i = t \\ (i-1)/i, & \text{if } O_{i-1} = t \text{ and } W_i = f \\ 1/i, & \text{if } O_{i-1} = f \text{ and } W_i = t \\ 0, & \text{if } O_{i-1} = W_i = f \end{cases};$$

3. *Add a value variable V with $pa(V) = \{O_n\}$, $U(pa(V) = t) = n\overline{U}$ and $U(pa(V) = f) = n\underline{U}$.*

Figure 3 illustrates the application of Transformation 2.

Remark 5. The transformation takes time polynomial in the size of the input influence diagram.

Remark 6. The transformation might introduce loops.

The following three results were proved in [22, Theorem 1].

Lemma 5. *Let $\mathcal{I}'$ be the result of applying Transformation 2 on an influence diagram $\mathcal{I}$. There is a polynomial-time computable bijection between strategies of $\mathcal{I}$ and $\mathcal{I}'$ that preserves expected utility.*

Corollary 3. *Let $\mathcal{I}'$ be the result of applying Transformation 2 on an influence diagram $\mathcal{I}$. The MEU of $\mathcal{I}'$ and $\mathcal{I}$ are equal.*

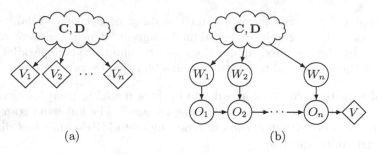

Fig. 3. (a) Influence diagram with multiple value variables. (b) Its equivalent influence diagram obtained by Transformation 2.

Lemma 6. *Transformation 2 increases the treewidth by at most three.*

We are now ready to describe the reduction from MEU to MAP problems.

1. While there is a decision variable with at least one parent, apply Transformation 1;
2. If there is more than a value variable, apply Transformation 2;
3. Transform each (parentless) decision variable D into a chance variable M taking on the same values, and with $P(M) = 1/v$, where v is the cardinality of D;
4. Replace the (single) value variable V by a chance variable E taking values t and f, and with

$$P(E = t | \text{pa}(V)) = \frac{U(\text{pa}(V)) - \underline{U}}{\overline{U} - \underline{U}},$$

where $\underline{U} = \min_\pi U(\text{pa}(V) = \pi)$ and $\overline{U} = \max_\pi U(\text{pa}(V) = \pi)$.

Let $\mathcal{N}$ be the Bayesian network obtained by the reduction above, and denote by $\mathbf{M}$ the set of variables introduced in step 3.

Theorem 3. *Let* MAP *be the value of the MAP problem with Bayesian network* $\mathcal{N}$*, MAP variables* $\mathbf{M}$ *and evidence* $E = t$*, and* MEU *be the value of the MEU problem with input* $\mathcal{I}$*. For any configuration* $\mathbf{m}$ *of* $\mathbf{M}$ *we have that*

$$\mathsf{MAP} = \frac{P(\mathbf{M} = \mathbf{m})}{\overline{U} - \underline{U}} \mathsf{MEU} - \underline{U},$$

where $P(\mathbf{M}) = \prod_{M \in \mathbf{M}} P(M)$*, and* $\underline{U}$ *and* $\overline{U}$ *are, respectively, the minimum and the maximum of the utility function defined by* $\mathcal{I}$*.*

Proof. Let $\mathbf{X}$ denote the variables in $\mathcal{N}$, and $\mathbf{Y} = \mathbf{X} \setminus (\mathbf{M} \cup \{E\})$. Since the set $\mathbf{M}$ contains only root variables associated to uniform probability distributions, $P(\mathbf{M} = \mathbf{m})$ equals some constant C for any configuration $\mathbf{m}$. Hence,

$$\mathsf{MAP} = \max_{\mathbf{m}} \sum_{\mathbf{Y}} P(E{=}t|\mathrm{pa}(E)) P(\mathbf{Y}|\mathbf{M}{=}\mathbf{m}) P(\mathbf{M}{=}\mathbf{m})$$

$$= C \max_{\mathbf{m}} \sum_{\mathbf{Y}} P(E{=}t|\mathrm{pa}(E)) P(\mathbf{Y}|\mathbf{M}{=}\mathbf{m})$$

$$= C \max_{\mathbf{m}} \sum_{\mathbf{Y}} \frac{U(\mathrm{pa}(V)) - \underline{U}}{\overline{U} - \underline{U}} P(\mathbf{Y}|\mathbf{M}{=}\mathbf{m}) = \frac{C}{\overline{U} - \underline{U}} \mathsf{MEU} - \underline{U},$$

which proves the result. □

The following result shows that the reduction maps NP-complete instances of MEU into NP-complete instances of MAP.

Corollary 4. *Let κ denote the treewidth of an influence diagram. Then the Bayesian network generated by the reduction has treewidth at most $\kappa + 5$.*

Proof. It follows from Lemmas 4 and 6. □

7   Conclusions

Computing the maximum posterior probability of a subset of variables in a Bayesian network and calculating the maximum expected utility of strategies in an influence diagrams are common tasks in probabilistic reasoning. Despite their similarities, these two problems have hitherto been investigated independently. In this work, we showed constructively that these two problems are computationally equivalent in that one problem can be reduced to the other in polynomial time. Hence, any algorithm designed for one problem can be immediately used for the other with a small overhead. Future work should evaluate the benefits and drawbacks of applying algorithms designed for one problem to solve the other, by means of the reductions presented here.

A common limitation of the correspondences devised here is that they map problems with polytree-shaped graph structure into problems with loopy graph structure. This reduces some tractable instances of one problem into apparently intractable instances of the other problem. For instance, MEU is tractable in polytree-shaped diagrams with binary variables and a single value node, but the reduction shown here creates a MAP problem in a loopy Bayesian network, for which no efficient algorithm exists. A similar problem appears if we try to use the reductions developed here to prove the hardness of instances with simple structure. For instance, the complexity of MAP in tree-shaped Bayesian networks with binary variables is not known, and it cannot be characterized by the reduction from MAP to MEU presented here because tree-shaped Bayesian networks are mapped into loopy influence diagrams. It would be interesting to devise reductions that preserve the topology of the graph structure.

Acknowledgments. The author thanks Fabio G. Cozman and the reviewers for their valuable comments and suggestions. This work was partially supported by the São Paulo Research Foundation (FAPESP) grant no. 2013/23197-4.

References

1. Antonucci, A., Piatti, A.: Modeling unreliable observations in Bayesian networks by credal networks. In: Godo, L., Pugliese, A. (eds.) SUM 2009. LNCS, vol. 5785, pp. 28–39. Springer, Heidelberg (2009)
2. Bodlaender, H., Koster, A., van den Eijkhof, F., van der Gaag, L.: Pre-processing for triangulation of probabilistic networks. In: Proceedings of the 17th Conference on Uncertainty in Artificial Intelligence (UAI), pp. 32–39 (2001)
3. de Campos, C.P.: New complexity results for MAP in Bayesian networks. In: Proceedings of the 22nd International Joint Conference on Artificial Intelligence (IJCAI), pp. 2100–2106 (2011)
4. de Campos, L.M., Gámez, J.A., Moral, S.: Partial abductive inference in Bayesian networks by using probability trees. In: 5th International Conference on Enterprise Information Systems (ICEIS), pp. 83–91 (2003)
5. de Campos, C.P., Ji, Q.: Strategy selection in influence diagrams using imprecise probabilities. In: Proceedings of the 24th Conference in Uncertainty in Artificial Intelligence (UAI), pp. 121–128 (2008)
6. de Paz, R.C., Gómez-Olmedo, M., Cano, A.: Approximate inference in influence diagrams using binary trees. In: Proceedings of the 6th European Workshop on Probabilistic Graphical Models (PGM), pp. 43–50 (2012)
7. Dechter, R.: An anytime approximation for optimizing policies under uncertainty. In: Workshop of Decision Theoretic Planning, AIPS (2000)
8. Dechter, R., Rish, I.: Mini-buckets: A general scheme for bounded inference. Journal of the ACM 50(2), 107–153 (2003)
9. Detwarasiti, A., Shachter, R.D.: Influence diagrams for team decision analysis. Decision Analysis 2(4), 207–228 (2005)
10. Howard, R.A., Matheson, J.E.: Influence diagrams. In: Readings on the Principles and Applications of Decision Analysis, pp. 721–762. Strategic Decisions Group (1984)
11. Huang, J., Chavira, M., Darwiche, A.: Solving MAP exactly by searching on compiled arithmetic circuits. In: Proceedings of the 21st National Conference on Artificial Intelligence (NCAI), pp. 1143–1148 (2006)
12. Jensen, F.V., Nielsen, T.D.: Bayesian Networks and Decision Graphs, 2nd edn. Information Science and Statistics. Springer (2007)
13. Jiang, J., Rai, P., Daume III, H.: Message-passing for approximate MAP inference with latent variables. In: Advances in Neural Information Processing Systems 24 (NIPS), pp. 1197–1205 (2011)
14. Khaled, A., Yuan, C., Hansen, E.: Solving limited memory influence diagrams using branch-and-bound search. In: Proceedings of the 29th Conference on Uncertainty in Artificial Intelligence, UAI (2013)
15. Koller, D., Friedman, N.: Probabilistic Graphical Models. MIT Press (2009)
16. Lauritzen, S.L., Nilsson, D.: Representing and solving decision problems with limited information. Management Science 47, 1235–1251 (2001)
17. Lim, H., Yuan, C., Hansen, E.: Scaling up MAP search in Bayesian networks using external memory. In: Proceedings of the 5th European Workshop on Probabilistic Graphical Models (PGM), pp. 177–184 (2010)
18. Liu, Q., Ihler, A.: Variational algorithms for marginal MAP. In: Proceedings of the 27th Conference on Uncertainty in Artificial Intelligence (UAI), pp. 453–462 (2011)

19. Liu, Q., Ihler, A.: Belief propagation for structured decision making. In: Proceedings of the 28th Conference on Uncertainty in Artificial Intelligence (UAI), pp. 523–532 (2012)
20. Madsen, A.L., Nilsson, D.: Solving influence diagrams using HUGIN, Shafer-Shenoy and lazy propagation. In: Proceedings of the 17th Conference in Uncertainty in Artificial Intelligence (UAI), pp. 337–345 (2001)
21. Mauá, D.D., de Campos, C.P.: Anytime marginal MAP inference. In: Proceedings of the 28th International Conference on Machine Learning (ICML), pp. 1471–1478 (2012)
22. Mauá, D.D., de Campos, C.P., Zaffalon, M.: The complexity of approximately solving influence diagrams. In: Proceedings of the 28th Conference on Uncertainty in Artificial Intelligence (UAI), pp. 604–613 (2012)
23. Mauá, D.D., de Campos, C.P., Zaffalon, M.: On the complexity of solving polytree-shaped limited memory influence diagrams with binary variables. Artificial Intelligence 205, 30–38 (2013)
24. Mauá, D.D., de Campos, C.P.: Solving decision problems with limited information. In: Advances in Neural Information Processing Systems 24 (NIPS), pp. 603–611 (2011)
25. Mauá, D.D., de Campos, C.P., Zaffalon, M.: Solving limited memory influence diagrams. Journal of Artificial Intelligence Research 44, 97–140 (2012)
26. Meek, C., Wexler, Y.: Approximating max-sum-product problems using multiplicative error bounds. Bayesian Statistics 9, 439–472 (2011)
27. Nilsson, D., Höhle, M.: Computing bounds on expected utilities for optimal policies based on limited information. Research Report 94, Dina (2001)
28. Nilsson, D., Lauritzen, S.L.: Evaluating influence diagrams using LIMIDs. In: Proceedings of the 16th Conference in Uncertainty in Artificial Intelligence (UAI), pp. 436–445 (2000)
29. Park, J.D., Darwiche, A.: Solving MAP exactly using systematic search. In: Proceedings of the 19th Conference on Uncertainty in Artificial Intelligence (UAI), pp. 459–468 (2003)
30. Park, J.D., Darwiche, A.: Complexity results and approximation strategies for MAP explanations. Journal of Artificial Intelligence Research 21, 101–133 (2004)
31. Shenoy, P.P.: Binary join trees for computing marginals in the Shenoy-Shafer architecture. International Journal of Approximate Reasoning 17(2-3), 239–263 (1997)
32. Yuan, C., Wu, X., Hansen, E.A.: Solving multistage influence diagrams using branch-and-bound search. In: Proceedings of the 26th Conference on Uncertainty in Artificial Intelligence (UAI), pp. 691–700 (2010)
33. Zhang, N.L., Qi, R., Poole, D.: A computational theory of decision networks. International Journal of Approximate Reasoning 11(2), 83–158 (1994)

Speeding Up k-Neighborhood Local Search in Limited Memory Influence Diagrams

Denis D. Mauá and Fabio G. Cozman

Universidade de São Paulo,
São Paulo, Brazil
{denis.maua,fgcozman}@usp.br

Abstract. Limited memory influence diagrams are graph-based models that describe decision problems with limited information, as in the case of teams and agents with imperfect recall. Solving a (limited memory) influence diagram is an NP-hard problem, often approached through local search. In this paper we investigate algorithms for k-neighborhood local search. We show that finding a k-neighbor that improves on the current solution is W[1]-hard and hence unlikely to be polynomial-time tractable. We then develop fast schema to perform approximate k-local search; experiments show that our methods improve on current local search algorithms both with respect to time and to accuracy.

1 Introduction

Limited memory influence diagrams (LIMIDs) are graph-based probabilistic decision making models particularly suited for teams and limited-resource agents [4,13]. LIMIDs relax the *perfect recall* requirement (a.k.a. no-forgetting assumption) of traditional influence diagrams [9], and by doing so, require considerably more computational effort in the search for optimal policies.

Finding an optimal strategy for polytree-shaped LIMIDs is NP-hard even if variables are ternary and the utility function is univariate [18], or if variables are binary and the utility function is multivariate [16]. Similar negative results hold for approximating the problem within any fixed constant when either variable cardinality or treewidth is unbounded [18]. And even though there are polynomial-time approximations when cardinalities and treewidth are bounded [17], constants in such solutions are so big as to prevent practical use. Currently the state-of-art algorithm for solving LIMIDs exactly is Multiple Policy Updating (MPU) [15], that works by verifying a dominance criterion between partial strategies. MPU has worst-case exponential cost but often finishes in reasonable time. There are also anytime solvers based on branch-and-bound that can trade off accuracy for efficiency [3,2,10].

In practice, local search methods are the most widely used algorithms for approximately solving LIMIDs. Lauritzen and Nilsson [13] developed the Single Policy Updating (SPU) algorithm for computing locally optimum strategies in arbitrary LIMIDs. Their algorithm remains the most referenced and probably used algorithm for solving medium and large LIMIDs. SPU iteratively seeks

L.C. van der Gaag and A.J. Feelders (Eds.): PGM 2014, LNAI 8754, pp. 334–349, 2014.

for a variable that can improve the incumbent global strategy by modifying its associated actions. If no such variable is found the algorithm halts at a local optimum. Detwarasiti and Shachter [4] extended SPU to allow larger moves in the search space. Roughly speaking, their approach can be seen as a k-neighbor local search in the space of strategies. In practice, exhaustive k-local search can only be applied to networks with say less than $n = 100$ decision variables and with very small values of k (say, $k = 2$), as every step requires exploration of $O(n^k)$ candidates. There have also been proposals based on message-passing algorithms to cope with high treewidth diagrams [14].

In this paper we focus on the local search that runs within the methods in the last paragraph. First, we prove that k-local search is W[1]-hard, which suggests that algorithms that run in time $O(f(k)n^\alpha)$ are unlikely to exist, where f is an arbitrary computable function and α is a constant independent of k (Section 3). We take such a result as an indication that approximating local-search is necessary for large values of k.

We thus investigate the use of MPU's pruning to speed up SPU and related k-local search schema (Section 4). We propose a relaxed and approximate version of MPU's pruning, with worst-case polynomial-time complexity (Section 5). This approximate pruning method is used in each k-local search, leading to very efficient versions of local search methods for LIMIDs. We prove that when k is the number of action variables, our approximate pruning provides an additive fully polynomial-time approximation scheme for LIMIDs of bounded treewidth and bounded variable cardinality. Finally, we show by experiments with random networks that our local search algorithms, both exact and approximate, outperform existing local search methods (Section 6).

2   Limited Memory Influence Diagrams

LIMIDs are graphical representations of structured decision problems [8]. Variables in a decision problem can be partitioned into state (or chance) variables $\mathcal{S}$, which represent quantities unknown at planning stage, action (or decision) variables $\mathcal{A}$, which enumerate alternative courses of action, and value variables $\mathcal{V}$, which assess the quality of decisions for every configuration of state variables. We assume here that variables take on finitely many values. Each variable in a decision problem represented as a LIMID is equated with a node in a directed acyclic graph; in particular, value variables are equated to leaf nodes. There is a (conditional) probability distribution $P(S|\mathcal{P}_S)$ for every state variable $S \in \mathcal{S}$, where the notation $\mathcal{P}_X$ denotes the parents of a variable X in the graph. There is also a utility function $U(\mathcal{P}_V)$ for every value variable. The overall utility U is assumed to decompose additively in terms of the value variables [22], that is, $U(\mathcal{S}, \mathcal{A}) = \sum_{V \in \mathcal{V}} U(\mathcal{P}_V)$. State variables are assumed to satisfy the Markov condition, which states that any (state) variable is independent of its non-descendant non-parents conditional on its parents. Consequently, the joint distribution of state variables conditioned on a configuration $\mathcal{A} = a$ of the action variables factorizes as $P(\mathcal{S}|\mathcal{A}=a) = \prod_{S \in \mathcal{S}} P(S|\mathcal{P}_S, \mathcal{A}=a)$.

A strategy $\delta = \{\delta_A : A \in \mathcal{A}\}$ is a multiset of local decision rules, or policies, one for each action variable in the problem. Each policy δ_A is a mapping from the configurations of the values of the parents $\mathcal{P}_A$ of A to values of A. We denote by Δ_A the set of all policies for variable A. A policy for an action variable with no parents is simply an assignment of a value to that variable. We assume that policies are encoded as tables. Hence, the size of a policy is exponential in the number of parents of the corresponding action variable, which in real scenarios forces us to constrain the maximum number of parents of an action node lest the implementation of a policy be not practicable.

The *perfect recall* condition (a.k.a. no forgetting) assumes that all decisions and observations are "remembered". Graphically, it entails that if A and A' are two action nodes such that A is a parent of A', then all parents of A are also parents of A'. We assume that when perfect recall is satisfied the "remembered" arcs are explicitly represented in the diagram.

The construction of an optimal strategy is harder for LIMIDs that do not satisfy the perfect recall requirement, exactly due to the absence of links between actions.

Given an action variable A and a policy δ_A, we let $P(A|\mathcal{P}_A, \delta_A)$ be the collection of degenerate conditional probability distributions that assign all mass to $a = \delta_A(\mathcal{P}_A)$ (or the degenerate marginal distribution $P(A|\delta_A)$ that places all mass on δ_A in case A has no parents). With this correspondence between policies and (conditional) probability distributions, we can define a joint probability distribution over the state and action variables for any given strategy δ as

$$P(\mathcal{S}, \mathcal{A}|\delta) = \prod_{S \in \mathcal{S}} P(S|\mathcal{P}_S) \prod_{A \in \mathcal{A}} P(A|\mathcal{P}_A, \delta).$$

The expected utility of a strategy δ, $E(U|\delta)$, is then $\sum_{\mathcal{S}, \mathcal{A}} U(\mathcal{S}, \mathcal{A}) P(\mathcal{S}, \mathcal{A}|\delta)$.

Given a strategy δ, computing $E(U|\delta)$ can be reduced to a marginal inference in a Bayesian network [1]. Conversely, marginal inference in Bayesian networks, a #P-complete problem [20], can be reduced to the computation of an expected utility by using a $\{0, 1\}$-valued utility and making the conditional probabilities of the children of action nodes numerically independent of strategies. Hence, those two problems are computationally equivalent. Marginal inference can be performed in time exponential in the treewidth of the underlying graph by e.g. variable elimination. This entails a polynomial-time algorithm for networks of small treewidth (with treewidth considered constant in the complexity analysis). Kwisthout et al. [12] showed that under the widely believed hypothesis that SAT is not subexponential-time solvable, variable elimination's performance is optimal. Thus it seems necessary to constrain LIMIDs to bounded treewidth diagrams if worst-case efficient computations are sought (it is possible that the average cost of marginal inference is polynomial; we do not pursue this possibility here).

An important task with LIMIDs is that of finding the

Maximum Expected Utility (MEU)
Input: A LIMID and a rational k
Question: Is there a strategy δ such that $E(U|\delta) \geq k$?

MEU is NP^{PP}-complete, and NP-complete for diagrams of bounded treewidth [2]. The problem is NP-complete on LIMIDs of bounded treewidth even when all variables are binary [16], and when all variables are ternary and there is a single value node [18].

Provided that expected utilities can be succinctly encoded (i.e., represented in space $O(b^\alpha)$, where b is the size of the encoding and α is a constant), we can compute the value of the maximum expected utility by binary search in polynomial time if MEU can be solved in polynomial time. Similarly, if the in-degrees of action nodes are bounded we can use a polynomial-time algorithm M that solves MEU to obtain an optimal strategy in polynomial time as follows. First, perform a binary search using M to compute the maximum expected utility and use that value as k. Select an action variable A and for every policy δ_A build a new LIMID where A is a state node with conditional probability $P(A|\delta_A)$. Now run the algorithm M on those LIMIDs: the algorithm will certainly return a yes answer on some of them; any policy δ_A corresponding to an affirmative answer is part of the optimal strategy, and we can repeat the procedure for another action variable until no action variables remains. Conversely, assuming the same requirements on the representation of expected utilities and strategies, MEU can be efficiently solved by any polynomial-time algorithm that computes the maximum expected utility. Finally, if the treewidth of the diagrams is bounded, MEU can trivially be solved in polynomial-time by any polynomial-time algorithm that finds optimal strategies. Thus, MEU is largely equivalent to selecting an optimal strategy and computing the value of the maximum expected utility.

We make extensive use of the following result that follows immediately from the results in [17] and [18].

Proposition 1. *Given a LIMID $\mathcal{L}$ of treewidth w we can construct in time polynomial in its size a LIMID $\mathcal{L}'$ and a function f such that (i) $\mathcal{L}'$ has a single value variable V such that $0 \leq U(\mathcal{P}_V) \leq 1$, and treewidth at most $w + 3$; (ii) the action nodes in $\mathcal{L}'$ have no parents; (iii) f maps strategies δ' of $\mathcal{L}'$ into strategies δ of $\mathcal{L}$ in linear time; (iv) if δ' is such that $E(U'|\delta') > 0$ and $\delta = f(\delta')$ then $E(U|\delta) \propto E(U'|\delta')$, where U and U' denote the utility functions of $\mathcal{L}$ and $\mathcal{L}'$, respectively.*

A corollary of the above result is that the MEU of $\mathcal{L}'$ equals the MEU of $\mathcal{L}$ up to a constant, and the optimal strategy for $\mathcal{L}$ can be obtained from the optimal strategy for $\mathcal{L}'$. Hence we assume in the rest of the paper that LIMIDs have a single value node taking its values in $[0, 1]$, and that action nodes are parentless.

3   The Complexity of k-Neighborhood Local Search

Consider a strategy δ and a set $\mathcal{N} \subseteq \mathcal{A}$. The $\mathcal{N}$-neighborhood of δ is the set of strategies δ' that coincide with δ on the policies of variables $A \in \mathcal{A} \setminus \mathcal{N}$. A k-neighbor of δ is any strategy in a $\mathcal{N}$-neighborhood of δ with $|\mathcal{N}| = k$. The k-neighborhood of δ is the set of its k-neighbors. Arguably, the most widely used scheme for selecting strategies is as follows.

> k-**Policy Updating** (k**PU**) Take a LIMID, a strategy δ_0 and a positive integer M: for $i = 1 \ldots M$ find a strategy δ_i in the k-neighborhood of δ_{i-1}; at the end, return δ_M.

For large enough and finite M the procedure converges to a local optimum. In particular, if k equals the number of action variables, a global optimum is found in one iteration. The main bottleneck of kPU is the k-local search step, where an improving solution is searched for; this can be formalized as

> k-**Policy Improvement** (k**PI**)
> *Input:* A LIMID $\mathcal{L}$ and a strategy δ
> *Parameter:* A positive integer k
> *Question:* Is there a k-neighbor δ' of δ such that $E(U|\delta') > E(U|\delta)$?

The same argument used when discussing MEU can be used here to show that the problem of finding the maximum expected utility in the k-neighborhood of a strategy and the problem of selecting a k-neighbor with higher expected utility (if it exists) are largely equivalent to kPI in the sense that (under the same requirements) a polynomial-time algorithm for one problem can be used to solve another.

For a LIMID whose action variables are parentless, kPI can be solved by exhaustive search in the k-neighborhood in time $O(n^k c^k)$, where $n = |\mathcal{A}|$ is the number of action variables and c is the maximum cardinality of an action variable. Such an approach is prohibitive for large values of n or c and moderate values of k. It is thus interesting to look for faster methods for searching the k-neighborhood of a strategy. In particular, we should ask whether there is an algorithm that runs in time $O(f(k)b^\alpha)$, where f is an arbitrary computable function, b is the size of the LIMID (encoded as a bitstring), and α is a constant independent of k. In other words, we are interested in knowing whether it is possible to scale up k-local search to diagrams with hundreds of variables if k is kept small. We now show that finding such an algorithm implies that FPT=W[1], and is therefore unlikely. To this aim, we need to introduce some background in the rich field of parameterized complexity.

Parameterized complexity investigates the runtime behavior of inputs (x, k) that can be decomposed into two parts, its main part x and a parameter k. Many interesting NP-hard problems are polynomial-time solvable for fixed values of the parameters, that is, when the parameter is not taken to be part of the input. There are essentially two kinds of polynomial running time for fixed parameters. A (decision) problem is said to be *fixed-parameter tractable* if there is an algorithm that solves any parameterized instance (x, k) of the problem in time $O(f(k)b^\alpha)$, where f is an arbitrary computable function, b is the size of the input and α is a constant that does not depend on k [5]. The class of all fixed-parameter tractable decision problems is denoted FPT. Note that NP-complete problems can be either fixed-parameter tractable or intractable.

Similar to the polynomial hierarchy in the NP-completeness framework, the family W[t] defines a hierarchy of nested and increasingly more complex parameterized problems for $t = 1, 2, \ldots,$. Roughly speaking, W[t] is the class of parameterized problems that can compute Boolean circuits of depth at most t. We have that FPT $\subseteq$ W[1] $\subseteq$ W[2] $\subseteq \cdots$. It is unknown whether any of these inclusions is proper, but there are good reasons to believe that at least the first inclusion (i.e., FPT $\subseteq$ W[1]) is proper [6]: if FPT=W[1] then NP-complete problems can be solved in subexponential time [7].

Instead of polynomial-time reductions, which one uses to show NP-hardness of non-parameterized problems, fixed-parameter intractability is usually shown by many-one parameterized reductions to W[t]-hard problems. A many-one parameterized reduction from a problem A to problem B takes an instance (x, k) of A and produces an instance $(x', g(k))$ of B in time $O(f(k)b^\alpha)$, where g and f are arbitrary computable functions, b is the length of x and α is a constant.

The next result shows that if a fixed-parameter tractable algorithm that performs k-local search on the space of strategies existed we would prove k-FLIP MAX SAT to be fixed-parameter tractable.

Theorem 1. *Unless W[1]=FPT, there is no algorithm that solves k-POLICY IMPROVEMENT in time $O(f(k)b^\alpha)$, where b is the size of bitstring encoding of the LIMID, f is an arbitrary computable function and α is a constant independent of k, even for polytree-shaped LIMIDs of bounded treewidth.*

Proof. We use a parameterized reduction from k-FLIP MAX SAT to prove the result; that is, we consider the following variant of MAX SAT:

> k-**Flip Max Sat**
> *Input:* A CNF formula F and a truth-value assignment τ_0
> *Parameter:* A positive integer k
> *Question:* Is there a k-flip of τ_0 satisfying more clauses of F?

A k-flip of an assignment τ is another assignment that differs from τ in the values assigned to at most k variables. Szeider [21] showed that the above problem is W[1]-hard.

Consider formula F, truth assignment τ and parameter k, and let $X_1, \ldots, X_n$ be the variables in F, and $C_1, \ldots, C_m$ be its clauses. We build a corresponding LIMID with graph structure as in Figure 1 and numerical parameters specified as follows (this construction is similar to the construction used by [19] to show NP-hardness of MAP inference in polytree-shaped Bayesian networks). The variables $S_1, \ldots, S_n$ take values in $\{0, 1, \ldots, m\}$. The variable S_0 takes values in $\{1, \ldots, m\}$, and the action variables are binary and take on values 0 and 1. The conditional probabilities of the chance variables are specified as follows:

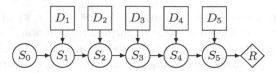

Fig. 1. LIMID used to prove Theorem 1

$$P(S_i=1|S_{i-1}, D_i) = \begin{cases} 1 & \text{if } S_i = S_{i-1} = 0, \\ 1 & \text{if } S_i = 0 \text{ and } S_{i-1} = k \geq 1 \text{ and } D_i \text{ satisfies } C_k, \\ 1 & \text{if } S_i = S_{i-1} = k \geq 1 \text{ and } D_i \text{ does not satisfy } C_k, \\ 0 & \text{otherwise}, \end{cases}$$

and $P(S_0 = s_0) = 1/m$ for all s_0. The utility is defined as $U(S_n = 0) = 1$ and $U(S_n = s_n) = 0$ for all $s_n \neq 0$. The variable S_0 serves as a clause selector: $S_0 = i$ denotes that clause i is being selected. Variable S_i, $i = 1, \ldots, n$, indicates whether the clause selected by S_0 is satisfied by some of $D_1, \ldots, D_i$. We then have that $E(U|\delta) = \#\mathsf{SAT}(\delta)/m$, where $\#\mathsf{SAT}$ is the number of clauses satisfied by a truth-value assignment corresponding to δ. Consider an arbitrary strategy δ corresponding to the truth-value assignment τ. A k-neighbor of δ is a strategy δ' differing from δ in at most k coordinates. Hence, there is a truth-value assignment satisfying more clauses in F than τ if and only there is k-policy improvement of δ, and the result is proved.   □

4   Improving k-Policy Updating: DkPU

The result in the previous section indicates that local search becomes difficult once we try to refine search by increasing its width (through k). Thus we must focus on approximate ways that allows us to climb up to reasonably large k (say, 10) for large values of n.

Assuming (w.l.o.g.) that a LIMID has parentless action variables of cardinality c, a brute-force approach to k-local search can be accomplished by examining the n^k subsets $\mathcal{N} \subseteq \mathcal{A}$ of cardinality k, and for each $\mathcal{N}$ examining all the c^k joint configurations of variables $\mathcal{N}$. Hence, there are two sources of inefficiency in this approach: finding $\mathcal{N}$ and selecting an $\mathcal{N}$-neighbor. We tackle the first problem by randomly sampling sets $\mathcal{N}$, which guarantees uniform coverage. The search for $\mathcal{N}$-neighbors is more intricate. In this section we develop a fast procedure for selecting the *optimal* $\mathcal{N}$-neighbor of an incumbent strategy for a fixed $\mathcal{N}$ (an optimal neighbor is one that maximizes the expected utility among all neighbors).

4.1   Dominance Pruning

We start with some basic concepts; first, the notion of a potential.

Definition 1. *A potential ϕ with scope $\mathcal{X}$ is a nonnegative real-valued mapping of configurations x of $\mathcal{X}$.*

We assume the usual algebra of potentials: the product $\phi \cdot \psi$ of potentials ϕ and ψ returns a potential γ on $z \sim \mathcal{Z} = \mathcal{X} \cup \mathcal{Y}$ such that $\gamma(z) = \phi(x) \cdot \psi(y)$, where x and y are the projections of z onto $\mathcal{X}$ and $\mathcal{Y}$, respectively. The $\mathcal{Y}$-marginal $\sum_{\mathcal{X} \setminus \mathcal{Y}} \phi$ of a potential ϕ with scope $\mathcal{X}$, where $\mathcal{Y} \subseteq X$, is the potential ψ with scope $\mathcal{Y}$ such that $\psi(y) = \sum \{\phi(x) : x \sim y\}$. The system of potentials with product and marginalization forms a valuation algebra [11]. This entails that a marginal $\sum_{\mathcal{X}} \prod_{\psi \in \Gamma} \psi$ can be computed by variable elimination: for each variable $X \in \mathcal{X}$ in some ordering, remove from Γ all potentials whose scope contain X, compute the marginal of those products which sums over X and add the result to Γ.

Since by definition the probability and utility functions in a LIMID are potentials, the expected utility of a given strategy can be computed by variable elimination. The potentials produced during variable elimination satisfy the following property, which we use later on:

Proposition 2. *Consider a LIMID with a single value variable V and a strategy δ, and let $\Gamma = \{P(S|\mathcal{P}_S), P(A|\delta_A), U(\mathcal{P}_V) : S \in \mathcal{S}, A \in \mathcal{A}\}$. If $0 \leq U(\mathcal{P}_V) \leq 1$, then every potential ϕ generated during variable elimination satisfies $0 \leq \phi \leq 1$.*

Proof. Consider the step in variable elimination where all potentials containing a variable X are collected, and let $\mathcal{Y}$ be all variables $Y < X$. Let also F_Y be the set of Y and its children. By design, the only potentials in the initial Γ whose scope include a variable Y are $P(Y|\mathcal{P}_Y)$ (or $U(\mathcal{P}_Y)$ if $Y = V$) and $P(Z|\mathcal{P}_Z)$ for $Z \in F_Y$. By the properties of a valuation algebra, it follows that $\psi_X = \sum_X P(X|\mathcal{P}_X) \sum_{\mathcal{Y}} \prod_{Z \in F_Y : Z \neq X, Y \mathcal{Y}} P(Z|\mathcal{P}_Z)$, where $P(Z|\mathcal{P}_Z)$ is $U(\mathcal{P}_Z)$ if $Z = V$. The right-hand side of the equality is a convex combination of $\sum_{\mathcal{Y}} \prod_{Z \in F_Y : Z \neq X, Y \in \mathcal{Y}} P(Z|\mathcal{P}_Z)$, and therefore is a function not greater or smaller than that in every coordinate. The result follows by induction. $\square$

The algebra of potentials can be extended to set-valued objects, so as to obtain maximum expected utility and hence solve MEU [15]. To do so, we define:

Definition 2. *A set-potential $\Phi(\mathcal{X})$ is a set of potentials ϕ with scope $\mathcal{X}$.*

Definition 3. *The product of two set-potentials $\Phi(\mathcal{X})$ and $\Psi(\mathcal{Y})$ is the set-potential $[\Phi \cdot \Psi](\mathcal{X} \cup \mathcal{Y}) = \{\phi \cdot \psi : \phi \in \Phi, \psi \in \Psi\}$.*

Definition 4. *The $\mathcal{Y}$-marginal of a set-potential $\Phi(\mathcal{X})$ with respect to a variable set $\mathcal{Y} \subseteq \mathcal{X}$ is the set-potential $[\sum_{\mathcal{X} \setminus \mathcal{Y}} \Phi](\mathcal{Y}) = \{\sum_{\mathcal{X} \setminus \mathcal{Y}} \phi : \phi \in \Phi\}$.*

Mauá et al. [18] proved that the algebra of set-potentials is a valuation algebra [11], and thus marginal inference with set-potentials can also be computed by variable elimination (with potentials and their operations replaced by their set counterpart).

The product of set-potentials may create exponentially larger set-potentials. Our interest in the algebra of set-potentials is, as we show later on, to select one single table in a set-potential produced by marginalization and product of many set-potentials. As most of the tables produced are irrelevant, we can save computations by pruning them from set-potentials generated during variable elimination. One way of doing this is by defining a dominance criterion between potentials:

Definition 5. *Consider two potentials $\phi(\mathcal{X})$ and $\psi(\mathcal{X})$ with the same scope. We say that ϕ dominates (resp., is dominated by) ψ if $\phi(x) \geq \psi(x)$ (resp., $\phi(x) \leq \psi(x)$) for all x.*

We can define the set of non-dominated potentials:

Definition 6. *The dominance-pruning of a set-potential $\Phi(\mathcal{X})$ is the set-potential $\mathsf{nd}[\Phi]$ of non-dominated potentials in Φ.*

Note that if nd is applied on a set-potential with empty scope, it produces a single real number. Mauá et al. [18] showed that dominance-pruning satisfies

$$\mathsf{nd}[\Phi(\mathcal{X})\Psi(\mathcal{Y})] = \mathsf{nd}\left[\mathsf{nd}[\Phi(\mathcal{X})]\mathsf{nd}[\Psi(\mathcal{Y})]\right], \quad \mathsf{nd}\left[\sum_{\mathcal{X}\setminus\mathcal{Z}} \Phi(\mathcal{X})\right] = \mathsf{nd}\left[\sum_{\mathcal{X}\setminus\mathcal{Z}} \mathsf{nd}[\Phi(\mathcal{X})]\right].$$

Those properties guarantee the correctness of computation of non-dominated marginals by *dominance-pruned variable elimination*; that is, by a version of variable elimination in which dominance-pruning is applied after every operation (product or marginalization). If dominance-pruned variable elimination is applied on a multi-set Γ of set potentials whose joint scopes are $\mathcal{X}$, by the properties above, we have at the end of the computation a real number $r = \mathsf{nd}\left[\sum_{\mathcal{X}} \prod_{\Psi \in \Gamma} \Psi\right]$. Note that while direct computation of the right-hand side of this equality above takes time exponential in the size of the set-potentials in Γ, applying dominance-pruning after each operation can enormously decrease the overall cost of computation.

4.2 Local Search With Dominance Pruning

The main idea here is to use dominance pruning as in the MPU algorithm, but to run efficient k-local search with the space of strategies. Our proposal is as follows.

Dominance-Based $\mathcal{N}$-Policy Updating (D$\mathcal{N}$PU). Let $\mathcal{N}$ be a subset of the action variables, δ be an arbitrary strategy, and Γ be an initially empty set.
1. For each state variable S add a set-potential $\Phi_S = \{P(S|\mathcal{P}_S)\}$ to Γ,
2. for each action variable A in $\mathcal{N}$ add a set-potential $\Phi_A = \{P(A|\delta'_A) : \delta'_A \in \Delta_A\}$ to Γ,
3. for each action variable A *not* in $\mathcal{N}$ add a set-potential $\Phi_A = \{P(A|\delta_A) : \delta_A\}$ to Γ, where δ_A is the policy of A in δ.

4. Add the set-potential $\Phi_V = \{U(\mathcal{P}_V)\}$, where V is the value node,
5. run dominance-pruned variable elimination and return result.

Theorem 2. $D\mathcal{N}PU$ *outputs a strategy* δ' *such that* $\delta'_A = \delta_A$ *for all* $A \notin \mathcal{N}$ *and* $E(U|\delta') \geq E(U|\delta)$.

Proof. Let $\Delta(\delta, \mathcal{N})$ be the set of all strategies that agree on $\mathcal{N}$ with δ, that is, all $\mathcal{N}$-neighbors of δ. Note that δ is an element of $\Delta(\delta, \mathcal{N})$. By design, we have that

$$\sum_{\mathcal{X}} \prod_{\Psi \in \Gamma} \Psi = \{E(U|\delta') : \delta' \in \Delta(\delta, \mathcal{N})\}.$$

Hence, the result follows from the properties of the algebra of set-potential with dominance pruning. □

If we set $\mathcal{N} = \mathcal{A}$, $D\mathcal{N}PU$ collapses to the MPU algorithm [15], and hence produces exact solution of LIMIDs (of moderate size). As with MPU, the worst-case running time of $D\mathcal{N}PU$ is exponential in $|\mathcal{N}|$, but dominance pruning can significantly decrease that complexity, as our experiments in Section 6 show.

We call $DkPU$ the method that randomly samples a fixed number of sets $\mathcal{N}$ and on each set run $D\mathcal{N}PU$. Importantly, D1PU offers an algorithm that produces exactly the same result as the popular SPU, but only quicker. As we show in Section 6, the gain in speed is not dramatic for D1PU, but it is very significant for $DkPU$ with larger values of k. The fact that $DkPU$ allows us to try larger values of k in practice is valuable as it leads to superior solutions through local search; depending on the application, even marginal gains can be important, and as such the move from kPU to $DkPU$ is always recommended.

Note that additional computational savings could be gained by structuring computations in a junction tree (as in SPU), and avoiding redundant computations among different runs of $DkPU$ (i.e., with different sets $\mathcal{N}$). We do not study such implementation techniques in this paper.

5 Approximate Policy Updating: AkPU

Even though dominance pruning often largely reduces the size of set-potentials, there are cases where pruning is ineffective, as the following example shows.

Example 1. Consider a LIMID with action variables $D_1, \ldots, D_n$, state variables A, B, C and utility node V. A and B have either all action variables as parents. C has A and B as parents and V as child. All variables are binary and take values in $\{0, 1\}$. The CPTs are $P(A=1|D_1, \ldots, D_n) = \sum_i 2^{-i} D_i$, $P(B=1|D_1, \ldots, D_n) = \sum_i 2^{-i}(1 - D_i)$, and $P(C=1|A, B)$ is 1 if $A = B = 1$, $1/2$ if $A \neq B$, and 0 if $A=B=0$. The utility is $U(C) = 2^{n+1}C$. Suppose we eliminate variables in order $D_1, \ldots, D_n, C$ and produce $\Psi(A, B) = \{\sum_C U(C)P(C|A, B)P(A|d)P(B|d) : d \in \{0, 1\}^n\}$. We have that $\sum_{A,B} \Psi(A, B) = \{2^n P(A|d) + 2^n P(B|d) : d \in \{0, 1\}^n\} = \{2^n - 1\}$. Hence, there are 2^n non-dominated tables in $\Psi(A, B)$ (as two potentials whose values add to the same constant cannot one dominate each other).

Bucketing, which we describe next, gives us a way of bounding the growth of tables in such case at the expense of producing approximate inferences.

Let M be an upper bound we wish to impose over the number of tables in a set-potential during variable elimination. Consider a set-potential Φ with dimension d. Partition the hyperrectangle $[0,1]^d$ into a lattice of smaller M hypercubes called buckets. Let $s \stackrel{\text{def}}{=} \lfloor M^{-1/d} \rfloor$. The bucket index of an arbitrary table $\phi = [\phi(x_1), \ldots, \phi(x_d)]$ in Φ is given by $[\lfloor \phi(x_1)/s \rfloor, \ldots, \lfloor \phi(x_d)/s \rfloor]$. Any two points assigned to the same bucket are less than a distance of s of each other in any coordinate. Thus, by keeping one table per non-empty bucket we are guaranteed not to introduce a local error of more than s. We call this approach AkPU. This solution can be improved by any greedy algorithm for clustering under absolute-norm or Euclidean norm (e.g., k-means).

Theorem 3. *Consider a LIMID of treewidth w and maximum variable cardinality c, and let r be the value computed by AkPU (in fact, with or without dominance pruning) on that LIMID, choosing M at every step in a way that $s \stackrel{\text{def}}{=} \lfloor M^{-1/d} \rfloor$ is bounded from above by a constant m, where $d = c^w$. Denoting by n the number of (action, state, and value) variables, we have*

$$\left| r - \max_{\delta} E(U|\delta) \right| \leq 4n^2 \cdot [c+1]m.$$

Proof. Consider set-potentials Φ' and Ψ' obtained by bucketing of set-potentials Φ and Ψ, respectively. Now consider an element $\gamma = \phi \cdot \psi$ in $\Gamma = \Phi \cdot \Psi$, and let $\gamma' = \phi' \cdot \psi' \in \Gamma' = \Phi' \cdot \Psi'$, where ϕ' and ψ' are in the same buckets as ϕ and ψ, respectively. That is, $|\phi - \phi'| \leq m$ and $|\psi - \psi'| \leq m$. Suppose that $\gamma(x) \geq \gamma'(x)$ at some coordinate x. Then

$$\gamma(x) - \gamma'(x) \leq \phi(x)\psi(x) - [\phi(x) + s][\psi(x) - s] = [\phi(x) - \psi(x)] \cdot m - m^2 \leq 2m,$$

where in the last passage we assumed that $\phi(x) \geq \psi(x)$ (otherwise $\gamma(x) - \gamma'(x) \leq 0$, contradicting our initial claim) and used Proposition 2 to bound expression $\phi(x) - \psi(x)$ in one. Similarly, suppose that $\gamma'(x) \geq \gamma(x)$. Then

$$\gamma(x) - \gamma'(x) \leq [\phi(x) + s][\psi(x) + s] - \phi(x)\psi(x) = [\phi(x) + \psi(x)] \cdot m + m^2 \leq 3m,$$

where in the last passage we used Proposition 2 to bound expression $\phi(x) + \psi(x)$ in two. Hence, for any γ in Γ there is γ' in Γ' such that $|\gamma(x) - \gamma'(x)| \leq 3m$. Moreover, if Γ'' is a set-potential produced by bucketing of Γ' that for any γ in Γ there is γ'' in Γ'' such that $|\gamma(x) - \gamma'(x)| \leq 4m$. This implies that bucketing after every product introduces an error of at most $4m$. Consider now a set-potential Γ produced by Y-marginalization of a set-potential $\Phi(\mathcal{X} \cup \{Y\})$, and let Φ' be the output of bucketing Φ. For any $\gamma = \sum_Y \phi$ in Γ there is $\gamma' = \sum_Y \phi'$ in Γ' such that

$$|\gamma(x) - \gamma'(x)| = \left| \sum_y \phi(x, y) - \sum_y \phi'(x, y) \right| \leq \sum_y |\phi(x, y) - \phi'(x, y)| \leq c \cdot m.$$

Thus, bucketing after every marginalization introduces an error of at most $[c+1]m$. Variable elimination performs $n-1$ products and $n-1$ marginalizations. Hence the overall error introduced by bucketing is at most $[n-1]\cdot4\cdot[n-1]\cdot[c+1]m$, and the result follows. □

This result leads to a conservative estimate of the maximum number of buckets M we should use if we want to guarantee before runtime a maximum error on the output. The rationale in the proof above can be used to obtain an estimate of the overall error in the output when we fix the value of M at every step of variable elimination (adjusting it according to the dimension of the tables). We simply need to compute the actual worst-case induced error introduced by bucketing in a given step, accounting for the propagated errors as in the proof: products increase the current error by four, marginalization by c. This way, the algorithm can provide bounds on its quality at the end of the computation. When the values of the treewidth and the maximum variable cardinality are bounded by constants, a similar approach serves to prove the existence of an additive fully polynomial-time approximation scheme:

Theorem 4. *Given a LIMID of treewidth bounded by a constant w and whose variables have cardinalities bounded by a constant c, and $\epsilon > 0$, AkPU returns a strategy δ_ϵ such that $|E(U|\delta_\epsilon) - \max_\delta E(U|\delta)| \leq \epsilon$ in time polynomial in the size of the input and in $1/\epsilon$.*

Proof. Let n be the number of (action, state and value) variables in a LIMID, and M be the maximum number of tables in a set potential produced during a run of variable elimination with dominance pruning on that LIMID. Then MPU takes time $O(c^w \cdot M \cdot n)$, which is $O(M \cdot n)$, as c^w is considered constant. Since bucketing can be computed in time polynomial in M and c^w, it follows that AkPU takes time polynomial in M. Choose M such that $M \geq [4n^2(c+1)]^{c^w}[1/\epsilon]^{c^w} = O(n^\alpha \cdot 1/\epsilon^\beta)$, where α and β are some integer constants. Hence, AkPU runs in time polynomial in the size of the input (which is at least linear in the number of variables), and in $1/\epsilon$. Let r be the result of AkPU and $s \stackrel{\text{def}}{=} \lfloor M^{-1/d} \rfloor$. By Theorem 3, it follows that $|r - \max_\delta E(U|\delta)| \leq \epsilon$. □

For even moderately large values of n or c the estimate M obtained by applying Theorem 3 is prohibitively high, which implies that the above theorem is mostly of theoretical interest except for small diagrams with binary or ternary variables. For example, for a LIMID with structure as in Figure 1, $n = 100$ and $c = 10$, we have that $m = 1/[4 \cdot 10^5]$. The maximum dimension of a set-potential produced during variable elimination for that LIMID (assuming a perfect elimination order) is $d = 10^3$. Hence, $M \geq s^d \geq 2^{2560}$. A more realistic estimate of the required number of tables at every step can be obtained during runtime by computation of the actual error introduced after every bucketing operation, and consideration of the propagated error estimates. This way, M can be adjusted adaptively, demanding much less computational resources than in the proof of Theorem 4, but still guaranteeing a maximum error in the output in fully polynomial time.

6   Experiments

We compared the performance of DkPU and AkPU in a large set of randomly sampled LIMIDs with graph structure as in Fig. 1. While the choice of a fixed structure might seem restrictive, we note that any diagram can be transformed into a diagram like that of Fig. 1 by merging and adding variables [18]. The algorithms were implemented in Python and ran using the Pypy interpreter.[1] We performed experiments varying the cardinality c of the variables and the number n of variables. For each configuration of c and n, we compared running times and expected value of the best strategy found by the algorithms in a set of 30 LIMIDs, whose conditional probability distributions were independently sampled from a symmetric Dirichlet distribution with parameter $1/c$, and whose utilities were independently sampled from a uniform distribution in $[0, 1]$.

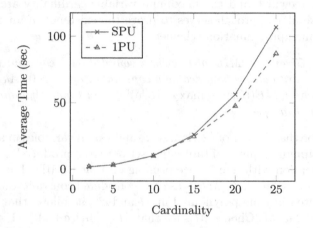

Fig. 2. Comparison between SPU and D1PU

Bucketing was performed with $M = 2^{20}/c^2$, thus keeping the size of set-potentials (number of tables times their dimension) below 2^{20}, as the treewidth of the LIMIDs we generate is 2. Local search was initialized with a uniform strategy, but while 1-local search was ran until convergence, k-local search with $k > 1$ ran for 1000 iterations.

Note first that SPU is by far the most popular algorithm for LIMIDs, and any gain in SPU's speed is welcome. Figure 2 shows that the overhead of dominance verification pays off in terms of speed. Gains are not dramatic, staying at about 20%, but these gains are obtained without any penalty in the quality of policies (as both SPU and D1PU produce identical runs).

Gains in speed with respect to kPU are important because they allow one to move up to higher values of k, hopefully searching deeper to produce higher expected values. Indeed, experiments summarized by Figures 3 and 4 show that

---

[1] The code and diagrams are available at http://github.com/denismaua/kpu.

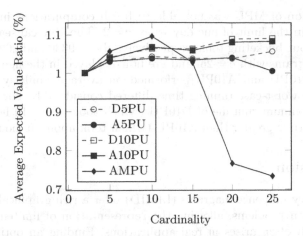

Fig. 3. Average accuracy relative to SPU

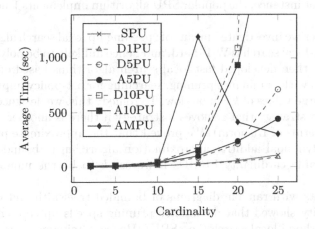

Fig. 4. Average time performance

DkPU can be used up to relatively high k with gains in expected value that reach 10%; also, the move to AkPU does control computing time while still leading to gains in expected value.

Figure 3 shows the average of the ratio between expected value obtained with k-local search schema to expected value obtained with 1-local search. Points higher than one indicate that the corresponding method outperforms 1-local search on average. Dashed curves report the performance of DkPU, whereas the solid curves report the performance of AkPU. We see that the bucketing does not decrease performance significantly, except for the approximate MPU variant, whose accuracy decays considerably with the increase of variable cardinality.

Figure 4 shows average running times. We also see that bucketing adds little overhead to computation for small values of k, but that it is crucial for effective tractability for large k. Indeed, differently than the approximate MPU (AMPU), the exact version of MPU was not able to finish computations in most of the diagrams within the limit of one day when $c = 2$. The 10-local search methods with and without bucketing took, respectively, about 9330 and 2600 seconds on average on diagrams with $c = 25$, and are not displayed in the figure for clarity.

Although D10PU and A10PU performed on average similarly in time and accuracy, their worst-case running time differed considerably. For example, for $c = 25$, the maximum runtime of D10PU (which took about 10 hour) was one order of magnitude greater than A10PU (which took about an hour).

7   Conclusion

Limited memory influence diagrams (LIMID) offer a rich graphical language to describe decision problems, allowing the representation of limited information scenarios which often arises in real applications. Finding an optimal strategy for a LIMID is an NP-hard problem, and practitioners resort to local search algorithms. For instance, the popular SPU algorithm implements 1-neighborhood local search.

In this paper we investigated means of speeding up local search algorithms. We showed that k-local search is W[1]-hard, and hence unlikely to be polynomial-time tractable. We then developed fast local search algorithms based on dominance pruning. Even with dominance pruning, searching for a k-policy improvement for moderately large values of k can be slow. To remedy this, we designed an approximate pruning strategy that removes a strategy if there is another close enough strategy (in terms of L1-norm). We proved that the approximate pruning leads to a fully polynomial additive approximation algorithm in bounded-treewidth bounded-variable cardinality LIMIDs if we set k to be the number of action variables.

Experiments with random diagrams of bounded treewidth and varying variable cardinality showed that dominance pruning speeds up computations even for 1-neighborhood local search (i.e., SPU). Hence dominance pruning is *always* useful when one wishes to resort to local search. Moreover, dominance pruning becomes essential for k-local search with $k > 5$. We also empirically showed that the quality of approximate pruning decays quickly with the increase of variable cardinalities, being only useful for variable cardinalities up to 15 or so.

Acknowledgements. The first author is partially supported by the FAPESP grant no. 13/23197-4. The second author is partially supported by CNPq.

References

1. Cooper, G.F.: A method for using belief networks as influence diagrams. In: Fourth Workshop on Uncertainty in Artificial Intelligence (1988)
2. de Campos, C.P., Ji, Q.: Strategy selection in influence diagrams using imprecise probabilities. In: Proceedings of the 24th Conference in Uncertainty in Artificial Intelligence (UAI), pp. 121–128 (2008)

3. Dechter, R.: An anytime approximation for optimizing policies under uncertainty. In: Workshop of Decision Theoretic Planning, AIPS (2000)
4. Detwarasiti, A., Shachter, R.D.: Influence diagrams for team decision analysis. Decision Analysis 2(4), 207–228 (2005)
5. Downey, R., Fellows, M.R.: Fixed-parameter tractability and completeness I: Basic theory. SIAM Journal of Computing 24, 873–921 (1995)
6. Downey, R., Fellows, M.R.: Fixed-parameter tractability and completeness II: Completeness for W[1]. Theoretical Computer Science A 141, 109–131 (1995)
7. Downey, R.G., Fellows, M.R.: Parameterized Complexity. Springer (1999)
8. Howard, R.A., Matheson, J.E.: Influence diagrams. In: Readings on the Principles and Applications of Decision Analysis, pp. 721–762. Strategic Decisions Group (1984)s
9. Jensen, F.V., Nielsen, T.D.: Bayesian Networks and Decision Graphs, 2nd edn. Information Science and Statistics. Springer (2007)
10. Khaled, A., Yuan, C., Hansen, E.: Solving limited memory influence diagrams using branch-and-bound search. In: Proceedings of the 29th Conference on Uncertainty in Artificial Intelligence, UAI (2013)
11. Kohlas, J.: Information Algebras: Generic Structures for Inference. Springer, New York (2003)
12. Kwisthout, J.H.P., Bodlaender, H.L., van der Gaag, L.C.: The Necessity of Bounded Treewidth for Efficient Inference in Bayesian Networks. In: Proceedings of the 19th European Conference on Artificial Intelligence, pp. 237–242 (2010)
13. Lauritzen, S.L., Nilsson, D.: Representing and solving decision problems with limited information. Management Science 47, 1235–1251 (2001)
14. Liu, Q., Ihler, A.: Belief propagation for structured decision making. In: Proceedings of the 28th Conference on Uncertainty in Artificial Intelligence (UAI), pp. 523–532 (2012)
15. Mauá, D.D., de Campos, C.P.: Solving decision problems with limited information. In: Advances in Neural Information Processing Systems 24 (NIPS), pp. 603–611 (2011)
16. Mauá, D.D., de Campos, C.P., Zaffalon, M.: On the complexity of solving polytree-shaped limited memory influence diagrams with binary variables. Artificial Intelligence 205, 30–38 (2013)
17. Mauá, D.D., de Campos, C.P., Zaffalon, M.: The complexity of approximately solving influence diagrams. In: Proceedings of the 28th Conference on Uncertainty in Artificial Intelligence (UAI), pp. 604–613 (2012)
18. Mauá, D.D., de Campos, C.P., Zaffalon, M.: Solving limited memory influence diagrams. Journal of Artificial Intelligence Research 44, 97–140 (2012)
19. Park, J.D., Darwiche, A.: Complexity results and approximation strategies for MAP explanations. Journal of Artificial Intelligence Research 21, 101–133 (2004)
20. Roth, D.: On the hardness of approximate reasoning. Artificial Intelligence 82(1-2), 273–302 (1996)
21. Szeider, S.: The parameterized complexity of k-flip local search for SAT and MAX SAT. Discrete Optimization 8(1), 139–145 (2011)
22. Tatman, J.A., Shachter, R.D.: Dynamic programming and influence diagrams. IEEE Transactions on Systems, Man and Cybernetics 20(2), 365–379 (1990)

Inhibited Effects in CP-Logic

Wannes Meert and Joost Vennekens

Dept. Computer Science – Campus De Nayer,
KU Leuven, Belgium
{wannes.meert,joost.vennekens}@cs.kuleuven.be

Abstract. An important goal of statistical relational learning formalisms is to develop representations that are compact and expressive but also easy to read and maintain. This is can be achieved by exploiting the modularity of rule-based structures and is related to the noisy-or structure where parents independently influence a joint effect. Typically, these rules are combined in an additive manner where a new rule increases the probability of the effect. In this paper, we present a new language feature for CP-logic, where we allow negation in the head of rules to express the inhibition of an effect in a modular manner. This is a generalization of the inhibited noisy-or structure that can deal with cycles and, foremost, is non-conflicting. We introduce syntax and semantics for this feature and show how this is a natural counterpart to the standard noisy-or. Experimentally, we illustrate that in practice there is no additional cost when performing inference compared to a noisy-or structure.

Keywords: Statistical Relational Learning, Bayesian networks, Noisy-or, Inhibited noisy-or, CP-logic.

1 Introduction

Statistical Relational Learning (SRL) [7] and probabilistic logic learning [2] are concerned with representations that combine the benefits of probabilistic models, such as Bayesian networks, with those of logic representations, such as first-order logic. In this work we focus on the family of SRL formalisms that associate probabilities to directed logic programming rules and can be interpreted as cause-effect pairs (e.g. CP-logic [17], ProbLog [6] or PRISM [14]).

An important goal of these formalism is to develop representations that are easy to read and maintain. One way in which they attempt to achieve this is by exploiting the inherent modularity of the rule-based structure of logic programs. We will illustrate this using CP-logic because of its intuitive, causal interpretation but the results are generally applicable. A CP-logic theory consists of a set of rules, and each rule is viewed as an independent causal mechanism. This makes it easy to update an existing theory by adding a (newly discovered) causal mechanism, since none of the existing rules have to be touched. In certain restricted cases, a theory in CP-logic can be translated into a Bayesian network in a very straightforward way. The translation may preserve this modularity property by using *noisy-or* nodes to represent the joint effect of different rules with the same

L.C. van der Gaag and A.J. Feelders (Eds.): PGM 2014, LNAI 8754, pp. 350–365, 2014.
© Springer International Publishing Switzerland 2014

head (i.e., different rules that may independently cause the same effect). It has been shown that the use of such noisy-or nodes in Bayesian networks makes it easier for human experts to supply probabilities and build more accurate models [18]. This is further evidence for the importance of this modularity property in probabilistic logics.

Currently, however, CP-logic's modularity is limited, in the sense that each new rule that is added for a given effect can only *increase* its probability. In practice, it occurs just as often that an existing theory has to be modified because a previously unknown mechanism makes some effect *less* likely in certain cases. CP-logic currently offers no modular way of adding such a new mechanism to an existing theory—it always requires changes to existing rules. In this paper, we present a new language feature, where we allow *negation in the head* of rules. We develop a syntax and semantics for this feature, and demonstrate that it indeed extends the modularity property to the discovery of new mechanisms that *decrease* the probability of existing events. In the special case where the CP-logic theory can easily be translated to a Bayesian network, we show that this feature of negation in the head reduces to an inhibited noisy-or structure [4]. While this is a little known kind of node in the literature on probabilistic graphical models, our analysis shows that it is a natural counterpart to the standard noisy-or. Additionally, we show experimentally that this intuitive structure exhibits the same advantageous properties as a noisy-or structure such as a linear number of parameters and inference that is polynomial in the number of parents.

2   Preliminaries and Motivation

CP-logic offers a compact and robust way of specifying certain kinds of probability distributions. This is due to three interacting properties:

- Different causes for the same effect can be represented as separate rules, each with their own probabilities, which are combined with a *noisy-or* when necessary. This leads to a modular representation.
- Logical variables may be used to write down first-order rules that serve as templates for sets of propositional rules. In this way, very compact representations can be achieved.
- The semantics of CP-logic is defined in a robust way, allowing, in particular, also cycles in the possible cause-effect relations.

To illustrate these properties, consider the following example.

Example 1. An infectious disease spreads through a population as follows: whenever two people are in regular contact with each other and one is infected, there is a probability of 0.6 of the infection spreading also to the other person. Given a set of initially infected people and a graph of connections between individuals in the population, the goal is to predict the spread of the disease.

In CP-logic, this can be represented by a set of two rules:

$$(Inf(x) : 1.0) \leftarrow InitialInf(x). \tag{1}$$
$$(Inf(x) : 0.6) \leftarrow Contact(x,y) \land Inf(y). \tag{2}$$

Given any set of individuals and any interpretation for the exogenous predicates *InitialInf* and *Contact*, this CP-theory defines the probability with which each individual will be infected. In particular, no restrictions (such as acyclicity) are imposed on the *Contact*-relation.

In addition to representing probability distributions in a compact way, CP-logic also aims at being *elaboration tolerant*: once a CP-theory for a given domain has been constructed, it should be easy to adapt this theory when we learn new facts about the domain. Ideally, new knowledge should be incorporated in a way which respects the inherent modularity of CP-logic, in the sense that it may involve adding or removing rules, but not changing existing rules.

One such operation for which CP-logic is obviously well-suited is when a new cause for some effect is discovered. For instance, suppose we learn that, in addition to being among the initially infected and having contact with infected individuals from the population, people may also contract the disease by travelling to particular locations (e.g., with probability 0.2). We can update our CP-logic model accordingly, by simply adding an additional rule:

$$(Inf(x) : 1.0) \leftarrow InitialInf(x).$$
$$(Inf(x) : 0.6) \leftarrow Contact(x,y) \land Inf(y).$$
$$(Inf(x) : 0.2) \leftarrow RiskyTravel(x).$$

Importantly, there is no need to change our existing rules.

A second operation is discovering that certain parts of the population form an exception to the general rules. For instance, suppose that certain people are discovered to be especially susceptible (e.g., probability 0.8) to contracting the disease through contact with an already infected person. We can represent this by "case splitting" rule (2) into the following two rules:

$$(Inf(x) : 0.6) \leftarrow Contact(x,y) \land Inf(y) \land \neg Susceptible(x).$$
$$(Inf(x) : 0.8) \leftarrow Contact(x,y) \land Inf(y) \land Susceptible(x).$$

However, this solution has the downside that it forces us to change an existing rule. A better alternative is to exploit the additive nature of different causes for the same effect in CP-logic:

$$(Inf(x) : 0.6) \leftarrow Contact(x,y) \land Inf(y).$$
$$(Inf(x) : 0.5) \leftarrow Contact(x,y) \land Inf(y) \land Susceptible(x).$$

For non-susceptible individuals, only the first rule is applicable, so they still get infected with the same probability of 0.6 as before. The same rule of course also applies to susceptible individuals, whom the second rule then gives an *additional* probability of getting infected *because* they are susceptible. This brings their total probability of being infected up to $0.6 + (1 - 0.6) \cdot 0.5 = 0.8$. When compared to the "case splitting" theory, this representation has the advantage that it allows the "default" rule for normal people to remain unchanged.

In addition to discovering that certain parts of the population are especially susceptible to the infection, it is equally possible to discover that certain people tend to be more resistant to it. Again, this can be solved by case splitting:

$$(Inf(x) : 0.6) \leftarrow Contact(x, y) \wedge Inf(y) \wedge \neg Resistant(x).$$
$$(Inf(x) : 0.4) \leftarrow Contact(x, y) \wedge Inf(y) \wedge Resistant(x).$$

A solution in which we can keep our original "default" rule unchanged is not possible using noisy-or or is not intuitive to impossible in current probabilistic logics. Indeed, this is an obvious consequence of the fact that adding additional rules can only *increase* probabilities. In this paper, we introduce the new feature of negation in the head of rules, which will allow us to represent also a *decrease* in probabilities. In particular, we will be able to represent our example as:

$$(Inf(x) : 0.6) \leftarrow Contact(x, y) \wedge Inf(y).$$
$$(\neg Inf(x) : 1/3) \leftarrow Resistant(x).$$

3 Preliminaries: Formal Semantics of CP-Logic

A theory in CP-logic [17] consists of a set of *CP-laws* of the form: $\forall \boldsymbol{x} \; (A_1 : \alpha_1) \vee \cdots \vee (A_n : \alpha_n) \leftarrow \phi$. Here, ϕ is a conjunction of literals and the A_i are atoms, such that the tuple of logic variables $\boldsymbol{x}$ contains all free logic variables in ϕ and the A_i. The α_i are non-zero probabilities with $\sum \alpha_i \leq 1$. Such a rule expresses that ϕ causes some (implicit) non-deterministic event, of which each A_i is a possible outcome with probability α_i. If $\sum_i \alpha_i = 1$, then at least one of the possible effects A_i must result if the event caused by ϕ happens; otherwise, the event may happen without any (visible) effect on the state of the world. For a CP-law r, we refer to ϕ as $body(r)$, and to the sequence $(A_i, \alpha_i)_{i=1}^n$ as $head(r)$.

The semantics of a theory in CP-logic is defined in terms of its grounding, so from now on we will restrict attention to ground theories, in which each tuple of logic variables $\boldsymbol{x}$ is empty. Any theory can be made ground by replacing the logic variables by constants. A ground atom can be considered as a binary random variable.

Example 2. Suzy and Billy may each decide to throw a rock at a bottle. Suzy throws with probability 0.5 and if she does, her rock breaks the bottle with probability 0.8. Billy always throws and his rock hits with probability 0.6.

$$(Throws(Suzy) : 0.5). \qquad (Broken : 0.8) \leftarrow Throws(Suzy).$$
$$(Throws(Billy) : 1). \qquad (Broken : 0.6) \leftarrow Throws(Billy).$$

The semantics of CP-logic is defined using the concept of an *execution model*. This is a probability tree in which each node s is labeled with a set of partial truth value assignments to atoms, which we denote as an *interpretation $\mathcal{I}(s)$*. Such trees are constructed, starting from a root node in which all atoms are false, by "firing" rules whose body holds. The following is an execution model for Example 2. States s in which the bottle is broken (i.e., $\mathcal{I}(s) \models Broken$) are represented by an empty circle, and those in which it is still whole by a full one.

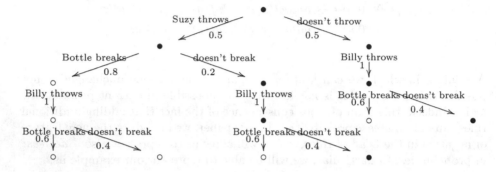

Each such tree defines a probability distribution over its leaves, which induces a probability distribution over the interpretations $\mathcal{I}(s)$ that are associated to these leaves. A CP-theory may have many execution models, which differ in the order in which they fire rules. The differences between these trees are irrelevant, in the sense that they all produce the same probability distribution π_T in the end [17].

The above example can easily be represented as a Bayesian network, where *Broken* is a noisy-or node with $Throws(Suzy)$ and $Throws(Billy)$ as its parents. This is in general the case for CP-theories that are acyclic [12]. Naively translating a CP-theory that is not acyclic to a Bayesian network would produce a cyclic graph.

The execution model semantics of CP-logic elegantly handles such cycles. As an example, we consider the following small instantiation of the previous example:

$$Inf(Alice) \leftarrow InitialInf(Alice).$$
$$Inf(Bob) \leftarrow InitialInf(Bob).$$
$$(Inf(Alice) : 0.2) \leftarrow RiskyTravel(Alice).$$
$$(Inf(Bob) : 0.2) \leftarrow RiskyTravel(Bob).$$
$$(Inf(Bob) : 0.6) \leftarrow Inf(Alice).$$
$$(Inf(Alice) : 0.6) \leftarrow Inf(Bob).$$

In the root of the execution model of this theory, $Inf(x)$ is still false for all x. It is only by applying the different rules that *Alice* and *Bob* may get infected. This ensure that the causal cycle between $Inf(Alice)$ and $Inf(Bob)$ is interpreted correctly and that, in particular, they cannot each cause the other to be infected unless at least one of them was also initially infected or infected by risky travel. However, this same property also makes it tricky to interpret negation in rule bodies. For instance, suppose we also have a rule:

$$Quarantine(x) \leftarrow \neg Inf(x).$$

In the root of the tree, $\neg Inf(x)$ still holds for all x — including those for which $InitialInf(x)$ holds! Naive application of this rule could therefore lead us to conclude that also initially infected people need to be quarantined, which is clearly not intended. To solve this problem, each node s in an execution model not only keeps track of an interpretation $\mathcal{I}(s)$ that represents the actual state of the world in that node, but also of an overestimate $\mathcal{U}(s)$ that looks ahead in the causal process to see which atoms could potentially still be caused. As long as an atom $A \in \mathcal{U}(s)$, it is still possible that A will be caused further on in the tree, even if at the current node it is still the case that $\mathcal{I}(s) \not\models A$. While $A \in \mathcal{U}(s)$, a rule that depends on the negative literal $\neg A$ will therefore be prevented from firing. We omit the formal details of how this $\mathcal{U}(s)$ is computed, but they can be found in [17].

To translate cyclic CP-theories into Bayesian networks, it is typically necessary to introduce additional nodes:

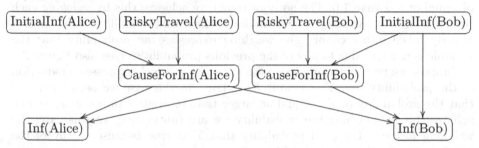

In general, this translation requires the addition of n^2 of such new nodes in order to eliminate a cycle between n nodes. For large CP-theories, such a blow-up may render inference intractable. Moreover, because all of these new nodes are latent, they also make the network harder to interpret or learn. Finally, because this translation needs to consider the cycle as a whole, it may no longer be possible to update the resulting network in a modular way.

4   Bayesian Net Interpretation for Negation in the Head

We now investigate how the semantics of CP-logic can be extended to accomodate negative literals in the head. Before addressing this question in general, we first focus on a fragment of CP-logic that can be trivially translated into a Bayesian net, namely, that of ground, acyclic CP-theories in which each rule has only one atom in the head. We first show how the idea behind noisy-or can be extended to accomodate negative literals in this simple fragment, before investigating—in the next section—how this result can be extended to the whole of CP-logic.

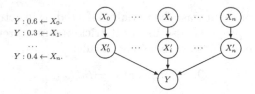

$$Y : 0.6 \leftarrow X_0.$$
$$Y : 0.3 \leftarrow X_1.$$
$$\ldots$$
$$Y : 0.4 \leftarrow X_n.$$

Fig. 1. CP-logic theory and the equivalent noisy-or Bayesian network

When adding a rule $(Y : \theta_i) \leftarrow X_i$ to a CP-logic theory, we increase the probability of Y being true given that condition X_i is true. This is equivalent to adding an additional parent X_i to a noisy-or construct (see Figure 1) and the probability of Y given the parents X_i can be calculated with:

$$Pr(Y = \top \mid X_{0:n}) = 1 - \prod_{\substack{i \in [0,n] \\ X_i = \top}} (1 - \theta_i) = \sum_{\substack{i \in [0,n] \\ X_i = \top}} \overbrace{\prod_{\substack{j \in [0,i[\\ X_j = \top}} (1 - \theta_j)}^{\text{weighing}} \cdot \theta_i$$

To adhere to the laws of probability, the total probability that Y is true should be equal or less than 1.0. The noisy-or structure achieves this by *weighing* each contribution of a parent by the remainder of the total probability of the parents already taken into account. The weighing expresses the probability that the variable is not true due to any of the previous probabilities (see also Figure 2).

Suppose we now add a rule $(\neg Y : \theta_{n+1}) \leftarrow X_{n+1}$, which expresses a reduction of the probability that Y is true if X_i is true. In this case we need to ensure that the probability of Y is equal or larger than 0. Similar to noisy-or, we can achieve this by weighing the probability we are subtracting. In this case, the weighing factor is the total probability that Y is true because of any of the previous (positive) parents (see Figure 2).

$$Pr(Y = \top \mid X_{0:n}, X_{n+1} = \top) = Pr(Y = \top \mid X_{0:n}) - \overbrace{Pr(Y = \top \mid X_{0:n})}^{\text{weighing}} \cdot \theta_{n+1}$$
$$= Pr(Y = \top \mid X_{0:n}) \cdot (1 - \theta_{n+1})$$

When adding multiple rules with negative literals in the head $\neg Y : \theta_{n+1} \leftarrow X_{n+1}., \ldots, \neg Y : \theta_m \leftarrow X_m$, the computation of the probability of Y can be generalized to:

$$Pr(Y = \top \mid X_{0:m}) = Pr(Y = \top \mid X_{0:n}) \cdot \left(1 - \sum_{\substack{i \in]n,m] \\ X_i = \top}} \prod_{\substack{j \in]n,i[\\ X_j = \top}} (1 - \theta_j) \cdot \theta_i \right) \quad (3)$$
$$= Pr(Y = \top \mid X_{0:n}) \cdot (1 - Pr(Y = \bot \mid X_{n+1:m})) \quad (4)$$

To represent Formula 4 as a Bayesian net it must be expressed as a set of conditional probability tables. Given two auxiliary variables P and N with

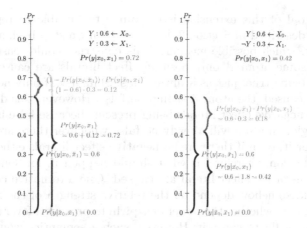

Fig. 2. Interpretation for probability scaling. For brevity we write $X = \top$ as x and $X = \bot$ as $\bar{x}$.

respectively the noisy-or structures $Pr(Y \mid X_{0:n})$ and $Pr(Y \mid X_{n+1:m})$, the formula $Pr(Y \mid X_{0:m})$ can now be written as $Pr(Y \mid P, N)$. The conditional probability table for this last conditional probability distribution is equivalent to $Y \Leftrightarrow P \wedge \neg N$ (see Figure 3).

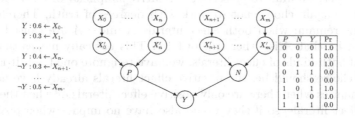

Fig. 3. Bayesian network that efficiently encodes an inhibited noisy-or. P is a noisy-or for $X_0, \ldots, X_n$ and N for $X_{n+1}, \ldots, X_m$.

5   Generalization to CP-Logic Programs

We now examine how we can incorporate the intuitions of the previous section into the general setting of CP-logic. To be more precise, from now on, we allow rules of the form:

$$\forall \boldsymbol{x} \quad (L_1 : \alpha_1) \vee \cdots \vee (L_n : \alpha_n) \leftarrow \phi.$$

Here, ϕ is again a first-order logic formula with $\boldsymbol{x}$ as free logic variables and the $\alpha_i \in [0, 1]$ are again such that $\Sigma \alpha_i \leq 1$. Each of the L_i is now either a *positive effect literal* A (i.e., an atom) or a *negative effect literal* $\neg A$.

While the goal of this extension is of course to be able to represent such phenomena as described in Section 2, let us first take a step back and consider, in the abstract, which possible meanings this construct could reasonably have. Clearly, if for some atom A only positive effect literals are caused, the atom should end up being true, just as it always has. Similarly, if only negative effect literals $\neg A$ are caused, the atom A should be false. However, this does not even depend on the negative effect literals being present: because false is the default value in CP-logic, an atom will already be false whenever there are no positive effect literals for it, even if there are no negative effect literals either.

The only question, therefore, is what should happen if, for some A, both a positive and a negative effect literal are caused. One alternative could be that the result would somehow depend on the relative strength of the negative and positive effects, e.g., whether the power of aspirin to prevent a fever is "stronger" than the power of flu to cause it. However, such a semantics would be a considerable departure from the original version of CP-logic, in which cumulative effects (synergy, interference, ...) are strictly ignored. In other words, CP-logic currently makes no distinction whatsoever between a headache that is simultaneously caused by five different conditions and a headache that has just a single cause. This design decision was made to avoid a logic that, in addition to probabilities, would also need to keep track of the degree to which a property holds. A logic combining probabilities with such fuzzy truth degrees would, in our opinion, become quite complex and hard to understand.

In this paper, we want to preserve the relative simplicity of CP-logic, and we will therefore again choose not to work with degrees of truth. Therefore, only two options remain: when both effect literals A and $\neg A$ are caused, the end result must be that A is either true of false. This basically means that, in the presence of both kinds of effect literals, we have to ignore one kind. It is obvious what this choice should be: the negative effect literals already have no impact on the semantics when there are only positive effect literals or when there are no positive effect literals, so if they would also have no impact when positive and negative effect literals are both present, then they would have never have any impact at all and we would have introduced a completely superfluous language construct. Therefore, the only reasonable choice is to give negative effect literals precedence over positive ones, that is, an atom A will be true if and only if it is caused at least once and no negative effect literal $\neg A$ is caused.

This can be formally defined by a minor change to the existing semantics of CP-logic. Recall that, in the current semantics, each node s of an execution model has an associated interpretation $\mathcal{I}(s)$, representing the current state of the world, and an associated three-valued interpretation $\mathcal{U}(s)$, representing an overestimate of all that could still be caused in s. We now add to this a third set, namely a set of atoms $\mathcal{N}(s)$, containing all atoms for which a negative effect literal has already been caused. The sets $\mathcal{I}(s)$ and $\mathcal{N}(s)$ evolve throughout an execution model as follows:

- In the root of the tree, $\mathcal{I}(s) = \mathcal{N}(s) = \{\}$

- When a *negative* effect literal $\neg A$ is caused in a node s, the execution model adds a child s' to s such that:
  - $\mathcal{N}(s') = \mathcal{N}(s) \cup \{A\}$;
  - $\mathcal{I}(s') = \mathcal{I}(s) \setminus \{A\}$.
- When a *positive* effect literal A is caused in a node s, the execution model adds a child s' to s such that:
  - $\mathcal{N}(s') = \mathcal{N}(s)$;
  - if $A \in \mathcal{N}(s)$, then $\mathcal{I}(s') = \mathcal{I}(s)$, else $\mathcal{I}(s') = \mathcal{I}(s) \cup \{A\}$.

Note that, throughout the execution model, we maintain the property that $\mathcal{N}(s) \cap \mathcal{I}(s) = \{\}$.

The overestimate $\mathcal{U}(s)$ is still constructed in the usual way (see [17]), with the exception that atoms from $\mathcal{N}(s)$ may no longer be added to it.

To illustrate, let us consider the following simple example, where we assume that *Alice* belongs to both exogenous predicates *RiskyTravel* and *Resistant*:

$$(Inf(Alice) : 0.2) \leftarrow RiskyTravel(Alice). \tag{5}$$

$$(\neg Inf(Alice) : \frac{1}{3}) \leftarrow Resistant(Alice). \tag{6}$$

Representing nodes in which *Alice* is infected by a full circle, these two rules may produce either of the following two execution models.

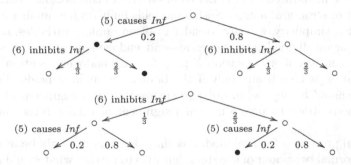

Again, the differences between these two execution models are irrelevant, because they both produce a distribution over final states in which $P(Inf(Alice)) = \frac{2}{3}{\cdot}0.2$, which is of course the same probability as we obtain with formula 4 of the previous section. Note that, in order to obtain this property, it is important that inhibition always trumps causation, regardless of which happens first. Unlike formula 4, the execution model semantics is equally applicable to cases with cyclic causation.

To implement this feature of negation-in-the-head, a simple transformation to regular CP-logic may be used. This transformation is based on the way in which [3] encode causal ramifications in their inductive definition modelling of the situation calculus.

For a CP-theory T in vocabulary Σ, let $\Sigma_\neg$ consist of all atoms A for which a negative effect literal $\neg A$ appears in T. For each atom $A \in \Sigma_\neg$, we introduce two new atoms, C_A and $C_{\neg A}$. Intuitively, C_A means that there is a cause for A, and $C_{\neg A}$ means that there is a cause for $\neg A$. Let τ_A be the following transformation:

- Replace all positive effect literals A in the heads of rules by C_A
- Replace all negative effect literals $\neg A$ in the heads of rules by $C_{\neg A}$
- Add this rule: $A \leftarrow C_A \wedge \neg C_{\neg A}$

Let $\tau_\neg(T)$ denote the result of applying to T, in any order, all the transformations τ_A for which $A \in \Sigma_\neg$. It is clear that $\tau_\neg(T)$ is a regular CP-theory, i.e., one without negation-in-the-head. As the following theorem shows, this reduction preserves the semantics of the theory.

Theorem 1. *For each interpretation X for the exogenous predicates, the projection of $\pi^X_{\tau_\neg(T)}$ onto the original vocabulary Σ of T is equal to π^X_T.*

· When comparing the transformed theory $\pi_{\tau_\neg}(T)$ to the original theory T, we see that the main benefit of having negation-in-the-head lies in its *elaboration tolerance*: there is no need to know before-hand for which atoms we later might wish to add negative effect literals, since we can always add these later, without having to change to original rules.

6 Application: Encoding Interventions

One of the interesting uses of negation-in-the-head is related to the concept of interventions, introduced by [13]. Let us briefly recall this notion. Pearl works in the context of *structural models*. Such a model is built from a number of random variables. For simplicity, we only consider Boolean random variables, i.e., atoms. These are again divided into exogenous and endogenous atoms. A structural model now consists of one equation $X := \varphi$ for each endogenous atom X, which defines that X is true if and only if the boolean formula φ holds. This set of equations should be acyclic, in order to ensure that an assignment of values to the exogenous atoms induces a unique assignment of values to the endogenous ones.

A crucial property of causal models is that they can not only be used to predicts the normal behaviour of a system, but also to predict what would happen if outside factors unexpectedly intervene with its normal operation. For instance, consider the following simple model of which students must repeat a class:

$$Fail := \neg Smart \wedge \neg Effort. \qquad Repeat := Fail \wedge Required.$$

Under the normal operation of this "system", only students who are not smart can fail classes and be forced to repeat them. Suppose now that we catch a student cheating on an assignment and decide to fail him for the class. This action was not foreseen by the causal model, so it does not follow from the normal behaviour. In particular, failing the student may cause him to have to repeat the class, but if the student is actually smart, then failing him will not make him stupid. Pearl shows that we can model our action of failing the student by means of an *intervention*, denoted $do(Fail = \top)$. This is a simple syntactic transformation, which removes and replaces the original equation for $Fail$:

$$Fail := \top. \qquad Repeat := Fail \wedge Required.$$

According to this updated set of equations, the student fails and may have to repeat the class, but he has not been made less smart.

In the context of CP-logic, let us consider the following simple medical theory:

$$(HighBloodPressure : 0.6) \leftarrow BadLifeStyle. \tag{7}$$

$$(HighBloodPressure : 0.9) \leftarrow Genetics. \tag{8}$$

$$(Fatigue : 0.3) \leftarrow HighBloodPressure. \tag{9}$$

Here, *BadLifeStyle* and *Genetics* are two exogenous predicates, which are both possible causes for *HighBloodPressure*. Suppose now that we observe a patient who suffers from *Fatigue*. Given our limited theory, this patient must be suffering from *HighBloodPressure*, caused by at least one of its two possible causes.

Now, suppose that a doctor is wondering whether it is a good idea to prescribe this patient some pills that lowers high blood pressure. Again, the proper way to answer such a question is by means of an *intervention*, that first prevents the causal mechanisms that normally determine someone's blood pressure and then substitutes a new "mechanism" that just makes *HighBloodPressure* false. This can be achieved by simply removing the two rules (7) and (8) from the theory. This is an instance of a general method, developed in [16], of performing Pearl-style interventions in CP-logic. The result is that probability of *Fatigue* drops to zero, i.e., $P(Fatigue \mid do(\neg HighBloodPressure)) = 0$.

In this way, we can evaluate the effect of prescribing the pills *without* actually having these pills in our model. This is a substantial difference to the way in which reasoning about actions is typically done in the field of knowledge representation, where formalisms such as situation or event calculus require an explicit enumeration of all available actions and their effects. Using an intervention, by contrast, we can envisage the effects of actions that we never even considered when writing our model.

Eventually, however, we may want to transform the above *descriptive* theory into a *prescriptive* one that tells doctors how to best treat a patient, given his or her symptoms. In this case, we would need rules such as this:

$$BPMedicine \leftarrow Fatigue. \tag{10}$$

Obviously, this requires us to introduce the action *BPMedicine* of prescribing the medicine, which previously was implicit in our intervention, as an explicit action in our vocabulary. Negation-in-the-head allows us to syntactically express the effect of this new action: $\neg HighBloodPressure \leftarrow BPMedicine$.

This transformation can be applied in general, as the following theorem shows.

Theorem 2. *Let T be a CP-theory over a propositional vocabulary Σ. For an atom $A \in \Sigma$, let T' be the theory $T \cup \{r\}$ with r the rule $\neg A \leftarrow B$ and B an exogenous atom not in Σ. For each interpretation X for the exogenous atoms of T', if $B \in X$, then $\pi_{T'}^X = \pi_{do(T, \neg A)}^X$ and if $B \notin X$, then $\pi_{T'}^X = \pi_T^X$.*

This theorem shows that negation-in-the-head allows CP-theories to "internalize" the intervention of *doing* ¬*A*. The result is a theory T' in which the intervention can be switched on or off by simply choosing the appropriate interpretation for the exogenous predicate that now explicitly represents this intervention. Once the intervention has been syntactically added to the theory in this way, additional rules such as (10) may of course be added to turn it from an exogenous to an endogenous property.

It is important to note that this is a fully modular and elaboration tolerant encoding of the intervention, i.e., the original CP-theory is left untouched and the rules that describe the effect of the intervention-turned-action are simply added to it. This is something that we can only achieve using negation-in-the-head.

7   Experiments

We have presented an intuitive and modular approach to express an inhibition structure. In this section, we evaluate the computational cost associated with this alternative structure. For this we perform inference on three theories: (i) the inhibited noisy-or structure from Fig. 3, (ii) the inhibited noisy-or structure translated to case splitting, and (iii) the infection example from Section 2. Inference was performed using ProbLog[1], an SRL system to which CP-logic can be compiled, for all three theories and using SMILE[2], a state-of-the-art PGM toolbox for the first two acyclic theories. All experiments are run on a 3GHz Intel Core2 Duo CPU with 2GB memory and timings are averaged over 3 runs.

For the inhibited noisy-or structure, the inference can be linear depending on the encoding of the noisy-or substructures [15,5]. The results (fig. 4a) show that the use of negative effect literals, implemented by means of their noisy-or encoding, is always more efficient than case splitting. Surprisingly, when using SMILE the inference has exponential complexity with a growing number of parents. This indicates that, although we used noisy-max encodings, noisy-or is not fully exploited. ProbLog is able to exploit the local structure more efficiently and performs inference for the inhibited noisy-or in time polynomial in the number of parents.

The infection example contains cycles and can therefore only be processed by ProbLog. For this theory, we let the number of people increase, while keeping the number of contacts per person fixed (Fig. 4b). This increases the number of inhibited noisy-or structures but, contrary to the previous theory, not the number of parents. We see that the version using case splitting is slower with approximately a constant factor.

We can conclude that the overhead introduced by the encoding for negative literals in the head is marginal compared to normal noisy-or combinations and inference can be performed efficiently.

---

[1] http://dtai.cs.kuleuven.be/problog
[2] http://genie.sis.pitt.edu

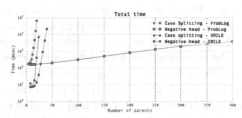
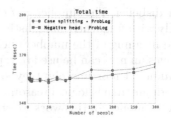

(a) Inference for an inhibited noisy-or structure.   (b) Inference for the infection example.

Fig. 4. Runtime of inference

8   Related Work

8.1   Inhibited Recursive Noisy-or

The structure we obtain is related to the inhibited recursive noisy-or structure [10] which states that:

$$Pr(Y|\mathbf{X}) = Pr(Y \text{ caused by } \mathbf{X}) \cdot Pr(Y \text{ not inhibited by } \mathbf{X})$$

The two parts are recursive noisy-or models, a generalisation of noisy-or that relaxes the ICI assumption. It allows to encode synergies, the combination of two causes to have a stronger effect than expected, and interferences, the combination to have a softer effect. A problem, however, with recursive noisy-or models is that the parametrisation may be asymmetric. As this causes confusion and conflicts, this model does not allow for a modular representation and is not popular in common use [4]. Different in CP-logic is that concepts like synergy and interference are not represented using a recursive parametric probability distribution but directly in the program using the conditions in the body and positive and negative literals. As such, CP-logic, offers a modular and non-conflicting alternative to inhibited recursive noisy-or models.

8.2   The Certainty Factor Model

Rule-based systems are popular for expert and diagnostics systems because they offer an intuitive syntax to human experts. In this setting, the concept of weighing the level of uncertainty of inhibiting factors has been proposed for *certainty factors* used in the MYCIN system [1,11]. The weighing, however, is performed independently for the measures of belief and disbelief and are joined only afterwards to define the certainty factor. These notions of uncertainty are not well-founded from a probabilistic point of view but are used in practice because they are computationally simple and behave satisfactorily. It was argued that the Bayesian framework was unsatisfactory because it would require too many conditional probability parameters that have to be filled in by an expert. This was a motivation to use the two different measures, one for belief and one for disbelief. The simplicity of the certainty factor model, however, was achieved only with frequently unrealistic assumptions and with persistent confusion about the meaning of the numbers being used [8]. Heckerman and Shortlife show how Bayesian nets can be used to represent the certainty factor model in a principled manner. Unfortunately, they show that "uncertain reasoning is inherently less modular than is logical reasoning", which is an attractive feature of the certainty factor model. In this work we show that both concepts of belief and disbelief can be represented in one rule-based framework with a strong foundation in probability theory and with the modularity properties of logical reasoning.

8.3   Interaction Rules in Probabilistic Logic

Negation in the head can be interpreted as a modification of the noisy-or interaction rule that is common among probabilistic logics. *Probabilistic interaction logic* [9] is a framework that generalizes languages like CP-logic and ProbLog to allow custom encodings of the interaction rules. This is achieved by building on top of default logic instead of logic programming. Part of the example in Section 2 can be expressed as:

$$D = \{\frac{RiskyTravel(x) \wedge p(x) \; : \; Inf(x)}{Inf(x)}\}, \quad W = \{Resistant(x) \wedge q(x) \rightarrow \neg Inf(x)\}$$

with $P(p(x)) = 0.2$ and $P(q(x)) = 1/3$. Here, the single default in D expresses that, if x has done risky travel, this will cause her to be infected with probability 0.2, *unless we know otherwise*. The implication in W then gives precisely such a reason for knowing otherwise, namely, the fact that x might be resistant.

This logic is obviously quite general, allowing many more interaction patterns to be expressed than just the simple inhibited effects we have considered here. However, it does depend on the user to correctly encode these patterns in first-order logic: for instance, adding the inhibiting effect of being resistant will require a change to the original theory, unless the user had the foresight to already include the justification $Inf(x)$ in his original default.

9   Conclusion

In this paper, we have presented the new language feature of negative effect literals. We have shown this for the case of CP-logic where it offers a natural extension the capacity to represent causal models in a modular way. In the particular case of theories that correspond to a Bayesian net, such negative effect literals correspond to an inhibited noisy-or structure. Additionally, we show that this new language feature can be encoded in such a manner that inference can be performed with a complexity similar to standard noisy-or.

References

1. Buchanan, B.G., Shortlife, E.H.: Rule Based Expert Systems: The MYCIN Exper-imehts of the Stanford Heuristic Programming Project. Addison-Wesley (1984)
2. De Raedt, L., Frasconi, P., Kersting, K., Muggleton, S. (eds.): Probabilistic Induc-tive Logic Programming. LNCS (LNAI), vol. 4911. Springer, Heidelberg (2008)
3. Denecker, M., Ternovska, E.: Inductive situation calculus. Artificial Intelli-gence 171(5-6), 332–360 (2007)
4. Díez, F.J., Druzdzel, M.: Canonical probabilistic models for knowledge engineering. Technical report cisiad-06-01, UNED, Madrid, Spain (2006)
5. Díez, F.J., Galán, S.F.: Efficient computation for the noisy max. International Journal of Intelligent Systems 18(2), 165–177 (2003)
6. Fierens, D., Van den Broeck, G., Thon, I., Gutmann, B., De Raedt, L.: Inference in probabilistic logic programs using weighted CNFs. arXiv preprint arXiv:1202.3719 (2012)
7. Getoor, L., Taskar, B. (eds.): Statistical Relational Learning. MIT Press (2007)
8. Heckerman, D.E., Shortliffe, E.H.: From certainty factors to belief networks. Arti-ficial Intelligence in Medicine 4(1), 35–52 (1992)
9. Hommersom, A., Lucas, P.J.F.: Generalising the interaction rules in probabilistic logic. In: Proceedings of the Twenty-Second International Joint Conference on Artificial Intelligence, vol. 2, pp. 912–917 (2011)
10. Kuter, U., Nau, D., Gossink, D., Lemmer, J.F.: Interactive course-of-action plan-ning using causal models. In: Third International Conference on Knowledge Sys-tems for Coalition Operations, KSCO 2004 (2004)
11. Lucas, P., Van Der Gaag, L.: Principles of expert systems. Addison-Wesley Longman Publishing Co., Inc. (1991)
12. Meert, W., Struyf, J., Blockeel, H.: Learning ground CP-logic theories by leveraging Bayesian network learning techniques. Fundamenta Informaticae 89(1), 131–160 (2008)
13. Pearl, J.: Causality: Models, Reasoning, and Inference. Cambridge University Press (2000)
14. Sato, T., Kameya, Y.: New advances in logic-based probabilistic modeling by PRISM. In: De Raedt, L., Frasconi, P., Kersting, K., Muggleton, S.H. (eds.) Prob-abilistic Inductive Logic Programming. LNCS (LNAI), vol. 4911, pp. 118–155. Springer, Heidelberg (2008)
15. Takikawa, M., D'Ambrosio, B.: Multiplicative factorization of noisy-max. In: Laskey, K.B., Prade, H. (eds.) Proceedings of the 15th Conference on Uncertainty in Artificial Intelligence (UAI), pp. 622–630 (1999)
16. Vennekens, J., Bruynooghe, M., Denecker, M.: Embracing events in causal mod-elling: Interventions and counterfactuals in CP-logic. In: Janhunen, T., Niemelä, I. (eds.) JELIA 2010. LNCS (LNAI), vol. 6341, pp. 313–325. Springer, Heidelberg (2010)
17. Vennekens, J., Denecker, M., Bruynooghe, M.: CP-logic: A language of causal prob-abilistic events and its relation to logic programming. Theory and Practice of Logic Programming (TPLP) 9(3), 245–308 (2009)
18. Zagorecki, A., Druzdzel, M.J.: An empirical study of probability elicitation under noisy-or assumption. In: FLAIRS Conference, pp. 880–886 (2004)

Learning Parameters in Canonical Models
Using Weighted Least Squares

Krzysztof Nowak[1,3] and Marek J. Druzdzel[1,2]

[1] Białystok University of Technology, Białystok, Poland
[2] School of Information Sciences, Pittsburgh, USA
[3] European Space Agency, Noordwijk, The Netherlands
knowak.ai@gmail.com, marek@sis.pitt.edu

Abstract. We propose a novel approach to learning parameters of canonical models from small data sets using a concept employed in regression analysis: weighted least squares method. We assess the performance of our method experimentally and show that it typically outperforms simple methods used in the literature in terms of accuracy of the learned conditional probability distributions.

Keywords: Bayesian networks, canonical models, noisy–MAX gates, parameter learning, weighted least squares.

1 Introduction

Methodologies for extracting information from data have been one of the key factors for the success of modern artificial intelligence. Bayesian networks — one of the prime examples among probabilistic modeling techniques — are widely acclaimed and used by the scientific communities as well as the industry. Nowadays, data for many problem domains are freely accessible and in many cases growing at an exponential rate. Nonetheless, in some fields the amount of data is small, usually due to the cost of acquisition or high complexity of the problem. The latter increases the number of parameters required for the accurate modeling of the problem, which in turn calls for learning samples of large size. A class of interactions within Bayesian networks, the so called ICI (*Independence of Causal Influence*) models, find their applications in problems where obtaining an adequately sampled dataset is infeasible. In this paper we focus on learning parameters for the ICI models by framing the problem in terms of linear algebra and then calculating the values of parameters by means of the weighted least squares. We follow this up by an empirical test for learning accuracy of the proposed method and highlight the cases in which it outperforms the two common approaches to this problem: Expectation-maximization approach and the method proposed by Onińsko and Druźdżel [1].

2 ICI Models

ICI models [2,3,4,5] are based on the assumption of independence of causal influences. An effect variable, along with its causes, may fit an ICI model if the

L.C. van der Gaag and A.J. Feelders (Eds.): PGM 2014, LNAI 8754, pp. 366–381, 2014.

mechanisms through which the causes impact the effect do not interact among each other. This simple restriction greatly simplifies elicitation of parameters from data and experts. Because some conditional probabilities can now be expressed as a function of a far smaller set of parameters, the number of independent parameters required to define the CPT (Conditional Probability Table) of the child node is reduced from exponential to linear in the number of parents. Figure 1 shows an example of such model.

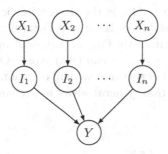

Fig. 1. Structure of an ICI model: additional auxiliary nodes I_1 - I_n are called *inhibitors*. The independence of mechanisms is represented by a lack of edges among the inhibitor nodes I_1 to I_n.

2.1   Noisy–OR/MAX

Deterministic models usually rely on a function that takes a set of input signals which determine the state of the child node Y. The deterministic OR function makes it impossible for the child to be activated if none of the parent nodes is present. Noisy–OR [2,3] and Noisy–MAX [6] models are cases of deterministic OR and deterministic MAX functions applied to the ICI framework [5]. Every variable has a special state indicating that the phenomenon that it represents is "absent," we call such state the "distinguished state." The term "activated state" will then refer to any of the non-distinguished states. Uncertainty is introduced to the model by adding a separate layer of inhibitor nodes (I_1 through I_n in Figure 1), which are activated based on their corresponding states of parents (nodes X_1 through X_n) with a certain probability. Once that is done, the states of inhibitor nodes are used as input signals for the deterministic function. As it turns out, all we need to provide to the model are the probabilities of every parent variable activating the child independently, that is when every variable other than the one in question is in its distinguished state. Thus, a Noisy–OR parameter p_i describes the probability of i-th parent activating the child:[1]

$$p_i = P(+y|\neg x_1, \neg x_2, \ldots, +x_i, \ldots, \neg x_n) . \tag{1}$$

---

[1] We will use lowercase to denote the states of variables. Distinguished states are preceded with a negation sign, while activated states with a plus sign, e.g., $+x_1$ and $\neg x_1$ are activated and distinguished states of variable X_1 respectively.

Obtaining conditional probabilities for the combination of states of nodes X_i other than the ones already described by the Noisy–MAX parameters can be derived using the following equation:

$$P(+y|\boldsymbol{x}) = 1 - \prod_{i\in I(\boldsymbol{x})} (1 - p_i) \,, \qquad (2)$$

where $I(\boldsymbol{x})$ is a set of indices of parents in combination $\boldsymbol{x}$ which are in their activated states. In our case, OR is the deterministic function — the child is active when any of the inhibitors is active. Deterministic MAX function is a generalization of the deterministic OR as proposed independently by Díez [6] and Srinivas [7]. For that reason, we can treat Noisy–OR as a binary case of the Noisy–MAX model. In further sections, we will occasionally use Noisy–OR/MAX interchangeably, knowing that general ideas apply equally well to binary and non-binary variables.

2.2   Leaky Noisy–OR/MAX

Deterministic OR function activates the child node only when at least one of the parents is in its non-distinguished state. This is not always the case in the real world — absence of any signal from the parental causes may still activate the child variable. Sometimes a problem does not allow for explicit modeling of each possible cause, either because it is not well understood, or because it would require a large number of additional variables that would have minimal impact on the child. In that case the unmodelled causes can be aggregated into a single node called *leak* [2] (see Figure 2). Since leak is modeling mechanisms that we do not control or actively observe, but that are present, we add the probability of leak (p_L) to the product on the right hand side of Equation 3:

$$P(+y|\boldsymbol{x}) = 1 - (1 - p_L) \prod_{i\in I(\boldsymbol{x})} (1 - p_i) \,. \qquad (3)$$

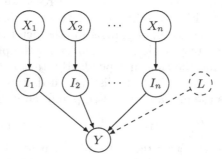

Fig. 2. Structure of a Leaky ICI model: L is the implicit leak node

2.3   Eliciting Parameters of ICI Models from Data

The problem of learning Noisy/Leaky–MAX parameters from data was addressed previously by Oniśko and Drużdżel [1], Zagórecki et al. [8], and others (e.g., [4,9]). The main problem with defining and learning the CPT is the exponential growth of the table in the number of parents. The number of independent parameters in a CPT (I_{CPT}) can be calculated using Equation 4:

$$I_{\mathrm{CPT}} = (|Y| - 1) \prod_{i=1}^{n} |X_i| \,, \tag{4}$$

where $|Y|$ and $|X_i|$ denote the number of states of the child node and the i-th parent respectively. The corresponding number of independent parameters for the Leaky–MAX gate (I_{MAX}) is:

$$I_{\mathrm{MAX}} = (|Y| - 1) \sum_{i=1}^{n} (|X_i| - 1) + |Y| - 1 \,. \tag{5}$$

We can see that the ICI models reduce the number of parameters required logarithmically. In real–life problems, it is not uncommon to find models consisting of thousands of variables, some having many parent nodes. In such cases, the size of the model's CPTs can become huge. Not only is the computational complexity of inference within such network high, but constructing such model in the first place can be daunting. What we would have to ask the expert are specific questions about conditional probabilities for every possible scenario that can occur within the problem domain. Assuming n binary parents, this involves eliciting 2^n numerical values from experts (this task becomes already cumbersome for even small values of n). This effort can be eased by learning the parameters from data, which in its basic form revolves around estimating the required conditional probabilities according to the distributions observed in the data. This, however, does not resolve the problem completely. Given a dataset of m records we obtain an average of $\frac{m}{2^n}$ records per conditional probability distribution within a CPT. When the probability distribution over various combinations of parent states is skewed, some parameters will have no corresponding records within the data at all. Most likely we will not be able to supply a sufficiently large dataset to maintain a reasonable records per parameter ratio, which is inherently associated with a large error. Thus, reducing the number of required parameters which define the model not only simplifies the storage and inference, but also the learning from both data and experts.

Oniśko and Drużdżel [1] propose a simple method of learning Noisy–MAX parameters from data by limiting learning to only those records that describe the parameters directly, namely the cases where only one of the parent nodes is in its activated state. Another approach by Zagórecki and Drużdżel [10] aims at learning the full CPT first and then fitting the Noisy/Leaky–MAX parameters which are the closest (in Euclidian distance) to the original CPT distribution. In this research, we propose yet another approach which somewhat resembles the previous two: Learning Leaky–MAX parameters directly from the data (yet using the information from the full dataset), while also minimizing the sum of distances towards each of the probabilistic scenarios in data using the weighted least squares method.

3   Least Squares Approximation

Least Squares (also known as "Ordinary" or "Linear" Least Squares) is a parameter estimation method, commonly used in regression analysis [11]. As we will show in this section, this approach resonates well with the constraints of our problem and can be employed for the purpose of learning parameters of canonical models from data. In this research we focus on the Leaky–OR/MAX gates, yet the same approach can be employed to the Leaky–AND/MIN models and other canonical gates, as long as it is possible to frame the problem in terms of a system of linear equations.

3.1   Expressing Probabilistic Information as a System of Linear Equations

Following Equation 3 we can express any conditional probability within the CPT using the Leaky–MAX parameters. Thus, given an arbitrary observation of parent states $\boldsymbol{x}$ and child state y we have the following:

$$1 - \hat{P}(+y|\boldsymbol{x}) = (1 - \hat{p_L}) \prod_{i \in I(\boldsymbol{x})} (1 - \hat{p}_i) , \tag{6}$$

where $\hat{P}(+y|\boldsymbol{x})$, $\hat{p_L}$ and $\hat{p}_i$ are the estimations of $P(+y|\boldsymbol{x})$, p_L and p_i from data. Taking the logarithm of both sides gives us:

$$\log(1 - \hat{P}(+y|\boldsymbol{x})) = \log(1 - \hat{p_L}) + \sum_{i \in I(\boldsymbol{x})} \log(1 - \hat{p}_i) . \tag{7}$$

By introducing substitutions $\hat{q} = \log(1 - \hat{P}(+y|\boldsymbol{x}))$, $\hat{q_L} = \log(1 - \hat{p_L})$ and $\hat{q}_i = \log(1 - \hat{p}_i)$ we obtain a linear equation:

$$\hat{q} = \hat{q_L} + \sum_{i \in I(\boldsymbol{x})} \hat{q}_i , \tag{8}$$

with $\hat{q}$ as a constant, and $\hat{q_L}, \hat{q}_1, \ldots, \hat{q}_n$ as the unknowns. Repeating the steps above for each combination $\boldsymbol{x}$ observed within the data gives us a system of linear equations. Solving for $\hat{q}_i$ and $\hat{q_L}$ gives us the corresponding values of $\log(1 - \hat{p}_i)$ and $\log(1 - \hat{p_L})$, from which we can obtain the original Leaky–MAX parameters $\hat{p}_i$ and $\hat{p_L}$:

$$\hat{p_k} = 1 - \exp(\hat{q_k}) \quad \text{for } k \in \{1, \ldots, n, L\} . \tag{9}$$

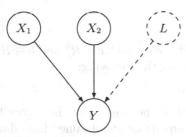

$x = x_1, x_2$ and y	# records	$\hat{P}(+y\|x)$
$+x_1, +x_2, +y$	40	$\dfrac{40}{52}$
$+x_1, +x_2, \neg y$	12	
$\neg x_1, +x_2, +y$	1	$\dfrac{1}{3}$
$\neg x_1, +x_2, \neg y$	2	
$+x_1, \neg x_2, +y$	16	$\dfrac{16}{25}$
$+x_1, \neg x_2, \neg y$	9	
$\neg x_1, \neg x_2, +y$	1	$\dfrac{1}{20}$
$\neg x_1, \neg x_2, \neg y$	19	

(a) Structure of the problem   (b) Sampled dataset

Fig. 3. Leaky–OR structure (left) along with a sample of 100 records (right)

Example 1. Let the binary nodes X_1, X_2 and Y form a Leaky–OR model, giving us 8 possible observations in the sample (Figure 3a). Let us assume a sample of 100 records reflecting the structure (Table 3b). Expressing the observations from the sample using the Equation 6, gives us the following system of equations:

$$
\begin{bmatrix}
(1-\hat{p}_1)\cdot(1-\hat{p}_2)\cdot(1-\hat{p}_L) \\
(1-\hat{p}_2)\cdot(1-\hat{p}_L) \\
(1-\hat{p}_1)\phantom{\cdot(1-\hat{p}_2)}\cdot(1-\hat{p}_L) \\
(1-\hat{p}_L)
\end{bmatrix}
=
\begin{bmatrix}
1-\hat{P}(+y|+x_1,+x_2) \\
1-\hat{P}(+y|\neg x_1,+x_2) \\
1-\hat{P}(+y|+x_1,\neg x_2) \\
1-\hat{P}(+y|\neg x_1,\neg x_2)
\end{bmatrix}. \tag{10}
$$

We take logarithm of both sides of each of the equations:

$$
\begin{bmatrix}
\log(1-\hat{p}_1)+\log(1-\hat{p}_2)+\log(1-\hat{p}_L) \\
\log(1-\hat{p}_2)+\log(1-\hat{p}_L) \\
\log(1-\hat{p}_1)\phantom{+\log(1-\hat{p}_2)}+\log(1-\hat{p}_L) \\
\log(1-\hat{p}_L)
\end{bmatrix}
=
\begin{bmatrix}
\log(1-{}^{40}/_{52}) \\
\log(1-{}^{1}/_{3}) \\
\log(1-{}^{16}/_{25}) \\
\log(1-{}^{1}/_{20})
\end{bmatrix}, \tag{11}
$$

and apply the substitutions $\hat{q}_L = \log(1-\hat{p}_L)$ and $\hat{q}_i = \log(1-\hat{p}_i)$, which gives us the following system of linear equations:

$$
\begin{bmatrix}
1 & 1 & 1 \\
0 & 1 & 1 \\
1 & 0 & 1 \\
0 & 0 & 1
\end{bmatrix}
\cdot
\begin{bmatrix}
\hat{q}_1 \\
\hat{q}_2 \\
\hat{q}_L
\end{bmatrix}
=
\begin{bmatrix}
\log({}^{12}/_{52}) \\
\log({}^{2}/_{3}) \\
\log({}^{9}/_{25}) \\
\log({}^{19}/_{20})
\end{bmatrix}
=
\begin{bmatrix}
-1.466\ldots \\
-0.405\ldots \\
-1.021\ldots \\
-0.051\ldots
\end{bmatrix}. \tag{12}
$$

■

3.2 Weighted Least Squares Method

In the previous section, we have shown a transition from discrete data, obeying the assumptions of the Leaky-MAX model, to a system of linear equations. The input in our case is an overdetermined system of linear equations, which can be thought of as a set of hyperplanes. Least squares method allows for finding a solution that minimizes the sum of vertical distances to each of the planes.

Given an overdetermined system:

$$Ax = b,$$ (13)

where A is an $m \times n$ matrix of real–valued coefficients, and $x \in R^n$ and $b \in R^m$, we can obtain an approximate solution x' solving the following:

$$A^T A x' = A^T b.$$ (14)

Thus obtained solution x' minimizes the sum of the squares of the errors for each of the m equations. Since in our case the equations are obtained from data, least squares method has no means of distinguishing between these equations that were represented by a larger fraction of the dataset (potential lower error) and those equations that were represented by only a handful of records (higher error). For that reason we employ the weighted variant of the method [11], thus allowing to promote the more significant equations.

Given the system of equations (13) and a diagonal matrix W with weights on its main diagonal,[2] we solve for the weight–induced approximate solution x'' using the following:

$$A^T W A x'' = A^T W b.$$ (15)

Example 2. Let us continue the previous example and assume an overdetermined system in Equation 12. By assuming

$$A = \begin{bmatrix} 1 & 1 & 1 \\ 0 & 1 & 1 \\ 1 & 0 & 1 \\ 0 & 0 & 1 \end{bmatrix}, b = \begin{bmatrix} -1.466\ldots \\ -0.405\ldots \\ -1.021\ldots \\ -0.051\ldots \end{bmatrix}, A^T A = \begin{bmatrix} 2 & 1 & 2 \\ 1 & 2 & 2 \\ 2 & 2 & 4 \end{bmatrix}, A^T b = \begin{bmatrix} -1.871\ldots \\ -2.487\ldots \\ -2.944\ldots \end{bmatrix}, x = \begin{bmatrix} \hat{q}_1 \\ \hat{q}_2 \\ \hat{q}_L \end{bmatrix},$$ (16)

we can solve for an ordinary least squares solution x'

$$A^T A x' = A^T b \Rightarrow \begin{bmatrix} 2 & 1 & 2 \\ 1 & 2 & 2 \\ 2 & 2 & 4 \end{bmatrix} x' = \begin{bmatrix} -1.821\ldots \\ -2.555\ldots \\ -2.894\ldots \end{bmatrix} \Rightarrow x' = \begin{bmatrix} -0.399\ldots \\ -1.015\ldots \\ -0.028\ldots \end{bmatrix}.$$ (17)

Since x' is a vector of approximate parameters q_1', q_2' and q_L', we apply Equation 9 to obtain the corresponding parameters p_1', p_2' and p_L':

$$\begin{bmatrix} p_1' \\ p_2' \\ p_L' \end{bmatrix} = \begin{bmatrix} 1 - \exp(q_1') \\ 1 - \exp(q_2') \\ 1 - \exp(q_L') \end{bmatrix} = \begin{bmatrix} 0.329\ldots \\ 0.638\ldots \\ 0.028\ldots \end{bmatrix}.$$ (18)

Alternatively, we can apply weights:[3]

$$W = \begin{bmatrix} 52 & 0 & 0 & 0 \\ 0 & 3 & 0 & 0 \\ 0 & 0 & 25 & 0 \\ 0 & 0 & 0 & 20 \end{bmatrix}, A^T W A = \begin{bmatrix} 55 & 52 & 55 \\ 52 & 77 & 77 \\ 55 & 77 & 100 \end{bmatrix}, A^T W b = \begin{bmatrix} -77.465\ldots \\ -101.790\ldots \\ -104.033\ldots \end{bmatrix},$$ (19)

---

[2] $w_{ii} \in W$ describes the weight of the i-th equation. Normalization of the weights is not necessary.

[3] For simplicity we assume $w_{ii} \in W$ to be the number of records describing i-th equation (see Table 3b).

and solve for a weighted least squares solution x'':

$$A^T W A x'' = A^T W b \Rightarrow \begin{bmatrix} 55 & 52 & 55 \\ 52 & 77 & 77 \\ 55 & 77 & 100 \end{bmatrix} x'' = \begin{bmatrix} -77.465\ldots \\ -101.790\ldots \\ -104.033\ldots \end{bmatrix} \Rightarrow x'' = \begin{bmatrix} -0.432\ldots \\ -0.988\ldots \\ -0.041\ldots \end{bmatrix}. \tag{20}$$

As previously highlighted, in order to obtain the final parameters p_1'', p_2'' and p_L'', we employ the Equation 9:

$$\begin{bmatrix} p_1'' \\ p_2'' \\ p_L'' \end{bmatrix} = \begin{bmatrix} 1 - \exp(q_1'') \\ 1 - \exp(q_2'') \\ 1 - \exp(q_L'') \end{bmatrix} = \begin{bmatrix} 0.351\ldots \\ 0.628\ldots \\ 0.040\ldots \end{bmatrix}. \tag{21}$$

■

4   Empirical Performance

This section describes experiments performed to test our approach in practice. We perform two types of experiments: learning parameters using datasets generated from an ideal Leaky–MAX definition and a definition "distorted" by a varying noise parameter κ. The latter models the situation in which we are fitting a Leaky–MAX gate to a distribution that is not exactly a Leaky–MAX.

4.1   Data Generation

Data for the experiment were prepared and generated using the GeNIe and SMILE© software packages.[4] We focus on learning networks composed of only one child and the number of binary parents varying between 2 and 6 plus leak. After testing the accuracy of our method on non-binary parents, we have found its performance to be correlated with the number of possible equations for given test case. For that reason we model the "hardness" of the problem simply by a varying number of binary parents. In order to provide statistical results, we generate a thousand randomized networks for each number of parents. The prior probabilities of parents, as well as the Leaky–MAX definition of the child are randomized uniformly each time. For each of the 5000 networks, we generate 20 random datasets consisting of 100, 200, ..., 1900 and 2000 records, which gives us 100,000 datasets total.

Noise–Induced Leaky–MAX: In order to test the accuracy of both methods in cases when the original distribution differs from the ideal Leaky–MAX, we introduce the "distortion" to the definition. We expand the Leaky–MAX parameters to a full CPT and introduce the noise to each of the probability

---

[4] Available at http://genie.sis.pitt.edu/.

distributions p_k (columns in CPT), by sampling a variate x from the Dirichlet distribution, as defined by the following probability density function:

$$\text{Dir}(x_1, \ldots, x_{n-1}; \alpha_1, \ldots, \alpha_n) = \frac{1}{B(\alpha)} \prod_{i=1}^{n} x_i^{\alpha_i - 1}, \tag{22}$$

where $\alpha = \kappa \cdot p_k$ and $B(\alpha)$ is the multinomial Beta function.

Figure 4 shows a visualization of the distortion of the distribution $q = [0.2, 0.3, 0.5]$ for decreasing values of κ. For the noise–induced data, we generate similar groups of 100,000 learning datasets for $\kappa = 100$, 25 and 10.

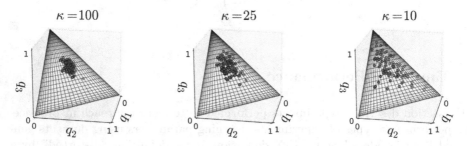

Fig. 4. Distortion of original distribution $q = [0.2, 0.3, 0.5]$ with a varying parameter κ visualized by sampling a 100 new variates from the Dirichlet distribution

4.2   Tested Learning Algorithms

We compare three approaches to learning the Leaky–MAX parameters: (1) Simple learning method by Onińsko and Drużdżel [1], (2) Leaky–MAX fitting to CPT by Zagórecki and Drużdżel [10], and (3) Weighted Least Squares method proposed in this paper. Small datasets may miss some of the information necessary for the estimation of the parameters: if m records describe a given combination of parent states, and the effect variable is activated in k of those records, we can safely assume that the approximate conditional probability that the variable will be activated is then equal to $\frac{k}{m}$. However, when none or all of the m records contain the child in an activated state, we end up with probabilities 0 and 1 respectively, which are best avoided when defining a probabilistic model [12]. Our method of handling such extreme cases (for both methods) is as follows:

− If none of the observed m records contains the child in an activated state, assume probability $\frac{1}{m+1}$.
− If all of the observed m records contain the child in an activated state, assume probability $\frac{m}{m+1}$.

- If a given scenario is nonexistent in the data, i.e., the joint probability of a given combination of parents states was very small, assume the uniform distribution for the child variable (Simple learning only).

In the remainder of this section, we will describe the three methods in detail:

Simple Learning

1. Perform a sweep through the dataset in search of records with only one parent being in its activated state and compute the conditional probabilities for the child node.
2. Since leak (denoted as L) is always taken into account, we are also interested in the conditional probability of the child being present, when all of the parents are in their corresponding distinguished states.
3. So far, the obtained values are the so–called *compound parameters* p_c [2], as they reflect the fact that implicit causes (leak node) might have acted upon the effect variable. In order to obtain each of the corresponding *net parameters* p_n [6], we compute the following:

$$p_n = 1 - \frac{1 - p_c}{1 - p_L}. \tag{23}$$

In case of a small dataset, it might occur that $p_c < p_L$, leading to p_n being a negative value, in which case we assume a uniform distribution.

Weighted Least Squares Method

1. Express the data as system of linear equations (Equations 6 through 8).
2. To reduce the potential error in the estimations, remove the equations which were represented by fewer than 10 records. While more relaxed and flexible rules can be considered here, they impact the general performance of the algorithm minimally, and only for the smallest datasets.
3. Assume the following weight for each of the equations:

$$w = (m \cdot 0.5^\delta)^2 , \tag{24}$$

where m is the total number of records describing a given equation, and δ is the number of parents that were in their activated state in a given probabilistic relationship. We want to promote the equations which were represented by a larger share of the sample, but at the same time we want to penalize complex equations with many activated parents. Higher prevalence of the latter results in more complex linear combination for the final parameter, increasing the error propagation. We have experimented with the degree of penalty that could be applied to each equation, and found that weight decreasing exponentially with the number of activated parents led to highest accuracy[5]. Proposed constants achieved the best results in preliminary experiments.

---

[5] Linear and polynomial penalty functions were also considered.

4. Apply the Weighted Least Squares method as described in the previous section. If any errors occur during learning (i.e., computed probability is negative or larger than 1), compute given parameter using the Simple learning method described above.

Leaky–MAX Fitting to CPT

1. Initialize the CPT and the prior probabilities with uniform distributions.
2. Perform the Expectation-maximization learning of the full CPT.
3. Perform fitting of the Leaky–MAX parameters to thus obtained CPT using the algorithm of Zagórecki and Drużdżel [10].[6]

4.3   Performance Assessment

This section describes our methodology for the performance evaluation of the tested algorithms (see Figure 5):

1. The first step differs for each of the experiments:
    (a) Leaky–MAX experiment: Parameters of the network (parents' prior probabilities and CPT) are randomized uniformly.
    (b) Noise–induced experiment: Parameters of the network are randomized uniformly and then subjected to a Dirichlet noising with parameter κ.
2. Thus obtained reference network is used for sampling of the datasets.
3. Tested algorithms learn the Leaky–MAX parameters from the data and expand them to full CPT.
4. Learned CPTs are compared against the original CPT using the Hellinger distance [13].

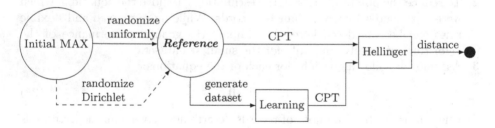

Fig. 5. Experiment flowchart. Dashed line is an noise–induced variant to the uniform randomization of the reference Leaky–MAX parameters.

We use two metrics based on the Hellinger distance to measure the average and maximal errors within the learned CPT:

$$H_{\mathrm{MAX}}(\mathbb{P}, \mathbb{Q}) = \max\{H(\mathbb{P}, \mathbb{Q}, j) \mid j \in J\}, \tag{25}$$

$$H_{\mathrm{AVG}}(\mathbb{P}, \mathbb{Q}) = \mathrm{avg}\{H(\mathbb{P}, \mathbb{Q}, j) \mid j \in J\}, \tag{26}$$

---

[6] Implementation of the Leaky–MAX fitting algorithm is available in GeNIe/SMILE©software packages at http://genie.sis.pitt.edu/.

where $H(\mathbb{P}, \mathbb{Q}, j)$ is the Hellinger distance between the j-th columns (out of J total) of the CPTs $\mathbb{P}$ and $\mathbb{Q}$:

$$H(\mathbb{P}, \mathbb{Q}, j) = \frac{1}{\sqrt{2}} \sqrt{\sum_{i=1}^{I} \left(\sqrt{\mathbb{P}^i_j} - \sqrt{\mathbb{Q}^i_j} \right)^2}, \qquad (27)$$

and I is the number of states of the child node.

4.4 Results

Figure 6 presents the results of the first experiment. We can observe almost unanimous tendency for the Least Squares–based learning to achieve better accuracy than the Simple learning method. However, for 2 and 3 parents Leaky–MAX fitting seems to achieve the best performance in both average and maximal error. The situation changes as we increase the number of parents to 5 and 6, in which case the accuracy of the Leaky–MAX fitting deteriorates drastically.

Figure 7 presents the results in more detail for 100, 1000 and 2000 records. Each of the boxplots represents 1000 experiment repetitions, with the center box describing the 50% of cases between the two quartiles (also known as the *interquartile range* – IQR). The whiskers extend further by a distances of 1.5·IQR, with the outliers marked in red. Above each of the boxplots we gather four statistical measures (rounded to 4 decimal points): mean (μ), median ($\tilde{x}$), standard deviation (σ) and a *win ratio* (ω), last of which was also shown on a separate plot on the right. Win ratio is simply the fraction of the 1000 experiments in which given method achieved the minimal distance among other algorithms (including ties). The best value for each of the statistics among the tested algorithms is signified with the boldface font.

Leaky–MAX fitting achieves the best overall performance for 2 parents by almost all statistical measures across the experiment. Similar scenario (although not as apparent) occurs for 3 parents, with Least Squares method performing akin to Leaky–MAX fitting in mean, median and win ratio. However, when we

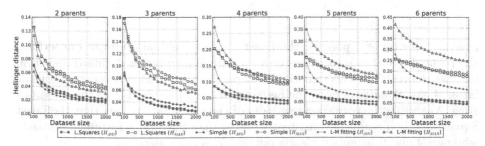

Fig. 6. Simple learning, Least Squares method and Leaky–MAX fitting approach. Average over 1000 repetitions.

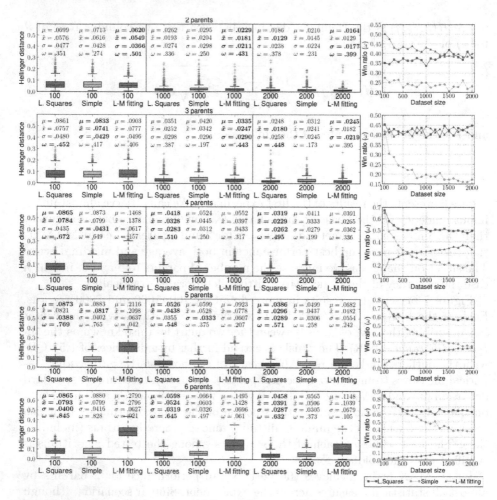

Fig. 7. Statistical performance of Least Squares (*L.Squares*), Simple learning (*Simple*) and Leaky–MAX fitting (*L-M fitting*) methods. Mean (μ), median ($\tilde{x}$), standard deviation (σ) and win ratio (ω) over 1000 experiment repetitions (H_{AVG} metric).

look at the case of 4 parents, the Least Squares method starts to outperform the remaining methods and loses only to Simple learning in standard deviation for 100 records. The effect continues to persist for 5 and 6 parents — for the latter, Least Squares obtains the best statistical indicators for all three dataset sizes. Analyzing the corresponding plots of the win ratio suggests that the Least Squares method is almost consistently better than the Simple learning approach, while for 4 and 6 parents it clearly outperforms both methods.

We show the results of the second experiment for 3, 4, and 5 parents in Figure 8. Since all methods assume certain unmet properties about the data, it is expected that the learning quality will deteriorate with lower values of κ. For

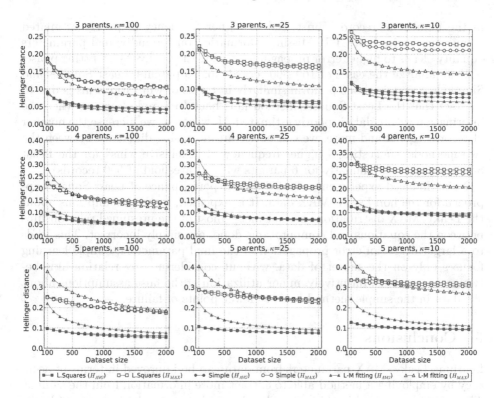

Fig. 8. Average accuracy of the Least Squres, Simple learning and Leaky–MAX fitting methods for various degrees of noise introduced to the learning data

relatively small deviation from the ideal Leaky–MAX ($\kappa = 100$), we observe that the Least Squares method achieves the accuracy as good as the Simple learning method, and not clearly different from the previous experiment. As we divert away from the perfect scenario ($\kappa = 25$ and 10) a degeneration in accuracy starts to occur, especially for the Least Squares method when assessed by the H_{MAX} metric. The Leaky–MAX fitting method seems to be more resistant to distortion, especially when one considers maximal error – the degeneration is not as severe as it is for the remaining methods.

4.5 Discussion

The degrading performance of the Leaky–MAX fitting for 4 and more parents is not surprising, as the main drawback of the method is the lack of information on the accuracy of the estimated parameters in the CPT after the Expectation-maximization algorithm. When the number of parents is small (i.e., 2 or 3 nodes, giving 4 and 8 parameters respectively), individual parameters in the CPT are estimated based on a larger share of the sample. However, exponential increase in the number of CPT parameters, without a corresponding exponential increase in the number of records, must lead to higher frequency of poorly estimated or missing entries in a CPT. Missing information on the individual parameters "quality" leads to higher error propagation during the Leaky–MAX fitting phase.

If we analyze the cases of 100–300 records and 5 parents, Least Squares and Simple learning methods perform similarly in terms of win ratio. This is due to the fact that the Least Squares method falls back to Simple learning when the information within data is poorly estimated. However, as we increase the number of records to 400 and more, we can notice an increasing difference between the two methods, suggesting a higher frequency of "exclusive wins" for the Least Squares approach (similarly for 6 parents with a threshold of 500 records).

Noise–induced experiment revealed a poor performance of Least Squares approach for the non-Leaky–MAX data. It is expected, as the assumptions about the Leaky–MAX distributions are not met – most of the presumed relationships between the equations and computations done, are simply incorrect. This leads to a larger degree of error propagation in the final parameters. Simple learning focuses only on the subset of data, without considering any relationships between observations. Intuitively, minor utilization of the false assumptions about the data in the simple method leads to fewer points of failure.

5   Conclusions

Learning Leaky–MAX parameters from data can be improved in terms of accuracy by employing a method able to extract more information from the sample. The method proposed in this research achieves better results in terms of learning precision than the simple approach and the Leaky–MAX fitting approach, assuming that the data fits the Leaky–MAX definition. By employing the assumptions of canonical models, we show that it is possible to render a connection between the probabilistic relationships and linear algebra. This opens up room for future research, as we believe that the approach thus presented resonates equally well with other canonical models. Besides a possible wider application of our solution, the core method itself can be improved as the problem of quasi linear regression presented in this research was given much attention in the field of regression analysis.

Acknowledgments. We acknowledge the support the National Institute of Health under grant number U01HL101066-01. Implementation of this work is based on SMILE[☺], a Bayesian inference engine developed at the Decision Systems Laboratory and available at http://genie.sis.pitt.edu/.

References

1. Oniśko, A., Druzdzel, M.J., Wasyluk, H.: Learning Bayesian network parameters from small data sets: Application of Noisy-OR gates. In: Working Notes on the European Conference on Artificial Intelligence (ECAI) Workshop Bayesian and Causal Networks: From Inference to Data Mining, August 22 (2000)

2. Henrion, M.: Some practical issues in constructing belief networks. In: Artificial Intelligence 3 Annual Conference on Uncertainty in Artificial Intelligence (UAI 1987), pp. 161–173. Elsevier Science, Amsterdam (1987)
3. Pearl, J.: Probabilistic Reasoning in Intelligent Systems: Networks of Plausible Inference. Morgan Kaufmann Publishers Inc., San Francisco (1988)
4. Heckerman, D.: A tutorial on learning with Bayesian networks. In: Jordan, M.I. (ed.) Learning in Graphical Models. The MIT Press, Cambridge (1998), Available on web at ftp://ftp.research.microsoft.com/pub/tr/TR-95-06.ps
5. Díez, F.J., Druzdzel, M.J.: Canonical probabilistic models for knowledge engineering. Technical report, UNED, Madrid, Spain (2006)
6. Diez, F.: Parameter adjustment in Bayes networks. the generalized noisy OR-gate. In: Proceedings of the Ninth Conference Annual Conference on Uncertainty in Artificial Intelligence (UAI 1993), pp. 99–105. Morgan Kaufmann, San Francisco (1993)
7. Srinivas, S.: A generalization of the noisy-OR model. In: Proceedings of the Ninth Conference Annual Conference on Uncertainty in Artificial Intelligence (UAI 1993), pp. 208–215. Morgan Kaufmann, San Francisco (1993)
8. Zagorecki, A., Voortman, M., Druzdzel, M.J.: Decomposing local probability distributions in Bayesian networks for improved inference and parameter learning. In: Sutcliffe, G., Goebel, R. (eds.) Recent Advances in Artificial Intelligence: Proceedings of the Nineteenth International Florida Artificial Intelligence Research Society Conference (FLAIRS 2006), pp. 860–865. AAAI Press, Menlo Park (2006)
9. Friedman, N., Goldszmidt, M.: Learning Bayesian networks with local structure. In: Jordan, M.I. (ed.) Learning and Inference in Graphical Models, pp. 421–459. The MIT Press, Cambridge (1999)
10. Zagorecki, A., Druzdzel, M.J.: Knowledge engineering for Bayesian networks: How common are noisy-MAX distributions in practice? IEEE Transactions on Systems, Man, and Cybernetics: Systems 43(1), 186–195 (2013)
11. Wasserman, L.: All of nonparametric statistics. Springer (2006)
12. Oniśko, A., Druzdzel, M.J.: Impact of precision of Bayesian network parameters on accuracy of medical diagnostic systems. Artificial Intelligence in Medicine 57(3), 197–206 (2013)
13. Gibbs, A.L., Su, F.E.: On choosing and bounding probability metrics. International Statistical Review 70(3), 419–435 (2002)

Learning Marginal AMP Chain Graphs under Faithfulness

Jose M. Peña

ADIT, IDA, Linköping University, SE-58183 Linköping, Sweden
jose.m.pena@liu.se

Abstract. Marginal AMP chain graphs are a recently introduced family of models that is based on graphs that may have undirected, directed and bidirected edges. They unify and generalize the AMP and the multivariate regression interpretations of chain graphs. In this paper, we present a constraint based algorithm for learning a marginal AMP chain graph from a probability distribution which is faithful to it. We also show that the extension of Meek's conjecture to marginal AMP chain graphs does not hold, which compromises the development of efficient and correct score+search learning algorithms under assumptions weaker than faithfulness.

Keywords: Chain graphs, marginal AMP chain graphs, learning graphical models.

1 Introduction

Chain graphs (CGs) are graphs with possibly directed and undirected edges, and no semidirected cycle. They have been extensively studied as a formalism to represent independence models, because they can model symmetric and asymmetric relationships between the random variables of interest. However, there are three different interpretations of CGs as independence models: The Lauritzen-Wermuth-Frydenberg (LWF) interpretation [6], the multivariate regression (MVR) interpretation [4], and the Andersson-Madigan-Perlman (AMP) interpretation [1]. It is worth mentioning that no interpretation subsumes another: There are many independence models that can be represented by a CG under one interpretation but that cannot be represented by any CG under the other interpretations [1,17]. Moreover, although MVR CGs were originally represented using dashed directed and undirected edges, we like other authors prefer to represent them using solid directed and bidirected edges.

Recently, a new family of models has been proposed to unify and generalize the AMP and MVR interpretations of CGs [11]. This new family, named marginal AMP (MAMP) CGs, is based on graphs that may have undirected, directed and bidirected edges. In this paper, we extend [11] by presenting an algorithm for learning an MAMP CG from a probability distribution which is faithful to it. Our algorithm is constraint based and builds upon those developed in [16] and [10] for learning, respectively, MVR and AMP CGs under the

L.C. van der Gaag and A.J. Feelders (Eds.): PGM 2014, LNAI 8754, pp. 382–395, 2014.

faithfulness assumption. Finally, note that there also exist algorithms for learning LWF CGs under the faithfulness assumption [7,19] and under the milder composition property assumption [13]. In this paper, we also show that the extension of Meek's conjecture to MAMP CGs does not hold, which compromises the development of efficient and correct score+search learning algorithms under assumptions weaker than faithfulness.

The rest of this paper is organized as follows. We start with some preliminaries in Section 2. Then, we introduce MAMP CGs in Section 3, followed by the algorithm for learning them in Section 4. We close the paper with some discussion in Section 5.

2   Preliminaries

In this section, we introduce some concepts of models based on graphs, i.e. graphical models. Most of these concepts have a unique definition in the literature. However, a few concepts have more than one and we opt for the most suitable in this work. All the graphs and probability distributions in this paper are defined over a finite set V. All the graphs in this paper are simple, i.e. they contain at most one edge between any pair of nodes. The elements of V are not distinguished from singletons.

If a graph G contains an undirected, directed or bidirected edge between two nodes V_1 and V_2, then we write that $V_1 - V_2$, $V_1 \to V_2$ or $V_1 \leftrightarrow V_2$ is in G. We represent with a circle, such as in $\circ\!\!\to$ or $\circ\!\!-\!\!\circ$, that the end of an edge is unspecified, i.e. it may be an arrow tip or nothing. The parents of a set of nodes X of G is the set $pa_G(X) = \{V_1 | V_1 \to V_2$ is in G, $V_1 \notin X$ and $V_2 \in X\}$. The children of X is the set $ch_G(X) = \{V_1 | V_1 \leftarrow V_2$ is in G, $V_1 \notin X$ and $V_2 \in X\}$. The neighbors of X is the set $ne_G(X) = \{V_1 | V_1 - V_2$ is in G, $V_1 \notin X$ and $V_2 \in X\}$. The spouses of X is the set $sp_G(X) = \{V_1 | V_1 \leftrightarrow V_2$ is in G, $V_1 \notin X$ and $V_2 \in X\}$. The adjacents of X is the set $ad_G(X) = ne_G(X) \cup pa_G(X) \cup ch_G(X) \cup sp_G(X)$. A route between a node V_1 and a node V_n in G is a sequence of (not necessarily distinct) nodes $V_1, \ldots, V_n$ such that $V_i \in ad_G(V_{i+1})$ for all $1 \le i < n$. If the nodes in the route are all distinct, then the route is called a path. The length of a route is the number of (not necessarily distinct) edges in the route, e.g. the length of the route $V_1, \ldots, V_n$ is $n - 1$. A route is called descending if $V_i \to V_{i+1}$ or $V_i - V_{i+1}$ is in G for all $1 \le i < n$. A route is called strictly descending if $V_i \to V_{i+1}$ is in G for all $1 \le i < n$. The descendants of a set of nodes X of G is the set $de_G(X) = \{V_n |$ there is a descending route from V_1 to V_n in G, $V_1 \in X$ and $V_n \notin X\}$. The strict ascendants of X is the set $san_G(X) = \{V_1 |$ there is a strictly descending route from V_1 to V_n in G, $V_1 \notin X$ and $V_n \in X\}$. A route $V_1, \ldots, V_n$ in G is called a cycle if $V_n = V_1$. Moreover, it is called a semidirected cycle if $V_n = V_1$, $V_1 \to V_2$ is in G and $V_i \to V_{i+1}$, $V_i \leftrightarrow V_{i+1}$ or $V_i - V_{i+1}$ is in G for all $1 < i < n$. A cycle has a chord if two non-consecutive nodes of the cycle are adjacent in G. An AMP chain graph (AMP CG) is a graph whose every edge is directed or undirected such that it has no semidirected cycles. A MVR chain graph (MVR CG) is a graph whose every edge is directed or bidirected such that it has no semidirected cycles. The subgraph of G induced by a set of its nodes X is the graph over X that has all and only the edges in G whose both ends are in X.

We now recall the semantics of AMP and MVR CGs. A node B in a path ρ in an AMP CG G is called a triplex node in ρ if $A \rightarrow B \leftarrow C$, $A \rightarrow B - C$, or $A - B \leftarrow C$ is a subpath of ρ. Moreover, ρ is said to be Z-open with $Z \subseteq V$ when

- every triplex node in ρ is in $Z \cup san_G(Z)$, and
- every non-triplex node B in ρ is outside Z, unless $A - B - C$ is a subpath of ρ and $pa_G(B) \setminus Z \neq \varnothing$.

A node B in a path ρ in an MVR CG G is called a triplex node in ρ if $A \rightarrowtail B \leftarrowtail C$ is a subpath of ρ. Moreover, ρ is said to be Z-open with $Z \subseteq V$ when

- every triplex node in ρ is in $Z \cup san_G(Z)$, and
- every non-triplex node B in ρ is outside Z.

Let X, Y and Z denote three disjoint subsets of V. When there is no Z-open path in an AMP or MVR CG G between a node in X and a node in Y, we say that X is separated from Y given Z in G and denote it as $X \perp_G Y | Z$. The independence model represented by G, denoted as $I(G)$, is the set of separations $X \perp_G Y | Z$. In general, $I(G)$ is different whether G is an AMP or MVR CG.

3   MAMP CGs

In this section, we review marginal AMP (MAMP) CGs. We refer the reader to [11] for more details. Specifically, a graph G containing possibly directed, bidirected and undirected edges is an MAMP CG if

C1. G has no semidirected cycle,
C2. G has no cycle $V_1, \ldots, V_n = V_1$ such that $V_1 \leftrightarrow V_2$ is in G and $V_i - V_{i+1}$ is in G for all $1 < i < n$, and
C3. if $V_1 - V_2 - V_3$ is in G and $sp_G(V_2) \neq \varnothing$, then $V_1 - V_3$ is in G too.

The semantics of MAMP CGs is as follows. A node B in a path ρ in an MAMP CG G is called a triplex node in ρ if $A \rightarrowtail B \leftarrowtail C$, $A \rightarrowtail B - C$, or $A - B \leftarrowtail C$ is a subpath of ρ. Moreover, ρ is said to be Z-open with $Z \subseteq V$ when

- every triplex node in ρ is in $Z \cup san_G(Z)$, and
- every non-triplex node B in ρ is outside Z, unless $A - B - C$ is a subpath of ρ and $sp_G(B) \neq \varnothing$ or $pa_G(B) \setminus Z \neq \varnothing$.

Let X, Y and Z denote three disjoint subsets of V. When there is no Z-open path in G between a node in X and a node in Y, we say that X is separated from Y given Z in G and denote it as $X \perp_G Y | Z$. The independence model represented by G, denoted as $I(G)$, is the set of separations $X \perp_G Y | Z$. We denote by $X \perp_p Y | Z$ (respectively $X \not\perp_p Y | Z$) that X is independent (respectively

dependent) of Y given Z in a probability distribution p. We say that p is faithful
to G when $X \perp_p Y|Z$ iff $X \perp_G Y|Z$ for all X, Y and Z disjoint subsets of V.
We say that two MAMP CGs are Markov equivalent if they represent the same
independence model. In an MAMP CG, a triplex $(\{A,C\},B)$ is an induced
subgraph of the form $A \multimap B \leftarrow\!\circ\, C$, $A \multimap B - C$, or $A - B \leftarrow\!\circ\, C$. We say that
two MAMP CGs are triplex equivalent if they have the same adjacencies and
the same triplexes. Two MAMP CGs are Markov equivalent iff they are triplex
equivalent [11, Theorem 7].

Clearly, the union of AMP and MVR CGs is a subfamily of MAMP CGs. The
following example shows that it is a proper subfamily.

Example 1. The independence model represented by the MAMP CG G below
cannot be represented by any AMP or MVR CG.

$$
\begin{array}{ccc}
A \longrightarrow B & \!\!-\!\! & C \\
\downarrow & & \uparrow \\
D & \longleftrightarrow & E
\end{array}
$$

To see it, assume to the contrary that it can be represented by an AMP
CG H. Note that H is an MAMP CG too. Then, G and H must have the
same triplexes. Then, H must have triplexes $(\{A,D\},B)$ and $(\{A,C\},B)$ but
no triplex $(\{C,D\},B)$. So, $C - B - D$ must be in H. Moreover, H must have
a triplex $(\{B,E\},C)$. So, $C \leftarrow E$ must be in H. However, this implies that H
does not have a triplex $(\{C,D\},E)$, which is a contradiction because G has
such a triplex. To see that no MVR CG can represent the independence model
represented by G, simply note that no MVR CG can have triplexes $(\{A,D\},B)$
and $(\{A,C\},B)$ but no triplex $(\{C,D\},B)$.

Finally, other families of models that are based on graphs that may contain
undirected, directed and bidirected edges are summary graphs after replacing
the dashed undirected edges with bidirected edges [4], MC graphs [5], maximal
ancestral graphs [14], and loopless mixed graphs [15]. However, the separation
criteria for these families are identical to that of MVR CGs. Then, MVR CGs
are a subfamily of these families but AMP CGs are not. See also [14, p. 1025]
and [15, Sections 4.1-4.3]. Therefore, MAMP CGs are the only graphical models
in the literature that generalize both AMP and MVR CGs.

4 Algorithm for Learning MAMP CGs

In this section, we present our algorithm for learning an MAMP CG from a
probability distribution which is faithful to it. The algorithm builds upon those
developed in [16] and [10] for learning, respectively, MVR and AMP CGs under the faithfulness assumption. The algorithm, which can be seen in Table 1,
resembles the well-known PC algorithm developed in [18] for learning Bayesian
networks under the faithfulness assumption, in the sense that it consists of two

Table 1. Algorithm for learning MAMP CGs

Input: A probability distribution p that is faithful to an unknown MAMP CG G.
Output: An MAMP CG H that is triplex equivalent to G.

1 Let H denote the complete undirected graph
2 Set $l = 0$
3 Repeat while $l \leq |V| - 2$
4   For each ordered pair of nodes A and B in H st $A \in ad_H(B)$ and
     $|[ad_H(A) \cup ad_H(ad_H(A))] \smallsetminus \{A, B\}| \geq l$
5   If there is some $S \subseteq [ad_H(A) \cup ad_H(ad_H(A))] \smallsetminus \{A, B\}$ st $|S| = l$ and $A \perp_p B|S$
       then
6   Set $S_{AB} = S_{BA} = S$
7   Remove the edge $A - B$ from H
8   Set $l = l + 1$
9 Apply the rules R1-R4 to H while possible
10 Replace every edge $A \vdash B$ in H with $A \to B$
11 Replace every edge $A - B$ or $A \vdash\!\!\dashv B$ in H with $A \leftrightarrow B$
12 Replace every induced subgraph $A \leftrightarrow B \leftrightarrow C$ in H st $B \in S_{AC}$ with $A - B - C$
13 If H has an induced subgraph $A \overset{\frown}{\longleftrightarrow} B \longrightarrow C$ then
14   Replace the edge $A \leftrightarrow B$ in H with $A - B$
15   Go to line 13
16 Return H

Table 2. Rules R1-R4 in the algorithm for learning MAMP CGs

R1: $A \circ\!\!\!-\!\!\!-\!\!\!\circ B \circ\!\!\!-\!\!\!-\!\!\!\circ C \Rightarrow A \vdash\!\!\!-\!\!\!-\!\!\!\circ B \circ\!\!\!-\!\!\!-\!\!\dashv C$
     $\wedge\ B \notin S_{AC}$

R2: $A \vdash\!\!\!-\!\!\!-\!\!\!\circ B \circ\!\!\!-\!\!\!-\!\!\!\circ C \Rightarrow A \vdash\!\!\!-\!\!\!-\!\!\!\circ B \vdash\!\!\!-\!\!\!-\!\!\!\circ C$
     $\wedge\ B \in S_{AC}$

R3: $A \overset{\frown}{\vdash\!\!-\!\!\circ \cdots \vdash\!\!-\!\!\dashv} B \Rightarrow A \overset{\frown}{\vdash\!\!-\!\!\circ \cdots \vdash\!\!-\!\!\dashv} B$

R4:
     $\wedge\ A \in S_{CD}$

---

phases: The first phase (lines 1-8) aims at learning adjacencies, whereas the second phase (lines 9-15) aims at directing some of the adjacencies learnt. Specifically, the first phase declares that two nodes are adjacent if and only if they are not separated by any set of nodes. Note that the algorithm does not test every possible separator (see line 5). Note also that the separators tested are tested in increasing order of size (see lines 2, 5 and 8). The second phase consists of two steps. In the first step (line 9), the ends of some of the edges learnt in the first phase are blocked according to the rules R1-R4 in Table 2. A block is represented by a perpendicular line at the edge end such as in $\vdash$ or $\vdash\!\dashv$, and it means that the edge cannot be a directed edge pointing in the direction of the block. Note that $\vdash\!\dashv$ does not mean that the edge must be undirected: It means that the edge cannot be a directed edge in either direction and, thus, it must be a bidirected or undirected edge. In the second step (lines 10-15), some edges get directed. Specifically, the edges with exactly one unblocked end get directed in the direction of the unblocked end. The rest of the edges get bidirected (see line 11), unless this produces a false triplex (see line 12) or violates the constraint C2 (see lines 13-15). Note that only cycles of length three are checked for the violation of the constraint C2.

The rules R1-R4 in Table 2 work as follows: If the conditions in the antecedent of a rule are satisfied, then the modifications in the consequent of the rule are applied. Note that the ends of some of the edges in the rules are labeled with a circle such as in $\vdash\!\!\circ$ or $\circ\!\!-\!\!\circ$. The circle represents an unspecified end, i.e. a block or nothing. The modifications in the consequents of the rules consist in adding some blocks. Note that only the blocks that appear in the consequents are added, i.e. the circled ends do not get modified. The conditions in the antecedents of R1, R2 and R4 consist of an induced subgraph of H and the fact that some of its nodes are or are not in some separators found in line 6. The condition in the antecedent of R3 consists of just an induced subgraph of H. Specifically, the antecedent says that there is a cycle in H whose edges have certain blocks. Note that the cycle must be chordless.

The rest of this section is devoted to prove that our algorithm is correct, i.e. it returns an MAMP CG the given probability distribution is faithful to. We start by proving some auxiliary results.

Lemma 1. *After having executed line 8, G and H have the same adjacencies.*

Proof. Consider any pair of nodes A and B in G. If $A \in ad_G(B)$, then $A \not\perp_p B | S$ for all $S \subseteq V \smallsetminus \{A, B\}$ by the faithfulness assumption. Consequently, $A \in ad_H(B)$ at all times. On the other hand, if $A \notin ad_G(B)$, then consider the following cases.

Case 1. Assume that $A \notin de_G(B)$ or $B \notin de_G(A)$. Assume without loss of generality that $B \notin de_G(A)$. Then, $A \perp_p B | pa_G(A)$ [11, Theorem 5]. Note that, as shown above, $pa_G(A) \subseteq ad_H(A) \smallsetminus B$ at all times.

Case 2. Assume that $A \in de_G(B)$ and $B \in de_G(A)$. Then, $A \perp_p B | ne_G(A) \cup pa_G(A \cup ne_G(A))$ [11, Theorem 5]. Note that, as shown above, $ne_G(A) \cup pa_G(A \cup ne_G(A)) \subseteq [ad_H(A) \cup ad_H(ad_H(A))] \smallsetminus \{A, B\}$ at all times.

Therefore, in either case, there will exist some S in line 5 such that $A \perp_p B | S$ and, thus, the edge $A - B$ will be removed from H in line 7. Consequently, $A \notin ad_H(B)$ after line 8.

Lemma 2. *The rules R1-R4 block the end of an edge only if the edge is not a directed edge in G pointing in the direction of the block.*

Proof. According to the antecedent of R1, G has a triplex $(\{A, C\}, B)$. Then, G has an induced subgraph of the form $A \dashrightarrow B \leftrightarrow C$, $A \dashrightarrow B - C$ or $A - B \leftrightarrow C$. In either case, the consequent of R1 holds.

According to the antecedent of R2, (i) G does not have a triplex $(\{A, C\}, B)$, (ii) $A \dashrightarrow B$ or $A - B$ is in G, (iii) $B \in ad_G(C)$, and (iv) $A \notin ad_G(C)$. Then, $B \to C$ or $B - C$ is in G. In either case, the consequent of R2 holds.

According to the antecedent of R3, (i) G has a path from A to B with no directed edge pointing in the direction of A, and (ii) $A \in ad_G(B)$. Then, $A \leftarrow B$ cannot be in G because G has no semidirected cycle. Then, the consequent of R3 holds.

According to the antecedent of R4, neither $B \to C$ nor $B \to D$ are in G. Assume to the contrary that $A \leftarrow B$ is in G. Then, G must have an induced subgraph that is consistent with

because, otherwise, G has a semidirected cycle. However, this contradicts that $A \in S_{CD}$.

Lemma 3. *At line 16, all the undirected edges in H are in G.*

Proof. Note that lines 10-11 imply that H has no undirected edge when line 12 is to be executed. Note also that any undirected edges $A - B$ and $B - C$ added to H in line 12 must exist in G, because this implies that H has an induced subgraph $A \longmapsto B \longmapsto C$ with $B \in S_{AC}$ when line 11 is to be executed, which implies that (i) A and B as well as B and C are adjacent in G whereas A and C are not adjacent in G by Lemma 1, and (ii) G has no directed edge between A and B or B and C. Then, $A - B - C$ must be in G by Lemma 2 and the fact that $B \in S_{AC}$.

The paragraph above implies that all the undirected edges in H are in G when lines 13-15 are to be executed for the first time, which implies that the undirected edge added to H in the first execution of lines 13-15 must also be in G due to the constraints C1 and C2. By repeatedly applying this argument, all the undirected edges in H at line 16 must be in G.

Lemma 4. *At line 16, G and H have the same triplexes.*

Proof. We first prove that any triplex in H at line 16 is in G. Assume to the contrary that H at line 16 has a triplex $(\{A,C\},B)$ that is not in G. This is possible if and only if H has an induced subgraph of one of the following forms when lines 10-11 are to be executed:

$$A \longrightarrow B \longrightarrow C \quad A \longrightarrow B \circ\!\!-\!\!\!\mid C \quad A \mid\!\!-\!\!\!\leftarrow B \longrightarrow C$$
$$A \mid\!\!-\!\!\!\leftarrow B \circ\!\!-\!\!\!\mid C \quad A \mid\!\!-\!\!\!\mid B \longrightarrow\!\!\circ C \quad A \mid\!\!-\!\!\!\mid B \mid\!\!-\!\!\!\mid C$$

Note that the induced subgraphs above together with Lemma 1 imply that A is adjacent to B in G, B is adjacent to C in G, and A is not adjacent to C in G. This together with the assumption made above that G has no triplex $(\{A,C\},B)$ implies that $B \in S_{AC}$. Now, note that the second and fourth induced subgraphs above are impossible because, otherwise, $A \circ\!\!-\!\!\!\mid B$ would be in H by R2. Likewise, the third and fifth induced subgraphs above are impossible because, otherwise, $B \mid\!\!-\!\!\!\circ C$ would be in H by R2. Now, note that any triplex that is added to H in line 11 due to the first and sixth induced subgraphs above is removed from H in line 12 because, as shown above, $B \in S_{AC}$. Finally, note that no triplex is added to H in lines 13-15.

We now prove that any triplex $(\{A,C\},B)$ in G is in H at line 16. Note that $B \notin S_{AC}$. Consider the following cases.

Case 1. Assume that the triplex in G is of the form $A \to B \circ\!\!-\!\!\!\circ C$ (respectively $A \circ\!\!-\!\!\!\circ B \leftarrow C$). Then, when lines 10-11 are to be executed, $A \vdash B \circ\!\!-\!\!\!\mid C$ (respectively $A \mid\!\!-\!\!\!\circ B \dashv C$) is in H by R1 and Lemmas 1 and 2. Then, the triplex is added to H in lines 10-11. Moreover, the triplex added is of the form $A \to B \circ\!\!-\!\!\!\circ C$ (respectively $A \circ\!\!-\!\!\!\circ B \leftarrow C$) and, thus, it does not get removed from H in lines 12-15.

Case 2. Assume that the triplex in G is of the form $A \leftrightarrow B \circ\!\!-\!\!\!\circ C$ or $A \circ\!\!-\!\!\!\circ B \leftrightarrow C$. Then, when lines 10-11 are to be executed, $A \mid\!\!-\!\!\!\circ B \circ\!\!-\!\!\!\mid C$ is in H by R1 and Lemmas 1 and 2. Then, the triplex is added to H in lines 10-11. Moreover, the triplex cannot get removed from H in lines 12-15. To see it, assume the contrary. Note that all lines 12-15 do is replacing bidirected edges in H with undirected edges. Thus, the triplex cannot get removed from H unless it is of the form $A \leftrightarrow B \leftrightarrow C$, $A \leftrightarrow B - C$, or $A - B \leftrightarrow C$. Consider the following cases.

Case 2.1. Assume that the triplex gets removed from H in line 12. Assume that the triplex is of the form $A \leftrightarrow B \leftrightarrow C$. The proofs for the forms $A \leftrightarrow B - C$ and $A - B \leftrightarrow C$ are similar. Note that the triplex cannot get removed from H by applying line 12 to $A \leftrightarrow B \leftrightarrow C$ because, as shown above, $B \notin S_{AC}$. Then, for the triplex to get removed from H in line 12, H must have two induced subgraphs $A' \leftrightarrow A \leftrightarrow B$ and $B \leftrightarrow C \leftrightarrow C'$ with $A \in S_{A'B}$ and $C \in S_{BC'}$ when line 12 is to be executed. This implies that $A \mid\!\!-\!\!\!\mid B \mid\!\!-\!\!\!\mid C$ is in H when lines 10-11 are to be executed because, as shown above, $A \mid\!\!-\!\!\!\circ B \circ\!\!-\!\!\!\mid C$ is in H when lines 10-11 are to be executed. Therefore, H has two induced subgraphs $A' \mid\!\!-\!\!\!\mid A \mid\!\!-\!\!\!\mid B$ and $B \mid\!\!-\!\!\!\mid C \mid\!\!-\!\!\!\mid C'$ by R2 when lines 10-11 are to be executed. Then, $A' - A$ or $A' \leftrightarrow A$ must be in G by Lemmas 1 and 2. Then, $A - B$ or $A \to B$ must be in G by Lemmas 1 and 2 and the fact that $A \in S_{A'B}$. However, $A \to B$ cannot be in G because $A \mapsto B$ is in H when line 11 is to be executed. Then, $A - B$ must be in G. Likewise, $B - C$ must be in G. However, this contradicts the assumption that G has a triplex $(\{A, C\}, B)$.

Case 2.2. Assume that the triplex gets removed from H in lines 13-15. Recall that all the undirected edges in H at line 16 are in G by Lemma 3. Therefore, any triplex that gets removed from H in lines 13-15 cannot exist in G.

The proofs of the following two lemmas can be found in [10, Lemmas 5 and 6]. The fact that G is an AMP CG in that work whereas it is an MAMP CG in this work is irrelevant for the proofs. What matters is that both works use the same rules R1-R4.

Lemma 5. *After having executed line 9, H does not have any induced subgraph of the form* $A \overset{\frown}{\mathrel{\mkern-3mu\vdash}} \!\!\circ\, B \,\text{---}\, C$.

Lemma 6. *After having executed line 9, every chordless cycle $\rho : V_1, \ldots, V_n = V_1$ in H that has an edge $V_i \vdash V_{i+1}$ also has an edge $V_j \dashv V_{j+1}$.*

Lemma 5 is used in the proof of Lemma 6. It is worth noting that one may think that Lemma 5 implies that H does not have any induced subgraph of the form $A \overset{\frown}{\longleftrightarrow} B \,\text{---}\, C$ after having executed line 12 and, thus, that lines 13-15 are not needed. However, this is wrong as the following example illustrates.

Example 2. The MAMP CG G below shows that lines 13-15 are necessary.

$$G \qquad\qquad H \text{ after line 9}$$

$$H \text{ after line 12}$$

We can now prove the correctness of our algorithm.

Theorem 1. *At line 16, H is an MAMP CG that is triplex equivalent to G.*

Proof. Lemma 1 implies that H at line 16 has the same adjacencies as G. Lemma 4 implies that H at line 16 has the same triplexes as G. Lemma 6 implies that H has no semidirected chordless cycle after having executed line 11. This implies that H has no semidirected chordless cycle at line 16, because all lines 12-15 do is replacing bidirected edges in H with undirected edges. To see that this in turn implies that H has no semidirected cycle at line 16, assume to the contrary that H has no semidirected chordless cycle but it has a semidirected cycle $\rho : V_1, \ldots, V_n = V_1$ with a chord between V_i and V_j with $i < j$. Then, divide ρ into the cycles $\rho_L : V_1, \ldots, V_i, V_j, \ldots, V_n = V_1$ and $\rho_R : V_i, \ldots, V_j, V_i$. Note that ρ_L or ρ_R is a semidirected cycle. Then, H has a semidirected cycle that is shorter than ρ. By repeated application of this reasoning, we can conclude that H has a semidirected chordless cycle, which is a contradiction. Therefore, H at line 16 satisfies the constraint C1.

We now show that H at line 16 satisfies the constraint C2. Assume to the contrary that H has a cycle $\rho : V_1, \ldots, V_n = V_1$ such that $V_1 \leftrightarrow V_2$ is in H and $V_i - V_{i+1}$ is in H for all $1 < i < n$. Note that ρ must be of length greater than three by lines 13-15, i.e. $n > 3$. Note also that $V_i - V_{i+1}$ must be in G for all $1 < i < n$ by Lemma 3, which implies that $V_1 - V_2$ is also in G by the constraints C1 and C2. This implies that V_1 and V_3 are adjacent in G because, otherwise, G and H have not the same triplexes, which contradicts Lemma 4. Then, V_1 and V_3 are adjacent in H by Lemma 1. In fact, $V_1 \leftrightarrow V_3$ must be in H because, otherwise, H has a cycle of length three that violates the constraint C1 or C2 which, as shown above, is a contradiction. Then, H has a cycle that violates the constraint C2 and that is shorter than ρ, namely $V_1, V_3, \ldots, V_n = V_1$. By repeated application of this reasoning, we can conclude that H has a cycle of length three that violates the constraint C2 which, as shown above, is a contradiction.

We finally show that H at line 16 satisfies the constraint C3. Assume to the contrary that H at line 16 has a subgraph of the form $V_1 - V_2 - V_3$, and $V_2 \leftrightarrow V_4$ is in H but $V_1 - V_3$ is not in H. We show below that G (respectively H at line 16) has the graph to the left (respectively right) below as an induced subgraph.

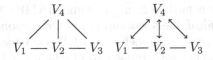

That $V_1 - V_2 - V_3$ is in H at line 16 but $V_1 - V_3$ is not implies that V_1 and V_3 cannot be adjacent in H because, otherwise, H violates the constraint C1 or C2 which, as shown above, is a contradiction. This implies that V_1 and V_3 are not adjacent in G either by Lemma 1. That $V_1 - V_2 - V_3$ is in H at line 16 implies that $V_1 - V_2 - V_3$ is also in G by Lemma 3. That $V_2 \leftrightarrow V_4$ is in H at line 16 implies that $V_2 \mapsto V_4$ is in H after having executed line 9, which implies that $V_2 - V_4$ or $V_2 \leftrightarrow V_4$ is in G by Lemmas 1 and 2. In fact, $V_2 - V_4$ must be in G because, otherwise, G violates the constraint C3 since, as shown above, $V_1 - V_2 - V_3$ is in G but $V_1 - V_3$ is not. Finally, note that V_1 and V_4 as well as V_3 and V_4 must be adjacent in G and H because, otherwise, H at line 16 does not have the same triplexes as G, which contradicts Lemma 4. Specifically, $V_1 - V_4 - V_3$ must be in G and $V_1 \leftrightarrow V_4 \leftrightarrow V_3$ must be in H at line 16 because, otherwise, G or H violates the constraint C1 or C2 which, as shown above, is a contradiction.

However, that G (respectively H at line 16) has the graph to the left (respectively right) above as an induced subgraph implies that H has a triplex ($\{V_1, V_3\}, V_4$) that G has not, which contradicts Lemma 4. Then, V_1 and V_3 must be adjacent in H which, as shown above, is a contradiction.

5 Discussion

MAMP CGs are a recently introduced family of models that is based on graphs that may have undirected, directed and bidirected edges. They unify and generalize AMP and MVR CGs. In this paper, we have presented an algorithm for learning an MAMP CG from a probability distribution p which is faithful to it. In practice, we do not usually have access to p but to a finite sample from it. Our algorithm can easily be modified to deal with this situation: Replace $A \perp_p B | S$ in line 5 with a hypothesis test, preferably one that is consistent so that the resulting algorithm is asymptotically correct. We are currently working in the implementation and empirical evaluation of our algorithm. It is worth mentioning that, whereas R1, R2 and R4 only involve three or four nodes, R3 may involve more. Unfortunately, we have not succeeded so far in proving the correctness of our algorithm with a simpler R3. Note that the output of our algorithm would be the same. The only benefit might be a decrease in running time.

The correctness of our algorithm relies upon the assumption that p is faithful to some MAMP CG. This is a strong requirement that we would like to weaken, e.g. by replacing it with the milder assumption that p satisfies the composition property. Specifically, p satisfies the composition property when $X \perp_p Y | Z \wedge X \perp_p W | Z \Rightarrow X \perp_p Y \cup W | Z$ for all X, Y, Z and W pairwise disjoint subsets of V. Note that if p is a Gaussian distribution, then it satisfies the composition property regardless of whether it is faithful or not to some MAMP CG [20, Corollary 2.4].

When making the faithfulness assumption is not reasonable, the correctness of a learning algorithm may be redefined as follows. Given an MAMP CG G, we say that p is Markovian with respect to G when $X \perp_p Y | Z$ if $X \perp_G Y | Z$ for all X, Y and Z pairwise disjoint subsets of V. We say that a learning algorithm is correct when it returns an MAMP CG H such that p is Markovian with respect to H and p is not Markovian with respect to any MAMP CG F such that $I(H) \subsetneq I(F)$.

Correct algorithms for learning Bayesian networks and LWF CGs under the composition property assumption exist [3,9,13]. The way in which these algorithms proceed (a.k.a. score+search based approach) is rather different from that of the algorithm presented in this paper (a.k.a. constraint based approach). In a nutshell, they can be seen as consisting of two phases: A first phase that starts from the empty graph H and adds single edges to it until p is Markovian with respect to H, and a second phase that removes single edges from H until

p is Markovian with respect to H and p is not Markovian with respect to any graph F such that $I(H) \subsetneq I(F)$. The success of the first phase is guaranteed by the composition property assumption, whereas the success of the second phase is guaranteed by the so-called Meek's conjecture [8]. Specifically, given two directed and acyclic graphs F and H such that $I(H) \subseteq I(F)$, Meek's conjecture states that we can transform F into H by a sequence of operations such that, after each operation, F is a directed and acyclic graph and $I(H) \subseteq I(F)$. The operations consist in adding a single edge to F, or replacing F with a triplex equivalent directed and acyclic graph. Meek's conjecture was proven to be true in [2, Theorem 4]. The extension of Meek's conjecture to LWF CGs was proven to be true in [13, Theorem 1]. The extension of Meek's conjecture to AMP and MVR CGs was proven to be false in [10, Example 1] and [12], respectively. Unfortunately, the extension of Meek's conjecture to MAMP CGs does not hold either, as the following example illustrates.

Example 3. The MAMP CGs F and H below show that the extension of Meek's conjecture to MAMP CGs does not hold.

$$
\begin{array}{ccc}
\begin{matrix} A & B \\ \downarrow & \downarrow \\ C - D - E \\ F \end{matrix}
&
\begin{matrix} A & B \\ \downarrow \nearrow & \downarrow \\ C - D - E \\ H \end{matrix}
&
\begin{matrix} A & B \\ \downarrow & \uparrow \\ C - D - E \\ F' \end{matrix}
\end{array}
$$

We can describe $I(F)$ and $I(H)$ by listing all the separators between any pair of distinct nodes. We indicate whether the separators correspond to F or H with a superscript. Specifically,

- $\mathcal{S}_{AD}^F = \mathcal{S}_{BE}^F = \mathcal{S}_{CD}^F = \mathcal{S}_{DE}^F = \varnothing$,
- $\mathcal{S}_{AB}^F = \{\varnothing, \{C\}, \{D\}, \{E\}, \{C, D\}, \{C, E\}\}$,
- $\mathcal{S}_{AC}^F = \{\varnothing, \{B\}, \{E\}, \{B, E\}\}$,
- $\mathcal{S}_{AE}^F = \{\varnothing, \{B\}, \{C\}, \{B, C\}\}$,
- $\mathcal{S}_{BC}^F = \{\varnothing, \{A\}, \{D\}, \{A, D\}, \{A, D, E\}\}$,
- $\mathcal{S}_{BD}^F = \{\varnothing, \{A\}, \{C\}, \{A, C\}\}$, and
- $\mathcal{S}_{CE}^F = \{\{A, D\}, \{A, B, D\}\}$.

Likewise,

- $\mathcal{S}_{AD}^H = \mathcal{S}_{BD}^H = \mathcal{S}_{BE}^H = \mathcal{S}_{CD}^H = \mathcal{S}_{DE}^H = \varnothing$,
- $\mathcal{S}_{AB}^H = \{\varnothing, \{C\}, \{E\}, \{C, E\}\}$,
- $\mathcal{S}_{AC}^H = \{\varnothing, \{B\}, \{E\}, \{B, E\}\}$,
- $\mathcal{S}_{AE}^H = \{\varnothing, \{B\}, \{C\}, \{B, C\}\}$,
- $\mathcal{S}_{BC}^H = \{\{A, D\}, \{A, D, E\}\}$, and
- $\mathcal{S}_{CE}^H = \{\{A, D\}, \{A, B, D\}\}$.

Then, $I(H) \subseteq I(F)$ because $\mathcal{S}_{XY}^H \subseteq \mathcal{S}_{XY}^F$ for all $X, Y \in \{A, B, C, D, E\}$ with $X \neq Y$. Moreover, the MAMP CG F' above is the only MAMP CG that is triplex equivalent to F, whereas there is no MAMP CG that is triplex equivalent to H. Obviously, one cannot transform F or F' into H by adding a single edge.

While the example above compromises the development of score+search learning algorithms that are correct and efficient under the composition property assumption, it is not clear to us whether it also does it for constraint based algorithms. This is something we plan to study.

Acknowledgments. We thank the Reviewers for their thorough reviews. This work is funded by the Center for Industrial Information Technology (CENIIT) and a so-called career contract at Linköping University, and by the Swedish Research Council (ref. 2010-4808).

References

1. Andersson, S.A., Madigan, D., Perlman, M.D.: Alternative Markov Properties for Chain Graphs. Scandinavian Journal of Statistics 28, 33–85 (2001)
2. Chickering, D.M.: Optimal Structure Identification with Greedy Search. Journal of Machine Learning Research 3, 507–554 (2002)
3. Chickering, D.M., Meek, C.: Finding Optimal Bayesian Networks. In: Proceedings of 18th Conference on Uncertainty in Artificial Intelligence, pp. 94–102 (2002)
4. Cox, D.R., Wermuth, N.: Multivariate Dependencies - Models, Analysis and Interpretation. Chapman & Hall (1996)
5. Koster, J.T.A.: Marginalizing and Conditioning in Graphical Models. Bernoulli 8, 817–840 (2002)
6. Lauritzen, S.L.: Graphical Models. Oxford University Press (1996)
7. Ma, Z., Xie, X., Geng, Z.: Structural Learning of Chain Graphs via Decomposition. Journal of Machine Learning Research 9, 2847–2880 (2008)
8. Meek, C.: Graphical Models: Selecting Causal and Statistical Models. PhD thesis, Carnegie Mellon University (1997)
9. Nielsen, J.D., Kočka, T., Peña, J.M.: On Local Optima in Learning Bayesian Networks. In: Proceedings of the 19th Conference on Uncertainty in Artificial Intelligence, pp. 435–442 (2003)
10. Peña, J.M.: Learning AMP Chain Graphs and some Marginal Models Thereof under Faithfulness. International Journal of Approximate Reasoning 55, 1011–1021 (2014)
11. Peña, J.M.: Marginal AMP Chain Graphs. International Journal of Approximate Reasoning 55, 1185–1206 (2014)
12. Peña, J.M.: Learning Multivariate Regression Chain Graphs under Faithfulness: Addendum. Available at the author's website (2014)
13. Peña, J.M., Sonntag, D., Nielsen, J.D.: An Inclusion Optimal Algorithm for Chain Graph Structure Learning. In: Proceedings of the 17th International Conference on Artificial Intelligence and Statistics, pp. 778–786 (2014)
14. Richardson, T., Spirtes, P.: Ancestral Graph Markov Models. The Annals of Statistics 30, 962–1030 (2002)
15. Sadeghi, K., Lauritzen, S.L.: Markov Properties for Mixed Graphs. Bernoulli 20, 676–696 (2014)
16. Sonntag, D., Peña, J.M.: Learning Multivariate Regression Chain Graphs under Faithfulness. In: Proceedings of the 6th European Workshop on Probabilistic Graphical Models, pp. 299–306 (2012)

17. Sonntag, D., Peña, J.M.: Chain graph interpretations and their relations. In: van der Gaag, L.C. (ed.) ECSQARU 2013. LNCS, vol. 7958, pp. 510–521. Springer, Heidelberg (2013)
18. Spirtes, P., Glymour, C., Scheines, R.: Causation, Prediction, and Search. Springer (1993)
19. Studený, M.: A Recovery Algorithm for Chain Graphs. International Journal of Approximate Reasoning 17, 265–293 (1997a)
20. Studený, M.: Probabilistic Conditional Independence Structures. Springer (2005)

Learning Maximum Weighted (k+1)-Order Decomposable Graphs by Integer Linear Programming

Aritz Pérez[1], Christian Blum[1,2], and Jose A. Lozano[1]

[1] University of the Basque Country UPV/EHU, San Sebastian, Spain
aritz.perez@ehu.es, christian.blum@ehu.es, ja.lozano@ehu.es
[2] IKERBASQUE, Basque Foundation for Science, Bilbao, Spain

Abstract. This work is focused on learning maximum weighted graphs subject to three structural constraints: (1) the graph is decomposable, (2) it has a maximum clique size of $k + 1$, and (3) it is coarser than a given maximum k-order decomposable graph. After proving that the problem is NP-hard we give a formulation of the problem based on integer linear programming. The approach has shown competitive experimental results in artificial domains. The proposed formulation has important applications in the field of probabilistic graphical models, such as learning decomposable models based on decomposable scores (e.g. log-likelihood, BDe, MDL, just to name a few).

Keywords: Maximum weighted graph problem, decomposable graph, bounded clique size, integer linear programming.

1 Introduction

The learning of probabilistic graphical models can be divided into two parts, the structural learning and the parametric learning [4]. Usually, the structural learning consists of obtaining a graph that maximizes a score function. This problem can be viewed from the more general perspective of Graph Theory [1] through the maximum weighted graph problems. Maximum weighted graph problems consist of learning a graph by maximizing a weight function associated to the constructed structure. Most of these problems restrict the space of solutions by imposing a set of structural constraints. Additionally, in some problems, the weight functions can be additively decomposed in terms of the structural components of the graph, e.g. the edges or the complete sets of vertices. The structural constraints and the decomposability of the weight function allow the design of efficient algorithms to deal with maximum weighted graph problems.

An illustrative example of this kind of problems is the maximum weighted tree problem. The solutions to this problem are constrained to the set of graphs with tree structure. In this problem the weight function is additively decomposable in terms of the edges included in the solution. This problem can be solved using learning algorithms with a polynomial computational complexity in the number

L.C. van der Gaag and A.J. Feelders (Eds.): PGM 2014, LNAI 8754, pp. 396–408, 2014.

of vertices [5]. The proposed algorithms have been applied to the problem of learning a maximum likelihood model with tree structure [2].

The maximum weighted k-order decomposable graph (MWk) problem was proposed in [9] as a natural generalization of the maximum weighted tree problem. Solutions to this problem are constrained to be k-order decomposable graphs (see Sect. 2). The weight function is additively decomposable in terms of complete sets of edges up to size k. For $k = 2$ the problem reduces to the maximum weighted tree problem. Unfortunately, MWk is an NP-hard problem for $k > 2$, even for $0 - 1$ weights associated to the sets of size two [9]. The author of [9,10] provides an approximate algorithm with theoretical guarantees for a version of the problem where the weights are constrained to be monotone increasing, i.e. the addition of a (valid) edge increases (or maintains) the weight. This algorithm has been applied to the problem of learning maximum likelihood k-order decomposable models [9,10].

In this work we propose a problem related to MWk with the following structural constraints (see Problem 1 in Section 3):

– The solution is a $(k + 1)$-order decomposable graph, and
– the solution is coarser than a given maximal k-order decomposable graph.

In this problem the weight function is also additively decomposable in terms of complete sets of vertices. We show that this problem remains being NP-hard and we provide a formulation based on Integer Linear Programming that can be used to solve it. The provided formulation could be used as the building block for constructing algorithms to approach the maximum weighted graph problem proposed in [9]. Hence, the formulation can be also used learning decomposable models [6] with a bounded clique size using decomposable scores, such as log-likelihood, Bayesian Dirichlet equivalent score or Minimum Description Length [4] (see Sect. 5 for further details).

This work is organized as follows. In Sect. 2 we introduce the theoretical concepts required for the correct understanding of the work. Section 3 defines our maximum weighted graph problem. In Sect. 4 we provide a set of properties to deal with this problem. Section 5 presents a formulation of the problem based on Integer Linear Programming. Section 6 provides a set of experimental results in a set of artificial domains using the Integer Linear Programming approach. In Sect. 7 we propose an approach to the MWk problem [9] which can be used to learn decomposable models with a bounded clique size. Finally, Sect. 8 summarizes the main contributions of this work and indicates how to avoid the drawbacks of the proposed formulation.

2   Background

This section provides a set of theoretical concepts related to decomposable graphs. The theoretical concepts are illustrated using the undirected graphs G_3 and G_3^+ presented in Figures 1 and 2, respectively.

Let $G = (V, E)$ be an undirected graph, where $V = \{1, ..., n\}$ is a set of indexes called the vertices and E is a set of pairs of vertices $\{u, v\}$ called edges. In the graphical representation of undirected graphs shown in Figures 1a and 2a the vertices are represented by circles and edges by lines. Let $G^+ = (V^+, E^+)$ be an undirected graph. G^+ is **coarser** than G (or equivalently, G is **thinner** than G^+) when $V = V^+$ and $E \subsetneq E^+$, and it is denoted as $G \prec G^+$. For example, G_3^+ is coarser than G_3.

A graph is said to be complete if it contains any possible edge, i.e. $E = \{\{u, v\} : \{u, v\} \subset V\}$. An empty graph is a graph without edges, i.e. $E = \emptyset$. The subgraph induced by $R \subseteq V$, $G(R) = (R, E(R))$, is a graph with the vertex set R, where the set of edges is given by $E(R) = \{\{u, v\} : \{u, v\} \subset R, \{u, v\} \in E\}$. When a set $R \subset V$ induces a complete subgraph for G, it is called a complete set. The set of all complete sets of G is denoted as $\mathbb{C}(G)$. Any set $C \subseteq V$ is called a **clique** for G if it is a complete set and there is no proper superset which is complete. For example, the set $\{1, 2, 4\}$ is a clique for G_3 but it is not for G_3^+ because the superset $\{1, 2, 4, 9\}$ is also a complete set for G_3^+.

A path of length l from u to v is a sequence $u = w_0, ..., w_l = v$ of distinct vertices such that $\{w_{i-1}, w_i\} \in E$ for $i = 1, ..., l$. A cycle of length l is a path $w_0, ..., w_l$ of length l with the modification that $w_0 = w_l$, i.e. it begins and ends in the same point. A chord of a cycle is an edge between two vertices not adjacent in the cycle. For example, the sequence $1, 2, 3, 8, 1$ is a cycle for G_3 and it has the chord $\{1, 3\}$. Two vertices u and v are said to be connected if there is a path from u to v. A connected component of a graph is a subgraph induced by $R \subseteq V$ such that there exists a path between any two vertices and, in the original graph, the vertices of the component are not connected to vertices from $V \setminus R$. For example, G_3 and G_3^+ have a single connected component.

The neighborhood of u in G is the set of vertices connected by an edge to u, $\mathbb{N}(u|G) = \{v \in V : \{u, v\} \in E\}$. We define the neighborhood of a set of vertices S in G as the set of vertices connected by edges to all the vertices in S, $\mathbb{N}(S|G) = \{v \in V : \forall u \in S, \{u, v\} \in E\} = \bigcap_{u \in S} \mathbb{N}(u|G)$. The neighborhood is denoted simply by $\mathbb{N}(S)$, when the graph is clear from the context. Note that this definition of neighborhood is the intersection of the neighborhood of each vertex instead of the union. For example, $\mathbb{N}(\{1, 2\}|G_3) = \{3, 4, 5\}$ and $\mathbb{N}(\{1, 2\}|G_3^+) = \{3, 4, 5, 8, 9\}$.

Given two non-adjacent indexes u and v, a subset $S \subseteq V \setminus \{u, v\}$ separates u and v, when the graph induced by $V \setminus S$ separates u and v into two different connected components. We denote by $u \perp_G v | S$ that S separates u and v in G. If no proper subset of S is a separator for u and v, then S is a minimal separator for u and v. The set of the minimal separators of G is denoted as $\mathbb{S}(G)$. For example, $\mathbb{S}(G_3) = \{\{1, 2\}, \{1, 3\}, \{1, 4\}, \{2, 3\}\}$ and $\mathbb{S}(G_3^+) = \{\{1, 2\}, \{2, 3\}, \{1, 3\}\}$. Henceforth, we call separator to a minimal separator, for the sake of brevity. The set of vertices associated to each connected component obtained by the removal of a separator S from G will be denoted by $comp(S|G)$. For example, $comp(\{1, 3\}|G_3) = \{\{2, 4, 5, 6, 9\}, \{7\}, \{8\}\}$ and $comp(\{1, 3\}|G_3^+) = \{\{2, 4, 5, 6, 8, 9\}, \{7\}\}$.

We say that an undirected graph is a **decomposable graph** (**DG**) if for any cycle of length greater than 3 there exists a chord. A remarkable property of DGs is that their minimal separators are complete sets and they are included in a minimum of two of its cliques [6]. For example, G_3 and G_3^+ are decomposable graphs because any cycle of length greater than 3 has a chord.

Next, we introduce a theoretical characterization, adapted from [3], of the set of the edges that can be added to a DG maintaining its decomposability. These edges are henceforth called **candidate edges**.

Theorem 1. *[3] Given a DG $G = (V, E)$, an edge $\{u, v\} \notin E$ is a **candidate edge** if and only if exists $S \subseteq V \setminus \{u, v\}$ such that*

- *S is the (minimal) separator for u and v in G, and*
- *$\{u, v\} \subset \mathbb{N}(S|G)$.*

We say that $\{u, v\}$ is a candidate edge due to S.

For example, $\{3, 4\}$ is a candidate edge due to $\{1, 2\}$ for G_3 because $3 \perp_{G_3} 4|\{1, 2\}$ and $\{3, 4\} \subset \mathbb{N}(\{1, 2\}|G_3) = \{3, 4, 5\}$. $\{3, 9\}$ is not a candidate edge due to $\{1, 2\}$ for G_3 because $\{3, 9\} \not\subset \mathbb{N}(\{1, 2\}|G_3)$. On the contrary, $\{3, 9\}$ is a candidate edge due to $\{1, 2\}$ for (the coarser) G_3^+ because $3 \perp_{G_3^+} 4|\{1, 2\}$ and $\{3, 9\} \subset \mathbb{N}(\{1, 2\}|G_3^+)$. Clearly, the addition of a set of edges can change the neighborhood of a separator S and, in consequence, the set of candidate edges due to S.

Corollary 1. *Let G be a DG. The addition of a candidate edge $\{u, v\}$ due to $S \in \mathbb{S}(G)$ to G creates the clique $\{u, v\} \cup S$.*

The addition of the candidate edge $\{2, 9\}$ due to $\{1, 4\}$ to G_3 creates the clique $\{1, 2, 4, 9\}$.

In this work we are particularly interested in two types of decomposable graphs which control explicitly the maximum clique size.

Definition 1. *A **k-order decomposable graph** (**kDG**) is a DG for which the maximum clique size is k. A **maximal k-order decomposable graph** (**MkDG**) is a kDG for which all the cliques are of size k and the addition of a candidate edge creates a clique of size $k + 1$.*

For example, a forest is a 2DG and a tree is an M2DG, G_3 is an M3DG and G_3^+ is a 4DG. MkDGs are also known in the literature as $(k - 1)$-hypertrees [9]. We are interested in kDGs and MkDGs because Problem 1 (see Sect. 3) is defined in terms of these types of DGs. An MkDG G_k has the following interesting structural properties [7], among others [9]:

- the set of cliques is of size $n - k + 1$, and
- the size of all the (minimal) separators in $\mathbb{S}(G_k)$ is $k - 1$.

MkDGs are maximal in the sense that the addition of any candidate edge creates a clique of size greater than k. For example, the addition of $\{2, 9\}$ and $\{2, 8\}$ to G_3, generates G_3^+, which has the cliques $\{1, 2, 4, 9\}$ and $\{1, 2, 3, 8\}$ of size four.

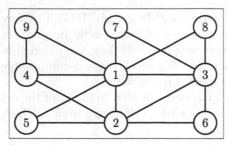 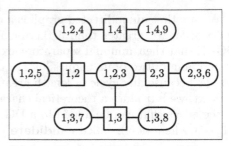

(a) Undirected graph. The edges of G_3 are represented by lines and vertices by circles.

(b) Junction tree. The cliques of G_3 are represented by ellipses and separators by squares.

Fig. 1. This figure shows two alternative representations of the M3DG G_3. The undirected graph representation provides the fine details of G_3 and the junction tree is a high level representation which highlights the cliques and separators of G_3.

3   Maximum Weighted $(k + 1)$DG Problem

In this section we define the maximum weighted graph problem to be studied.

Problem 1. *Let G be an MkDG and let $W = \{w_R : R \subset V, \text{ where } |R| \leq k+1\}$ be a set of weights associated to the sets of vertices of size lower than or equal to $k + 1$. Find a maximum weighted $(k + 1)$DG coarser than G:*

$$G^* = \arg_{G^+ \in \mathcal{G}_{k+1}} w(G^+) \tag{1}$$

where $\mathcal{G}_{k+1}$ is the set of $(k + 1)$DGs coarser than G and $w(G^+)$ is the weight function given by

$$w(G^+) = \sum_{R \in \mathbb{C}(G^+)} w_R \tag{2}$$

Note that $\mathbb{C}(G^+)$ denotes the entire set of complete subgraphs of G^+, not only the cliques. This weight function includes the decomposable scores such as log-likelihood, Bayesian Dirichlet equivalent and Minimum Description Length [4] (see Sect. 7).

   As we will see in Sect. 4, many of the weights included in W are irrelevant for solving Problem 1 (see Corollary 2). However, we have decided to include the entire set W in the definition of Problem 1, to highlight the similarities with the MWk problem defined in [9].

Theorem 2. *Problem 1 is NP-hard for $k > 2$.*

Proof. The proof is equivalent to the proof provided in [9], Theorem 4.1. We can reduce 3SAT to Problem 1, by constructing an MkDG coarser than the core structure given in Theorem 4.1 of [9].

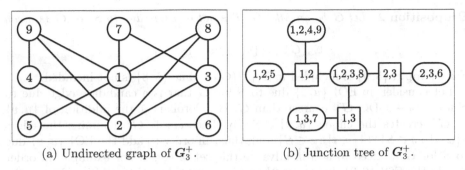

(a) Undirected graph of G_3^+.   (b) Junction tree of G_3^+.

Fig. 2. This figure shows two alternative representations of the 4DG G_3^+. G_3^+ is coarser than G_3. The set of candidate edges has changed with respect to G_3, due to the addition of $\{2,8\}$ and $\{2,9\}$: now $\{3,9\}, \{4,8\}, \{5,8\}, \{5,9\}$, and $\{8,9\}$ are candidate edges for G_3^+.

4   Results

In this section we illustrate the main ideas behind the Integer Linear Programming approach to Problem 1, which will be presented in Section 5. These intuitions make use of the particular structural features of MkDGs and coarser $(k+1)$DGs.

Proposition 1. *(Lemma 2.21, [6]) Let G be a decomposable graph and let G^+ be a coarser decomposable graph with exactly m edges more. Then, there is a sequence of coarser decomposable graphs $G = G^0 \prec G^1 \prec ... \prec G^m = G^+$, where G^i is obtained from G^{i-1} by adding a candidate edge for G^{i-1}, for $i = 1, ..., m$.*

Clearly, any $(k+1)$DG coarser than a given MkDG G can be obtained starting from G by adding candidate edges sequentially. Therefore, Problem 1 can be solved by considering the addition of candidate edges of successive DGs, only. We would like to highlight that, as we noted in Sect. 2, the neighborhood of a separator can change by the addition of edges and, in consequence, the set of candidate edges of the obtained structure can be different (see Figure 2). In other words, the candidate edge added to G^i does not need to be a candidate edge for G^{i-1}.

Next, we define the set of all candidate edges due to S that create cliques of size $k+1$ for an MkDG G and for all the possibles $(k+1)$DG coarser than G.

Definition 2. *Let G be an MkDG. An edge is called **edge of interest** (EOI) due to S for G when it is a candidate edge for G or for a $(k+1)$DG coarser than G, and its addition creates a clique of size $k+1$.*

For example, the edge $\{6,8\}$ is an EOI due to $\{2,3\}$ for G_3 because it is a candidate edge due to $\{2,3\}$ for G_3^+, $G_3 \prec G_3^+$ and its addition to G_3^+ creates the clique $\{2,3,6,8\}$ of size 4.

Proposition 2. *Let G be an MkDG. The set of EOIs due to S for G is given by*

$$\mathbb{E}_S(G) = \{\{u,v\} : u \perp_G v|S\}$$

For example, $\{5,9\}$ is included in $\mathbb{E}_{\{1,4\}}(G_3)$ while $\{5,8\}$ is not included.

Let consider an EOI $\{u,v\}$ due to S for G that is a candidate edge due to S for a $(k+1)$DG G^+ coarser than G. By Corollary 1 the addition of $\{u,v\}$ to G^+ creates the clique $\{u,v\} \cup S$. By Theorem 1, G^+ contains the edges $\{\{u,s\} : s \in S\} \cup \{\{v,s\} : s \in S\}$ and, thus, in order to add the EOI $\{u,v\}$ due to S for G we need to add in advance this set of edges. For example, in order to add the EOI $\{6,8\}$ due to $\{2,3\}$ to G_3 we would need to add in advance the edge $\{2,8\}$ which creates the clique $\{2,3,6,8\}$. Taking into account the edges that must be added in advance to G, we say that the EOI $\{u,v\}$ due to S for G is associated to the clique $\{u,v\} \cup S$. Note that the addition of the same EOI due to a different separator has associated a different clique. For example, the addition of the EOI $\{5,6\}$ due to $\{2,3\}$ for G_3 requires to add in advance $\{3,5\}$ and creates the clique $\{2,3,5,6\}$, while its addition due to $\{1,2\}$ requires to add in advance of $\{1,6\}$ and creates the clique $\{1,2,5,6\}$. Thus the addition of the same edge $\{u,v\}$ can create different cliques and, therefore, the separator must be stated explicitly in the EOIs.

In order to characterize the set of cliques that can form part of the solution to Problem 1, we need to determine the set of separators of size $k-1$ for G and for any $(k+1)$DG coarser than G. The next result states that this set is the set of separators of G.

Proposition 3. *[8] Let G^+ be a $(k+1)$DG coarser than the MkDG G. The set of the separators of size $k-1$ of G^+ is contained in the set of separators of G.*

This result reduces the set of separators to be considered for solving Problem 1 to the separators of the given MkDG. Note that, as already indicated in Sect. 2, all the separators of an MkDG are of size $k-1$. The following result characterizes the set of cliques of size $k+1$ that can be contained in the solution to Problem 1.

Proposition 4. *Let G be an MkDG. The set of cliques of size $k+1$ of any kDG coarser than G is given by*

$$\{\{u,v\} \cup S : \{u,v\} \in \mathbb{E}_S(G), \text{ for an } S \in \mathbb{S}(G)\}$$

The cliques are associated to an EOI due to a separator for G. By Propositions 2 and 3 the EOIs correspond to $\mathbb{E}_S(G)$ for $S \in \mathbb{S}(G)$. By Theorem 1 in order to add the EOI $\{u,v\}$ due to S for G we need to add in advance the edges $\{\{u,s\} : s \in S\} \cup \{\{v,s\} : s \in S\}$, which generates G^+. Finally, by Corollary 1 the addition of $\{u,v\}$ due to S to G^+ generates the clique $\{u,v\} \cup S$. For example, the clique $\{1,2,8,9\}$ can be created from G_3 because $\{8,9\} \in \mathbb{E}_{\{1,2\}}(G_3)$. But the clique $\{1,3,4,5\}$ can not be created from G_3 because $\{4,5\} \notin \mathbb{E}_{\{1,3\}}(G_3)$ and $\{3,5\} \notin \mathbb{E}_{\{1,4\}}(G_3)$. Thus, only a subset of cliques can be part of the solution structure.

The next result quantifies the contribution of an EOI due to S to the weight of a structure.

Proposition 5. *Let G be an $MkDG$ and let $\{u,v\}$ be an EOI due to S, $\{u,v\} \in \mathbb{E}_S(G)$. The contribution of $\{u,v\}$ due to S to the total weight of a solution is $w_{u,v|S} = \sum_{R \subseteq S} w_{\{u,v\} \cup R}$.*

By Theorem 1, the addition of the EOI $\{u,v\}$ due to S requires to add in advance the edges $\{\{u,s\} : s \in S\} \cup \{\{v,s\} : s \in S\}$, which generates G^+. Therefore, the addition of $\{u,v\}$ to G^+ creates the new complete sets $\{\{u,v\} \cup R : R \subseteq S\}$, only. For example the EOI $\{5,9\}$ due to $\{1,4\}$ for G_3 requires the addition of $\{4,5\}$ in advance. After the inclusion of $\{4,5\}$, the addition of $\{5,9\}$ creates the new complete sets $\{5,9\}, \{1,5,9\}, \{4,5,9\}$ and $\{1,4,5,9\}$, only. Thus the EOI $\{5,9\}$ due to $\{1,4\}$ has a weight $w_{5,9|\{1,4\}} = w_{\{5,9\}} + w_{\{1,5,9\}} + w_{\{4,5,9\}} + w_{\{1,4,5,9\}}$.

Clearly, the set of weights that have to be considered to solve Problem 1 is a subset of $\mathcal{W}$.

Corollary 2. *Let G be an $MkDG$. The set of weights required to solve Problem 1 is limited to*

$$\mathcal{W}(G) = \{w_{\{u,v\} \cup R} : R \subseteq S \text{ and } \{u,v\} \in \mathbb{E}_S(G), \text{ where } S \in \mathbb{S}(G)\}$$

5 An Integer Linear Programming Formulation of Problem 1

This section presents a formulation based on Integer Linear Programing for solving Problem 1, using the results presented in Sect. 4.

Let $G = (V, E)$ be the input MkDG. The **decision variables** of the problem are binary and correspond to the next set (see Proposition 2):

$$\{X_{u,v|S} : S \in \mathbb{S}(G), \{u,v\} \in \mathbb{E}_S(G)\} \tag{3}$$

where

$$X_{u,v|S} = \begin{cases} 1, & \text{if the EOI } \{u,v\} \text{ due to } S \text{ is included in the solution.} \\ 0, & \text{if the EOI } \{u,v\} \text{ due to } S \text{ is NOT included in the solution.} \end{cases} \tag{4}$$

An edge $\{u,v\}$ is included in the solution if $X_{u,v|S} = 1$ for any $S \in \mathbb{S}(G)$, where $\{u,v\} \in \mathbb{E}_S(G)$, while $\{u,v\}$ is not included in the solution if $X_{u,v|S} = 0$ for all $S \in \mathbb{S}(G)$, where $\{u,v\} \in \mathbb{E}_S(G)$. The $(k+1)DG$ codified by the decision variables is $G^+ = (V, E^+)$, where $E^+ = E \cup \{\{u,v\} : X_{u,v|S} = 1 \text{ for some } S \in \mathbb{S}(G)\}$. As we noted in Sect. 4, by Proposition 5 the addition of an EOI $\{u,v\}$ due to S contributes to the total weight of the obtained $(k+1)DG$ with $w_{u,v|S}$ and, thus, the assignment $X_{u,v|S} = 1$ has the associated weight $w_{u,v|S}$ to the solution.

Problem 1 can be rewritten as the next **Integer Linear Programming** (**ILP**) problem:

$$\max \sum_{S \in \mathbb{S}(G), \{u,v\} \in \mathbb{E}_S(G)} w_{u,v|S} X_{u,v|S} \tag{5}$$

subject to the following set of **constraints**:

1. $\sum_{S\in\mathbb{S}(G):\{u,v\}\in\mathbb{E}_S(G)} X_{u,v|S} \leq 1$ for $\{u,v\} \notin E$
2. $[\sum_{R\in\mathbb{S}(G):\{u,v\}\in\mathbb{E}_R(G)} \sum_{s\in S} X_{u,s|R}] - (k-1) \cdot X_{u,v|S} \geq -\sum_{s\in S} 1_{u,s}$
   $[\sum_{R\in\mathbb{S}(G):\{u,v\}\in\mathbb{E}_R(G)} \sum_{s\in S} X_{v,s|R}] - (k-1) \cdot X_{u,v|S} \geq -\sum_{s\in S} 1_{v,s}$
   for $X_{u,v|S}$.
3. $\sum_{T\neq U\in\mathcal{C}} \sum_{u\in T} \sum_{v\in U} X_{u,v|S} \leq |\mathcal{C}| - 1$ for $S \in \mathbb{S}(G)$ and $\mathcal{C} \subset comp(S|G)$

The constants $1_{u,v}$ of constraints (2) take the value 1 if $\{u,v\} \in E$, and 0 otherwise. In constraints (3), $\mathcal{C}$ represents a set of sets of vertices, and T and U are sets of vertices associated to (distinct) connected components of $G(V \setminus S)$.

Constraints (1) guarantee that each EOI is considered by one separator, which guarantees that each edge can be added only once in the solution. Constraints (2) guarantee that $\{u,v\}$ is in the neighborhood of S in the solution (see Theorem 1 and Figure 3a). In Sect. 4 we indicate that Problem 1 can be solved by adding sequentially a set of candidate edges (see Proposition 1). The proposed ILP formulation does not define an explicit ordering of the candidate edges added, but, it is possible to define the next partial ordering among the EOIs: if $\{u,v\}$ is an EOI due to S, then the edges $\{u,s\}$ and $\{v,s\}$ for $s \in S$ not included in the MkDG G have to be added in advance. Constraints (3) guarantee that the vertices u and v were separated due to S before the edge was added. For example, in Figure 3b, it is possible to add an edge from the component of v to the component of w because they are separated due to S. On the contrary, an edge from the component of v to the component of u can not be added because they are not separated due to S. Note that constraints (3) guarantee that we add a forest among the connected components of each separator.

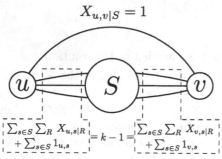 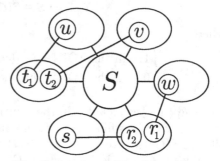

(a) Constraints (2): These constraints ensure that if $\{u,v\}$ is added to the structure creating the clique $\{u,v\} \cup S$ then u and v are in the neighborhood of S.

(b) Constraints (3): These constraints ensure that the edges between two vertices in different connected components form a forest.

Fig. 3. This figure illustrates the constraints of the Integer Linear Programming formulation to Problem 1. Small circles represent vertices, big circles separators, ellipses connected components and lines edges.

In the worst case, the number of decision variables of the ILP formulation is $\mathcal{O}(n^3)$ because the set of separators can be of size $\mathcal{O}(n)$ and the number of edges can be $\mathcal{O}(n^2)$. The number of constraints (1) correspond to the number of edges which can be $\mathcal{O}(n^2)$. The number of constraints (2) is proportional to the number of decision variables and is $\mathcal{O}(n^3)$. In the worst case, the bottleneck of this approach are the constraints (3), which are $\mathcal{O}(\sum_{S \in \mathbb{S}(G)} 2^{d_S})$. Therefore, they are exponential in d^{max}, the maximum degree of the separators of G. The worst case, for example, corresponds to an MkDG with a single separator because there are $\mathcal{O}(2^{n-k})$ constraints (3). This fact limits the range of applications of the proposed approach to deal with MkDGs to the ones with moderate values of d^{max}. However, in the performed experimentation, we have observed a linear growth in the number of constraints (3) with respect to n (see Sect. refsec:experiments).

6 Experiments

The experiments have been restricted to generating maximum weighted 3DGs from M2DGs (Problem 1 for $k = 2$). Further experimentation is left to future work.

The experimental results outlined in this section were obtained on a cluster of PCs with "Intel(R) Xeon(R) CPU E5450" CPUs of 8 nuclii of 3000 MHz and 32 Gigabyte of RAM. Moreover, the ILP formulation was solved with IBM ILOG CPLEX V12.1 [12]. A run of CPLEX was stopped once (at least) 3600 CPU seconds had passed. The output (result) of CPLEX is the value of the best feasible integer solution found within the CPU time limit. The results obtained by CPLEX are compared against the ones of a greedy approach where, at each step, the algorithm considers the addition of the candidate edge $\{u, v\}$ due to S with the maximum weight $w_{u,v|S}$ among the set of candidate edges.

The instances of Problem 1 have been randomly generated as follows:

- An M2DG (a tree) with n vertices G is generated at random using the Kruskall algorithm for maximization where the weights associated to the edges are randomly generated.
- The set of weights $\mathcal{W}(G)$ (see Corollary 2) has been uniformly sampled from the set $\{-50, 49, ..., -1, 0, 1, ..., 149, 150\}$.

100 domains have been generated for each $n \in \{25, 50, 75, 100\}$. The obtained experimental results are summarized in Table 1.

The proposed ILP approach has obtained the best average results for all the values of n (see Table 1). ILP obtains the optimum for all the domains with $n = 25$ and, for $n = 50$, has reached the optimum in 91 domains out of 100. Moreover, the time required to approach Problem 1 for $n = 50$ is very low (less than 10 seconds) and for $n = 25$ is even lower (less than 1 second). However, as n increases CPLEX is generally not able to provide (provenly) optimal solutions within the given CPU time limit (see column Gap). For instance, with $n = 75$ the optimum solution has been obtained for only one problem instance.

Table 1. Summary of the results obtained for the generated artificial domains. The columns Vars., Cons-1, Cons-2 and Cons-3 show the average number of decision variables and constraints (1), (2) and (3), respectively, for the ILP formulation of the problem. The column Weight shows the average objective function values obtained by Greedy and ILP. The column Edge provides the average weight of each edge added to the solution by both approaches. The column Gap provides information about the average optimality gap (in percent), which refers to the gap between the value of the best valid solution that was found and the current lower bound at the time of stopping a run. The column APD is the average percentage deviation of ILP with respect to Greedy.

	Problem features				Greedy		ILP			
n	Vars.	Cons-1	Cons-2	Cons-3	Weight	Edge	Weight	Edge	Gap	APD
25	1173.9	276	1795.8	55.9	2449.0	98.9	2718.3	118.2	0.0	11.8
50	7311.5	1176	12271.0	131.2	5195.1	105.1	5802.9	116.8	0.4	12.0
75	21262.0	2701	37122.1	201.5	78911.1	109.8	8733.7	121.0	6.0	11.0
100	43458.7	4851	77215.5	277.2	10663.8	107.9	11053.1	109.8	15.7	3.7

There is a severe drop in the gain with respect to Greedy from $n = 75$ to $n = 100$ (see column APD) which indicates that CPLEX is not appropriate— with a CPU time limit of 3600 seconds—for $n > 100$. Besides, in two instances CPLEX is not able to provide any feasible integer solution within the CPU time limit. These facts suggests the use of the solution of the greedy approach as starting solution for the ILP approach. This is called a *warm start* in CPLEX terminology.

As we noted in Sect. 5, the number of constraints (3) grows exponentially with the maximum size of the neighborhoods of the separators. However, the experimental results show a linear increase in the number of constraints (3) with respect to n (see column Cons-3 in Table 1). The reason for the linear growth is because, on average, the number of vertices of degree d in a random tree (M2DG) decreases exponentially with d, $\mathcal{O}(\frac{n}{2^d})$. This observation can be generalized to other values of k. The obtained experimental results suggest that the bottleneck of the proposed formulation in order to deal with MkDGs with many vertices is the number of decision variables and constraints (2). Actually, we are developing strategies to alleviate this problem.

7   An Approach to the MWk Problem [9]

In this section we propose an approach to the MWk problem [9] (see Sect. 1) based on the proposed ILP formulation. The intuition is to construct a sequence of coarser M$(i + 1)$DGs for $i = 1, ..., k - 1$, where M$(i + 1)$DG is obtained from the given MiDG using the ILP formulation proposed in Sect. 7. In order to guarantee that an M$(i + 1)$DG is constructed from a given MiDG using the ILP formulation the inequality ($\leq$) in constraints (3) is replaced by an equality ($=$). This slight modification ensures that $n - i$ edges are added to the given

MiDG, which produces an M(i + 1)DG. Based on this modified formulation we can approach the MWk problem [9] as follows:

- Growth stage: Construct a sequence of coarser M(i+1)DGs from $i = 1, ..., k-1$ starting from the empty graph (i.e. M1DG). At the end of this stage an MkDG is generated.
- Removal stage: Obtain a maximum weighted DG thinner than the constructed MkDG using a procedure similar to [11].

As we noted before, the modification of constraints (3) forces the addition of $n - i$ edges at step i of the growth stage. Consequently, some spurious edges could be added to the structure during the growing stage and, if the score is not monotone increasing, the spurious edges could degrease the weight of the obtained graph. The removal stage tries to eliminate this set of spurious edges.

This procedure can be used to learn decomposable models with a bounded clique size using decomposable scores such as log-likelihood, Bayesian Dirichlet equivalent or Minimum Description Length. The decomposable scores can be additively decomposed as follows:

$$s(\boldsymbol{G}; \mathcal{D}) = \sum_{u \in V} s(u|Pa(u); \mathcal{D})$$

where $\mathcal{D}$ is the available data and $Pa(u)$ is the set of parents of u fixed an ancestral ordering for the vertices V. Using a decomposable score, the weight of the EOI $\{u, v\}$ by S in the ILP formulation corresponds to $w_{u,v|S} = s(u|v, S; \mathcal{D}) - s(u|S; \mathcal{D})$. For example, using the log-likelihood score the weight $w_{u,v|S}$ is the empirical conditional mutual information $\hat{I}(u; v|S)$.

8   Conlusions

In this work we have formally defined the problem of learning a maximum weighted (k + 1)-order decomposable graph coarser than a given maximum k-order decomposable graph. We have provided a set of intuitions for the design of learning algorithms to deal with this problem. Using the presented results we have proposed a formulation based on Integer Linear Programming for solving the problem. The proposed approach has obtained competitive results, especially for graphs with moderate values of n ($n < 100$). Due to the general nature of the weights of Problem 1 and the fact that it considers decomposable graphs only, the proposed formulation is an interesting building block in order to design algorithms for learning decomposable models with a bounded clique size (see Sect. 7).

The main drawback of the presented formulation based on Integer Linear Programming is related with the number of decision variables and constraints (2), $\mathcal{O}(n^3)$. This bottleneck implies practical difficulties to deal with structures with many vertices. Actually, we are developing strategies to alleviate this problem by limiting the set of edges of interest considered: we only consider the edge of interest $\{u, v\}$ due to S if the length of the minimum path from u (and v) to $s \in S$

is lower than a predefined threshold l. With this additional constraint we can effectively control the number of decision variables and constraints (constraints (1) and (2)) considered for obtaining an approximate solution to Problem 1.

Acknowledgments. A. Pérez and J. A. Lozano were partially supported by the Saiotek and IT609-13 programs (Basque Government), TIN2010-14931 (Spanish Ministry of Science and Innovation), COMBIOMED network in computational bio-medicine (Carlos III Health Institute). C. Blum was supported by project TIN2012-37930 of the Spanish Government. In addition, support is acknowledged from IKERBASQUE (Basque Foundation for Science).

References

1. Boundy, J.A., Murty, U.S.R.: Graph Theory with Applications. The Macmillan Press LTD (1976)
2. Chow, C.K., Liu, C.: Approximating discrete probability distributions with dependence trees. IEEE Transactions on Information Theory 14, 462–467 (1968)
3. Desphande, A., Garofalakis, M., Jordan, M.I.: Efficient stepwise selection in decomposable models. In: Proceedings of UAI, pp. 128–135 (2001)
4. Koller, D., Friedman, N.: Probabilistic Graphical Models. Principles and Techniques. The MIT Press, Cambridge (2009)
5. Kruskal, J.B.: On the shortest spanning subtree of a graph and the traveling salesman problem. Proceedings of the American Mathematical Society 7(1), 48–50 (1968)
6. Lauritzen, S.L.: Graphical Models. Oxford University Press, New York (1996)
7. Malvestuto, F.M.: Approximating discrete probability distributions with decomposable models. IEEE Trans on SMC 21(5), 1287–1294 (1991)
8. Pérez, A., Inza, I., Lozano, J.A.: Efficient learning of decomposable models with a bounded clique size. Technical Report, University of the Basque Country, EHU-KZAA-TR-2014-07 (2014)
9. Srebro, N.: Maximum likelihood markov networks: An algorithmic approach. Master thesis, MIT (2000)
10. Srebro, N.: Maximum likelihood bounded tree-width Markov networks. Artificial Intelligence 143, 123–138 (2003)
11. Vats, D., Robert, R.N.: A junction tree framework for undirected graphical model selection. Journal of Machine Learning Research 15, 147–191 (2014)
12. IBM ILOG CPLEX VI2.1: User's Manual for CPLEX, International Business Machines Corporation (2009)

Multi-label Classification for Tree and Directed Acyclic Graphs Hierarchies

Mallinali Ramírez-Corona, L. Enrique Sucar, and Eduardo F. Morales

Instituto Nacional de Astrofísica, Óptica y Electrónica,
Luis Enrique Erro No. 1, Sta. Ma. Tonantzintla, Puebla, 72840, México
{mallinali.ramirez,esucar,emorales}@inaoep.mx

Abstract. Hierarchical Multi-label Classification (HMC) is the task of assigning a set of classes to a single instance with the peculiarity that the classes are ordered in a predefined structure. We propose a novel HMC method for tree and Directed Acyclic Graphs (DAG) hierarchies. Using the combined predictions of locals classifiers and a weighting scheme according to the level in the hierarchy, we select the "best" single path for tree hierarchies, and multiple paths for DAG hierarchies. We developed a method that returns paths from the root down to a leaf node (Mandatory Leaf Node Prediction or MLNP) and an extension for Non Mandatory Leaf Node Prediction (NMLNP). For NMLNP we compared several pruning approaches varying the pruning direction, pruning time and pruning condition. Additionally, we propose a new evaluation metric for hierarchical classifiers, that avoids the bias of current measures which favor conservative approaches when using NMLNP. The proposed approach was experimentally evaluated with 10 tree and 8 DAG hierarchical datasets in the domain of protein function prediction. We concluded that our method works better for deep, DAG hierarchies and in general NMLNP improves MLNP.

1 Introduction

The traditional classification task deals with problems where each example e is associated with a single label $y \in L$, where L is the set of classes. However, some classification problems are more complex and multiple labels are needed. This is called multi-label classification. A multi-label dataset D is composed of N instances $(x_1, J_1), (x_2, J_2), ..., (x_N, J_N)$, where $J \subset L$. The task is called Hierarchical Multi-label Classification (HMC) when the labels are ordered in a predefined structure, typically a tree or a DAG (Direct Acyclic Graph), the main difference between them is that in the DAG a node can have more than one parent node.

In hierarchical classification, an example that belongs to certain class automatically belongs to all its superclasses (hierarchy constraint), e.g., in Figure 1b an instance that belongs to class node 3 also belongs to nodes 1, 4 and root.

Some major applications of HMC can be found in the fields of text categorization [10], protein function prediction [13], music genre classification [12], phoneme classification [6], etc.

L.C. van der Gaag and A.J. Feelders (Eds.): PGM 2014, LNAI 8754, pp. 409–425, 2014.

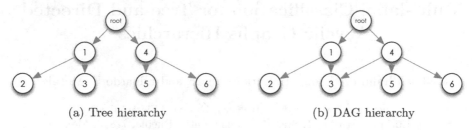

(a) Tree hierarchy   (b) DAG hierarchy

Fig. 1. An example of a tree and a DAG structure

Two general approaches can be distinguished in HMC [14]. The first is a global approach that builds a single classification model, taking into account the class hierarchy as a whole. These methods are incapable of handling large scale datasets because the models become too complex and thus time consuming. The second is a local approach that divides the problem in several subproblems according to a strategy (can be a local classifier per level, per node or per non leaf node). The main problem of this approach is that it does not incorporate the relations (underlying structure) in the local classification.

We propose a novel HMC approach, Chained Path Evaluation (CPE). CPE belongs to the local approaches so it can work effciently with large scale datasets; a local classifier is trained for each non-leaf node in the hierarchy. To include the relations between the classes, and diminish the limitation of local approaches, an extra attribute is added to the instances in each node which corresponds to the parent node class according to the hierarchy. We also incorporated a weighting scheme to value more the predictions of the more general classes than the more particular ones. CPE scores all the paths in the hierarchy to select the best one. CPE predicts single paths from the root down to a leaf node for tree hierarchies (e.g., in Figure 1a 2, 1, root) and multiple paths for DAG hierarchies (e.g., in Figure 1b 3, 1, 4, root).

We developed an extension of the base method for Non Mandatory Leaf Node Prediction (NMLNP); in which a pruning phase is performed to select the best path. We compared several pruning approaches, the best approach for the task was to prune and then choose the optimal path in a top-down fashion, using the most probable child as the condition to prune a node. Additionally, we proposed a new evaluation metric for hierarchical classifiers, that avoids the bias of current measures which favor conservative predictions (predictions of short paths that only predict the most general classes) when using NMLNP.

The proposed approach was experimentally evaluated with 10 tree and 8 DAG hierarchical datasets in the domain of protein function prediction. We concluded that our method in both versions, MLNP and NMLNP, is competitive with other methods in the state of the art, and performs better in deeper, DAG hierarchies.

The document is organized as follows. Section 2 reviews the relevant work in the area, Section 3 describes the method in detail, Section 4 outlines the framework for the experiments, Section 5 evaluates experimentally our approach, and Section 6 summarizes the paper and suggests possible future work.

2   Related Work

When the labels in a multi-label classification problems are ordered in a pre-defined structure, typically a tree or a Direct Acyclic Graph (DAG), the task is called Hierarchical Multi-label Classification (HMC). The class structure represent an "IS-A" relationship, these relations in the structure are asymmetric (e.g., all cats are animals, but not all animals are cats) and transitive (e.g., all Siameses are cats, and all cats are animals; therefore all Siameses are animals). In hierarchical classification, there are basically two types of classifiers: global classifiers and local classifiers.

Global classifiers construct a global model and train it to predict all the classes of an instance at once. Vens et al. [15] present a global method that applies a Predicting Clustering Tree (PCT) to hierarchical multi-label classification, transforms the problem in a hierarchy of clusters with reduced intra-cluster variance. One problem of global classifiers is that the computational complexity grows exponentially with the number of labels in the hierarchy.

Local classifiers can be trained in three different ways: a Local Classifier per hierarchy Level (LCL), that trains one multi-class classifier for each level of the class hierarchy; training a Local binary Classifier per Node (LCN), where each classifier decides if a node is predicted or not; the third way is training a Local Classifier per Parent Node (LCPN), where a multi-class classifier is trained to predict its child nodes.

Cerri et al. [5] propose a method that incrementally trains a multilayer perceptron for each level of the classification hierarchy (LCL). Predictions made by a neural network at a given level are used as inputs to the network of the next level. The labels are predicted using a threshold value. Finally, a post processing phase is used to correct inconsistencies (when a subclass is predicted but its superclass is not). Some difficulties of this approach are the selection of a correct threshold and the need of a post-processing phase.

Alaydie et al. [1] developed HiBLADE (Hierarchical multi-label Boosting with LAbel DEpendency), an LCN algorithm that takes advantage of not only the predefined hierarchical structure of the labels, but also exploits the hidden correlation among the classes that is not shown through the hierarchy. This algorithm attaches the predictions of the parent nodes as well as the related classes. However, appending multiple attributes can create models that over-fit the data.

Silla et al. [12] propose an LCPN algorithm combined with two selective methods for training. The first method selects the best features to train the classifiers, the second selects both the best classifier and the best subset of features simultaneously, showing that selecting a classifier and features improves the classification performance. A drawback of this approach is that the selection of the best features and the best classifier for each node can be a time-consuming process.

Bi et al. [3,4] propose HIROM, a method that uses the local predictions to search for the optimal consistent multi-label classification using a greedy strategy. Using Bayesian decision theory, they derive the optimal prediction rule by minimizing the conditional risk. The limitations of this approach is that it optimizes a function that does not necessarily maximizes the performance in other measures.

The approach of Hernandez et al. [7], used for tree structured taxonomies, learns an LCPN. In the classification phase, it classifies a new instance with the local classifier at each node, and combines the results of all of them to obtain a score for each path from the root to a leaf-node. Two fusion rules were used to achieve this: product rule and sum rule. Finally it returns the path with the highest score. One limitation of this method is that it favors shorter (product rule) or longer paths (sum rule) depending on which combination rule is used. Another limitation is that it does not take into account the relations between nodes when classifying an instance.

Extending the work of Hernandez et al., our method (Chained Path Evaluation or CPE), changes the way the classifiers are trained to include the relations between the labels, specifically of the parent nodes of the labels, to boost the prediction. The score for each path is computed using a fusion rule that takes into account the level in the hierarchy, thus minimizing the effect that the length of the path has in the score. We also extended the method to work with DAG structured hierarchies.

To include the relations of the parent nodes we used the idea of chain classifiers proposed by Read et al. [9] and further extended by Zaragoza et al. [16]. The chain classifiers proposed by Read link the classifiers along a chain where each classifier deals with the binary classification problem associated with a label. The feature space of each classifier in the chain is extended with the 0/1 label of all the previous classifiers in the chain. Zaragoza et al. propose a Bayesian Chain Classifier where they obtain a dependency structure out of the data. This structure determines the order of the chain, so that the order of the class variables in the chain is consistent with the structure found in the first stage. We adapt this idea to a hierarchical classifier, such that the chain structure is determined by the hierarchy.

3   Chained Path Evaluation

Let D be a training set with N examples, $e_k = (x_k, J_k)$, where x_k is a d-dimensional feature vector and $J \subset L$, $L = \{l_1, l_2, ..., l_M\}$ a finite set of M possible labels. These labels are represented as $Y \in \{0, 1\}^M$, where $y_i = 1$ iff $y_i \in J_k$ else $y_i = 0$. The parent of label y_i in the hierarchy is represented as $pa(y_i)$, the children nodes as $child(y_i)$ and the siblings as $sib(y_i)$, the siblings include all the children nodes of $pa(y_i)$ except y_i. Our method exploits the correlation of the labels with its ancestors in the hierarchy and evaluates each possible path from the root to a leaf node, taking into account the level of the predicted labels to give a score to each path and finally return the one with the best score. The method is composed of two phases: training and classification. There is an additional optional phase, pruning, which can be applied for non-mandatory leaf node prediction.

3.1   Training

The method trains local classifiers per parent node (LCPN). A multi-class classifier C_i is trained for each non leaf node y_i. The classes in C_i are the labels in the set of $child(y_i)$ plus an "*unknown*" label that corresponds to the instances that do not belong to any $child(y_i)$.

The training set for C_i is composed of two sets. The positive training set $(Tr^+(C_i))$ consists of the instances where $child(y_i) = 1$. Each instance in this set will be labeled with the corresponding $child(y_i)$ label. The negative training set $(Tr^-(C_i))$ consists of instances in $sib(y_i)$, in case y_i has no siblings this set will include the uncle nodes, these instances are labeled as "*unknown*". The number of instances on $Tr^-(C_i)$ is proportional to the average of the training examples for each $child(y_i)$ to create a balanced training set. The idea behind $Tr^-(C_i)$ is to include instances where the associated label of the parent has the value zero. The intuition is that the instances that has the parent label set as zero, will have less probability to be predicted as true that the ones that have the parent predicted as one.

As in multidimensional classification, the class of each node in the hierarchy is not independent from the other nodes. To incorporate these relations, inspired by chain classifiers, we include the class predicted by the parent node(s) as an additional attribute in the LCPN classifier. That is, the feature space of each node in the hierarchy is extended with the 0/1 label association of the parent (tree structure) or parents (DAG structure) of the node, as in a Bayesian Chain Classifier [16].

3.2   Classification

The classification phase consists in calculating for each new instance with feature vector x_e, the probability of a node i to occur given the feature vector and the prediction of the parents at each label $P(y_i = 1 | x_e, pa(y_i))$. When the structure of the dataset is a DAG it is possible to obtain more than one prediction for one class, then the associated prediction is the average of the prediction of all the parents for that class. After computing a probability for each node, the predictions are merged using a rule to obtain a score for each path.

Merging Rule. The rule that merges the predictions of each local classifier into one score considers the level in the hierarchy of the node to determine the weight that this node will have in the overall score. Misclassifications at the upper hierarchy levels (which correspond to more generic concepts) are more expensive than those at the lower levels (which correspond to more specific concepts).

To achieve this task, the weight of a node $(w(y_i))$ is defined in Equation (2) and depicted on Figure 2, where $level(y_i)$ is the level at which the node y_i is placed in the hierarchy (Equation (1)). For a tree structure it is simply the weight of its parent plus one, and for DAG structures it is computed as the mean of the levels of the m parents $(pa(y_i))$ of the node (y_i) plus one. Finally, $maxLevel$ is the length of the longest path in the hierarchy. This way of computing the weight of each node assures that the weights are well distributed along the hierarchy; so that the weights of the lower levels do not tend rapidly to zero, as in other approaches [3,15].

$$level(y_i) = 1 + \frac{\sum_{j=1}^{m} level(pa(y_i)_j)}{|pa(y_i)|} \tag{1}$$

$$w_i = 1 - \frac{level(y_i)}{maxLevel + 1} \tag{2}$$

Equation (3) describes the merging rule which is the sum of the logarithms of the probabilities on the nodes along the path or paths (when it is a DAG), where n is the number of nodes in the path, h_i is the ith node in the path and $P(h_i = 1|x_e, pa(h_i))$ is the probability of the node h_i to be predicted as true by the local classifier. Taking the sum of logarithms is used to ensure numerical stability when computing the probability for long paths. Figure 2 depicts the classification procedure.

$$score = \sum_{i=1}^{n} w_{h_i} * log(P(h_i|x_e, pa(h_i))) \tag{3}$$

This scheme assumes independence between the labels, although in an indirect way the dependencies with the parent nodes are considered by incorporating them as additional attributes. As in chain classifiers, this scheme looks for a balance between classification accuracy and computational complexity.

For DAG structures there might be numerous paths from the root to one leaf node. In that case, all the paths that end in that leaf node are returned.

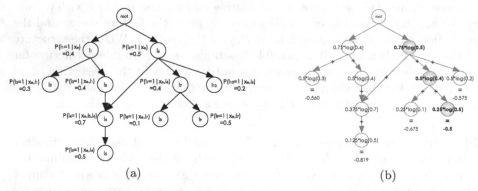

(a)   (b)

Fig. 2. Example of the application of the merging rule. (a) Each node has an associated probability. (b) The local probabilities are combined to obtain a score for each path. The path with highest score is highlighted.

3.3 Pruning

Sometimes the information available is not sufficient to estimate the class of an instance at the lower levels in the hierarchy, so it could be better to truncate the predicted path at some level, this is known as non-mandatory leaf node prediction (NMLNP). We introduce a pruning phase to obtain NMLNPs. We consider three decisions that need to be taken into account for pruning: pruning direction, pruning time and pruning condition.

Pruning direction. Determines the way the hierarchy is traversed to prune:

1. Top-Down. The hierarchy is traversed starting from the root node, when the pruning condition is met in one node, the traversing is stopped and the descendants of the node are pruned.
2. Bottom-Up. The hierarchy is traversed starting from the leaf nodes, when the pruning condition is met in one node, the traversing is stopped and the node and its descendants are pruned.

Pruning time. Determines when to perform the pruning stage:

1. Prune & Choose. Prune the hierarchy before the classification phase.
2. Choose & Prune. Prune the path that the classification phase selected.

Pruning condition. Establishes the condition to fulfill to prune a node:

1. Sum children probabilities (SUM). Prunes if the sum of the probabilities of the children is less than the probability of the 'unknown' label.
2. Most probable child (BEST). Prunes if the probability of the most probable child is less than the probability of the 'unknown' label.
3. Information Gain (IG). Prunes if there is not information gain when including the child in the prediction.

In the experiments we compared the different pruning strategies.

4 Experimental Setup

The proposed method, Chained Path Evaluation (CPE), was evaluated experimentally with several tree and DAG structured hierarchies, using five different evaluation metrics and compared with several state of the art hierarchical classification techniques.

4.1 Databases

Eighteen datasets were used in the tests, these datasets are from the field of functional genomics[1]. Ten of them (tree structured) are labeled using the FunCat annotation scheme [11]. The remaining eight datasets (DAG structured) are labeled using the Gene Ontology vocabulary [2]. From the set of paths that each instance owned we selected only the first to get instances with just one path. We prunned the hierarchy to obtain nodes with enough instances (more than 50) to train. The two tables in Table 1 represent two datasets: the first used in

---

[1] http://dtai.cs.kuleuven.be/clus/hmcdatasets/

MLNP experiments (all the class paths end on a leaf node) and the second used in NMLNP experiments (the class paths does not always end on a leaf node); they have the same names and number of attributes but not the same number of labels and instances, because the prunning was more exhaustive in MLNP case.

Table 1. Description of the datasets for the experiments. M=Number of Labels, A=Number of Attributes, N=Number of Instances and D=Maximum Depth.

(a) MLNP experiments

Dataset	M	A	N	D
Tree Hierarchies				
cellcylcle_FUN	36	77	2339	4
church_FUN	36	29	2340	4
derisi_FUN	37	65	2381	4
eisen_FUN	25	81	1681	3
expr_FUN	36	553	2346	4
gasch1_FUN	36	175	2356	4
gasch2_FUN	36	54	2356	4
pheno_FUN	17	71	1162	3
seq_FUN	39	480	2466	4
spo_FUN	36	82	2302	4
DAG Hierarchies				
cellcycle_GO	53	77	1708	11
church_GO	53	29	1711	11
derisi_GO	54	65	1746	11
expr_GO	53	553	1720	11
gasch1_GO	53	175	1716	11
gasch2_GO	53	54	1720	11
seq_GO	52	480	1711	11
spo_GO	53	82	1685	11

(b) NMLNP experiments

Dataset	M	A	N	D
Tree Hierarchies				
cellcylcle_FUN	49	77	3602	4
church_FUN	49	29	3603	4
derisi_FUN	49	65	3675	4
eisen_FUN	35	81	2335	4
expr_FUN	49	553	3624	4
gasch1_FUN	49	175	3611	4
gasch2_FUN	49	54	3624	4
pheno_FUN	22	71	1462	3
seq_FUN	51	480	3765	4
spo_FUN	49	82	3553	4
DAG Hierarchies				
cellcycle_GO	56	77	3516	11
church_GO	56	29	3515	11
derisi_GO	57	65	3485	11
expr_GO	56	553	3537	11
gasch1_GO	56	175	3524	11
gasch2_GO	56	54	3537	11
seq_GO	59	480	3659	11
spo_GO	56	82	3466	11

4.2 Evaluation Metrics

Measures for conventional classification are not adequate for hierarchical multi-label classification, for that reason specific measures for HMC have been proposed. In our work we use four of the most common evaluation metrics and propose a new metric.

Let M be the number of labels in L, N the number of instances in the training set, y_i the real set of labels and $\hat{y}_i$ the predicted set of labels. The labels of an instance are represented in a $0/1$ vector of size M where the predicted/real labels are set to 1 and the rest with 0.

Accuracy. Is the ratio of the size of the union and intersection of the predicted and actual label sets, taken for each example, and averaged over the number of examples.

$$Accuracy = \frac{1}{N} \sum_{i=1}^{N} \frac{|y_i \wedge \hat{y}_i|}{|y_i \vee \hat{y}_i|} \qquad (4)$$

Exact Match. The exact match represents the proportion of the real label sets that were predicted.

$$ExactMatch = \frac{1}{N} \sum_{i=1}^{N} 1_{y_i = \hat{y}_i} \tag{5}$$

F1-measure. F1-measure (F1) is calculated as in Equation (6) but redefining precision and recall as:

precision   as the fraction of predicted labels which are actually true $\frac{|z_i \wedge \hat{z}_i|}{|\hat{z}_i|}$.

recall   as the fraction of true labels which are also predicted $\frac{|z_i \wedge \hat{z}_i|}{|z_i|}$.

$$F1 = \frac{2 \times precision \times recall}{precision + recall} \tag{6}$$

We have specified a vector z instead of the y_i vector, because in the multi-label context there are two ways to average this measure.

F1-macro D (Equation (7)) is averaged by instances; we obtain N vectors of $z_i \equiv y_i$.

$$F1macro^{\times D}(D) = \frac{1}{N} \sum_{i=0}^{N} F1(z_i, \hat{z}_i) \tag{7}$$

F1-macro L (Equation (8)) is averaged by labels; we obtain M vectors of $z_i \equiv [y_i^1, ..., y_i^N]$.

$$F1macro^{\times L}(D) = \frac{1}{M} \sum_{i=0}^{M} F1(z_i, \hat{z}_i) \tag{8}$$

Gain-Loose Balance. In this paper we propose a new evaluation measure for hierarchical classifieres that avoids conservative predictions when using NMLNP. Gain-Loose Balance (GLB) is a measure that rewards the nodes that are correct and penalizes the ones that are incorrect. The rewards and penalties are determined using the number of siblings of the node and the depth of the node in the hierarchy.

Based on the notion that discriminating few categories is much easier than discriminating many of them, a correctly classified node with few siblings has a minor impact on the rewards than one with many. On the contrary, a misclassified node with few sibling has a mayor impact on the penalty than one with many.

A correctly classified node that belongs to a deep level in the hierarchy has more impact on the rewards than one in shallow levels, because reaching the most specific node is the goal of the prediction. In contrast, a deeper misclassified node in a deep level of the hierarchy has less impact in the penalty than one in shallow levels, while the predicted classification is near to the real it become less expensive.

Equation (9) describes the GLB measure, where n_p is the number of correct classified labels, n_{fp} is the number of false positive errors, n_{fn} is the number of false negative errors, n_t is the number of true labels; N represents the number of siblings plus one (the node that is being evaluated).

$$\frac{\sum_{i=0}^{n_p}(1 - \frac{1}{N})(1 - w_i)}{\sum_{i=0}^{n_t}(1 - \frac{1}{N})(1 - w_i)} - \left(\sum_{i=0}^{n_{fp}} \frac{1}{N}w_i + \sum_{i=0}^{n_{fn}} \frac{1}{N}w_i \right) \tag{9}$$

Gain-Loose Balance ranges from 1 (when the predicted path is equal to the real path) to $-\frac{maxL}{2}$ (see Equation (10)), where $maxL$ is the maximum number of levels in the hierarchy (see Equation (1)). In the worst case scenario the node has just two sibling and $N = 2$. w_i is defined in Equation (2).

As we know the maximum and minimum values of the GLB measure we transformed it into a (0, 1) range maintaining the ratio.

$$minValue = -2\sum_{i=1}^{maxL} \frac{1}{N}w_i = -2\sum_{i=1}^{maxL} \frac{1}{2}\left(1 - \frac{i}{maxL + 1}\right) \tag{10}$$

$$= -2\left(\sum_{i=1}^{maxL} \frac{1}{2} - \frac{1}{2(maxL + 1)} \sum_{i=1}^{maxL} i \right) \tag{11}$$

$$= -maxL + \frac{maxL}{2} \tag{12}$$

5   Experiments

The proposed method, Chained Path Evaluation (CPE), was evaluated experimentally with a number of tree and DAG structured hierarchies and compared with various hierarchical classification techniques. We performed three sets of experiments to: (i) compare our method using MLNP with other state of the art techniques, (ii) evaluate the different pruning strategies, (iii) analyze the NMLNP alternative, comparing it with MLNP and other method.

For MLNP related experiments we used the datasets in Table 1a and considered the first four evaluation metrics, for NMLNP related we used Table 1b and considered the five metrics, including GLB. The results were obtained by a stratified 10-fold cross-validation. The best results are marked in bold. Random Forest was used as base classifier for CPE in all the experiments because is the base classifier that best suits the data in our method.

5.1   Comparison of CPE-MLNP against other Methods

Results are summarized in Tables 2 and 3, the complete set of tables is presented in Appendix A.1.

Tree Structured Datasets. For tree structured hierarchies, we compared CPE against three HMC methods:

1. Top-Down LCPN (TD). Proposed by Koller et al. [8], this method trains a LCPN and selects at each level the most probable node. Only the children of this node are explored to preserve the consistency of the prediction. Random Forest was used as base classifier. This is the most typical approach for MHC.
2. Multidimensional Hierarchical Classifier (MHC). Proposed by Hernandez et al. [7]. Random Forest was used as base classifier as reported by the authors. This is the method that we are extending.
3. HIROM. Proposed by Bi et al. [4]. The used base classifier was Support Vector Machines as reported by the authors. This is a recent work in local based HMC.

Table 2. Comparing CPE against other methods in tree structured datasets

Metric	CPE	TD	MHC	HIROM
Accuracy	**23.63**	20.67	19.93	3.10
Exact Match	**18.33**	16.63	9.79	3.05
F1-macro D	**26.30**	22.76	25.07	3.12
F1-macro L	13.79	**14.38**	2.44	0.86

DAG Structured Datasets. For DAG structured datasets, we compared CPE against tree HMC methods:

1. Top-Down LCPN (TD).
2. Top-Down LCPN Corrected (TD-C). The only difference between this method and TD is that when a leaf node is reached, all the paths to that node are appended to the final prediction. TD returns a single path.
3. HIROM. The variant for DAG structures.

Table 3. Comparing CPE against other methods in DAG structured datasets

Metric	CPE	TD	TD-C	HIROM
Accuracy	**38.74**	36.48	36.42	19.02
Exact Match	**21.87**	18.59	19.22	0.0
F1-macro D	**48.86**	46.76	46.57	29.28
F1-macro L	13.68	16.18	**16.94**	2.93

Discussion. For tree structured datasets we observe that the proposed method is superior in terms of accuracy and exact match to other methods, and in most cases the difference is significant. In the case of F1-macro D and F1-macro L, it is superior to HIROM and competitive with MHC and TD. For DAG hierarchies, our method is clearly superior for the first three measures and for the fourth measure it is beaten by TD-C. One possible explanation is that it is difficult to design a method that is better in all measures, however the proposed approach is overall competitive in all measures and superior in some.

F1-macro L metric is the exception where CPE never obtains the best results. In this metric the results are averaged by labels, that means that there is probably a label(s) which does not have many instances classified, where our method could fail.

5.2   Selection of the Best NMLNP Approach

The pruning approaches described in Subsection 3.3 for the NMLNP version of CPE were tested to select the best one. In the case of the DAG structured datasets the root of the hierarchy has only one child and this child (l_0) is parent of the rest of the nodes of the hierarchy. The problem in this kind of hierarchies is that most measures score the conservative classifications as the better ones. In this case, the method *"Top-Down, Select & Prune, IG"* predict just the l_0 for every new instance, this classification is useless due to the fact that every instance belongs to l_0 and nevertheless is the one that is better scored.

Gain-Loose Balance deals with this problem and gives better scores to other methods that return relevant predictions. For that reason, and the fact that the other measures were inconsistent along the databases and the different structures, the methods were compared using Gain-Loose Balance. The results for the NMLNP approaches are depicted on Table 4, the results are averaged along the datasets.

Table 4. Comparison in terms of GLB (%) of the different approaches for NMLNP in tree and DAG structures

Dataset	Top-Down						Bottom-Up					
	Prune & Select			Select & Prune			Prune & Select			Select & Prune		
	SUM	BEST	IG	SUM	BEST	IG	SUM	BEST	IG	SUM	BEST	IG
Tree/DAG	71.02	**71.53**	68.71	69.79	70.12	69.31	70.34	70.61	68.68	69.66	69.83	68.58

Discussion. We observe that *"Top-Down, Prune & Select, BEST"* method obtains in most of the cases the better score in both, tree and DAG structures. The datasets have approximately 16% percent of instances which real label set is just the label l_0 . Since the average number of labels per instance is three, when a method predict only l_0 it already has 1/3 of the correct answer. This can be one of the reasons why the rest of the metrics give high scores with the *"Top-Down, Select & Prune, IG"* method.

5.3   Comparison of CPE NMLNP-version against MLNP-version

We compared the best NMLNP method (Top-Down, Prune & Select, BEST) against the MLNP to determine if it is worth to prune the hierarchies. Table 5 depicts the results. The datasets in Table 1b were split by hierarchy structure and the results averaged along the datasets.

Table 5. Gain-Loose Balance Metric (%) comparing NMLNP against MLNP

Dataset structure	NMLNP	MLNP
Tree	**61.688**	59.419
DAG	**83.826**	80.262

Discussion. In every dataset the results obtained by NMLNP version were significantly superior compared to the MLNP version for NMLNP datasets. Thus, pruning obtains better scores than returning complete paths to a leaf node according to the GLB metric.

5.4 Comparison of CPE-NMLNP against Other Methods

We compared CPE-NMLNP against HIROM, a method proposed by Bi et al. [4] which has a variant for DAG structures. The base classifier used for HIROM was SVM as reported by the authors. The results are depicted on Table 6.

Table 6. Comparison of MLNP methods

(a) Tree structured datasets

Metric	CPE	HIROM
Accuracy	**17.10**	3.98
Exact Match	**11.16**	3.89
F1-macro D	**19.87**	4.02
F1-macro L	**11.51**	0.76
GLB	**59.41**	41.35

(b) DAG structured datasets

Metric	CPE	HIROM
Accuracy	**32.29**	14.92
Exact Match	**9.31**	0.02
F1-macro D	**44.68**	24.81
F1-macro L	**8.56**	3.33
GLB	**80.18**	64.03

Discussion. CPE obtains better results than HIROM in most of the datasets along all the metrics, most of them with statistical relevance. This can be due to the fact that HIROM optimizes a loss function that does not necessarily improves its score in other metrics. Other fact is that HIROM is designed to return multiple paths instead of just one.

6   Conclusions and Future Work

We presented a novel approach for hierarchical multi-label classification for tree and DAG structures. The method estimates the probability of each path by combining LCPNs, including a pruning phase for NMLNP. A new metric was introduced that avoids conservative classifications. Experiments with 18 tree and DAG hierarchies show that: (i) the proposed method is competitive compared against other state-of-the-art methods for tree hierarchies and superior for DAGs, (ii) the best pruning strategy is top-down, prune first and based on the most probable child; (iii) NMLNP improves mandatory leaf-node prediction.

As future work we plan to extend the proposed method for multiple path prediction.

References

1. Alaydie, N., Reddy, C.K., Fotouhi, F.: Exploiting label dependency for hierarchical multi-label classification. In: Tan, P.-N., Chawla, S., Ho, C.K., Bailey, J. (eds.) PAKDD 2012, Part I. LNCS, vol. 7301, pp. 294–305. Springer, Heidelberg (2012)
2. Ashburner, M., Ball, C.A., Blake, J.A.: Gene Ontology: Tool for the unification of biology. Nature Genetics 25, 25–29 (2000)
3. Bi, W., Kwok, J.T.: Multilabel classification on tree- and dag-structured hierarchies. In: Proc. of the 28th Inter. Conf. on ML (ICML), pp. 17–24. Omnipress (2011)
4. Bi, W., Kwok, J.T.: Hierarchical multilabel classification with minimum bayes risk. In: IEEE Intl. Conf, on Data Mining (ICDM), pp. 101–110. IEEE Computer Society (2012)
5. Cerri, R., Barros, R.C., de Carvalho, A.C.P.L.F.: Hierarchical multi-label classification using local neural networks. J. Comput. System Sci. 1, 1–18 (2013)
6. Dekel, O., Keshet, J., Singer, Y.: An online algorithm for hierarchical phoneme classification. In: Bengio, S., Bourlard, H. (eds.) MLMI 2004. LNCS, vol. 3361, pp. 146–158. Springer, Heidelberg (2005)
7. Hernandez, J.N., Sucar, L.E., Morales, E.F.: A hybrid global-local approach for hierarchical classification. In: Boonthum-Denecke, C., Youngblood, G.M. (eds.) Proceedings FLAIRS Conference, pp. 432–437. AAAI Press (2013)
8. Koller, D., Sahami, M.: Hierarchically classifying documents using very few words. In: Proceedings of the 14th International Conference on Machine Learning, ICML 1997, pp. 170–178. Morgan Kaufmann Publishers Inc., San Francisco (1997)
9. Read, J., Pfahringer, B., Holmes, G., Frank, E.: Classifier chains for multi-label classification. Machine Learning, 254–269 (2011)
10. Rousu, J., Saunders, C., Szedmak, S., Shawe-Taylor, J.: Kernel-based learning of hierarchical multilabel classification models. J Mach. Learn. Res. 7, 1601–1626 (2006)
11. Ruepp, A., Zollner, A., Maier, D., Albermann, K., Hani, J., Mokrejs, M., Tetko, I., Güldener, U., Mannhaupt, G., Münsterkötter, M., Mewes, H.W.: The FunCat, a functional annotation scheme for systematic classification of proteins from whole genomes. Nucleic Acids Research 32(18), 5539–5545 (2004)
12. Silla Jr., C.N., Freitas, A.A.: Novel top-down approaches for hierarchical classification and their application to automatic music genre classification. In: IEEE Inter. Conf. on Systems, Man, and Cybernetics, pp. 3499–3504 (2009)
13. Silla Jr., C.N., Freitas, A.A.: A global-model naive bayes approach to the hierarchical prediction of protein functions. In: Proceedings of the 2009 9th IEEE International Conference on Data Mining, ICDM 2009, pp. 992–997. IEEE Computer Society, Washington, DC (2009)
14. Tsoumakas, G., Katakis, I.: Multi-Label Classification: An Overview. Int. J. Data Warehouse. Min. 3, 1–13 (2007)
15. Vens, C., Struyf, J., Schietgat, L., Džeroski, S., Blockeel, H.: Decision trees for hierarchical multi-label classification. Machine Learning 73(2), 185–214 (2008)
16. Zaragoza, J.H., Sucar, L.E., Morales, E.F., Bielza, C., Larrañaga, P.: Bayesian Chain Classifiers for Multidimensional Classification. Comp. Intelligence, 2192–2197 (2011)

A   Appendix

For the comparison of more than two methods (Appendix A.1) we performed a Friedman test and for the comparison of two methods (Appendix A.2) we performed a one tailed t-test; both with a confidence degree of 95%. Statistically inferior results against CPE are marked with ↓ and statistically superior results are marked with ↑.

A.1   Comparison of CPE-MLNP against Other Methods

Table 7. Tree structured datasets

(a) Accuracy (%)

Dataset	CPE	TD	MHC	HIROM
cellcycle_FUN	**22.76**	20.14	19.74	4.20↓
church_FUN	14.24	12.79	**19.71**	6.30↓
derisi_FUN	18.58	13.20	**19.43**	4.41↓
eisen_FUN	**31.07**	27.25	21.77	3.27↓
expr_FUN	**29.46**	23.36	19.63↓	1.96↓
gasch1_FUN	**28.39**	19.83	19.66↓	2.83↓
gasch2_FUN	**22.32**	19.57	19.63	3.94↓
pheno_FUN	18.39	**23.0**	22.49	6.63
seq_FUN	30.52	**32.36**	17.30	2.48↓
spo_FUN	**20.57**	15.22↓	19.92	3.82↓

(b) Exact Match (%)

Dataset	CPE	TD	MHC	HIROM
cellcycle_FUN	**17.53**	17.02	9.92	4.10↓
church_FUN	**11.20**	10.90	9.87	6.24↓
derisi_FUN	**13.27**	10.63	9.70	4.28↓
eisen_FUN	**23.32**	**23.32**	9.99↓	3.27↓
expr_FUN	**23.47**	20.20	9.85↓	1.78↓
gasch1_FUN	**22.12**	16.88	9.85↓	2.77↓
gasch2_FUN	**17.49**	15.58	9.85	3.78↓
pheno_FUN	**14.90**	11.62	9.47↓	6.63↓
seq_FUN	25.26	27.90	9.41	2.31↓
spo_FUN	**14.73**	12.29	9.95↓	3.69↓

(c) F1-macro D (%)

Dataset	CPE	TD	MHC	HIROM
cellcycle_FUN	**25.40**	21.71	24.74	4.24↓
church_FUN	15.77	13.87	**24.71**	6.32↓
derisi_FUN	21.27	14.73	**24.38**	4.45↓
eisen_FUN	**34.98**	29.33	27.66	3.27↓
expr_FUN	**32.47**	24.98	24.60↓	2.01↓
gasch1_FUN	**31.53**	21.33↓	24.64	2.85↓
gasch2_FUN	**24.76**	21.62	24.60	3.99↓
pheno_FUN	20.16	28.71	**29.00↑**	6.63
seq_FUN	33.17	34.64	21.40	2.53↓
spo_FUN	**23.49**	16.64	24.98	3.87↓

(d) F1-macro L (%)

Dataset	CPE	TD	MHC	HIROM
cellcycle_FUN	14.34	**18.05**	2.09	0.69↓
church_FUN	**7.72**	6.80	2.09	0.98↓
derisi_FUN	**11.54**	10.53	2.01↓	0.70↓
eisen_FUN	**19.59**	17.73	3.21↓	0.76↓
expr_FUN	18.45	**20.30**	2.08	0.57↓
gasch1_FUN	**17.24**	15.73	2.08↓	0.47↓
gasch2_FUN	8.74	**11.22**	2.08	0.61↓
pheno_FUN	**11.14**	9.83	4.86↓	1.46↓
seq_FUN	17.21	**21.05**	1.75	0.71↓
spo_FUN	11.97	**12.54**	2.11↓	0.63↓

Table 8. DAG structured datasets

(a) Accuracy (%)

Dataset	CPE	TD	TC-C	HIROM
cellcycle_GO	**36.60**	34.70	34.64	19.01↓
church_GO	**32.09**	31.11	30.86	19.11↓
derisi_GO	**33.42**	32.61	32.45	18.44↓
expr_GO	**42.80**	38.93	38.91↓	19.18↓
gasch1_GO	**42.03**	39.39	39.23	19.17↓
gasch2_GO	**39.53**	36.44	36.29↓	19.18↓
seq_GO	**48.99**	46.03	46.55	19.26↓
spo_GO	**34.45**	32.63	32.44↓	18.79↓

(b) Exact Match (%)

Dataset	CPE	TD	TD-C	HIROM
cellcycle_GO	**19.26**	16.74	17.27	0.00↓
church_GO	**13.79**	12.45	12.74	0.00↓
derisi_GO	**15.41**	14.09	14.43	0.00↓
expr_GO	**27.33**	21.86↓	22.50	0.00↓
gasch1_GO	**26.28**	22.09	22.73	0.00
gasch2_GO	**22.62**	19.01↓	19.30	0.00↓
seq_GO	**33.31**	28.46↓	30.51	0.00↓
spo_GO	**16.97**	14.01↓	14.24	0.00↓

(c) F1-macro D (%)

Dataset	CPE	TD	TD-C	HIROM
cellcycle_GO	**47.03**	45.19	44.98	29.28↓
church_GO	**43.24**	42.05	41.68	29.40↓
derisi_GO	**44.25**	43.40	43.14	28.59↓
expr_GO	**52.14**	48.77	48.64↓	29.47↓
gasch1_GO	**51.52**	49.27	48.97	29.47↓
gasch2_GO	**49.60**	46.69	46.41↓	29.47↓
seq_GO	**57.94**	55.28	55.55	29.54↓
spo_GO	**45.12**	43.45	43.15	29.03↓

(d) F1-macro L (%)

Dataset	CPE	TD	TD-C	HIROM
cellcycle_GO	10.45	14.32	**15.17↑**	2.92
church_GO	8.72	10.41	**10.51**	2.93↓
derisi_GO	9.50	13.17	**13.35↑**	2.81
expr_GO	17.45	17.82	**18.80**	2.95↓
gasch1_GO	16.70	18.24	**18.29**	2.94↓
gasch2_GO	12.52	15.02	**15.42**	2.95
seq_GO	24.28	28.21	**31.60↑**	3.02
spo_GO	9.79	12.21	**12.38**	2.89

A.2   Comparison of CPE-NMLNP against Other Methods

Table 9. Tree structured datasets

(a) Accuracy (%)   (b) Exact Match (%)   (c) F1-macro D

Dataset	CPE	HIROM	Dataset	CPE	HIROM	Dataset	CPE	HIROM
cellcycle_FUN	16.26	4.2↓	cellcycle_FUN	10.74	4.2↓	cellcycle_FUN	18.87	4.24↓
church_FUN	9.12	6.3↓	church_FUN	5.86	**6.3**	church_FUN	10.7	6.32↓
derisi_FUN	13.71	4.41↓	derisi_FUN	8.62	4.41↓	derisi_FUN	16.22	4.45↓
eisen_FUN	21.46	3.27↓	eisen_FUN	14.91	3.27↓	eisen_FUN	24.45	3.27↓
expr_FUN	21.44	1.96↓	expr_FUN	13.99	1.96↓	expr_FUN	24.81	2.01↓
gasch1_FUN	20.92	2.83↓	gasch1_FUN	13.87	2.83↓	gasch1_FUN	24.19	2.85↓
gasch2_FUN	16.44	3.94↓	gasch2_FUN	10.76	3.94↓	gasch2_FUN	19.11	3.99↓
pheno_FUN	13.36	6.63↓	pheno_FUN	8.35	6.63↓	pheno_FUN	15.63	6.63↓
seq_FUN	24.05	2.48↓	seq_FUN	15.64	2.48↓	seq_FUN	27.85	2.53↓
spo_FUN	14.26	3.82↓	spo_FUN	8.81	3.82↓	spo_FUN	16.84	3.87↓

(d) F1-macro L (%)   (e) GLB (%)

Dataset	CPE	HIROM	Dataset	CPE	HIROM
cellcycle_FUN	11.94	0.69↓	cellcycle_FUN	58.96	39.99↓
church_FUN	4.47	0.98↓	church_FUN	56.40	41.18↓
derisi_FUN	8.58	0.7↓	derisi_FUN	57.66	40.03↓
eisen_FUN	13.28	0.76↓	eisen_FUN	60.96	44.5↓
expr_FUN	16.21	0.57↓	expr_FUN	60.48	38.78↓
gasch1_FUN	15.82	0.47↓	gasch1_FUN	60.69	39.2↓
gasch2_FUN	8.28	0.61↓	gasch2_FUN	59.26	39.89↓
pheno_FUN	10.64	1.46↓	pheno_FUN	59.6	50.62↓
seq_FUN	16.36	0.71↓	seq_FUN	62.16	39.55↓
spo_FUN	9.52	0.63↓	spo_FUN	57.9	39.71↓

Table 10. DAG structured datasets

(a) Accuracy (%)   (b) Exact Match (%)   (c) F1-macro D (%)

Dataset	CPE	HIROM	Dataset	CPE	HIROM	Dataset	CPE	HIROM
cellcycle_GO	41.08	16.73↓	cellcycle_GO	3.61	0.0↓	cellcycle_GO	55.81	27.43↓
church_GO	40.41	18.49↓	church_GO	3.10	0.0↓	church_GO	55.27	29.90↓
derisi_GO	40.98	17.08↓	derisi_GO	4.10	0.09↓	derisi_GO	55.61	27.85↓
expr_GO	39.15	14.7↓	expr_GO	8.76	0.0↓	expr_GO	52.53	24.59↓
gasch1_GO	40.82	15.4↓	gasch1_GO	7.01	0.0↓	gasch1_GO	54.80	25.60↓
gasch2_GO	42.14	17.19↓	gasch2_GO	5.00	0.0↓	gasch2_GO	56.69	28.14↓
seq_GO	40.9	14.08↓	seq_GO	9.65	0.0↓	seq_GO	54.28	23.70↓
spo_GO	41.19	17.18↓	spo_GO	4.53	0.0↓	spo_GO	55.76	28.10↓

(d) F1-macro L (%)   (e) GLB (%)

Dataset	CPE	HIROM	Dataset	CPE	HIROM
cellcycle_GO	5.09	3.96↓	cellcycle_GO	83.83	65.51↓
church_GO	3.77	4.07	church_GO	83.57	67.32↓
derisi_GO	5.16	3.93↓	derisi_GO	83.73	65.65↓
expr_GO	10.34	3.34↓	expr_GO	83.55	63.76↓
gasch1_GO	8.87	3.55↓	gasch1_GO	83.96	64.35↓
gasch2_GO	4.64	3.87	gasch2_GO	84.12	66.27↓
seq_GO	11.44	2.96↓	seq_GO	84.01	63.24↓
spo_GO	5.37	4.12↓	spo_GO	83.84	65.87↓

Min-BDeu and Max-BDeu Scores
for Learning Bayesian Networks

Mauro Scanagatta, Cassio P. de Campos, and Marco Zaffalon

Istituto Dalle Molle di Studi sull'Intelligenza Artificiale (IDSIA), Switzerland
{mauro,cassio,zaffalon}@idsia.ch

Abstract. This work presents two new score functions based on the Bayesian Dirichlet equivalent uniform (BDeu) score for learning Bayesian network structures. They consider the sensitivity of BDeu to varying parameters of the Dirichlet prior. The scores take on the most adversary and the most beneficial priors among those within a contamination set around the symmetric one. We build these scores in such way that they are decomposable and can be computed efficiently. Because of that, they can be integrated into any state-of-the-art structure learning method that explores the space of directed acyclic graphs and allows decomposable scores. Empirical results suggest that our scores outperform the standard BDeu score in terms of the likelihood of unseen data and in terms of edge discovery with respect to the true network, at least when the training sample size is small. We discuss the relation between these new scores and the accuracy of inferred models. Moreover, our new criteria can be used to identify the amount of data after which learning is saturated, that is, additional data are of little help to improve the resulting model.

Keywords: Bayesian networks, structure learning, Bayesian Dirichlet score.

1 Introduction

A Bayesian network is a versatile and well-known probabilistic graphical model with applications in a variety of fields. It relies on a structured dependency among random variables to represent a joint probability distribution in a compact and efficient manner. These dependencies are encoded by an acyclic directed graph (DAG) where nodes are associated to random variables and conditional probability distributions are defined for variables given their parents in the graph. Learning the graph (or structure) of Bayesian networks from data is one of its most challenging problems.

The topic of Bayesian network learning has been extensively discussed in the literature and many different approaches are available. In general terms, the problem is to find the structure that maximizes a given score function that depends on the data [1]. The research on this topic is very active, with numerous methods and papers [2, 3, 4, 5, 6, 7, 8, 9, 10]. The main characteristic tying

L.C. van der Gaag and A.J. Feelders (Eds.): PGM 2014, LNAI 8754, pp. 426–441, 2014.

together all these methods is the score function. Arguably, the most commonly used score function is the Bayesian Dirichlet (likelihood) equivalent uniform (BDeu), which derives from BDe and BD [11, 12, 1] (other examples of score functions are the Bayesian Information Criterion [13], which is equivalent to *Minimum Description Length*, and the Akaike Information Criterion [14]). There are also more recent attempts to devise new score functions. For example, [15] presents a score that aims at having its maximization computationally facilitated as the amount of data increases.

The BDeu score aims at maximizing the posterior probability of the DAG given data, while assuming a uniform prior over possible DAGs. In this work we propose two new score functions, namely Min-BDeu and Max-BDeu. These scores are based on the BDeu score, but they consider all possible prior probability distributions inside an ε-contaminated set [16] of Dirichlet priors around the symmetric one (which is the one used by the original BDeu). Min-BDeu is the score obtained by choosing the most adversary prior distributions (that is, those minimizing the score) from the contaminated sets, while Max-BDeu is the score that uses the most beneficial priors to maximize the resulting value. We demonstrate that Min-BDeu and Max-BDeu can be efficiently calculated and are decomposable. Because of that, any structure learning solver can be used to find the best scoring DAG with them. We empirically show that Min-BDeu achieves better predictive accuracy (based on the likelihood of held-out data) than the original BDeu for small sample sizes, and performs similarly to BDeu when the amount of data is large. On the other hand, Max-BDeu achieves better edge accuracy (evaluated by the Hamming distance between the set of edges of true and learned moralized graphs).

A very important question regarding structure learning is whether the result is *accurate*, that is, whether it produces a network that will give accurate results on future unseen data. In this regard, we empirically show an interesting relation between accuracy obtained with a given training sample size and the gap between Max-BDeu and Min-BDeu. This relation might be used to identify the amount of data that is necessary to obtain an accurate network, as we will discuss later on.

The paper is divided as follows. Section 2 defines Bayesian networks, introduces our notation and the problem of structure learning. Section 3 presents our new score functions and demonstrates the existence of efficient algorithms to compute them. Section 4 describes our experimental setting and discusses two experiments regarding the accuracy of Min-BDeu and the use of Max-BDeu and Min-BDeu to help in predicting the amount of data needed to achieve a desired learning accuracy. Finally, Section 5 concludes the paper and discusses future work.

2   Learning Bayesian Networks

A Bayesian network represents a joint probability distribution over a collection of random variables, which we assume to be categorical. It can be defined as a triple $(\mathcal{G}, \mathcal{X}, \mathcal{P})$, where $\mathcal{G} \doteq (V_{\mathcal{G}}, E_{\mathcal{G}})$ is an acyclic directed graph (DAG) with $V_{\mathcal{G}}$ a collection of n nodes associated to random variables $\mathcal{X}$ (a node per variable, which might be used interchangeably to denote each other), and $E_{\mathcal{G}}$ a collection of arcs; $\mathcal{P}$ is a collection of conditional mass functions $p(X_i|\Pi_i)$ (one for each instantiation of Π_i), where Π_i denotes the parents of X_i in the graph (Π_i may be empty), respecting the relations of $E_{\mathcal{G}}$. In a Bayesian network every variable is conditionally independent of its non-descendant non-parent variables given its parent variables (Markov condition).

We use uppercase letters such as X_i, X_j to represent variables (or nodes of the graph), and x_i to represent a generic state of X_i, which has state space $\Omega_{X_i} \doteq \{x_{i1}, x_{i2}, \ldots, x_{ir_i}\}$, where $r_i \doteq |\Omega_{X_i}| \geq 2$ is the number of (finite) categories of X_i ($|\cdot|$ is the cardinality of a set or vector). Bold letters are used to emphasize sets or vectors. For example, $\mathbf{x} \in \Omega_{\mathbf{X}} \doteq \times_{X \in \mathbf{X}} \Omega_X$, for $\mathbf{X} \subseteq \mathcal{X}$, is an instantiation for all the variables in $\mathbf{X}$. $r_{\Pi_i} \doteq |\Omega_{\Pi_i}| = \prod_{X_t \in \Pi_i} r_t$ is the number of possible instantiations of the parent set Π_i of X_i, and $\boldsymbol{\theta} = (\theta_{ijk})_{\forall ijk}$ is the entire vector of parameters with elements $\theta_{ijk} = p(x_{ik}|\boldsymbol{\pi}_{ij})$, for $i \in \{1, \ldots, n\}$, $j \in \{1, \ldots, r_{\Pi_i}\}$, $k \in \{1, \ldots, r_i\}$, and $\boldsymbol{\pi}_{ij} \in \Omega_{\Pi_i}$. Because of the Markov condition, the Bayesian network represents a joint probability distribution by the expression $p(\mathbf{x}) = p(x_1, \ldots, x_n) = \prod_i p(x_i|\boldsymbol{\pi}_{ij})$, for every $\mathbf{x} \in \Omega_{\mathcal{X}}$, where every x_i and $\boldsymbol{\pi}_{ij}$ are consistent with $\mathbf{x}$.

Given a complete data set $D \doteq \{D_1, \ldots, D_N\}$ with N instances, where $D_u \doteq \mathbf{x}_u \in \Omega_{\mathcal{X}}$ is an instantiation of all the variables, the goal of structure learning is to find a DAG $\mathcal{G}$ that maximizes a given score function, that is, we look for $\mathcal{G}^* \doteq \operatorname{argmax}_{\mathcal{G} \in \mathbf{G}} s_D(\mathcal{G})$, with $\mathbf{G}$ the set of all DAGs with nodes $\mathcal{X}$, for a given score function s_D (the dependency on data is indicated by the subscript D).[1] In this paper, we consider the Bayesian Dirichlet equivalent uniform (BDeu) [11, 12, 1]. The BDeu score idea is to compute a function based on the posterior probability of the structure $p(\mathcal{G}|D)$. In this work we use $p(D|\mathcal{G})$, which is equivalent to the former (in the sense of yielding the same optimal graphs) if one assumes $p(\mathcal{G})$ to be uniform over DAGs:

$$s_D(\mathcal{G}) \doteq \log p(D|\mathcal{G}) = \log \int p(D|\mathcal{G}, \boldsymbol{\theta}) \cdot p(\boldsymbol{\theta}|\mathcal{G}) d\boldsymbol{\theta} \ ,$$

where the logarithm is used to simplify computations, $p(\boldsymbol{\theta}|\mathcal{G})$ is the prior of $\boldsymbol{\theta}$ for a given graph $\mathcal{G}$, assumed to be a Dirichlet with parameters $\boldsymbol{\alpha} \doteq (\boldsymbol{\alpha}_i)_{\forall i}$ with $\boldsymbol{\alpha}_i \doteq (\alpha_{ijk})_{\forall jk}$ (which are assumed to be strictly positive and whose dependence on $\mathcal{G}$, or more specifically on Π_i, is omitted unless necessary in the context):

$$p(\boldsymbol{\theta}|\mathcal{G}) = \prod_{i=1}^{n} \prod_{j=1}^{r_{\Pi_i}} \Gamma\left(\sum_k \alpha_{ijk}\right) \prod_{k=1}^{r_i} \frac{\theta_{ijk}^{\alpha_{ijk}-1}}{\Gamma(\alpha_{ijk})} \ .$$

From now on, we denote the Dirichlet prior by its defining parameter $\boldsymbol{\alpha}$. Under these assumptions, it has been shown [12] that

---

[1] In case of many optimal DAGs, then we assume to have no preference and argmax returns one of them.

$$s_D(\mathcal{G}) = \log \prod_{i=1}^{n} \prod_{j=1}^{r_{\Pi_i}} \frac{\Gamma(\sum_k \alpha_{ijk})}{\Gamma(\sum_k (\alpha_{ijk} + n_{ijk}))} \prod_{k=1}^{r_i} \frac{\Gamma(\alpha_{ijk} + n_{ijk})}{\Gamma(\alpha_{ijk})}, \quad (1)$$

where n_{ijk} indicates how many elements of D contain both x_{ik} and π_{ij} (the dependence of n_{ijk} on Π_i is omitted too). The BDe score assumes the prior $\boldsymbol{\alpha}$ to be such that $\alpha_{ijk} \doteq \alpha^* \cdot p(\theta_{ijk}|\mathcal{G})$, where α^* is the parameter known as the Equivalent Sample Size (or the prior strength), and $p(\theta_{ijk}|\mathcal{G})$ is the prior probability for $(x_{ik} \wedge \pi_{ij})$ given $\mathcal{G}$ (or simply given Π_i). The BDeu score assumes further that $p(\theta_{ijk}|\mathcal{G})$ is uniform and thus $\alpha_{ijk} \doteq \frac{\alpha^*}{r_{\Pi_i} r_i}$ and α^* becomes the only free parameter.

An important property of BDeu is that its function is decomposable and can be written in terms of the local nodes of the graph, that is, $s_D(\mathcal{G}) = \sum_{i=1}^{n} s_i(\Pi_i)$ (the subscript D is omitted from now on), such that

$$s_i(\Pi_i) = \sum_{j=1}^{r_{\Pi_i}} \left(\log \frac{\Gamma(\sum_k \alpha_{ijk})}{\Gamma(\sum_k (\alpha_{ijk} + n_{ijk}))} + \sum_{k=1}^{r_i} \log \frac{\Gamma(\alpha_{ijk} + n_{ijk})}{\Gamma(\alpha_{ijk})} \right). \quad (2)$$

3   Min-BDeu and Max-BDeu Scores

In order to study the sensitivity of the BDeu score to different choices of prior $\boldsymbol{\alpha}$, we define an ε-contaminated set of priors. Let β denote $\frac{\alpha^*}{r_{\Pi_i}}$, $\mathbf{1}$ denote the vector $[1, \ldots, 1]$ with length r_i, and $\mathcal{S}$ denote the set of the r_i distinct degenerate mass functions of dimension r_i. Then

$$\forall ij : A_{ij}^{\varepsilon} \doteq \mathrm{CH} \left\{ \beta \left(\frac{(1-\varepsilon)}{r_i} \mathbf{1} + \varepsilon \mathbf{v} \right) \mid \mathbf{v} \in \mathcal{S} \right\}$$

$$= \left\{ \forall k : \alpha_{ijk} \in \left[\beta \frac{(1-\varepsilon)}{r_i}, \beta \left(\varepsilon + \frac{(1-\varepsilon)}{r_i} \right) \right], \sum_{k=1}^{r_i} \alpha_{ijk} = \beta \right\}, \quad (3)$$

where CH means the convex hull operator. Equation (3) defines a set of priors for each i, j by allowing the Dirichlet parameters to vary "around" the symmetric Dirichlet with sum of parameters β. To accommodate possible different choices of priors, we rewrite the score function for each node to take into account the value of $\boldsymbol{\alpha}$:

$$s_i(\Pi_i, \boldsymbol{\alpha}_i) = \sum_{j=1}^{r_{\Pi_i}} s_{i,j}(\Pi_i, \boldsymbol{\alpha}_{ij}),$$

$$s_{i,j}(\Pi_i, \boldsymbol{\alpha}_{ij}) = \log \frac{\Gamma(\sum_k \alpha_{ijk})}{\Gamma(\sum_k (\alpha_{ijk} + n_{ijk}))} + s'_{i,j}(\Pi_i, \boldsymbol{\alpha}_{ij}), \text{ and}$$

$$s'_{i,j}(\Pi_i, \boldsymbol{\alpha}_{ij}) = \sum_{k=1}^{r_i} \log \frac{\Gamma(\alpha_{ijk} + n_{ijk})}{\Gamma(\alpha_{ijk})} = \sum_{\substack{k \in \{1, \ldots, r_i\}: \\ n_{ijk} \neq 0}} \log \frac{\Gamma(\alpha_{ijk} + n_{ijk})}{\Gamma(\alpha_{ijk})}. \quad (4)$$

where $\boldsymbol{\alpha}_{ij} \in A_{ij}^{\varepsilon}$ for a given $0 < \varepsilon \leq 1$. Using this extended parametrization of the score, we can define our new score functions Min-BDeu $\underline{s}$ and Max-BDeu $\overline{s}$.

$$\underline{s}(\mathcal{G}) \doteq \min_{\boldsymbol{\alpha}} \sum_{i=1}^{n} s_i(\Pi_i, \boldsymbol{\alpha}_i) \quad \text{and} \quad \overline{s}(\mathcal{G}) \doteq \max_{\boldsymbol{\alpha}} \sum_{i=1}^{n} s_i(\Pi_i, \boldsymbol{\alpha}_i) \ ,$$

where maximization and minimization are taken with respect to the sets A_{ij}^{ε}, for every i, j. The names Max-BDeu and Min-BDeu represent the fact that a maximization (or minimization) of the Bayesian Dirichlet equivalent uniform score over the ε-contamination set is performed. However, we note that Min-BDeu and Max-BDeu scores as defined here do not necessarily respect the *likelihood equivalence* property of BDeu. These scores can be seen as a sensitivity analysis of the structure under different prior distributions [17]. It is possible to maintain likelihood equivalence by enforcing constraints among the Dirichlet parameters, but decomposability would be lost and the score computation would become very expensive [18]. Arguments about varying priors and likelihood equivalence are given in [19].

We devote the final part of this section to demonstrate that Min-BDeu and Max-BDeu can be efficiently computed. The first important thing to note is that the maximization can be performed independently for each i, j (the same holds for the minimization):

$$\overline{s}(\mathcal{G}) = \max_{\boldsymbol{\alpha}} \sum_{i=1}^{n} s_i(\Pi_i, \boldsymbol{\alpha}_i) = \sum_{i=1}^{n} \max_{\boldsymbol{\alpha}_i} s_i(\Pi_i, \boldsymbol{\alpha}_i) = \sum_{i=1}^{n} \sum_{j=1}^{r_{\Pi_i}} \max_{\boldsymbol{\alpha}_{ij} \in A_{ij}^{\varepsilon}} s_{i,j}(\Pi_i, \boldsymbol{\alpha}_{ij}) \ ,$$

where

$$\max_{\boldsymbol{\alpha}_{ij} \in A_{ij}^{\varepsilon}} s_{i,j}(\Pi_i, \boldsymbol{\alpha}_{ij}) = \log \frac{\Gamma(\beta)}{\Gamma(\beta + \sum_k n_{ijk})} + \max_{\boldsymbol{\alpha}_{ij} \in A_{ij}^{\varepsilon}} s'_{i,j}(\Pi_i, \boldsymbol{\alpha}_{ij}) \ .$$

Expanding the Gamma functions of $s'_{i,j}(\Pi_i, \boldsymbol{\alpha}_{ij})$ by repeatedly using the equality $\Gamma(z+1) = z\Gamma(z)$, we obtain the following convex optimization problem with linear constraints:

$$\max_{(\alpha_{ijk}) \forall k} \sum_{\substack{k \in \{1, \dots, r_i\}: \\ n_{ijk} \neq 0}} \sum_{w=0}^{n_{ijk}-1} \log(\alpha_{ijk} + w)$$

$$\text{subject to} \sum_{k=1}^{r_i} \alpha_{ijk} = \beta \quad \text{and} \quad \forall k : \alpha_{ijk} \in \left[\beta \frac{(1-\varepsilon)}{r_i}, \beta \left(\varepsilon + \frac{(1-\varepsilon)}{r_i} \right) \right] \ .$$

Hence, the solution of $\max_{\boldsymbol{\alpha}_i} s_i(\Pi_i, \boldsymbol{\alpha}_i)$ to obtain the local score of parent set Π_i of X_i can be done with r_{Π_i} calls to a convex programming solver, each of which runs in worst-case time cubic in r_i [20].

The solution for the minimization $\min_{\boldsymbol{\alpha}_{ij} \in A_{ij}^{\varepsilon}} s'_{i,j}(\Pi_i, \boldsymbol{\alpha}_{ij})$ is even simpler: it is enough to find $k_* = \arg\min n_{ijk}$ and take as optimal solution the prior with

$$\alpha_{ijk_*} = \beta \left(\varepsilon + \frac{(1-\varepsilon)}{r_i} \right) \quad \text{and} \quad \forall k \neq k_* : \alpha_{ijk} = \beta \frac{(1-\varepsilon)}{r_i} \ .$$

In order to compute the score of parent set Π_i, we simply repeat this procedure for every j and compute the associated scores. While much easier to solve, the proof of correctness is slightly more intricate. We do it in three steps. The first step considers the case when at least one $n_{ijk} = 0$. In this case, we can safely choose any k_* such that $n_{ijk_*} = 0$ and the solution value is trivially minimal, because the corresponding term does not appear in the objective function defined by (4) and each function

$$g(\eta, \alpha_{ijk}) = \sum_{w=0}^{\eta-1} \log(\alpha_{ijk} + w)$$

appearing in (4) is monotonically increasing with $\eta > 0$, hence we cannot do better than choosing the minimum possible value (that is, $\beta \frac{(1-\varepsilon)}{r_i}$) for each α_{ijk} associated to non-zero n_{ijk}.

Now we can assume that $n_{ijk} \geq 1$ for every k. The second step of the proof is by contradiction and its goal is to show that only one α_{ijk} will be different from $\beta \frac{(1-\varepsilon)}{r_i}$. So, suppose that the optimal solution is attained at a point α_{ij} such that there are $k_1 \neq k_2$ with $\alpha_{ijk_1} > \frac{(1-\varepsilon)\beta}{r_i}$, $\alpha_{ijk_2} > \frac{(1-\varepsilon)\beta}{r_i}$, $n_{ijk_1} \geq 1$ and $n_{ijk_2} \geq 1$. Let $\mu = \alpha_{ijk_1} + \alpha_{ijk_2}$. Take the terms of the objective function in (4) that correspond to k_1 and k_2:

$$f(\mu, \alpha_{ijk_1}) = \sum_{w=0}^{n_{ijk_1}-1} \log(\alpha_{ijk_1} + w) + \sum_{w=0}^{n_{ijk_2}-1} \log(\mu - \alpha_{ijk_1} + w) \ .$$

While keeping μ constant, we can decrease α_{ijk_1} until $\beta \frac{(1-\varepsilon)}{r_i}$, or increase it until $\alpha_{ijk_2} = \beta \frac{(1-\varepsilon)}{r_i}$. The second derivative of $f(\mu, \alpha_{ijk_1})$ with respect to α_{ijk_1} is

$$-\sum_{w=0}^{n_{ijk_1}-1} \frac{1}{(\alpha_{ijk_1} + w)^2} - \sum_{w=0}^{n_{ijk_2}-1} \frac{1}{(\mu - \alpha_{ijk_1} + w)^2} < 0 \ ,$$

so the function is concave in α_{ijk_1}. Because of that, the minimum is attained at one of its extremes, that is, either at $\alpha_{ijk_1} = \beta \frac{(1-\varepsilon)}{r_i}$ or at $\alpha_{ijk_2} = \beta \frac{(1-\varepsilon)}{r_i}$. If we take such a new solution α'_{ij}, it achieves value strictly smaller than that of α_{ij}, which is a contradiction.

Hence we can assume that all $k \neq k_*$ must have $\alpha_{ijk} = \beta \frac{(1-\varepsilon)}{r_i}$ and we only need to choose the k_* whose α_{ijk_*} will be $\beta \left(\varepsilon + \frac{(1-\varepsilon)}{r_i} \right)$. The third step of the proof is straightforward: the best choice for k_* in order to minimize $s'_{i,j}(\Pi_i, \alpha_{ij})$ is such that

$$k_* = \arg\min_k g\left(n_{ijk}, \beta \left(\varepsilon + \frac{(1-\varepsilon)}{r_i} \right) \right) \ ,$$

that is, the one of smallest n_{ijk}, simply because the function g is monotonically increasing with η, as described previously, and also with α_{ijk}.

4   Experimental Setup

We begin this section by describing data and settings that are used in the experiments. Our goals are to assess the accuracy of the new proposed scores and to understand the relation between them and the quality of inferred networks. The experimental procedure is performed in steps, as explained in the following.

In order to allow for a comparison against the true model, we generate data from pre-defined Bayesian networks. We experiment both with networks from the literature and with random-generated networks. In the former case, we use the well-known networks named *child* (20 nodes) [21], *insurance* (27 nodes) [22], *water* (32 nodes) [23], and *alarm* (37 nodes) [24]. In the latter case, we employ the *BNGenerator* package v0.3 of [25] to obtain random Bayesian networks. The options passed to the program are the desired number of nodes in the randomly generated network and the maximum degree (sum of number of parents and children) allowed for each node in the graph (to avoid excessively dense graphs, which would require a too large amount of data for learning because of their complexity). The maximum degree (sum of incoming and outgoing arcs) is fixed to six in all the experiments, while the number of nodes varies from 20 to 50 nodes. The number of states is 2 for every node. For each different number of nodes, ten networks are randomly generated.

From each one of these networks, data sets are sampled with ten different sample sizes $N = 10 \cdot 2^i, \forall i \in \{1, \dots, 10\}$ using the R package *bnlearn* v3.5 [26]. These data sets are then used to compute the BDeu scores for each node, as well as Min-BDeu and Max-BDeu, with equivalent sample size α^* set to one, ε set to one half, and upper limit of three parents per node (because of decomposability, scores are always computed per node and stored in some data structure for later querying). We point out that we have not tuned these values, but instead we chose values that are common in the literature. In spite of that, the results presented in this work remain unaltered under small modifications of α^* around the chosen value (data not shown). We leave for future work a thorough analysis of different values for α^*. Limiting the number of parents per node is a common practice and has the purpose of avoiding a large computational cost of evaluating the scores of a great number of parent sets per node (this number increases exponentially with the limit). We discuss the implications of this decision later on. After the scores are computed and pruned [27], we call the structure learning solver *Gobnilp* v1.4.1 [2, 3] to infer the Bayesian network in an optimal way (that is, we wait until the solver finds the globally optimal structure for the given local scores). With that network, parameter estimation is performed using the same Dirichlet prior parameters and the same data as used for structure learning.

In order to check the accuracy of the inferred networks, from each true network we generate an additional data set with 10000 samples that is not available to the learning procedure. On these data we compute the log-likelihood function using both the true network and the inferred network, and to make the outputs comparable we take the percentage difference between them, which we call *likelihood accuracy*. Lower values of this measure indicate that the inferred network evaluates the log-likelihood of the held-out data in a more similar way to the evaluation done by the true network. The evaluation of the log-likelihood using the inferred network over unseen data sampled from the true network approximates the Kullback-Leibler (KL) divergence and converges to the latter when the amount of data goes to infinity. We refrained from a direct use of KL divergence or other measure of distance between the true distribution (represented by the true network) and the inferred one because of the computational cost of evaluating KL, given that true and inferred networks have (almost always) different structures.

In summary, each execution (i) generates a network (or picks a known one) of a given number of nodes; (ii) generates a data set for accuracy evaluation; (iii) generates a data set for training of a given number of samples; (iv) computes the local scores for each variable using the training data set, for a given decomposable score function; (v) calls the learning procedure using the just calculated local scores; (vi) evaluates the likelihood accuracy with learned/true networks using the held-out data.

4.1 Comparison among Scores

In this experiment, we compare the likelihood accuracy obtained by our proposed scores and the original BDeu. Figure 1 show the average results of the experiment

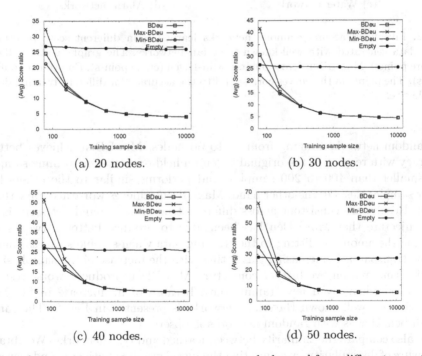

(a) 20 nodes.   (b) 30 nodes.

(c) 40 nodes.   (d) 50 nodes.

Fig. 1. Average likelihood accuracy among networks learned from different score functions. The graphs feature BDeu (square points), Max-BDeu (cross points), Min-BDeu (circle points), Empty net (bullet points). Each curve of the graph corresponds to the average over random networks used for sampling the data. The points in those curves correspond to the accuracy for different training data sample size.

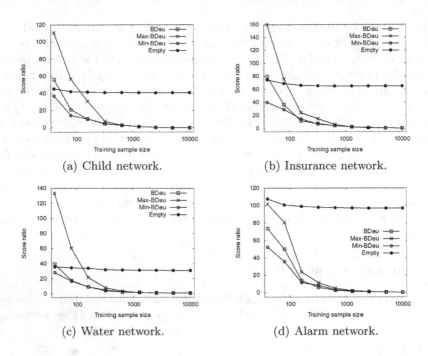

(a) Child network.   (b) Insurance network.

(c) Water network.   (d) Alarm network.

Fig. 2. Likelihood accuracy among networks learned from different score functions using data generated with well-known Bayesian networks. The graphs feature BDeu (square points), Max-BDeu (cross points), Min-BDeu (circle points), Empty net (bullet points). The points in the curves correspond to the accuracy for different training data sample size.

for random networks ranging from 20 to 50 nodes. Min-BDeu achieves better accuracy with respect to the original BDeu on held-out data for training sample size smaller than 100 to 200 samples, and performs similar to the others for larger sample sizes. On the other hand, Max-BDeu achieves worse accuracy than BDeu. Results are consistent across different true networks and network sizes. We conjecture that Min-BDeu has been able to produce better networks by reducing the amount of fitting to the training data when sample size was small, and by transparently increasing this fitting with the increase of the sample size. For the same reason, we further conjecture Max-BDeu produced worse results than BDeu when given few training data because of its increase in fit. The evaluation for well-known Bayesian networks is presented in Fig. 2. The same overall behavior as with random networks is observed.

We also compare the similarity between learned and true networks. We obtain a measure of dissimilarity by computing the moral graph of both true and learned networks, and by counting the total number of mismatches over all pairs of nodes. Figure 3 shows average results for random networks ranging from 20 to 50 nodes. Max-BDeu achieves a better similarity than others for training sample size smaller than 100 to 200 samples, and performs similarly for larger sample

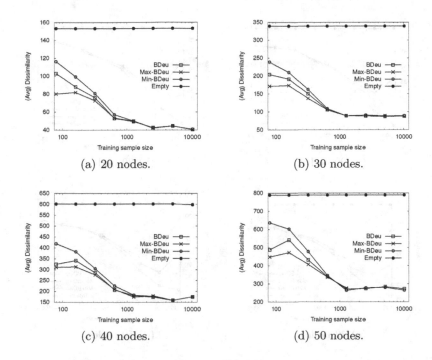

(a) 20 nodes.   (b) 30 nodes.

(c) 40 nodes.   (d) 50 nodes.

Fig. 3. Comparison on average computed dissimilarity between true and learned moral graphs using different score functions. The graphs feature BDeu (square points), Max-BDeu (cross points), Min-BDeu (circle points), Empty net (bullet points). The points in the curves correspond to the total number of mismatches between the moral graph of true and learned networks.

sizes. On the other hand, Min-BDeu achieves worse similarity than BDeu. Results are consistent across different network sizes. To analyze further these findings, we show the average total number of arcs produced by the different scores on the same experiments (Fig. 4). Notably, Max-BDeu yields denser inferred networks, while Min-BDeu prefers sparser networks.

The results obtained so far suggest that Min-BDeu and Max-BDeu aim at different goals. Min-BDeu has better likelihood accuracy, which is achieved by using sparser graphs than original BDeu's. Max-BDeu has better edge accuracy, computed as the learned graph similarity to the true one, which is achieved by using denser graphs than BDeu's. These results suggest that with small amount of data, better likelihood accuracy is obtained with networks that are simpler than the true network, probably because the data are not enough to learn good parameters if a denser network were used. This might explain the reason why Max-BDeu has poorer performance in terms of log-likelihood of held-out data. In fact, we further analyzed this situation by learning the parameters of inferred graphs using a large data set of 5000 samples (the graphs themselves were learned with the appropriate varying sample sizes). Figure 5 shows average results for random networks of 20 nodes and different training sample sizes for structure

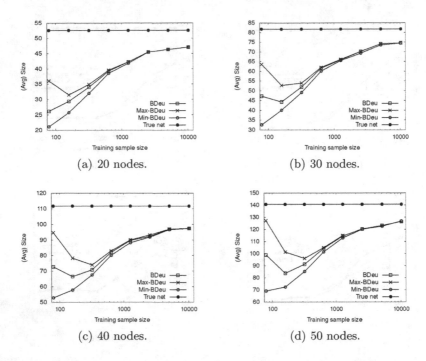

(a) 20 nodes.

(b) 30 nodes.

(c) 40 nodes.

(d) 50 nodes.

Fig. 4. Comparison on average total number of arcs in the learned networks from different score functions. The graphs feature BDeu (square points), Max-BDeu (cross points), Min-BDeu (circle points), true network (bullet points). The points in the curves correspond to the total number of arcs in the graphs of learned networks.

learning, while using 5000 samples for parameter learning. In this scenario, Max-BDeu performed better than the others did even in terms of likelihood accuracy, suggesting that its poorer performance was related to poorer parameter estimation. When training sample size becomes large enough, the behavior of all different methods becomes similar.

4.2   Relation between Scores and Learning Saturation

The proposed scores Min-BDeu and Max-BDeu have an interpretation as the pessimistic and optimistic scenarios with respect to unknown "ideal" prior Dirichlet distributions. The set of priors that we employ can be seen as a way to take into account the sensitivity of the score to variations of the local priors. Min-BDeu is aimed at finding the structure that maximizes the BDeu score under the most adversary prior, that is, learning with Min-BDeu is done by a maximin approach. Because of that, the fitting of Min-BDeu is less aggressive than that of BDeu. On the other hand, Max-BDeu is a maximax approach: it yields the structure that maximizes BDeu under the most beneficial prior, so it tends to fit more aggressively than BDeu.

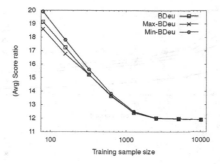

Fig. 5. Average likelihood accuracy among networks learned from different score functions. The graphs feature BDeu (square points), Max-BDeu (cross points), Min-BDeu (circle points) for networks with 20 nodes. The structure of the networks is learned with the given training data sample size, while the network parameters are learned with a training data set of 5000 samples.

With these considerations in mind, one may define a measure of "sensitivity" for an inferred graph $\mathcal{G}$ as the difference between $\overline{s}(\mathcal{G})$ and $\underline{s}(\mathcal{G})$ (note that these values are logarithms of probabilities, so imprecision here is defined as the ratio between upper and lower probabilities $\log(\overline{p}(D|\mathcal{G})/\underline{p}(D|\mathcal{G}))$). We have performed experiments using only the BDeu score for learning, but whose result is later evaluated in terms of this measure. That is, we execute experiments with only the BDeu score using randomly generated networks to sample training and held-out data. The curves in the upper graph of Fig. 6 show the average likelihood accuracy (over ten runs) obtained by the original BDeu for domains with different number of nodes, as done before. Such accuracy can only be computed because (for testing) we have available the true networks, thus we can generate samples from it to create the held-out test data. Such curves decrease with the amount of data used for learning, as expected. On the other hand, the same Fig. 6 (lower graph) displays the average sensitivity measure (score ratio) of the optimal networks that have been found by BDeu (because of scale differences, we divided them by the lower score so as to have them displayed in the same graph), which does not depend on knowing the true network and thus can be computed in the moment of learning. These curves of the lower graph of Fig. 6 also decrease as the amount of data increases, an expected phenomenon as the importance of the prior reduces with the increase in sample size. We can see in the figure a relation between likelihood accuracy and the sensitivity measure. This suggests that one can use the measure to determine (approximately) the amount of data after which structure learning will not benefit anymore (or will benefit very little) from additional data. In these tests, this saturation point happens around one thousand samples. Previous to the saturation point, the likelihood accuracy of the inferred networks greatly increases from the provision of more training data. Past this saturation point, this increase becomes small and might not be cost-effective in many situations. We note that the score alone could not be used for

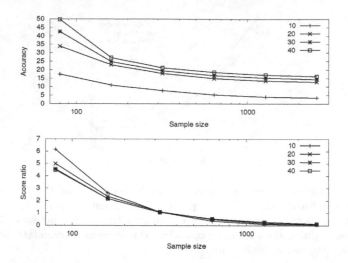

Fig. 6. Curves for likelihood accuracy (upper graph) and sensitivity measure as ratio of Max- and Min-BDeu (lower graph) for inferred networks learned from training data of different sizes using randomly generated Bayesian networks with 10 to 40 nodes.

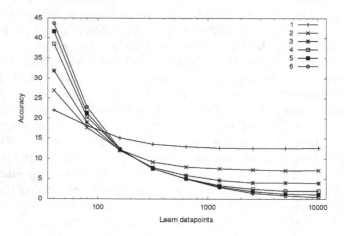

Fig. 7. Average likelihood accuracy curves of learned networks from data of randomly generated Bayesian networks with 20 nodes and maximum degree of six. Different curves represent the result of inferred networks under different parent set size limits during learning: maximum of one parent up to maximum of six parents.

this task, as it does not usually converge with the increase in the amount of data (recall that we use the logarithm of the probability of data given the structure).

One might have noticed that the accuracy curves in Fig. 6 do not converge to zero with the increase in the amount of data. This can be explained as an effect of the restricted maximum number of parents per node during learning (defined as three in our simulations). Figure 7 shows the likelihood accuracy of original BDeu for varying limits on the number of parents (from 1 to 6) for learned networks from data of randomly generated Bayesian networks with 20 nodes (true networks have no such limit in the number of parents, but have a total degree limit of six). It is also very interesting to note the shape of the accuracy curves. With small training sample size, it is seems better to force stronger limits, as the inferred networks with greater number of parents might be unreliable (recall that the number of training data and number of parents per node are very related in terms of learning accuracy). This indicates, at least in these experiments, that the original BDeu score was not able to control well the complexity of the learned network with small sample sizes, given that the limit on the number of parents was able to improve learning accuracy. This empirical result corroborates with our previous results about Min-BDeu outperforming the original BDeu when sample size is small.

5 Conclusions

In this paper we presented two new score functions for learning the structure of Bayesian networks from data. They are based on allowing the Dirichlet prior parameters of the Bayesian Dirichlet equivalent uniform (BDeu) to vary inside a contamination set around the symmetric Dirichlet priors, while keeping its strength fixed. Over this set of priors we choose the most adversary and the most beneficial priors to construct the Min-BDeu and the Max-BDeu, respectively. Learning with Min-BDeu is equivalent to a maximin approach, as one must find the graph that maximizes the minimum score (over BDeu scores with priors in the contaminated set). Min-BDeu prefers sparser graphs than the original BDeu does when sample size is small, and converges to the original BDeu as the amount of data increases. Max-BDeu is analogous, but using the most beneficial prior, that is, a maximax approach.

We demonstrate that these new score functions can be efficiently computed and are decomposable just like the original BDeu, so they can be used within most of the current state-of-the-art structure learning solvers. In our experiments, Min-BDeu has led to networks with higher accuracy than that of the original BDeu score, in terms of fitting the true model. On the other hand, Max-BDeu has led to better edge accuracy, that is, has fit the graph structure better. We also employ a combination of Max-BDeu and Min-BDeu as a measure of sensitivity. This measure visually correlates with the accuracy of inferred networks and might be used to identify the amount of data that saturates the learning, that is, the amount of data after which the accuracy of the inferred network does not considerably improve anymore if additional data were made available. Finally, an analysis of the BDeu accuracy with respect to the restriction on the maximum number of parents per node helps to explain the better accuracy of Min-BDeu against the original BDeu score when the training sample size is small. Scenarios of small sample size are particularly important in applied fields such as biomedicine and bioinformatics.

As future work we intend to expand our experiments, including the study of the sensitivity of the parameters that define the size of the contamination and the strength of the prior, as well as a deeper analysis about the consequences of limiting the number of parents of each variable, in order to better understand the properties of Min-BDeu and Max-BDeu. We also want to further study the characteristics of the original BDeu and to investigate other score functions, for instance using entropy as criterion to select priors.

Acknowledgments. Work partially supported by the Swiss NSF grants Nos. 200021_146606 / 1 and 200020_137680 / 1, and by the Swiss CTI project with Hoosh Technology.

References

1. Heckerman, D., Geiger, D., Chickering, D.M.: Learning Bayesian networks: The combination of knowledge and statistical data. Machine Learning 20, 197–243 (1995)
2. Barlett, M., Cussens, J.: Advances in Bayesian network learning using integer programming. In: Proceedings of the 29th Conference on Uncertainty in Artificial Intelligence, UAI 2013, pp. 182–191 (2013)
3. Cussens, J.: Bayesian network learning with cutting planes. In: Proceedings of the 27th Conference on Uncertainty in Artificial Intelligence, UAI 2011, pp. 153–160. AUAI Press, Barcelona (2011)
4. de Campos, C.P., Ji, Q.: Efficient structure learning of Bayesian networks using constraints. Journal of Machine Learning Research 12, 663–689 (2011)
5. de Campos, C.P., Zeng, Z., Ji, Q.: Structure learning of Bayesian networks using constraints. In: Proceedings of the 26th International Conference on Machine Learning, ICML 2009, pp. 113–120. Omnipress, Montreal (2009)
6. Jaakkola, T., Sontag, D., Globerson, A., Meila, M.: Learning Bayesian Network Structure using LP Relaxations. In: Proceedings of the 13th International Conference on Artificial Intelligence and Statistics, AISTATS 2010, pp. 358–365 (2010)
7. Niinimäki, T., Koivisto, M.: Annealed importance sampling for structure learning in Bayesian networks. In: Proceedings of the 23rd International Joint Conference on Artificial Intelligence, IJCAI 2013, pp. 1579–1585. AAAI Press (2013)
8. Parviainen, P., Koivisto, M.: Exact structure discovery in Bayesian networks with less space. In: Proceedings of the 25th Conference on Uncertainty in Artificial Intelligence, UAI 2009, pp. 436–443. AUAI Press (2009)
9. Parviainen, P., Koivisto, M.: Finding optimal Bayesian networks using precedence constraints. Journal of Machine Learning Research 14, 1387–1415 (2013)
10. Yuan, C., Malone, B.: Learning optimal Bayesian networks: A shortest path perspective. Journal of Artificial Intelligence Research 48, 23–65 (2013)
11. Buntine, W.: Theory refinement on Bayesian networks. In: Proceedings of the 8th Conference on Uncertainty in Artificial Intelligence, UAI 1992, pp. 52–60. Morgan Kaufmann, San Francisco (1991)
12. Cooper, G.F., Herskovits, E.: A Bayesian method for the induction of probabilistic networks from data. Machine Learning 9, 309–347 (1992)
13. Schwarz, G.: Estimating the dimension of a model. The Annals of Statistics 6(2), 461–464 (1978)
14. Akaike, H.: A new look at the statistical model identification. IEEE Transactions on Automatic Control 19(6), 716–723 (1974)

15. Brenner, E., Sontag, D.: Sparsityboost: A new scoring function for learning Bayesian network structure. In: Proceedings of the 29th Conference on Uncertainty in Artificial Intelligence, UAI 2013, pp. 112–121. AUAI Press, Corvallis (2013)
16. Walley, P.: Statistical Reasoning with Imprecise Probabilities. Chapman and Hall, London (1991)
17. Silander, T., Kontkanen, P., Myllymäki, P.: On Sensitivity of the MAP Bayesian Network Structure to the Equivalent Sample Size Parameter. In: Proceedings of the 23rd Conference on Uncertainty in Artificial Intelligence, UAI 2007, pp. 360–367 (2007)
18. Abellan, J., Moral, S.: New score for independence based on the imprecise Dirichlet model. In: International Symposium on Imprecise Probability: Theory and Applications, ISIPTA 2005, SIPTA, pp. 1–10 (2005)
19. Cano, A., Gómez-Olmedo, M., Masegosa, A.R., Moral, S.: Locally averaged Bayesian Dirichlet metrics for learning the structure and the parameters of Bayesian networks. International Journal of Approximate Reasoning 54(4), 526–540 (2013)
20. Ben-Tal, A., Nemirovski, A.: Lectures on Modern Convex Optimization: Analysis, Algorithms, and Engineering Applications. MPS/SIAM Series on Optimization. SIAM (2001)
21. Spiegelhalter, D.J., Cowell, R.G.: In: Learning in probabilistic expert systems, pp. 447–466. Clarendon Press, Oxford (1992)
22. Binder, J., Koller, D., Russell, S., Kanazawa, K.: Adaptive probabilistic networks with hidden variables. Machine Learning 29 (1997)
23. Jensen, F.V., Kjærulff, U., Olesen, K.G., Pedersen, J.: Et forprojekt til et ekspertsystem for drift af spildevandsrensning (an expert system for control of waste water treatment — a pilot project). Technical report, Judex Datasystemer A/S, Aalborg, Denmark (1989) (in Danish)
24. Beinlich, I.A., Suermondt, H.J., Chavez, R.M., Cooper, G.F.: The ALARM Monitoring System: A Case Study with Two Probabilistic Inference Techniques for Belief Networks. In: Proceedings of the 2nd European Conference on Artificial Intelligence in Medicine. Lecture Notes in Medical Informatics, vol. 38, pp. 247–256. Springer, Heidelberg (1989)
25. Ide, J.S., Cozman, F.G.: Random generation of Bayesian networks. In: Bittencourt, G., Ramalho, G.L. (eds.) SBIA 2002. LNCS (LNAI), vol. 2507, pp. 366–375. Springer, Heidelberg (2002)
26. Nagarajan, R., Scutari, M., Lèbre, S.: Bayesian Networks in R with Applications in Systems Biology. Use R! series. Springer (2013)
27. de Campos, C.P., Ji, Q.: Properties of Bayesian Dirichlet scores to learn Bayesian network structures. In: AAAI Conference on Artificial Intelligence, pp. 431–436. AAAI Press (2010)

Causal Discovery from Databases
with Discrete and Continuous Variables

Elena Sokolova, Perry Groot, Tom Claassen, and Tom Heskes

Radboud University, Faculty of Science,
Postbus 9010, 6500 GL Nijmegen, The Netherlands

Abstract. Bayesian Constraint-based Causal Discovery (BCCD) is a state-of-the-art method for robust causal discovery in the presence of latent variables. It combines probabilistic estimation of Bayesian networks over subsets of variables with a causal logic to infer causal statements. Currently BCCD is limited to discrete or Gaussian variables. Most of the real-world data, however, contain a mixture of discrete and continuous variables. We here extend BCCD to be able to handle combinations of discrete and continuous variables, under the assumption that the relations between the variables are monotonic. To this end, we propose a novel method for the efficient computation of BIC scores for hybrid Bayesian networks. We demonstrate the accuracy and efficiency of our approach for causal discovery on simulated data as well as on real-world data from the ADHD-200 competition.

Keywords: Causal discovery, hybrid data, structure learning.

1 Introduction

Causal discovery is widely used for analysis of experimental data focusing on the exploratory analysis and suggesting probable causal dependencies. There is a variety of causal discovery algorithms in the literature. Some of these algorithms rely on the assumption that there are no latent variables in the model; others do not provide a scoring metric to easily compare the reliability of two candidate models. Bayesian Constraint-based Causal Discovery (BCCD) [6] is a state-of-the-art-algorithm for causal discovery that tries to combine the strength of the best algorithms in the field. BCCD is able to detect latent variables in the model and determines the reliability of the edges between variables that makes it very easy to compare alternative models.

The idea of BCCD is to estimate the reliability of causal relations by scoring Directed Acyclic Graphs (DAGs) for a smaller subset of variables using a Bayesian score and then to combine these statements to infer a final causal model. The Bayesian score has a closed form solution for discrete variables that makes the scoring of causal relations fast and efficient. The BCCD algorithm is currently limited to discrete or Gaussian variables as there is no closed form solution for the Bayesian score for a mixture of discrete and continuous variables.

L.C. van der Gaag and A.J. Feelders (Eds.): PGM 2014, LNAI 8754, pp. 442–457, 2014.

To extend BCCD, we need a new scoring method to estimate the reliability of causal relations.

There are several scoring methods in the literature for mixtures of discrete and continuous variables. Most of these methods either rely on strict assumptions about the structure of the network that do not apply in practice, such as forbidding structures in the network with a continuous variable as a parent having a discrete variable as child, or are time consuming or/and memory inefficient. In this paper we propose a fast and memory efficient method to score DAGs with both discrete and continuous variables, under the assumption that the relationships between these variables are monotonic. This appears to be a reasonable assumption for many real-world data sets.

The scoring method proposed in this paper estimates the Bayesian information criterion (BIC) score by approximating the mutual information for a combination of discrete and continuous variables. Through simulations we will show that the BCCD algorithm with this scoring method can accurately estimate the structure of the Bayesian network for both simulated and real-world data.

The rest of the paper is organized as follows. Section 2 describes background information about causal discovery and graphical models. Section 3 describes algorithms for structure learning. Section 4 briefly summarizes the idea of the BCCD algorithm. Section 5 explains the scoring method for a mixture of discrete and continuous variables. Section 6 presents the results of the experiments of the BCCD algorithm on simulated data and real-world data. Section 7 provides our conclusion and future work.

2   Background

A Bayesian network is a pair $(\mathcal{G}, \Theta)$ where $\mathcal{G} = (\mathbf{X}, \mathbf{E})$ is a Directed Acyclic Graph (DAG) with a set of nodes $\mathbf{X}$ representing domain variables and a set of arcs $\mathbf{E}$; $\theta_{X_i} \subset \Theta$ is a set of parameters representing the conditional probability of variable $X_i \subset \mathbf{X}$ given its parents Pa_i in a graph $\mathcal{G}$. Using Bayesian networks we can model causal relationships between variables. In that case an edge $A \rightarrow B$ between variables represents a direct causal link from A to B. This means that A influences the values of B, but not the other way around.

Saying that two variables A and B are conditionally independent given C, means that if we know C, learning B would not change our belief in A. Two DAGs are called equivalent to one another, if they entail the same conditional (in)dependencies. All DAGs that are equivalent to a graph $\mathcal{G}$ form an equivalence class of a graph $\mathcal{G}$, where all members are indistinguishable in terms of implied independencies. To represent the members of this equivalence class, a different type of structure is used, known as a partially directed acyclic graph (PDAG).

The three main assumptions that are often used when learning the structure of causal networks are the following.

1. Causal Markov Condition: each variable is independent of its non-descendant conditioned on all its direct causes.

2. Faithfulness assumption: there are no independencies between variables that are not implied by the Causal Markov Condition.
3. Causal sufficiency assumption: there are no common confounders of the observed variables in $\mathcal{G}$.

In this paper we do not rely on the causal sufficiency assumption, i.e. we do allow for latent variables. One can represent the structure of a Bayesian network with latent variables using a so-called Maximal Ancestral Graph (MAG) on only the observed variables. In contrast to DAGs, MAGs can also contain bi-directed $X \leftrightarrow Y$ arcs (indicating that there is a common confounder) and undirected arcs $X - Y$. The equivalence class for MAGs is a partial ancestral graph (PAG). Edge directions are marked with " $-$ " and ">" if the direction is the same for all graphs belonging to the PAG and with "o" otherwise.

3   Structure Learning of Causal Networks

There is a variety of methods that can be used to learn the structure of a causal network. A broad description of methods can be found in [8]. In general, methods are divided into two approaches: constraint-based and score-based. The constraint-based approach works with statistical independence tests. Firstly, this approach finds a skeleton of a graph by starting from the complete graph and excluding edges between variables that are conditionally independent, given some other set of variables (possibly empty). Secondly, the edges are oriented to arrive at an output graph. The constraint-based approach learns the equivalence class of DAGs and outputs a PDAG. Examples of the constraint-based approach are the IC algorithm [23], PC-FCI [27], and TC [24].

A score-based approach uses a scoring metric. It measures the data goodness of fit given a particular graph structure and accounts for the complexity of the network. There are many different scoring metrics, where the Bayesian score [9] and the BIC score [26] are among the most common. The goal is to find the graph that has the highest score. Unfortunately, this optimization problem is NP-hard, so different heuristics are used in practice. These methods are divided in local search methods, such as greedy search [5], greedy equivalence search [4], and global search methods, such as simulated annealing [10] and genetic algorithms [17].

An advantage of the constraint-based approach is that it does not have to rely on the causal sufficiency assumption, which means that the algorithm can detect common causes of the observed variables. A disadvantage of the constraint-based approach is that it is sensitive to propagating mistakes in the resulting graph. A standard approach makes use of independence tests, whose results for borderline independencies/dependencies sometimes can be incorrect. The outcome of learning a network can be sensitive to such errors. In particular, one such error can produce multiple errors in the resulting graph. A set of conservative methods such as CPC [25] and CFCI [28] tackles the problem of lack of robustness, outperforming standard constraint-based methods such as PC.

An advantage of the score-based approach is that it indicates the measure of reliability of inferred causal relations. This makes the interpretation of the results easier and prevents incorrect categorical decisions. A main drawback of the approach is that it relies on the causal sufficiency assumption and as a result cannot detect latent confounders.

4 Bayesian Constraint-Based Causal Discovery

One of the state-of-the-art algorithms in causal discovery is Bayesian Constraint-based Causal Discovery (BCCD). Claassen and Heskes [6] showed that BCCD outperforms reference algorithms in the field, such as FCI and Conservative PC. Moreover, it provides an indication of the reliability of the causal links that makes it easier to interpret the results and compare alternative models. The advantage of the BCCD algorithm is that it combines the strength of constraint-based and score-based approaches. We here describe only the basic idea of the method. A more detailed description can be found in [6]. The main two steps of BCCD are the following:

Step 1. Start with a fully connected graph and perform adjacency search, estimating the reliability of causal relations, for example $X \to Y$. If a causal relation declares a variable conditionally independent with a reliability higher than a predefined threshold, delete an edge from the graph between these variables.

Step 2. Rank all causal relations in decreasing order of reliability and orient edges in the graph starting from the most reliable relations. If there is a conflict, pick the causal relation that has a higher reliability.

Based on the score of the causal relations, we can rank these relations and avoid propagating unreliable decisions giving preference to more confident ones. This can solve the drawback of a standard constraint-based method that can end up with an unreliable result. Moreover, using a Bayesian score we get a reliability measure of the final output, which makes it easier to interpret the results and compare with other alternative models. The BCCD algorithm does not rely on the causal sufficiency assumption, thus it can detect latent variables in the model.

The first step of the algorithm requires estimating the reliability of a causal relations $L : `X \to Y'$ given a data set $\mathbf{D}$, which is done using a Bayesian score:

$$p(L : `X \to Y'|\mathbf{D}) = \frac{\sum_{\mathcal{M} \in \mathbf{M}(L)} p(\mathbf{D}|\mathcal{M})p(\mathcal{M})}{\sum_{\mathcal{M} \in \mathbf{M}} p(\mathbf{D}|\mathcal{M})p(\mathcal{M})} \ , \tag{1}$$

where $p(\mathbf{D}|\mathcal{M})$ denotes the probability of data $\mathbf{D}$ given structure $\mathcal{M}$, $p(\mathcal{M})$ represents the prior distribution over structures and $\mathbf{M}(L)$ is the set of structures containing the relation L. This reliability measure (1) gives a conservative estimate of the probability of a causal relation. Claassen and Heskes approximate the probability $p(\mathbf{D}|\mathcal{M})$ by $p(\mathbf{D}|\mathcal{G})$, the marginal likelihood of the data given graph $\mathcal{G}$ that has a closed form solution for discrete variables, known as the Bayesian Dirichlet (BD) metric [16]. There is also a closed-form solution when all variables have a Gaussian distribution, called the BGe metric [16]. To estimate (1), the algorithm requires calculating the marginal likelihood over all possible graphs for each causal relation that we infer. For speed and efficiency of the algorithm, the set of possible graphs is limited to the graphs with at most five vertices, which gives a list of at most 29,281 DAGs. In theory, limiting the number of vertices to five may lead to loss of information. In practice, however, the accuracy of the BCCD algorithm is hardly affected and it still outperforms standard algorithms that perform conditional independence tests for more than five vertices [6].

The use of BD/BGe metrics to score DAGs estimating the marginal likelihood, limits the BCCD algorithm to work only with discrete variables or only with Gaussian variables. However, many real-world data sets contain a mixture of discrete and continuous variables. One of the possible solutions to this problem would be to discretize continuous variables [12,21]. This, however, can lead to a loss of information comprised in continuous variables, spurious dependencies and misleading results of the method [27]. Another solution to this problem is to define a new scoring method that can estimate the marginal likelihood when the data is a mixture of discrete and continuous variables. Since the estimation of the marginal likelihood is repeated many times for a large number of possible graphs, such a new scoring method must be fast and memory efficient.

5   The Scoring Metric for Discrete and Continuous Variables

In this section we discuss methods that can estimate the Bayesian score (1) for a mixture of discrete and continuous variables. In Sect. 5.1 we provide an overview of existing scoring algorithms and in Sect. 5.2 we propose our scoring method.

5.1   Alternative Approaches

There are few score-based methods that can work with a mixture of discrete and continuous variables. Geiger and Heckerman [13] proposed a closed-form solution for the Bayesian score of a mixture of discrete and continuous variables, but this solution only works in case a number of assumptions are met. These assumptions imply that the data are drawn from a conditional Gaussian distribution and forbid structures in the network with a continuous variable having a discrete variable as a child.

An alternative method is described in [11] which uses a multiple regression framework for scoring structures. However, the method is applicable only for time-series data. Bach and Jordan [1] use Mercer kernels to estimate the structure of causal models, but calculation of a Gramm matrix requires significant computational costs ($O(N^3)$, where N is the sample size) and may be inefficient for data sets with large sample sizes. Monti and Cooper [20] use neural networks to represent the density function for a mixture of discrete and continuous variables. Estimation of the neural network parameters to calculate (1) requires significant computational costs which should be repeated multiple times and becomes too slow to be applicable in the BCCD algorithm.

5.2   Proposed Scoring Method

Existing scoring methods for a mixture of discrete and continuous variables are either computationally or/and memory expensive or rely on strict assumptions that limit the types of data that can be used considerably. In this section we propose a fast and memory efficient method for scoring DAGs that relies only on the assumption that relationships between variables are monotonic. This is a reasonable assumption for many domains.

We consider the BIC score, which is an approximation of the Bayesian score, to estimate the marginal likelihood $p(\mathbf{D}|\mathcal{G})$. The BIC score can be decomposed into the sum of two components, the mutual information $I(X_i, Pa_i)$ with Pa_i the parents of node X_i and $Dim[\mathcal{G}]$ the number of parameters necessary to estimate the model. The first component measures the goodness of fit, and the second penalizes the complexity of the model:

$$BICscore(\mathbf{D}|\mathcal{G}) = M \sum_{i=1}^{n} I(X_i, Pa_i) - \frac{\log M}{2} Dim[\mathcal{G}] \ , \qquad (2)$$

where n is the number of variables and M is a sample size.

To estimate (2) for a mixture of discrete and continuous variables we need to estimate the mutual information and the complexity penalty. We propose to approximate the mutual information based on the formula for continuous variables drawn from a Gaussian distribution [7]:

$$I(X_i, Pa_i) = -\frac{1}{2} \log \frac{|R|}{|R_{Pa_i}|} \ , \qquad (3)$$

where R is a correlation matrix between all variables and R_{Pa_i} is a correlation matrix between the parents of variable X_i.

Estimation of the complexity penalty for the model containing a mixture of variables is reduced to calculation of the number of parameters in the model:

$$Dim[\mathcal{G}] = \sum_{i=1}^{n} d_i d_{Pa_i}, \text{with } d_{Pa_i} = \sum_{j \in Pa_i(\mathcal{G})} d_j, \text{ and}$$

$$d_j = \begin{cases} 1, & \text{if } X_j \text{ is continuous} \\ s - 1, & \text{if } X_j \text{ is discrete} \ , \end{cases} \qquad (4)$$

where s is the number of states of the discrete variable X_j. In case the variable X_i does not have parents, we assign (3) and (4) a zero value. Due to that assignment,

(4) represents the difference in the number of parameters, when there is an edge between X_i and Pa_i and when there are no edges between them.

We propose a scoring method that approximates the BIC score for a mixture of discrete and continuous variables using (3) and (4) and substituting Pearson correlation with Spearman in (3). We assume hereby explicitly that the variables obey a so-called non-paranormal distribution. For univariate monotone functions $f_1, ..., f_d$ and a positive-definite correlation matrix $\Sigma^0 \in \mathbb{R}^{d \times d}$ we say that a d-dimensional random variable $X = (X_1, ..., X_d)^T$ has a non-paranormal distribution $X \backsim \mathrm{NPN}_d(f, \Sigma^0)$, if $f(X) = (f_1(X_1), ..., f_d(X_d)) \backsim N_d(0, \Sigma^0)$. As shown in [15], conditional independence tests for non-paranormal data based on Spearman correlations are more accurate than those based on Pearson correlations for non-Gaussian continuous data. Ignoring the discreteness of the discrete variables does introduce some bias in the approximation of the BIC score, however, as we will argue and show in the next section, this bias hardly affects the scoring of the network structures. In case of categorical variables we propose to perform so-called dummy coding, binarising the variable into several variables when calculating the correlation matrix and then adjusting for the correlation between these variables when calculating the mutual information.

The most computationally expensive part of the proposed scoring method is the calculation of the correlation matrix. However, one can compute the full correlation matrix once beforehand, which can then be stored and used to efficiently construct the correlation matrices for any subset of variables. The proposed method is thus computationally and memory efficient.

In this section we described the scoring method for a mixture of discrete and continuous variables. Now we can use this scoring method to estimate marginal likelihood in the loop of the BCCD algorithm.

6   Experiments

In this section we describe the results of our experiments. Section 6.1 describes the accuracy of estimating mutual information using (3). In Sect. 6.2 we describe the results of the experiments on simulated data, where the ground truth about the structure of the network is known. In Sect. 6.3 we describe the results of the experiments on real-world data, where the ground truth is unknown.

6.1   Testing Mutual Information

In the previous section we proposed a method to score DAGs for a mixture of discrete and continuous variables that approximates mutual information in the BIC score using (3). To estimate the accuracy of this approximation, we performed a series of simulations. In these simulations we randomly generated two variables with particular distributions and a sample size of 1000 and then compared the exact value of mutual information with the approximated value. Since the strength of association between variables can influence the accuracy of the estimation we performed the simulations changing the correlation between the variables from 0 to 1. Simulations were performed for three types of data: continuous, discrete, and a mixture of discrete and continuous.

Figure 1 (a) shows the difference between the exact mutual information and mutual information estimated using (3) between two normally distributed random variables. There is a slight difference between the exact and approximated mutual information, since we use Spearman correlation instead of Pearson in (3) in order to capture monotonic relations between variables that may be non-normally distributed.

Figure 1 (b) shows the difference between the exact mutual information and mutual information estimated using (3) between two discrete binary variables. The difference between the estimates of mutual information arises when the correlation between two variables is close to one. This happens because our mutual information estimate is based upon the incorrect assumption that discrete variables can have an infinite amount of values.

In practice, however, the approximated mutual information for discrete variables will not strongly affect the selection of the correct structure. For the range of correlations when the mutual information is off, the probability of the edge between two variables is always close to one. Thus, higher levels of mutual information cannot overestimate the probability of an edge, since it has already achieved its maximum. As a result, overestimation of the mutual information for discrete variables hardly affects the scoring of the structures.

Figure 1 (c) shows the difference between the exact mutual information and mutual information estimated using (3) between one binary parent and one normally distributed child with its mean depending on the value of the binary parent. Figure 1 (c) shows that the approximated value of mutual information stays very close to the actual value.

6.2 Application on Simulated Data

To test our algorithm on simulated data we chose two widely used Bayesian networks: the Asia Network [18] and the Waste Incinerator Network [19] to test the algorithm for discrete variables and a mixture of discrete and continuous variables, respectively. The case of only continuous variables was already discussed in [15]. We randomly generated data for four different sample sizes: 100, 500, 1000, and 1500 and repeated our experiments 20 times. Performance is measured by PAG accuracy measure, that evaluates how many edges were oriented correctly in the output PAG. We also estimated the correctness of the skeleton by calculating the amount of correct, missing, and spurious edges of the resulting graph.

The Asia data set describes the effect of visiting Asia and smoking behavior on the probability of contracting tuberculosis, cancer or bronchitis. The network contains eight binary variables that are connected by eight arcs as can be seen in Figure 2. For this network we compared the results of the BCCD using BD as a scoring method, following [16] and using our approximated BIC score.

Figure 3 compares the accuracy of the BCCD algorithm for the Asia Network when the BIC score is estimated using BD metric (BIC original) or correlation

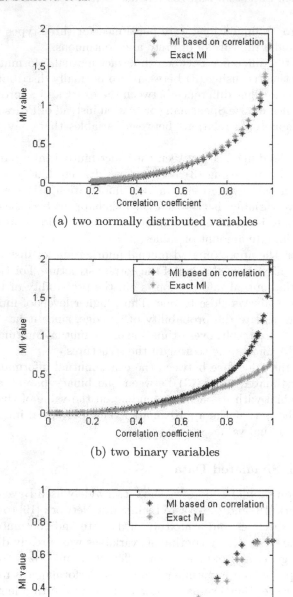

(a) two normally distributed variables

(b) two binary variables

(c) one binary and one normally distributed variable

Fig. 1. The exact mutual information (MI) and mutual information calculated using Spearman correlation for different values of the correlation coefficient between: (a) two normally distributed variables, (b) two binary variables, (c) one binary and one normally distributed variable.

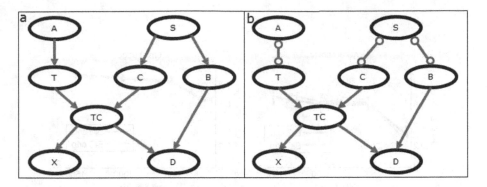

Fig. 2. Asia Network represented as (a) DAG and (b) PAG. The node names are abbreviated as follows: Visit to Asia? (A) Smoker? (S) Has tuberculosis? (T) Has lung cancer? (C) Has bronchitis? (B) Tuberculosis or cancer (TC) Positive X-ray? (X) Dyspnoea? (D)

matrix (BIC Corr). Figure 3 suggests that there is no significant difference in PAG accuracy and in the skeleton of the graph between the two methods which is also confirmed by a Students t-test. The larger the sample size, the higher is the PAG accuracy and the more accurate the skeleton. Based on these results, we suggest that estimating mutual information using the correlation matrix for discrete variables gives an accurate estimation of the causal structure.

The Waste Incinerator Network describes the emission from a waste incinerator depending on filter efficiency, waste type, burning regime, and other factors. The network contains nine variables that are connected by ten arcs as can be seen in Fig. 4.

The network contains three discrete (B, F, W) and six continuous variables (C, E, MW, ME, D, L). The original parameters of Waste Incinerator Network have a very low variance that results in an almost one-to-one deterministic relationship between the variables, which violates the faithfulness assumption. To avoid this problem, we increased the variance between variables E, F, and V to 2.5 and between other variables to 0.2.

Figure 5 represents the accuracy of the BCCD algorithm when the BIC score is estimated using the correlation matrix for the Waste Incinerator Network. Figure 5 shows that the PAG and skeleton accuracy increases with the increase of sample size and becomes close to one. That suggests that the BCCD algorithm is able to estimate the structure of a Bayesian network for a mixture of discrete and continuous variables. Based on this conclusion we now can apply the BCCD algorithm on real-world data.

6.3   Application on Real-World Data: ADHD

To test the BCCD algorithm on real-world data, we use the data set collected by [3] that contains a mixture of discrete and continuous variables. This data set

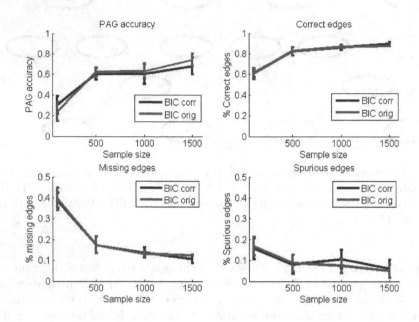

Fig. 3. The accuracy of the BCCD algorithm for the Asia Network in which mutual information for the BIC score is estimated using: frequency counts (BIC orig) and correlation matrices (BIC corr). (a) PAG accuracy. (b) Percentage of correct edges. (c) Percentage of missing edges. (d) Percentage of spurious edges.

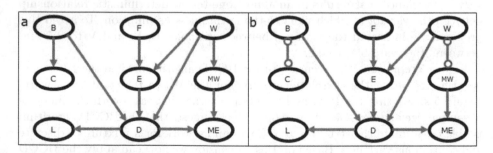

Fig. 4. Waste Incinerator Network represented as (a) DAG and (b) PAG. The node names are abbreviated as follows: Burning regime (B) Filter state (F) Waste type(W) CO2 concentration (C) Filter efficiency (E) Metal in waste (MW) Light penetrability (L) Dust emission (D) Metals emission (ME)

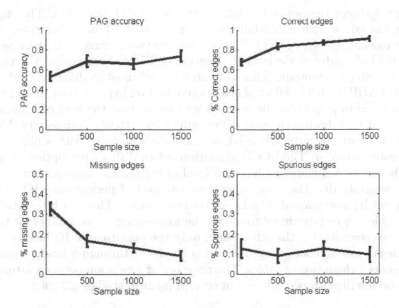

Fig. 5. The accuracy of the BCCD algorithm for the Waste Incinerator Network (a) PAG accuracy. (b) Percentage of correct edges. (c) Percentage of missing edges. (d) Percentage of spurious edges.

describes phenotypic information about children with Attention Deficit Hyperactivity Disorder (ADHD) and is publicly available as a part of the ADHD-200 competition. The ADHD data set contains 23 variables for 245 subjects. We excluded subjects that had missing information, so our final data set contained 223 subjects. We also excluded eleven variables that either did not have enough data or were considered irrelevant, leaving only the following nine variables:

1. Gender (binary: male/female)
2. Attention deficit score (continuous)
3. Hyperactivity/impulsivity score (continuous)
4. Verbal IQ (continuous)
5. Full IQ (continuous)
6. Performance IQ (continuous)
7. Aggressive behavior (binary: yes/no)
8. Medication status (binary: naïve/not naïve)
9. Handedness (binary: right/left)

The BCCD algorithm using Spearman correlation was applied to the ADHD data set. Due to the small sample size the BCCD algorithm inferred only the skeleton of the network, but not the direction of the edges for the resulting network. However, including prior knowledge about the domain that no variable in the network can cause gender, BCCD inferred the direction of several edges.

Figure 6 shows the network inferred by the BCCD algorithm. This figure includes the edges with a reliability of a direct causal link higher than 50%, that are calculated based on (1). The resulting network suggests that there is a strong effect of gender on the level of attention deficit and consequently hyperactivity/impulsivity symptoms. This statement is confirmed by different studies in the field of ADHD [2]. The effect of attention deficit on hyperactivity/impulsivity was also found in [30]. From the network we can see that the level of aggression is associated with both: attention deficit and hyperactivity/impulsivity. Moreover, the network suggests that left-handedness is associated with a higher risk of aggressive behavior. The BCCD algorithm inferred that prescription of medication is mainly associated with a high level of inattention and aggression. The network suggests that the association between level of performance IQ, verbal IQ, and full IQ is explained by a latent common cause. Thus, an IQ-related latent variable can be introduced to model the association between different types of IQ measurements. On the other hand, only the performance IQ has a causal link with attention deficit symptoms either direct or through a latent common cause between these two variables. The presence of a relation between attention deficit and intelligence is confirmed in several medical studies [22,29].

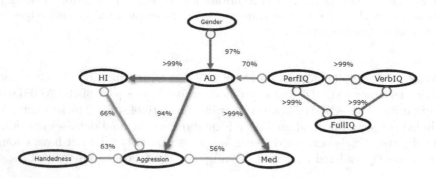

Fig. 6. The causal graph representing causal relationships between variables for the ADHD data set. The graph represents a PAG, where edge directions are marked with " − " and " >" for invariant edge directions and with "o" for non-invariant edge directions. The reliability of an edge between two variables is depicted with a percentage value near each edge.

7   Conclusion and Future Work

In this paper we extended the state-of-the-art BCCD algorithm for causal discovery to mixtures of discrete and continuous variables. In order to extend BCCD, we developed a method to estimate the score of the DAG for data sets with a mixture of discrete and continuous variables, under the assumption that relationships between variables are monotonic. The developed method approximates mutual information using Spearman correlation to calculate the BIC score. Simulation studies on different types of data show that this appears to be a reasonable approximation of mutual information. The small bias introduced to it hardly influences the outcome of the BCCD algorithm.

The most computationally expensive part of the proposed scoring method is the calculation of the correlation matrix. However, to estimate the BIC score for all possible subsets of variables it is not necessary to recalculate correlation matrix for each subset. The correlation matrix between all variables can be calculated beforehand and the required elements of the matrix can be selected depending on a subset of variables used to calculate the BIC score. As a result, the proposed scoring method does not require expensive calculations and requires to store only the correlation matrix, which makes it fast and memory efficient.

Our extended version of BCCD is now able to learn causal relationships between a mixture of discrete and continuous variables, assuming that the relations between them are monotonic. For future work we will relax this assumption for discrete variables and learn non-monotonic relations, by introducing a preliminary transformation to combine the parents of the variables. Another possible extension of the method is to handle missing values. This is a quite common problem in many fields, while standard solutions such as data deletion or data imputation lead to loss of information or introduction of bias into the results.

An alternative approach for structure learning is through conditional independence tests. Several methods [31,14] employing kernels have been proposed to measure conditional independence without assuming a functional form of dependency between the variables as well as the data distributions. It would be interesting to consider how we can use these tests to obtain alternative scoring methods for the BCCD algorithms.

Acknowledgments. The research leading to these results has received funding from the European Community's Seventh Framework Programme (FP7/2007-2013) under grant agreement n° 278948.

References

1. Bach, F.R., Jordan, M.I.: Learning graphical models with Mercer kernels. In: Proceedings of the NIPS Conference, pp. 1009–1016 (2002)
2. Bauermeister, J.J., Shrout, P.E., Chávez, L., Rubio-Stipec, M., Ramírez, R., Padilla, L., Anderson, A., García, P., Canino, G.: ADHD and Gender: Are risks and sequela of ADHD the same for boys and girls? Journal of Child Psychology and Psychiatry 48(8), 831–839 (2007)
3. Cao, Q., Zang, Y., Sun, L., Sui, M., Long, X., Zou, Q., Wang, Y.: Abnormal neural activity in children with attention deficit hyperactivity disorder: a resting-state functional magnetic resonance imaging study. Neuroreport 17(10), 1033–1036 (2006)
4. Chickering, D.M.: Optimal structure identification with greedy search. Journal of Machine Learning Research 3, 507–554 (2002)

5. Chickering, D.M., Geiger, D., Heckerman, D.: Learning Bayesian networks: Search methods and experimental results. In: Proceedings of the Fifth International Workshop on Artificial Intelligence and Statistics, pp. 112–128 (January 1995)
6. Claassen, T., Heskes, T.: A Bayesian approach to constraint based causal inference. In: Proceedings of the UAI Conference, pp. 207–216. AUAI Press (2012)
7. Cover, T.M., Thomas, J.A.: Elements of Information Theory. Wiley Series in Telecommunications and Signal Processing. Wiley-Interscience (2006)
8. Daly, R., Shen, Q., Aitken, J.S.: Learning Bayesian networks: approaches and issues. Knowledge Eng. Review 26(2), 99–157 (2011)
9. Dawid, A.P.: Statistical theory: the prequential approach (with discussion). J. R. Statist. Soc. A 147, 278–292 (1984)
10. de Campos, L.M., Huete, J.F.: Approximating causal orderings for Bayesian networks using genetic algorithms and simulated annealing. In: Proceedings of the Eighth IPMU Conference, pp. 333–340 (2000)
11. de Santana, Á.L., Francês, C.R.L., Costa, J.C.W.: Algorithm for graphical Bayesian modeling based on multiple regressions. In: Gelbukh, A., Kuri Morales, Á.F. (eds.) MICAI 2007. LNCS (LNAI), vol. 4827, pp. 496–506. Springer, Heidelberg (2007)
12. Friedman, N., Goldszmidt, M.: Discretizing continuous attributes while learning Bayesian networks. In: Proceedings of the ICML Conference, pp. 157–165 (1996)
13. Geiger, D., Heckerman, D.: Learning Gaussian networks. In: Proceedings of the UAI Conference, pp. 235–243. Morgan Kaufmann (1994)
14. Gretton, A., Fukumizu, K., Teo, C.H., Song, L., Schölkopf, B., Smola, A.J.: A kernel statistical test of independence. In: NIPS. Curran Associates, Inc. (2007)
15. Harris, N., Drton, M.: PC algorithm for nonparanormal graphical models. Journal of Machine Learning Research 14, 3365–3383 (2013)
16. Heckerman, D., Geiger, D., Chickering, D.M.: Learning Bayesian networks: The combination of knowledge and statistical data. In: Machine Learning, pp. 197–243 (1995)
17. Larrañaga, P., Kuijpers, C.M.H., Murga, R.H., Yurramendi, Y.: Learning Bayesian network structures by searching for the best ordering with genetic algorithms. IEEE Transactions on Systems, Man and Cybernetics 26, 487–493 (1996)
18. Lauritzen, S., Spiegelhalter, D.J.: Local computations with probabilities on graphical structures and their application to expert systems (with discussion). Journal of the Royal Statistical Society Series B 50, 157–224 (1988)
19. Lauritzen, S.L., Lauritzen, S.L.: Propagation of probabilities, means and variances in mixed graphical association models. Journal of the American Statistical Association 87, 1098–1108 (1992)
20. Monti, S., Cooper, G.F.: Learning hybrid Bayesian networks from data. Technical Report ISSP-97-01, Intelligent Systems Program, University of Pittsburgh (1997)
21. Monti, S., Cooper, G.F.: A multivariate discretization method for learning Bayesian networks from mixed data. In: Cooper, G.F., Moral, S. (eds.) Proceedings of the UAI Conference, pp. 404–413. Morgan Kaufmann (1998)
22. Paloyelis, Y., Rijsdijk, F., Wood, A., Asherson, P., Kuntsi, J.: The genetic association between adhd symptoms and reading difficulties: the role of inattentiveness and IQ. J. Abnorm. Child Psychol. 38, 1083–1095 (2010)
23. Pearl, J., Verma, T.: A theory of inferred causation. In: Proceedings of the KR Conference, pp. 441–452. Morgan Kaufmann (1991)
24. Pellet, J.P., Elisseeff, A.: Using Markov blankets for causal structure learning. J. Mach. Learn. Res. 9, 1295–1342 (2008)
25. Ramsey, J., Zhang, J., Spirtes, P.: Adjacency-faithfulness and conservative causal inference. In: Proceedings of the UAI Conference, pp. 401–408. AUAI Press (2006)

26. Schwarz, G.: Estimating the dimension of a model. The Annals of Statistics 6, 461–464 (1978)
27. Spirtes, P., Glymour, C., Scheines, R.: Causation, Prediction, and Search, 2nd edn. MIT Press (2000)
28. Spirtes, P., Glymour, C., Scheines, R.: The TETRAD Project: Causal Models and Statistical Data. Carnegie Mellon University Department of Philosophy, Pittsburgh (2004)
29. Vaida, N., Mattoo, N.H., Wood, A., Madhosh, A.: Intelligence among attention deficit hyperactivity disordered (adhd) children (aged 5-9). J. Psychology 4(1), 9–12 (2013)
30. Willcutt, E., Pennington, B., DeFries, J.: Etiology of inattention and hyperactivity/impulsivity in a community sample of twins with learning difficulties. J. Abnorm. Child Psychol. 28(2), 149–159 (2000)
31. Zhang, K., Peters, J., Janzing, D., Schölkopf, B.: Kernel-based conditional independence test and application in causal discovery. CoRR 1202.3775 (2012)

On Expressiveness of the AMP Chain Graph Interpretation

Dag Sonntag

ADIT, IDA, Linköping University, Sweden
dag.sonntag@liu.se

Abstract. In this paper we study the expressiveness of the Andersson-Madigan-Perlman interpretation of chain graphs. It is well known that all independence models that can be represented by Bayesian networks also can be perfectly represented by chain graphs of the Andersson-Madigan-Perlman interpretation but it has so far not been studied how much more expressive this second class of models is. In this paper we calculate the exact number of representable independence models for the two classes, and the ratio between them, for up to five nodes. For more than five nodes the explosive growth of chain graph models does however make such enumeration infeasible. Hence we instead present, and prove the correctness of, a Markov chain Monte Carlo approach for sampling chain graph models uniformly for the Andersson-Madigan-Perlman interpretation. This allows us to approximate the ratio between the numbers of independence models representable by the two classes as well as the average number of chain graphs per chain graph model for up to 20 nodes. The results show that the ratio between the numbers of representable independence models for the two classes grows exponentially as the number of nodes increases. This indicates that only a very small fraction of all independence models representable by chain graphs of the Andersson-Madigan-Perlman interpretation also can be represented by Bayesian networks.

Keywords: Chain graphs, Andersson-Madigan-Perlman interpretation, MCMC sampling.

1 Introduction

Chain graphs (CGs) is a class of probabilistic graphical models (PGMs) that can contain two types of edges, representing symmetric resp. asymmetric relationships between the variables in question. This allows CGs to represent more independence models than for example Bayesian networks (BNs) and thereby represent systems more accurately than this less expressive class of models. Today there do however exist several different ways of interpreting CGs and what conditional independences they encode, giving rise to different so called CG interpretations. The most researched CG interpretations are the Lauritzen-Wermuth-Frydenberg (LWF) interpretation [6, 9], the Andersson-Madigan-Perlman (AMP)

L.C. van der Gaag and A.J. Feelders (Eds.): PGM 2014, LNAI 8754, pp. 458–470, 2014.

interpretation [1] and the multivariate regression (MVR) interpretation [3, 4]. Each interpretation has its own way of determining conditional independences in a CG and it can be noted that no interpretation subsumes another in terms of representable independence models [5, 15].

Although it is well known that any independence model that can be represented by BNs also can be represented by CGs, and that the opposite does not hold, it is unclear how much more expressive the latter class of models is. This question has previously been studied for CGs of the LWF interpretation (LWF CGs) and MVR interpretation (MVR CGs) [11, 16]. The results from these studies show that the ratio of the number of BN models (independence models representable by BNs) compared to the number of CG models (independence models representable by CGs) decreases exponentially as the number of nodes in the graphs increases. In this article we carry out a similar study for the AMP CG interpretation and investigate if the previous results also hold for this interpretation of CGs.

Measuring the ratio of representable independence models should in principle be easy to do. We only have to enumerate every independence model representable by CGs of the AMP interpretation (AMP CGs) and check if they also can be perfectly represented by BNs. The problem here is that the number of AMP CG models grows superexponentially as the number of nodes in the graphs increases and it is infeasible to enumerate all of them for more than five nodes. Hence we instead use a Markov chain Monte Carlo (MCMC) approach that previously has been shown to be successful when studying BN models, LWF CG models and MVR CG models [11, 16]. The approach consists of creating a Markov chain whose states are all possible AMP CG models with a certain number of nodes and whose stationary distribution is the uniform distribution over these models. This does then allow us to sample the AMP CG models from approximately the uniform distribution over all AMP CG models by transitioning through the Markov chain. With such a set of models we can thereafter approximate the ratio of BN models to AMP CG models for up to 20 nodes. Moreover, since there exists an equation for calculating the exact number of AMP CGs for a set number of nodes [17], we can also estimate the average number of AMP CGs per AMP CG model. Finally we also make the AMP CG models publicly available online and it is our intent that they can be used for further studies of AMP CG models and to evaluate for example learning algorithms to get more accurate evaluations than what is achieved with the randomly generated CGs used today [10, 12].

The rest of the article is organised as follows. In the next section we will cover the definitions and notations used in this article. This is then followed by Section 3 where we discuss the theory of the MCMC sampling algorithm used and also prove that its unique stationary distribution is the uniform distribution of all AMP CG models. In Section 4 we then discuss the results found and in Section 5 we give a short conclusion.

2   Notation

All graphs are defined over a finite set of variables V. If a graph G contains an edge between two nodes V_1 and V_2, we denote with $V_1{\rightarrow}V_2$ a *directed edge* and with $V_1{-}V_2$ an *undirected edge*. A set of nodes is said to be *complete* if there exist edges between all pairs of nodes in the set.

The *parents* of a set of nodes X of G is the set $pa_G(X) = \{V_1|V_1{\rightarrow}V_2$ is in G, $V_1 \notin X$ and $V_2 \in X\}$. The *children* of X is the set $ch_G(X) = \{V_1|V_2{\rightarrow}V_1$ is in G, $V_1 \notin X$ and $V_2 \in X\}$. The *neighbours* of X is the set $nb_G(X) = \{V_1|V_1{-}V_2$ is in G, $V_1 \notin X$ and $V_2 \in X\}$. The *boundary* of X is the set $bd_G(X) = pa_G(X) \cup nb_G(X)$. The *adjacents* of X is the set $ad_G(X) = \{V_1|V_1{\rightarrow}V_2, V_1{\leftarrow}V_2$ or $V_1{-}V_2$ is in G, $V_1 \notin X$ and $V_2 \in X\}$.

A *route* from a node V_1 to a node V_n in G is a sequence of nodes $V_1, \ldots, V_n$ such that $V_i \in ad_G(V_{i+1})$ for all $1 \leq i < n$. A *path* is a route containing only distinct nodes. The length of a path is the number of edges in the path. A path is called a *cycle* if $V_n = V_1$. A path is *descending* if $V_i \in pa_G(V_{i+1}) \cup nb_G(V_{i+1})$ for all $1 \leq i < n$. The *descendants* of a set of nodes X of G is the set $de_G(X) = \{V_n|$ there is a descending path from V_1 to V_n in G, $V_1 \in X$ and $V_n \notin X\}$. A path is *strictly descending* if $V_i \in pa_G(V_{i+1})$ for all $1 \leq i < n$. The *strict descendants* of a set of nodes X of G is the set $sde_G(X) = \{V_n|$ there is a strict descending path from V_1 to V_n in G, $V_1 \in X$ and $V_n \notin X\}$. The *ancestors* (resp. *strict ancestors*) of X is the set $an_G(X) = \{V_1|V_n \in de_G(V_1), V_1 \notin X, V_n \in X\}$ (resp. $san_G(X) = \{V_1|V_n \in sde_G(V_1), V_1 \notin X, V_n \in X\}$). A cycle is called a *semi-directed cycle* if it is descending and $V_i{\rightarrow}V_{i+1}$ is in G for some $1 \leq i < n$.

A Bayesian network (BN) is a graph containing only directed edges and no semi-directed cycles while a Markov network (MN) is graph containing only undirected edges. A CG under the Andersson-Madigan-Perlman (AMP) interpretation, denoted AMP CG, contains only directed and undirected edges but no semi-directed cycles. Note that we in this article only will study the AMP CG interpretation, but that we will use the term CG when the results or notion generalizes to all CG interpretations. A *connectivity component* C of an AMP CG is a maximal (wrt set inclusion) set of nodes such that there exists a path between every pair of nodes in C containing only undirected edges. A *subgraph* of G is a subset of nodes and edges in G. A subgraph of G induced by a set of its nodes X is the graph over X that has all and only the edges in G whose both ends are in X.

To illustrate these concepts we can study the AMP CG G with five nodes shown in Fig. 1a. In the graph we can for example see that the only child of A is B and that D is a neighbour of E. D is also a strict descendant of A due to the strictly descending path $A{\rightarrow}B{\rightarrow}D$, while E is not. E is however in the descendants of A together with B and D. A is therefore an ancestor of all nodes except itself and C. We can also see that G contains no semi-directed cycles since it contains no cycle at all. Moreover we can see that G contains four connectivity components: $\{A\}, \{B\}, \{C\}$ and $\{D, E\}$. In Fig. 1b we can see a subgraph of G

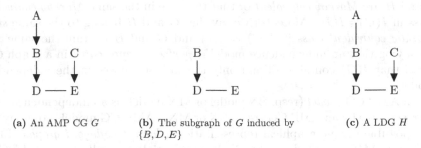

(a) An AMP CG G

(b) The subgraph of G induced by $\{B, D, E\}$

(c) A LDG H

Fig. 1. Three different AMP CGs

with the nodes B, D and E. This is also the induced subgraph of G with these nodes since it contains all edges between the nodes B, D and E in G.

Let X, Y and Z denote three disjoint subsets of V. We say that X is *conditionally independent* from Y given Z if the value of X does not influence the value of Y when the values of the variables in Z are known, i.e. $pr(X, Y|Z) = pr(X|Z)pr(Y|Z)$ holds and $pr(Z) > 0$. We say that X is *separated* from Y given Z, denoted $X \perp_G Y|Z$, in an AMP CG, BN or MN G iff there exists no S-open path between X and Y. A path is said to be *S-open* iff every non-head-no-tail node on the path is not in Z and every head-no-tail node on the path is in Z or $san_G(Z)$. A node B is said to be a *head-no-tail* in an AMP CG, BN or MN G between two nodes A and C on a path if one of the following configurations exists in G: $A{\rightarrow}B{\leftarrow}C$, $A{\rightarrow}B{-}C$ or $A{-}B{\leftarrow}C$. Moreover G is also said to contain a *triplex* $(\{A, C\}, B)$ iff one such configuration exists in G and A and C are not adjacent in G. A triplex $(\{A, C\}, B)$ is said to be a *flag* (resp. a *collider*) in an AMP CG or BN G iff G contains one following subgraphs induced by A, B and C: $A{\rightarrow}B{-}C$ or $A{-}B{\leftarrow}C$ (resp. $A{\rightarrow}B{\leftarrow}C$). If an AMP CG G is said to contain a *biflag* we mean that G contains the induced subgraph $A{\rightarrow}B{-}C{\leftarrow}D$ where A and D might be adjacent, for four nodes A, B, C and D. Given a graph G we mean with $G \cup \{A{\rightarrow}B\}$ the graph H with the same structure as G but where H also contains the directed edge $A{\rightarrow}B$ in addition to any other edges in G. Similarly we mean with $G \smallsetminus \{A{\rightarrow}B\}$ the graph H with the same structure as G but where H does not contain the directed edge $A{\rightarrow}B$. Note that if G did not contain the directed edge $A{\rightarrow}B$ then H is not a valid graph.

To illustrate these concepts we can once again look at the graph G shown in Fig. 1a. G contains no colliders but two flags, $B{\rightarrow}D{-}E$ and $C{\rightarrow}E{-}D$, that together form the biflag $B{\rightarrow}D{-}E{\leftarrow}C$. This means that $B \perp_G E$ holds but that $B \perp_G E|D$ does not hold since D is a head-no-tail node on the path $B{\rightarrow}D{-}E$.

The *independence model* M induced by a graph G, denoted as $I(G)$, is the set of separation statements $X \perp_G Y|Z$ that hold in G. We say that two graphs

G and H are *Markov equivalent* or that they are in the same *Markov equivalence class* iff $I(G) = I(H)$. Moreover we say that G and H belong to the same *strong Markov equivalent class* iff $I(G) = I(H)$ and G and H contain the same flags. By saying that an independence model is *perfectly represented* in a graph G we mean that $I(G)$ contains all and only the independences in the independence model.

An AMP CG *model* (resp. BN model or MN model) is an independence model representable by an AMP CG (resp. BN or MN). AMP CG models do today have two possible unique graphical representations, *largest deflagged graphs* (LDGs) [14] and *AMP essential graphs* [2]. In this article we will only use LDGs. A LDG is the AMP CG of a Markov equivalence class that has the minimum number number of flags while at the same time contains the maximum number of undirected edges for that strong Markov equivalence class.

For AMP CGs there exists a set of operations that allows for changing the structure of edges within an AMP CG without altering the Markov equivalence class it belongs to. In this article we use the *feasible split* operation [15] and the *legal merging* [14] operation. A split is said to be feasible for a connectivity component C of an AMP CG G iff it can be divided into two disjoint sets U and L such that $U \cup L = C$ and if replacing all undirected edges between U and L with directed edges orientated towards L results in an AMP CG H such that $I(G) = I(H)$. It has been shown that if the following conditions hold in G then such a split is possible: [15] (1) $\forall A \in ne_G(L) \cap U, L \subseteq ne_G(A)$, (2) $ne_G(L) \cap U$ is complete and (3) $\forall B \in L, pa_G(ne_G(L) \cap U) \subseteq pa_G(B)$.

A merging is on the other hand said to be legal for two connectivity components U and L in an AMP CG G, such that $U \in pa_G(L)$, iff replacing all directed edges between U and L with undirected edges results in an AMP CG H such that G and H belongs to the same strong Markov equivalence class. It has been shown that if the following conditions hold in G then such a merging is possible [14] (1) $pa_G(L) \cap U$ is complete in G, (2) $\forall B \in pa_G(L) \cap U, pa_G(L) \setminus U = pa_G(B)$ and (3) $\forall A \in L, pa_G(L) = pa_G(A)$. Note that a legal merging is not the reverse operator of a feasible split since a feasible split handles Markov equivalence classes, while a legal merging handles strong Markov equivalence classes.

If we once again look at the CGs in Fig. 1 we can see that G, shown in Fig. 1a, and H, shown in Fig. 1c, are Markov equivalent since $I(G) = I(H)$. Moreover we can note that G and H must belong to the same strong Markov equivalence class since they contain the same flags. This means that G cannot be a LDG since H is larger than G. In G it exists no feasible split, but one legal merging which is the merging of the connectivity components $\{A\}$ and $\{B\}$ that results in a CG with the same structure as H. In H it does on the other hand exist no legal merging but one feasible split which is the split of the connectivity component $\{A, B\}$ into either $A \rightarrow B$ or $B \rightarrow A$. We can finally note that this feasible split would result in no additional flags and hence, since no legal mergings are possible in H, that H must be a LDG.

3   Markov Chain Monte Carlo Approach

In this section we cover the theory behind the MCMC approach used in this article for sampling LDGs from the uniform distribution. Note that we only cover the theory of the MCMC sampling method very briefly and for a more complete introduction of the sampling method we instead refer the reader to the work by Häggström [7].

The MCMC sampling approach consists of creating a Markov chain whose unique stationary distribution is the desired distribution and then sample this Markov chain after a number of transitions. The Markov chain is defined by a set of operators that allows us to transition from one state to another. It can then be shown that if these operators have certain properties and the number of transitions goes to infinity then all states have the same probability of being sampled [7]. Moreover, in practice it has also been shown that the number of transitions can be relatively small and a good approximation of the uniform distribution can still be achieved.

In our case the possible states of the Markov chain are all possible independence models representable by AMP CGs, i.e. AMP CG models, represented by LDGs. The operators then add and remove certain edges in in these LDGs, allowing the MCMC method to transition between all possible LDGs. For the stationary distribution to be the unique uniform distribution of all possible LDGs we only have to prove that the operators fulfill the following properties [7]: aperiodicity, irreducibility and reversibility. Aperiodicity, i.e. that the Markov chain does not end up in the same state periodicly, and irreducibility, i.e. that any LDG can be reached from any other LDG using only the defined operators, proves that the Markov chain has a unique distribution. Reversibility then proves that this distribution also is the stationary distribution.

The operators used are defined in Definition 1 and it follows from Lemma 1, 2 and 3 that they fulfill the properties described above for LDGs with at least two nodes.

Definition 1. *Markov Chain Operators*
Choose uniformly and perform one of the following six operators to transition from a LDG G to the next LDG H in the Markov chain.

- **Add directed edge.** *Choose two nodes X, Y in G uniformly and with replacement. If $X \notin ad_G(Y)$ and $G \cup \{X \rightarrow Y\}$ is a LDG let $H = G \cup \{X \rightarrow Y\}$, otherwise let $H = G$.*
- **Remove directed edge.** *Choose two nodes X, Y in G uniformly and with replacement. If $X \rightarrow Y$ is in G and $G \setminus \{X \rightarrow Y\}$ is a LDG let $H = G \setminus \{X \rightarrow Y\}$, otherwise let $H = G$.*
- **Add undirected edge.** *Choose two nodes X, Y in G uniformly and with replacement. If $X \notin ad_G(Y)$ and $G \cup \{X - Y\}$ is a LDG let $H = G \cup \{X - Y\}$, otherwise let $H = G$.*

- **Remove undirected edge.** *Choose two nodes X, Y in G uniformly and with replacement. If X–Y is in G and $G \smallsetminus \{X$–$Y\}$ is a LDG let $H = G \smallsetminus \{X$–$Y\}$, otherwise let $H = G$.*
- **Add two directed edges.** *Choose four nodes X, Y, Z, W in G uniformly and with replacement. If $X \notin ad_G(Y)$, $Z \notin ad_G(W)$ and $G \cup \{X{\to}Y, Z{\to}W\}$ is a LDG let $H = G \cup \{X{\to}Y, Z{\to}W\}$, otherwise let $H = G$. Note that Y might be equal to W in this operation.*
- **Remove two directed edges.** *Choose four nodes X, Y, Z, W in G uniformly and with replacement. If $X{\to}Y$ and $Z{\to}W$ are in G and $G \smallsetminus \{X{\to}Y, Z{\to}W\}$ is a LDG let $H = G \smallsetminus \{X{\to}Y, Z{\to}W\}$, otherwise let $H = G$. Note that Y might be equal to W in this operation.*

Lemma 1. *The operators in Definition 1 fulfill the aperiodicity property when G contains at least two nodes.*

Proof. Here it is enough to show that there exists at least one operator such that H is equal to G, and at least one operator such that it is not, for any possible G with at least two nodes. The latter follows directly from Lemma 3 since there exist more than one possible state when G contains at least two nodes. To see that the former must hold note that if the add directed edge operation results in a LDG for some nodes X and Y in a LDG G, then clearly the remove directed edge $X{\to}Y$ operation must result in a LDG H equal to G for that G.   □

Lemma 2. *The operators in Definition 1 fulfill the reversibility property for the uniform distribution.*

Proof. Since the desired distribution for the Markov chain is the uniform distribution proving that reversibility holds for the operators simplifies to proving that symmetry holds for them. This means that we need to show that for any LDG G the probability to transition to any LDG H is equal to the probability to transition from H to G. Here it is simple to see that each operator has a reverse operator such as remove directed edge for add directed edge etc. and that the "forward" operator and reverse operator are chosen with equal probability for a certain set of nodes. Moreover we can also see that if H is G with one operator performed upon it, such that $H \neq G$, then clearly there can exist no other operator that transition G to H. Hence the operators fulfill the symmetry property and thereby the reversibility property.   □

Lemma 3. *Given a LDG G any other LDG H can be reached using the operators described in Definition 1 such that all intermediate graphs are LDGs.*

Proof. Here it is enough to prove that any LDG H can be reached from the empty graph $G_\varnothing$ since we, due to the reversibility property, then also know that $G_\varnothing$ can be reached from H (or G). That such a procedure exists follows from Lemma 4.   □

Lemma 4. *There exists an algorithm that, given a LDG H, constructs G from the empty graph using the operators in Definition 1 such that G = H when the algorithm terminates and all intermediate graphs are LDGs.*

The algorithm and the proof of its correctness are rather technical and we omit these due to page limitations. In short the algorithm works by iteratively adding the connectivity components to G one by one until $G = H$. The complete algorithm, and the proof of its correctness, can be found in an extended version of this paper available at $http://www.ida.liu.se/\sim dagso62/PGM2014Extended.pdf$.

Finally we do also have to prove what the conditions are for when the independence model of a LDG can be represented as a BN resp. a MN.

Lemma 5. *Given a LDG G there exists a BN H such that I(G) = I(H) iff G contains no flags and G contains no chordless undirected cycles.*

Proof. Since H is a BN and BNs is a subclass of AMP CGs we know that for $I(G) = I(H)$ to hold H must be in the same Markov equivalence class as G if it is interpreted as an AMP CG. However, since G is a LDG and contains a flag we know that that flag must exist in every AMP CG in the Markov equivalence class of G. Hence $I(G) = I(H)$ cannot hold if G contains a flag. Moreover it is also well known that the independence model of a MN containing chordless cycles cannot be represented as a BN. On the other hand, if G contains no flag and no chordless undirected cycles, then clearly all triplexes in G must be unshielded colliders. Hence, together with the fact that G can contain no semi-directed cycles, we can orient all undirected edges in G to directed edges without creating any semi-directed cycles as shown by Koller and Friedman [8, Theorem 4.13]. This means that the resulting graph must be a BN H such that $I(G) = I(H)$.   □

Lemma 6. *Given a LDG G there exists a MN H such that I(G) = I(H) iff G contains no directed edges.*

Proof. This follows directly from the fact that G contains a triplex iff it also contains a directed edge and MNs cannot represent any triplexes. It is also clear that if G contains no directed edges than it can only contain undirected edges and hence be interpreted as a MN.   □

4   Results

Using the Markov chain presented in the previous section we were able to sample AMP CG models uniformly for a set number of nodes for up to 20 nodes. For each number of nodes 10^5 samples were sampled with 10^5 transitions between each sampled sample. The success rate for the transitions, i.e. the percentage of transitions such that $H \neq G$, was for 20 nodes approximately 10% and this corresponds well with earlier studies for LWF and MVR CGs [16]. To check if a graph G was a LDG it was first checked whether it was an AMP CG. If so then the LDG H in

Table 1. Exact and approximate ratios of AMP CG models whose independence mode can be represented as BNs, MNs, neither (in that order)

NODES	EXACT			APPROXIMATE		
2	1.0000	1.0000	0.0000	1.0000	1.0000	0.0000
3	1.0000	0.7273	0.0000	1.0000	0.7275	0.0000
4	0.8259	0.2857	0.1607	0.8253	0.2839	0.1609
5	0.5987	0.0689	0.3958	0.6019	0.0632	0.3931
6				0.4292	0.0112	0.5694
7				0.2987	0.0019	0.7010
8				0.2021	0.0002	0.7979
9				0.1373	0.0000	0.8627
10				0.0924	0.0000	0.9076
11				0.0604	0.0000	0.9396
12				0.0394	0.0000	0.9606
13				0.0251	0.0000	0.9749
14				0.0152	0.0000	0.9849
15				0.0106	0.0000	0.9894
16				0.0065	0.0000	0.9935
17				0.0044	0.0000	0.9956
18				0.0028	0.0000	0.9972
19				0.0017	0.0000	0.9983
20				0.0010	0.0000	0.9990

Table 2. Exact and approximate ratios of AMP CG models to AMP CGs

NODES	EXACT	APPROXIMATE
2	0.5000	0.5074
3	0.2200	0.2235
4	0.1327	0.1312
5	0.1043	0.1008
6		0.0860
7		0.0775
8		0.0701
9		0.0668
10		0.0616
11		0.0594
12		0.0595
13		0.0608
14		0.0635
15		0.0559
16		0.0598
17		0.0557
18		0.0567
19		0.0596
20		0.0632

the Markov equivalence class of G was created whereafter it was checked whether H and G had the same structure. To construct the LDG of G the algorithm defined by Roverato and Studený was used [14]. The implementation was carried out in C++, run on an Intel Core i5 processor and it took approximately one week to complete the sampling of all sample sets. Moreover we also enumerated all LDGs for up to five nodes to allow a comparison between the approximated and exact values. The sampled graphs and code are available for public access at $http://www.ida.liu.se/divisions/adit/data/graphs/CGSamplingResources$.

The results are shown in Tables 1 and 2. If we start with Table 1 we can here see the ratio of the number of BN models resp. MN models compared to number of AMP CG models. We can note that the approximation seems to be very accurate for up to five nodes but for more than five nodes we do not have any exact values to compare the approximations to. We can however plot the ratios of the number of BN models compared to the number of AMP CG models in a graph with logarithmic scales, as seen in Fig. 2. We can then see that the ratio follows the equation $R = 5 * 0.664^n$ very closely, where n is the number of nodes and R the ratio. This means that the ratio decreases exponentially as the number of nodes increases and hence that the previous results seen for LWF and MVR CGs also hold for AMP CGs [16].

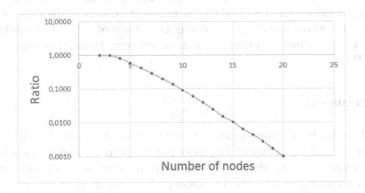

Fig. 2. The ratios of the number of BN models compared to the number of AMP CG models for different number of nodes. (Displayed with a logarithmic scale)

If we go back to the table we can also note that the ratio of the number of MN models compared to the number of AMP CG models is almost zero for more than eight nodes similarly as seen in previous studies for LWF and MVR CGs. Moreover, if we compare the ratio of the number BN models to the number of CG models for the different CG interpretations we can see that the ratio is approximately 0.0017 for LWF CGs, 0.0011 for MVR CGs and 0.0010 for AMP

CGs for 20 nodes [16]. This means that LWF CGs can represent the smallest amount of independence models while MVR and AMP CGs can represent about the same amount. Another interesting observation that can be made here is that the difference between the ratios of the number of BN models compared to the number of AMP CG models resp. MVR CG models is very small for any number of nodes.

If we continue to Table 2 we can here see the exact and approximate ratios of AMP CG models to AMP CGs. This ratio can be calculated using the equation

$$\frac{\#CGmodels}{\#CGs} = \frac{\#BNs}{\#CGs} * \frac{\#BNmodels}{\#BNs} * \frac{\#CGmodels}{\#BNmodels} \tag{1}$$

where $\#CGmodels$ represents the number of AMP CG models etc. The ratio $\frac{\#BNs}{\#CGs}$ can then be found using the iterative equations by Robinsson [13] and Steinsky [17] while $\frac{\#BNmodels}{\#BNs}$ can be found in previous studies for BN models [11]. Finally we can also get the ratio $\frac{\#CGmodels}{\#BNmodels}$ by inverting the ratio found in Table 1. If we study the values in Table 2 we can see that the ratio seems to converge to somewhere around 0.06 similarly as seen for the average number of BNs per BN model [16]. In the previous study estimating the average number of LWF CGs per LWF CG model no such convergence has been seen [11]. We can however note that this study only goes up to 13 nodes but also that the approximated values for the LWF CG interpretation is considerably lower than 0.06. The ratio also shows that traversing the space of AMP CG models when learning CG structures is considerably more efficient than traversing the space of all AMP CGs.

5   Conclusion

In this article we have approximated the ratio of the number of AMP CG models compared to the number of BN models for up to 20 nodes. This has been achieved by creating a Markov chain whose unique stationary distribution is the uniform distribution of all AMP CG models. The results show that the ratio of the number of independence models representable by AMP CGs compared to the corresponding number for BNs grows exponentially as the number of nodes increases. This confirms previous results for LWF and MVR CGs but a new, and unexpected, result is also that AMP and MVR CGs seem to be able to represent almost the same amount of independence models. It has previously been shown that there exist some independence models perfectly representable by both MVR and AMP CGs, but not BNs [15], and it would now be interesting to see how large this set is compared to the set of all AMP resp. MVR CG models.

In addition to the comparison of the number of AMP CG models to the number of BN models we have also approximated the average number of AMP CGs per CG model. The results indicate that the average ratio converges to somewhere around 0.06. Such a convergence has not previously been seen for any other CG interpretation but it has been found to be ≈ 0.25 for BNs [11].

With these new results sampling methods for CG models of all CG interpretations are now available [11, 16]. This opens up for MCMC based structure learning algorithms for the different CG interpretations that traverses the space of CG models instead of the space of CGs. More importantly it does also open up for further studies on what independence models and systems the different CG interpretations can represent, which is an important question where the answer is still unclear.

Acknowledgments. This work is funded by the Swedish Research Council (ref. 2010-4808).

References

[1] Andersson, S.A., Madigan, D., Perlman, M.D.: An alternative markov property for chain graphs. Scandinavian Journal of Statistics 28, 33–85 (2001)

[2] Andersson, S.A., Perlman, M.D.: Characterizing Markov Equivalence Classes For AMP Chain Graph Models. The Annals of Statistics 34, 939–972 (2006)

[3] Cox, D.R., Wermuth, N.: Linear Dependencies Represented by Chain Graphs. Statistical Science 8, 204–283 (1993)

[4] Cox, D.R., Wermuth, N.: Multivariate Dependencies: Models, Analysis and Interpretation. Chapman and Hall (1996)

[5] Drton, M.: Discrete Chain Graph Models. Bernoulli 15, 736–753 (2009)

[6] Frydenberg, M.: The Chain Graph Markov Property. Scandinavian Journal of Statistics 17, 333–353 (1990)

[7] Häggström, O.: Finite Markov Chains and Algorithmic Applications. Campbridge University Press (2002)

[8] Koller, D., Friedman, N.: Probabilistic Graphcal Models. MIT Press (2009)

[9] Lauritzen, S.L., Wermuth, N.: Graphical Models for Association Between Variables, Some of Which are Qualitative and Some Quantitative. The Annals of Statistics 17, 31–57 (1989)

[10] Ma, Z., Xie, X., Geng, Z.: Structural Learning of Chain Graphs via Decomposition. Journal of Machine Learning Research 9, 2847–2880 (2008)

[11] Peña, J.M.: Approximate Counting of Graphical Models via MCMC Revisited. In: Bielza, C., Salmerón, A., Alonso-Betanzos, A., Hidalgo, J.I., Martínez, L., Troncoso, A., Corchado, E., Corchado, J.M. (eds.) CAEPIA 2013. LNCS, vol. 8109, pp. 383–392. Springer, Heidelberg (2013)

[12] Peña, J.M., Sonntag, D., Nielsen, J.: An Inclusion Optimal Algorithm for Chain Graph Structure Learning. In: Proceedings of the 17th International Conference on Artificial Intelligence and Statistics, pp. 778–786 (2014)

[13] Robinson, R.W.: Counting Labeled Acyclic Digraphs. New Directions in the Theory of Graphs, pp. 239–273 (1973)

[14] Roverato, A., Studený, M.: A Graphical Representation of Equivalence Classes of AMP Chain Graphs. Journal of Machine Learning Research 7, 1045–1078 (2006)

[15] Sonntag, D., Peña, J.M.: Chain Graph Interpretations and Their Relations (under review, 2014)
[16] Sonntag, D., Peña, J.M., Gómez-Olmedo, M.: Approximate Counting of Graphical Models Via MCMC Revisited (under review, 2014)
[17] Steinsky, B.: Enumeration of Labelled Chain Graphs and Labelled Essential Directed Acyclic Graphs. Discrete Mathematics 270, 266–277 (2003)

Learning Bayesian Network Structures When Discrete and Continuous Variables Are Present

Joe Suzuki

Department of Mathematics, Osaka University,
Toyonaka, Osaka 560-0043, Japan

Abstract. In any database, some fields are discrete and others continuous in each record. We consider learning Bayesian network structures when discrete and continuous variables are present. Thus far, most of the previous results assumed that all the variables are either discrete or continuous. We propose to compute a new Bayesian score for each subset of discrete and continuous variables, and to obtain a structure that maximizes the posterior probability given examples. We evaluate the proposed algorithm and make experiments to see that the error probability and Kullback-Leibler divergence diminish as n grows whereas the computation increases linearly in the logarithm of the number of bins in the histograms that approximate the density.

Keywords: Learning Bayesian network structures, discrete and continuous variables, Kullback-Leibler divergence, density estimation, universality.

1 Introduction

We consider a learning Bayesian network structure from examples.

Suppose we have three random variables X, Y, Z and wish to express the distribution by one of the eleven factorizations:

$$P(X)P(Y)P(Z), P(X)P(Y,Z), P(Y)P(Z,X),$$

$$P(Z)P(X,Y), \frac{P(X,Y)P(X,Z)}{P(X)}, \frac{P(X,Y)P(Y,Z)}{P(Y)},$$

$$\frac{P(X,Z)P(Y,Z)}{P(Z)}, \frac{P(Y)P(Z)P(X,Y,Z)}{P(Y,Z)}, \frac{P(Z)P(X)P(X,Y,Z)}{P(Z,X)},$$

$$\frac{P(X)P(Y)P(X,Y,Z)}{P(X,Y)} \text{ , and } P(X,Y,Z) \text{ ,}$$

where hereafter we number the eleven equations as (1) through (11). Then, we can express the corresponding Bayesian networks as in Figure 1. In this paper, we do not distinguish structures that share the same distribution (Markov equivalent structures). For examples, although the three structures can be considered for (5) as in Figure 2, they express the same distribution. On the other hand, only one structure exists for (8).

L.C. van der Gaag and A.J. Feelders (Eds.): PGM 2014, LNAI 8754, pp. 471–486, 2014.

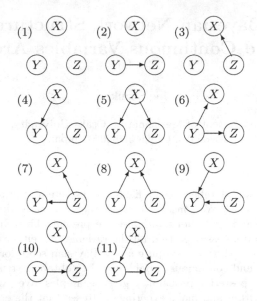

Fig. 1. The eleven Bayesian networks with three variables X, Y, Z

The problem we consider in this paper is to identify the structure from examples $(X = x_1, Y = y_1, Z = z_1), \cdots, (X = x_n, Y = y_n, Z = z_n)$. We assume that the n examples without missing values are independently emitted.

There are several approaches to tackle this problem. For example, we may apply statistical tests on conditional independence w.r.t. a positive error probability α many times to choose one structure such as the PC Algorithm [22]. In this paper, we solve the problem based on the Bayes principle: specifying the prior probabilities over parameters and models, we choose a model that maximizes the posterior probability.

Suppose that $P(X), P(Y), P(X, Y)$ are expressed in terms of some parameter θ, and denote them by $P(X|\theta), P(Y|\theta)$, and $P(X, Y|\theta)$, respectively. For $x^n = (x_1, \cdots, x_n)$ and $y^n = (y_1, \cdots, y_n)$, let $P^n(x^n|\theta) := \prod_{i=1}^n P(x_i|\theta)$, $P^n(y^n|\theta) := \prod_{i=1}^n P(y_i|\theta)$, $P^n(x^n, y^n|\theta) := \prod_{i=1}^n P(x_i, y_i|\theta)$,

$$Q^n(x^n) := \int P^n(x^n|\theta)w(\theta)d\theta \ , \quad Q^n(y^n) := \int P^n(y^n|\theta)w(\theta)d\theta \ ,$$

and

$$Q^n(x^n, y^n) := \int P^n(x^n, y^n|\theta)w(\theta)d\theta \ ,$$

respectively, where $w(\cdot)$ is the prior probability over the candidate parameters. If the prior probability p of $X \perp\!\!\!\perp Y$ (X, Y are independent) is available, the posterior probability of $X \perp\!\!\!\perp Y$ given examples (x^n, y^n) can be expressed by

$$P(X \perp\!\!\!\perp Y|x^n, y^n) = \frac{pQ^n(x^n)Q^n(y^n)}{pQ^n(x^n)Q^n(y^n) + (1-p)Q^n(x^n, y^n)} \ ,$$

so that we can decide whether $X \perp\!\!\!\perp Y$ or $X \not\!\perp\!\!\!\perp Y$ by

$$X \perp\!\!\!\perp Y \iff pQ^n(x^n)Q^n(y^n) \geq (1-p)Q^n(x^n, y^n) .$$

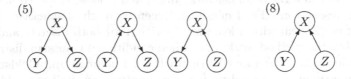

Fig. 2. Markov equivalent Bayesian networks: (5) and (8) have three and one Markov equivalent structures, respectively

The idea can be applied to learning Bayesian networks: let $p_1, \cdots, p_{11}$ be the prior probabilities of the eleven structures above. If we are given examples z^n as well as x^n, y^n, we can compute $Q^n(z^n), Q^n(x^n, z^n), Q^n(y^n, z^n), Q^n(x^n, y^n, z^n)$ as well. Then, we obtain the eleven quantities

$$p_1 Q^n(x^n)Q^n(y^n)Q(z^n), p_2 Q^n(x^n)Q^n(y^n, z^n), p_3 Q^n(y^n)Q^n(z^n, x^n),$$

$$p_4 Q^n(z^n)Q^n(x^n, y^n), p_5 \frac{Q^n(x^n, y^n)Q^n(x^n, z^n)}{Q^n(x^n)}, p_6 \frac{Q^n(x^n, y^n)Q^n(y^n, z^n)}{Q^n(y^n)},$$

$$p_7 \frac{Q^n(x^n, z^n)Q^n(y^n, z^n)}{Q^n(z^n)}, p_8 \frac{Q^n(y^n)Q^n(z^n)Q(x^n, y^n, z^n)}{Q^n(y^n, z^n)}, p_9 \frac{Q^n(z^n)Q^n(x^n)Q^n(x^n, y^n, z^n)}{Q^n(z^n, x^n)},$$

$$p_{10} \frac{Q^n(x^n)Q^n(y^n)Q^n(x^n, y^n, z^n)}{Q^n(x^n, y^n)}, p_{11} Q^n(x^n, y^n, z^n) ,$$

so that we can select a structure that maximizes the posterior probability.

The same idea with three variables can be extended into the cases with N variables in a straightforward manner.

The idea of replacing the true P by such some Q to obtain a Bayesian solution has been used since the early 1990's. Wray Buntine considered its application to construction of classification trees [1], Cooper and Herskovits estimated Bayesian network structures using such a Q [5], and Suzuki [24] used the MDL principle to the Bayesian network structure learning problem and proposed a modified version of the Chow-Liu algorithm. Since then, many authors reported applications using similar techniques thus far.

However, if some attributes take continuous values, it is hard to construct such a Q in order to identify the structure given $\{(x_i, y_i, z_i)\}_{i=1}^n$. For example, suppose that X is continuous, that Y is finite, and that Z is countable and infinite, such as $X \in [0, 1)$, $Y \in \{0, 1\}$, and $Z \in \{1, 2, \cdots\}$.

Suppose that all the variables are continuous and that a marginal density function exists. Then, constructing a kernel function [18] to which the training data fit may show better performance in some cases. However, and in order to compute scores for each structure, multivariate correlation should be obtained.

So, it seems to be hard for the kernel methods to obtain the structure that maximizes the posterior probability.

Most previous works have either solved the problem by discretization, or assumed that the data are generated by a Gaussian distribution. Heckerman and Geiger [8] considered structure estimation when both Gaussian and finite variables are present in a Bayesian network. Monti and Cooper [15] used neural networks to represent conditional densities. Recently, in the R-package, Bottcher and Dethlefsen [2] deal with a learning algorithm for both discrete and continuous data, but the method works only when conditional Gaussian distributions can be assumed (Lauritzen and Wermuth [14]). Friedman and Goldszmidt [7] proposed an estimation method to discretize continuous attributes while learning Bayesian networks.

However, if we quantize the continuous data to take care of only discrete data, it is not easy to optimize the number of bins in the histogram for each n. For example, if we discretize $[0, 1)$ for X to $[0, 2^{-k}), [2^{-k}, 2 \cdot 2^{-k}), \cdots, [(2^k - 1) \cdot 2^{-k}, 1)$, how should we decide $k := k(n)$? What quantization is the best in $\{1, 2, \cdots\}$ for Y ? The previous works ([12], for example) do not give any answer to those questions.

Recently, several authors take approaches such as mixtures of truncated exponentials, mixtures of polynomials and mixtures of truncated basis functions. However, there is no guarantee for any required property such as consistency [21].

In this paper, we propose how to choose an optimal structure given $\{(x_i, y_i, z_i)\}_{i=1}^n$ in the sense of maximizing the posterior probability without assuming that each variable is either discrete or continuous. The main issue is what Q is qualified to be an alternative to P in more general settings. We will give an answer to the problem by generalizing the aforementioned idea. The theory developed in this paper is partially due to Boris Ryabko's density estimation [20]. Based on the generalized principle, we propose a method to estimate Bayesian network structures when the random variables are arbitrary.

Section 2 introduces the notion of universality for sequences consisting of elements in a finite set, and the idea of density function developed by Boris Ryabko [20]; Section 3 generalizes the notions of universality and density estimation based on histogram sequences; Section 4 gives a way of learning Bayesian network structures based on the generalized method; Section 5 makes several experiments to see the error probability and Kullback-Leibler divergence diminish; and Section 6 summarizes the contributions and state future works.

2   Preliminaries

2.1   Universality

Suppose random variable X takes binary values, and that the prior probability of $\theta = P(X = 1)$ is expressed by

$$w(\theta) = \frac{1}{C} \cdot \theta^{a-1} (1 - \theta)^{b-1}$$

in terms of $a, b > 0$ and $C := \int \theta^{a-1}(1 - \theta)^{b-1}d\theta$ (the Dirichlet distribution). For example, if $a = b = 1$, then the prior distribution w is uniform. In particular, it is known that Q^n takes the closed form:

$$Q^n(x^n) = \int_0^1 \theta^c(1 - \theta)^{n-c}dw(\theta)d\theta = \prod_{i=1}^n \frac{c_i + a_i}{i - 1 + a + b} , \tag{12}$$

where c_i is the number of x_i in x^{i-1}, and $a_i = a, b$ for $x_i = 1, 0$, respectively, and c is the number of ones in x^n.

Proposition 1 (Ryabko[19]). For any θ,

$$\frac{1}{n} \log \frac{P^n(x^n|\theta)}{Q^n(x^n)} \to 0$$

as $n \to \infty$ with probability one.

Notice that $a, b > 0$ are arbitrary, even if our prior knowledge on the values of a, b is not correct, the estimation is asymptotically correct although the convergence may be slower.

It is known that Proposition 1 is true even if X takes finite values rather than binary values. Let A, B be finite sets, and suppose X, Y takes values in A, B, respectively. Then, $A \times B = \{(a, b)|a \in A, b \in B\}$ is apparently a finite set, so that we have for any θ,

$$\frac{1}{n} \log \frac{P^n(y^n|\theta)}{Q^n(y^n)} \to 0 \quad \text{and} \quad \frac{1}{n} \log \frac{P^n(x^n, y^n|\theta)}{Q^n(x^n, y^n)} \to 0 .$$

Thus, from $\frac{1}{n} \log \frac{p}{1-p} \to 0$, we have

$$\frac{1}{n} \log \frac{pQ^n(x^n)Q^n(y^n)}{(1 - p)Q^n(x^n, y^n)} - \frac{1}{n} \log \frac{P^n(x^n|\theta)P^n(y^n|\theta)}{P^n(x^n, y^n|\theta)} \to 0 ,$$

which means that the decision does not depend on the prior probability p asymptotically although more correct prior information makes estimation converge faster for finite n.

The idea with two variables can be extended into the ones with N variables in a straightforward manner.

In summary, we can asymptotically obtain a Bayesian network structure with the maximum posterior probability even if the prior probabilities w and $\{p_i\}$ over parameters and structures, respectively, are not available.

Hereafter, we refer Q^n in Proposition 1 to a universal probability w.r.t. finite set A.

2.2 Learning Bayesian Networks with Continuous Data

Let $\{A_j\}$ be such that $A_0 = \{A\}$, and that A_{j+1} is a refinement of A_j, where A is the range of X. For example, suppose that random variable X takes values in

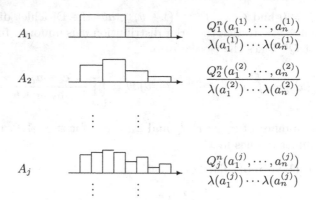

A_1 $\dfrac{Q_1^n(a_1^{(1)}, \cdots, a_n^{(1)})}{\lambda(a_1^{(1)}) \cdots \lambda(a_n^{(1)})}$

A_2 $\dfrac{Q_2^n(a_1^{(2)}, \cdots, a_n^{(2)})}{\lambda(a_1^{(2)}) \cdots \lambda(a_n^{(2)})}$

A_j $\dfrac{Q_j^n(a_1^{(j)}, \cdots, a_n^{(j)})}{\lambda(a_1^{(j)}) \cdots \lambda(a_n^{(j)})}$

Fig. 3. Density estimation based on weighting histograms

$A = [0, 1]$ and generate the sequence as follows:
$A_1 = \{[0, \frac{1}{2}), [\frac{1}{2}, 1)\}$
$A_2 = \{[0, \frac{1}{4}), [\frac{1}{4}, \frac{1}{2}), [\frac{1}{2}, \frac{3}{4}), [\frac{3}{4}, 1)\}$
$\cdots$
$A_j = \{[0, 2^{-(j-1)}), [2^{-(j-1)}, 2 \cdot 2^{-(j-1)}), \cdots, [(2^{j-1} - 1)2^{-(j-1)}, 1)\}$.
$\cdots$

For each j, we quantize each $x \in [0, 1]$ into the $a \in A_j$ such that $x \in a$. For example, if $x = 0.4$ and $j = 2$, then $a = [\frac{1}{4}, \frac{1}{2}) \in A_2$. Let λ be the Lebesgue measure (width of the interval). For example, $\lambda([\frac{1}{4}, \frac{1}{2})) = \frac{1}{4}$ and $\lambda(\{\frac{1}{2}\}) = 0$.

Notice that each A_j is a finite set. As we constructed Q^n in (12) for $A = \{0, 1\}$, we can construct a universal probability Q_j^n w.r.t. finite set A_j for each j. Given $x^n = (x_1, \cdots, x_n) \in [0, 1]^n$, we obtain a quantized sequence $(a_1^{(j)}, \cdots, a_n^{(j)}) \in A_j^n$ for each j, so that we can compute the quantity

$$g_j^n(x^n) := \frac{Q_j^n(a_1^{(j)}, \cdots, a_n^{(j)})}{\lambda(a_1^{(j)}) \cdots \lambda(a_n^{(j)})}$$

for each j. If we prepare a sequence of positive reals $w_1, w_2, \cdots$ such that $\sum_j w_j = 1$, we can compute the quantity

$$g^n(x^n) := \sum_{j=1}^{\infty} w_j g_j^n(x^n) .$$

On the other hand, let f be the true density function, and $f_j(x) := P(X \in a)/\lambda(a)$ for $a \in A_j$ and $j = 1, 2, \cdots$, if $x \in a$. We may regard f_j as an approximated density functions assuming histograms $\{A_j\}$ (Figure 3). For the given x^n, we define $f^n(x^n) = f(x_1) \cdots f(x_n)$ and $f_j^n(x^n) := f_j(x_1) \cdots f_j(x_n)$.

Thus, we have the following proposition, which is a continuous version of Proposition 1.

Proposition 2 (Ryabko 2009). For any density function f such that $D(f\|f_j) \to 0$ as $j \to \infty$,

$$\frac{1}{n} \log \frac{f^n(x^n)}{g^n(x^n)} \to 0$$

as $n \to \infty$ with probability one, where $D(f\|f_j)$ is the Kullback-Leibler divergence between f and f_j.

For example, for three variables X, Y, Z, if we compute the quantities $g^n(y^n)$, $g^n(z^n)$, $g^n(x^n, y^n)$, $g^n(x^n, z^n)$, $g^n(y^n, z^n)$, $g^n(x^n, y^n, z^n)$ as well as $g^n(x^n)$, we obtain the eleven quantities

$$p_1 g^n(x^n) g^n(y^n) g^n(z^n), p_2 g^n(x^n) g^n(y^n, z^n), p_3 g^n(y^n) g^n(z^n, x^n),$$

$$p_4 g^n(z^n) g^n(x^n, y^n), p_5 \frac{g^n(x^n, y^n) g^n(X, z^n)}{g^n(x^n)}, p_6 \frac{g^n(x^n, y^n) g^n(y^n, z^n)}{g^n(y^n)},$$

$$p_7 \frac{g^n(x^n, z^n) g^n(y^n, z^n)}{g^n(z^n)}, p_8 \frac{g^n(y^n) g^n(z^n) g^n(x^n, y^n, z^n)}{g^n(y^n, z^n)}, p_9 \frac{g^n(z^n) g^n(x^n) g^n(x^n, y^n, z^n)}{g^n(z^n, x^n)},$$

$$p_{10} \frac{g^n(x^n) g^n(y^n) g^n(x^n, y^n, z^n)}{g^n(x^n, y^n)}, p_{11} g^n(x^n, y^n, z^n) \ ,$$

so that we can select a structure that maximizes the posterior probability. The idea with three variables can be extended into the ones with N variables in a straightforward manner.

3   General Case

It is easy to verify that Propositions 1 and 2 can be applied to the Bayesian networks with only discrete variables and only continuous variables, respectively, which is considered to be rather unrealistic. In fact, in our daily life, we experience that some fields are discrete and others continuous in any database. If X is discrete and Y is continuous, the joint random variable (X, Y) is neither discrete nor continuous, and none of Propositions 1 and 2 can be applied.

Besides, from Proposition 2, we may not obtain universality without the assumption $D(f\|f_j) \to 0$ as $j \to \infty$. For example, if $\int_{\frac{1}{2}}^{1} f(x)dx > 0$ and $\{A_j\}$ is given by the following sequence, we cannot obtain universality because the range $[\frac{1}{2}, 1)$ will not be refined:
$A_1 = \{[0, \frac{1}{2}), [\frac{1}{2}, 1)\}$
$A_2 = \{[0, \frac{1}{4}), [\frac{1}{4}, \frac{1}{2}), [\frac{1}{2}, 1)\}$
$\cdots$
$A_j = \{[0, 2^{-(j-1)}), [2^{-(j-1)}, 2 \cdot 2^{-(j-1)}), \cdots, [\frac{1}{2} - 2^{-(j-1)}, \frac{1}{2}), [(\frac{1}{2}, 1)\} \ .$
$\cdots$

3.1   For the Random Variables that Are Neither Discrete nor Continuous

Suppose we wish to estimate the distribution over the positive integers $\mathbb{N}$. Apparently, $\mathbb{N}$ is not a finite set and has no density function. We consider the histogram sequence $\{B_k\}$: $B_0 = \{\mathbb{N}\}$,

$B_1 := \{\{1\}, \{2, 3, \cdots\}\}$

$B_2 := \{\{1\}, \{2\}, \{3, 4, \cdots\}\}$

$\cdots$

$B_k := \{\{1\}, \{2\}, \cdots, \{k\}, \{k+1, k+2, \cdots\}\}$

$\cdots$

For each k, we quantize each $y \in \mathbb{N}$ into the $b \in B_k$ such that $y \in b$. For example, if $y = 4$ and $k = 2$, then $b = \{3, 4, \cdots\} \in B_2$. Let η be the measure such that

$$\eta(\{k\}) = \frac{1}{k} - \frac{1}{k+1} \ , \ k \in \mathbb{N} \ . \tag{13}$$

For example, $\eta(\{2\}) = \frac{1}{6}$ and $\eta(\{3, 4\}) = \frac{2}{15}$.

Notice that each B_k is a finite set, and we construct a universal probability Q_k^n w.r.t. finite set B_k for each k. Given $y^n = (y_1, \cdots, y_n) \in \mathbb{N}^n$, we obtain a quantized sequence $(b_1^{(k)}, \cdots, b_n^{(k)}) \in B_k^n$ for each k, so that we can compute the quantity

$$g_k^n(y^n) := \frac{Q_k^n(b_1^{(k)}, \cdots, b_n^{(k)})}{\eta(b_1^{(k)}) \cdots \eta(b_n^{(k)})}$$

for each k. If we prepare a sequence of positive reals $w_1, w_2, \cdots$ such that $\sum_k w_k = 1$, we can compute the quantity

$$g^n(y^n) := \sum_{k=1}^{\infty} w_k g_k^n(y^n) \ .$$

On the other hand, although no density function (w.r.t. λ) exists over $\mathbb{N}$ because Y is a discrete random variable, however, there does exist a density function in a generalized sense. In this case, the generalized density function w.r.t. η is obtained via

$$f(y) = \frac{P(Y = y)}{\eta(\{y\})} \tag{14}$$

for $y \in \mathbb{N}$ (f takes arbitrary values for $y \notin \mathbb{N}$). Even if η is not given by (13), if $\eta(\{y\}) \neq 0$ for all $y \in \mathbb{N}$ such that $P(Y = y) \neq 0$, (14) will become the density function. The readers who wants to prove (14) might want to consult the theory of Radon-Nikodym.

In general, we obtain the following result:

Theorem 1. For any generalized density function f such that $D(f \| f_k) \to 0$ as $k \to \infty$,

$$\frac{1}{n} \log \frac{f^n(y^n)}{g^n(y^n)} \to 0$$

as $n \to \infty$ with probability one

Proof: see https://arxiv.org/submit/985408/view

Notice that Theorem 1 contains Proposition 1 as well as Proposition 2 as special cases. In fact, for random variable Z such that $P(Z = z) \neq 0 \Longleftrightarrow z \in C$ for a finite set C, the generalized density function can be expressed by (14) using $\eta(z) \neq 0$ for $z \in C$.

3.2   Universal Histogram Sequences

Theorem 1 assumes a specific $\{B_k\}$. We find a histogram sequence $\{B_k\}$ such that for any f, $D(f \| f_k) \to 0$ as $k \to \infty$, and remove the condition $D(f \| f_k)$ converge to zero as $k \to \infty$.

To this end, we choose any $\mu, \sigma \in \mathbb{R}$, where σ should be positive, and generate the sequence:

$$C_0 = \{(-\infty, \infty)\}$$

$$C_1 = \{(-\infty, \mu], (\mu, \infty)\}$$

Given $C_k = \{(-\infty, c_{k,1}], (c_{k,1}, c_{k,2}], \cdots, (c_{k,2^k-2}, c_{k,2^k-1}], (c_{k,2^k-1}, \infty)\}$ for $k \geq 1$, we define

$$C_{k+1} = \{(-\infty, c_{k+1,1}], (c_{k+1,1}, c_{k+1,2}], \cdots, (c_{k+1,2^{k+1}-2}, c_{k+1,2^{k+1}-1}], (c_{k+1,2^{k+1}-1}, \infty)\}$$

by

$$c_{k+1,1} = \mu - k\sigma \ , \ c_{k+1,2^{k+1}-1} = \mu + k\sigma$$

$$c_{k+1,2j} = c_{k,j}, \ j = 1, \cdots, 2^k - 1$$

$$c_{k+1,2j+1} = \frac{c_{k,j} + c_{k,j+1}}{2} \ , \ j = 1, \cdots, 2^k - 2$$

In this way, given the values of μ, σ, we obtain the sequence $\{C_k\}_{k=0}^{\infty}$.

Let B be the set in which random variable Y takes values, and define

$$B_k^* := \{B \cap c | c \in C_k\} \setminus \{\phi\} \ ,$$

where ϕ is the empty set.

For example, suppose that density function f exists over $[0, 1]$ as in Section 2.3. Then, there exists a unique sequence $\{(a_k, b_k]\}_{k=1}^{\infty}$ such that $0 \leq a_k \leq y \leq b_k \leq 1$, $k = 1, 2, \cdots$, so that the ratio

$$\frac{P(Y \in (a_k, b_k]))}{\lambda((a_k, b_k])} = \frac{F_X(b_k) - F_X(a_k)}{b_k - a_k}$$

converges to $f(x)$ as $k \to \infty$. Thus, $D(f \| f_k) \to 0$ as $k \to \infty$ for any f, where F_X is the distribution function of X.

For example, the sequence $\{B_k\}$ in Section 3.1 is obtained by $\mu = 1$ and $\sigma = 1$ in the histogram sequence $\{B_k^*\}$. Then, for each $y \in \mathbb{N} = \{1, 2, \cdots\}$, there exists $K \in \mathbb{N}$ and a unique $\{D_k\}_{k=1}^{\infty}$ such that $y \in D_k \in B_k^*$, $k = 1, 2, \cdots$ and $\{y\} = D_k \in B_k^*$ for $k = K, K + 1, \cdots$, so that

$$f_k(y) = \frac{P(Y \in D_k)}{\eta(D_k)} \to f(y) = \frac{P(Y = y)}{\eta(\{y\})}$$

for each $y \in B$ and $D(f||f_k) \to 0$ as $k \to \infty$ for any f.

The choice of μ, σ may be arbitrary, but we should take the prior knowledge into consideration in order to make the estimation correct even for small n. For $\{B_k^*\}$ and the supporting measure, we can compute $g^n(y^n)$ given examples y^n.

Theorem 2. For any generalized density function f,

$$\frac{1}{n} \log \frac{f^n(y^n)}{g^n(y^n)} \to 0$$

as $n \to \infty$ with probability one

Proof: see https://arxiv.org/submit/985408/view

3.3   Multivariable Universal Probabilities

Suppose we wish to estimate the distribution over $[0,1] \times \mathbb{N}$. Apparently, $[0,1] \times \mathbb{N}$ is not a finite set and has no density function. Theorem 1 can be applied to this situation.

Since $A_j \times B_k$ is a finite set, we can construct a universal probability $Q_{j,k}^n$ for $A_j \times B_k$

$$g_{jk}^n(x^n, y^n) := \frac{Q_{j,k}^n(a_1^{(j)}, \cdots, a_n^{(j)}, b_1^{(k)}, \cdots, b_n^{(k)})}{\lambda(a_1^{(j)}) \cdots \lambda(a_n^{(j)}) \eta(b_1^{(k)}) \cdots \eta(b_n^{(k)})}$$

If we prepare the sequence such that $\sum_{jk} \omega_{jk} = 1$, $\omega_{jk} > 0$, then we obtain the quantity

$$g^n(x^n, y^n) := \sum_{k=1}^{\infty} \omega_{jk} g_{jk}^n(x^n, y^n) .$$

In this case, the (generalized) density function is obtained via

$$f(x, y) = \frac{F_X(x|y)}{dx} \cdot \frac{P(Y = y)}{\eta(\{y\})}$$

for $y \in \mathbb{N}$ (f takes arbitrary values for $x \notin [0,1]$ and $y \notin \mathbb{N}$), where $F_X(\cdot|y)$ is the conditional distribution function of X given $Y = y$.

Theorem 3. For any generalized density function f,

$$\frac{1}{n} \log \frac{f^n(x^n, y^n)}{g^n(x^n, y^n)} \to 0$$

as $n \to \infty$ with probability one

Proof: see https://arxiv.org/submit/985408/view

It is straightforward to extend Theorem 2 into the N variable case.

4 Learning Bayesian Network Structures When Discrete and Continuous Variables Are Present

There are two stages before choosing a structure that minimizes the score. For example, if we are given $n = 100$ examples x^n, y^n, z^n and obtain the seven quantities of $-\log g^n(\cdot)$ in the upper values of Table 1, we compute the eleven scores w.r.t. the structures in the lower values of Table 1. For example, for (10), the score can be computed via

$$\frac{1}{n}\{-\log p_{10} - \log g^n(x^n) - \log g^n(y^n) - \log g^n(x^n, y^n, z^n) + \log g^n(x^n, y^n)\}. \quad (15)$$

Then, a structure that minimizes $-\log g^n(\cdot)$ will be chosen. In this case, structure (1) is chosen.

Let $M(N)$ be the number of the DAGs with N nodes when the structures within each Markov equivalent class are regarded to be identical. For example, $M(2) = 2$ and $M(3) = 11$ (see Figure 1). First, we compute 2^N values of $-\log g^n(\cdot)$, Then, we compare $M(N)$ values of the scores such as (15) for $N = 3$ and structure (10).

Table 1. The eleven scores obtained from the seven scores for X, Y, Z

	X	Y	(X,Y)	Z	(X,Z)	(Y,Z)	(X,Y,Z)
$-\log g^n(\cdot)$	1.617	1.533	3.249	1.647	3.318	3.290	4.943

(1)	(2)	(3)	(4)	(5)	(6)	(7)	(8)	(9)	(10)	(11)
4.799	4.908	4.852	4.897	4.950	5.006	4.962	4.833	4.890	4.845	4.943

For the first stage, we need to bound the length K of histogram sequence $\{A_k\}$ and specify the values of the weights $\{w_k\}_{k=1}^K$. In the experiments in Section 5, we set $w_k = 1/K$ although the convergence will be faster if we know the values clearly. For two variables, the histogram sequence $\{A_k \times B_k\}_{k=1}^K$ of length K rather than $\{A_j \times B_k\}_{j,k}$ of length K^2 should be specified. Even for the settings, we can obtain the asymptotically optimal property for large n if K is large enough.

For each $x^n = (x_1, \cdots, x_n) \in A^n$, we obtain $(a_1^{(k)}, \cdots, a_n^{(k)})$ for each $k = 1, \cdots, K$, where A is the range of random variable X. In the experiments in Section 5, we apply binary search to obtain $a_i^{(k+1)} \in A_{k+1}$ from $a_i^{(k)} \in A_k$ for each $i = 1, \cdots, n$, where $\{A_k\}$ contains 2^k elements.

On the other hand, if A is a finite set and $x^n \in A^n$, it is known that

$$Q^n(x^n) = \prod_{i=1}^{n} \frac{c(x_i, x^{i-1}) + 1/2}{i - 1 + |A|/2}$$

is a universal probability of $x^n \in A^n$, where $c(x, x^{i-1})$ is the number of occurrences of $x \in A$ in $x^{i-1} \in A^{i-1}$, and $|A|$ is the cardinality of A.

The resulting algorithm is shown in Figure 4, where η_X and η_Y are the supporting measures of X and Y, respectively. Those procedure can be applied to the N variable case.

(A) Input $x^n \in A^n$, Output $g^n(x^n)$

1. For each $k = 1, \cdots, K$, $g_k^n(x^n) := 0$
2. For each $k = 1, \cdots, K$ and each $a \in A_k$, $c_k(a) := 0$
3. For each $i = 1, \cdots, n$, for each $k = 1, \cdots, K$
   (a) Find $a \in A_k$ from $x_i \in A$
   (b) $g_k^n(x^n) := g_k^n(x^n) - \log \frac{c_k(a)+1/2}{i-1+|A_k|/2} + \log(\eta_X(a))$
   (c) $c_k(a) := c_k(a) + 1$
4. $g^n(x^n) := \sum_{k=1}^{K} \frac{1}{K} g_k^n(x^n)$

(B) Input $x^n \in A^n$ and $y^n \in B^n$, Output $g^n(x^n, y^n)$

1. For each $k = 1, \cdots, K$, $g_k^n(x^n, y^n) := 0$
2. For each $k = 1, \cdots, K$ and each $a \in A_k$ and $b \in B_k$, $c_k(a, b) := 0$
3. For each $i = 1, \cdots, n$, for each $k = 1, \cdots, K$
   (a) Find $a \in A_k$ and $b \in B_k$ from $x_i \in A$ and $y_i \in B$
   (b) $g_k^n(x^n, y^n) := g_k^n(x^n, y^n) - \log \frac{c_k(a,b)+1/2}{i-1+|A_k||B_k|/2} + \log(\eta_X(a)\eta_Y(b))$
   (c) $c_k(a, b) := c_k(a, b) + 1$
4. $g^n(x^n, y^n) := \sum_{k=1}^{K} \frac{1}{K} g_k^n(x^n, y^n)$

Fig. 4. The Algorithm calculating g^n: for given supporting measures η_X, η_Y etc., the score g^n is calculated. In Step 3(a) for (A) and (B), binary search is applied, so that the computations in those steps complete in constant times.

For the first stage, its computational complexity is at most $O(nK)$ for obtaining the value of $g(x^n)$ for one variable X and even for obtaining the values of $g(x^n, y^n, \cdots)$ w.r.t. variables $X, Y, \cdots$, whereas even if all the variables are discrete, $O(n)$ computation is required. For the second stage, since we compare $M(N)$ structures, we need $O(M(N))$ time (we do not care about whether the original values are either discrete or continuous). Thus, the total time is $O(\max\{nK2^N, M(N)\})$.

Another issue is its space complexity. We require memory space for storing $c_k(a)$ for $a \in A$ and $k = 1, \cdots, K$ with size $O(2^K)$. For the score with m variables, the size is $O(2^{mK})$, which is is more severe than the discrete only case. However, if we fix the total number n of examples, the number of examples in each bin dramatically reduces as K grows, so we specify the value of K for the scores with m variables to be les than $\frac{1}{n} \log M(N)$, so that the expected number of examples in each bin is uniform over all the scores. On the other hand, for the second stage, we need to keep $M(N)$ values. Thus, the total space complexity is $O(\max\{2^{NK}, M(N)\}) = O(M(N))$ and we find that the space complexity remains the same even if the variable set contain continuous one.

5   Experiments

5.1   Convergence of Kullback-Leibler Divergence

First, we consider two Bayesian networks with four nodes to see convergence of the KL divergence of the score from the true distribution: we generate (X, Y, U, V) defined by

1. $X, Y \in \{0,1\}$, $X \perp\!\!\!\perp Y$, takes one with probability 0.5, and $U \sim N(x+y,1)$ and $V \sim N(x-y,1)$, and
2. $X, Y \sim N(0,1)$, $X \perp\!\!\!\perp Y$, and $U, V \in \{0,1\}$ such that

$$P(U=1|X+Y=z) = P(V=1|X-Y=z) = \begin{cases} 0, & z < -1 \\ (z+1)/2, & -1 \le z \le 1 \\ 1, & z > 1 \end{cases}$$

in Figures 4(a) and 4(b), respectively, where $N(\mu, \sigma^2)$ is the normal distribution with mean μ and variance σ^2.

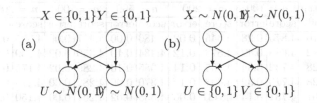

Fig. 5. Experiment 1: Bayesian Networks with 2 Discrete and 2 Gaussian Nodes

We obtain the values of arithmetic average over 100 trials of $-\log g^n(\cdot)$ and their execution times in Table 2. As can be seen from Table 2, $-\frac{1}{n} \log g^n(x^n, y^n, u^n, v^n)$ converges to $h(X, Y, U, V)$ as n grows, which means that the K-L divergence goes to zero. This property (universality) is the key to learning Bayesian network structures. The closer score g^n to differential entropy h for each subset of the N variables, the more correct structure obtained. The computation is almost linear in n.

Table 2. Experiment 1: the results for the Bayesian networks Figures 5(a) and 5(b), where $h(X, Y, U, V)$ is the differential entropy of X, Y, U, V

Bayesian Networks	n	100	200	500	1000	2000
Figure 5(a)	$g^n(x^n, y^n, u^n, v^n)$	5.009	4.858	4.626	4.616	4.552
$h(X, Y, U, V) = 4.224$	KL divergence	0.785	0.634	0.402	0.392	0.328
	execution time (sec)	1.079	1.276	1.939	4.596	7.047
Figure 5(b)	$g^n(x^n, y^n, u^n, v^n)$	4.435	4.191	4.002	3.867	3.771
$h(X, Y, U, V) = 3.372$	KL divergence	1.063	0.819	0.630	0.495	0.399
	execution time (sec)	0.601	0.849	1.721	2.582	4.619

5.2   Structure Learning of Bayesian Networks

Secondly, we made experiments using Bayesian networks with three variables to see its basic performance.

In the experiment (Experiment 2), we prepare eleven Bayesian networks with three Gaussian variables as in Figure 1. In particular, we generate n tuples of random numbers as follows, where the values of $-1 \leq \rho, \rho_a, \rho_b \leq 1$ in (2) through (10) are 0.6, and $-1 \leq \rho_a, \rho_b, \rho_c \leq 1$ in (11) are 0.5.

(1) $X, Y, Z \sim N(0,1)$.
(2)(3)(4) $X, U \sim N(0,1)$, $Y = \rho X + \sqrt{1 - \rho^2} U$, $Z \sim N(0,1)$.
(5)(6)(7) $X, U \sim N(0,1)$, $Y = \rho_a X + \sqrt{1 - \rho_a^2} U$, $Z = \rho X + \sqrt{1 - \rho_b^2} U$.
(8)(9)(10) $X, U, V \sim N(0,1)$, $Y = \rho_a X + \sqrt{1 - \rho_a^2} U$, $Z = \rho_b X + \sqrt{1 - \rho_b^2} V$.
(11) $X, U, V \sim N(0,1)$, $Y = \rho_a X + \sqrt{1 - \rho_a^2} U$, $Z = \rho_b X + \rho_c Y + \sqrt{1 - \rho_b^2 - \rho_c^2} V$.

Table 3. Experiment 2: the KL divergence and error probabilities for Bayesian network structure learning

true structure	differential entropy	$n = 100$ score	error	$n = 200$ score	error	$n = 500$ score	error	$n = 1000$ score	error	$n = 2000$ score	error
(1)	4.256816	4.875	0.28	4.645	0.02	4.480	0.00	4.417	0.00	4.355	0.00
(2)(3)(4)	4.033672	4.699	0.42	4.573	0.12	4.434	0.10	4.350	0.02	4.269	0.00
(5)(6)(7)	3.810528	4.732	0.34	4.565	0.14	4.385	0.10	4.289	0.02	4.175	0.00
(8)(9)(10)	3.810528	4.710	0.32	4.498	0.12	4.370	0.06	4.282	0.00	4.178	0.00
(11)	3.766401	4.731	0.14	4.5431	0.06	4.335	0.02	4.261	0.00	4.150	0.00

For each of the five structure classes, given x^n, y^n, z^n, if the true and chosen structures do not coincide in any edge, we say that an error occurs. We repeat $L = 50$ times to generate x^n, y^n, z^n and choose a structure that minimizes the score such as (15). We compute the KL divergence between the true and chosen Bayesian networks, where we define the error rate to be the number of errors in L trials divided by L. We obtained data as in Table 3, and find that both the error rate and KL divergence diminish to zero for all the cases.

Next, We apply the proposed algorithm for actual data prepared in the R data set. For the second part, we do not care about whether the original values are either discrete or continuous. Thus, we evaluate the execution times only for the first part (the true Bayesian network structures are not known but only data sets are given). The five data sets faithful, quakes, attitude, longley, and USJudgeRatings in Table 4 are obtained in the R datasets package[1].

From Table 4, by comparing attitude and longley with seven discrete and continuous variables, respectively, we can see that computing $g^n(\cdot)$ for continuous variables does not take so much time compared with computing them for discrete variables. For the dataset USJudgeRatings, it took about 30 minutes but the total execution time divided by $2^N = 1024$ was not so large.

---

[1] http://stat.ethz.ch/R-manual/R-patched/library/datasets/html/00Index.html

Table 4. Experiment 3: using real datasets in the R

data.frame	N	data.type	n	time (sec)	time (sec)/2^N
faithful	2	c,d	272	6.08	3.04
quakes	5	c,c,d,d,c	1000	60.77	1.90
attitude	7	d,d,d,d,d,d,d	30	27.66	0.216
longley	7	c,c,c,c,c,c,d	16	44.63	0.349
USJudgeRatings	12	c,c,c,c,c,c,c,c,c,c,c,c	43	1946.63	0.4752

6   Concluding Remarks

We proposed the Bayesian network structure learning algorithm such that each variable in the dataset is either discrete or continuous. The posterior probability converges to the true value even if the prior probabilities over parameters and models are not correct. The computational and space complexities are at most $O(max\{nK2^N, M(N)\})$ and $O(M(N))$, and we find that they are not severe for implementation. In fact, the execution time is almost linear in K because binary search is applied for finding the quantized values.

Although the proposed algorithm is not the first algorithm to learn Bayesian network structures with discrete and continuous variables, it finds the structure that maximizes the posterior probability given examples, and can be applied to general situations.

Future work includes further experiments to find a way to obtain the depth K for n and N, histogram sequences such as $\{A_k\}, \{B_k\}$, and the supporting measures such as μ_X, μ_Y. They are not mathematically analyzed, and experience by data will be required.

References

1. Buntine, W.L.: Learning Classification Trees. Statistics and Computing 2, 63–73 (1991)
2. Boettcher, S.G., Dethlefsen, C.: A Package for Learning Bayesian Networks. Journal of Statistical Software 8(20), 1–40 (2003), http://www.jstatsoft.org/v08/i20/
3. Billingsley, P.: Probability & Measure, 3rd edn. Wiley, New York (1995)
4. Cai, H., Kulkarni, S., Verdú, S.: Universal divergence estimation for finite-alphabet sources. IEEE Trans. Information Theory 52(8), 3456–3475 (2006)
5. Cooper, G.F., Herskovits, E.: A Bayesian Method for the Induction of Probabilistic Networks from Data. Machine Learning 9, 309–347 (1992)
6. Cover, T.M., Thomas, J.A.: Elements of Information Theory, 2nd edn. Wiley, New York (1995)
7. Friedman, N., Goldszmidt, M.: Discretizing Continuous Attributes While Learning Bayesian Networks. In: International Conference on Machine Learning, pp. 157–165 (1996)

8. Heckerman, D., Geiger, D.: Learning Bayesian networks: A unification for discrete and Gaussian domains. In: Eleventh Conference on Uncertainty in Artificial Intelligence, pp. 274–284 (1995)

9. Hofmann, R., Tresp, V.: Discovering Structure in Continuous Variables Using Bayesian Networks. In: Advances in Neural Information Processing Systems, vol. 8. MIT Press, Cambridge (1996)

10. John, G., Langley, P.: Estimating Continuous Distributions in Bayesian Classifiers. In: Eleventh Conference on Uncertainty in Artificial Intelligence, pp. 338–345 (1995)

11. Kullback, S., Leibler, R.A.: On information and sufficiency. Ann. Math. Statistics 22(1), 79–86 (1951)

12. Kozlov, A.V., Koller, D.: Nonuniform Dynamic Discretization in Hybrid Networks. In: Uncertainty in Artificial Intelligence, pp. 314–325 (1997)

13. Krichevsky, R.E., Trofimov, V.K.: The Performance of Universal Encoding. IEEE Trans. Information Theory 27(2), 199–207 (1981)

14. Lauritzen, S.L., Wermuth, N.: Graphical models for associations between variables, some of which are quantitative and some qualitative. Annals of Statistics 17, 31–57 (1989)

15. Monti, S., Cooper, G.F.: Learning Bayesian Belief Networks with Neural Network Estimators. In: Advances in Neural Information Processing Systems, vol. 8, pp. 578–584 (1996)

16. Pearl, J.: Probabilistic reasoning in intelligent systems: networks of plausible inference. Morgan Kaufmann, San Francisco (1988)

17. Rissanen, J.: Modeling by shortest data description. Automatica 14, 465–471 (1978)

18. Romero, V., Rumi, R., Salmeron, A.: Learning hybrid Bayesian networks using mixtures of truncated exponentials. International Journal of Approximate Reasoning 42, 54–68 (2006)

19. Ryabko, B.: Prediction of random sequences and universal coding. Problems Inform. Transmission 24(2), 87–96 (1988)

20. Ryabko, B.: Compression-Based Methods for Nonparametric Prediction and Estimation of Some Characteristics of Time Series. IEEE Trans. on Inform. Theory 55(9), 4309–4315 (2009)

21. Shenoy, P.P.: Two Issues in Using Mixtures of Polynomials for Inference in Hybrid Bayesian Networks. International Journal of Approximate Reasoning 53(5), 847–866 (2012)

22. Spirtes, P., Glymour, C., Scheines, R.: Causation, Prediction, and Search, 2nd edn. MIT Press (2000)

23. Suzuki, J.: On Strong Consistency of Model Selection in Classification. IEEE Trans. on Information Theory 52(11), 4767–4774 (2006)

24. Suzuki, J.: A Construction of Bayesian Networks from Databases on an MDL Principle. In: The Ninth Conference on Uncertainty in Artificial Intelligence, Washington D.C., pp. 266–273 (1993)

25. Suzuki, J.: The Universal Measure for General Sources and its Application to MDL/Bayesian Criteria. In: Data Compression Conference (one page abstract), Snowbird, Utah, p. 478 (2011)

26. Suzuki, J.: MDL/Bayesian Criteria based on Universal Coding/Measure. In: Dowe, D.L. (ed.) Solomonoff Festschrift. LNCS, vol. 7070, pp. 399–410. Springer, Heidelberg (2013)

Learning Neighborhoods of High Confidence in Constraint-Based Causal Discovery

Sofia Triantafillou[1,2], Ioannis Tsamardinos[1,2], and Anna Roumpelaki[1,2]

[1] Institute of Computer Science,
Foundation for Research and Technology - Hellas (FORTH),
N. Plastira 100 Vassilika Vouton, GR-700 13 Heraklion, Crete, Greece
[2] Computer Science Department, University of Crete, Heraklion, Greece

Abstract. Constraint-based causal discovery algorithms use conditional independence tests to identify the skeleton and invariant orientations of a causal network. Two major disadvantages of constraint-based methods are that (a) they are sensitive to error propagation and (b) the results of the conditional independence tests are binarized by being compared to a hard threshold; thus, the resulting networks are not easily evaluated in terms of reliability. We present PROPeR, a method for estimating posterior probabilities of pairwise relations (adjacencies and non-adjacencies) of a network skeleton as a function of the corresponding p-values. This novel approach has no significant computational overhead and can scale up to the same number of variables as the constraint-based algorithm of choice. We also present BiND, an algorithm that identifies neighborhoods of high structural confidence on causal networks learnt with constraint-based algorithms. The algorithm uses PROPeR to estimate the confidence of all pairwise relations. Maximal neighborhoods of the skeleton with minimum confidence above a user-defined threshold are then identified using the Bron-Kerbosch algorithm for identifying maximal cliques. In our empirical evaluation, we demonstrate that (a) the posterior probability estimates for pairwise relations are reasonable and comparable with estimates obtained using more expensive Bayesian methods and (b) BiND identifies sub-networks with higher structural precision and recall than the output of the constraint-based algorithm.

Keywords: Posterior probabilities, causal networks, constraint-based causal discovery.

1 Introduction

Constraint-based algorithms are a popular choice for learning causal models; they are fast, scalable, and usually guarantee soundness and completeness in the sample limit. However, for smaller sample sizes, identification of false constraints poses a challenge: An erroneous identification of a conditional independence can propagate through the network and lead to erroneous edge identifications or conflicting orientations even in seemingly unrelated parts of the network. Particularly for networks with many variables and small sample sizes, error propagation can result in unreliable networks.

L.C. van der Gaag and A.J. Feelders (Eds.): PGM 2014, LNAI 8754, pp. 487–502, 2014.
© Springer International Publishing Switzerland 2014

Constraint-based algorithms query the data for conditional independencies and then use the results to constrain the search space of possible causal models. Failure to identify which parts of the output of a constraint-based algorithm are reliable is partly due to the nature of conditional independence tests: The test returns a p-value, which stands for the probability of getting a test statistic at least as extreme as the the one actually observed in the data, given that the null hypothesis (conditional independence) is true. If this probability is lower than a chosen significance threshold (typically 5-10%), the null hypothesis is rejected, and the alternative hypothesis is implicitly accepted. While lower p-values indicate higher confidence conditional dependencies, the p-value can not be interpreted as the probability of a conditional independence, and it therefore cannot be used to compare a conditional dependence to a conditional independence in terms of belief. Thus, the decisions made by the constraint-based algorithm (accept or reject a conditional independence) cannot be evaluated in terms of confidence.

We propose Posterior RatiO PRobability (PROPeR), a method for identifying posterior probabilities for all (non) adjacencies of a causal network learnt with a constraint-based algorithm. We use the term **pairwise relations** to denote adjacencies and non-adjacencies in a causal graph (ignoring orientations).

For each pair of variables, a constraint-based algorithm tries a number of conditional tests of independence. We use the maximum p-value obtained for every pair of variables as a representative of the corresponding pairwise relation. Posterior probabilities are then estimated as a function of these representative p-values. The method has no significant computational overhead, and can therefore scale up to the same number of variables as the algorithm of choice. Moreover, it does not depend on any additional assumptions (e.g. acyclicity, causal sufficiency, parametric assumptions) and can therefore be used with any constraint-based algorithm equipped with an appropriate test of conditional independence.

Notice that PROPeR *is not used to improve the algorithm per se, but to produce confidence estimates for pairwise relations learnt from the algorithm.* Identifying which parts of the learnt network are reliable is of great importance for practitioners who use causal discovery methods, and are often interested in high-confidence pairwise connections among variables or in avoiding a specific type of error (e.g. false positive or false negative edges). It can also be useful for selecting subsequent experiments for a system under study, by pointing out relationships that are uncertain.

We use the estimates obtained by PROPeR, to identify neighborhoods of high structural confidence in causal networks. The proposed method, called BiND (β-NeighborhooDs), takes as input a causal graph $\mathcal{G}$ along with representative p-values for every pairwise relation in $\mathcal{G}$ and a desired threshold of confidence β. The algorithm outputs all neighborhoods in $\mathcal{G}$ for which all pairwise relations have confidence estimates above β. Internally, BiND uses PROPeR to obtain probability estimates for each pairwise relation, creates a graph $\mathcal{H}_\beta$ where edges correspond to pairwise relations with confidence above β, and then uses the Bron-Kerbosch algorithm to identify all maximal cliques in graph $\mathcal{H}_\beta$.

In our empirical evaluation, we use simulated data to test the calibration of PROPeR's probability estimates, and compare against two Bayesian methods that can be used to obtain similar estimates [1][2]. Results indicate that PROPeR produces reasonable probability estimates, while being significantly faster than other approaches. The behavior of BiND was also examined using simulated data sets. Results indicate that BiND identifies neighborhoods that include a smaller proportion of false positive and false negative edges, compared to the original induced network.

2 Background

We use V to denote random variables (interchangeably nodes of a causal graph), and bold upper-case letters to denote sets of variables. We use the notation $X \perp\!\!\!\perp Y | \mathbf{Z}$ to denote the independence of variables X and Y given the set of variables $\mathbf{Z}$. We use $\mathcal{G} = (\mathbf{V}, \mathcal{E})$ to denote a graph over variables $\mathbf{V}$ with edges $\mathcal{E}$. For the scope of this work, we only deal with undirected edges, thus, members of $\mathcal{E}$ are unordered tuples of $\mathbf{V}$.

Bayesian networks consist of a Directed Acyclic Graph (DAG) $\mathcal{G}$ over a set of variables $\mathbf{V}$ and a joint probability distribution $\mathcal{P}$ over the same variables. A directed edge in $\mathcal{G}$ denotes a direct causal relation (in the context of measured variables). The DAG $\mathcal{G}$ and the distribution $\mathcal{P}$ are connected by the Causal Markov condition (CMC): Every variable is independent of its non-descendants given its parents. The graph in conjunction with the CMC entails a set of conditional independencies that hold in $\mathcal{P}$. The faithfulness condition (FC) states that all the conditional independencies that hold in $\mathcal{P}$ stem from $\mathcal{G}$ and the CMC, instead of being accidental parametric properties of the distribution.

Under CMC and FC, the conditional (in)dependencies that hold in $\mathcal{P}$ can be identified from the graph $\mathcal{G}$ according to a graphical criterion, namely d-separation. For graphs and distributions that are faithful to each other, we say that $\mathcal{P}$ satisfies the global Markov property with respect to $\mathcal{G}$: $X \perp\!\!\!\perp Y | \mathbf{Z}$ in $\mathcal{P}$ if and only if X and Y are d-separated given $\mathbf{Z}$ in $\mathcal{G}$. Constraint-based methods for learning Bayesian Networks use independence relations present in the data to constrain the search space of possible underlying causal graphs. The following theorem is the cornerstone of constraint-based causal learning:

Theorem 1. *[3] If $\langle \mathcal{G}, \mathcal{P} \rangle$ is a Bayesian network over $\mathbf{V}$ and $\mathcal{G}$ is faithful to $\mathcal{P}$, then the following holds: For every pair of variables $X, Y \in \mathbf{V}$: X and Y are not adjacent in $\mathcal{G} \leftrightarrow \exists \mathbf{Z} \subseteq V \setminus \{X, Y\}$ s.t. $X \perp\!\!\!\perp Y | \mathbf{Z}$.*

The theorem states that every missing edge in $\mathcal{G}$ corresponds to a conditional independence in $\mathcal{P}$. This is also known as the *pairwise Markov property*. Essentially, the theorem matches the skeleton of the causal graph to a kernel of conditional independencies (one for every missing edge). Thus, to identify the network skeleton, constraint-based algorithms use a search strategy to iterate over all pairs of variables in $\mathbf{V}$. For each such pair (X, Y), the algorithm tries to identify a set of variables $\mathbf{Z}$ that renders X and Y independent. If no such set exists, X and Y are adjacent in the resulting causal graph $\mathcal{G}$, otherwise the edge between them is removed and $\mathbf{Z}$ is reported as the separating set of X and Y. The set of all conditional independencies that hold in a probability distribution is called the **independence model** $\mathcal{J}$ of the distribution $\mathcal{P}$. Under CMC and FC, the minimal set of independencies identified by a (sound and complete) constraint-based algorithm are sufficient to entail *all* conditional independencies in $\mathcal{J}$.

Apart from CMC and FC, Bayesian networks rely on the assumption of causal sufficiency: Pairs of variables in a Bayesian network cannot be confounded, i.e. they cannot be effects of the same unmeasured common cause. This assumption is very restrictive and likely to be violated in many applications. Maximal Ancestral Graphs (MAGs) are extensions of Bayesian networks that can handle possible hidden confounders. In faithful MAGs, the graph $\mathcal{G}$ and the distribution $\mathcal{P}$ are connected through a graphical criterion similar to d-separation, called m-separation. MAGs also satisfy the pairwise Markov property: a missing edge in $\mathcal{G}$ corresponds to a conditional independence in $\mathcal{P}$. Edges and orientations in MAGs, however, have slightly different causal semantics than in Bayesian networks.

Methods presented in this work do not depend on the assumption of causal sufficiency, and can therefore work for both DAGs and MAGs. They do require, however, that the causal graph and the distribution satisfy the pairwise Markov property: every missing edge must correspond to a conditional independence. This holds for DAGs and MAGs, but is not true for all graphical models.

Typically, for a joint probability distribution $\mathcal{P}$ over a set of variables $\mathbf{V}$, there exists a class (instead of a single) of causal graphs (DAGs or MAGs) that entail all and only the conditional independencies that hold in $\mathcal{P}$. Causal graphs that belong to the same class, and cannot be distinguished based on conditional independencies alone, called Markov Equivalent. For both DAGs and MAGs, Markov Equivalent graphs share the same skeleton, and vary in some of the orientations.

For the scope of this work, we only attempt to quantify our belief to the adjacency or non-adjacency of each pair of variables, regardless of orientations. Thus, we only need to take into account the output of *the skeleton identification step of a constraint-based algorithm*. In the remainder of this paper, we use $\mathcal{G}$ $=(\mathbf{V}, \mathcal{E})$ to denote the output such an algorithm, thus, the skeleton of a BN or a MAG *without orientations*.

3   Posterior Probabilities for Pairwise Relations

In this section, we present the PROPeR algorithm for estimating posterior probabilities of pairwise relations in causal networks. PROPeR takes as input the causal skeleton $\mathcal{G}$ returned by a constraint-based algorithm and a set of representative p-values and outputs a posterior probability estimate for every adjacency and non-adjacency in $\mathcal{G}$. We use $P(X\!-\!Y)$ and $P(\neg X\!-\!Y)$ to denote the posterior probability of the adjacency and non-adjacency of X and Y in $\mathcal{G}$, respectively.

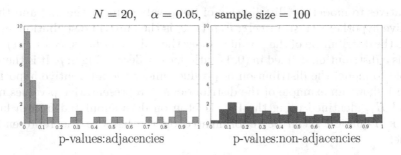

$$N = 20, \quad \alpha = 0.05, \quad \text{sample size} = 100$$

Fig. 1. Representative p-values for adjacencies and non-adjacencies. Normalized histograms of 190 representative p-values identified by the PC skeleton algorithm for a random network of 20 variables. p-values corresponding to adjacencies in the data-generating network (left) follow a distribution with decreasing density. p-values corresponding to non-adjacencies in the data-generating network (right) follow a uniform distribution in the interval $[\alpha, 1]$. A smaller number of predictions fall in the $[0, \alpha]$ interval. This bias is introduced due to constraint-base search strategy: while the representative p-value is below the threshold, the algorithm performs more tests. Naturally, in real scenarios, we do not know which p-values come from which distribution.

According to the pairwise Markov condition, a non-adjacency in a causal graph $\mathcal{G}$ over variables $\mathbf{V}$ corresponds to a conditional independence given a subset of $\mathbf{V}$. In contrast, an adjacency in $\mathcal{G}$ corresponds to the lack of such a subset: If X and Y are adjacent in $\mathcal{G}$, there exists no subset $\mathbf{Z}$ of observed variables such that $X \perp\!\!\!\perp Y | \mathbf{Z}$. Thus, edge X—Y will be present in $\mathcal{P}$ if the data support the null hypothesis

$$H_0 : \exists \mathbf{Z} \subset \mathbf{V} : X \perp\!\!\!\perp Y | \mathbf{Z} \ \textit{less} \text{ than the alternative } H_1 : \forall \mathbf{Z} \subset \mathbf{V} : X \not\!\perp\!\!\!\perp Y | \mathbf{Z} \quad (1)$$

For a network with N variables, this complex set of hypotheses involves $|2^{N-2}|$ conditional independencies. To simplify Equation 1, we use a surrogate conditioning set. For each pair of variables, during the skeleton search, a constraint-based algorithm performs a number of tests, each for a different conditioning set. To avoid performing all possible tests, most algorithms avoid conditioning sets that are theoretically not likely to be d-separating the variables, and also use a threshold on the cardinality of attempted conditioning sets. Let p_{XY} be the maximum p-value of any attempted test of conditional independence between X and Y, and let $\mathbf{Z}_{XY}$ be the corresponding conditioning set. p_{XY} is used in constraint-based algorithms to determine whether X and Y are adjacent. If p_{XY} is lower than the threshold α, the edge is present in $\mathcal{G}$. Otherwise, the edge is absent in $\mathcal{G}$. We approximate Equation 1 with the following set of hypotheses:

$$H_0 : Ind(X, Y | \mathbf{Z}_{XY}) \text{ against the alternative } H_1 : \neg Ind(X, Y | \mathbf{Z}_{XY}), \quad (2)$$

Under H_0, the p-values follow a uniform distribution. Under H_1, the p-values follow a distribution with decreasing density. Sellke et al. [4] propose using Beta alternatives to model the distribution of the p-values under the null and the alternative hypotheses, respectively: $Beta(1,1)$ is the uniform distribution and describes the distribution of the p-values under the null hypothesis. $Beta(\xi, 1)$, $0 < \xi < 1$ is a distribution defined in $(0,1)$ with density decreasing in p. It is therefore suitable to model the distribution of p-values under the alternative hypothesis. Figure 1 shows an example of the distributions of representative p-values under H_0 and H_1, identified using the PC skeleton on data simulated from a known network. Equation 2 can be re-formulated on the basis of the representative p-value:

$$H_0 : p_{XY} \sim Beta(1,1) \text{ against } H_1 : p_{XY} \sim Beta(\xi, 1) \text{ for some } \xi \in (0,1). \quad (3)$$

We can now estimate whether adjacency is more probable than non-adjacency for a given representative p-value p, *by estimating which of the Beta alternatives it is most likely to follow*. We use $\mathbf{V}^2 = \{(X, Y), X, Y \in \mathbf{V}, X \neq Y\}$ to denote the set of unordered pairs of $\mathbf{V}$, i.e. the set of pairwise relations in a causal skeleton $\mathcal{G}$. Let $\mathbf{p} = \{p_{XY} : (X, Y) \in \mathbf{V}^2\}$ be the set of the representative p-values for each pairwise relation. We assume that this population of p-values follows a mixture of $Beta(\xi, 1)$ and $Beta(1,1)$ distributions. If π_0 is the proportion of p-values following $Beta(1,1)$, then the corresponding probability density function is:

$$f(p|\xi, \pi_0) = \pi_0 + (1 - \pi_0)\xi p^{\xi - 1}$$

For given estimates $\hat{\pi}_0$ and $\hat{\xi}$, the posterior odds of H_0 against H_1 for variables X, Y is

$$
\begin{aligned}
PO(p_{XY}) &= \frac{P(p_{XY}|H_0)P(H_0)}{P(p_{XY}|H_1)P(H_1)} = \\
&\frac{P(p_{XY}|p_{XY} \sim Beta(1,1))P(p_{XY} \sim Beta(1,1))}{P(p_{XY}|p_{XY} \sim Beta(\hat{\xi},1))P(p_{XY} \sim Beta(\hat{\xi},1))} = \frac{\hat{\pi}_0}{\hat{\xi}p_{XY}^{\hat{\xi}-1}(1 - \hat{\pi}_0)}.
\end{aligned}
\quad (4)
$$

Obviously, if $PO(p_{XY}) > 1$, non-adjacency is more probable than adjacency for the pair of variables X, Y. Notice that for some $\hat{\xi}$ and $\hat{\pi}_0$, it is possible that $PO(p_{XY}) > 1$, while X and Y are adjacent in $\mathcal{G}$.

Based on the ratios in Equation 4, we can obtain the probability estimates:

$$P(X\!-\!Y) = \frac{1}{1 + PO(p_{XY})}, \quad P(\neg X\!-\!Y) = \frac{PO(p_{XY})}{1 + PO(p_{XY})} \quad (5)$$

To estimate the probabilities in Equation 5, we need to obtain estimates for $\hat{\pi}_0$ and $\hat{\xi}$. To estimate π_0, we use the method described in [5]. The authors propose fitting a natural cubic spline to the distribution of the p-values to estimate the proportion of p-values that come from the null hypothesis.

The method requires that the p-values are i.i.d., an assumption that is clearly violated for the sample of p-values obtained during a skeleton identification algorithm: Typically, the tests of independence attempted by constraint-based network learning algorithms depend on the results of previously attempted tests.

Algorithm 1. PROPeR

---

    input : causal network $\mathcal{G}$ over $\mathbf{V}$, representative p-values $\{p_{XY}\}$
    output: Probability estimates $P(X\!-\!Y), P(\neg X\!-\!Y)$

1 Estimate $\hat{\pi}_0$ from $\{p_{XY}\}$ using the method described in [5];
2 Find $\hat{\xi}$ that minimizes $-\sum_{(X,Y)\in\mathbf{V}^2} log(\hat{\pi}_0 + (1-\hat{\pi}_0)\xi p_{XY}^{\xi-1})$;
3 **foreach** $(X,Y)\in\mathbf{V}^2$ *with representative p-value* p_{XY} **do**
4    $PO(p_{XY}) \leftarrow \frac{\hat{\pi}_0}{\hat{\xi} p_{XY}^{\xi-1}(1-\hat{\pi}_0)}$;
5    $P(X\!-\!Y) \leftarrow \frac{1}{PO(p_{XY})+1}, \quad P(\neg X\!-\!Y) \leftarrow \frac{PO(p_{XY})}{PO(p_{XY})+1}$;
6 **end**

Moreover, each p_{XY} is the maximum among many attempted tests. Finally, the p-values coming from the null hypothesis are not uniform, since independence is only accepted if $p > \alpha$. Thus, the obtained estimate $\hat{\pi}_0$ may be biased. Nevertheless, we believe that the estimates produced using this method are reasonable approximations. An example of the distribution of representative p-values coming from H_0 and H_1 is illustrated in Figure 1.

For a given $\hat{\pi}_0$, the likelihood for a set of representative p-values $\{p_{XY}\}$ is

$$L(\xi) = \prod_{(X,Y)\in\mathbf{V}^2} (\hat{\pi}_0 + (1-\hat{\pi}_0)\xi p_{XY}^{\xi-1}).$$

The respective negative log likelihood is

$$-LL(\xi) = - \sum_{(X,Y)\in\mathbf{V}^2} log(\hat{\pi}_0 + (1-\hat{\pi}_0)\xi p_{XY}^{\xi-1}). \tag{6}$$

Equation 6 can easily be optimized for ξ. Algorithm 1 describes how to obtain probability estimates for all pairwise relations given their representative p-values.

4 Identifying Neighborhoods of High Structural Confidence

Algorithm 2 takes as input a causal skeleton $\mathcal{G}$, confidence estimates on $\mathcal{G}$'s pairwise relations and a confidence threshold β and outputs the set of all β-neighborhoods in $\mathcal{G}$. In the previous section we presented a method for obtaining posterior probability estimates for all pairwise relations in a causal skeleton. In this section, we will use these estimates to identify neighborhoods of high structural confidence on the same skeleton. We define a neighborhood of structural confidence β as follows:

Definition 1 (β-neighborhood). *Let* $\mathcal{G} = (\mathbf{V},\mathcal{E})$ *be a causal skeleton, and* $\{P_{XY}, (X,Y)\in\mathbf{V}^2\}$ *the set of probability estimates:*

$$P_{XY} = \begin{cases} P(X\!-\!Y), & \text{if } (X,Y) \text{ adjacent in } \mathcal{G} \\ P(\neg X\!-\!Y), & \text{if } (X,Y) \text{ not adjacent in } \mathcal{G} \end{cases}$$

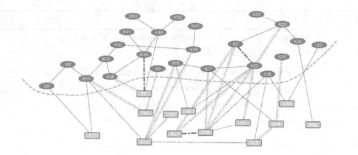

Fig. 2. An example maximum 0.8-neighborhood identified using Algorithm 2. We used the DAG of the Alarm network coupled with random parameters to simulate 100 samples. PC-skeleton was used to obtain the network skeleton $\mathcal{G}$, consisting of 34 edges: 31 true positive edges (solid lines in the figure) and 3 false positive edges ($- \cdot -$ lines). 15 edges were not identified by the algorithm, even though they are present in the data-generating graph (false negative edges, depicted as ⫼ lines). Algorithm 2 was used to identify the maximum 0.8-neighborhoods of $\mathcal{G}$. One of the maximum 0.8-neighborhoods, consisting of 24 variables that share 17 adjacencies, is noted: elliptical blue nodes denote variables in the neighborhood, while the remaining variables are shown as rectangular grey nodes (the neighborhood is also separated from the rest of the network with a dashed grey line). The proportion of false inferences within the clique is far lower than the overall proportion of false inferences: The clique includes only two false negative edges and only one false positive. Most of the false inferences are pairwise relations between members and non-members of the neighborhood.

*A subgraph $\mathcal{G}' = (\mathbf{V}', \mathcal{E}')$ of G is a β-**neighborhood** iff: $\forall X, Y \in \mathbf{V}' : P_{XY} > \beta$ The size of a β-neighborhood $\mathcal{G}' = (\mathbf{V}', \mathcal{E}')$ is $|\mathbf{V}'|$.*

Thus, a neighborhood of confidence β is a subgraph of the causal network in which the posterior probability of every pairwise relation is above a given threshold β. For a causal skeleton and a set of confidence estimates on all pairwise relations, finding a β - neighborhood can be reformulated as a graph theoretical problem: Let $\mathcal{H} = (\mathbf{V}, \mathcal{E}_\beta)$ be an undirected graph with edges defined as follows:

$$(X, Y) \in \mathcal{E}_\beta \text{ if } P_{XY} \geq \beta, \quad (X, Y) \notin \mathcal{E}_\beta \text{ if } P_{XY} < \beta \qquad (7)$$

Variables X and Y are adjacent in $\mathcal{H}_\beta$ only if the probability of their respective pairwise relation in $\mathcal{G}$ is above the confidence threshold β. Finding β-neighborhoods in $\mathcal{G}$ is equivalent to identifying cliques in $\mathcal{H}_\beta$.

Naturally, a causal skeleton can have many β-neighborhoods. Moreover, if a subgraph $\mathcal{G}' = (\mathbf{V}', \mathcal{E}')$ of $\mathcal{G}$ is a β-neighborhood, then every subgraph of $\mathcal{G}'$ is a β-neighborhood. More interesting inferences may be made by identifying all **maximal** β-neighborhoods on a graph:

Definition 2. *Let $\mathcal{G}=(\mathbf{V}, \mathcal{E})$ be a causal skeleton and $\mathcal{G}' = (\mathbf{V}', \mathcal{E}')$ be a β-neighborhood. $\mathcal{G}$ is a **maximal** β-neighborhood if $\nexists \mathbf{V}'' \supset \mathbf{V}'$ such that the subgraph $\mathcal{G}'' = (\mathbf{V}'', \mathcal{E}'')$ is a β-neighborhood.*

Algorithm 2. BiND

---

input : causal network $\mathcal{G}$ over $\mathbf{V}$, pairwise confidence estimates P_{XY},
          confidence threshold β
output: β-neighborhoods $\{\mathcal{G}'\}$
1   $\mathcal{H}_\beta \leftarrow$ empty graph;
2   **foreach** $(X, Y), X, Y \in \mathbf{V}$ **do**
3 | **if** $P_{XY} \geq \beta$ **then** add (X, Y) to $\mathcal{H}_\beta$
4   **end**
5   $\{\mathbf{V}'\} \leftarrow$ Bron-Kerbosch$(\mathcal{H}_\beta)$;
6   $\{\mathcal{G}'\} \leftarrow$ subgraphs of $\mathcal{G}$ over $\{\mathbf{V}'\}$;

Thus, a maximal β-neighborhood is a β-neighborhood that is not part of a larger neighborhood. Identifying all maximal β-neighborhoods in $\mathcal{G}$ can be solved by finding all maximal cliques in the corresponding $\mathcal{H}_\beta$. Identifying maximal cliques is NP-hard [6], but algorithms that run in exponential time or identify approximate solutions are available. We use the Bron-Kerbosch algorithm [7].

Maximal cliques can often be very small; for example, if no larger cliques exist, all adjacencies and all non-adjacencies with $P_{XY} > \beta$ are (trivial) maximal cliques of size 2. Another interesting problem that could be solved using Algorithm 2 is to identify the **maximum** β-neighborhoods of a causal skeleton, i.e. the maximal β-neighborhoods with the maximum possible number of variables. This is equivalent to identifying all maximum cliques in $\mathcal{H}_\beta$, and can be easily obtained from the output of Algorithm 2. Figure 2 shows an example maximum clique, identified using Algorithm 2 on simulated data. The neighborhood includes 24 out of 37 variables. While the neighborhood includes more than half of the total variables and edges of $\mathcal{G}$, the number of false positive and false negative edges within the neighborhood is much lower than the corresponding number in the entire skeleton.

5 Related Work

Friedman et al. [8] propose a method for estimating probabilities on features of Bayesian networks. They use bootstrap to resample the data and learn a Bayesian network from each sampled data set. The probability of a structural feature is then estimated as the proportion of appearances of the feature in the resulting networks. Friedman and Koller [9] present a Bayesian method for estimating probabilities of features using MCMC samples over variable orderings. The methods are evaluated in terms of the classification performance (i.e. how accurately they accept or reject a feature), but not in terms of the calibration of predicted probability estimates.

Koivisto and Sood [10] and Koivisto [11] present algorithms for identifying exact posterior probabilities of edges in Bayesian networks. The methods use a dynamic programming strategy and constrain the search space of candidate causal models by bounding the number of possible parents per variable. The algorithms require a special type of non-uniform prior that does not respect Markov equivalence. Thus, resulting probabilities may be biased. Subsequent methods try to fix this problem by using MCMC simulations to compute network priors [2] or exploiting special types of nodes [12]. All methods in this category scale up to about 25 variables, since the minimum time and space requirement of these algorithms is $\mathcal{O}(n2^n)$.

Claasen and Heskes [1] propose a method for estimating Bayesian probabilities of a feature as a normalized sum of the posterior probabilities of all networks that entail this feature. The method requires exhaustive search of the space of possible networks, and is therefore not applicable for networks with more than 5-6 variables. The authors propose using this method as a standalone test of conditional independence, and also use it to decide on features inside a constraint-based algorithm. Pena, Kocka and Nielsen [13] estimate the confidence of a feature as the fraction of models containing the feature out of the different locally optimal models.

6 Experimental Evaluation

We performed a series of experiments to characterize the behavior of the proposed algorithms.

6.1 Calibration of Estimated Probabilities

We initially used simulated data to examine if the returned probability estimates are calibrated. We generated random DAGs with 10 and 20 variables, where each variable had 0 to 5 parents (randomly selected). The networks were then coupled with random parameters to create linear gaussian networks (continuous data) or discrete Bayesian networks (binary data). For continuous variables, a minimum correlation coefficient of 0.2 was imposed on the parameters to avoid weak interactions. We then simulated networks of various sample sizes, to test the method's behavior in different settings.

We used the PC skeleton identification step [3] with significance threshold $\alpha = 0.05$ and maximum conditioning set size 3 (explained below), modified to additionally return the maximum p-value encountered for each pair of variables. The set of maximum p-values was then used as input in Algorithm 1 to produce probability estimates for all pairwise relations. We compared our method against two alternative approaches:

1. **BCCD-P**: A method based on the BCCD algorithm presented in [1]. As mentioned above, the method estimates the posterior probability of a feature as a normalized sum of the posterior probabilities of DAGs that entail this feature. The algorithm scores all possible DAGs, and the authors use it to estimate probabilities for networks of at most 5 variables. To estimate the probabilities of pairwise relations, we scored the DAGs over variables X, Y and $\mathbf{Z}_{XY}$, where $\mathbf{Z}_{XY}$ is the conditioning set maximizing the p-value of

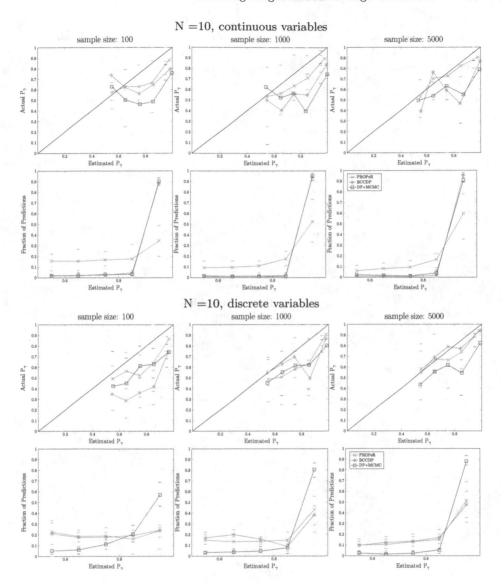

Fig. 3. Probability calibration plots for PROPeR, BCCD-P and MCMC+DP for networks of 10 variables. Bars indicate the quartiles. All methods tend to overestimate probabilities. Bayesian scoring methods are often very confident: For continuous variables, most of the probability estimates predicted by BCCD-P or MCMC+DP lie in the interval [0.9, 1], while MCMC+DP exhibits similar behavior for discrete variables also.

the tests $X \perp\!\!\!\perp Y | \mathbf{Z}$ performed by PC. This means that the cardinality of $\mathbf{Z}_{XY}$ cannot exceed 3. For a fair comparison, we used 3 as the maximum conditioning set of PC in all experiments. The probability of an adjacency

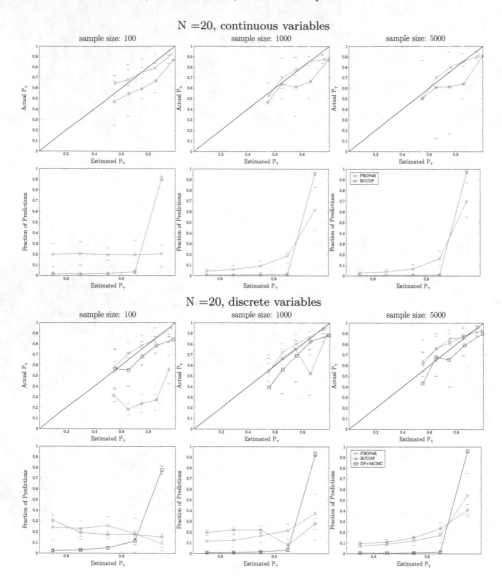

Fig. 4. Probability calibration plots for PROPeR, BCCD-P and MCMC+DP for networks of 20 variables. Bars indicate the quartiles. Similar to the results in Figure 3, Bayesian scoring methods tend to overestimate probabilities. MCMC+DP produced memory errors and failed to complete in all iterations for the BGE score, and is therefore not inlcuded in the corresponding plot.

was estimated as: $P(X{-}Y) = \sum_{\mathcal{G}\vdash X{-}Y} P(\mathbf{D}|\mathcal{G})P(\mathcal{G})$. Consistent priors described in [1] were pre-calculated and cached. To speed up the algorithm, we only scored one DAG per Markov equivalence class. For both approaches,

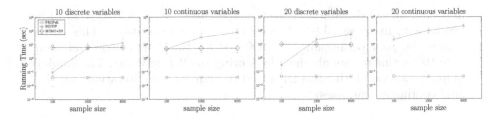

Fig. 5. Running times for PROPeR, BCCD-P and MCMC+DP

we used the BDe metric for discrete data and the BGe metric for gaussian data. Both metrics are score-equivalent.

2. **DP+ MCMC**: The method presented in [2] for identifying exact probabilities for edges in Bayesian networks. The method uses a combination of the DP algorithm [11] and MCMC sampling to correct the bias from the modular priors. We used the implementation provided by the authors in the BDAGL package. Maximum parents was set to 5, and the default parameters suggested by the authors in the package documentation were used. The method estimates probability estimates for directed edges, so we used $P(X—Y) = P(X{\rightarrow}Y) + P(Y{\rightarrow}X)$, $P(\neg X—Y) = 1 - P(X—Y)$.

To produce the probability calibration plots, the resulting predicted probabilities in [0.5, 1] were binned in 5 intervals. For every pair of variables, $P(X—Y)$ $=1-P(\neg X—Y)$. Thus, to consider each estimate once, we only need to consider half of the interval [0, 1]. If N pairwise relations have probability estimates $\{\hat{P}_i\}_{i=1}^{N}$ that lie in interval $[\gamma, \gamma + 0.1]$, we expect that $\bar{\hat{P}}_i \times N$ of the corresponding relations will be true. The actual probability P_γ for each interval is the fraction of relations with probability estimates in the given interval that are actually true in the data-generating graph. Figures 3 and 4 illustrate the mean estimated versus the mean actual probability for each bin, as well as the fraction of predictions in each bin for networks with 10 and 20 variables. Running times for all methods are shown in Figure 5.

Overall, results indicate that:

– PROPeR produces reasonable probability estimates, particularly in comparison to the more expensive BCCD-P and MCMC+DP approaches.
– MCMC+DP tends to identify very high (resp. very low) probabilities for the pairwise relations, even for small sample sizes for both metrics (BGE and BDE). BCCD has similar behavior for the BGE score, but not for the BDE score. This could explain the large deviations (and the seemingly unpredictable behavior) observed for these algorithms in the first four bins ([0.5 0.9]), since the means are computed over very few data points.
– As far as running times are concerned, both BCCD-P and MCMC+DP algorithms have (theoretically) exponential complexity with respect to the number of variables. BCCD-P also increases exponentially with sample size, but this is probably due to an increase in maximum conditioning set sizes

reported by PC skeleton for larger sample sizes. i.e., BCCD-P iterates networks with many variables (4-5) for most pairwise relations. This also explains the poor performance of BCCD-P for the BDE metric and sample size 100: estimates are obtained by scoring smaller networks. The employed implementation of MCMC+DP failed to complete any iterations for N=20 and continuous variables.

We must point out that the calibration of the probability estimates *is not necessarily related to the predictive power of the respective approaches*, which depends more on the relative ranking of probabilities among pairwise relations, rather than the actual estimates. For example, MCMC+DP has been shown to produce rankings of edges with very high AUC [2]. For the purposes of this work, however, obtaining estimates that are calibrated is important for identifying neighborhoods of a user-defined confidence. For example, using MCMC+DP estimates in Algorithm 2 would result in an almost fully connected $\mathcal{H}_{0.9}$, since most of the pairwise relations have probability estimates above this threshold.

6.2   Evaluation of Neighborhoods Identified with BiND

To demonstrate the value of BiND, we simulated data of 100 and 1000 samples from random networks with 20 and 50 variables, as described above. For the causal skeletons identified with the PC skeleton algorithm and the posterior probability estimates produced by PROPeR, all maximal β-neighborhoods for β=0.6, 0.7 and 0.9 were identified using Algorithm 2.

We examined the structural precision ($\frac{\text{\# edges in } \mathcal{G}' \text{ and the ground truth}}{\text{\# edges in } \mathcal{G}'}$) and recall ($\frac{\text{\# edges in } \mathcal{G}' \text{ and the ground truth}}{\text{\# edges in the ground truth}}$) of the resulting neighborhoods, compared to the baseline precision and recall for $\mathcal{G}$. As mentioned above,the maximal cliques can be very small and uninformative, particularly for high confidence thresholds. We are more interested in identifying large parts of the networks that we are confident about, and therefore focused in the maximum β-neighborhoods. Figure 6 illustrates the precision, recall and size of maximum β-neighborhoods for networks of 20 and 50 variables, for both discrete and continuous data. The algorithm took 85.56 seconds on average to identify the maximum 0.6-neighborhoods for 50 variables and 1000 samples (the most expensive case). Detailed time results are omitted due to space limitations.

Results indicate the following:

- The method identifies subgraphs with lower ratios of false inferences compared to the entire skeleton.
- For high confidence thresholds and small sample sizes, the algorithm cannot identify large neighborhoods.
- The algorithm is particularly useful in small sample sizes, where the overall recall is very low.

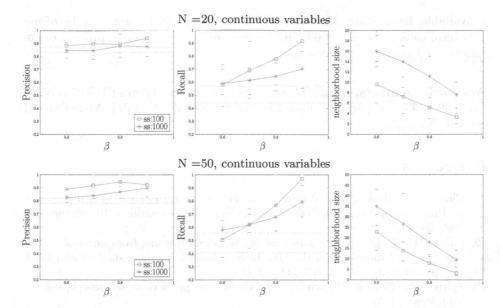

Fig. 6. Precision, recall, number and size of maximum cliques identified using BiND in networks of 20 continuous variables. Bars indicate quartiles. Dashed horizontal lines show the mean baseline precision and recall (mean precision and recall of the output of PC skeleton. BiND identifies neighborhoods of higher structural precision and recall than the corresponding baseline, particularly for small sample sizes.

7 Discussion

Equipping constraint-based causal discovery algorithms with a method that can provide some measure of confidence on their output improves their usability. Bayesian scoring and bootstrapping methods can be employed for this purpose, but are computationally expensive and do not scale up to the number of variables constraint-based algorithms can handle.

We have presented PROPeR, an algorithm for estimating posterior probabilities of adjacencies and non-adjacencies in networks learnt using constraint-based methods. The algorithm has no significant computational overhead and is scalable to practically any input size: increasing the number of variables processed by the constraint-based algorithm merely increases the sample size of p-values on which PROPeR fits a probability density function. PROPeR is shown to produce calibrated probability estimates, while being significantly faster than other state of the art algorithms. We have also presented BiND, an algorithm that identifies the maximal (or maximum) neighborhoods of high confidence on a causal network. In simulated scenarios, the algorithm is able to identify neighborhoods that are indeed more reliable.

PROPeR and BiND can easily accompany any constraint-based algorithm on any type of data, provided an appropriate test of conditional independence is available. Estimating posterior probabilities based on p-values can be of use in several causal discovery tasks, including conflict resolution, improving orientations, and experiment selection.

Acknowledgements. ST and IT were funded by the STATegra EU FP7 project, No 306000. IT was partially funded by the EPILOGEAS GSRT ARISTEIA II project, No 3446.

References

1. Claassen, T., Heskes, T.: A Bayesian approach to constraint based causal inference. In: Proceedings of the 28th Conference on Uncertainty in Artificial Intelligence, pp. 2992–2996 (2012)
2. Eaton, D., Murphy, K.P.: Exact bayesian structure learning from uncertain interventions. In: Proceedings of the 11th International Conference on Artificial Intelligence and Statistics, pp. 107–114 (2007)
3. Spirtes, P., Glymour, C., Scheines, R.: Causation, prediction, and search, vol. 81. MIT Press (2000)
4. Sellke, T., Bayarri, M., Berger, J.: Calibration of ρ values for testing precise null hypotheses. The American Statistician 55(1), 62–71 (2001)
5. Storey, J., Tibshirani, R.: Statistical significance for genomewide studies. PNAS 100(16), 9440 (2003)
6. Karp, R.: Reducibility Among Combinatorial Problems. In: Complexity of Computer Computations, pp. 85–103. Plenum Press (1972)
7. Bron, C., Kerbosch, J.: Algorithm 457: finding all cliques of an undirected graph. Communications of the ACM 16(9), 575–577 (1973)
8. Friedman, N., Goldszmidt, M., Wyner, A.: On the application of the bootstrap for computing confidence measures on features of induced Bayesian networks. In: Proceedings of the 7th International Workshop on Artificial Intelligence and Statistics, pp. 196–205 (1999)
9. Friedman, N., Koller, D.: Being Bayesian about network structure. A Bayesian approach to structure discovery in Bayesian networks. Machine Learning 50(1-2), 95–125 (2003)
10. Koivisto, M., Sood, K.: Exact Bayesian structure discovery in Bayesian networks. JMLR 5, 549–573 (2004)
11. Koivisto, M.: Advances in exact Bayesian structure discovery in Bayesian networks. In: Proceedings of the 22nd Conference on Uncertainty in Artificial Intelligence, pp. 241–248 (2006)
12. Tian, J., He, R.: Computing posterior probabilities of structural features in Bayesian networks. In: Proceedings of the 25th Conference on Uncertainty in Artificial Intelligence, pp. 538–547 (2009)
13. Pena, J., Kocka, T., Nielsen, J.: Featuring multiple local optima to assist the user in the interpretation of induced Bayesian Network models. In: Proceedings of the 10th International Conference on Information Processing and Management of Uncertainty in Knowledge-Based Systems, pp. 1683–1690 (2004)

Causal Independence Models
for Continuous Time Bayesian Networks

Maarten van der Heijden and Arjen Hommersom*

Institute for Computing and Information Sciences,
Radboud University Nijmegen, The Netherlands
{m.vanderheijden,arjenh}@cs.ru.nl

Abstract. The theory of causal independence is frequently used to facilitate the assessment of the probabilistic parameters of probability distributions of Bayesian networks. Continuous time Bayesian networks are a related type of graphical probabilistic models that describe stochastic processes that evolve over continuous time. Similar to Bayesian networks, the number of parameters of continuous time Bayesian networks grows exponentially in the number of parents of nodes. In this paper, we study causal independence in continuous time Bayesian networks. This new theory can be used to significantly reduce the number of parameters that need to be estimated for such models as well. We show that the noisy-OR model has a natural interpretation in the parameters of these models, and can be exploited during inference. Furthermore, we generalise this model to include synergistic and anti-synergistic effects, leading to noisy-OR synergy models.

1 Introduction

During the past two decades, probabilistic graphical models, and in particular Bayesian networks [17], have become popular methods for building applications involving uncertainty in many real-world domains. One of the key challenges in building probabilistic graphical models is that the number of parameters needed to assess a family of conditional probability distributions for a variable E grows exponentially with the number of its causes. This is clearly problematic when developing such models manually, e.g., by acquiring relevant knowledge from experts in a domain, but also when learning these models from data.

The theory of causal independence is frequently used in such situations, basically to decompose a probability table in terms of a small number of factors [10,17,8,12]. A well-known example of this is the noisy-OR model, where the number of required parameters is only linear in the number of parents. Later, the theory of causal independence was also called *intercausal independence* or *independence of causal influence* [2]. These names emphasise that in these models the causes are not independent, but the effects of causes are modelled independently, which leads to the reduction in parameters.

---

* The second author of this paper is supported by the ITEA2 MoSHCA project (ITEA2 ip11027).

L.C. van der Gaag and A.J. Feelders (Eds.): PGM 2014, LNAI 8754, pp. 503–518, 2014.
© Springer International Publishing Switzerland 2014

Several generalisations of Bayesian networks that incorporate time have been proposed. One of the most popular models is *dynamic Bayesian networks* (DBNs), where random variables are indexed by elements from a set of discrete time points. One could look upon such networks as a regular Bayesian network, where causal independence principles can be applied. Another type of temporal network are temporal-nodes Bayesian networks [1] and networks of probabilistic events in discrete time [5], where time intervals are modelled as part of the values of random variables. Also for these type of temporal models, causal independence models have been proposed [2]. Finally, for models in continuous time, continuous time Bayesian networks (CTBNs) [14] have been proposed. CTBNs represent a Markov process over continuous time, using a factorisation to obtain a more compact representation, similar to Bayesian networks. However, also similar to Bayesian networks, the number of parameters in CTBNs, which are represented as intensity matrices rather than conditional probability tables, increases exponentially in the number of parents as well. Therefore, a further decomposition of the these matrices is required if the number of parents is large.

This paper introduces the concept of causal independence models for CTBNs. In particular, we show that the noisy-OR applied to CTBNs has a natural interpretation in terms of the parameters of this model. To obtain more flexibility, we further generalise this noisy-OR, using the theory of probabilistic independence of causal influence (PICI) [22] such that known synergies and anti-synergies between the parents of a node can be modelled in an intuitive manner. We illustrate this concept with an example of antibiotic treatment where there are several synergistic and anti-synergistic effects between the treatments.

This paper is organised as follows. In the next section, we will introduce the required preliminaries with respect to causal independence modelling, as well as CTBNs. Then, in Section 3, we introduce one particular manner to incorporate causal independence models into CTBNs, in particular the CTBN noisy-OR model. This acts as a basis for Section 4, where this model is further generalised to the case where there are synergistic and anti-synergistic relationships between parents on the effect. Then, in Section 5, we discuss related work, in particular with respect to causal independence modelling. Finally, in Section 6, we conclude the paper and discuss directions for further research.

2 Preliminaries

In the following, we will denote random variables by upper case, e.g., X, Y, etc. In this paper, we will mainly consider binary variables where $X = x$ indicates that X is 'true' and $X = \bar{x}$ indicates that X has the value 'false'. Instead of $X = x$ we will frequently write simply x. We will also use capital letters X to vary over values in summations and products.

2.1 Causal Independence Models

As explained, a popular way to specify interactions among statistical variables in a compact fashion is offered by the notion of *causal independence* [8]

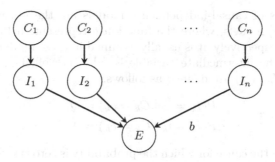

Fig. 1. Causal-independence model

or *independence of causal influences* (ICI). The global structure of a causal-independence model is shown in Fig. 1; it expresses the idea that causes $C = \{C_1, \ldots, C_n\}$ influence a given common effect E through intermediate variables $I = \{I_1, \ldots, I_n\}$ and a Boolean-valued function b. The influence of each cause C_k on the common effect E is independent of each other cause C_j, $j \neq k$. The function b represents in which way the intermediate effects I_k, and indirectly also the causes C_k, interact to yield the final effect E. Hence, the function b is defined in such a way that when a relationship, as modelled by the function b, between I_k, $k = 1, \ldots, n$, and $E = true$ is satisfied, then it holds that $b(I_1, \ldots, I_n) = true$.

In terms of probability theory, the notion of causal independence can be formalised for the occurrence of the effect E as follows. By standard probability theory:

$$P(E \mid C_1, \ldots, C_n) = \sum_{I_1, \ldots, I_n} P(E \mid I_1, \ldots, I_n, C_1, \ldots, C_n)$$

$$P(I_1, \ldots, I_n \mid C_1, \ldots, C_n) \qquad (1)$$

meaning that the causes $C = \{C_1, \ldots, C_n\}$ influence the common effect E through the intermediate effects $I_1, \ldots, I_n$. The deterministic probability distribution $P(E \mid I_1, \ldots, I_n)$ corresponds to the Boolean function b, such that $b(I_1, \ldots, I_n) = true$ if $P(e \mid I_1, \ldots, I_n) = 1$; otherwise, $b(I_1, \ldots, I_n) = false$ if $P(e \mid I_1, \ldots, I_n) = 0$. Note that the effect variable E is conditionally independent of $C_1, \ldots, C_n$ given the intermediate variables $I_1, \ldots, I_n$, and that each variable I_k is only dependent on its associated variable C_k; hence, it holds that

$$P(e \mid I_1, \ldots, I_n, C_1, \ldots, C_n) = P(e \mid I_1, \ldots, I_n)$$

and

$$P(I_1, \ldots, I_n \mid C_1, \ldots, C_n) = \prod_{k=1}^{n} P(I_k \mid C_k)$$

Formula (1) can now be simplified to:

$$P_b(e \mid C) = \sum_{b(I_1, \ldots, I_n)} \prod_{k=1}^{n} P(I_k \mid C_k) \qquad (2)$$

Typical examples of causal-independence models are the noisy-OR [17,8] and noisy-MAX [3,10] models, where the function b represents a logical OR and a MAX function, respectively. It is usually assumed in these models that if a cause is absent, then the intermediate variable is false as well, i.e., the probability distributions $P(I_k \mid C_k)$ are defined as follows:

$$P(I_k = i_k \mid C_k = c_k) = Q_i$$
$$P(I_k = i_k \mid C_k = \bar{c}_k) = 0$$

where the state of the cause for which the probability is zero is sometimes referred to as the *distinguished state*.

These causal independence or ICI models were later generalised to a more general model, called *probabilistic ICI* [22]. As said, in a normal causal independence model, we have a Boolean function b such that $E = b(I)$. Similar to ICI models, a PICI model assumes intermediate variables, but instead of a function b defined in I, it assumes a function f such that $P(E) = f(Q, I)$, where Q are the parameters defined for the conditional probabilities $P(I_k = i_k \mid C_k = c_k)$. This allows, for example, the modelling of the average probability, leading to a *noisy-average* model [22].

2.2 Continuous Time Bayesian Networks

Let $V(t) = \{V_1(t), \ldots, V_n(t)\}$ be a set of random variables at time t. A CTBN models a continuous time Markov process over $\{V(t) \mid t \geq 0\}$. Let $G = (V, E)$ be a directed graph and Q_V an intensity matrix that models rates $q \geq 0$ for moving between states over time, where each state is described by a configuration of random variables V. Then, in a CTBN, the joint intensity Q_V is factorised according to G, i.e., it can be written as

$$Q_V = \prod_{V_i \in V} Q_{V_i \mid \mathrm{pa}(V_i)},$$

where multiplication is defined as *amalgamation* [14] and $Q_{V_i \mid \mathrm{pa}(V_i)}$ is the conditional intensity matrix of V_i given the parents of V_i in G. For example, consider a set of binary variables $V = \{C, D, E\}$ with $\mathrm{pa}(E) = \{C, D\}$ and the following conditional intensity matrices:

$$Q_C = \begin{bmatrix} -q^c & q^c \\ q^{\bar{c}} & -q^{\bar{c}} \end{bmatrix} \qquad Q_D = \begin{bmatrix} -q^d & q^d \\ q^{\bar{d}} & -q^{\bar{d}} \end{bmatrix}$$

$$Q_{E \mid \bar{c}, \bar{d}} = \begin{bmatrix} -q^e_{\bar{c}, \bar{d}} & q^e_{\bar{c}, \bar{d}} \\ q^{\bar{e}}_{\bar{c}, \bar{d}} & -q^{\bar{e}}_{\bar{c}, \bar{d}} \end{bmatrix} \qquad Q_{E \mid c, \bar{d}} = \begin{bmatrix} -q^e_{c, \bar{d}} & q^e_{c, \bar{d}} \\ q^{\bar{e}}_{c, \bar{d}} & -q^{\bar{e}}_{c, \bar{d}} \end{bmatrix}$$

$$Q_{E \mid \bar{c}, d} = \begin{bmatrix} -q^e_{\bar{c}, d} & q^e_{\bar{c}, d} \\ q^{\bar{e}}_{\bar{c}, d} & -q^{\bar{e}}_{\bar{c}, d} \end{bmatrix} \qquad Q_{E \mid c, d} = \begin{bmatrix} -q^e_{c, d} & q^e_{c, d} \\ q^{\bar{e}}_{c, d} & -q^{\bar{e}}_{c, d} \end{bmatrix}$$

where, for each variable $X \in V$, q^x models the transition rate from $\bar{x}$ to x and $q^{\bar{x}}$ models the transition rate from x to $\bar{x}$, given the state of their parents.

The joint intensity Q_V is then given by:

$$Q_V = \begin{bmatrix} -z_1 & q^c & q^d & 0 & q^e_{\bar{c},\bar{d}} & 0 & 0 & 0 \\ q^{\bar{c}} & -z_2 & 0 & q^d & 0 & q^e_{c,\bar{d}} & 0 & 0 \\ q^{\bar{d}} & 0 & -z_3 & q^c & 0 & 0 & q^e_{\bar{c},d} & 0 \\ 0 & q^{\bar{d}} & q^{\bar{c}} & -z_4 & 0 & 0 & 0 & q^e_{c,d} \\ q^{\bar{e}}_{\bar{c},\bar{d}} & 0 & 0 & 0 & -z_5 & q^c & q^d & 0 \\ 0 & q^{\bar{e}}_{c,\bar{d}} & 0 & 0 & q^{\bar{c}} & -z_6 & 0 & q^d \\ 0 & 0 & q^{\bar{e}}_{\bar{c},d} & 0 & q^{\bar{d}} & 0 & -z_7 & q^c \\ 0 & 0 & 0 & q^{\bar{e}}_{c,d} & 0 & q^{\bar{d}} & q^{\bar{c}} & -z_8 \end{bmatrix}$$

where the states in Q_V are ordered by

$$\{(\bar{c},\bar{d},\bar{e}),(c,\bar{d},\bar{e}),(\bar{c},d,\bar{e}),(c,d,\bar{e}),(\bar{c},\bar{d},e),(c,\bar{d},e),(\bar{c},d,e),(c,d,e)\}$$

For example, the $(1,2)$-entry in this matrix indicates that the transition rate from $C = \bar{c}, D = \bar{d}, E = \bar{e}$ to $C = c, D = \bar{d}, E = \bar{e}$ is q^c, which follows directly from the specification of Q_C. Note that two variables cannot transition at the same time, which results in the zeros in Q_V. The values for z_i ensure that each row in the intensity matrix sums to 0, e.g., $z_1 = q^c + q^d + q^e_{\bar{c},\bar{d}}$. These parameters can be interpreted as the parameters that govern the time until the next state change, while the q parameters indicate to which state the transition leads.

Given a joint intensity matrix Q_V, we can describe the transient behaviour of V as follows. Each row and column refers to a particular state, modelled by a particular configuration of V. Let z_i be at the i'th position of the diagonal of Q_V. The time of transitioning from a state i to another state is then modelled by some random variable $T_i \sim \text{Exp}(z_i)$. We will denote its probability density function by f and its cumulative distribution function by F, i.e., $F(t) = P(T_i \leq t)$. The function G is the tail distribution of T, i.e., $G(t) = P(T_i > t) = 1 - F(t)$. The expected time of transitioning, i.e., the expected value of T_i is $1/z_i$. It shifts to state j with probability q_{ij}/z_i. Given this transient behaviour, we can describe the probability of not transitioning as follows:

$$P(V(t + \Delta t) = v_i \mid V(t) = v_i) = G(\Delta t) = \exp(-z_i \Delta t)$$

Usually, we are interested in the instantaneous transitions, i.e., when Δt is small. In this case, this probability can be given a straightforward interpretation.

Proposition 1. *For $\Delta t \to 0$, it holds that the* instantaneous transition probability *is:*

$$P(V(t + \Delta t) = \bar{v}_i \mid V(t) = v_i) = 1 - G(\Delta t) \approx z_i \Delta t$$

Proof. The Taylor series of $G(t) = \exp(-z_i t)$ at $t_0 = 0$ is:

$$\sum_{n=0}^{\infty} \frac{G^{(n)}(t_0)}{n!}(t - t_0)^n = \sum_{n=0}^{\infty} \frac{(-z_i)^n}{n!}t^n$$

$$= 1 - z_i t + \frac{z_i^2 t^2}{2} - \frac{z_i^3 t^3}{3!} + \frac{z_i^4 t^4}{4!} - \cdots$$

$$\approx 1 - z_i t$$

$\square$

In other words, the probability of transitioning from the i'th state is approximately proportional to both z_i and Δt.

3   Causal Independence in CTBNs

In this section, we will introduce a general methodology for specifying causal independence models in CTBNs. First, we discuss causal independence in general. Then we study the case where we have noisy-OR models. This will be taken as a starting point for the next section, where this model is generalised so that it can include synergistic and anti-synergistic effects.

3.1   General Causal Independence Models

While in general Bayesian networks the goal is to model the distribution $P(E \mid C_1, \ldots, C_n)$, in CTBNs we aim to model the distribution:

$$P(E(t + \Delta t) = e \mid E(t) = \bar{e}, C_1, \ldots, C_n)$$

i.e., the probability of transitioning from a state where $\bar{e}$ at t to some other state where e holds within Δt time, depending on the causes $C_1, \ldots, C_n$. While these causes are also time-dependent, whether E transitions always depends on a particular configuration of causes that hold between t and $t+\Delta$, so time is ignored in the notation of this conditional probability distribution. As illustrated in the preliminaries, for each configuration of C, a different parameter is introduced to model this transition.

Similar to the general causal independence model, we introduce additional variables $I_1, \ldots, I_n$ as intermediate nodes with intensity matrices $Q_{I_i \mid C_i}$, so that the number of parameters is only linear in n. Furthermore, we define:

$$P(E(t + \Delta t) = e \mid E(t) = \bar{e}, C_1, \ldots, C_n)$$
$$= \sum_{b(I_1, \ldots, I_n)} \prod_{k=1}^{n} P(I_k(t + \Delta t) = I_k \mid I_k(t) = \bar{i}_k, C_k)$$

where b is a Boolean interaction function. In other words, it defines that a transition occurs if a Boolean combination of intermediate nodes I transitions from $\bar{i}$ to i. Note that such interactions are not necessarily representable in a CTBN, however, as we will show in the remainder of this paper, in some cases it is.

3.2   Continuous Time Noisy-OR Models

A particularly useful causal independence model for Bayesian networks is the noisy-OR model. As mentioned in the preliminaries, it is usually assumed in causal independence models that there is a distinguished state (say *false* in the binary case) such that the probability of $P(\bar{i}_j \mid \bar{c}_j) = 1$, which implies that E can only become true if at least one of its causes is true, i.e., it implies

$P(\bar{e} \mid \bar{c}_1, \ldots, \bar{c}_n) = 1$. In the CTBN case, we have a similar assumption: suppose E has parents C and D, then:

$$P(E(t + \Delta t) = \bar{e} \mid E(t) = \bar{e}, C = \bar{c}, D = \bar{d}) = 1$$

for all Δt. Since:

$$P(E(t + \Delta t) = \bar{e} \mid E(t) = \bar{e}, C = \bar{c}, D = \bar{d}) = \exp(-q^e_{\bar{c}, \bar{d}} \Delta t)$$

this assumption implies that $q^e_{\bar{c}, \bar{d}} = 0$.

The noisy-OR effect of independent influences turns out to result in additive parameters, as shown in the following theorem.

Theorem 1. *Given a CTBN with effect variable E and parents $C = \{C_1, \ldots, C_n\}$. Suppose E is determined by a noisy-OR, i.e., a causal independence model where b is the logical OR:*

$$P(E(t + \Delta t) = e \mid E(t) = \bar{e}, C_1, \ldots, C_n) =$$

$$\sum_{I_1 \vee \cdots \vee I_n} \prod_{k=1}^{n} P(I_k(t + \Delta t) = I_k \mid I_k(t + \Delta t) = \bar{\imath}_k, C_k)$$

Let q^e_k be the intensity of transitioning from $\bar{\imath}_k$ to i_k given C_k for all k. Then the intensity of transitioning from $\bar{e}$ to e is given by $q^e_C = \sum_{k=1}^{n} q^e_k$.

Proof. The noisy-OR model can be rephrased as:

$$P(E(t + \Delta t) = \bar{e} \mid E(t) = \bar{e}, C_1, \ldots, C_n) =$$

$$\prod_{k=1}^{n} P(I_k(t + \Delta t) = \bar{\imath}_k \mid I_k(t + \Delta t) = \bar{\imath}_k, C_k)$$

Let G_k be the tail distribution of the transition function for I_k given C_k, i.e, $G_k(t) = \exp(-q^e_k t)$. So then:

$$\begin{aligned} P(E(t + \Delta t) = \bar{e} \mid E(t) = \bar{e}, C_1, \ldots, C_n) &= \prod_{k=1}^{n} G_k(\Delta t) \\ &= \prod_{k=1}^{n} \exp(-q^e_k \Delta t) \\ &= \exp(-(\sum_{k=1}^{n} q^e_k) \Delta t) \end{aligned}$$

and hence,

$$P(E(t + \Delta t) = e \mid E(t) = \bar{e}, C_1, \ldots, C_n) = 1 - \exp\left(-\left(\sum_{k=1}^{n} q^e_k\right) \Delta t\right)$$

which implies that the intensity of transitioning from $\bar{e}$ to e is $\sum_{k=1}^{n} q^e_k$.   □

Therefore, the noisy-OR results into an additive effect on the transition probability:

$$P(E(t + \Delta t) = e \mid E(t) = \bar{e}, C_1, \ldots, C_n) \approx \sum_{k=1}^{n} q^e_k \Delta t$$

when $\Delta t \rightarrow 0$.

Another way to look at this is to consider a causal independence model for the probability of staying in the same state, i.e.,

$$P(E(t + \Delta t) = \bar{e} \mid E(t) = \bar{e}, C_1, \ldots, C_n)$$

$$= \sum_{b(I_1,\ldots,I_n)} \prod_{k=1}^{n} P(I_k(t + \Delta t) = \bar{I}_k \mid I_k(t + \Delta t) = \bar{\iota}_k, C_k)$$

where b is a Boolean function. If we take b to be a logical-AND, i.e., we have a noisy-AND model, it follows that:

$$P(E(t + \Delta t) = \bar{e} \mid E(t) = \bar{e}, C_1, \ldots, C_n)$$

$$= \prod_{k=1}^{n} P(I_k(t + \Delta t) = \bar{\iota}_k \mid I_k(t + \Delta t) = \bar{\iota}_k, C_k)$$

$$= \prod_{k=1}^{n} G_k(\Delta t) = \exp\left(-\left(\sum_{k=1}^{n} q_k^e\right)\Delta t\right)$$

From now, we will describe this interaction between causes on the effect by an *interaction function* h, which maps the state of causes $C_1, \ldots, C_n$ to a linear function over the intensities defined for each of the causes separately. As the discussion above shows, the noisy-OR model may be described by a simple sum of intensities, i.e., $h(C_1, \ldots, C_n) = \sum_{k=1}^{n} q_k^e$.

Having such a simple interaction function, this noisy-OR formalisation can also be exploited during inference. We illustrate this by means of elimination of the parents from the conditional probability distribution of E, which is normally a summation exponential in the number of parents.

Theorem 2. *Let* $P(E(t + \Delta t) = e \mid E(t) = \bar{e}, C_1, \ldots, C_n)$ *be defined in terms of a noisy-OR CTBN model, and let* $P(C_k)$ *be given for all* $1 \leq k \leq n$. *Then* $P(E(t + \Delta t) = e \mid E(t) = \bar{e})$ *can be computed in* $O(n)$.

Proof. Consider computing $P(E(t + \Delta t) = \bar{e} \mid E(t) = \bar{e})$, which is $1 - P(E(t + \Delta t) = e \mid E(t) = \bar{e})$. Let G_{C_k} be the tail distribution of the transition function for I_k given C_k, i.e., $G_{C_k} = \exp(-q_k^e \Delta t)$ and $G_{\bar{C}_k} = 1$. By basic probability theory, we have:

$$P(E(t + \Delta t) = \bar{e} \mid E(t) = \bar{e})$$

$$= \sum_{C_1,\ldots,C_n} P(E(t + \Delta t) = \bar{e} \mid E(t) = \bar{e}, C_1, \ldots, C_n) \prod_{k=1}^{n} P(C_k)$$

$$= \sum_{C_1,\ldots,C_n} \prod_{k=1}^{n} G_{C_k}(\Delta t) \prod_{k=1}^{n} P(C_k)$$

$$= \sum_{C_1} G_{C_1}(\Delta t) P(C_1) \cdots \sum_{C_n} G_{C_n}(\Delta t) P(C_n)$$

$$= \prod_{k=1}^{n} (\exp(-q_k^e \Delta t) P(c_k) + P(\bar{c}_k)) \qquad \qquad \square$$

4 Noisy-OR Synergy Models

The previous section shows that the noisy-OR has an additive effect in terms of intensities of parents. In practice, however, there are often synergies of parents on their common children, e.g., parents may be reinforcing or inhibiting each other. Some of these synergies can be modelled using Boolean functions [12]; however, most of these functions do not have a straightforward interpretation in terms of transition intensities in CTBNs.

In this section, we introduce a new causal-independence synergy model, which is closely related to the additive nature of intensities in parameters. The main advantage is that this synergistic model can be easily exploited using existing CTBN methods. Moreover, it has a natural interpretation in CTBN models.

4.1 Synergies

Synergy has been studied in context of qualitative probabilistic networks [19], where an *additive synergy* expresses how the interaction between two variables influences the probability of observing the values of a third variable.

Definition 1. *Let C_i and C_j be two parents of E and Z the set of other parents of E. We say there is a* positive additive synergy *of C_i and C_j on E, if*

$$P(e \mid c_i, c_j, Z) - P(e \mid \bar{c}_i, c_j, Z) \geq P(e \mid c_i, \bar{c}_j, Z) - P(e \mid \bar{c}_i, \bar{c}_j, Z)$$

Negative and zero additive synergies are defined similarly by replacing the $\geq$ sign with $\leq$ or $=$, respectively. The intuition behind this definition is that the influence of C_i, i.e., the difference of the probability of e when C_i is true and when C_i is false, is larger when C_j is also true compared to when C_j is false. Also note that this definition is symmetric in C_i and C_j. Similarly, we can say that in a CTBN a positive synergy for the *transition* from $\bar{e}$ to e occurs if:

$$P(E(t + \Delta t) = e \mid E(t) = \bar{e}, c_i, c_j, Z) - P(E(t + \Delta t) = e \mid E(t) = \bar{e}, \bar{c}_i, c_j, Z)$$
$$\geq P(E(t + \Delta t) = e \mid E(t) = \bar{e}, c_i, \bar{c}_j, Z) - P(E(t + \Delta t) = e \mid E(t) = \bar{e}, \bar{c}_i, \bar{c}_j, Z)$$

i.e., C_i and C_j have a synergistic effect on the influence of the transition probability of e. Again, negative and zero synergies can be defined similarly.

Wellman [19] also proved that the noisy-OR model results in a negative synergy. For example, suppose e has two parents C_1 and C_2, then:

$$P(e \mid \bar{c}_1, \bar{c}_2) = 0 \qquad\qquad P(e \mid c_1, \bar{c}_2) = P(i_1 \mid c_1)$$
$$P(e \mid \bar{c}_1, c_2) = P(i_2 \mid c_2) \qquad\qquad P(e \mid c_1, c_2) = 1 - P(\bar{i}_1 \mid c_1) P(\bar{i}_2 \mid c_2)$$

The negative synergy is easy to verify as:

$$(1 - (1 - P(i_1 \mid c_1))(1 - P(i_2 \mid c_2))) - P(i_2 \mid c_2)$$
$$= P(i_1 \mid c_1) + P(i_2 \mid c_2) - P(i_1 \mid c_1)P(i_2 \mid c_2) - P(i_2 \mid c_2) \leq P(i_1 \mid c_1) - 0$$

While in a noisy-OR the probability of e monotonically increases by the number of true causes, the negative synergy holds because causes have relatively less effect if other causes are already true, especially when the effects of individual causes are relatively large.

4.2 Basic Noisy-OR Synergies

Given the discussion in Section 3, it is clear to see that the noisy-OR can also be seen as a PICI model defined on the parameters of the intermediate variables:

$$P(E(t + \Delta t) = e \mid E(t) = \bar{e}, c_1, \ldots, c_n) = f(G_1(\Delta t), \ldots, G_n(\Delta t))$$

The restriction that there are no further interactions between different causes is a fairly strong assumption. We can relax the assumption by allowing some interaction between causes, where the goal is to increase the expressivity without resorting to the full (exponential) model. We define an interaction function on the intensities of the causes with a parameter to model synergistic or anti-synergistic effects between causes. This leads to the following noisy-OR synergy model:

$$P(E(t + \Delta t) = \bar{e} \mid E(t) = \bar{e}, C_1, \ldots, C_n) = \begin{cases} G_{C_k}(\Delta t) & \text{if one } C_k \text{ is true} \\ \left(\prod_{k=1}^{n} G_{C_k}(\Delta t) \right)^s & \text{otherwise} \end{cases}$$

with s an additional synergistic parameter.

In the full model with binary variables given in the introduction, we have four parameters for the intensities $Q_{e|\mathrm{pa}(E)}$, one for each parent configuration. A synergistic effect implies that the intensity of the effect increases when multiple causes are present.

Similar to the noisy-OR, this model can also be written as an interaction function $h(C)$ which models the intensity of transitioning from $\bar{e}$ to e. Suppose all causes are true, then:

$$P(E(t + \Delta t) = e \mid E(t) = \bar{e}, c_1, \ldots, c_n) = 1 - \exp\left(- s\left(\sum_{k=1}^{n} q_k^e \right) \Delta t \right)$$

so $h(c_1, \ldots, c_n) = s \sum_{k=1}^{n} q_k^e$. For example, let h be an interaction function for the intensity of E given only two causes C, D. Now, when we let the base rates for the single causes be $q_{c,\bar{d}}^e$ and $q_{\bar{c},d}^e$ we obtain:

$$h(\bar{c}, \bar{d}) = 0 \qquad\qquad h(\bar{c}, d) = q_{\bar{c},d}^e$$
$$h(c, \bar{d}) = q_{c,\bar{d}}^e \qquad\qquad h(c, d) = s \cdot (q_{\bar{c},d}^e + q_{\bar{c},d}^e)$$

where s is the synergy parameter that applies when both causes are true. Note that we regain the noisy-OR as a special case by choosing $s = 1$; we call it a synergistic model when $s > 1$ and an anti-synergistic model when $s < 1$. The motivation for this is given by the following theorem.

Theorem 3. *Given a CTBN over the variables $\{C, D, E\}$, let E be the effect variable and h an interaction function with synergy parameter s between causes C and D. If*

$$s = \frac{\log(\exp(-q_C^e \Delta t) + \exp(-q_D^e \Delta t) - 1)}{-(q_C^e + q_D^e)\Delta t}$$

then the synergy between C and D is zero on E at Δt. Furthermore, if $\Delta t \to 0$, then $s \approx 1$.

Proof. In this model, it is given that:

$$P(E(t + \Delta t) = e \mid E(t) = \bar{e}, \bar{c}, \bar{d}) = 0$$
$$P(E(t + \Delta t) = e \mid E(t) = \bar{e}, c, \bar{d}) = 1 - \exp(-q_C^e \Delta t)$$
$$P(E(t + \Delta t) = e \mid E(t) = \bar{e}, \bar{c}, d) = 1 - \exp(-q_D^e \Delta t)$$
$$P(E(t + \Delta t) = e \mid E(t) = \bar{e}, c, d) = 1 - \exp(-s(q_C^e + q_D^e)\Delta t)$$

So a zero synergy implies:

$$(1 - \exp(-s(q_C^e + q_D^e)\Delta t)) - (1 - \exp(-q_D^e \Delta t)) = 1 - \exp(-q_C^e \Delta t) - 0$$

which implies the formula given for s at Δt. It can be shown that the series expansion of s at $\Delta t = 0$ is:

$$s = 1 + \frac{q_C^e \cdot q_D^e \cdot \Delta t}{q_C^e + q_D^e} + \frac{1}{2} \cdot q_C^e \cdot q_D^e \cdot (\Delta t)^2 + O((\Delta t)^3)$$

so $s \approx 1$ if $\Delta t \to 0$. Alternatively, one can also consider the instantaneous transition probabilities directly, where a zero synergy implies that:

$$P(E(t + \Delta t) = e \mid E(t) = \bar{e}, c, d) - P(E(t + \Delta t) = e \mid E(t) = \bar{e}, \bar{c}, d)$$
$$\approx s(q_C^e + q_D^e)\Delta t - q_D^e \Delta t = q_C^e \Delta t - 0\Delta t \approx$$
$$P(E(t + \Delta t) = e \mid E(t) = \bar{e}, c, \bar{d}) - P(E(t + \Delta t) = e \mid E(t) = \bar{e}, \bar{c}, \bar{d})$$

if $\Delta t \to 0$. Obviously, this implies $s = 1$. $\square$

This theorem may seem to contradict the earlier statement that a noisy-OR interaction is a negative synergy and therefore $s < 1$, but this is not the case. Recall that the noisy-OR is a negative synergy because

$$P(i_1 \mid c_1) - P(i_1 \mid c_1)P(i_2 \mid c_2) \le P(i_1 \mid c_1).$$

The second term on the left hand side can be written as

$$P(i_1 \mid c_1)P(i_2 \mid c_2) = (1 - \exp(-q_{c_1}^e \Delta t))(1 - \exp(-q_{c_2}^e \Delta t)),$$

which will vanish as Δt goes to zero, resulting in the statement in the theorem that $s = 1$.

4.3 Noisy-OR Synergies with Local Interactions

It may sometimes be appropriate to allow local synergy effects instead of synergy over all causes C. This implies that particular subsets of causes can have different synergy parameters. Let h_i be a *local interaction function*, i.e. a synergistic noisy function over i arguments. In order to allow arbitrary synergy effects we can assign a synergy parameter to all subsets of causes, such that $h(C_1, \ldots, C_n)$ is defined in terms of 2^n synergy parameters. We then obtain the probability

$$P(E(t + \Delta t) = \bar{e} \mid E(t) = \bar{e}, C_1, \ldots, C_n) = \prod_{k=1}^{n} G_{C_k}(\Delta t)^{s_{C_1, \ldots, C_n}}$$

where $s_{C_1, \ldots, C_n}$ indicates that s depends on the configuration of causes. In practice we are interested in particular restricted classes of local interaction models, using just enough synergy parameters to represent certain effects of interest. The most restricted case with a single s is the situation described above. It is easier to work with the interaction function directly in terms of combinations of local interaction functions. Let s_i be the synergy parameter corresponding to the interaction function h_i. Then, for example, a three-way synergy could be $h_3(c_1, c_2, c_3) = s_3 \cdot (q_{c_1}^e + q_{c_2}^e + q_{c_3}^e)$, while e.g. $h(c_1, \ldots, c_6) = h_3(c_1, c_2, c_3) + h_3(c_4, c_5, c_6)$.

Since arbitrary synergy interaction functions can be constructed from local interaction functions, it is also possible to combine synergistic and anti-synergistic local interactions. In order to specify synergistic and anti-synergistic effects that involve the same cause, a local interaction h_1 can be used to apply a particular synergy parameter to a single cause. This is illustrated in an example in Section 4.4.

The idea of a noisy-OR synergy easily generalises to multi-valued variables. Consider n parent variables where variable i has r_i possible values. The number of parameters without CI is then $\prod_{i=1}^{n} r_i$. After introducing a noisy-OR this reduces to $\sum_{i=1}^{n} r_i$. If we then introduce synergy parameters s_j, with j the number of variables not in their distinguished state, which implies that the values except the distinguished state do not influence the amount of synergy, the number of parameters is $(n-1) + \sum_{i=1}^{n} r_i$. Allowing independent anti-synergy would add another $n-1$ parameters, which is still linear in the number of variables and values. However, a single parameter for synergy and anti-synergy can also be defined by taking $s^- = 1/s^+$, leading to a further reduction in parameters.

4.4 Drug Interaction Example

Consider a patient with an infection who has to be treated with antibiotics. Different antibiotics each target particular types of bacteria; however, there might be interaction effects between the different drugs. Say we have three drugs, $\{A, B, C\}$. A and B have a synergistic effect, because when both are present it is more likely that the infection is removed than when used separately. C targets a different type of bacteria and therefore does not have a positive synergistic effect; the effect is independent of A. Unfortunately, the presence of B and

C results in an adverse chemical reaction that reduces the effectiveness of both drugs. Now assume that we have three variables corresponding to the presence of drugs A, B, C and an effect variable E that models the infection, all binary for ease of exposition. The rate parameters for E making a transition $e \to \bar{e}$ – from infection to no infection – can be specified by the interaction function h as follows:

$$h(\bar{a}, \bar{b}, \bar{c}) = 0 \qquad\qquad h(a, \bar{b}, \bar{c}) = q_a$$
$$h(\bar{a}, \bar{b}, c) = q_c \qquad\qquad h(a, \bar{b}, c) = h_2^0(q_a, q_c)$$
$$h(\bar{a}, b, \bar{c}) = q_b \qquad\qquad h(a, b, \bar{c}) = h_2^+(q_a, q_b)$$
$$h(\bar{a}, b, c) = h_2^-(q_b, q_c) \quad h(a, b, c) = h_2^0(h_2^+(q_a, h_1^-(q_b)), h_1^-(q_c))$$

where h^+, h^- are synergistic and anti-synergistic local interaction functions respectively, and h^0 a zero synergy, i.e. a noisy-OR. Note that here we have a synergy and anti-synergy on B, which only has an effect if $s^+ \neq s^-$. The specification in terms of local interaction functions is not unique as $h(a, b, c) = h_2^0(h_1^+(q_a), h_2^-(h_1^+(q_b), q_c))$, however, it results in simple linear statements for the parameters of E that are straightforward to implement. For example, $h(a, b, c) = s^+ q_a + s^+ s^- q_b + s^- q_c$.

Now consider the situation where an infection is observed at $t = 0$. Antibiotic A is applied at $t = 0.5$; B is added at $t = 1$ and C at $t = 2$. The effect of the synergy parameters is visualised for this situation in Fig. 2 for different values of s^+ and s^-. Observe that the positive synergy decreases the probability of an infection, the anti-synergy increases the probability and the combination results in the expected intermediate scenario.

5 Related Work

General causal independence models for Bayesian networks were first described in [7] and further refined in [9]. It discusses the noisy-OR, noisy-MAX, and a noisy-ADD for discrete variables as special cases. The last paper also considers a continuous version of causal independence, namely the linear Gaussian model.

For general Bayesian networks, there have been two approaches to exploit causal independence for speeding up inference by changing the network structure, namely the *parent-divorcing method* [16] and the *temporal transformation method* [7]. Other approaches use the insight that efficient probabilistic inference is made possible by working with a factorisation of the joint probability distribution, rather than working with the joint probability distribution itself. As causal independence models allow decomposition of the probability distribution beyond the factorisation implied by the conditional independences derived from the associated graph, this insight can be exploited in algorithms that work with these factorisations directly such as variable elimination [23]. The results in this paper suggests that similar principles apply for CTBNs.

Right from the beginning, causal-independence models were not only used to improve probabilistic inference (by approximating the actual model), but also to

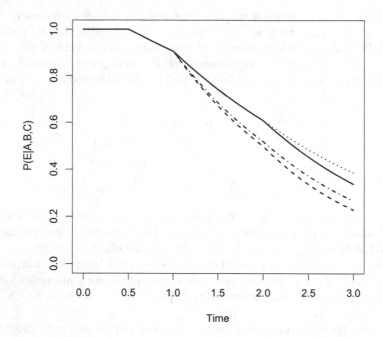

Fig. 2. Probability of infection given the presence of drugs A, B, C from $t = 0.5$, $t = 1$, $t = 2$ respectively, for $(s^+, s^-) = (1, 1)$ (solid line); $(s^+, s^-) = (3/2, 1)$ (dashed line); $(s^+, s^-) = (1, 2/3)$ (dotted line); $(s^+, s^-) = (7/5, 5/6)$ (dash-dot line)

facilitate the manual construction of Bayesian networks, as the number of parameters in, e.g., noisy-OR models that has to be estimated is proportional rather than exponential in the number of parents [10,16,15]. This is the main reason why causal-independence models are considered as important canonical models for knowledge engineering. Particularly related are a number of approaches specifically designed to model *undermining* effects, where the presence of more causes decreases the probability of the effect. Undermining can be modelled with anti-synergies, however, anti-synergy and undermining are not equivalent, cf. [20]. In noisy-OR models this undermining cannot be represented, because the causes collectively increase the probability of the effect. In general causal independence, however, this can be modelled by a Boolean expression that incorporates negation. The recursive noisy-OR model [11] is an approach to represent positive and negative influences, but these cannot be combined within the same model. A more general approach related to this work is the non-impeding noisy-AND tree (NIN-AND) [20], which can be seen as a noisy-AND with negations. A similar approach is [13], where gates are modelled by a conjunctive normal form.

Related work in context of CTBNs is done by Weiss et al. [18], where a different parameterisation is given for CTBNs which allows arbitrary parameter tying between states. Although this approach is very general, the parameter tying is not related to the graph structure, and generally no interpretation of such models can be given.

6 Conclusions

CTBNs are a widely applicable class of graphical models to reason about temporal processes, with applications ranging from social network dynamics [4], intrusion detection [21], and medicine [6]. However, similar to Bayesian networks, the number of parameters grows exponentially with the maximal in-degree of the graph. In this paper we have studied whether causal independence models can be applied to CTBNs to overcome this problem.

This paper shows that noisy-OR models are a particularly suitable framework for defining parameters in CTBNs. Similar to Bayesian network noisy-OR models, this leads to exponential savings in the number of parameters and can lead to exponential savings in inference time when computing transition probabilities. As a second contribution, we have introduced synergy models that allow a limited kind of interaction between different causes, while remaining linear in the number of parameters. In particular, we introduced the noisy-OR synergy model which allows the modelling of synergistic relationships between parents on their effect explicitly, where synergy here has a natural interpretation derived from the qualitative reasoning literature [19].

In conclusion, the paper shows that causal independence models are a useful tool to simplify model construction, not only in static probabilistic graphical models, or discrete-time temporal models, but also in models with continuous time. An open question is which noisy-OR synergy models can be used to speed up inference. In the noisy-OR case, the speedup is related to the decomposability of the local interaction function; however, it appears most noisy-OR synergy models are not decomposable. In future research, we will further study the relationship between properties of the interaction function and tractable inference.

References

1. Arroyo-Figueroa, G., Sucar, L.: A temporal Bayesian network for diagnosis and prediction. In: Laskey, K., Prade, H. (eds.) UAI 1999: Proceedings of the 15th Conference on Uncertainty in Artificial Intelligence, pp. 13–20. Morgan Kaufmann, San Francisco (1999)
2. Díez, F.J., Druzdzel, M.J.: Canonical probabilistic models for knowledge engineering. Technical Report CISIAD-06-01, UNED, Madrid, Spain (2006)
3. Díez, F.: Parameter adjustment in Bayes networks: the generalized noisy OR-gate. In: UAI 1993: Proceedings of the 9th Conference on Uncertainty in Artificial Intelligence, pp. 99–105 (1993)
4. Fan, Y., Shelton, C.R.: Learning continuous-time social network dynamics. In: UAI 1993: Proceedings of the Twenty-Fifth Conference on Uncertainty in Artificial Intelligence, pp. 161–168. AUAI Press (2009)
5. Galán, S.F., Díez, F.J.: Networks of probabilistic events in discrete time. International Journal of Approximate Reasoning 30(3), 181–202 (2002)

6. Gatti, E., Luciani, D., Stella, F.: A continuous time bayesian network model for cardiogenic heart failure. Flexible Services and Manufacturing Journal 24(4), 496–515 (2012)
7. Heckerman, D.: Causal independence for knowledge acquisition and inference. In: UAI 1993: Proceedings of the 9th Conference on Uncertainty in Artificial Intelligence (1993)
8. Heckerman, D., Breese, J.: Causal independence for probability assessment and inference using Bayesian networks. IEEE Transactions on Systems, Man and Cybernetics 26, 826–831 (1996)
9. Heckerman, D., Breese, J.: A new look at causal independence. In: UAI 1994: Proceedings of the 10th Conference on Uncertainty in Artificial Intelligence, pp. 286–292 (1994)
10. Henrion, M.: Practical issues in constructing a Bayes belief network. In: Lemmer, J., Levitt, T., Kanal, L. (eds.) UAI 1987: Proceedings of the 3rd Conference on Uncertainty in Artificial Intelligence, pp. 132–139 (1987)
11. Lemmer, J., Gossink, D.: Recursive noisy OR-A rule for estimating complex probabilistic causal interactions. IEEE Transactions on Systems, Man and Cybernetics - Part B: Cybernetics 34(6), 2252–2261 (2004)
12. Lucas, P.J.: Bayesian network modelling through qualitative patterns. Artificial Intelligence 163(2), 233–263 (2005)
13. Maaskant, P., Druzdzel, M.: An independence of causal interactions model for opposing influences. In: PGM 2008: Proceedings of the 4th European Workshop on Probabilistic Graphical Models, pp. 185–192 (2008)
14. Nodelman, U., Shelton, C., Koller, D.: Continuous-time Bayesian networks. In: UAI 2002: Proceedings of the 18th Conference on Uncertainty in Artificial Intelligence (2002)
15. Olesen, K., Andreassen, S.: Specification of models in large expert systems based on causal probabilistic networks. Artificial Intelligence in Medicine 5(3), 269–281 (1993)
16. Olesen, K., Kjærulff, U., Jensen, F., Jensen, F., Falck, B., Andreassen, S., Andersen, S.: A MUNIN network for the median nerve - a case study on loops. Applied Artificial Intelligence 3(2-3), 385–403 (1989)
17. Pearl, J.: Probabilistic reasoning in intelligent systems: networks of plausible inference. Morgan Kaufmann Publishers Inc., San Francisco (1988)
18. Weiss, J., Natarajan, S., Page, D.: Multiplicative forests for continuous-time processes. In: NIPS 2012: Proceedings of the Conference on Neural Information Processing Systems, pp. 467–475 (2012)
19. Wellman, M.: Fundamental concepts of qualitative probabilistic networks. Artificial Intelligence 44(3), 257–303 (1990)
20. Xiang, Y., Jia, N.: Modeling causal reinforcement and undermining for efficient CPT elicitation. IEEE Trans. on Knowl. and Data Eng. 19(12), 1708–1718 (2007)
21. Xu, J., Shelton, C.R.: Intrusion detection using continuous time bayesian networks. Journal of Artificial Intelligence Research 39(1), 745–774 (2010)
22. Zagorecki, A., Druzdzel, M.: Probabilistic independence of causal influence. In: PGM 2006: Proceedings of the 3rd European Workshop on Probabilistic Graphical Models, pp. 325–332 (2006)
23. Zhang, N., Poole, D.: Exploiting causal independence in Bayesian network inference. Journal of Artificial Intelligence Research 5, 301–328 (1996)

Expressive Power of Binary Relevance and Chain Classifiers Based on Bayesian Networks for Multi-label Classification

Gherardo Varando, Concha Bielza, and Pedro Larrañaga

Departamento de Inteligencia Artificial,
Universidad Politécnica de Madrid,
Campus de Montegancedo, 28660 Boadilla del Monte, Madrid, Spain
gherardo.varando@upm.es, {mcbielza,pedro.larranaga}@fi.upm.es
http://cig.fi.upm.es

Abstract. Bayesian network classifiers are widely used in machine learning because they intuitively represent causal relations. Multi-label classification problems require each instance to be assigned a subset of a defined set of h labels. This problem is equivalent to finding a multi-valued decision function that predicts a vector of h binary classes. In this paper we obtain the decision boundaries of two widely used Bayesian network approaches for building multi-label classifiers: Multi-label Bayesian network classifiers built using the *binary relevance method* and Bayesian network *chain classifiers*. We extend our previous single-label results to multi-label chain classifiers, and we prove that, as expected, chain classifiers provide a more expressive model than the binary relevance method.

Keywords: Bayesian network classifier, multi-label classification, expressive power, chain classifier, binary relevance.

1 Introduction

We consider a multi-label classification problem [19] over categorical predictors, that is, mapping every instance $\mathbf{x} = (x_1, \ldots, x_n)$ to a subset of h labels:

$$\Omega_1 \times \cdots \times \Omega_n \to Y \subset \mathcal{Y} = \{y_1, \ldots, y_h\},$$

where $\Omega_i \subset \mathbb{R}$, $|\Omega_i| = m_i < \infty$. This could be transformed into a multi-dimensional binary classification problem, that is, finding an h-valued decision function $\mathbf{f}$ that maps every instance of n predictor variables $\mathbf{x}$ to a vector of h binary values $\mathbf{c} = (c_1, \ldots, c_h) \in \{-1, +1\}^h$:

$$\mathbf{f}: \quad \Omega = \Omega_1 \times \cdots \times \Omega_n \to \{-1, +1\}^h$$

$$(x_1, \ldots, x_n) \quad \mapsto (c_1, \ldots, c_h),$$

where $c_i = +1$ (-1) means that the ith label is present (absent) in the predicted label subset. Moreover, we consider the predictor variables $X_1, \ldots, X_n$ and the

L.C. van der Gaag and A.J. Feelders (Eds.): PGM 2014, LNAI 8754, pp. 519–534, 2014.
© Springer International Publishing Switzerland 2014

binary classes $C_i \in \{-1, +1\}$ as categorical random variables. Real examples include classification of texts into different categories by counting selected words, diagnosis of multiple diseases from common symptoms and identification of multiple biological gene functions.

The simplest method to build a multi-label classifier is to consider h single-label binary classifiers, one for each class variable C_i. Each classifier f_i is learned from predictor variables and C_i data, and the results are combined to form multi-label prediction. This method, called *binary relevance* [6], is easily implementable, has low computational complexity and is fully parallelizable. Hence it is scalable to a large number of classes. However, it completely ignores dependencies among labels and generally it does not represent the most likely set of labels.

Chain classifiers [14] relax the independence assumption by iteratively adding class dependencies in the binary relevance scheme, that is, the kth classifier in the chain predicts class C_k from $X_1, \ldots, X_n, C_1, \ldots, C_{k-1}$.

We study differences in the expressive power of these two methods when Bayesian network (BN) classifiers [1] are used. Sucar *et al.* [15] employed naive Bayes within chain classifiers. We use the results on the decision boundaries and expressive power of one-dimensional BN classifiers. (a) For naive Bayes classifiers, Minsky [9] proved that the decision boundaries are hyperplanes if binary predictors are used. (b) Peot [11] observed that Minsky's results could be extended to categorical predictors. (c) Recently, we have developed a method [18] to compute decision boundaries for a broad class of BN classifiers. In this paper we extend these results to multi-label classifiers. Moreover, we suggest some theoretical reasons why the binary relevance method performs poorly and prove that chain classifiers provide more expressive models.

The paper is organized as follows. In Sect. 2 we give some definitions and report our results on one-label classifiers. We describe the binary relevance method in Sect. 3 and chain classifiers in Sect. 4. In Sect. 5 we compare the decision boundaries, and expressive power of the two methods. In Sect. 6 we present our conclusions and some ideas for future research.

2   Expressive Power of One-Dimensional BN Classifiers

We first report some results on the decision boundary and expressive power of one-label, or equivalently one-dimensional binary, BN classifiers [18]. In particular we look at Bayesian network-augmented naive Bayes (BAN) classifiers [5].

BAN classifiers are Bayesian network classifiers where the class variable C is assumed to be a parent of every predictor and the predictor sub-graph can be a general BN. The decision function induced by the BAN classifier is

$$f_{\mathcal{G}}^{BAN}(x_1, \ldots, x_n) = \arg \max_{c \in \{-1, +1\}} P(C = c, X_1 = x_1, \ldots, X_n = x_n),$$

where $P(C = c, X_1 = x_1, \ldots, X_n = x_n)$ could be factorized according to BN theory [10] as

$$P(C = c) \prod_{i=1}^{n} P\left(X_i = x_i | C = c, \mathbf{X}_{\mathbf{pa}(i)} = \mathbf{x}_{\mathbf{pa}(i)}\right),$$

where $\mathbf{X}_{\mathbf{pa}(i)}$ stands for the vector of parents of X_i in the predictor sub-graph $\mathcal{G}$. Moreover, $\mathbf{pa}(i)$ denotes the set of indexes defining the parents of X_i that are not C and $\mathbb{M}_i = \times_{s \in \mathbf{pa}(i)} \{1, \ldots, m_s\}$, the set of possible configurations of the parents of X_i.

Let us recall that the sign function $sgn(t)$ is defined as

$$sgn(t) = \begin{cases} +1 & \text{if } t > 0 \\ 0 & \text{if } t = 0 \\ -1 & \text{if } t < 0. \end{cases}$$

We define [18]:

Definition 1. *Given a decision function* $f : \Omega \to \{-1, +1\}$, *where* $\Omega \subset \mathbb{R}^n$, $|\Omega| < \infty$ *and* $r : \mathbb{R}^n \mapsto \mathbb{R}$ *is a polynomial, we say that* r *sign-represents* f *if*

$$f(\mathbf{x}) = sgn(r(\mathbf{x})) \text{ for every } \mathbf{x} \in \Omega.$$

Moreover, given a set of polynomials $\mathcal{P}$, *we denote by* $sgn(\mathcal{P})$ *the set of decision functions that are sign-representable by polynomials in* $\mathcal{P}$ *and by* $\{-1, +1\}^{|\Omega|}$, *the set of all* $2^{|\Omega|}$ *decision functions over* Ω.

Example 1. We consider $\Omega = \{0, 2\} \times \{-3, 1\}$ and the decision function over Ω

$$f(x_1, x_2) = \begin{cases} +1 & \text{if } (x_1, x_2) = (0, -3), (2, -3), (0, 1) \\ -1 & \text{if } (x_1, x_2) = (2, 1). \end{cases}$$

We have that the polynomial $r(x_1, x_2) = -x_1^2 - x_2 + 3$ sign-represents f over Ω, precisely:

$$r(0, -3) = 6 > 0, \quad r(2, -3) = 2 > 0, \quad r(0, 1) = 2 > 0 \text{ and } r(2, 1) = -2 < 0.$$

Next let us recall the definition of the Vapnik-Chervonenkis (VC) dimension [17].

Definition 2. *Given a subset of decision functions* $\mathcal{F} \subset \{-1, +1\}^{|\Omega|}$, *we say that* $\mathcal{F}$ *shatters* $\Omega_0 \subset \Omega$ *if for every* $g \in \{-1, +1\}^{|\Omega_0|}$ *there exists a decision function* $f \in \mathcal{F}$ *such that* $f_{|\Omega_0} = g$, *where* $f_{|\Omega_0}$ *indicates the restriction of* f *over* Ω_0.

That is, $\mathcal{F}$ shatters Ω_0 if every decision over Ω_0 is representable by some elements of $\mathcal{F}$. The cardinality of the largest subset shattered by $\mathcal{F}$ is called the VC dimension of $\mathcal{F}$. It indicates the maximum number of points that can be discriminated by $\mathcal{F}$.

Definition 3. *The VC dimension of* $\mathcal{F} \subset \{-1, +1\}^{|\Omega|}$, *denoted by* $d_{VC}(\mathcal{F})$, *is defined by*

$$d_{VC}(\mathcal{F}) = \max\{|\Omega_0| \ s.t. \ \Omega_0 \ is \ shattered \ by \ \mathcal{F}\}.$$

For every predictor variable $X_i \in \Omega_i = \{\xi_i^1, \ldots, \xi_i^{m_i}\}$, we define the Lagrange basis polynomials over Ω_i

$$\ell_j^{\Omega_i}(x) = \prod_{k \neq j} \frac{(x - \xi_i^k)}{(\xi_i^j - \xi_i^k)} \ \text{for every } j = 1, \ldots, m_i \text{ and } x \in \mathbb{R}. \tag{1}$$

Then we have [18]:

Lemma 1. *If f is the decision function induced by a BAN classifier for a classification problem with n categorical predictor variables $\{X_i \in \Omega_i \subset \mathbb{R}, |\Omega_i| = m_i\}_{i=1}^n$, then there exists a polynomial of the form*

$$\sum_{i=1}^n \sum_{j=1}^{m_i} \ell_j^{\Omega_i}(x_i) \sum_{\mathbf{k} \in \mathbb{M}_i} \beta_i(j|\mathbf{k}) \prod_{s \in \mathbf{pa}(i)} \ell_{k_s}^{\Omega_s}(x_s)$$

that sign-represents f, where we write $\sum_{\mathbf{k} \in \mathbb{M}_i} \beta_i(j|\mathbf{k}) \prod_{s \in \mathbf{pa}(i)} \ell_{k_s}^{\Omega_s}(x_s) = \beta_i(j)$ when a variable does not have parents different from C, that is, $\mathbf{pa}(i) = \emptyset$.

The proof of Lemma 1 [18] is constructive and the coefficients $\beta_i(j|\mathbf{k})$ of the built polynomial are related to the conditional probability tables of the BAN. Precisely we have that

$$\beta_i(j|\mathbf{k}) = \ln \frac{P(X_i = \xi_i^j | X_s(i) = \xi_s^{k_s}, \forall s \in \mathbf{pa}(i), C = +1)}{P(X_i = \xi_i^j | X_s(i) = \xi_s^{k_s}, \forall s \in \mathbf{pa}(i), C = -1)}, \tag{2}$$

where $\mathbf{k} = (k_s)_{s \in \mathbf{pa}(i)}$, $k_s \in \{1, \ldots, m_s\}$.

When the predictor sub-graph $\mathcal{G}$ does not contain V-structures, the inverse implication of Lemma 1 is provable and thus the following theorem [18] holds.

Theorem 1. *Let $\mathcal{G}$ be a directed acyclic graph with nodes X_i for $i \in \{1, 2, \ldots, n\}$ and f, a decision function over predictor variables $X_i \in \Omega_i = \{\xi_i^1, \ldots, \xi_i^{m_i}\}$. Suppose that $\mathcal{G}$ does not contain V-structures, then we have that f is sign-represented by a polynomial of the form*

$$r(\mathbf{x}) = \sum_{i=1}^n \sum_{j=1}^{m_i} \ell_j^{\Omega_i}(x_i) \sum_{\mathbf{k} \in \mathbb{M}_i} \beta_i(j|\mathbf{k}) \prod_{s \in \mathbf{pa}(i)} \ell_{k_s}^{\Omega_s}(x_s),$$

if and only if f is induced by a BAN classifier whose predictor sub-graph is $\mathcal{G}$.

The above result applies in a lot of practical cases as naive Bayes (NB) classifier [9], tree augmented naive Bayes (TAN) classifier [5] and super-parent one-dependence-estimator (SPODE) classifier [8], because the corresponding predictor sub-graphs do not contain V-structures. Moreover, Theorem 1 implies that when $\mathcal{G}$ does not contain V-structures the family of polynomials

$$\mathcal{P}_{\mathcal{G}} = \left\{ r(\mathbf{x}) = \sum_{i=1}^{n} \sum_{j=1}^{m_i} \ell_j^{\Omega_i}(x_i) \sum_{\mathbf{k} \in \mathbb{M}_i} \beta_i(j|\mathbf{k}) \prod_{s \in \mathbf{pa}(i)} \ell_{k_s}^{\Omega_s}(x_s) \text{ s.t. } \beta_i(j|\mathbf{k}) \in \mathbb{R} \right\}$$

(3)

completely represents the set of decision functions induced by BAN classifiers, that is, $sgn(\mathcal{P}_{\mathcal{G}})$ is exactly the set of decision functions induced by BAN classifiers whose predictor sub-graph is $\mathcal{G}$.

Remark 1. In the simplest NB classifier case, that is, when the predictor subgraph $\mathcal{G}$ is an empty graph, we have that

$$\mathcal{P}_{\mathcal{G}} \equiv \mathcal{P}_{NB} = \left\{ r(\mathbf{x}) = \sum_{i=1}^{n} \sum_{j=1}^{m_i} \alpha_i(j) \ell_j^{\Omega_i}(x_i) \text{ s.t. } \alpha_i(j) \in \mathbb{R} \right\}$$

is exactly the set of polynomials that sign-represent the decision function induced by NB classifiers.

We can prove that the set $\mathcal{P}_{\mathcal{G}}$ is a vector space of dimension

$$d = \sum_{i=1}^{n} \left((m_i - 1) \prod_{s \in \mathbf{pa}(i)} m_s \right) + 1$$

and that the VC dimension of $sgn(\mathcal{P}_{\mathcal{G}})$ is precisely d. Theorem 1 also places an upper bound on the number of decision functions representable by BAN classifiers without V-structures [18].

Corollary 1. *Consider a BAN classifier over predictor variables $X_i \in \Omega_i$, $|\Omega_i| = m_i$ for every $i = 1, \ldots, n$. Moreover suppose that the predictor sub-graph $\mathcal{G}$ does not contain V-structures. Then we have*

$$|sgn(\mathcal{P}_{\mathcal{G}}^{BAN})| \leq C(M, d) = 2 \sum_{k=0}^{d-1} \binom{M-1}{k},$$

where $d = \sum_{i=1}^{n} \left((m_i - 1) \prod_{s \in \mathbf{pa}(i)} m_s \right) + 1$ and $M = \prod_{i=1}^{n} m_i$.

Remark 2. If $\Omega = \Omega_1 \times \cdots \times \Omega_n$, we observe that $|\{-1, +1\}^{|\Omega|}| = 2^{|\Omega|} = 2^M$. Thus Corollary 1 implies that in the case of the NB classifier the quotient of decision functions representable by NB classifiers over 2^M becomes vanishingly small as the number of predictors increase. Figure 1 shows the number of total decision functions $(2^{|\Omega|})$ and the bounding of Corollary 1 for NB classifiers with n binary predictors, $C(M, d)$. Observing that the scale of the graph is logarithmic, the graph shows that the number of decision functions induced by NB classifiers is *small* compared with all possible decision functions over Ω.

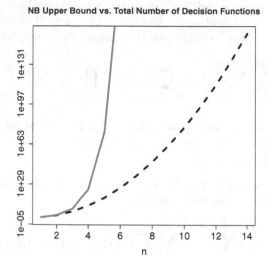

NB Upper Bound vs. Total Number of Decision Functions

Fig. 1. Total number of decision functions over n binary predictors (gray) and the bounding $C(M, d)$ of Corollary 1 (dashed black) for NB classifiers

Remark 2 could be extended to every type of BAN classifier, such that for every variable the number of parents is bounded (Corollary 17 in Varando *et al.* [18]), that is, $|\mathbf{pa}(i)| < K$.

Remark 3. When the predictor sub-graph $\mathcal{G}$ of a BAN classifiers contains V-structures, Lemma 1 is still valid and exists a polynomial that sign-represents the induced decision function. The problem is that the associated family of polynomials is not a linear space as in (3), thus is not possible to employ the same techniques as in Varando *et al.* [18] to prove the bounding in Corollary 1.

3   Binary Relevance Method

We consider the binary relevance method with BAN classifiers, that is, for every class C_i we build a BAN classifier with predictor sub-graph $\mathcal{G}$. Thus every one-dimensional classifier has the same predictor structure and differs with respect to the values of the conditional probability tables that define the BAN models. From a practical point of view, the advantages of this method are that the structure of the predictor sub-graph has only to be learned once and the parameters of the BN are then fitted to the different data sets related to each class.

From Lemma 1 it follows that if $\mathbf{f} = (f_1(\mathbf{x}), f_2(\mathbf{x}), \ldots, f_h(\mathbf{x}))$ is the multi-valued decision function induced by the h BAN classifiers, then there exist

$$p_1(\mathbf{x}), \ldots, p_h(\mathbf{x}) \in \mathcal{P}_{\mathcal{G}},$$

such that $f_i(\mathbf{x}) = sgn(p_i(\mathbf{x}))$ for every $i \in \{1, \ldots, h\}$. Thus, in Lemma 2, we bound the number of multi-valued decision functions representable by the BAN binary relevance method, when the predictor sub-graph does not contain V-structures.

Lemma 2. *Consider h BAN classifiers, whose predictor sub-graph $\mathcal{G}$ contains no V-structures, to predict h binary classes. We have that $N(\mathcal{G}, h)$, the number of h-valued decision functions representable by the BAN binary relevance method, satisfies*

$$N(\mathcal{G}, h) \leq C(M, d)^h,$$

where $C(M, d) = 2 \sum_{k=0}^{d-1} \binom{M-1}{k}$, $d = \sum_{i=1}^{n} \left((m_i - 1) \prod_{s \in \mathbf{pa}(i)} m_s \right) + 1$ and $M = \prod_{i=1}^{n} m_i$.

Proof. The proof is a straightforward application of Corollary 1.   □

Remark 4. The total number of h-valued decision functions over n categorical predictors is $2^{h \prod m_i} = 2^{hM}$. Then the fraction of h-valued decision functions representable by the BAN binary relevance method is bounded by

$$\frac{N(\mathcal{G}, h)}{2^{hM}} \leq \left(\frac{C(M, d)}{2^M} \right)^h.$$

Thus, as in Remark 2, we have that if we fix the structure of the predictor sub-graph, and it does not contain V-structures, the number of representable multi-valued decision functions becomes vanishingly small as the number of predictors increase. Moreover, using the binary relevance method, the *speed* at which the ratio between representable multi-valued decision functions and the total number of multi-valued decision functions drops to zero, is exponential in h, the number of classes.

The above bound could also be computed when each of the h BAN classifiers is built with different structures, that is, the kth classifier to predict class C_k is a BAN classifier whose predictor sub-graph $\mathcal{G}_k$ does not contain V-structures. Then if we denote $N(\mathcal{G}_1, \ldots, \mathcal{G}_h)$ the number of h-valued decision functions built with h BAN classifiers whose predictor sub-graph is $\mathcal{G}_1, \ldots, \mathcal{G}_h$ respectively, we have that

$$N(\mathcal{G}_1, \ldots, \mathcal{G}_h) \leq \prod_{k=1}^{h} C(M, d_k),$$

where $d_k = \sum_{i=1}^{n} \left((m_i - 1) \prod_{s \in \mathbf{pa}_k(i)} m_s \right) + 1$, $\mathbf{pa}_k(i)$ is the set of X_i parents in $\mathcal{G}_k$ and $M = \prod_{i=1}^{n} m_i$.

Example 2. We consider two binary classes C_1, C_2 and two predictor variables $X_1 \in \{0, 1\}$ and $X_2 \in \{2, 3, 4\}$. Using the binary relevance method we build two independent NB classifiers, see Fig. 2.

Next, we list the conditional probability tables for both classifiers (Tables 1a and 1b).

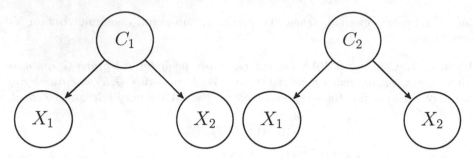

Fig. 2. Two NB classifiers in Example 2

Table 1. Conditional probability tables in Example 2 for the two NB classifiers

(a) NB for C_1

X_1	$C_1 = +1$	$C_1 = -1$
0	0.5	0.25
1	0.5	0.75

X_2	$C_1 = +1$	$C_1 = -1$
2	0.3	0.1
3	0.5	0.7
4	0.2	0.2

(b) NB for C_2

X_1	$C_2 = +1$	$C_2 = -1$
0	0.7	0.4
1	0.3	0.6

X_2	$C_2 = +1$	$C_2 = -1$
2	0.1	0.6
3	0.1	0.2
4	0.8	0.2

From the representation of Theorem 1 we have that there exist two polynomials p_1, p_2 that sign-represent the decision functions induced by the two NB classifiers

$$p_1(x_1, x_2) = \ln\left(\frac{0.5}{0.25}\right)\frac{x_1 - 1}{-1} + \ln\left(\frac{0.5}{0.75}\right)\frac{x_1}{1}$$
$$+ \ln\left(\frac{0.3}{0.1}\right)\frac{(x_2 - 3)(x_2 - 4)}{2} + \ln\left(\frac{0.5}{0.7}\right)\frac{(x_2 - 2)(x_2 - 4)}{-1}$$
$$+ \ln\left(\frac{0.2}{0.2}\right)\frac{(x_2 - 2)(x_2 - 3)}{2}$$

and

$$p_2(x_1, x_2) = \ln\left(\frac{0.7}{0.4}\right)\frac{x_1 - 1}{-1} + \ln\left(\frac{0.3}{0.6}\right)\frac{x_1}{1}$$
$$+ \ln\left(\frac{0.1}{0.6}\right)\frac{(x_2 - 3)(x_2 - 4)}{2} + \ln\left(\frac{0.1}{0.2}\right)\frac{(x_2 - 2)(x_2 - 4)}{-1}$$
$$+ \ln\left(\frac{0.8}{0.2}\right)\frac{(x_2 - 2)(x_2 - 3)}{2}.$$

We have that

$$\mathbf{f}(\mathbf{x}) = \Big(sgn\big(p_1(\mathbf{x})\big), sgn\big(p_2(\mathbf{x})\big) \Big)$$

is the bi-valued decision function that predicts C_1, C_2 from X_1, X_2. Figure 3 shows the decision boundaries of the two classifiers (black for C_1 and gray for C_2). We observe that the predictor space $\Omega = \{0,1\} \times \{2,3,4\}$ is partitioned into four subsets corresponding to the four different predictions of the two binary classes. Moreover, the value of the respective predicted class changes when one of the decision boundaries is crossed.

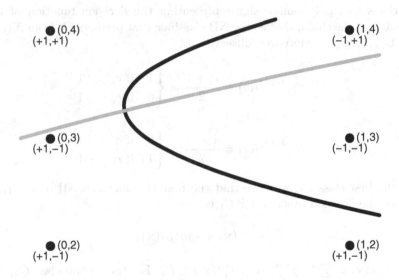

Fig. 3. Decision boundaries for the two NB classifiers in Example 2. The value of the predicted classes and the coordinates of the points are reported.

4   BN Chain Classifiers

The easiest way to relax the strong independence assumption of the binary relevance method is to gradually add the predicted classes to the predictors. Specifically, suppose that we have to predict h binary classes $C_1, \ldots, C_h$ from n predictor variables $X_1, \ldots, X_n$. We consider h BAN classifiers such that the kth BAN classifier predicts C_k from the variables

$$X_1, \ldots, X_n, C_1, \ldots, C_{k-1}.$$

From Lemma 1 we have that there exist h polynomials $p_1, \ldots, p_h$ such that

$$p_k(\mathbf{x}, c_1, \ldots, c_{k-1}) : \mathbb{R}^{n+k-1} \to \mathbb{R}$$

$$p_k \in \mathcal{P}_{\mathcal{G}_k},$$

where $\mathcal{G}_k$ is the predictor sub-graph related to the kth BAN classifier over $X_1, \ldots, X_n, C_1, \ldots, C_{k-1}$.

If we consider only naive Bayes classifiers, we state

$$\mathcal{P}_k = \left\{ r(\mathbf{x}) = \sum_{i=1}^{n} \sum_{j=1}^{m_i} \alpha_i(j) \ell_j^{\Omega_i}(x_i) + \sum_{i=1}^{k-1} \beta_i(+1) \ell_{+1}^{\{-1,+1\}}(c_i) + \beta_i(-1) \ell_{-1}^{\{-1,+1\}}(c_i) \atop \text{s.t. } \alpha_i(j), \beta_i(+1), \beta_i(-1) \in \mathbb{R} \right\},$$

$$(4)$$

for the set of polynomials sign-representing the decision function of the kth classifier in the chain, that is, the NB classifier that predicts C_k from $X_1, \ldots, X_n$ and $C_1, \ldots, C_{k-1}$. Moreover, observe that

$$\ell_{+1}^{\{-1,+1\}}(c_i) = \frac{c_i + 1}{2} = \begin{cases} 1 & \text{if } c_i = +1 \\ 0 & \text{if } c_i = -1 \end{cases}$$

$$\ell_{-1}^{\{-1,+1\}}(c_i) = \frac{1 - c_i}{2} = \begin{cases} 0 & \text{if } c_i = +1 \\ 1 & \text{if } c_i = -1 \end{cases}$$

For the first class C_1, we have that the first classifier is a NB over $X_1, \ldots, X_n$ and so the decision function for C_1 is

$$f_1(\mathbf{x}) = sgn\big(p_1(\mathbf{x})\big), \tag{5}$$

where $p_1(\mathbf{x}) = \sum_{i=1}^{n} \sum_{j=1}^{m_i} \alpha_i(j) \ell_j^{\Omega_i}(x_i) \in \mathcal{P}_1$. For the second class C_2, we have a NB classifier over $X_1, \ldots, X_n, C_1$. Thus $f_2(\mathbf{x})$, the decision function for C_2, is

$$f_2(\mathbf{x}) = sgn\Big(p_2\big(\mathbf{x}, c_1\big) \Big), \tag{6}$$

where $p_2 \in \mathcal{P}_2$ and $c_1 = f_1(\mathbf{x})$. Substituting (5) in (6), we obtain

$$f_2(\mathbf{x}) = sgn\Big(p_2\big(\mathbf{x}, sgn(p_1(\mathbf{x}))\big) \Big).$$

This chain classifier over two classes is equivalent to the bi-valued decision function

$$\mathbf{f} = \big(f_1(\mathbf{x}), f_2(\mathbf{x})\big).$$

Iterating the above computations, we have that the kth decision function that predicts class C_k is given by

$$f_k(\mathbf{x}) = sgn\Big(p_k\big(\mathbf{x}, f_1(\mathbf{x}), \ldots, f_{k-1}(\mathbf{x})\big) \Big),$$

where $p_k \in \mathcal{P}_k$. More explicitly, we have that

$$
f_k(\mathbf{x}) = \begin{cases} sgn\Big(q_k(\mathbf{x}) + \gamma(+1,+1,\ldots,+1)\Big) & \text{if } f_1(\mathbf{x}) = +1,\ldots,f_{k-1}(\mathbf{x}) = +1 \\ \quad\vdots & \qquad\vdots \\ sgn\Big(q_k(\mathbf{x}) + \gamma(\sigma_1,\sigma_2\ldots,\sigma_{k-1})\Big) & \text{if } f_1(\mathbf{x}) = \sigma_1,\ldots,f_{l-1}(\mathbf{x}) = \sigma_{k-1} \quad (7) \\ \quad\vdots & \qquad\vdots \\ sgn\Big(q_k(\mathbf{x}) + \gamma(-1,-1,\ldots,-1)\Big) & \text{if } f_1(\mathbf{x}) = -1,\ldots,f_{k-1}(\mathbf{x}) = -1 \end{cases}
$$

where $q_k(\mathbf{x}) \in \mathcal{P}_1$ and $\gamma(\sigma_1,\ldots,\sigma_{k-1}) \in \mathbb{R}$ for every $(\sigma_1,\ldots,\sigma_{k-1}) \in \{-1,+1\}^{k-1}$. In other words, the kth decision function, in every subset of Ω defined by the previous $k-1$ decision functions, is sign-represented by a polynomial in $\mathcal{P}_1$ or equivalently by a NB classifier over the original predictors. The only difference between these polynomials is the additive coefficients. Precisely the additive coefficients $\gamma(\sigma_1,\ldots,\sigma_{k-1})$ are obtained from the representation in (4) as follows:

$$
\gamma(\sigma_1,\ldots,\sigma_{k-1}) = \sum_{i=1}^{k-1} \beta_i(\sigma_i),
$$

where

$$
\beta_i(\sigma_i) = \ln \frac{P(C_i = \sigma_i | C_k = +1)}{P(C_i = \sigma_i | C_k = -1)}.
$$

Figure 4 shows two examples of decision boundaries of a NB chain classifier for two classes. The predictor domain in both examples is $\{0,1,2,3\} \times \{0,1,2,3\}$. We observe that the decision boundaries related to the second class in the chain C_2 (dashed black line) are dependent on the decision boundaries of the first class C_1 (gray line).

Remark 5. For simplicity's sake, we have presented the computation of the decision boundaries in the NB case. The same arguments as used above could be applied to a broader class of chain classifiers, specifically to every model where a BAN classifier with predictor sub-graph $\mathcal{G}_k$ is built in the kth step of the chain. If the previously predicted classes $C_1,\ldots,C_{k-1}$ are added in a *naive* way, that is, they have only one parent, C_k and they have no children, we have that the form of the kth decision function is similar to (7), where the previously predicted classes contribute in the form of additive constants.

Example 3. We use a chain NB classifier over the prediction problems of Example 2. The NB classifier for predicting class C_1 is the same as in Example 2 (see Fig. 2 left and Table 1a). The predictors of the NB classifier for predicting C_2 now include C_1. We consider the same conditional probability tables as in Example 2 (Tables 1a and 1b). Moreover we have to specify the conditional probabilities of C_1 given C_2 in the NB that predicts C_2. We set

$$
P(C_1 = +1 | C_2 = +1) = 0.3 \text{ and } P(C_1 = -1 | C_2 = +1) = 0.7
$$

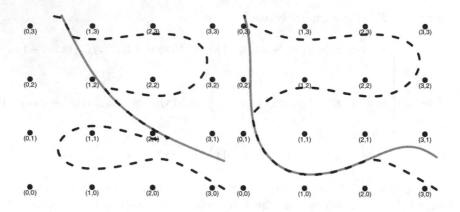

Fig. 4. Decision boundaries for NB chain classifiers with two predictor variables

$$P(C_1 = +1|C_2 = -1) = 0.9 \text{ and } P(C_1 = -1|C_2 = -1) = 0.1$$

And, thus, coefficients $\beta_1(+1)$ and $\beta_1(-1)$ as defined in (4) are given by

$$\beta_1(+1) = \ln\left(\frac{0.3}{0.9}\right) \text{ and } \beta_1(-1) = \ln\left(\frac{0.7}{0.1}\right).$$

We have that the decision function to predict C_2 is given by

$$f_2(x_1, x_2) = \begin{cases} sgn\bigg(p_2(x_1, x_2) + \beta_1(+1)\bigg) & \text{if } p_1(x_1, x_2) > 0 \\ sgn\bigg(p_2(x_1, x_2) + \beta_1(-1)\bigg) & \text{if } p_1(x_1, x_2) < 0 \end{cases}$$

where p_1 and p_2 are the polynomials defined in Example 2. The decision boundaries of the two classes are shown in Fig. 5. We observe that the two boundaries are no longer independent; the decision boundary for the second class C_2 (dashed black line) depends on the predicted value of the first class C_1.

5   Binary Relevance vs. Chain Classifier

We denote the set of multi-valued decision functions representable by a NB chain classifier over $X_1, \ldots, X_n$ and by a multiple independent NB classifiers built as in the binary relevance method by $\mathcal{F}$ and $\mathcal{D}$, respectively. We can prove the following lemma.

Lemma 3.
$$|\mathcal{F}| > |\mathcal{D}|.$$

In other words, NB chain classifiers are more expressive than the NB binary relevance method.

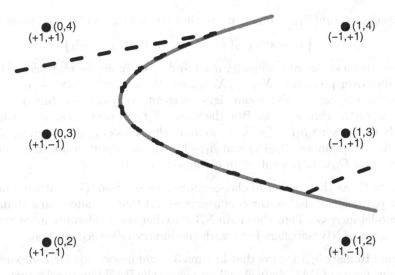

●(0,4)
(+1,+1)

●(1,4)
(−1,+1)

●(0,3)
(+1,−1)

●(1,3)
(−1,+1)

●(0,2)
(+1,−1)

●(1,2)
(+1,−1)

Fig. 5. Decision boundaries for the chain NB classifier in Example 3. The value of the predicted classes and the coordinates of the points are reported

Proof. We need only consider two class variables, since the result in the general case is proved analogously. If we define $\mathcal{P}_k$ for $k = 1, 2$ as in (4), we have that

$$\mathcal{D} = sgn(\mathcal{P}_1) \times sgn(\mathcal{P}_1) \subset sgn(\mathcal{P}_1) \times sgn(\mathcal{P}_2) = \mathcal{F}.$$

So, obviously, $|\mathcal{F}| \geq |\mathcal{D}|$. Thus to prove the lemma we just have to disprove the equality. Moreover, the VC dimension of $sgn(\mathcal{P}_1)$ (the cardinality of the maximum shattered subset) is equal to

$$d = \sum_{i=1}^{n} m_i - n + 1 < |\Omega| = \prod_{i=1}^{n} m_i.$$

Then, by the definition of VC dimension, there exists $\Omega_0 \subset \Omega$ such that $|\Omega_0| = d$ which is shattered by $sgn(\mathcal{P}_1)$. We now choose $\omega \in \Omega \setminus \Omega_0$ and find that there exists $p_0(\mathbf{x}) \in \mathcal{P}_1$ such that

$$p_0(\omega) < 0$$

and

$$p_0(\mathbf{x}) > 0 \text{ for every } \mathbf{x} \neq \omega.$$

Consider the bi-valued decision function $\mathbf{f} \in \mathcal{F}$ with the form

$$\mathbf{f} = \Big(sgn\big(p_0(\mathbf{x})\big), sgn\big(p_2(\mathbf{x}, sgn(p_0(\mathbf{x})))\big) \Big).$$

We observe from (4) that we have

$$p_2\big(\mathbf{x}, sgn(p_0(\mathbf{x}))\big) = \begin{cases} q(\mathbf{x}) + \beta_1(+1) & \text{if } p_0(\mathbf{x}) > 0 \\ q(\mathbf{x}) + \beta_1(-1) & \text{if } p_0(\mathbf{x}) < 0, \end{cases}$$

where $q(\mathbf{x}) \in sgn(\mathcal{P}_1)$. We now prove that the set of decision functions

$$\left\{ f_2 = sgn\left(p_2\left(\mathbf{x}, sgn(p_0(\mathbf{x}))\right)\right) \text{ s.t. } p_2 \in \mathcal{P}_2 \right\}$$

can shatter a subset of cardinality $d+1$ and thus cannot be represented by a NB classifier over predictors $X_1, \ldots, X_n$ alone. We have that $q(\mathbf{x}) + \beta_1(+1) \in \mathcal{P}_1$. Thus, by varying $q \in \mathcal{P}_1$, it can sign-represent every decision function over $\boldsymbol{\Omega}_0$ because of the choice of $\boldsymbol{\Omega}_0$. But the value of $f_2(\mathbf{x})$ over ω can be set independently by choosing $\beta_1(-1) \in \mathbb{R}$. So we have that choosing the polynomial $q \in \mathcal{P}_1$ and the real numbers $\beta_1(+1)$ and $\beta_1(-1)$, the defined decision functions $f_2(\mathbf{x})$ can shatter $\boldsymbol{\Omega}_0 \cup \{\omega\}$, a subset of cardinality $d+1$. $\qquad\square$

Remark 6. As the number of classes grows, we see from (7) that the number of extra parameters, that is, the coefficients $\gamma(\ldots)$ that are added in a chain classifier model increase. Thus the chain NB classifier is considerably more expressive than a set of NB classifiers built with the binary relevance method.

From Remark 5, it follows that Lemma 3 could be extended to compare the expressive power of BAN chain classifiers versus the BAN binary relevance method, proving that BAN chain classifiers are in general more expressive than classifiers built using binary relevance.

Moreover we observe that changing the order of classes in which the classifier is built implies a change in the expressive power of the resulting multi-label classifier. In fact we find that the first class in NB chain classifiers is predicted as in the binary relevance method, and from Lemma 3, we get that the chain classifier is more expressive than binary relevance over the second variable. In general it is possible to prove that if the chain classifier for classes $C_1, \ldots, C_h$, is built with the class ordering $j_1, \ldots, j_h$, we have that the kth classifier for C_{j_k} is more expressive than all the previous classifiers in the chain. So, by changing the order of the classes, we obtain a multi-label classifier with different expressive power. This last observation led us to formulate an easy expressiveness-based heuristic to select an ordering for the chain classifier. We built h classifiers, one for each class as in the binary relevance method. We sorted the classifiers according to some evaluation metric and we used the resulting order to build a chain classifier. Precisely we started with the classifier with the best prediction performance and we ended with the worst predicted classes. In other words, we tried to employ the more expressive classifiers in the chain for the classes that were predicted worst by the binary relevance model. Moreover, if the BAN chain classifier is built as suggested in Remark 5, that is, by adding the previously predicted classes in a naive way, we find that the above heuristic introduces a low computational complexity: once the binary relevance model is built we have only to compute the additive coefficient, corresponding to the previously predicted classes to build the chain classifier. In real problems, where the coefficient of the models have to be estimated, overfitting could be an issue, specially with a limited number of observations available. In those cases we have to check that the increased expressive power of the chain model does not increase the classification errors. This could be achieved estimating the errors with cross-validation techniques [7] or using structural risk minimization [16].

6   Conclusions and Future Work

In this paper we have extended our previous results on the decision boundaries and expressive power of one-label BN classifiers to two types of BN multi-label classifiers: BAN classifiers built with binary relevance method and BAN chain classifiers. We have given theoretical grounds for why the binary relevance method provides models with poor expressive power and why this gets worst for larger numbers of classes. In both models we have expressed the multi-label decision boundaries in polynomial forms, and we have proved that chain classifiers provide more expressive models than the binary relevance method when the same type of BAN classifier is used as the base classifier.

As possible future research, we would like to extend our results to general multi-dimensional BN classifiers [4,12,2,13]. Multi-dimensional BN classifiers permit BN structures between classes and predictors, and so the multi-valued decision functions have to be found by a global maximum search over the possible class values. This fact does not permit to employ the same arguments used in this work. *Class-Bridge* decomposable multi-dimensional BN classifiers [2,3] could be easier to study due to the factorization of the maximization problem into a number of maximization problems in lower dimensional spaces.

Acknowledgments. This work has been partially supported by the Spanish Ministry of Economy and Competitiveness through the Cajal Blue Brain (C080020-09) and TIN2013-41592-P projects.

References

1. Bielza, C., Larrañaga, P.: Discrete Bayesian network classifier: A survey. ACM Computing Surveys 47(1) (in press, 2015)
2. Bielza, C., Li, G., Larrañaga, P.: Multi-dimensional classification with Bayesian networks. International Journal of Approximate Reasoning 52, 705–727 (2011)
3. Borchani, H., Bielza, C., Larrañaga, P.: Learning CB-decomposable multi-dimensional Bayesian network classifiers. In: Petri, M., Teemu, R., Tommi, J. (eds.) Proceedings of the 5th European Workshop on Probabilistic Graphical Models (PGM 2010), pp. 25–32. HIIT Publications (2010)
4. van der Gaag, L.C., de Waal, P.R.: Multi-dimensional Bayesian network classifiers. In: Studený, M., Vomlel, J. (eds.) Third European Workshop on Probabilistic Graphical Models, pp. 107–114 (2006)
5. Friedman, N., Geiger, D., Goldszmidt, M.: Bayesian network classifiers. Machine Learning 29(2-3), 131–163 (1997)
6. Godbole, S., Sarawagi Discriminative, S.: methods for multi-labeled classification. In: Advances in Knowledge Discovery and Data Mining, pp. 22–30. Springer (2004)
7. Kelner, R., Lerner, B.: Learning bayesian network classifiers by risk minimization. International Journal of Approximate Reasoning 53(2), 248–272 (2012)
8. Keogh, E.J., Pazzani Learning, M.J.: the structure of augmented Bayesian classifiers. International Journal on Artificial Intelligence Tools 11(04), 587–601 (2002)
9. Minsky, M.: Steps toward artificial intelligence. In: Computers and Thought, pp. 406–450. McGraw-Hill (1961)

10. Pearl, J.: Probabilistic Reasoning in Intelligent Systems: Networks of Plausible Inference. Morgan Kaufmann Publishers Inc. (1988)
11. Peot, M.A.: Geometric implications of the naive Bayes assumption. In: Eric, H., Finn, J. (eds.) Proceedings of the Twelfth International Conference on Uncertainty in Artificial Intelligence, pp. 414–419. Morgan Kaufmann Publishers Inc (1996)
12. de Waal, P.R., van der Gaag, L.C.: Inference and Learning in Multi-dimensional Bayesian Network Classifiers. In: Mellouli, K. (ed.) ECSQARU 2007. LNCS (LNAI), vol. 4724, pp. 501–511. Springer, Heidelberg (2007)
13. Read, J., Bielza, C., Larrañaga, P.: Multi-dimensional classification with super-classes. IEEE Transactions on Knowledge and Data Engineering (2013)
14. Read, J., Pfahringer, B., Holmes, G., Frank, E.: Classifier Chains for Multi-label Classification. In: Buntine, W., Grobelnik, M., Mladenić, D., Shawe-Taylor, J. (eds.) ECML PKDD 2009, Part II. LNCS, vol. 5782, pp. 254–269. Springer, Heidelberg (2009)
15. Sucar, L.E., Bielza, C., Morales, E.F., Hernandez-Leal, P., Zaragoza, J.H., Larrañaga, P.: Multi-label classification with Bayesian network-based chain classifiers. Pattern Recognition Letter 41, 14–22 (2014)
16. Vapnik, V.N.: The Nature of Statistical Learning Theory. Springer (1995)
17. Vapnik, V.N., Chervonenkis, A.: On the uniform convergence of relative frequencies of events to their probabilities. Theory of Probability and Its Applications 16(2), 264–280 (1971)
18. Varando, G., Bielza, C., Larrañaga, P.: Decision boundary for discrete Bayesian network classifiers. Technical Report UPM-ETSIINF/DIA/2014-1, Universidad Politecnica de Madrid (2014), http://oa.upm.es/26003/
19. Zhang, M.-L., Zhou, Z.-H.: A review on multi-label learning algorithms. IEEE Transactions on Knowledge and Data Engineering (in press, 2014)

An Approximate Tensor-Based Inference Method Applied to the Game of Minesweeper*

Jiří Vomlel and Petr Tichavský

Institute of Information Theory and Automation of the AS CR,
Pod vodárenskou věží 4,
Prague, 182 08, Czech Republic
http://www.utia.cas.cz/

Abstract. We propose an approximate probabilistic inference method based on the CP-tensor decomposition and apply it to the well known computer game of Minesweeper. In the method we view conditional probability tables of the exactly ℓ-out-of-k functions as tensors and approximate them by a sum of rank-one tensors. The number of the summands is $\min\{l + 1, k - l + 1\}$, which is lower than their exact symmetric tensor rank, which is k. Accuracy of the approximation can be tuned by single scalar parameter. The computer game serves as a prototype for applications of inference mechanisms in Bayesian networks, which are not always tractable due to the dimensionality of the problem, but the tensor decomposition may significantly help.

Keywords: Bayesian Networks, Probabilistic Inference, CP Tensor Decomposition, Symmetric Tensor Rank.

1 Introduction

In many applications of Bayesian networks [1,2,3], conditional probability tables (CPTs) have a certain local structure. Canonical models [4] form a class of CPTs with their local structure being defined either by:

- a deterministic function of the values of the parents (*deterministic models*),
- a combination of the deterministic model with independent probabilistic influence on each parent variable (*ICI models*), or
- a combination of the deterministic model with probabilistic influence on a child of the deterministic model (*simple canonical models*).

In this paper we will pay special attention to deterministic models.

A common task solved efficiently with the help of Bayesian networks is probabilistic inference, which is the computation of marginal conditional probabilities of all unobserved variables given observations of other variables. During the inference the special local structure of deterministic CPTs can be exploited. Díez

---

* This work was supported by the Czech Science Foundation through projects 13–20012S and 14–13713S.

L.C. van der Gaag and A.J. Feelders (Eds.): PGM 2014, LNAI 8754, pp. 535–550, 2014.

and Galán [5] suggested to rewrite each CPT of a noisy-max model as a product of two-dimensional potentials $\psi_i, i = 1, \ldots, k$. Later, Savický and Vomlel [6] generalized the method to any CPT. Assume a CPT $P(Y = y | X_1, \ldots, X_k)$ with the state y of variable Y being observed, then we can write

$$P(Y = y | X_1, \ldots, X_k) = \sum_B \prod_{i=1}^k \psi(B, X_i) \; , \qquad (1)$$

where B is an auxiliary variable and the summation proceeds over all its values.

The above equality can be always satisfied if the number of states of B is the product of the number of states of variables $X_1, \ldots, X_k$. The transformation becomes computationally advantageous if the number of states of B is low. It was observed in [6] that each CPT can be understood as a tensor and the minimum number of states of B equals the rank of tensor A defined as

$$\mathcal{A}_{i_1, \ldots, i_k} = P(Y = y | X_1 = x_{i_1}, \ldots, X_k = x_{i_k}),$$

for all combinations of states $(x_{i_1}, \ldots, x_{i_k})$ of variables $X_1, \ldots, X_k$. The decomposition of tensors into the form corresponding to the right hand side of formula (1) is known now as Canonical Polyadic (CP) or CANDECOMP-PARAFAC (CP) decomposition [7,8].

In [9] we have shown how the CP decomposition can be applied to the noisy threshold model of the probabilistic tables. We have presented exact CP decomposition of these tensors, which have rank k if the table size is $2 \times 2 \times \ldots \times 2$ ($k \times$) in real domain, and slightly lower rank in complex domain. Similar decompositions were derived for the probabilistic tables that represent deterministic exact ℓ-out-of-k functions. The tensor rank is about the same. It was shown that using the CP decomposition approach it is possible to perform probabilistic inference also in cases where the classical method cannot be applied because of a large dimensionality of the probabilistic tables. Next, it was shown that the complexity reduction using CP decomposition is better than in the popular parent divorcing method. Finally, it was shown that the tensor decomposition approach can be combined with another alternative mechanism for Bayesian inference, which is Weighted Model Counting (WMC) [9].

In this paper we take a closer look at tensors representing one specific type of a canonical model – deterministic exact ℓ-out-of-k functions. An ℓ-out-of-k function is a function of k binary arguments that takes the value one if exactly ℓ out of its k arguments take value one – otherwise the function value is zero. These tensors appear naturally in Bayesian network models with CPTs $P(y | X_1, \ldots, X_k)$ representing the addition of binary parent variables $X_1, \ldots, X_k$ and with evidence $Y = y$ on the child variable. We suggest a new approximation by a sum of rank-one tensors, where the number of the summands is $\min\{l+1, k-l+1\}$ and the approximation error can be tuned by a single scalar parameter. This means that we propose less complex (lower rank) approximations, which are computationally simpler, but they approach the desired probabilistic table (tensor) quite accurately, with an arbitrarily small error. The main advantage is the lower rank of the approximation, which is much lower than the true rank (k), if ℓ is low or ℓ is close to k.

The paper is organized as follows. In Section 2 we introduce the necessary tensor notation, define tensors of the exact ℓ-out-of-k functions, and present their basic properties. Section 3 represents the main original contribution of this paper. We propose two approximate CP decompositions of tensors of the ℓ-out-of-k functions based on the symmetric border rank of these tensors. We present a comparison of the CP decomposition with the parent divorcing method in Section 4. In Section 5 we introduce our Bayesian network model for the game of Minesweeper. In Section 6 we apply the suggested decomposition to Minesweeper and compare the computational efficiency and the approximation error of the suggested approximate CP decompositions, the exact CP decomposition, and the standard inference approach based on moralization of parent variables.

2 Preliminaries

Tensor is a mapping $\mathcal{A} : \mathbb{I} \to \mathbb{X}$, where $\mathbb{X} = \mathbb{R}$ or $\mathbb{X} = \mathbb{C}$, $\mathbb{I} = I_1 \times \ldots \times I_k$, k is a natural number called the order of tensor $\mathcal{A}$, and $I_j, j = 1, \ldots, k$ are index sets. Typically, I_j are sets of integers of cardinality n_j. Then we can say that tensor $\mathcal{A}$ has dimensions $n_1, \ldots, n_k$. In this paper all index sets will be $\{0, 1\}$.

Example 1. A visualization of a tensor of order $k = 4$ and dimensions $n_1 = n_2 = n_3 = n_4 = 2$ with successive dimensions alternating between rows and columns[1]:

$$\mathcal{A} = \begin{pmatrix} \begin{pmatrix} 0 & 1 \\ 1 & 0 \end{pmatrix} & \begin{pmatrix} 1 & 0 \\ 0 & 0 \end{pmatrix} \\ \begin{pmatrix} 1 & 0 \\ 0 & 0 \end{pmatrix} & \begin{pmatrix} 0 & 0 \\ 0 & 0 \end{pmatrix} \end{pmatrix}$$

Tensor $\mathcal{A}$ has rank one if it can be written as an outer product of vectors:

$$\mathcal{A} = \boldsymbol{a}_1 \otimes \ldots \otimes \boldsymbol{a}_k ,$$

with the outer product being defined for all $(i_1, \ldots, i_k) \in I_1 \times \ldots \times I_k$ as

$$\mathcal{A}_{i_1, \ldots, i_k} = a_{1, i_1} \cdot \ldots \cdot a_{k, i_k} ,$$

where $\boldsymbol{a}_j = (a_{j,i})_{i \in I_j}$, $j = 1, \ldots, k$ are real or complex valued vectors.

Each tensor can be decomposed as a linear combination of rank-one tensors:

$$\mathcal{A} = \sum_{i=1}^{r} b_i \cdot \boldsymbol{a}_{i,1} \otimes \ldots \otimes \boldsymbol{a}_{i,k} , \tag{2}$$

The rank of a tensor $\mathcal{A}$, denoted $rank(\mathcal{A})$, is the minimal r over all such decompositions. The decomposition of a tensor $\mathcal{A}$ to tensors of rank one that sum up to $\mathcal{A}$ is called CP tensor decomposition.

---

[1] The first dimension is the row of the outer matrix, the second is the column of the outer matrix, the third is the row of the inner matrix, and the fourth is the column of the inner matrix.

Example 2. The tensor $\mathcal{A}$ from Example 1 can be written as:

$$
\begin{aligned}
\mathcal{A} = \ & (0,1) \otimes (1,0) \otimes (1,0) \otimes (1,0) \\
& +(1,0) \otimes (0,1) \otimes (1,0) \otimes (1,0) \\
& +(1,0) \otimes (1,0) \otimes (0,1) \otimes (1,0) \\
& +(1,0) \otimes (1,0) \otimes (1,0) \otimes (0,1) \ .
\end{aligned}
$$

This implies that its rank is at most 4.

The tensors studied in this paper are symmetric.

Definition 1. *Let $\mathbb{X}$ be either $\mathbb{R}$ or $\mathbb{C}$. Tensor $\mathcal{A} : \{0,1\}^k \to \mathbb{X}$ is symmetric if for $(i_1, \ldots, i_k) \in \{0,1\}^k$ it holds that*

$$
\mathcal{A}_{i_1, \ldots, i_k} = \mathcal{A}_{i_{\sigma(1)}, \ldots, i_{\sigma(k)}} \ ,
$$

for any permutation σ of $\{1, \ldots, k\}$.

Example 3. The tensor $\mathcal{A}$ from Example 1 is symmetric.

Definition 2. *Let $\mathbb{X}$ be either $\mathbb{R}$ or $\mathbb{C}$. The symmetric rank $\mathrm{srank}_{\mathbb{X}}(\mathcal{A})$ of a tensor $\mathcal{A}$ is the minimum number of symmetric rank-one tensors taking values from $\mathbb{X}$ such that their linear combination is equal to $\mathcal{A}$, i.e.,*

$$
\mathcal{A} = \sum_{i=1}^{r} b_i \cdot \boldsymbol{a}_i^{\otimes k} \ , \tag{3}
$$

where $\boldsymbol{a}_i, i = 1, \ldots, r$ are vectors of length equal to dimensions of $\mathcal{A}$ taking values form $\mathbb{X}$, $b_i \in \mathbb{X}, i = 1, \ldots, r$, and $\boldsymbol{a}_i^{\otimes k}$ is used to denote $\underbrace{\boldsymbol{a}_i \otimes \ldots \otimes \boldsymbol{a}_i}_{k \ copies}$.

As we will discuss later some tensors $\mathcal{A}$ can be approximated with arbitrarily small error by tensors of lower rank than their rank. This can be formalized using the notion of border rank.

Definition 3. *The border rank of $\mathcal{A} : \{0,1\}^k \to \mathbb{R}$ is*

$$
brank(\mathcal{A}) = \min\{r : \forall \varepsilon > 0 \ \exists \mathcal{E} : \{0,1\}^k \to \mathbb{R}, \|\mathcal{E}\| < \varepsilon, rank(\mathcal{A} + \mathcal{E}) = r\} \ ,
$$

where $\| \cdot \|$ is any norm.

Next we give an example of a tensor that has its border rank at most two. The example is a specialization of Example 4.2 from [10].

Example 4. Let $k = 4$. Then for $q > 0$ tensor

$$
\begin{aligned}
\mathcal{B}(q) = \ & \frac{1}{2q} \cdot (1,q) \otimes (1,q) \otimes (1,q) \otimes (1,q) \\
& -\frac{1}{2q} \cdot (1,-q) \otimes (1,-q) \otimes (1,-q) \otimes (1,-q)
\end{aligned}
$$

has rank at most two. Note that

$$\lim_{q \to 0} \mathcal{B}(q) = \mathcal{A}$$

where $\mathcal{A}$ is the tensor from Example 1. This implies that $brank(\mathcal{A}) \leq 2$.

A class of tensors that appear in the real applications are tensors representing functions. In this paper we pay special attention to tensors representing the exact ℓ-out-of-k functions, i.e. a Boolean function taking value 1 if and only if exactly ℓ of its k inputs have value 1.

Definition 4. *Tensor* $\mathcal{S}(\ell, k) : \{0,1\}^k \to \{0,1\}$ *represents an exact ℓ-out-of-k function if it holds for* $(i_1, \ldots, i_k) \in \{0,1\}^k$:

$$\mathcal{S}_{i_1,\ldots,i_k}(\ell, k) = \delta(i_1 + \ldots + i_k = \ell)$$

$$\delta(i = \ell) = \begin{cases} 1 \ if \ i = \ell \\ 0 \ otherwise. \end{cases}$$

Example 5. The tensor $\mathcal{A}$ presented in Examples 1– 4 is tensor $\mathcal{S}(1,4)$. It follows from Example 2 it has rank at most 4 and border rank at most 2 (Example 4).

3   Approximate Tensor Decompositions

Tensors $\mathcal{S}(\ell, k)$ were studied in [9]. It was shown that their symmetric rank in the real domain is equal to k for all integer k, ℓ [9, Proposition 1 and Proposition 3], except for the trivial cases $\ell \in \{0, k\}$, where the rank is one [9, Lemma 2], symbolically:

$$srank_{\mathbb{R}}(\mathcal{S}(\ell, k)) = \begin{cases} k & for \ 1 \leq \ell \leq (k-1) \\ 1 & for \ \ell \in \{0, k\}. \end{cases} \tag{4}$$

In the complex domain the tensor rank is slightly smaller for ℓ in vicinity of $k/2$:

$$srank_{\mathbb{C}}(\mathcal{S}(\ell, k)) = \max\{\ell + 1, k - \ell + 1\} \quad for \ 1 \leq \ell \leq (k-1), \tag{5}$$

see [9, Proposition 3]. The proofs in [9] are constructive.

In practical applications, the tensors with ℓ near zero, $\ell = 1, 2, 3, 4$ and with ℓ near k, i.e. $\ell = k - 1, k - 2, k - 3, k - 4$, seem to be more common that those with ℓ around $k/2$. For example, in Section 5 we discuss an application to the computer game of Minesweeper where CPTs with values of ℓ around $k/2$ appear rarely. For ℓ near $k/2$ we recommend decomposition in the complex domain [9], which has the rank specified in formula (5).

Earlier it was shown in [10, Theorem 4.3] that the symmetric border rank of the tensor $\mathcal{S}(\ell, k)$ can be bounded as

$$brank(\mathcal{S}(\ell, k)) \leq \min\{\ell + 1, k - \ell + 1\} .$$

It means that the tensor can be expressed as a limit of a series of tensors having the displayed rank. Unfortunately, the CP decomposition of the approximating tensors are such that some elements of the factor matrices converge to zero and some other converge to infinity. For practical applications it is indeed possible to work with an inaccurate decomposition, provided that the approximation error is sufficiently low, and the corresponding factor matrices do not have too large Frobenius norm, so that there are no serious numerical issues with these factors.

The paper [10, Section 6] contains a general construction of series of tensors of rank $\min\{\ell + 1, k - \ell + 1\}$ that converge to $\mathcal{S}(\ell, k)$ for a general pair (k, ℓ). Convergence of the series is relatively slow with respect to the Frobenius norm of the factor matrices, except for the special case $\ell = 1$.

1-out-of-k

The tensor $\mathcal{S}(1, k)$ can be written as a limit

$$\mathcal{S}(1, k) = \lim_{x \to \infty} \mathcal{S}_a(1, k, x)$$

where

$$\mathcal{S}_a(1, k, x) = (x, y)^{\otimes k} - (x, -y)^{\otimes k}$$
$$y = y(x, k) \quad = \quad \frac{1}{2x^k} \ .$$

Obviously, rank of $\mathcal{S}_a(1, k, x)$ is 2. The error of the approximation is

$$E(1, k, x) = \|\mathcal{S}(1, k) - \mathcal{S}_a(1, k, x)\|_\infty \quad = \quad \frac{1}{4x^{2k}} \ .$$

Note that Example 4 represents a special case for $k = 4$.

A Method for Tensor Approximations

In this paper we extend the above result for the cases $\ell = 2, 3, 4$ and a general k. In other words we write the tensor $\mathcal{S}(\ell, k)$ as a limit of an appropriately parameterized tensor $\mathcal{S}_a(\ell, k, x)$ of a low rank,

$$\mathcal{S}(\ell, k) = \lim_{x \to \infty} \mathcal{S}_a(\ell, k, x) \ .$$

We have conducted a series of numerical experiments attempting to decompose the tensors numerically, using Levenberg-Marquardt method [11] starting from different random starting points, which allowed us to guess a functional form of suitable approximations.

Once the functional form of the approximation was found, we used symbolic matlab tool to evaluate the assumed tensor decomposition as a function of 2 to 5 designed parameters. These parameters were selected to approximate the exact tensor of interest to the maximum possible extent, with a single parameter left. This parameter allows one to control the quality of the approximation, possibly at the expense of numerical stability.

2-out-of-k

For $\ell = 2$ we get

$$\mathcal{S}_a(2, k, x) = (x, y)^{\otimes k} + (x, -y)^{\otimes k} - 2x^k(1, 0)^{\otimes k}$$

$$y = y(x, k) = \frac{1}{\sqrt{2}x^{(k-2)/2}} \; .$$

The error of the approximation is

$$E(2, k, x) = \|\mathcal{S}(2, k) - \mathcal{S}_a(2, k, x)\|_\infty = \frac{1}{2x^k} \; .$$

3-out-of-k

For $\ell = 3$ we get

$$\mathcal{S}_a(3, k, x) = (x, y)^{\otimes k} - (x, -y)^{\otimes k} - (z, w)^{\otimes k} + (z, -w)^{\otimes k}$$

with

$$y = y(x, k) = -\frac{1}{2}x^{1-k/3}$$

$$z = z(x, y, k) = \left(\frac{2x^{3k-3}y^3}{2x^{k-3}y^3 - 1} \right)^{1/(2k)}$$

$$w = w(x, y, z, k) = y\left(\frac{x}{z} \right)^{k-1} \; .$$

The error of the approximation is

$$E(3, k, x) = \|\mathcal{S}(3, k) - \mathcal{S}_a(3, k, x)\|_\infty = \frac{3}{2x^{2k/3}} \; .$$

4-out-of-k

Finally, for $\ell = 4$ we get

$$\mathcal{S}_a(4, k, x) = (x, y)^{\otimes k} + (x, -y)^{\otimes k} - (z, w)^{\otimes k} - (z, -w)^{\otimes k} - 2(x^k - z^k)(1, 0)^{\otimes k}$$

with

$$y = y(x, k) = x^{1-k/4}$$

$$z = z(x, y, k) = \left(\frac{2x^{2k-4}y^4}{2x^{k-4}y^4 - 1} \right)^{1/k}$$

$$w = w(x, y, z, k) = y\left(\frac{x}{z} \right)^{k/2-1} \; .$$

The error of the approximation is

$$E(4, k, x) = \|\mathcal{S}(4, k) - \mathcal{S}_a(4, k, x)\|_\infty = \frac{3}{2x^{k/2}} \; .$$

$(k - \ell)$-out-of-k

Approximations for $\ell = k - 1, k - 2, k - 3, k - 4$ can be constructed from $\ell = 1, 2, 3, 4$, respectively, by swapping values of all vectors in the CP decompositions, i.e., from

$$\mathcal{S}_a(\ell, k) = \sum_{i=1}^{r} b_i \cdot \boldsymbol{a}_i^{\otimes k} \,,$$

we get

$$\mathcal{S}_a(k - \ell, k) = \sum_{i=1}^{r} b_i \cdot \overline{\boldsymbol{a}}_i^{\otimes k} \,,$$

where vector $\overline{\boldsymbol{a}}_i = (y_i, x_i)$ is obtained from $\boldsymbol{a}_i = (x_i, y_i)$ by swapping its values.

Approximate Decompositions of Threshold Tensors

Similar functional forms can be derived also for approximate decompositions of threshold tensors discussed in [9]. For tensors $\mathcal{T}(\ell, k)$ with ℓ near zero ($\ell = 1, 2, 3, 4$) and with ℓ near k ($\ell = k - 1, k - 2, k - 3, k - 4$) we can use already derived expressions for $\mathcal{S}(\ell, k)$ and combine them using the following identity:

$$\mathcal{T}(\ell, k) = \begin{cases} \sum_{m=\ell}^{k} \mathcal{S}(m, k) & \text{for } \ell = k - 1, k - 2, \ldots \\ (1, 1)^{\otimes k} - \sum_{m=1}^{\ell - 1} \mathcal{S}(m, k) & \text{for } \ell = 1, 2, \ldots. \end{cases}$$

Complex Valued Decompositions

It is worth noting that if complex-valued factors in the decomposition are allowed, the approximation is possible with higher accuracy for the same variable x. We consider the same functional form as in the real-valued decomposition.

In particular, for $\ell = 3$ the dependence of the variable y on x can be taken as $y = (2^{-2/3} - 1/x^2)\, x^{1-k/3}$, and for $\ell = 4$ we propose the choice $y = (2^{-1/2} - 1/x^2)\, x^{1-k/4}$. With these choices, the variables z and w become complex-valued, but the decomposition remains valid and the total approximation error is reduced.

Approximation Errors

In Table 1 we present maximum[2] approximation error for approximate decompositions of ℓ-out-of-k tensors. These errors were obtained for variable $x = 10$. With higher variable x, the approximation errors could be still lower, but for the price of a risk of numerical issues.

---

[2] The maximum is taken over all absolute values of differences of all corresponding pairs of tensor values.

The first two column present errors of the complex-valued and real-valued decomposition described above. For $\ell = 1$ and $\ell = 2$ there is no difference, no increase of accuracy can be attained in the complex domain. For $\ell = 3$ and $\ell = 4$ the former decomposition is more accurate, but for the price of involving arithmetic with complex numbers. The third column contains the error obtained by a rank-k approximation in the real domain suggested in [9]. The error is effectively zero.

Table 1. Maximum approximation error for approximate and exact decompositions of ℓ-out-of-k tensors. The errors for decompositions of $(k - \ell)$-out-of-k tensors are the same as of ℓ-out-of-k by their construction.

CPT	complex approx.	real approx.	real exact
1-out-of-4	2.5e-09	2.5e-09	1.465e-14
2-out-of-4	5,00e-05	5,00e-05	1.908e-15
1-out-of-5	2.5e-11	2.5e-11	1.399e-14
2-out-of-5	5,00e-06	5,00e-06	5.995e-15
1-out-of-6	2.5e-13	2.5e-13	1.654e-14
2-out-of-6	5,00e-07	5,00e-07	6.573e-14
3-out-of-6	3.78e-06	6.3e-05	1.248e-13
1-out-of-7	2.5e-15	2.5e-15	7.472e-14
2-out-of-7	5,00e-08	5,00e-08	2.315e-12
3-out-of-7	8.144e-07	4.454e-05	6.625e-14
1-out-of-8	1.11e-16	1.11e-16	1.798e-12
2-out-of-8	1.192e-07	1.192e-07	1.396e-12
3-out-of-8	1.755e-07	2.18e-05	2.376e-12
4-out-of-8	5.698e-06	0.00015	1.239e-12

Remark 1. For the three cases 2-out-of-4, 3-out-of-6 and 4-out-of-8 presented in Table 1 we can get exact CP complex decomposition of the same symmetric rank as the approximate one – see formula (5).

4   A Comparison with the Parent Divorcing Method

A different transformation that can be applied to CPTs of ℓ-out-of-k functions is the parent-divorcing method [12].

On the right hand side of Figure 1 we present the graph after parent divorcing and consequent moralization for $k = 5$ and $\ell = 1$. First, we add $k - 2$ auxiliary variables and connect each of them with two parents. The CPT of each auxiliary node is for $\ell \leq k/2$, $j = 2, \ldots, k - 1$ and $y = 0, 1, \ldots, \min\{j, \ell + 1\}$ defined as

$$P(Y_j = y | Y_{j-1} = y', X_j = x) = \begin{cases} 1 & \text{if either } y = y' + x \text{ or} \\ & y' + x \geq \ell \text{ and } y = \ell + 1 \\ 0 & \text{otherwise,} \end{cases}$$

where for $j = 2$ variable Y_{j-1} is replaced by X_1. Note that we need not consider the values of Y_j greater than $\ell + 1$ since in these cases the exact ℓ-out-of-k function is already ensured to be zero and by adding the values of the remaining variables $X_j, \ldots, X_k$ the sum cannot decrease. For $\ell > k/2$ the CPTs are defined similarly but with values of y swapped.

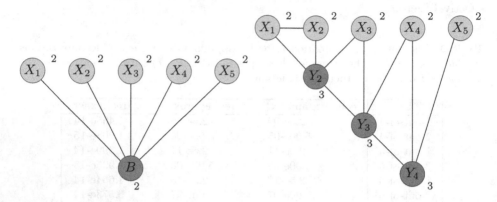

Fig. 1. The graph after the CP decomposition (left) and the parent divorcing method with consequent moralization (right) for $k = 5$ and $\ell = 1$. The small numbers attached to nodes represent number of states of corresponding variables.

In the moralization step all parents of each node are pairwise connected by an undirected edge and directions of edges are removed. Finally, the last auxiliary node is to the last parent node by an undirected edge. The table corresponding to clique $\{Y_{k-1}, X_k\}$ is for the observed value y of Y defined as

$$P(Y = y | Y_{j-1} = y', X_j = x) = \begin{cases} 1 & \text{if either } \ell = y' + x \text{ and } y = 1 \text{ or} \\ & \ell \neq y' + x \text{ and } y = 0 \\ 0 & \text{otherwise.} \end{cases}$$

Table Size

The table size is the number of numerical values (memory units) that are needed to represent all tables of a CPT. In Table 2 we compare the table size of CPT after the real approximate (ts_{CPa}) and real exact CP decompositions (ts_{CPe}) of ℓ-out-of-k tensors compared with the parent divorcing (PD) method (ts_{PD}) and the full table size (ts_f). The table sizes were for given $k \geq 4$ and $(k-1) \geq \ell \geq 1$ computed by following formulas:

$$ts_{CPa} = 2k \min\{k - \ell + 1, \ell + 1\}$$
$$ts_{CPe} = 2k^2$$
$$ts_{PD} = \sum_{j=2}^{k-1} 2 \min\{j, k - \ell + 2, \ell + 2\} \cdot \min\{j + 1, k - \ell + 2, \ell + 2\}$$
$$+ 2 \min\{k - \ell + 2, \ell + 2\}$$
$$ts_f = 2^k$$

Table 2. Table size for the real approximate and real exact CP decompositions compared with the parent divorcing (PD) method and the full table size. On the right we present a plot of table sizes for $k = 8$. The table sizes for decompositions of $(k - \ell)$-out-of-k tensors are the same as of ℓ-out-of-k by their construction.

CPT	ts_{CPa}	ts_{CPe}	ts_{PD}	ts_f
1-out-of-4	16	32	36	16
2-out-of-4	24	32	44	16
1-out-of-5	20	50	54	32
2-out-of-5	30	50	76	32
1-out-of-6	24	72	72	64
2-out-of-6	36	72	108	64
3-out-of-6	48	72	136	64
1-out-of-7	28	98	90	128
2-out-of-7	42	98	140	128
3-out-of-7	56	98	186	128
1-out-of-8	32	128	108	256
2-out-of-8	48	128	172	256
3-out-of-8	64	128	236	256
4-out-of-8	80	128	292	256

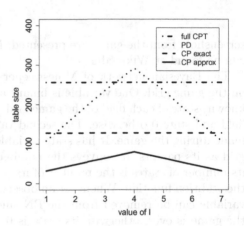

5 The Game of Minesweeper

In [13] the computer game of Minesweeper was used to illustrate a few modeling tricks utilized when applying Bayesian networks in real applications. In [10] this game was used to illustrate the benefits of CP tensor decompositions of CPTs of noisy exact ℓ-out-of-k functions. In this paper we will use the Bayesian network model of this game to compare exact and approximate CP tensor decompositions of deterministic CPTs of exact ℓ-out-of-k functions.

Minesweeper is a one-player game. The game starts with a grid of $n \times m$ blank fields. During the game the player clicks on different fields. If the player clicks on a field containing a mine the game is over. Otherwise the player gets information on how many fields in the neighborhood of the selected field contain a mine. The goal of the game is to find all mines without clicking on them. In Figure 2 two

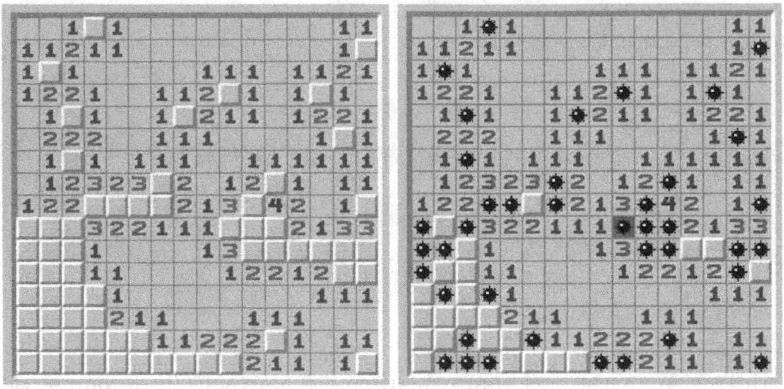

Fig. 2. Two screenshots from the game of Minesweeper. The screenshot on the right hand side is taken after the player stepped on a mine and it shows the actual position of mines.

screenshots from the game are presented. More information about Minesweeper can be found at Wikipedia [14].

The Bayesian network of Minesweeper contains two variables for each field on the game grid. One variable is binary and corresponds to the (originally unknown) state of each field of the game grid. It has state 1 if there is a mine on this field and state 0 otherwise. The second variable corresponds to the observation made during the game. It has state variables on the neighboring positions in the grid as its parents. It conveys the number of its neighbors with a mine. Thus, its number of states is the number of its parents plus one. Its CPT is defined by the addition function. Whenever an observation is made the corresponding state variable can be removed from the BN since its state is known. If its state is 1 the game is over, otherwise its state is 0. When evidence from an observation is distributed to its neighbors the node corresponding to the observation can be removed. By entering evidence to a CPT of addition a table of exact ℓ-out-of-k function is created. Variables from the second set that were not observed are not included in the BN model since they are barren variables [2, Section 5.5.1]. The above considerations implies that in every moment of the game we will have at most one node for each field of the grid and all tables in the BN are either one dimensional priors that are the same for each position or tables of exact ℓ-out-of-k function. Thus, the BN of Minesweeper represent a good test bed for inference algorithms exploiting the local structure of tables of ℓ-out-of-k functions. This paragraph is a digest of a more detailed description of the BN of Minesweeper in [10, Section 7.1].

In Figure 3 we present an example of the game grid after 175 random observations of fields without a mine from the point of view of a game oracle. The players do not see the positions of mines.

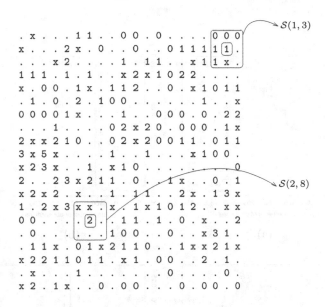

Fig. 3. The game grid after 175 random observations. The points ".." represent covered fields without a mine, crosses "x" represent covered fields with a mine. The numbers correspond to uncovered fields and give the number of mines in the neighborhood. The neighborhoods of 2 out of 175 observed fields are denoted by rectangles. In the corresponding steps of the game the CPTs $\mathcal{S}(1,3)$ and $\mathcal{S}(2,8)$ are added to the Bayesian network. Note that nodes of observed fields are connected to uncovered fields only – therefore we add CPT $\mathcal{S}(1,3)$ instead of $\mathcal{S}(1,8)$.

6 Numerical Experiments

We performed experiments with Minesweeper of 20×20 grid size. We implemented all algorithms in the R language [15]. In each of 350 steps of the game the oracle randomly selected a field to be observed from those 350 not containing a mine and we created a Bayesian network corresponding to that step. We compared two transformations:

- the standard method consisting of moralization and triangulation steps and
- the CP tensor decomposition applied to CPTs with higher number of parents[3] – for other CPTs we used the moralization followed by the triangulation step.

In both networks we used the lazy propagation method [16] which is junction tree based methods where the computations are performed with messages that are kept as long as possible as lists of tables.

---

[3] We applied CP tensor decomposition only when the total size of created tables was less than the size of the table after moralization. Roughly speaking, this happened when the number of parents was higher than three for the approximate methods and higher than six for the exact method.

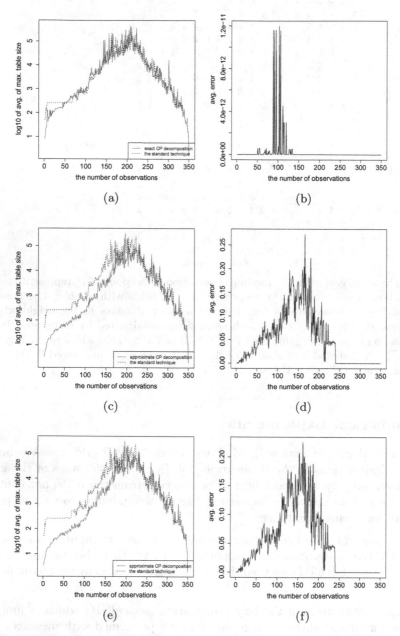

Fig. 4. Results of the experiments for the real exact decompositions – (a) and (b), the real approximate decompositions – (c) and (d), and the complex approximate decompositions – (e) and (f).

At each step of the game we recorded (1) decadic logarithm of the size of the largest table created during the lazy propagation and (2) the conditional marginal probabilities given current observations computed by (1) the standard method and three versions of methods exploiting the CP tensor decomposition: (2) the real exact decompositions, (3) the real approximate decompositions, and (4) the complex approximate decompositions. The value of parameter x was set to 10.

For the results of experiments see Figure 4. All values represent the average over ten different games. The plots in the first column present the decadic logarithm of the size of the largest table created at each step of the game. The plots in the second column present the average error (measured by the absolute value of difference) of the conditional marginal probabilities.

We can see that when using exact CP tensor decomposition the size of largest tables are not reduced but there is no approximation error (as expected). When the real or complex approximate CP tensor decomposition is used at some stages of the game the size of largest tables is reduced by an order of magnitude. But this is achieved at an expense of a certain loss of the accuracy. The loss is lower for the complex CP tensor decomposition (ranging from 0 to 0.2) than for the real CP tensor decomposition (ranging from 0 to 0.25). Unfortunately, for some particular configurations the approximation error is high. The numerical stability of probabilistic inference seems to be an important issue here – however, we did not study this issue in depth and leave it as a topic for our future research.

7 Conclusions

The reduction of maximal table size is important for applications where large tables imply memory requirements that forbid using standard probabilistic inference schemes based on moralization and triangulation. The game of Minesweeper is a prototype application, where the tensor decomposition approach can be really useful for reducing computational load of the inference mechanism. For this particular game, the computational savings are low, if any, because the maximum number of parents (order of the probability tables) is at most 8. However, we can imagine more complex situations, where the number of parents is higher. In such cases the computational advantage would be more apparent.

We can consider, for example, a 3D generalization of Minesweeper, where each field has not only 8 but 26 neighbors. The complexity of such problem would grow significantly. We believe that the tensor decompositions presented in this paper might be very suitable for the probabilistic inference in Bayesian networks of such a type.

References

1. Pearl, J.: Probabilistic Reasoning in Intelligent Systems: Networks of Plausible Inference. Morgan Kaufmann Publishers Inc., San Francisco (1988)
2. Jensen, F.V.: Bayesian Networks and Decision Graphs. Springer, New York (2001)

3. Jensen, F.V., Nielsen, T.D.: Bayesian Networks and Decision Graphs, 2nd edn. Springer (2007)
4. Díez, F.J., Druzdzel, M.J.: Canonical probabilistic models for knowledge engineering.Technical Report CISIAD-06-01, UNED, Madrid, Spain (2006)
5. Díez, F.J., Galán, S.F.: An efficient factorization for the noisy MAX. International Journal of Intelligent Systems 18, 165–177 (2003)
6. Savicky, P., Vomlel, J.: Exploiting tensor rank-one decomposition in probabilistic inference. Kybernetika 43(5), 747–764 (2007)
7. Carroll, J.D., Chang, J.J.: Analysis of individual differences in multidimensional scaling via an n-way generalization of Eckart-Young decomposition. Psychometrika 35, 283–319 (1970)
8. Harshman, R.A.: Foundations of the PARAFAC procedure: Models and conditions for an "explanatory" multi-mode factor analysis. UCLA Working Papers in Phonetics 16, 1–84 (1970)
9. Vomlel, J., Tichavský, P.: Probabilistic inference with noisy-threshold models based on a CP tensor decomposition. International Journal of Approximate Reasoning 55, 1072–1092 (2014)
10. Vomlel, J.: Rank of tensors of l-out-of-k functions: an application in probabilistic inference. Kybernetika 47(3), 317–336 (2011)
11. Phan, A.H., Tichavský, P., Cichocki, A.: Low complexity damped Gauss-Newton algorithms for CANDECOMP/PARAFAC. SIAM Journal on Matrix Analysis and Applications 34, 126–147 (2013)
12. Olesen, K.G., Kjærulff, U., Jensen, F., Jensen, F.V., Falck, B., Andreassen, S., Andersen, S.K.: A MUNIN network for the median nerve — a case study on loops. Applied Artificial Intelligence 3, 384–403 (1989); Special issue: Towards Causal AI Models in Practice
13. Vomlelová, M., Vomlel, J.: Applying Bayesian networks in the game of Minesweeper. In: Proceedings of the Twelfth Czech-Japan Seminar on Data Analysis and Decision Making under Uncertainty, pp. 153–162 (2009)
14. Wikipedia: Minesweeper (computer game), http://en.wikipedia.org/wiki/Minesweeper_(computer_game)
15. R Development Core Team: R: A Language and Environment for Statistical Computing. R Foundation for Statistical Computing, Vienna, Austria (2008) ISBN 3-900051-07-0
16. Madsen, A.L., Jensen, F.V.: Lazy propagation: A junction tree inference algorithm based on lazy evaluation. Artificial Intelligence 113(1–2), 203–245 (1999)

Compression of Bayesian Networks with NIN-AND Tree Modeling

Yang Xiang and Qing Liu

School of Computer Science, University of Guelph, Canada

Abstract. We propose to compress Bayesian networks (BNs), reducing the space complexity to being fully linear, in order to widely deploy in low-resource platforms, such as smart mobile devices. We present a novel method that compresses each conditional probability table (CPT) of an arbitrary binary BN into a Non-impeding noisy-And Tree (NAT) model. It achieves the above goal in the binary case. Experiments demonstrate that the accuracy is reasonably high in the general case and higher in tested real BNs. We show advantages of the method over alternatives on expressiveness, generality, space reduction and online efficiency.

Keywords: Bayesian networks, causal models, local distributions, approximation of CPTs, NIN-AND tree models.

1 Introduction

The space complexity of a discrete BN of μ variables, each with up to n parents and up to κ possible values, is $O(\mu \, \kappa^n)$. Consider a BN that has at least 25 variables (among others) whose CPTs satisfy $\kappa = 6$ (perhaps due to discretizing continuous variables) and $n = 8$. With 4 bytes per probability, this BN takes a space of about $25 * 6^{8+1} * 4 \approx 1$ gigabytes, more than the full memory of iPhone 5s, making it undeployable in such a device. We propose to compress BNs so that they can be widely deployed in low-resource platforms, such as smart mobile devices and intelligent sensors. Specifically, the reported research has the ultimate goal to reduce the BN space complexity from $O(\mu \, \kappa^n)$ to $O(\mu \, \kappa \, n)$ (fully linear). An important contribution of this paper is the accomplishment of this goal for the case $\kappa = 2$.

A compression method must maintain the expressiveness of the BN reasonably well and enable more efficient inference. The combination of space and time reduction is the key to broader applications of BNs in low-resource platforms. Many existing techniques aimed at reducing parameter complexity in BNs to ease knowledge acquisition can be viewed from the perspective of space reduction. QPNs [1] represent qualitative influences by signs without using numerical parameters. They are limited in expressiveness and cannot resolve parallel influences of opposite signs, although they have been extended to alleviate the limitation, e.g., [2]. Coarsening [3] reduces κ to $\tau < \kappa$, but the space complexity is still exponential on n (i.e., $O(\mu \, \tau^n)$). Divorcing [4] cuts down the value of n, reducing the space exponentially, but is not always applicable. The noisy-OR

L.C. van der Gaag and A.J. Feelders (Eds.): PGM 2014, LNAI 8754, pp. 551–566, 2014.
© Springer International Publishing Switzerland 2014

[5] and several causal independence models, e.g., [6,7], reduce space complexity to being linear on n. They capture only reinforcing causal interactions and are limited in expressiveness. A NAT model [8] can also express undermining and is more expressive. Assuming that a BN family (a child plus its parents) satisfies a NAT model, it can be recovered by exhaustively evaluating alternative NATs [9]. The space of the junction tree of a BN is reduced in lazy propagation [10].

We present a novel method, denoted TDPN (Tree-Directed, Pci-based Nat search), to compress an arbitrary binary ($\kappa = 2$) BN CPT into a NAT model. It overcomes limitations of several alternatives discussed above on expressiveness, generality and efficiency. The compression process consists of an offline and an online stage. A novel search tree is constructed offline and is used online to reduce the search space of NAT structures. Numerical search for parameters is performed over the subspace to yield a compressed model.

Section 2 reviews the background on NAT models. The main idea of TDPN is described in Section 3. Its key components are elaborated in Sections 4 through 8. Section 9 reports empirical evaluations. We analyze the advantages of TDPN over alternatives in Section 10.

2   Background on NIN-AND Tree Models

We briefly introduce terminologies on NAT models and more details are found in [9]. A NAT model is defined over an effect variable e and a set $C = \{c_1, ..., c_n\}$ of uncertain causes, where $e \in \{e^-, e^+\}$, e^+ denotes $e = true$, $n \geq 2$, and $c_i \in \{c_i^-, c_i^+\}$. The probability of a causal success, where c_i caused e to occur when other causes are false, is denoted $P(e^+ \leftarrow c_i^+)$ and is referred to as a single-causal (probability). Similarly, $P(e^+ \leftarrow c_1^+, c_2^+)$ is a double-causal or multi-causal. Causal and conditional probabilities are related. If $C = \{c_1, c_2, c_3\}$, then $P(e^+|c_1^+, c_2^+, c_3^-) = P(e^+ \leftarrow c_1^+, c_2^+)$. With $X = \{c_1, c_2\} \subset C$, the multi-causal is also written as $P(e^+ \leftarrow \underline{x}^+)$. The probability of the corresponding causal failure is $P(e^- \leftarrow \underline{x}^+) = 1 - P(e^+ \leftarrow \underline{x}^+)$. When all causes are false, the effect is false, i.e., the leak probability is $P(e^+|c_1^-, c_2^-, c_3^-) = 0$.

Causes reinforce each other if they are collectively at least as effective as when only some act. They undermine each other if collectively they are less effective. A NAT expresses reinforcing and undermining between individual causes as well disjoint subsets. Fig. 1 (a) shows a NAT over $C = \{c_1, ..., c_4\}$, where black nodes are labeled by causal events. Undermining between c_1 and c_2 is encoded by a direct Non-Impeding Noisy-AND (NIN-AND) gate g_1. The gate dictates $P(e^+ \leftarrow c_1^+, c_2^+) = P(e^+ \leftarrow c_1^+)P(e^+ \leftarrow c_2^+)$. Hence, $P(e^+ \leftarrow c_1^+, c_2^+) < P(e^+ \leftarrow c_i^+)$ for $i = 1, 2$ (undermining). The similar holds for c_3 and c_4. Subsets $\{c_1, c_2\}$ and $\{c_3, c_4\}$ reinforce each other, encoded by a dual NIN-AND gate g_3. The left input event to g_3 is causal failure $e^- \leftarrow c_1^+, c_2^+$, where the white oval negates an event. Gate g_3 dictates $P(e^- \leftarrow c_1^+, c_2^+, c_3^+, c_4^+) = P(e^- \leftarrow c_1^+, c_2^+)P(e^- \leftarrow c_3^+, c_4^+)$. Hence, we have $P(e^+ \leftarrow c_1^+, c_2^+, c_3^+, c_4^+) \geq P(e^+ \leftarrow c_1^+, c_2^+)$ (reinforcing).

A NAT model is a tuple $M = (e, C, T, Sp)$, where $|C| = n$, T is a NAT over e and C, and Sp is a set of n single-causals. M uniquely defines a CPT $P_M(e|C)$. Hence, a NAT modeled-CPT has a space complexity linear on n.

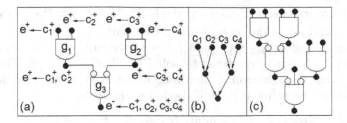

Fig. 1. (a) A NAT of 4 root events, (b) its root-labeled tree of 4 roots, and (c) a NAT of $n = 7$ causes with labels omitted

A NAT can be concisely represented as a root-labeled tree by omitting gates and non-root labels, and simplifying root labels, as shown in Fig. 1 (b). From the root-labeled tree and the type of leaf gate g_3, the NAT in (a) is uniquely determined. Hence, each root-labeled tree encodes two distinct NATs.

A NAT defines a pairwise causal interaction (PCI) function from pairs of causes to the set $\{u, r\}$ (undermining or reinforcing). For instance, the NAT in Fig. 1 (a) defines $pci(c_1, c_2) = u$ and $pci(c_1, c_4) = r$. Given an order of cause pairs, denoting $\{u, r\}$ by $\{0, 1\}$, a PCI pattern (bit string) is derived from the function. Using the order $(\langle c_1, c_2 \rangle, \langle c_1, c_3 \rangle, \langle c_1, c_4 \rangle, ...)$, the PCI pattern from Fig. 1 (a) is $(u, r, r, ...)$ or $(0, 1, 1, ...)$. Each NAT has a unique PCI pattern [11].

3   Tree-Directed, PCI-Based NAT Search

We consider how to compress a target CPT P_T into a NAT model M. In other words, we approximate P_T by P_M (the CPT defined by M). The accuracy of approximation is measured by the *average Euclidean distance* (denoted by ED) below, where $K = 2^n$. The range of ED is $[0, 1]$ and matches that of probability.

$$ED(P_T, P_M) = \sqrt{\frac{1}{K} \sum_{i=1}^{K}(P_T(e^+|\underline{c}_i^+) - P_M(e^+|\underline{c}_i^+))^2}$$

When $ED(P_T, P_M) = 0$, the approximation is perfect. This is not possible for every P_T in general. Hence, M^* that minimizes ED is deemed optimal.

A related problem was solved [9] where e and C are assumed to observe an underlying NAT model. Single-causals and double-causals are estimated from observed frequencies, from which a PCI pattern is derived. The pattern is compared with that of every alternative NAT, and a best matching NAT plus the single-causals define the output model. Although the method works well, it does not solve the current problem.

First, P_T does not always yield a well-defined PCI pattern. A bit $pci(c_i, c_j) = u$ in a PCI pattern is well-defined if $P(e^+ \leftarrow c_i^+, c_j^+) < P(e^+ \leftarrow c_k^+)$ $(k = i, j)$, and $pci(c_i, c_j) = r$ is well-defined if $P(e^+ \leftarrow c_i^+, c_j^+) \geq P(e^+ \leftarrow c_k^+)$. The PCI bit

$pci(c_i, c_j)$ is not well-defined when P_T yields $P(e^+ \leftarrow c_i^+) < P(e^+ \leftarrow c_i^+, c_j^+) < P(e^+ \leftarrow c_j^+)$, and nor is the corresponding PCI pattern.

Assuming an underlying NAT model, the above case may still occur due to sampling errors. It can be corrected by soft PCI identification [9]. That is, if c_i and c_j are closer to undermining than reinforcing, treat them as undermining. In particular, assign $pci(c_1, c_2) = u$ if the following holds,

$$|P(e^+ \leftarrow c_1^+, c_2^+) - \min(P(e^+ \leftarrow c_1^+), P(e^+ \leftarrow c_2^+))|$$
$$< |P(e^+ \leftarrow c_1^+, c_2^+) - \max(P(e^+ \leftarrow c_1^+), P(e^+ \leftarrow c_2^+))|,$$

and assign $pci(c_1, c_2) = r$ otherwise. This heuristics often recovers correct PCI bits.

Given an arbitrary P_T, interaction between a pair of its causes may be neither undermining nor reinforcing. Hence, softly identified PCI bits do not always lead to an accurate P_M. On the other hand, requiring well-defined PCI bits leads to a partial PCI pattern with missing bits.

Second, single-causals estimated from frequencies are directly used in the output model. This works well when the underlying model is a NAT and the NAT is recovered correctly. For an arbitrary P_T (not a NAT model in general), no matter which NAT is used, its combination with the single-causals directly from P_T is unlikely to yield an accurate P_M. We demonstrate this in Section 9.

In additional to these fundamental limitations, the method evaluates NATs exhaustively. The number of alternative NATs is $O(n! \, 0.386^{-n} \, n^{-3/2})$ [12]. Even though n is not unbounded in BN CPTs, the exhaustive NAT evaluation can be costly for larger n values.

To overcome these limitations, we propose a new method, TDPN, for approximating an arbitrary P_T with a NAT model. The basic idea is to organize seemingly incomparable NATs in a search tree based on their PCI patterns. Through the search tree, a partial pattern retrieves a small number of promising candidate NATs. Only these NATs are evaluated, which greatly improves efficiency. Rather than using the single-causals directly from P_T, they are searched during evaluation of the candidate NATs.

TDPN consists of an offline and an online phase. The offline phase is conducted before P_T is given. It enumerates root-labeled trees of n roots and constructs a PCI pattern based search tree (PST). Each non-root node of PST is assigned a value of a PCI bit. Each path from the root to a leaf is the PCI pattern of a NAT. For each n value, a PST is constructed offline and is reused online to process any P_T of n causes. Sections 4 and 5 elaborate on PST and its construction. The online phase below starts when a P_T is given.

1. Identify a well-defined partial PCI pattern Pat from P_T.
2. Use Pat to retrieve a set of candidate NATs from the PST.
3. For each candidate, search for single-causals so that the resultant NAT model M minimizes $ED(P_T, P_M)$.
4. From the above candidate NAT models, select M^* of the minimum ED as the approximate NAT model of P_T.

Steps 2 and 3 are elaborated in Sections 6 and 7.

4   PCI Pattern Based Search Trees

A PST is used to retrieve candidate NATs according to a given PCI pattern. First, we specify the PST for retrieving a single NAT from a full PCI pattern.

A NAT of n roots defines a PCI pattern of $N = C(n, 2)$ bits. A PST for a given n has $N + 1$ levels, indexed from 0 to N, with the root t placed at level 0. Each level $k > 0$ is associated with a PCI bit, and each node at the level is labeled by a bit value u or r. All leaf nodes occur at level N. A leaf z exists iff the path from t to z forms the PCI pattern of a NAT, and the NAT is assigned to z. For each node y at level $k < N$, y has a child node x iff bit values at y and x are part of a well-defined PCI pattern.

A PST for $n = 3$ is shown in Fig. 2. It has 4 levels ($N = 3$). Levels 1 to 3 are labeled by PCI bit values. The NAT assigned to each leaf is indicated by an arrow. This PST ($n = 3$) is the only balanced one. For instance, when $n = 4$, we have $N = 6$. A balanced binary tree of 7 levels has $2^N = 64$ leaf nodes, but the PST has only 52 leaf nodes (for 52 NATs). From a PCI pattern, e.g., (r, u, r), a path in the PST can be followed to retrieve a NAT, e.g., T_6. It takes $O(n^2)$ time (since N is quadratic in n).

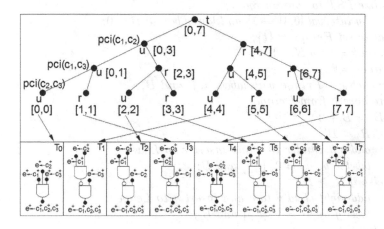

Fig. 2. A PST for $n = 3$ with the NATs at the bottom

If P_T yields a partial PCI pattern, all NATs whose PCI patterns are consistent with the partial pattern need to be retrieved. A partial pattern missing $m > 0$ bits has up to 2^m consistent NATs. Following these paths takes $O(2^m n^2)$ time. We enhance the PST below so that the retrieval from a pattern of m missing bits takes $O(n^2 - m)$ time at best (see below) and $O(2^m n^2)$ time at worst.

After the above PST is formed, index NATs assigned to leaves from left to right in ascending order. Thus, for any node x, the NATs assigned to leaves under x are consecutively indexed. These indexes can be specified by an interval $[i, j]$, which is used to label x. For leaf x, specify $i = j$.

In Fig. 2, NATs assigned to leaves are indexed from left to right at 0 through 7, e.g., T_7 is indexed at 6. If the m missing bits are located at the deepest levels, the search for NATs follows a single path to the level $N - m$ only, and the interval at the last node contains all NATs consistent with the partial pattern. It takes $O(n^2 - m)$ time. For example, the pattern $(r, u, ?)$ leads to the node with the interval $[4, 5]$ and retrieves T_1 and T_6.

5   PST Construction

Let W be the number of distinct NATs of n causes. Before building a PST for n, the W NATs are enumerated [13]. For each NAT, treat its PCI pattern as a base-2 number $R = (a_0, ..., a_j, ..., a_{N-1})$, where $a_j \in \{0, 1\} = \{u, r\}$ and its position has weight 2^{N-1-j}. We refer to PCI patterns and PCI numbers interchangeably.

Algorithm 1. *SetPST*
Input: the number n of causes; the set of W NATs;
1   for each NAT, compute $R = (a_0, ..., a_j, ..., a_{N-1})$;
2   sort PCI numbers ascending into $(R_0, ..., R_{W-1})$;
3   sequence corresponding NATs as $(T_0, ..., T_{W-1})$;
4   initialize PST to contain root t;
5   label t by interval $[0, W - 1]$ and lower bound $B = 0$;
6   initialize set $Fringe = \{t\}$;
7   for level $k = 0$ to $N - 1$,
8   Leaves $= \emptyset$;
9   for each $v \in Fringe$ with labels $[i, j]$ and B,
10   remove v from Fringe;
11 $B' = B + 2^{N-(k+1)}$;
12   if $R_j < B'$,
13   add left child x to v and insert x to Leaves;
14   label x by $a_k = 0$, $[i, j]$ and B;
15   else if $R_i \geq B'$,
16   add right child y to v and insert y to Leaves;
17   label y by $a_k = 1$, $[i, j]$ and B';
18   else
19   add left child x and right child y to v;
20   insert x and y to Leaves;
21   search index d in $(R_i, ..., R_j)$ such that $R_{d-1} < B'$ and $R_d \geq B'$;
22   label x by $a_k = 0$, $[i, d - 1]$ and B;
23   label y by $a_k = 1$, $[d, j]$ and B';
24   Fringe $=$ Leaves;
25 return PST;

To compute the interval $[i, j]$ for each PST node x, associate x with a lower bound of PCI numbers for NATs assigned to leaves below x. The lower bound is discarded once the PST is completed.

Construction of a PST starts at level 0 and proceeds to each deeper level. Current leaf nodes are maintained in a set *Fringe*. Newly generated leaf nodes are recorded in a set *Leaves*. SetPST specifies details of construction.

Lines 12 to 23 expand a current leaf by one child or both, depending on whether each leads to a NAT. Hence, a PST is generally imbalanced. Its size is about $2W$. W has the complexity of $O(n!\ 0.386^{-n}\ n^{-3/2})$ [12] and so does SetPST. It grows faster than $n!$. Fortunately, PSTs are reusable and can be constructed offline once for all.

6 Search for Candidate NATs

At online time, after a well-defined partial PCI pattern *Pat* is obtained from P_T, candidate NATs whose PCI patterns are consistent with *Pat* are retrieved using a PST. For a given *Pat*, some deep PST levels may not be involved, and need not be loaded. Each unloaded deep level reduces the loading time and space by half. Given the $O(n!\ 0.386^{-n}\ n^{-3/2})$ space complexity of PST, such saving is worthwhile.

Denote the set of bits in *Pat* by *Bits*. The partially loaded PST includes only top levels of the full PST such that all bits in *Pat* are covered. Denote the PCI bit for level $i > 0$ by b_i. We assume that the PST has $K + 1$ levels ($K \leq N$) and $b_K \in Bits$.

The retrieval starts from the root t. Each path consistent with *Pat* is followed to a node at level K, where the interval specifies candidate NATs. If $b_1 \in Bits$, one child of t is followed, according to the value of b_1 in *Pat*. Otherwise, both child nodes of t are followed. In general, when the loaded PST includes PCI bits absent from *Bits*, multiple nodes at a given level may be followed. They are maintained in a set *Front*. GetCandidateNAT specifies details of the retrieval.

A *Pat* extracted from P_T may be invalid (no defining NATs). This condition is captured in line 10, where no PST path is consistent with *Pat* and hence an empty candidate set is returned. To handle such cases, one option is to continue as if the error causing bit is a missing bit. It will guarantee an non-empty candidate set in the end.

From uniqueness of PCI pattern [11] and construction of PST by SetPST, it can be proven that, given any well-defined *Pat*, GetCandidateNAT will return exactly the set of NATs whose PCI patterns are consistent with *Pat*. As shown in Section 4, the time complexity is $O(2^m n^2)$.

A refinement to SetPST and GetCandidateNAT can be devised to improve efficiency for both. As mentioned before, each root-labeled tree corresponds to two NATs T_0 and T_1, differing in the type of leaf gate. Hence, their PCI patterns are bitwise complement of each other. Consider a PST built by SetPST. Let x be the leaf assigned with T_0, and y be the leaf assigned with T_1. On the path from root t to x, the PCI bit value at each level is the complement of the corresponding bit value on the path from t to y. That is, the path from t to x is the bitwise complement of the path from t to y.

Algorithm 2. *GetCandidateNAT*
*Input: a partial pattern Pat over a set Bits of PCI bits; a PST of $K + 1$ levels
($K \leq N$) that covers Bits;*

1 *initialize Front = {t}, where t is the root of PST;*
2 *for level $i = 1$ to K,*
3   *Temp = $\emptyset$;*
4   *if $b_i \in Bits$,*
5   *retrieve value $b_i = v_i$ in Pat;*
6   *for each $x \in Front$,*
7   *remove x from Front;*
8   *if x has child y whose bit value is v_i,*
9   *add y to Temp;*
10   *if Temp = $\emptyset$, return $\emptyset$;*
11   *else*
12   *for each $x \in Front$,*
13   *remove x from Front;*
14   *add each child of x to Temp;*
15   *Front = Temp;*
16 *initialize Candidates = $\emptyset$;*
17 *for each $x \in Front$ with interval $[i, j]$,*
18   *Candidates = Candidates $\cup \{T_i, ..., T_j\}$;*
19 *return Candidates;*

This observation allows to refine SetPST and to construct a PST only for
NATs of direct leaf gates. The reduced PST relative to that in Fig. 2 is shown
in Fig. 3. With the reduced PST, GetCandidateNAT must be run twice and the
candidate set is the union of results from both runs. The 2nd run assumes the
bit value complement at each node and a dual leaf gate for each NAT.

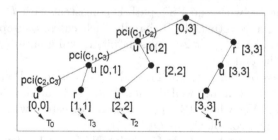

Fig. 3. A PST for NATs of direct leaf gates

This refinement reduces the time of SetPST to half and reduce the space of
the PST to half. For GetCandidateNAT, the total search time in both runs is
equivalent to the time used before. Since the half-sized PST needs to be loaded
only once, the load time is reduced to half.

7   Parameter Search by Steepest Descent

After candidate NATs are obtained, single-causals must be obtained to fully specify corresponding NAT models. It is possible to use the single-causals defined by the target CPT P_T. However, this option, though efficient, does not usually lead to accurate compressed NAT models, as we demonstrate in Section 9. Instead, TPDN searches single-causals for each NAT so that the resultant model M minimizes $ED(P_T, P_M)$. To this end, we apply the method of steepest descent, where a point moves along the surface of a multi-dimensional function, in the direction of the steepest descent, until the gradient is less than a threshold. Alternative techniques, e.g., simulated annealing, may be used instead.

Descent is guided by the function $ED(P_T, P_M)$ between the target CPT P_T and the CPT P_M of a candidate NAT model M. The search space has the dimension n, where each point p is a vector of n single-causals. Each single-causal is bounded in $(0, 1)$. To guard against finding a local extremum, multiple searches are randomly started for each given NAT. If they converge to a single point, the confidence increases that it is the global extremum.

8   Anytime Approximation

The TDPN presented above extracts a well-defined partial PCI pattern Pat from P_T and uses it to search for candidate NATs. As Pat is a heuristic guide to focus evaluation of alternative NATs, there is no guarantee that candidate NATs consistent with Pat include the best NAT. The number of bits in Pat directly affects the number of candidate NATs. The shorter Pat is, the larger the number of candidate NATs and the more accurate the resultant NAT model. On the other hand, evaluating more NATs is more costly, due primarily to time for steepest descent. To balance accuracy and efficiency, we present an anytime enhancement to TDPN that constrains the model approximation computation by a user specified time limit.

Let W be the number of NATs of n causes, rt be the average runtime for steepest descent per NAT, Rt be the user specified runtime, and m be the number of missing bits in Pat. The number of candidate NATs that can be processed in Rt time is about Rt/rt. The number of NATs consistent with a PCI pattern of m missing bits is about $2^{m-C(n,2)} W$. Hence, the largest m such that $2^{m-C(n,2)} W < Rt/rt$ best balances accuracy and efficiency, given the user time.

What remains is to decide the m missing bits. Suppose P_T yields $C(n, 2) - m'$ well-defined PCI bits. If $m > m'$, then $m - m'$ deepest bits in PST (Section 4) can be dropped from Pat before it is used for NAT retrieval. If $m < m'$, then $m' - m$ soft-identified PCI bits that are shallowest in PST can be added. This ensures a Pat of $C(n, 2) - m$ length, where well-defined PCI bits are used as much as possible.

9   Experimental Evaluation

To evaluate the effectiveness of TDPN, we conducted four groups of experiments. The 1st group compares approximation accuracy of TDPN with the optimal baseline, the 2nd compares TDPN with the well-known noisy-OR, and the 3rd demonstrates effectiveness of the anytime enhancement to TDPN. Five batches of target CPTs (a total of 353 CPTs) were used in these experiments. The 4th group examines the offline PST construction time.

Approximation Accuracy. The 1st group mainly evaluates the approximation accuracy of TDPN. The 1st batch of 99 target CPTs were randomly generated (the most general targets), where $n = 4$ and the leak probability is 0 (Section 2).

For comparison, the approximation accuracy by exhaustive search over all NAT models is used as the baseline. We refer to the method by OP. The choice of $n = 4$ (with $W = 52$) was made because running OP for a larger n value is much more costly, e.g., $W = 472$ for $n = 5$.

For each target CPT P_T, OP computes an optimal NAT model as follows. For each distinct NAT, the best set of single-causals is obtained by steepest descent and ED of the resultant NAT model is calculated. The model of the minimum ED is selected by OP. Note that exhaustive search over all NATs by OP is the same as an earlier method [9]. On the other hand, the earlier method uses single-causals from the target CPT, while OP produces them by steepest descent. Hence, the ED resultant from OP signifies the best accuracy that NAT models can achieve.

For each P_T, a NAT model is also obtained by TDPN, as well as its ED. If TDPN selects the same NAT as OP, then difference between the two EDs is zero, and the performance of TDPN for this P_T is deemed optimal.

The 1st box in Fig. 4 (left) depicts EDs obtained by OP, where ends of whiskers represent the minimum and maximum. The sample mean is 0.1787. It shows that, under the most general condition, NAT model approximation can achieve a reasonably high accuracy, while reducing the CPT complexity from being exponential to being linear on n.

The 2nd box in Fig. 4 (left) depicts EDs by TDPN. Out of 99 target CPTs, TDPN selected the same NAT as OP in 69 of them. The sample mean of EDs is 0.1855 and is fairly close to that of OP. Fig. 4 (right) shows the runtime comparison between OP (average 7.7 sec) and TDPN (average 0.8 sec). This result shows that PCI heuristics of TDPN works well. Using 10% of time of OP, TDPN either found the optimal NAT model or one very close to the optimal.

The runtime ratio of OP versus TDPN is expected to grow more than exponentially on n (in favour of TDPN). This is due to super-exponential growth of W on n and the same growth of OP runtime. On the other hand, the number of NATs evaluated by TDPN is determined by the length of the partial PCI pattern and can be well controlled through the anytime enhancement.

Each P_T was also run with a modified version (SOFT) of TDPN. Instead of using a well-defined, partial PCI pattern, SOFT uses soft PCI identification [9]

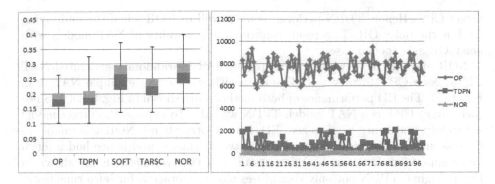

Fig. 4. Left: Boxplot of EDs of NAT models by OP, TDPN, SOFT, TARSC and NOR from the 1st batch of target CPTs. Right: Runtime by OP, TDPN and NOR in msec.

(introduced in Section 3) and a full, softly-defined PCI pattern. The 3rd box in Fig. 4 (left) depicts EDs by SOFT. Out of 99 target CPTs, it selected the same NAT as OP did in 30 of them (less than half compared to TDPN). Its sample mean of EDs is 0.2550 and much worse than TDPN, relative to the OP baseline. This comparison demonstrates the superiority of partial PCI patterns consisting of only well-defined PCI bits.

In addition, each P_T was run with an earlier method [9], referred to as TARSC. It can be viewed as a modified OP, where single-causals of the target CPT are used directly. The 4th box in Fig. 4 (left) depicts EDs by TARSC. It selected the same NAT as OP did in 54 of them and its sample mean of EDs is 0.2234. Even though TARSC searches NATs exhaustively, it performed worse than TPDN, relative to the OP baseline. This comparison demonstrates the benefit of single-causal search by steepest descent.

The results by SOFT and TARSC show that the existing technique [9] for recovering underlying NAT models does not work well for compressing general CPTs. Hence, the innovations in TDPN (well-defined PCI bits, partial PCI patterns, PST guided search, and single-causals by steepest descent) are both effective and necessary.

Comparison with Noisy-OR. The 2nd group of experiments compares TDPN with noisy-OR approximation. For each P_T, the latter method (NOR) searches single-causals of a noisy-OR model by steepest descent.

The result of running NOR on the 1st batch of target CPTs is shown by the 5th box in Fig. 4 (left). For two target CPTs, NOR selected the same NAT as OP did, compared with 69 by TDPN. The sample mean of EDs by NOR is 0.2662. In comparison, TDPN is much more accurate, relative to the OP baseline.

Both NOR and TPDN were run on a 2nd batch of 100 randomly generated noisy-OR target CPTs (each CPT follows a noisy-OR model) with $n = 7$. For each P_T, TDPN returned the same NAT as NOR did without exception. The EDs by both methods are about 0.0013, and both ran about 1.5 seconds per target CPT. Hence, TDPN performs equally well as NOR when the underlying CPT is the noisy-OR. This result confirms the generality of NAT models with noisy-OR as a special case.

NOR and TPDN were run on a 3rd batch of 99 randomly generated NAT CPTs (each CPT follows a NAT model) with $n = 7$. An example NAT is in Fig. 1 (c). The ED performance of both methods are shown in Fig. 5 (left). Since each target CPT is a NAT model, TPDN was able to express the target model accurately, and hence close to zero ED. On the other hand, NOR was unable to express undermining causal interaction between causes, and hence had a much lower approximation accuracy. The runtimes of the two methods are shown in Fig. 5 (right). TPDN not only compresses more accurately, but also runs faster than NOR. This is because TDPN selected the NAT that matched well with the target CPT, making subsequent search for single-causals converging quickly. On the other hand, causal interactions assumed by noisy-OR did not match well with the NAT modeled CPT, slowering down the search for single-causals.

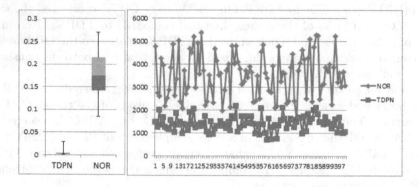

Fig. 5. Left: Euclidean distances obtained by TDPN and NOR from the 3rd batch of target CPTs. Right: Runtime by TDPN and NOR in msec.

The fourth batch was run on target CPTs from three real BNs: Alarm [14], Hailfinder [15], and HeparII [16]. Since all three BNs use multi-valued variables, they were coarsened equivalently into binary, and target CPTs were collected from all variables of 3 or more parents since a non-trivial NAT model has at least 3 causes. A total of 25 target CPTs were collected: 3 from Alarm, 6 from Hailfinder, and 16 from HeparII. Among them, 14 CPTs have $n = 3$, 7 CPTs have $n = 4$, 3 CPTs have $n = 5$, and one CPT has $n = 6$.

The ED performances of TPDN and NOR are shown in Fig. 6. For almost all target CPTs, the data points are above the $X = Y$ line, signifying smaller ED by TPDN. A Friedman test with $k = 2$ [17] resulted in the test statistic 8.8947, which is larger than the critical 0.01 χ^2 value 6.63. Therefore, TDPN compresses these real BN CPTs significantly more accurate than NOR.

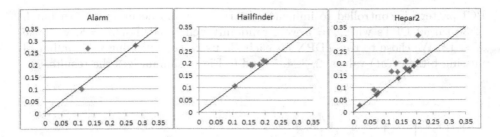

Fig. 6. Euclidean distances obtained by TDPN (X-axis) and NOR (Y-axis) from the 4th batch of target CPTs

The sample mean of EDs by TDPN over all 25 target CPTs is 0.1497. Compared with 0.1855 from the random target CPTs, it suggests that TDPN approximates real BN CPTs more accurately than the random CPTs. In other words, real BN CPTs are closer to NAT models than random CPTs.

The runtimes for the 4th batch are shown in Fig. 7. Because a single model structure was evaluated by NOR, while multiple NATs were evaluated by TDPN, NOR runs faster. The three most time consuming CPTs in HeparII (at top of chart) took TDPN between 16 to 50 seconds due to the need to evaluate a large number of alternative NATs.

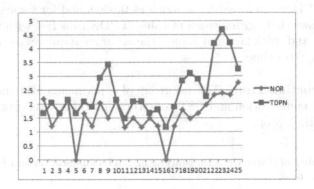

Fig. 7. Log10 runtimes from 4th batch in order of Alarm, Hailfinder and HeparII

By comparing TDPN runtimes between the 3rd batch (Fig. 5, mostly < 2 sec) and 4th (up to 50 sec), we see that TDPN is able to adapt its amount of computation according to the target CPT. When the target CPT is fairly close to a NAT model, very few alternative NATs are evaluated by steepest descent. More NATs are evaluated only when it is necessary.

Effectiveness of Anytime Enhancement. The 3rd group of experiment examines effectiveness of anytime enhancement to TDPN, where the length of a

PCI pattern is controlled to influence the runtime and accuracy. A 5th batch of 30 target CPTs with $n = 5$ were randomly generated. For $n = 5$, we have $N = 10$. We chose to run TDPN for each P_T using the following m (number of missing bits in Pat) values: 2, 3, 4, 5, and 6. Fig. 8 summarizes the results.

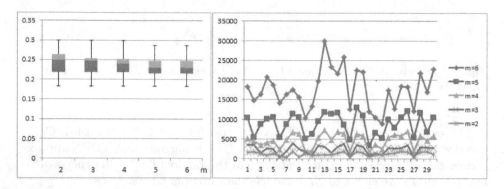

Fig. 8. Left: Euclidean distances obtained by TDPN from the 5th batch of target CPTs. Right: Runtimes in msec.

As m was increased from 2 to 6, the number of candidates NATs and runtime was increased by more than 10 times, though still less than 30 sec. For some P_T, the resultant ED was reduced by as much as 0.0609, and for some as little as 0. The result shows that, with a small m value, TPDN runs fast with a reasonably high accuracy and, with larger m values, the accuracy improves moderately with the controlled extra time.

PST Construction Time. The 4th group of experiments examines the offline time for NAT enumeration and PST construction. Table 1 reports the runtime using a 2.9 GHz laptop.

Table 1. Number of distinct NATs, runtime of NAT enumeration and PST construction, in relation to n

n	6	7	8	9
# NATs	5,504	78,416	1.32×10^6	2.56×10^7
NAT enu.	110ms	0.3s	4.6s	134.5s
PST con.	828ms	16.8s	1.8h	42.9h

Table 1 shows that for $n \leq 9$, construction of PSTs are practical, since it is performed offline once for all and the resultant PSTs are reusable. Due to conditional independence encoded in BN structures, the n value for BN CPTs are not unbounded and are unlikely to be much larger than 9. Furthermore, for target CPTs with $n \geq 10$, it is possible to decompose the BN family into two or more subsets where $n < 10$ and to apply TDPN to each. Constructing PSTs for $n \geq 10$ will then be unnecessary. We leave this to a sequel of the current work.

10   Conclusion

This paper presents three key contributions in order to compress BNs for deployment in low-resource platforms. First, we defined PCI pattern based search trees (PSTs) and developed an algorithm for their construction. PSTs organize seemingly incomparable NAT structures into a uniform searchable representation. PSTs enable exponential online complexity [9] to be shifted to offline and to be incurred once for all. Second, we presented an algorithm for searching highly promising NAT structures based on partial PCI patterns and proposed its combination with steepest descent. This combination enables efficient online compression of general target CPTs with reasonable accuracy. Third, our experimental study demonstrated several key results: (a) many general target CPTs can be approximated fairly well by NAT models, (b) NAT models lead to more accurate approximations than noisy-OR models, and (c) TDPN is superior in approximation accuracy than alternatives where either soft PCI identification or single-causals from target CPT are used.

TDPN overcomes limitations of several existing techniques. Relative to QPNs, a NAT-modeled BN retains the full range of probability. Relative to the noisy-OR, NAT-modeled BNs encode both reinforcing and undermining. Hence, TDPN leads to more expressive compression. Earlier method [9] can only recover a target CPT accurately if it is truly a NAT model. TDPN does not assume an underlying NAT model and still achieves a reasonably high accuracy. Hence, TDPN provides a general method to reduce the space complexity of BNs from $O(\mu \ \kappa^n)$ to $O(\mu \ \kappa \ n)$ for $\kappa = 2$. It can be viewed as complementing techniques such as divorcing by providing yet another alternative. Relative to coarsening whose space complexity is still exponential on n, complexity of NAT-modeled BNs is linear on n. Hence, TDPN has better space efficiency. The earlier method [9] takes an exponential online time, while online computation of TDPN is efficient ($O(2^m n^2)$ where m is a user-controllable, small integer). BNs compressed by TDPN support more efficient inference: lazy propagation in NAT-modeled BNs can be one-order of magnitude faster [18].

Extension of TDPN to multi-valued NAT models where $\kappa \geq 2$ remains the most important for further research. We have shown that TDPN can practically compress binary CPTs of up to at least $n = 9$. Although complexity of PST construction is exponential, the computation is offline, it is incurred once for all, and n is not unbounded in BNs. For target CPTs with $n \geq 10$, promising directions include decomposition and parallel PST construction. Relaxation to target CPTs of positive leak probabilities will also extend the generality of TPDN.

Acknowledgement. We thank anonymous reviewers for their helpful comments. Financial support through Discovery Grant from NSERC, Canada is acknowledged.

References

1. Wellman, M.: Graphical inference in qualitative probabilistic networks. Networks 20(5), 687–701 (1990)
2. Renooij, S., Parsons, S., van der Gaag, L.: Context-specific sign-propagation in qualitative probabilistic networks. Artificial Intelligence 140, 207–230 (2002)
3. Chang, K., Fung, R.: Refinement and coarsening of Bayesian networks. In: Proc. Conf. on Uncertainty in Artificial Intelligence, pp. 475–482 (1990)
4. Jensen, F., Nielsen, T.: Bayesian Networks and Decision Graphs, 2nd edn. Springer, New York (2007)
5. Pearl, J.: Probabilistic Reasoning in Intelligent Systems: Networks of Plausible Inference. Morgan Kaufmann (1988)
6. Galan, S., Diez, F.: Modeling dynamic causal interaction with Bayesian networks: temporal noisy gates. In: Proc. 2nd Inter. Workshop on Causal Networks, pp. 1–5 (2000)
7. Lemmer, J., Gossink, D.: Recursive noisy OR - a rule for estimating complex probabilistic interactions. IEEE Trans. on System, Man and Cybernetics, Part B 34(6), 2252–2261 (2004)
8. Xiang, Y., Jia, N.: Modeling causal reinforcement and undermining for efficient CPT elicitation. IEEE Trans. Knowledge and Data Engineering 19(12), 1708–1718 (2007)
9. Xiang, Y., Truong, M., Zhu, J., Stanley, D., Nonnecke, B.: Indirect elicitation of NIN-AND trees in causal model acquisition. In: Benferhat, S., Grant, J. (eds.) SUM 2011. LNCS, vol. 6929, pp. 261–274. Springer, Heidelberg (2011)
10. Madsen, A., Jensen, F.: Lazy propagation: A junction tree inference algorithm based on lazy evaluation. Artificial Intelligence 113(1-2), 203–245 (1999)
11. Xiang, Y., Li, Y., Zhu, Z.J.: Towards effective elicitation of NIN-AND tree causal models. In: Godo, L., Pugliese, A. (eds.) SUM 2009. LNCS, vol. 5785, pp. 282–296. Springer, Heidelberg (2009)
12. Schroeder, E.: Vier combinatorische probleme. Z. f. Math. Phys. 15, 361–376 (1870)
13. Xiang, Y., Zhu, Z.J., Li, Y.: Enumerating unlabeled and root labeled trees for causal model acquisition. In: Gao, Y., Japkowicz, N. (eds.) AI 2009. LNCS, vol. 5549, pp. 158–170. Springer, Heidelberg (2009)
14. Beinlich, I., Suermondt, H., Chavez, R., Cooper, G.: The alarm monitoring system: A case study with two probabilistic inference techniques for belief networks. In: Proc. 2nd European Conf. Artificial Intelligence in Medicine, pp. 247–256 (1989)
15. Abramson, B., Brown, J., Edwards, W., Murphy, A., Winkler, R.L.: Hailfinder: A Bayesian system for forecasting severe weather. Inter. J. Forecasting 12, 57–71 (1996)
16. Onisko, A.: Probabilistic Causal Models in Medicine: Application to Diagnosis of Liver Disorders. PhD thesis, Institute of Biocybernetics and Biomedical Engineering, Polish Academy of Science (2003)
17. Sheskin, D.: Handbook of Parametric and Nonparametric Statistical Procedures, 3rd edn. Chapman & Hall/CRC (2004)
18. Xiang, Y.: Bayesian network inference with NIN-AND tree models. In: Cano, A., Gomez-Olmedo, M., Nielsen, T. (eds.) Proc. 6th European Workshop on Probabilistic Graphical Models, Granada, pp. 363–370 (2012)

A Study of Recently Discovered Equalities about Latent Tree Models Using Inverse Edges

Nevin L. Zhang[1], Xiaofei Wang[2], and Peixian Chen[1]

[1] Department of Computer Science and Engineering,
The Hong Kong University of Science and Technology, Hong Kong
{lzhang,pchenac}@cse.ust.hk
[2] School of Mathematics and Statistics, Northeast Normal University, China
wangxf341@nenu.edu.cn

Abstract. Interesting equalities have recently been discovered about latent tree models. They relate distributions of two or three observed variables with joint distributions of four or more observed variables, and with model parameters that depend on latent variables. The equations are derived by using matrix and tensor decompositions. This paper sheds new light on the equalities by offering an alternative derivation in terms of variable elimination and structure manipulations. The key technique is the introduction of inverse edges.

Keywords: Matrix decomposition, parameter estimation, latent tree models.

1   Introduction

The Expectation-Maximization (EM) algorithm, introduced by Dempster et al. [4], is commonly used to estimate parameters of latent variable models. A well-known drawback of EM is that it tends to get trapped in local optima. New estimation techniques have recently been developed that do not share the shortcoming [7,5,2]. Those techniques are suitable for a class of latent variable models called latent tree models [11,8]. They are based on matrix decompositions and tensor computations.

Anandkumar et al. [2] further developed the techniques and provided a unified framework in terms of low order moments and tensor decompositions. Their works cover a number of latent variable models, including latent tree models, Gaussian mixture models, hidden Markov models, and Latent Dirichlet allocation.

At the heart of the techniques are equalities that related model parameters to quantities that can be directly estimated from data. Equalities of similar flavor are discovered by Parikh et al. [10] that relate joint distributions of four or more observed variables in a latent tree model with distributions of two or three observed variables. Those equalities allows one to estimate the joint probability of a particular assignment of all observed variables without estimating the model parameters.

L.C. van der Gaag and A.J. Feelders (Eds.): PGM 2014, LNAI 8754, pp. 567–580, 2014.

In this paper, we study the equalities in the context of latent tree models. We augment latent tree models with what we call inverse edges. The equalities are then derived by eliminating variables from the augmented models according to different orders. The derivations are insightful and give intuitive explanations as why the equalities hold.

The rest of this paper is organized as follows. In Section 2 we introduce preliminary concepts and notations, and in Section 3 we introduce the key concept of inverse edges. Equalities for joint probability estimation are derived in Section 4, and equalities for parameter estimation are derived in Section 5. Conclusions are provided in Section 6.

2   Preliminaries

We start by introducing several technical concepts.

2.1   Markov Random Fields

A Markov random field (MRF) over a set of discrete variables is defined by a list of potentials. Each potential is a non-negative function of some of the variables. The product of the potentials is the joint distribution of all the variables.

An example MRF is shown in Figure 1. There are four potentials $\phi_1(A, B)$, $\phi_2(B, C, D)$, $\phi_3(D, E)$ and $\phi_4(E, F)$. The figure shows the structure of the MRF. The edge between A and B indicates that there is a potential for them; the hyperedge consisting of B, C, and D indicates the same for those three variables; and so on.

2.2   Latent Tree Model

A *latent tree model (LTM)* is a tree structured MRF where the leaf nodes represent observed variables, while the internal nodes represent latent variables [11]. An example is shown in Figure 2. The variables A, B, C and D are observed, while H and G are latent. There are multiple ways to specify parameters for the LTM. One way is to give: $P(A|H)$, $P(B|H)$, $P(H, G)$, $P(C|G)$ and $P(D|G)$.

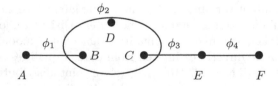

Fig. 1. An example on Markov random fields

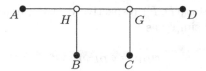

Fig. 2. An example on latent trees

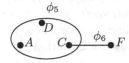

Fig. 3. Result of eliminating B and E from Figure 1

2.3 Variable Elimination in MRF

To *eliminate a variable* X from an MRF means to remove all the potentials that involve X, compute their product, marginalize out X from the product, and add the result to the MRF as a new potential [12].

Consider eliminating the variable B from the MRF shown in Figure 1. The potentials ϕ_1 and ϕ_2 are first removed and the following new potential is created:

$$\phi_5(A, C, D) = \sum_B \phi_1(A, B)\phi_2(B, C, D). \tag{1}$$

If we further eliminate E, then the potentials ϕ_3 and ϕ_4 are also removed and the following new potential is created:

$$\phi_6(D, F) = \sum_E \phi_3(D, E)\phi_4(E, F). \tag{2}$$

The new model is shown in Figure 3. The new structure is obtained from the old one by deleting the edges that involve B and E, and creating two new edges that consist of the neighbors of B and E respectively.

2.4 Matrix Representation of Potentials

For any variable X, use $|X|$ to denote its *cardinality*, i.e., the number of possible values. Denote the values of X as 1, 2, ..., $|X|$. A generic value of X will be referred to using the lower case letter x.

A potential of two variables can be represented using a matrix. Take $\phi_1(A, B)$ for example. The matrix representation of $\phi_2(A, B)$ is a $|A| \times |B|$ matrix. The value at the a-th row and b-th column is $\phi_2(A=a, B=b)$. We denote this matrix as ϕ_{AB}, and the value $\phi_2(A=a, B=b)$ as ϕ_{ab}.

Matrix representation allows us to write the results of variable elimination as matrix multiplications. Let ϕ_{DE}, ϕ_{EF}, ϕ_{DF} be the matrix representations of

$\phi_3(D,E)$, $\phi_4(E,F)$ and $\phi_6(D,F)$ respectively. Using matrix multiplication, we can write Equation (2) simply as

$$\phi_{DF} = \phi_{DE}\phi_{EF}. \tag{3}$$

For particular values a, b, c and d of the corresponding variables, denote the values $\phi_5(A{=}a, C{=}c, D{=}d)$ and $\phi_2(B{=}b, C{=}c, D{=}d)$ as ϕ_{acd} and ϕ_{bcd} respectively. Use ϕ_{Bcd} to denote the column vector where the value on the b-th row is ϕ_{bcd}, and use ϕ_{aB} to denote the column vector where the value on the b-th row is ϕ_{ab}. Those notations allow us to rewrite Equation (1) as:

$$\phi_{acd} = \phi_{aB}^T \phi_{Bcd}, \tag{4}$$

for any given value a, c and d of A, C and D,

For the probability distributions $P(A|H)$, $P(B|H)$, $P(H,G)$, $P(C|G)$ and $P(D|G)$ of the LTM shown in Figure 2, their matrix representations are denoted as $P_{A|H}$, $P_{B|H}$, P_{HG}, $P_{C|G}$ and $P_{D|G}$ respectively.

3   Identity and Inverse Edges

In an MRF, potentials are non-negative. In this paper, for introducing inverse matrices or inverse edges into the potential operations, we generalize the concept by allowing potentials to take negative values. This results in *generalized MRF*. In a generalized MRF, the product of all potentials is a *joint potential* on all the variables. It is not necessarily a probability distribution. *Marginal potentials* can be obtained from the joint through marginalization. Variable elimination and the corresponding manipulations with model structure are the same as in the case of MRF. LTMs are MRFs, and hence are generalized MRFs.

3.1   Identity Edges

For the rest of this section, we consider only generalized MRFs with tree structures. Let X and X' be two neighboring variables that have equal cardinality. Suppose the matrix representation $\phi_{XX'}$ of the potential $\phi(X,X')$ is an identity matrix I. Then we say that the edge (X,X') is an *identity edge*. The potential matrix is written as $I_{XX'}$.

Let X be a variable with two or more neighbors. *Splitting X into an identity edge* means to: (1) create another variable X' such that $|X'| = |X|$, (2) divide the neighbors of X between X' and X, (3) connect X' and X and make it an identity edge.

In the model of Figure 4 (a), splitting the variable B results in the model of (b). The identity edge (B, B') is introduced. The potential matrix $\phi_{B'C}$ equals ϕ_{BC}. It is the same matrix, except the rows are indexed by values of B', not of B.

The following theorem is obvious.

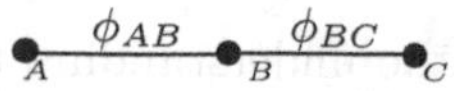

$$\bullet \xrightarrow{\phi_{AB}} \bullet \xrightarrow{\phi_{BC}} \bullet$$
$$A \qquad\quad B \qquad\quad C$$

(a) A simple MRF.

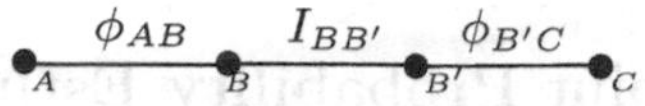

$$\bullet \xrightarrow{\phi_{AB}} \bullet \xrightarrow{I_{BB'}} \bullet \xrightarrow{\phi_{B'C}} \bullet$$
$$A \qquad\quad B \qquad\quad B' \qquad\quad C$$

(b) Obtained from (a) by splitting variable B into an identity edge.

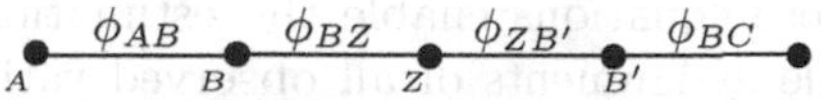

$$\bullet \xrightarrow{\phi_{AB}} \bullet \xrightarrow{\phi_{BZ}} \bullet \xrightarrow{\phi_{ZB'}} \bullet \xrightarrow{\phi_{BC}} \bullet$$
$$A \qquad\quad B \qquad\quad Z \qquad\quad B' \qquad\quad C$$

(c) Obtained from (b) by replacing the identity edge (B, B') with a pairs of edges (B, Z) and (Z, B') that are inverse of each others.

Fig. 4. Illustration of identity and inverse edge introduction

Theorem 1. *Let m be a generalized MRF with a tree structure, and X be a variable in m and $\mathbf{Y}$ be a subset of other variables. Suppose X has two or more neighbors. Let m' be a new model obtained from m by splitting X into an identity edge. Then the marginal potential of $\mathbf{Y}$ in the model m equals that in m'.*

Another way to state the theorem is that the elimination of X' from m' results in the model m.

3.2   Inverse Edges

Continue with the example shown in Figure 4 (b). Suppose there is another variable Z such that $|Z| \geq |B|$. Let ϕ_{BZ} be matrix representation for a potential of the two variables. Construct a new model by: (a) inserting Z between B and B', (2) setting the potential matrix of (B, Z) to be ϕ_{BZ}, and (3) setting the potential matrix $\phi_{ZB'}$ of (Z, B') to satisfy that $I_{BB'} = \phi_{BZ}\phi_{ZB'}$. This is well-defined because $|B'| = |B|$. The new model is shown in Figure 4 (c).

It is obvious that $\phi_{BZ} = \phi_{ZB'}^{-1}$ when $|Z| = |B|$. So, we say that (Z, B') is the *inverse edge* of (B, Z) and vice versa. Because of $I_{BB'} = \phi_{BZ}\phi_{ZB'}$, eliminating Z from (c) gives us (b).

In general, we have the following theorem.

Theorem 2. *Suppose (X, X') is an identity edge in a generalized MRF m whose model structure is a tree. Let $\mathbf{Y}$ be a subset of other variables. Constructing a new model m' by replacing the edge (X, X') with a pair of edges that are inverse of each other. Then the marginal potential of $\mathbf{Y}$ in the model m equals that in m'.*

Two remarks are in order. First, the matrix $\phi_{ZB'}$ might contain negative values even though the matrix ϕ_{BZ} does not. This is why we need to generalize the concept of MRF. Second, when $|Z| = |B|$, the notation $\phi_{ZB'}$ in Figure 4 (c)

can be replaced by ϕ_{BZ}^{-1}, with the understanding the columns of the matrix are indexed using values of B'.

4   Equalities for Joint Probability Estimation

Parikh *et al.* [10] have recently discovered equations about LTMs that relate distributions of four or more observed variables to distributions of two or three observed variables. Those equations enable the estimation of the joint probability of particular value assignments of all observed variables without having to estimate the model parameters. The equations were derived using matrix decomposition and tensor computation. In this section, we give an alternative derivation using inverse edges.

4.1   Quartet Trees

We start with model shown in Figure 2. It is called a *quartet model* and will be referred to as M1. We assume that all the variables have equal cardinality and all the probability matrices have full rank. In the following, we will derive an equation that relates $P(A, B, C, D)$ to $P(A, B, D)$, $P(B, D)$, and $P(B, C, D)$.

Starting from M1, we split the two latent nodes H and G into two identity edges (H', H) and (G, G'), resulting in the model M2 of Figure 5 (a). Next, we replace the edge (H', H) with a pairs of edges (H', D') and (D', H), where D' is new variable such that $|D'| = |D|$. The potential matrices for the two edges are set as: $\phi_{H'D'} = P_{HD}$ and $\phi_{D'H} = P_{HD}^{-1}$. Here P_{HD} is the matrix representation of $P(H, D)$. It is invertible because all the probability matrices in M1 have full rank. So, the two edges (H', D') and (D', H) are inverse of each other. Similarly, we replace the edge (G, G') with two edges (G, B') and (B', G') that are inverse of each other. The resulting model $M3$ is shown in Figure 5 (b). According to Theorems 1 and 2, the joint distribution $P(A, B, C, D)$ in M3 is the same as that in M1. In other words, eliminating D', B', H' and G' from $M3$ yields M1. Note that further eliminating H and G in M1 gives us $P(A, B, C, D)$.

Another way to compute $P(A, B, C, D)$ in M3 is to eliminate the variables in the following order: H', G', H, G, D' and B'. The model M4 shown in Figure 5 (c) is that we obtain after eliminating the first four variables. In the following, we explain how the result is obtained.

The elimination of H' involves three potentials. In function form, they are $P(A|H')$, $P(B|H')$, and $P(H', D')$. The elimination of H' gives us the potential

$$\sum_{H'} P(A|H')P(B|H')P(H', D')P(A, B, D').$$

Similarly, the elimination of G' gives us $P(B', C, D)$.

The elimination of H and G also involves three potentials. In matrix form, they are $\phi_{D'H} = P_{HD}^{-1}$, P_{HG} and $\phi_{GB'} = P_{BG}^{-1}$. Note that in M1 we have $P_{HD} = P_{HG}P_{D|G}^T$ and $P_{BG} = P_{B|H}P_{HG}$. So, the elimination of H and G gives us the following potential:

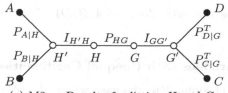

(a) M2 — Result of splitting H and G in the model of Figure 2.

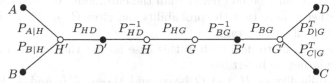

(b) M3 — Results of replacing the identity edges in M2 with pairs edges that are inverse of each other.

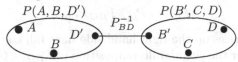

(c) M4 — Result of eliminating H', G', H and G from M3. Two of the potentials are given in function form, while the third in matrix form.

Fig. 5. Transforms applied on the model of Figure 2

$$\phi_{D'H}P_{HG}\phi_{GB'} = P_{HD}^{-1}P_{HG}P_{BG}^{-1}$$
$$= (P_{D|G}^T)^{-1}P_{HG}^{-1}P_{HG}P_{HG}^{-1}P_{B|H}^{-1}$$
$$= (P_{D|G}^T)^{-1}P_{HG}^{-1}(P_{B|H})^{-1}$$
$$= (P_{B|H}P_{HG}P_{D|G}^T)^{-1} = P_{BD}^{-1}.$$

This is the matrix representation of the potential for the edge (D', B'), i.e., $\phi_{D'B'} = P_{BD}^{-1}$.

Finally, the elimination of D' and B' in M4 gives us the distribution $P(A, B, C, D)$. For specific values a, b, c and d of the corresponding variables, denote the probability $P(A=a, B=b, C=c, D=d)$ as P_{abcd}. It is clear that

$$P_{abcd} = P_{abD}^T P_{BD}^{-1} P_{Bcd}, \tag{5}$$

where P_{abD} and P_{Bcd} are column vectors obtained from the joint distributions $P(A, B, D)$ and $P(B, C, D)$ in the way described in Section 2.4.

The following theorem summarizes the foregoing derivations, which construct inverse edges and potentials for variable elimination:

Theorem 3. *In the quartet model of Figure 2, suppose all the variables have equal cardinality and all probability matrices have full rank. Then the distribution*

$P(A, B, C, D)$ can be computed from $P(A, B, D)$, $P(B, D)$, $P(B, C, D)$ using Equation (5).

4.2   Observed Variables with Unequal Cardinalities

Next we generalize Theorem 3 to the case where the observed variable might have unequal cardinalities. The latent variables are still required to have equal cardinality and it must be no greater than the cardinality of any observed variable. We further require that the probability matrices $P_{A|H}$, $P_{B|H}$, P_{HG}, $P_{C|G}$ and $P_{D|G}$ have full column rank.

The technical issue to deal with in this case is that the matrices P_{HD}, P_{BG} and P_{BD} might not be invertible.

Let the cardinality of H and G be r, and those of B and D be s and t respectively. In model M1 we have

$$P_{BD} = P_{B|H} P_{HG} P_{D|G}^T.$$

Because all the matrices on the right hand side have full column rank, the rank of P_{BD} is r. Consider the singular decomposition of $P_{BD} = U \Lambda V^T$, where Λ is a $r \times r$ diagonal matrix, U and V are $s \times r$ and $t \times r$ column orthogonal matrices respectively. We have

$$\begin{aligned} P_{BD} &= P_{B|H} P_{HG} P_{D|G}^T \\ &= P_{B|H} P_{HG} P_{HG}^{-1} P_{HG} P_{D|G}^T \\ &= P_{BG} P_{HG}^{-1} P_{HD}. \end{aligned} \tag{6}$$

Consequently,

$$P_{BG} P_{HG}^{-1} P_{HD} = U \Lambda V^T,$$
$$U^T P_{BG} P_{HG}^{-1} P_{HD} V = \Lambda.$$

This implies that $U^T P_{BG}$ and $P_{HD} V$ are invertible.

Construct the model M3 as in the previous subsection, except that we set the potential matrices for the edges (D', H) and (G, B') as follows:

$$\phi_{D'H} = V(P_{HD}V)^{-1}, \phi_{GB'} = (U^T P_{BG})^{-1} U^T.$$

It is clear that the product of P_{HD} and $V(P_{HD}V)^{-1}$ is an identity matrix. So, the edge (D', H) is the inverse edge of (H', D'). It is also clear that the product of $(U^T P_{BG})^{-1} U^T$ and P_{BG} is an identity matrix. So, the edge (G, B') is the inverse edge of (B', G').

When we move from M3 to M4, everything is the same as in the previous subsection, except that the elimination of H and G now involves different potential matrices. The result is

$$\begin{aligned} \phi_{D'H} P_{HG} \phi_{GB'} &= V(P_{HD}V)^{-1} P_{HG} (U^T P_{BG})^{-1} U^T \\ &= V(U^T P_{BG} P_{HG}^{-1} P_{HD} V)^{-1} U^T \\ &= V(U^T P_{BD} V)^{-1} U^T, \end{aligned}$$

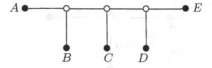

Fig. 6. A general latent tree model

where the last equality is due to Equation 6. This is the potential matrix for the edge (D', B').

Consequently, for any specific values a, b, c and d of the variables, we have

$$P_{abcd} = P_{abD}^T V (U^T P_{BD} V)^{-1} U^T P_{Bcd}. \tag{7}$$

Theorem 4. *In the quartet model of Figure 2, suppose the conditions specified in the first paragraph of this subsection hold. Then the distribution $P(A, B, C, D)$ can be computed from $P(A, B, D)$, $P(B, D)$, $P(B, C, D)$ using Equation (7), where U and V are from the singular decomposition of P_{BD}.*

Note that Equation (7) is same as Equation (5) except that the matrices U and V are used to deal with the issue that P_{BD} might not be invertible.

4.3   General Trees

To apply Theorems 3 and 4 in general trees, we divide all the variables into four groups S_1, S_2, S_3 and S_4 such that, when each group is viewed as a joint variable, the relationship among them is a quartet tree. By applying the theorems, we can compute $P(S_1, S_2, S_3, S_4)$ to from $P(S_1, S_2, S_4)$, $P(S_2, S_4)$, $P(S_2, S_3, S_4)$. In the process, we need to invert the matrix representation of $P(S_2, S_4)$. For computational efficiency, S_2 and S_4 should be singletons. The same strategy can then be repeated on $P(S_1, S_2, S_4)$ and $P(S_2, S_3, S_4)$ until all the distributions needed for the computation are for no more than 3 variables. Such distributions are directly estimated from data.

Let us illustrate the strategy using the model shown in Figure 6. First, partition the variables into four groups as follows: $S_1 = \{A\}$, $S_2 = \{B\}$, $S_3 = \{C, D\}$ and $S_4 = \{E\}$. This allows us to reduce $P(A, B, C, D, E)$ to $P(A, B, E)$, $P(B, E)$, and $P(B, C, D, E)$. The first two distributions involve no more than three variables and are directly estimated from data. To compute the last distribution, consider a restriction of the model onto the four variables involved. Let $S_1 = \{B\}$, $S_2 = \{C\}$, $S_3 = \{D\}$ and $S_4 = \{E\}$. This allows us to reduce $P(B, C, D, E)$ to $P(B, C, E)$, $P(C, D)$, and $P(C, D, E)$. All those distributions are estimated from data.

The description of a complete algorithm and the discussion of related issues are out the scope of this paper.

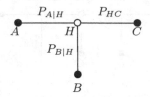

Fig. 7. A latent tree model with one latent variable

5   Equalities for Parameter Estimation

In this section, we derive a group of equations that are used to estimate parameters of LTMs [2].

5.1   Some Notations

Consider the LTM shown in Figure 7. Suppose all variables have equal cardinality. We parameterize the model with the distributions $P(A|H)$, $P(B|H)$ and $P(H,C)$. In matrix notation, they are $P_{A|H}$, $P_{B|H}$ and P_{HC}.

Let b be a particular value of B. For reasons to become clear later, we consider the joint probability $P(A=a, B=b, C=c)$ when variables $A = a$ and $C = c$. It is obvious that

$$P(A=a, B=b, C=c) = \sum_{H} P(A=a|H)P(B=b|H)P(H, C=c). \qquad (8)$$

Use P_{AbC} to denote the matrix where the element at the a-th row and c-th column is P_{abc}. Use $P_{b|H}$ to denote the column vector where the element at the h-row is $P(B=b|H=h)$, which in turn is denoted as $P_{b|h}$. Use $diag(P_{b|H})$ to denote the diagonal matrix with the elements of the vector $P_{b|H}$ as the diagonal elements. Equation (8) can be rewritten in matrix form as follows:

$$P_{AbC} = P_{A|H}diag(P_{b|H})P_{HC}. \qquad (9)$$

5.2   The Case of Equal Cardinality

Suppose all the probability matrices are invertible. Augment the model of Figure 7 with two edges (C, H') and (H', A'), where $|A'| = |A|$ and $|H'| = |H|$. Let the potential matrices for the two new edges be:

$$\phi_{CH'} = P_{HC}^{-1}, \phi_{H'A'} = P_{A|H}^{-1}.$$

Note that the edge (C, H') is the inverse edge of (H, C).

For the rest of this subsection, fix the value of B at b. Consider calculating the marginal potential $P(A, B=b, A')$ in the model of Figure 8 (a). In matrix form, it is $P_{AbA'}$.

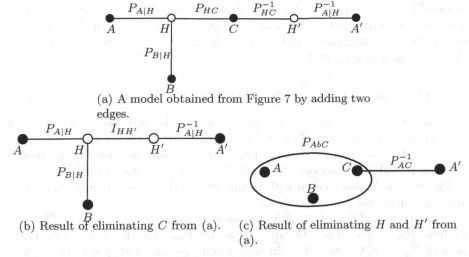

(a) A model obtained from Figure 7 by adding two edges.

(b) Result of eliminating C from (a).   (c) Result of eliminating H and H' from (a).

Fig. 8. Operations on a model obtained by augmenting the model of Figure 7

One way to compute $P_{AbA'}$ is to first eliminate C and then eliminate H' and H. Eliminating C from the model of (a) results in the model of (b). Because the edge (C, H') is the inverse edge of (H, C), the potential for the edge (H, H') is an identity matrix $I_{HH'}$. Further eliminating H' results in a model that has the structure shown in Figure 7 with the variable C replaced with A' (ref. Theorem 1). The potential matrix for the edge (H, A') is $P_{A|H}^{-1}$. According to Equation (9), we have

$$P_{AbA'} = P_{A|H} diag(P_{b|H}) P_{A|H}^{-1}. \tag{10}$$

Another way to compute $P_{AbA'}$ is to first eliminate H and H', and then eliminate C. The elimination of H and H' leads to the model of (c). According to Equation (9), the elimination of H gives us the potential matrix P_{AbC}. The elimination of H' gives us the following potential matrix:

$$\phi_{CH'} \phi_{H'A'} = P_{HC}^{-1} P_{A|H}^{-1} = (P_{A|H} P_{HC})^{-1} = P_{AC}^{-1}.$$

Further eliminating C from the model of (c), we get

$$P_{AbA'} = P_{AbC} P_{AC}^{-1}. \tag{11}$$

Putting the two equations (10) and (11) together, we get the following theorem.

Theorem 5. *Suppose that all the variables in the LTM shown Figure 7 have equal cardinality and that all the probability matrices are invertible. Then,*

$$P_{A|H} diag(P_{b|H}) P_{A|H}^{-1} = P_{AbC} P_{AC}^{-1}. \tag{12}$$

Note that the two terms on the right hand side involve only observed variables. They can be directly estimated from data. On the other hand, the left hand side involves model parameters, and particularly the diagonal elements of $P_{b|H}$ are the eigenvalues of $P_{AbC}P_{AC}^{-1}$.

5.3   The Case of Unequal Cardinalities

Next we generalize Theorem 5 to the case where the observed variables might have unequal cardinalities. The cardinality of H latent variables is required to be no greater than the cardinality of any observed variable. We further require that the probability matrices $P_{A|H}$ and P_{HC}^T have full column rank.

The technical issue to deal with in this case is that the matrices $P_{A|H}$, P_{HC} and P_{AC} might not be invertible.

Let the cardinalities of H, A and C be r, s and t. In the model of Figure 7, we have $P_{AC} = P_{A|H}P_{HC}$. Because $P_{A|H}$ and P_{HC}^T have full column rank, the rank of P_{AC} is r. Consider the singular decomposition of $P_{AC} = U\Lambda V^T$, where Λ is a $r \times r$ diagonal matrix, U and V are $s \times r$ and $t \times r$ column orthogonal matrices respectively. By following the same line of reasoning as in Section 4.2, we can conclude that $U^T P_{A|H}$ and $P_{HC}V$ are invertible.

Note that $(U^T P_{A|H})^{-1}U^T$ is a $r \times s$ matrix, and $V(P_{HC}V)^{-1}$ is a $t \times r$ matrix. Construct the model of Figure 8 (a) in the same way as in the previous subsection, except that we set potential matrices of the edges (H', A') and (C, H') as follows:

$$\phi_{H'A'} = (U^T P_{A|H})^{-1}U^T, \phi_{CH'} = V(P_{HC}V)^{-1}.$$

Now consider the marginal potential matrix $P_{AbA'}$. One way to compute it is to first eliminate C and then eliminate H' and H. Eliminating C from the model of (a) results in the model of (b). Because the product of P_{HC} and $V(P_{HC}V)^{-1}$ is an identity matrix, the potential for the edge (H, H') an identity edge. Further eliminating H' results in a model that has the structure shown in Figure 7 with the variable C replaced with A' (ref. Theorem 1). The potential matrix for the edge (H, A') is $(U^T P_{A|H})^{-1}U^T$. According to Equation (9), we have

$$P_{AbA'} = P_{A|H}diag(P_{b|H})(U^T P_{A|H})^{-1}U^T. \tag{13}$$

Another way to compute $P_{AbA'}$ is to first eliminate H and H', and then eliminate C. The elimination of H and H' lead to the model of (c). As in the previous section, the elimination of H gives us the potential matrix P_{AbC}. The elimination of H' gives us the following potential matrix:

$$\phi_{CH'}\phi_{H'A'} = V(P_{HC}V)^{-1}(U^T P_{A|H})^{-1}U^T$$
$$= V(U^T P_{A|H}P_{HC}V)^{-1}U^T$$
$$= V(U^T P_{AC}V)^{-1}U^T$$

Further eliminating C from the model of (c), we get

$$P_{AbA'} = P_{AbC}V(U^T P_{AC}V)^{-1}U^T. \tag{14}$$

Putting the two equations (13) and (14) together, we get

$$P_{A|H} diag(P_{b|H})(U^T P_{A|H})^{-1} U^T$$
$$= P_{AbC} V (U^T P_{AC} V)^{-1} U^T.$$

Consequently,

$$U^T P_{A|H} diag(P_{b|H})(U^T P_{A|H})^{-1}$$
$$= U^T P_{AbC} V (U^T P_{AC} V)^{-1}. \tag{15}$$

Theorem 6. *Suppose that, in the LTM shown Figure 7, the cardinality of H is no greater than that of any observed variable, and that the probability matrices $P_{A|H}$ and P_{HC}^T have full column rank. Then, Equation (15) holds.*

Note that Equation (15) is the same as Equation (12) except that the column orthogonal matrices U and V are used to deal the issue that $P_{A|H}$ and P_{AC} might not be invertible.

5.4   Parameter Estimation

To use Equation (15) for parameter estimation, observe that the eigenvalues of the matrix on the right hand side are the diagonal elements of $diag(P_{b|H})$, which in turn are the values for the conditional distribution $P(B=b|H)$. To estimate $P(B|H)$, we can: (1) estimate P_{ABC} and P_{AC} from data; (2) compute the singular decomposition of the of P_{AC} to obtain the matrices U and V; (3) for each value b for B, form the matrix on the right hand side; and (4) calculate the eigenvalues of the matrix. Those eigenvalues are the values for the distribution $P(B|H)$.

   If all the variables have equal cardinality, we can use Equation (12) instead. In this case, there is no need calculate the matrices U and V.

   To see how the strategy can be applied to LTMs with multiple latent variables, consider the model of Figure 2. For simplicity, we assume all the variables have equal cardinality. Restricting the model to the variable A, B, C and H, we get the model of Figure 7. Using the strategy, we can estimate $P(A|H)$, $P(B|H)$ and $P(C|H)$. In similar fashion, we can estimate $P(C|G)$, $P(D|G)$ and $P(A|G)$. Then $P(G|H)$ can be calculated from $P(H|C)$ and $P(C|G)$ using the relationship $P_{G|H} = P_{C|G}^{-1} P_{C|H}$.

   The description of a complete algorithm and the discussion of related issues are out the scope of this paper.

6   Conclusions

Starting from a latent tree model, we introduce inverse edges to obtain a generalized MRFs. Variables are then eliminated from the generalized MRFs in different orders to obtain equalities about the model. Two groups of equalities are obtained. One group allows us to calculate the joint probability one of particular assignment of all observed variables without estimating the model parameters. The other group of equalities gives us a new method for estimating the model parameters, which has advantages over the commonly used EM algorithm. For example, it does not have the difficulty of getting trapped in local maxima.

References

1. Anandkumar, A., Ge, R., Hsu, D., Kakade, S.M., Telgarsky, M.: Tensor Decompositions for Learning Latent Variable Models (2012) (Preprint)
2. Anandkumar, A., Hsu, D., Kakade, S.M.: A Method of Moments for Mixture Models and Hidden Markov Models. In: An Abridged Version Appears in the Proc. of COLT (2012)
3. Bartholomew, D.J., Knott, M., Moustaki, I.: Latent variable models and factor analysis, 2nd edn. Wiley (1999)
4. Dempster, A.P., Laird, N.M., Rubin, D.B.: Maximum Likelihood from Incomplete Data via the EM Algorithm. Journal of the Royal Statistical Society, Series B 39, 1–38 (1977)
5. Hsu, D., Kakade, S., Zhang, T.: A spectral algorithm for learning hidden markov models. In: COLT (2009)
6. Lauritzen, S.L.: Graphical Models. Clarendon Press (1996)
7. Mossel, E., Roch, S.: Learning nonsingular phylogenies and hidden markov models. AOAP 16, 583–614 (2006)
8. Mourad, R., Sinoquet, C., Zhang, N.L., Liu, T.F., Leray, P.: A survey on latent tree models and applications. Journal of Artificial Intelligence Research 47, 157–203 (2013)
9. Murphy, K.P.: Machine learning: a probabilistic perspective. MIT Press (2012)
10. Parikh, A.P., Song, L., Xing, E.P.: A Spectral Algorithm for Latent Tree Graphical Models. In: ICML (2011)
11. Zhang, N.L.: Hierarchical latent class models for cluster analysis. Journal of Machine Learning Research 5, 697–723 (2004)
12. Zhang, N.L., Poole, D.: Exploiting causal independence in Bayesian network inference. Journal of Artificial Intelligence Research 5, 301–328 (1996)

An Extended MPL-C Model
for Bayesian Network Parameter Learning
with Exterior Constraints

Yun Zhou[1,2,*], Norman Fenton[1], and Martin Neil[1]

[1] Risk and Information Management (RIM) Research Group,
Queen Mary University of London, United Kingdom
[2] Science and Technology on Information Systems Engineering Laboratory,
National University of Defense Technology, PR China
{yun.zhou,n.fenton,m.neil}@qmul.ac.uk

Abstract. Lack of relevant data is a major challenge for learning Bayesian networks (BNs) in real-world applications. Knowledge engineering techniques attempt to address this by incorporating domain knowledge from experts. The paper focuses on learning node probability tables using both expert judgment and limited data. To reduce the massive burden of eliciting individual probability table entries (parameters) it is often easier to elicit *constraints* on the parameters from experts. Constraints can be interior (between entries of the same probability table column) or exterior (between entries of different columns). In this paper we introduce the first auxiliary BN method (called MPL-EC) to tackle parameter learning with exterior constraints. The MPL-EC itself is a BN, whose nodes encode the data observations, exterior constraints and parameters in the original BN. Also, MPL-EC addresses (i) how to estimate target parameters with both data and constraints, and (ii) how to fuse the weights from different causal relationships in a robust way. Experimental results demonstrate the superiority of MPL-EC at various sparsity levels compared to conventional parameter learning algorithms and other state-of-the-art parameter learning algorithms with constraints. Moreover, we demonstrate the successful application to learn a real-world software defects BN with sparse data.

Keywords: BN parameter learning, Monotonic causality, Exterior constraints, MPL-EC model.

1 Introduction

Bayesian networks have proven valuable in modeling uncertainty and supporting decision making in practice [1]. However, in many applications there is extremely

* The authors would like to thank the three anonymous reviewers for their valuable comments and suggestions. This work was supported by European Research Council (grant no. ERC-2013-AdG339182-BAYES-KNOWLEDGE). The first author was supported by China Scholarship Council (CSC)/Queen Mary Joint PhD scholarships and National Natural Science Foundation of China (grant no. 61273322).

L.C. van der Gaag and A.J. Feelders (Eds.): PGM 2014, LNAI 8754, pp. 581–596, 2014.

limited data available to learn either the BN structure or probability tables. In such situations we have to use qualitative knowledge from domain experts in addition to any quantitative data available [2]. There are numerous recent real-world applications in which BN models incorporate significant expert judgment – for example, in medical diagnostics [3,4], traffic incident detection [5] and facial action recognition [6]. However, eliciting expert judgment remains a major challenge.

Directly asking experts to provide quantitative parameter values is time consuming and error-prone because the number of parameters increase exponentially with the number of nodes in the BN. For example, for a node X with 3 states that has 5 parents (each with 2-states), the probability table for X has 32 columns and 3 rows, i.e., 96 probability values to be elicited. Since the columns sum to 1, each column requires only 2 probability values to be elicited, so we consider these as 'parameters' and there are 64 in total. Recent study [7] shows exploring qualitative relationships and their generated constraints would greatly reduce the elicitation burden. However, in applying this method, central challenges include *how to estimate* parameters with both data and constraints [8], *how to optimally perform* expert judgments elicitation [9], and *how to fuse* different weights from different causal relationships and different parent state configurations. These are crucial to ensure that parameter learning is accurate and effective. Despite the finding of qualitative relationships published more than twenty years ago, only limited work [8,10,6,11] has been done on addressing these challenges.

In this paper we assume the BN structure is already defined and only investigate elicited *constraints* on parameters to help learn a target BN with sparse data. The paper extends earlier work [12] in which we introduced an auxiliary BN method (multinomial parameter learning with constraints, which is also referred as MPL-C) for learning parameters given expert constraints and limited data. In that work we considered only parameters constraints restricted to a single probability table column; for example:

"$P(\text{cancer} = \text{true}|\text{smoker} = \text{true}) > 0.01$" or

"$P(\text{cancer} = \text{true}|\text{smoker} = \text{true}) > P(\text{cancer} = \text{false}|\text{smoker} = \text{true})$"

In this paper we extend this to exterior parameter constraints (across columns) like:

"$P(\text{cancer} = \text{true}|\text{smoker} = \text{true}) > P(\text{cancer} = \text{true}|\text{smoker} = \text{false})$"

This kind of exterior parameter constraints are encoded in monotonic causality between two BN variables [13,14,15]. Parameter learning with this constraints normally is solved via establishing a constrained optimization problem [6,11], and is restricted to assumptions of binary nodes and convex constraints.

Our contribution in this paper is to extend the original MPL-C model (now refered to as MPL-EC) to support parameter learning with both data observations and exterior constraints. In MPL-EC the original parameter estimation problem converts to a BN inference problem. In this way, our model supports either convex or non-convex exterior constraints. Because the MPL-EC is a hybrid BN (contains continuous, as well as discrete, nodes) the inference is achieved via a dynamic discretization junction tree (DDJT) algorithm [16]. Some other works

[17,18,19] also support inference in hybrid BNs with deterministic conditional distributions. In this paper, we mainly focus on building the hybrid BN model to support the parameter learning with exterior constraints. Hence, we will not compare the DDJT with other inference algorithms. In our model, different exterior constraints have different strengths (added as the margin in each inequality [13]), which has a generative equation that encodes the weights from different causal relationships and the weights from different parent state configurations. This is itself an important output for modeling the constraints in a more precise way. To evaluate the algorithm, we conduct experiments on three standard networks, i.e., Weather, Cancer and Asia BNs, comparing against three baselines and prior learning with constraints methods. Finally, we apply our method to parameter learning in a real-world software defects BN.

2   Bayesian Networks Parameter Learning

2.1   Preliminaries

A BN consists of a directed acyclic graph (DAG) $G = (U, E)$ (whose nodes $U = \{X_1, X_2, X_3, \ldots, X_n\}$ correspond to a set of random variables, and whose arcs E represent the direct dependencies between these variables), together with a set of probability distributions associated with each variable. For discrete variables[1] the probability distribution is described by a node probability table (NPT) that contains the probability of each value of the variable given each instantiation of its parent values in G. We write this as $P(X_i | pa(X_i))$ where $pa(X_i)$ denotes the set of parents of variable X_i in DAG G. Thus, the BN defines a simplified joint probability distribution over U given by:

$$P(X_1, X_2, \ldots, X_n) = \prod_{i=1}^{n} P(X_i | pa(X_i)) \tag{1}$$

Let r_i denote the cardinality of the space of X_i, and q_i represent the cardinality of the space of parent configurations of X_i. The k-th probability value of a conditional probability distribution $P(X_i | pa(X_i) = j)$ can be represented as $\theta_{ijk} = P(X_i = k | pa(X_i) = j)$, where $\theta_{ijk} \in \theta$, $1 \leq i \leq n$, $1 \leq j \leq q_i$ and $1 \leq k \leq r_i$. Assuming $D = \{D_1, D_2, \ldots, D_N\}$ is a dataset of fully observable cases for a BN, then D_l is the l-th complete case of D, which is a vector of values of each variable. The classical maximum likelihood estimation (MLE) is to find the set of parameters that maximize the data loglikelihood $l(\theta | D) = \log \prod_l P(D_l | \theta)$. Let N_{ijk} be the number of data records in sample D for which X_i takes its k-th value and its parent $pa(X_i)$ takes its j-th value. Then $l(\theta | D)$ can be rewritten as $l(\theta | D) = \sum_{ijk} N_{ijk} \log \theta_{ijk}$. The MLE seeks to estimate θ by maximizing $l(\theta | D)$. In particular, we can get the estimation of each parameter as follows:

$$\theta_{ijk}^* = \frac{N_{ijk}}{N_{ij}} \tag{2}$$

---

[1] For continuous nodes we normally refer to a conditional probability distribution.

However, for several cases in the unified model, a certain parent-child state combination would seldom appear, and the MLE learning fails in this situation. Hence, another classical parameter learning algorithm (maximum a posteriori, MAP) can be used to mediate this problem via introducing *Dirichlet* prior: $\theta^* = \arg\max_\theta P(D|\theta)P(\theta)$. Therefore, we can derive the following equation for MAP:

$$\theta_{ijk}^* = \frac{N_{ijk} + \alpha_{ijk} - 1}{N_{ij} + \alpha_{ij} - 1} \tag{3}$$

Intuitively, one can think of the hyperparameter α_{ijk} in *Dirichlet* prior as an experts' guess of the virtual data counts for the parameter θ_{ijk}. When there is no related expert judgments, people usually use uniform prior or BDeu prior [2] in the MAP.

2.2   Constrained Optimization Approach

Although the *Dirichlet* prior is widely used, it is usually difficult to elicit the numerical hyperparameters from experts. Since the ultimate goal of MAP is to infer a posterior distribution, people directly introduce expert provided constraints to regularize the posterior estimation. As discussed above, some related work solves this problem via constrained optimization (CO). In CO, the expert judgments are encoded as convex constraints. For example, based on the previous definition, a convex constraint can be defined as $f(\theta_{ijk}) \leq \mu_{ijk}$, where $f : \Omega_{\theta_{ijk}} \to R$ is a convex function over θ_{ijk}, and $\mu_{ijk} \in [0, 1]$. Regarding parameter constraints, the scores are computed by a constrained optimization approach (i.e., gradient descent). In detail, for $\forall_{i,j,k} \theta_{ijk}$, we maximize the score function $l(\theta|D)$ subject to $g(\theta_{ijk}) = 0$ and $f(\theta_{ijk}) \leq \mu_{ijk}$, where the constraint $g(\theta_{ijk}) = -1 + \sum_{k=1}^{r_i} \theta_{ijk}$ ensures the sum of all the estimated parameters in a probability distribution is equal to one. To model the strength of the constraints, [6] introduced a confidence level λ_{ijk} for the penalty term in the objective function, i.e., let $f(\theta_{ijk}) = \theta_{ijk}$, and the penalty term is defined as $penalty(\theta_{ijk}) = [\mu_{ijk} - \theta_{ijk}]^-$, where $[x]^- = \max(0, -x)$. Therefore, the constrained maximization problem can be rewritten as follows:

$$\begin{array}{c} \arg\max_\theta \; l(\theta|D) - \frac{w}{2}\sum_{ijk}\lambda_{ijk} \cdot penalty(\theta_{ijk})^2 \\ s.t. \; \forall_{i,j,k} \; g(\theta_{ijk}) = 0 \end{array} \tag{4}$$

where w is the penalty weight, which is chosen empirically. Obviously, the penalty varies with the confidence level for each constraint λ_{ijk}. To ensure the solutions move towards the direction of reducing constraint violations (the maximal score), the score function must be convex, which limits the usage of constraints. Meanwhile, because the starting points are randomly generated in gradient descent, this may cause unacceptably poor parameter estimation results when learning with zero or limited data counts N_{ijk} in the score function.

2.3   Multinomial Parameter Learning with Constraints

Because the basic parameter learning method can be modeled with an auxiliary BN model, the constraints can be easily incorporated as the shared child of the nodes representing the constrained parameters. This auxiliary BN is a *hybrid* model (see Figure 1) containing a mixture of discrete and (non-normally distributed) continuous nodes. Therefore, the parameter estimation problem converts to a BN inference problem, where the data statistics and constraints are observed, and the target parameters are updated by a dynamic discretization inference algorithm [16].

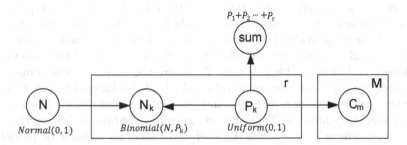

Fig. 1. The multinomial parameter learning model with constraints (MPL-C) and its associated distributions. C_m is a constraint node, which encodes constraints within a NPT column, i.e., $C_1 : P_1 > 0.5$ and $C_2 : P_2 > P_1$.

In Figure 1, for simplification, we use P_k ($k = 1$ to r) to represent the r parameters of a single column instead of θ_{ijk}. Similarly, the N (instead of N_{ij}) represents the data counts of a parent state configuration, and the N_k (instead of N_{ijk}) represent the data counts for its k-th state under this parent state configuration. Given the above model and its related observations, inference refers to the process of computing the discretized posterior marginal of unknown nodes P_k (these are the nodes without evidence). These nodes encode uniform priors, which prevents the problem of random initial values in constrained optimization. After inference, the mean value of P_k will be assigned as the parameter estimation (i.e., the corresponding NPT cell value). Full details can be found in [12].

3 The New Method

In this section we first describe (Section 3.1) the type of monotonic causality and its associated exterior constraints. In Section 3.2 we describe the extended version of the auxiliary BN model to incorporate new forms of exterior constraints provided from expert judgments in order to supplement the MPL-C. Because there is a state combination explosion problem in the extended BN model, we describe a novel alternative BN model which keeps the properties of the original extended BN but with fewer state combinations. In Section 3.3, a simple example is presented to show how to build and apply the MPL-EC model for parameter learning.

3.1   Parameter Constraints

There are two types of node parameter constraints that we consider: interior and exterior. An interior constraint, which is also called inter-relationship constraint, constrains two parameters that share the same node index i, and parent state configuration j (i.e., this is a constraint between values in the same column of a node probability table). An example of such a constraint is $\theta_{ijk} \geq \theta_{ijk'}$, where $k \neq k'$. Interior constraints, which can only be elicited from expert judgment, were studied extensively in our previous work [12]. We showed in [12] that significant improvements to table learning could be achieved from relatively small number of expert provided interior constraints. However, in many situations it is possible (and actually more efficient) to elicit constraints between parameters in different probability table columns. These are the exterior constraints.

Formally, an exterior constraint (also called inter-relationship constraint) is where two parameters in a relative relationship constraint share the same node index i, and state index k. Typically an exterior constraint will have the form: $\theta_{ijk} \geq \theta_{ij'k}$ where $j \neq j'$. This kind of constraint is encoded in monotonic causality which can greatly reduce the burden of expert judgment elicitation. Before we examine exterior constraints in detail, we need some definitions and notations:

The positive/negative monotonic causality: For the simplest single monotonic causal connection: X causes Y ($X \rightarrow Y$), the causality can either be positive or negative. Positive monotonic causality is represented by $X \xrightarrow{+} Y$ (increasing value of X leads to increases in Y). Negative monotonic causality is represented by $X \xrightarrow{-} Y$ (increasing value of X leads to decrease in Y); for example, if X is a particular medical treatment and Y is patient mortality.

Let $cdf(\cdot)$ denote the cumulative distribution function. The formal equation of these two kinds of monotonic causality can be formulated as exterior constraints as follows:

$$X \xrightarrow{+} Y: cdf(P(Y|pa(Y) = j)) \geq cdf(P(Y|pa(Y) = j'))$$
$$X \xrightarrow{-} Y: cdf(P(Y|pa(Y) = j)) \leq cdf(P(Y|pa(Y) = j'))$$

Here both X and Y are ordered categorical variables, j' and j are integers satisfying the inequality relationships $0 < j' < j < |pa(Y)|$, where the $|pa(Y)|$ represents the total number of state configurations in $pa(Y)$. In $X \rightarrow Y$, $pa(Y) = X$. As we can see, the negative causality represents the opposite causal relationship compared with positive causality. The model of introducing a single positive monotonic causality has been well discussed in previous work [7,13,20]. However, real-world BNs usually contain nodes whose parents provide a mixture of positive and negative causality, as synergistic interactions [11]. Previous work [11] has addressed this synergy problem at some point, where all the causalities should either be positive or negative (homogeneous synergies). Therefore, this work does not allow the synergy relationship have different types of monotonic causalities, which is referred as heterogeneous synergies. Recently, researchers [21,22] introduced a novel canonical gate (refered to as NIN-AND tree) to model different causal interactions: reinforcing and undermining. However, this work does not support learning with monotonic constraints and their margins. Actually, the synergies of different causalities are different when the causal weights (the confidences of the causal connections) are considered. Previous studies rarely discussed this problem, and no relevant model has tackled this issue.

In this paper, we introduce a generative form of the exterior constraint equation, which support homogeneous/heterogeneous synergies with different weights. Assume we have a BN with variables $U = \{Y, X_1, X_2, \ldots, X_n\}$ and the simple inverted naive structure, which means the variable Y is the shared child of $X_1, X_2, \ldots, X_n$. Then our generative exterior constraint is:

$$\begin{cases} cdf(P(Y|pa(Y) = j)) - cdf(P(Y|pa(Y) = j')) \geq M_{jj'} & if\ M_{jj'} > 0 \\ cdf(P(Y|pa(Y) = j)) - cdf(P(Y|pa(Y) = j')) \leq M_{jj'} & if\ M_{jj'} < 0 \end{cases} \tag{5}$$

where $M_{jj'} = \sum_{i=1}^{n} M_{jj'}^i = \sum_{i=1}^{n} w_i \cdot cl_i \cdot \varepsilon_{jj'}^i$ and $0 < j' < j < |pa(Y)|$. The $M_{jj'}$ represents the overall margin of the synergies, which is the summation of each single margin $M_{jj'}^i$. $M_{jj'}^i$ contains three terms: $w_i \geq 1$ represent the global weight (the subjective confidence) of the causal relationship $X_i \to Y$, its default value $w_i = 1$ indicates there is no subjective confidence on the causality; cl is the causality label ($cl_i = 1$ indicates the positive causality $X_i \overset{+}{\to} Y$; and $cl_i = -1$ represents the negative causality $X_i \overset{-}{\to} Y$); $\varepsilon_{jj'}^i$ is the term that describes the confidence of the inequality introduced by state configuration gap in a causality. That is to say, the $\varepsilon_{jj'}^i$ is a small positive value proportional to the state configuration distance in X_i under two indices j and j' in $pa(Y) = \{X_1, X_2, \ldots, X_n\}$.

To calculate $\varepsilon_{jj'}^i$, we need to find the subindices $(ind2sub_i(j)$ and $ind2sub_i(j'))$ of X_i from the single indices in $pa(Y)$. Thus we have: $\varepsilon_{jj'}^i = \frac{ind2sub_i(j) - ind2sub_i(j')}{\lambda \cdot |X_i|}$. Here the $\lambda > 1$ is the trade-off parameter that controls the effect of the confidence introduced by state configuration gap. Because size $|pa(Y)| = \prod_{i=1}^{n} |X_i|$ increases exponentially with an increase of parent nodes, it would be very expensive to find all combinations of two indices in $|pa(Y)|$. Therefore, in this paper, we only discuss a very simple way to get the combinations. For state configuration size $|pa(Y)|$, we generate two indices pairs iteratively ("$|pa(Y)|, 1$", "$(|pa(Y)| - 1), 2$", ...) until no more pairs can be found.

As shown in equation 5, the type ($\geq$ or $\leq$) of the exterior constraint is decided by the value of the margin. The margin is equal to zero ($M = 0$) only in the situation where the effects of different causalities are intermediate in the shared child node. Thus, there is no associated exterior constraints.

Next, we present a simple example of our model: we assume the target variable Y is binary, and it has two binary parents X_1 and X_2 with "T" and "F" states. Assume the first causality is positive $X_1 \overset{+}{\to} Y$, and the second causality is negative $X_2 \overset{-}{\to} Y$. Therefore, the exterior constraints induced by these two monotonic causalities can be represented as:

$$cdf(P(Y|pa(Y) = 4)) - cdf(P(Y|pa(Y) = 1)) \geq w_1 \cdot \varepsilon_{41}^1 - w_2 \cdot \varepsilon_{41}^2$$
$$cdf(P(Y|pa(Y) = 3)) - cdf(P(Y|pa(Y) = 2)) \geq w_1 \cdot \varepsilon_{32}^1 - w_2 \cdot \varepsilon_{32}^2$$

In addition, there is no subjective judgments on their weights, i.e. $w_1 = w_2 = 1$. Thus the margin of the first equation equal to zero ($w_1 \cdot \varepsilon_{41}^1 = w_2 \cdot \varepsilon_{41}^2$), and this equation is discarded. Also, because Y is binary, this means $y = \{y_T, y_F\}$. Therefore, we can have the following exterior constraints based on the above equation:

$$P(y_T|x_{1T}, x_{2F}) - P(y_T|x_{1F}, x_{2T}) \geq \tfrac{1}{\lambda}$$
$$P(y_T|x_{1T}, x_{2F}) + P(y_F|x_{1T}, x_{2F}) - P(y_T|x_{1F}, x_{2T}) - P(y_F|x_{1F}, x_{2T}) = 0$$

Note the equality only happens when y reaches the full range (the biggest) value in $cdf(P(y))$.

3.2   The Extended MPL-C Model

In this subsection, we present the extended MPL-C model (MPL-EC) to encode the constraints in equation 5. For any monotonic causality, we need to introduce a set of shared children nodes to model the introduced constraints $C_1, C_2, \ldots, C_r$ (see Figure 2). The size of the constraints set is equal to the number of states (ranges from 1 to r) in variable Y. In order to simplify the notation, we use P_k and $P_{k'}$ ($k/k' = 1$ to r/r') to represent parameters in Y under different state configurations of X_i. Therefore, for a single positive monotonic causality $X_i \overset{+}{\to} Y$, we have the following arithmetic constraints encoded in the MPL-EC model to constrain the parameters under two state configurations of X_i (j and j'):

$$\begin{cases} C_1 : P_1 - P_{1'} \geq w_i \cdot cl_i \cdot \varepsilon_{jj'}^i \\ C_2 : P_1 + P_2 - P_{1'} - P_{2'} \geq w_i \cdot cl_i \cdot \varepsilon_{jj'}^i \\ C_3 : P_1 + P_2 + P_3 - P_{1'} - P_{2'} - P_{3'} \geq w_i \cdot cl_i \cdot \varepsilon_{jj'}^i \\ \qquad\qquad \vdots \\ C_r : \sum_{k=1}^r P_k - \sum_{k'=1}^r P_{k'} = 0 \end{cases} \qquad (6)$$

In the last exterior constraint equation C_r, two sides of the relative relationship are equal to each other. As we can see there are additional $(1 + n) \cdot n$ edges when we introduce n constraint nodes. To reduce the model complexity it must be replaced by an equivalent model whose structure has a restricted number of parents.

Previous work has proposed a binary factorization algorithm [23] to improve the efficiency of the DDJT algorithm. This idea can also be applied here to produce an alternative model of the straightforward MPL-EC. The new model is called binary summation model, which introduces an additional $2 \cdot (n - 1)$ auxiliary nodes, which only encode the simple sum arithmetic equations to model the summations of its parents. This model has the same number of edges as the straightforward model, but the maximal number of parents is fixed as two in this model. This avoids the parent state combination explosion problem. The detail of its structure can be found in Figure 2(b).

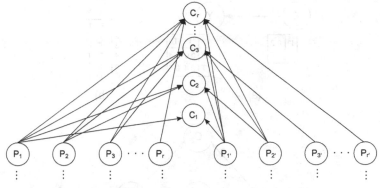

(a) The straightforward model of introducing exterior constraints

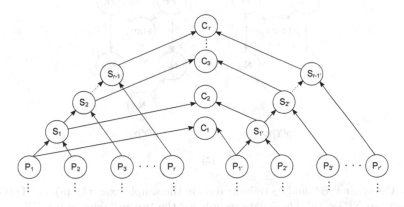

(b) The binary summation model of introducing exterior constraints

Fig. 2. The straightforward MPL-EC model and its alternative binary summation model. Due to the space limitation, the MPL-EC model presented here only display the part for modeling introduced constraints, the left part for modeling multinomial parameter learning is not displayed, which is the same as MPL-C in Section 2.3.

3.3   A Simple Example

In this subsection, we use a simple example to demonstrate the exterior constraints and its generated MPL-EC model. This example encodes the simplest single positive causal connection: $X \overset{+}{\to} Y$, where the two nodes involved are both binary with "T" and "F" states. Therefore, we have two parameter columns under two parent state instantiations to estimate in Y, which are $P(Y|x_T)$ and $P(Y|x_F)$. Its MPL-EC model is shown in Figure 3.

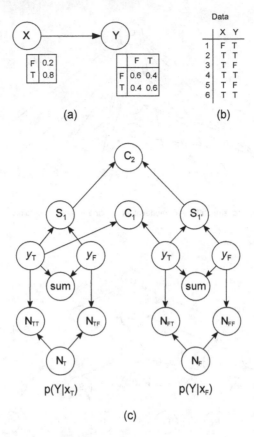

Fig. 3. The original BN and its training data in the simple example. (a) The DAG and its associated NPTs. (b) The 6 data records for the two variables in the BN. (c) The MPL-EC model for estimating the parameters in Y. The constraint nodes are modeled as binary (True/False) nodes with expressions that specify the constraint relationships between its parents. The auxiliary BN model is implemented in AgenaRisk [24], which supports hybrid BNs containing conditionally deterministic expressions. For example, the software statement for C_1 is: $if(P(y_T|x_T) - P(y_T|x_F) \geq w \cdot cl \cdot \varepsilon, "True", "False")$.

The detail of the exterior constraints encoded in the constraint nodes of MPL-EC is:

$$\begin{cases} S_1 : P(y_T|x_T) + P(y_F|x_T) \\ S_{1'} : P(y_T|x_F) + P(y_F|x_F) \\ C_1 : P(y_T|x_T) - P(y_T|x_F) \geq w \cdot cl \cdot \varepsilon \\ C_2 : S_1 - S_{1'} \geq w \cdot cl \cdot \varepsilon \end{cases} \quad (7)$$

where $cl = 1$, $\varepsilon = \frac{1}{2\lambda}$. according to above definition, and w represents the subjective confidence whose value can be chosen empirically from the domain knowledge.

Based on the statistics on the dataset (Figure 3(b)) and previous definition, we have: $N_T = 5$ ($N_{TF} = 2$, $N_{TT} = 3$) under the condition of $X = x_T$, and $N_F = 1$ ($N_{FF} = 0$, $N_{FT} = 1$) in the state initiation of $X = x_F$. Therefore, the MLE results of Y are $P(y_T|x_T) = 0.6$ and $P(y_T|x_F) = 1$. As we can see, the estimation of $P(y_T|x_F)$ is far away from the ground truth (0.6 and 0.4) due to the sparse data records under the $X = x_F$ condition.

With the above data observations, we now can set the evidence for certain nodes including constraint nodes (all are set as "True" observations), number of trials, total numbers, and the summation of all the estimated parameters. Based on these evidences, the inference in the MPL-EC is to compute the discretized posterior marginals of each of the unknown nodes $y_{F/T}$ (these are the nodes without evidence) via DDJT algorithm [16]. This algorithm alternates between two steps: 1) performing dynamic discretization, which searches and splits the regions with the highest relative entropy error determined by a bounded K-L divergence with the current approximated estimates of the marginals; 2) performing junction tree inference, which updates the posterior of the marginals. At convergence, the mean value of $y_{F/T}$ will be assigned as the final corresponding NPT cell values. After inference with the model in Figure 3(c), we have $P(y_T|x_T) = 0.67$ and $P(y_T|x_F) = 0.50$, which are much reasonable than the MLE results.

4   Experiments

The goal of the experiments is to demonstrate the benefits of our method and show the advantages of using elicited signs of causalities (either from ground truth or from expert judgment) and their generated exterior constraints to improve the parameter learning performance. We test the method against the conventional learning techniques (MLE and MAP) as well as against the competing method that incorporates exterior constraints (i.e., the constraint optimization method). Sections 4.1 and 4.2 describe the details of the experiments. The first (Section 4.1) uses the well-known Weather, Cancer and Asia BN (their signs are elicited from the ground truth), while the second (Section 4.2) uses a software defects BN, and its signs of causalities are elicited from a real expert.

In all cases, we assume that the structure of the model is known and that the 'true' NPTs that we are trying to learn are those that are provided as standard with the models. Obviously, for the purpose of the experiment we are not given these 'true' NPTs but instead are given a number of sample observations which are randomly generated based on the true NPTs. The experiments consider a range of sample sizes. In all case the resulting learnt NPTs are evaluated against the true NPTs by using the K-L divergence measure [25], which is recommended to measure the distance between distributions. The smaller the K-L divergence is, the closer the estimated NPT is to the true NPT. If frequency estimated values are zero in MLE, Laplace smoothing is applied to guarantee they can be computed. The global weights of all causal relationships are set as default value $w_i = 1$ in all experiment settings, and the trade-off value λ is set as 10.

4.1   Different Standard BNs Experiments

In the first set of experiments we use three standard models [26,27,28] that have been widely used for evaluating different learning algorithms. Based on these BNs and elicited signs, we compare the performance of different parameter learning algorithms: MLE, MAP, CO and MPL-EC.

Table 1 shows the structure of each BN and its associated parameter learning results. The BN structures are presented in the middle column of the table and annotated with positive/negative signs on their edges. The learning results in each setting are presented in the last column for each row. In each sub figure, the x-coordinate denotes the data sample size from 10 to 100, and the y-coordinate denotes the average K-L divergence for each parameter. For each data sample size, the experiments are repeated 5 times, and the results are presented with their mean and standard deviation.

As shown in the last column of Table 1, for all parameter learning methods, the K-L divergence decreases as expected when the sample size increases. Specifically, methods of learning with constraints, i.e., CO and MPL-EC always outperform the conventional MLE algorithm, especially in the sparse data situations. However, the CO failed to outperform MAP in all data settings of Cancer and Asia BNs, while the MPL-EC method always achieves the best performance in all cases for the three different BNs.

4.2   Software Defects BN Experiment

In this section, we consider a very well documented BN model that has been used by numerous technology companies worldwide [29] to address a real-world problem: the software defects prediction problem. The idea is to be able to predict the quality of software in terms of defects found in operation based on observations that may be possible during the software development (such as component complexity and defects found in testing). This BN contains eight nodes: "design process quality (DQ)", "component complexity (C)", "defects inserted (DI)", "testing quality (T)", "defects found in testing (DT)", "residual defects (R)", "operation usage (O)" and "defects found in operation (DO)". All of them are discrete, which have 3 ordered states: "Low", "Medium", and "High".

Figure 4(a) represents the structure of the BN, the signs on the edges indicate whether the associate monotonic causalities are positive or negative. These causalities are elicited from real expert judgments, i.e., as design process quality (DQ) goes from "Low" to "High", the defects inserted (DI) go from "High" to "Low", this encodes a negative monotonic causality.

Figure 4(b) shows the learning results, where the MPL-EC outperforms all other algorithms in every scenario. Compared with the state-of-art CO algorithm, our MPL-EC significantly improves the parameter learning performance, i.e., the MPL-EC outperforms the CO in all training sample sizes, with an overall 47.06% K-L divergence reduction.

Table 1. Learning results for MLE, MAP, CO and MPL-EC in Weather, Cancer and Asia BN learning problems. Four lines are presented in each sub figure, where the solid line with circle marker represents the learning results of baseline MLE algorithm, the dotted line with right-pointing triangle marker represents the learning results of MAP algorithm, the dotted line with square marker denotes the results of the CO algorithm, and the bold dash-dot line with diamond marker shows the learning results of the MPL-EC method.

Name	Directed acyclic graph (DAG)	Learning performance
Weather		
Cancer		
Asia		

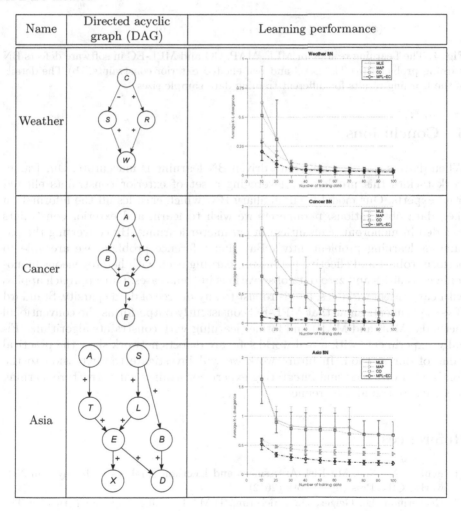

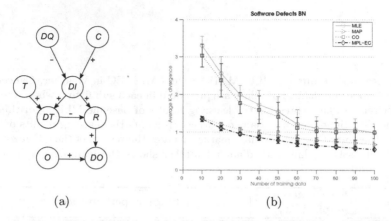

Fig. 4. The Learning results for MLE, MAP, CO and MPL-EC in software defects BN learning problem: (a) The DAG and real elicited exterior constraints; (b) The details of the learning results for different training data sample sizes

5 Conclusions

When data is sparse, purely data driven BN learning is inaccurate. Our framework tackles this problem by leveraging a set of exterior constraints elicited from experts. Our model is an auxiliary BN, which encodes all the information (i.e., data observations, parameters we wish to learn, and exterior constraints encoded in monotonic causalities) in parameter learning. By converting the parameter learning problem into a Bayesian inference problem, we are able to perform robust and effective parameter learning even with heterogeneous monotonic causalities and zero data observations in some cases. Our approach applies with categorical variables, and is robust to any degree of data sparsity. Standard BNs experiments show that MPL-EC consistently outperforms the conventional methods (MLE and MAP) and former learning with constraints algorithms. Finally, experiments with a real-world software defects network show the practical value of our method. In future work we will investigate the extension to the continuous variables, and integrating expert constraints with structure learning so structure can also be refined.

References

1. Fenton, N., Neil, M.: Risk Assessment and Decision Analysis with Bayesian Networks. CRC Press, New York (2012)
2. Heckerman, D., Geiger, D., Chickering, D.M.: Learning Bayesian networks: The combination of knowledge and statistical data. Mach. Learn. 20(3), 197–243 (1995)
3. Hutchinson, R.A., Niculescu, R.S., Keller, T.A., Rustandi, I., Mitchell, T.M.: Modeling fMRI data generated by overlapping cognitive processes with unknown onsets using hidden process models. NeuroImage 46(1), 87–104 (2009)

4. Yet, B., Perkins, Z., Fenton, N., Tai, N., Marsh, W.: Not just data: A method for improving prediction with knowledge. J. Biomed. Inform. 48, 28–37 (2014)
5. Šingliar, T., Hauskrecht, M.: Learning to detect incidents from noisily labeled data. Mach. Learn. 79(3), 335–354 (2010)
6. Liao, W., Ji, Q.: Learning Bayesian network parameters under incomplete data with domain knowledge. Pattern Recogn 42(11), 3046–3056 (2009)
7. Wellman, M.P.: Fundamental concepts of qualitative probabilistic networks. Artif. Intell. 44(3), 257–303 (1990)
8. Druzdzel, M.J., Van Der Gaag, L.C.: Elicitation of probabilities for belief networks: Combining qualitative and quantitative information. In: Proceedings of the Eleventh Conference on Uncertainty in Artificial Intelligence, Morgan Kaufmann Publishers Inc, pp. 141–148. Morgan Kaufmann, San Francisco (1995)
9. Cano, A., Masegosa, A.R., Moral, S.: A method for integrating expert knowledge when learning Bayesian networks from data. IEEE Trans. on Sys. Man Cyber. Part B 41(5), 1382–1394 (2011)
10. Niculescu, R.S., Mitchell, T., Rao, B.: Bayesian network learning with parameter constraints. J. Mach. Learn. Res. 7, 1357–1383 (2006)
11. Yang, S., Natarajan, S.: Knowledge intensive learning: Combining qualitative constraints with causal independence for parameter learning in probabilistic models. In: Blockeel, H., Kersting, K., Nijssen, S., Železný, F. (eds.) ECML PKDD 2013, Part II. LNCS, vol. 8189, pp. 580–595. Springer, Heidelberg (2013)
12. Zhou, Y., Fenton, N., Neil, M.: Bayesian network approach to multinomial parameter learning using data and expert judgments. Int. J. Approx. Reasoning 55(5), 1252–1268 (2014)
13. Altendorf, E.E.: Learning from sparse data by exploiting monotonicity constraints. In: Proceedings of the 21st Conference on Uncertainty in Artificial Intelligence, pp. 18–26. Morgan Kaufmann Publishers Inc., San Francisco (2005)
14. van der Gaag, L.C., Renooij, S., Geenen, P.L.: Lattices for studying monotonicity of Bayesian networks. In: Proceedings of the 3rd European Workshop on Probabilistic Graphical Models, Prague, Czech Republic, pp. 99–106 (2006)
15. van der Gaag, L.C., Tabachneck-Schijf, H.J.M., Geenen, P.L.: Verifying monotonicity of bayesian networks with domain experts. Int. J. Approx. Reasoning 50(3), 429–436 (2009)
16. Neil, M., Tailor, M., Marquez, D.: Inference in hybrid Bayesian networks using dynamic discretization. Stat. and Comput. 17(3), 219–233 (2007)
17. Lunn, D.J., Thomas, A., Best, N., Spiegelhalter, D.: WinBUGS-a Bayesian modelling framework: concepts, structure, and extensibility. Stat. and Comput. 10(4), 325–337 (2000)
18. Shenoy, P.P., West, J.C.: Inference in hybrid Bayesian networks using mixtures of polynomials. Int. J. Approx. Reasoning 52(5), 641–657 (2011)
19. Shenoy, P.P.: Two issues in using mixtures of polynomials for inference in hybrid Bayesian networks. Int. J. Approx. Reasoning 53(5), 847–866 (2012)
20. Feelders, A., van der Gaag, L.: Learning Bayesian network parameters under order constraints. Int. J. Approx. Reasoning 42(1), 37–53 (2006)
21. Xiang, Y., Jia, N.: Modeling causal reinforcement and undermining for efficient CPT elicitation. IEEE Trans. on Knowl. and Data Eng. 19(12), 1708–1718 (2007)
22. Xiang, Y., Truong, M.: Acquisition of causal models for local distributions in Bayesian networks. IEEE Trans. on Cyber (2013), doi:10.1109/TCYB. 2013.2290775

23. Neil, M., Chen, X., Fenton, N.: Optimizing the calculation of conditional probability tables in hybrid Bayesian networks using binary factorization. IEEE Trans. on Knowl. and Data Eng. 24(7), 1306–1312 (2012)
24. AgenaRisk (2014), http://www.agenarisk.com/
25. Cover, T.M., Thomas, J.: Entropy, relative entropy and mutual information. Elements of Information Theory, 12–49 (1991)
26. Lauritzen, S., Spiegelhalter, D.: Local computations with probabilities on graphical structures and their application to expert systems. J. Roy. Statist. Soc. Ser. B, 157–224 (1988)
27. Friedman, N., Geiger, D., Goldszmidt, M.: Bayesian network classifiers. Mach. Learn. 29(2-3), 131–163 (1997)
28. Korb, K.B., Nicholson, A.E.: Bayesian Artificial Intelligence. CRC Press, New York (2003)
29. Fenton, N., Neil, M., Marsh, W., Hearty, P., Radliński, Ł., Krause, P.: On the effectiveness of early life cycle defect prediction with Bayesian nets. Empirical Softw. Engg. 13(5), 499–537 (2008)

Author Index

Albrecht, David 1
Antal, Peter 222
Antonucci, Alessandro 145
Arias, Jacinto 17

Bendtsen, Marcus 49
Ben Mrad, Ali 33
Bielza, Concha 519
Blum, Christian 396
Bolt, Janneke H. 65
Butz, Cory J. 81

Cabañas, Rafael 97
Cano, Andrés 97, 113
Chen, Peixian 567
Claassen, Tom 442
Cobb, Barry R. 129
Corani, Giorgio 145, 176
Cozman, Fabio G. 334
Cuccu, Marco 176

De Bock, Jasper 160
De Boom, Cedric 160
de Campos, Cassio P. 176, 426
de Cooman, Gert 160
de Jongh, Martijn 190
de S. Oliveira, Jhonatan 81
Delcroix, Véronique 33
del Sagrado, José 206
Druzdzel, Marek J. 190, 238, 366

Fenton, Norman 581
Fernández, Antonio 206

Gabaglio, Sandra 145
Gámez, José A. 17
Gómez-Olmedo, Manuel 97, 113
Groot, Perry 442

Heskes, Tom 442
Hommersom, Arjen 503
Hullam, Gabor 222

Jensen, Frank 286, 302

Karlsen, Martin 286, 302
Kraisangka, Jidapa 238
Kwisthout, Johan 254, 271

Langseth, Helge 302
Larrañaga, Pedro 519
Leicester, Philip 33
Liu, Qing 551
Lozano, Jose A. 396

Madsen, Anders L. 81, 97, 286, 302
Mauá, Denis D. 145, 318, 334
Meert, Wannes 350
Moral, Serafín 113
Morales, Eduardo F. 409

Neil, Martin 581
Nicholson, Ann E. 1
Nielsen, Thomas D. 17, 302
Nowak, Krzysztof 366

Peña, Jose M. 49, 382
Pérez, Aritz 396
Pérez-Ariza, Cora B. 113
Piechowiak, Sylvain 33
Puerta, José M. 17

Ramírez-Corona, Mallinali 409
Renooij, Silja 65
Roumpelaki, Anna 487
Rumí, Rafael 206

Salmerón, Antonio 206, 302
Scanagatta, Mauro 426
Soendberg-Jeppesen, Nicolaj 286
Sokolova, Elena 442
Sonntag, Dag 458
Sucar, L. Enrique 409
Suzuki, Joe 471

Tichavský, Petr 535
Triantafillou, Sofia 487
Tsamardinos, Ioannis 487

Van Camp, Arthur 160
van der Heijden, Maarten 503
Varando, Gherardo 519

Vennekens, Joost   350
Vomlel, Jiří   535

Wang, Xiaofei   567
Whittle, Chris   1

Xiang, Yang   551

Zaffalon, Marco   176, 426
Zhang, Nevin L.   567
Zhou, Yun   581